GABA$_B$ RECEPTOR PHARMACOLOGY:
A Tribute to Norman Bowery

GABA$_B$ RECEPTOR PHARMACOLOGY: A TRIBUTE TO NORMAN BOWERY

Edited by

Thomas P. Blackburn

Translational Pharmacology BioVentures LLC
Hoboken, New Jersey, USA

Serial Editor

S. J. Enna

Department of Molecular and Integrative Physiology, Department of Pharmacology, Toxicology and Therapeutics, University of Kansas Medical Center, Kansas City, Kansas, USA

ADVANCES IN
PHARMACOLOGY

VOLUME 58

AMSTERDAM • BOSTON • HEIDELBERG • LONDON
NEW YORK • OXFORD • PARIS • SAN DIEGO
SAN FRANCISCO • SINGAPORE • SYDNEY • TOKYO
Academic Press is an imprint of Elsevier

Academic Press is an imprint of Elsevier
525 B Street, Suite 1900, San Diego, CA 92101-4495, USA
30 Corporate Drive, Suite 400, Burlington, MA 01803, USA
32 Jamestown Road, London NW1 7BY, UK
Radarweg 29, PO Box 211, 1000 AE Amsterdam, The Netherlands

First edition 2010

British Library Cataloguing in Publication Data
A catalogue record for this book is available from the British Library

Library of Congress Cataloging-in-Publication Data
A catalog record for this book is available from the Library of Congress

ISBN: 978-0-12-378647-0
ISSN: 1054-3589

Printed and bound in USA
10 11 12 10 9 8 7 6 5 4 3 2 1

Contents

Contributors xi
Preface xv

Historical Perspective and Emergence of the $GABA_B$ Receptor

N. G. Bowery

I. Introduction 2
II. $GABA_B$ Receptor Ligands 5
III. Receptor Distribution 6
IV. Allosteric Modulation 7
V. Receptor Function 8
VI. Potential Therapeutic Significance 8
VII. Conclusion 13
References 13

Chemistry and Pharmacology of $GABA_B$ Receptor Ligands

Wolfgang Froestl

I. Introduction 20
II. $GABA_B$ Receptor Agonists 20
III. $GABA_B$ Receptor Partial Agonists 27

IV. Positive Modulators of $GABA_B$ Receptors 29
V. $GABA_B$ Receptor Antagonists 32
VI. Conclusion 42
Acknowledgments 43
References 44

Heterodimerization of the $GABA_B$ Receptor—Implications for GPCR Signaling and Drug Discovery

Fiona H. Marshall and Steven M. Foord

I. Introduction 64
II. Identification of the $GABA_B$ Receptors 65
III. The Phylogeny of $GABA_B$ Receptors $GABA_B$ Receptors in Different Species 66
IV. Structure of $GABA_B$ Receptors 70
V. Function of the $GABA_B$ Receptor Heterodimer 73
VI. $GABA_B$ as a Model System for GPCR Dimerization 79
VII. Conclusions 83
References 85

Mechanisms of $GABA_B$ Receptor Exocytosis, Endocytosis, and Degradation

Dietmar Benke

I. Introduction 94
II. Cell-Surface Trafficking of $GABA_B$ Receptors 94
III. Endocytosis of $GABA_B$ Receptors 99
IV. Degradation of $GABA_B$ Receptors 104
V. Conclusions 105
Acknowledgments 107
References 108

Functional Modulation of $GABA_B$ Receptors by Protein Kinases and Receptor Trafficking

Miho Terunuma, Menelas N. Pangalos, and Stephen J. Moss

I. Introduction 114
II. Phosphorylation of $GABA_BR$ and Its Functional Modulation 115
III. Phosphorylation-Independent Desensitization of $GABA_BRs$ by Protein Kinases 117
IV. $GABA_BR$ Endocytic Sorting and the Control of Receptor Cell Surface Stability 118

V. $GABA_BR$ Phosphorylation and Diseases 118
VI. Conclusion 119
Acknowledgments 120
References 120

$GABA_B$ Receptor Coupling to G-proteins and Ion Channels

Claire L. Padgett and Paul A. Slesinger

I. Introduction 124
II. $GABA_B$ Receptor Structure 125
III. $GABA_B$ Receptor Protein Expression 128
IV. $GABA_B$ Receptor Effectors 129
V. Formation of a Macromolecular Signaling Heterocomplex 134
VI. $GABA_B$ Receptor-Dependent Desensitization 136
VII. Conclusion 139
References 139

$GABA_B$ Receptor-Mediated Modulation of Metabotropic Glutamate Signaling and Synaptic Plasticity in Central Neurons

Toshihide Tabata and Masanobu Kano

I. Introduction 150
II. Synaptic Organization Around Cerebellar Purkinje Cells 151
III. Possible Ligands for $GABA_BR$ in Cerebellar Purkinje Cells 152
IV. Functions of $GABA_BR$ Near the Excitatory Synapses 154
V. Possible Physiological Significance of $GABA_BR$-Mediated Effects on mGluR1 Signaling 162
VI. Conclusion 167
Acknowledgments 168
References 169

GABA Transporter GAT1: A Crucial Determinant of $GABA_B$ Receptor Activation in Cortical Circuits?

Guillermo Gonzalez-Burgos

I. Introduction 176
II. The Plasma Membrane GABA Transporter 1 176
III. Cellular and Subcellular Localization of $GABA_BRs$ and GAT1 180
IV. $GABA_BR$ Activation by GABA Released from Endogenous Sources 182

V. GAT1 Activity and $GABA_B$R Activation 188
VI. Conclusions 195
Acknowledgments 196
References 197

The Roles of $GABA_B$ Receptors in Cortical Network Activity

Michael M. Kohl and Ole Paulsen

I. Introduction 206
II. $GABA_B$ Receptors in Cortical Microcircuits 207
III. Control of Network Activity by $GABA_B$ Receptors 210
IV. Conclusion 221
References 222

$GABA_B$ Receptors: Physiological Functions and Mechanisms of Diversity

Audrée Pinard, Riad Seddik, and Bernhard Bettler

I. Introduction 232
II. Physiological Functions of $GABA_B$ Receptors 233
III. Heterogeneity of Native $GABA_B$ Responses 234
IV. Functions of the Cloned $GABA_B$ Receptor Subtypes 238
V. Additional Mechanisms of Diversity 246
VI. Conclusions 248
Acknowledgments 249
References 249

Role of $GABA_B$ Receptors in Autonomic Control of Systemic Blood Pressure

De-Pei Li and Hui-Lin Pan

I. Introduction 258
II. Overview of Autonomic Nervous System 260
III. Distribution of $GABA_B$ Receptors in CNS Autonomic Centers 261
IV. $GABA_B$ Receptor Function in the RVLM 261
V. $GABA_B$ Function in the NTS 263
VI. $GABA_B$ Receptor Function in the Hypothalamus 267
VII. Conclusion 274
Acknowledgments 275
References 276

$GABA_B$ Receptor Agonism as a Novel Therapeutic Modality in the Treatment of Gastroesophageal Reflux Disease

Anders Lehmann, Jörgen M. Jensen, and Guy E. Boeckxstaens

I. Introduction 288
II. Underlying Pathophysiological Mechanisms of GERD 289
III. The $GABA_B$ Receptor 290
IV. $GABA_B$ Receptor Agonists as Reflux Inhibitors 295
V. Conclusions 306
Acknowledgments 307
References 307

$GABA_B$ Receptors in Reward Processes

Styliani Vlachou and Athina Markou

I. Introduction 316
II. $GABA_B$ Receptor Agonists, $GABA_B$ Receptor Positive Modulators, and Reward: Effects on the Rewarding Properties of Food, Intracranial Self-Stimulation, and Drugs of Abuse 336
III. $GABA_B$ Receptor Agonists versus $GABA_B$ Receptor Positive Modulators 359
IV. Conclusion 360
Acknowledgments 360
References 362

$GABA_B$ Receptors in Addiction and Its Treatment

Robin J. Tyacke, Anne Lingford-Hughes, Laurence J. Reed, and David J. Nutt

I. Introduction 374
II. Alcohol Dependence 375
III. Cocaine 379
IV. Nicotine 384
V. Opiates 385
VI. Methamphetamine 387
VII. GHB 387
VIII. Discussion 388
IX. Conclusion 392
References 392

GABA$_B$–GIRK2-Mediated Signaling in Down Syndrome

Nathan P. Cramer, Tyler K. Best, Marcus Stoffel, Richard J. Siarey, and Zygmunt Galdzicki

I. Introduction 398
II. Mouse Models of DS 399
III. Evidence for Cognitive Impairment in Mouse Models of DS 400
IV. Hippocampal Deficit in DS and in DS Mouse Models 402
V. A Role of GABA$_B$–GIRK Coupling in Hippocampal Overinhibition in DS 404
VI. Potential Evidence for GABA$_B$-Mediated Developmental and Morphological Changes in DS 415
VII. Role of GIRK in DS Pain and Cerebellar Phenotypes 417
VIII. Conclusions 418
Acknowledgments 419
References 419

GABA$_B$ Receptors and Depression: Current Status

John F. Cryan and David A. Slattery

I. Introduction 428
II. GABA$_B$ Receptors 429
III. Role of GABA$_B$ Receptors in Animal Models of Antidepressant Action 430
IV. GABA$_B$ Receptors and Cognition 435
V. Chronic Antidepressants and GABA$_B$ Receptor Function 437
VI. GABA$_B$ Receptors and the Reward System 437
VII. GABA$_B$–5-HT Interactions 438
VIII. Clinical Evidence for a Role of GABA$_B$ Receptors in Depression 439
IX. Conclusion 441
Acknowledgments 441
References 442

GABA-B Receptors in *Drosophila*

Hari Manev and Svetlana Dzitoyeva

I. Introduction 454
II. *Drosophila* Model in Pharmacology 454
III. *Drosophila* GABA System 456
IV. *Drosophila* GABA-B Receptors 458
V. Conclusion 462
References 462

Index 465
Contents of Previous Volumes 485

Contributors

Numbers in parentheses indicate the pages on which the authors' contributions begin.

Dietmar Benke (93) Institute of Pharmacology and Toxicology, University of Zurich, Zurich, Switzerland

Tyler Best (397) Department of Anatomy, Physiology, and Genetics, School of Medicine, Uniformed Services University of the Health Sciences, Bethesda, Maryland, USA

Bernhard Bettler (231) Department of Biomedicine, Institute of Physiology, Pharmazentrum, University of Basel, Basel, Switzerland

Guy E. Boeckxstaens (287) Department of Gastroenterology, University Hospital Leuven, Catholic University of Leuven, Belgium; Department of Gastroenterology & Hepatology, Academic Medical Center, Amsterdam, The Netherlands

Norman Bowery (1) University of Birmingham Medical School, Edgbaston, Birmingham, UK

Nathan Cramer (397) Department of Anatomy, Physiology, and Genetics, School of Medicine, Uniformed Services University of the Health Sciences, Bethesda, Maryland, USA

John F. Cryan (427) School of Pharmacy, Department of Pharmacology & Therapeutics, Alimentary Pharmabiotic Centre, University College Cork, Cork, Ireland

Svetlana Dzitoyeva (453) The Psychiatric Institute, Department of Psychiatry, University of Illinois, Chicago, Illinois, USA

Steven M. Foord (63) Heptares Therapeutics, Biopark, Welwyn Garden City, Hertfordshire, UK

Wolfgang Froestl (19) AC Immune SA, Lausanne, Switzerland

Zygmunt Galdzicki (397) Department of Anatomy, Physiology, and Genetics, School of Medicine, Uniformed Services University of the Health Sciences, Bethesda, Maryland, USA

Guillermo Gonzalez-Burgos (175) Department of Psychiatry, Translational Neuroscience Program, University of Pittsburgh School of Medicine, Pittsburgh, Pennsylvania, USA

Jörgen Jensen (287) AstraZeneca R&D, Mölndal, Sweden

Masanobu Kano (149) Department of Neurophysiology, Graduate School of Medicine, University of Tokyo, Bunkyo-ku, Tokyo, Japan

Michael M. Kohl (205) Department of Physiology, Development and Neuroscience, University of Cambridge, Cambridge, UK

Anders Lehmann (287) AstraZeneca R&D, Mölndal, Sweden

De-Pei Li (257) Department of Critical Care, The University of Texas, M. D. Anderson Cancer Center, Houston, Texas, USA

Anne Lingford-Hughes (373) Neuropsychopharmacology Unit, Division of Experimental Medicine and Therapeutics, Imperial College London, London, UK

Hari Manev (453) The Psychiatric Institute, Department of Psychiatry, University of Illinois, Chicago, Illinois, USA

Athina Markou (315) Department of Psychiatry, School of Medicine, University of California San Diego, La Jolla, California, USA

Fiona H. Marshall (63) Heptares Therapeutics, Biopark, Welwyn Garden City, Hertfordshire, UK

Stephen J. Moss (113) Department of Neuroscience, Physiology and Pharmacology, University College London, London, UK; Department of Neuroscience, Tufts University School of Medicine, Boston, MA, USA

David J. Nutt (373) Neuropsychopharmacology Unit, Division of Experimental Medicine and Therapeutics, Imperial College London, London, UK

Claire L. Padgett (123) Peptide Biology Laboratory, The Salk Institute for Biological Studies, La Jolla, CA, USA

Hui-Lin Pan (257) Department of Anesthesiology and Perioperative Medicine, The University of Texas, M. D. Anderson Cancer Center, Houston, Texas, USA

Menelas N. Pangalos (113) AstraZeneca, Alderley Park, UK

Ole Paulsen (205) Department of Physiology, Development and Neuroscience, University of Cambridge, Cambridge, UK

Audrée Pinard (231) Department of Biomedicine, Institute of Physiology, Pharmazentrum, University of Basel, Basel, Switzerland

Laurence John Reed (373) Neuropsychopharmacology Unit, Division of Experimental Medicine and Therapeutics, Imperial College London, London, UK

Riad Seddik (231) Department of Biomedicine, Institute of Physiology, Pharmazentrum, University of Basel, Basel, Switzerland

Richard J. Siarey (397) Department of Anatomy, Physiology, and Genetics, School of Medicine, Uniformed Services University of the Health Sciences, Bethesda, Maryland, USA

David A. Slattery (427) Institute of Zoology, University of Regensburg, Regensburg, Germany

Paul A. Slesinger (123) Peptide Biology Laboratory, The Salk Institute for Biological Studies, La Jolla, CA, USA

Marcus Stoffel (397) ETH Zurich, Institute of Molecular Systems Biology, Zurich, Switzerland

Toshihide Tabata (149) Laboratory for Neural Information Technology, Graduate School of Sciences and Engineering, University of Toyama, Toyama, Japan

Miho Terunuma (113) Department of Neuroscience, Tufts University School of Medicine, Boston, MA, USA

Robin J. Tyacke (373) Neuropsychopharmacology Unit, Division of Experimental Medicine and Therapeutics, Imperial College London, London, UK

Styliani Vlachou (315) Department of Psychiatry, School of Medicine, University of California San Diego, La Jolla, California, USA

Preface

Studies on γ-aminobutyric acid (GABA) receptor systems have yielded an abundance of fundamental information on chemical neurotransmission. For example, characterization of the structure and function of the $GABA_A$ receptor provided much of the evidence that ligand-gated ion channels are composed of multiple, interacting subunits. The discovery that the benzodiazepines, one of the most widely used drug classes, enhance $GABA_A$ responsiveness by increasing the sensitivity of this site for endogenous GABA provided conclusive proof that allosteric agents can be safe and effective medications. Moreover, the revelation that the stoichiometry and subunit composition of $GABA_A$ receptors differ throughout the brain, and that the affinity of these sites for orthosteric and allosteric compounds is a function of these properties, points the way for the development of more selective ligand-gated channel receptor agonists and antagonists.

Likewise, the characterization of $GABA_B$ sites has led to a deeper understanding of seven transmembrane receptor systems. Of particular significance was the discovery that $GABA_B$ receptors are heterodimers. While second messenger receptor dimers had been suspected for some time, it was not until the heterodimeric nature of the $GABA_B$ site was identified that direct support was provided for their existence and functionality *in vivo*. This finding energized research in this area, with the results indicating that homo- and heterodimerization is a characteristic of a variety of seven transmembrane sites. The implications of these discoveries for most neurotransmitter systems, either in terms of biological abnormalities responsible for clinical conditions, or for drug discovery, remain to be fully exploited.

Because of the impact $GABA_B$ receptor research has had in the neurosciences it is noteworthy that the discovery and initial characterization of this site can be so easily traced to a single person, Norman Bowery. Rarely is it possible to identify one individual who single-handedly opens an entire new avenue of research. In this case, however, the historical record is clear. In 1980 Norman and his associates were the first to report on a bicuculline-insensitive response to GABA that is mediated by a novel receptor activated by baclofen. Subsequently they christened this site the $GABA_B$ receptor to distinguish it from the well-known bicuculline-sensitive, ligand-gated ion channel receptor, thereafter known as $GABA_A$ (see Chapter 1 by Bowery, this volume).

This monograph was assembled to honor Norman and his work on the 30th anniversary of his discovery. While it is impossible to include reports from all who have contributed significantly to characterizing this receptor, the topics covered provide a comprehensive review of the field and the current state of research in this area. Included are chapters on the chemistry of $GABA_B$ agonists and antagonists, on the molecular composition of the site, its regulation and trafficking, and its role in controlling cellular, autonomic, and behavioral function. There are also offerings describing the potential clinical utility of drugs regulating $GABA_B$ activity. The information contained in this text will be of interest to neuroscientists in general and neuropharmacologists in particular.

It has been our privilege and pleasure to assemble this volume. The task was greatly simplified by the eagerness and enthusiasm of the contributors to prepare their reports out of respect for Norman and his work. We are pleased this effort has yielded not only a fitting tribute to a good friend and colleague, but also an authoritative reference on a topic of contemporary interest and importance to biomedical scientists.

June 1st, 2010
S. J. Enna, Ph.D.
Thomas P. Blackburn, Ph.D.

N. G. Bowery
University of Birmingham Medical School, Edgbaston, Birmingham, UK

Historical Perspective and Emergence of the $GABA_B$ Receptor

Abstract

This chapter forms an introduction to the subsequent chapters in this volume which highlight the significance and potential therapeutic application of $GABA_B$ receptors. It is now 30 years since the $GABA_B$ site was first described in mammalian tissue. Since then much has emerged about its physiological role in the mammalian nervous system and its relationship to other neurotransmitter receptors. It appears to function at pre- and postsynaptic locations as both an auto- and a hetero-receptor where its activation modulates the membrane conductance of Ca^{2+} and K^+. The receptor is

Advances in Pharmacology, Volume 58

1054-3589/10 $35.00
10.1016/S1054-3589(10)58001-3

G-protein coupled and was the first to be shown to exist, possibly in multiple forms, as a heterodimer. The primary agonist for the receptor is baclofen and this continues to be used therapeutically as a centrally active muscle relaxant. Other potential applications for agonists are suggested and positive allosteric modulators may provide an alternative and more effective approach. One application of an agonist, for which there are strong positive clinical data, is in gastroesophageal reflux disease where the receptor target is outside the brain. Antagonists of the $GABA_B$ receptor may also have therapeutic applications such as in cognitive deficits, affective disorders, and absence seizures but robust clinical evidence remains to be demonstrated. Each of these applications is also discussed in the chapters that follow.

I. Introduction

It is truly an honor to be given the opportunity to introduce this volume on the $GABA_B$ receptor. The collection of chapters that have been assembled here provides a really impressive and extensive coverage of the current research status of this G-protein-coupled receptor.

The original research that led up to the discovery of the $GABA_B$ receptor began in the 1970s. This followed on from the many studies that had been performed during the previous decade, establishing GABA as the major inhibitory transmitter in higher centers of the brain (Curtis, 1975; Krnjevic, 1974; Krnjevic & Schwartz, 1967). GABA (γ-amino-*n*-butyric acid) is the single most important inhibitory neurotransmitter within the mammalian brain where it is estimated that 40% of all inhibitory synaptic activity is mediated through this neutral amino acid. It was initially shown to act by increasing the neuronal membrane permeability to Cl^-. Generally the neuronal membrane potential is such that when chloride ion channels are open these negative ions tend to flow in an inward direction hyperpolarizing the cell. But even when the flow is in the outward direction this can still stabilize the cell to reduce its excitability (see Curtis, 1977).

We had noticed that this effect of GABA was not confined to neurons from within the central nervous system (CNS) but also occurred in neurons in the periphery, in particular, neurons of sympathetic and dorsal root ganglia (Bowery & Brown, 1974; Desarmenien et al., 1984). The action of GABA on these neurons was comparable to that observed in the CNS with the production of a selective increase in chloride ion conductance (Adams & Brown, 1975). However, there appeared to be a major difference in the membrane reversal potential in sympathetic neurons when compared to that in neurons of the CNS. In ganglion neurons the reversal potential was observed to be –42 mV whereas in cortical neurons it is normally in the region of –75 mV. Thus, the opening of chloride ion channels in ganglion neurons produces an outward flux of anions whereas the same increase in membrane conductance in higher center neurons generally has the opposite effect. The net effect in ganglion neurons is to produce a membrane depolarization rather than a hyperpolarization.

The same depolarization produced by GABA in primary afferent terminals of the mammalian spinal cord is believed to provide the mechanism whereby inhibitory interneurons control the outflow of afferent transmitter within the dorsal horn (Curtis, 1977). It was this comparative action which stimulated our interest in using the sympathetic ganglion as a model for the action of GABA on primary afferent terminals. If GABA depolarized the terminals of sympathetic fibers in the same manner as the cell bodies, by increasing an efflux of chloride ions, this could be comparable to the endogenous action in the spinal cord.

The first problem that arose was how to examine any depolarization in sympathetic terminals. A direct approach was not possible so we took an indirect approach examining the effect of GABA on the evoked release of the sympathetic transmitter, noradrenaline. After all, it was the inhibition of transmitter release that we were attempting to model. At that time much interest had focused on the uptake and release of noradrenaline from heart tissue so we employed the isolated atrium, in which radiolabeled noradrenaline had been allowed to accumulate, and then examined the effect of GABA on its release evoked by transmural stimulation. As predicted GABA decreased the release without affecting basal release and the effect became more evident in the presence of yohimbine, the presynaptic α-adrenoceptor antagonist (Bowery et al., 1981). This inhibitory effect was subsequently observed in cerebrocortical slices of the rat (Bowery et al., 1980). It was only when we examined the pharmacology of this effect that we realized that a GABA receptor different from the classical type might be involved. The primary criteria which differentiated this novel site were that it was unaffected by the competitive GABA antagonist, bicuculline; it was activated by β-chlorophenyl GABA (baclofen) which is inactive at classical sites for GABA; and it was not mimicked by isoguvacine or muscimol which are both active at ionotropic receptors for GABA. Moreover, the action of GABA at this novel site was shown to be associated with a decrease in neuronal membrane Ca^{2+} conductance and not chloride ion related (Barral et al., 2000; Chen & Van der Pol, 1998). Of course, all of this work preceded any knowledge about receptor structure. It was only in the late 1980s that the structure of the ionotropic GABA receptor was discovered (Levitan et al., 1988; Schofield et al., 1987) and more than 10 years later that the structure of the novel GABA site was first described (Jones et al., 1998; Kaupmann et al., 1998; Kuner et al., 1999; White et al., 1998).

A popular method for characterizing receptors before the advent of the molecular biology approach was the use of high-affinity radiolabeled binding probes. This technique was developed during the 1960s and 1970s when suitable probes became available. The characterization of GABA receptor binding to neuronal membranes was first described by Zukin and colleagues in 1974 but further analysis of this binding failed to indicate the presence of any additional binding sites for the amino acid (Enna & Snyder, 1975, 1977). The only binding site that could be demonstrated was sensitive to bicuculline and isoguvacine but if there was a novel site for GABA this must surely be present as well. In 1980 I felt sure that radiolabeled baclofen must have been prepared in the laboratories of Ciba–Geigy and so I contacted them in Switzerland with a request for a sample for our studies. Eventually after some 9 months or more I was kindly sent a sample together with a much-appreciated letter from Dr Helmut Bittiger. This informed us that they were aware of our studies and this had led them to attempt to obtain a binding assay for ^{3}H-baclofen. Their attempts had failed and all the indications were that a binding site does not exist. They also kindly sent us a resume of the approaches they had employed in trying to obtain displaceable binding. A variety of buffer systems had been tried and this meant that we could immediately look elsewhere. It occurred to us that all the buffer solutions that had been tested were, as usual, nonphysiological so, as we knew we could measure the functional site in Kreb's physiological solution, perhaps we could do the same with the binding site. Even in our first study using neuronal membranes under these conditions, a displaceable binding component could be detected although this was less than 20% of total binding (Bowery et al., 1983; Hill & Bowery, 1981). This enabled us to show the presence of a site that had the same pharmacological characteristics as the functional site in isolated atria and brain slices (Bowery et al., 1980, 1981). It was at this point in 1981 that we designated the term "$GABA_B$" to distinguish this novel site from the classical GABA receptor which we classified as "$GABA_A$" (Hill & Bowery, 1981) and this nomenclature has been retained as the International Union of Basic and Clinical Pharmacology (IUPHAR) terminology (Bowery et al., 2000).

The primary factor, which enabled $GABA_B$ binding to be detected, was dependence on divalent cations such as Ca^{2+} and Mg^{2+} and this was supported by data indicating that functional $GABA_B$ responses were dependent on Ca^{2+} with no dependence on chloride ions, unlike $GABA_A$-mediated responses. The detection of a novel binding site enabled us not only to characterize the receptor but also to demonstrate its distribution within the brain and spinal cord (Price et al., 1984; Wilkin et al., 1981) and to make comparisons with that described for $GABA_A$ sites.

Many electrophysiological studies indicated that $GABA_B$ sites were located presynaptically to control the release of endogenous transmitter as well as on postsynaptic membranes to influence neuronal function directly (Colmers & Pittman, 1989; Dutar & Nicoll, 1988a, 1988b; Mott & Lewis, 1994; Newberry & Nicoll, 1984). The activation of $GABA_B$ receptors decreases membrane Ca^{2+} conductance (Doze et al., 1995; Isaacson, 1998; Wu & Saggau, 1995) but also increases K^{+} conductance. These effects tend, though not exclusively, to be site directed such that the decrease in Ca^{2+} is more associated with presynaptic receptors (Dunlap, 1981) while K^+ effects are predominantly postsynaptic (Luscher et al., 1997; see Deisz, 1997; Premkumar & Gage, 1994). As a consequence of these effects receptor activation can produce neuronal hyperpolarization (K^+ mediated) (Gage, 1992) or a decrease in evoked neurotransmitter release (Ca^{2+} effect).

Both of these events are mediated by G-proteins that are members of the pertussis toxin-sensitive family $G_{i\alpha}/G_{o\alpha}$ (Odagaki & Koyama, 2001; Odagaki et al., 2000). However, presynaptically mediated events, which are generally associated with reduced Ca^{2+} conductance, appear to be insensitive to pertussis toxin (Harrison et al., 1990). The predominant calcium channel linked to $GABA_B$ sites appears to be the "N" type, although "P"- and "Q"-type channels have also been implicated (Barral et al., 2000; Lambert & Wilson, 1996; Santos et al., 1995). Multiple types of K^+ channels seem to be associated with postsynaptic $GABA_B$ receptors (Wagner & Deakin, 1993). A third signaling system associated with $GABA_B$ sites is adenylyl cyclase which is normally inhibited by receptor activation (Xu & Woczik, 1986). However, if the enzyme has been activated by G_s-coupled receptor agonists such as isoprenaline, the β-adrenoceptor ligand, $GABA_B$ receptor activation enhances the formation of cyclic adenosine monophosphate (cAMP) above the level achieved with isoprenaline alone (Hill, 1985; Karbon et al., 1984). This observation prompted a search for receptor subtypes by comparing the abilities of different $GABA_B$ receptor agonists to inhibit or enhance cAMP production. However, no absolute separation has been established (Cunningham & Enna, 1997; Knight & Bowery, 1996).

In 1998 a major advance was made with the discovery of the structure of the $GABA_B$ receptor by three independent groups (Jones et al., 1998; Kaupmann et al., 1998; White et al., 1998) some 10 years after the $GABA_A$ receptor structure was first defined. Naturally there are major differences between the metabotropic ($GABA_B$) and ionotropic ($GABA_A$) receptors and not least of these is that the metabotropic receptor exists as a heterodimer with seven membrane-spanning regions in each of the two components. This contrasts with the five-subunit structure of the $GABA_A$ receptor which forms a pentamer to act as an ion channel selectively allowing the passage of chloride ions.

Information on the characteristics, G-protein coupling and trafficking of the $GABA_B$ receptor heterodimer are provided in many of contributions of this volume. All of these experts have made major contributions to our understanding of the molecular biology of this receptor.

II. $GABA_B$ Receptor Ligands

Prior to the emergence of the detailed structure of the receptor, the primary criterion for establishing the presence of $GABA_B$ receptors was to demonstrate that β-[4-chlorophenyl] GABA (baclofen) was a stereospecific agonist. Subsequently, other selective agonists such as 3-aminopropylphosphinic acid (3-APPA) and 3-aminopropyl-methylphosphinic acid (3-APMPA) were developed which were ~10-fold more potent than *R*-(−)-baclofen. The structure/activity relationship of compounds with affinity for $GABA_B$ sites is discussed in the subsequent presentation by Wolfgang Froestl who has been responsible for the discovery of the majority of the agonists and antagonists at the $GABA_B$ receptor. More recently, Alstermark et al. (2008) have described novel agonists with affinity for $GABA_B$ receptors which do not readily penetrate the CNS. This, they suggest, will have considerable importance in the treatment of gastroesophageal disease. One of these compounds, lesogaberan, has already been successfully tested in human gastroesophageal reflux disease (Bredenoord, 2009). Details of this series of compounds are presented in the contribution by Lehmann, Jensen, and Boeckxstaens.

The first selective $GABA_B$ antagonists to be described were phaclofen, saclofen, and 2-hydroxy saclofen (Kerr & Ong, 1995; Kerr et al., 1987). While these have only low affinity for the receptor, with pK_i values in the micromolar range, they were useful for obtaining the first clear evidence of a physiological role of $GABA_B$ receptors in synaptic transmission within the rat hippocampus (Dutar & Nicoll, 1988a). Subsequent compounds were introduced that were able to enter the brain: CGP 35348, CGP 36742, and SCH50911. However, these too have low receptor affinities. A crucial breakthrough came with the attachment of 3,4-dichlorobenzyl or 3-carboxybenzyl substituents to the existing molecules to produce a variety of compounds with affinities in the low nanometer range such as CGP 55845 and many others, all of which contain a phosphinic acid moiety. These antagonists, when radiolabeled with ^{125}I, provided photoaffinity ligands suitable for the elucidation of the structure of the $GABA_B$ receptor. The group responsible for producing these important tools was also headed by Wolfgang Froestl.

III. Receptor Distribution

$GABA_B$ receptors are distributed throughout the mammalian system and are not confined to nervous tissue (see Erdo & Bowery, 1986). But, of course, while the receptors might be present, their significance only emerges when the agonist is also present. Their presence on autonomic nerve terminals provides an example of this. When activated by GABA, a decrease in the evoked release of transmitter from the terminal occurs. In the enteric nervous system, where GABA-releasing neurons are present and impinge on autonomic cholinergic nerve fibers, $GABA_B$ receptors probably contribute to the control of intestinal movement and sphincters (e.g., Kleinrok & Kilbinger, 1983; Ong & Kerr, 1983, 1990).

$GABA_B$ sites in the mammalian brain appear to be quite widely distributed, although there are regional variations. Receptor autoradiography revealed that the highest receptor densities are in the interpeduncular nucleus, dorsal horn of the spinal cord, the thalamic nuclei, cerebellum (molecular layer), and cerebral cortex (Bowery et al., 1987; Chu et al., 1990). However, moderate to low levels in other brain regions do not necessarily reflect a lack of physiological importance of the receptor. For example, in the hippocampus the overall density of receptors does not match the undoubted significance of $GABA_B$ receptors in neural transmission.

The distribution of the individual components of the receptor heterodimer, $GABA_{B1}$ and $GABA_{B2}$, when detected by immunocytochemistry generally match each other (Charles et al., 2001; Durkin et al., 1999; Margeta-Metrovic et al., 1999). However, in the caudate–putamen, $GABA_{B2}$ is not detectable even though $GABA_{B1}$ and the native receptor are present (Durkin et al., 1999). $GABA_{B1}$ and $GABA_{B2}$ subunits have been detected in peripheral tissue (Calver et al., 2000), but their distribution does not always appear to be coincident, for example, in the uterus and spleen (Calver et al., 2000). In general, there appears to be a paucity of $GABA_{B2}$, suggesting that another protein may form a dimer with $GABA_{B1}$ to facilitate its function.

Perhaps, not surprisingly, $GABA_B$ sites have been demonstrated in lower species as illustrated in the contribution by Manev and Dzitoyeva, indicating that from an evolutionary perspective it is an "old" receptor.

IV. Allosteric Modulation

Allosteric modulation of $GABA_B$ receptors was first described by Urwyler et al. (2001). The location of the modulatory site within the receptor is the heptahelical domain of $GABA_{B2}$. A number of compounds have been reported to be positive modulators. Included are CGP7930 and GS39783 (Urwyler et al., 2003) and BHF177 (Maccioni et al., 2009). While none of these has any direct agonist activity at $GABA_B$ sites, they all accentuate the responses to GABA and baclofen. This effect was firstly demonstrated *in vitro* and was subsequently shown in *in vivo* models of addiction (e.g., Smith et al., 2004 and see present contributions by Vlachou and Markou and Nutt and colleagues) where, for example, both modulators reduce the self-administration of cocaine without the need to administer a receptor agonist. This suggests that positive allosteric modulation may provide an effective therapy avoiding the possible adverse effects of an agonist like baclofen as well as the possible desensitization that frequently accompanies the use of direct acting receptor agonists.

V. Receptor Function

The predominant cellular location of $GABA_B$ receptors appears to be presynaptic where they function as autoreceptors and heteroreceptors to limit the release of a variety of neurotransmitters. Receptor activation produces a reduction in presynaptic membrane Ca^{2+} conductance, which inhibits transmitter release. While the physiological significance of autoreceptors (Davies et al., 1990; Nicoll, 2004) appears obvious because of the local availability of neurotransmitter, the role of heteroreceptors (Nicoll, 2004) on non-GABAergic terminals is not apparent as evidence for axo-axonic contacts at nerve terminals is lacking. Primary afferent terminals in the spinal cord, where GABA interneurons innervate the terminal regions of the primary afferent fibers, are the exception (see Price et al., 1987). Despite the apparent lack of innervation elsewhere, there is evidence for GABA acting in a paracrine manner to activate $GABA_B$ heteroreceptors. GABA released from GABAergic terminals appears to access adjacent terminals where $GABA_B$ sites are present (Isaacson et al., 1993). As the estimated concentration of GABA in the synaptic cleft, following its release, is in the millimolar range, and the affinity of GABA for heteroreceptors is in the nanomolar range, there is sufficient GABA to interact with any sites in close proximity.

VI. Potential Therapeutic Significance

A. Analgesia

$GABA_B$ receptor agonists, including baclofen, are antinociceptive in acute pain models, such as the tail flick and hot plate tests in rodents, at doses below the threshold for muscle relaxation (e.g., Aley & Kulkarni, 1991; Wilson & Yaksh, 1978). Similarly, baclofen and the positive allosteric modulator CGP7930 inhibit visceral pain-related responses to colorectal distension in rats (Brusberg et al., 2009). While $GABA_B$ receptor activation reduces the release of nociceptive transmitter from primary afferent fibers within the dorsal horn of the spinal cord (Malcangio & Bowery, 1996; Riley et al., 2001), the antinociceptive effect might also result from an action within higher centers, in particular the thalamus (Liebman & Pastor, 1980). For example, focal injections of baclofen into the thalamic ventrobasal complex suppress nociceptive processing in chronic inflammation. However, the clinical use of baclofen as an analgesic is limited because of the rapid tolerance that develops as well as the adverse effects of the compound when it is administered systemically. The antinociceptive effect of the GABA uptake inhibitor, tiagabine, in rodents, has been attributed to indirect $GABA_B$ receptor activation produced by an increase in GABA levels within the thalamus (Ipponi et al., 1999). In the contribution by Gonzalez-Burgos, the author considers the possibility that GAT-1, which is inhibited by tiagabine, may be an important determinant in the activation of $GABA_B$ receptors in higher centers.

The significance of $GABA_B$ receptors in pain mechanisms is supported by studies with "knock-out" mice. In these animals, functional $GABA_B$ receptors are not formed, as the mice fail to produce either $GABA_{B1}$ or $GABA_{B2}$ subunits (Schuler et al., 2001). In both forms of mutant mice, hyperalgesia was exhibited in acute nociceptive tests suggesting that functional heteromeric $GABA_B$ receptors are required to increase or maintain pain thresholds.

B. Skeletal Muscle Relaxation

The well-established muscle relaxant properties of baclofen, which are centrally mediated, make it an effective drug in the treatment of spasticity associated with cerebral palsy, multiple sclerosis, stiff-man syndrome, and tetanus. But the side effects associated with its use in such patients, which include seizures, nausea, drowsiness, dizziness, hypotension, muscle weakness, hallucinations, and mental confusion, are often poorly tolerated.

These adverse effects appear due to the need for the administration of high doses because of poor brain penetration. Intrathecal administration via an indwelling cannula connected to a pump inserted into the peritoneal cavity is one approach for addressing this problem. Direct administration to the site of action within the spinal cord requires only small amounts of baclofen, minimizing adverse effects (e.g., Ochs et al., 1989; Orsnes et al., 2000; Penn et al., 1989).

C. Affective Disorders

An upregulation in $GABA_B$-binding sites, which occurs in rat frontal cortex after chronic administration of a variety of antidepressants, was first observed more than 20 years ago (Lloyd et al., 1985; see Enna & Bowery, 2004). While these findings were disputed at the time, there is now solid evidence suggesting that $GABA_B$ mechanisms are associated with depression. Antagonism of $GABA_B$ receptors produces a reversal of depressant-like behavior in animal models of this condition, such as the rodent forced swim test and learned helplessness models (Cryan & Kaupmann, 2005; Nakagawa et al., 1999). Moreover, mice lacking $GABA_{B1}$ or $GABA_{B2}$ receptor subunits exhibit antidepressant-like behavior and anxiety. So $GABA_B$ receptor activation appears to produce anxiolytic activity while a loss or blockade of $GABA_B$ receptor function produces antidepressant-like effects (Mombereau et al., 2004).

D. Drug Addiction

A successful therapeutic treatment for drug dependence is still a major clinical goal, with a number of potential targets under active consideration. Included in this group are dopamine and glutamate, and $GABA_B$ receptors. While baclofen was initially shown to reduce the reinforcing effects of cocaine in rats, it soon became clear that the addictive behavior associated with nicotine, morphine-related agents, and ethanol can also be attenuated by $GABA_B$ agonists (Corrigall et al., 2000; Roberts & Andrews, 1997; Xi & Stein, 1999). This suggests that there might be an underlying common mechanism for the $GABA_B$ agonist, possibly in the ventral tegmental area within the reward center of the mesolimbic system. If this is true, raising endogenous GABA levels within the mesolimbic system should have the same effect as administering a $GABA_B$ agonist. Indeed, central administration of vigabatrin, an inhibitor of GABA metabolism, or NO-711, a GABA uptake inhibitor, into this region attenuates heroin and cocaine self-administration and prevents cocaine-induced increases in dopamine in this brain region (Ashby et al., 1999; Xi & Stein, 1999).

Clinical data indicate that baclofen is also effective against cocaine and alcohol craving in man (Ling et al., 1998). However, as noted above, the administration of baclofen can produce muscle relaxation and other unwanted side effects which could compromise potential benefits. Again, positive allosteric modulators might be a better approach for treating this condition (Smith et al., 2004; Urwyler, 2006). The contribution by Nutt and colleagues elegantly addresses this possible approach for the treatment of drug addiction.

E. Memory

A problem that might arise from the use of a receptor agonist, or even a positive allosteric modulator, is that $GABA_B$ receptor activation suppresses cognitive behavior in animals (DeSousa et al., 1994; Stackman & Walsh, 1994). This action is reversed by $GABA_B$ antagonists. In fact, even basal cognitive activity is enhanced by these antagonists. While this action might limit the use of receptor agonists, it does raise the possibility that $GABA_B$ antagonists might be useful for the clinical treatment of cognitive impairment (Froestl et al., 2004; Genkova-Papazova et al., 2000; John et al., 2009; Lasarge et al., 2009; Sunyer et al., 2008). Evidence suggests that the hippocampus may be the site of action of $GABA_B$ antagonists in relation to cognition. Indeed, an increase in long-term potentiation (LTP) has been implicated in response to $GABA_B$ antagonists, although the nature of this modification appears to depend on the frequency of the stimulation employed to produce LTP (Mott & Lewis, 1994; Olpe & Karlsson, 1990).

F. Absence Epilepsy

Mechanisms underlying the production of epileptic seizures have long been associated with aberrations in the GABA system. Facilitation of $GABA_A$ receptor function using, for example, a benzodiazepine reduces seizure activity, while antagonism of the action of $GABA_A$ receptors enhances it (for overview, see Upton & Blackburn, 1997).

In addition, $GABA_B$ mechanisms might play a role in the extent of seizure production. In temporal lobe epilepsy, for example, impairment of $GABA_B$ receptor function has been noted in cerebrocortical slices obtained from patients undergoing surgery for drug-resistant epilepsy (Teichgräber et al., 2009). Nonetheless, the evidence of a role for $GABA_B$ is greatest in the generation of absence seizures. The occurrence of these seizures is primarily associated with juveniles and often disappears during the late teens. The characteristic electroencephalography (EEG) activity of a 3-Hz spike and wave, which is manifest during an absence attack, stems from discharges in the thalamic nuclei. It has been believed for many years that the thalamus is the site from which these discharges originate and, while an intact thalamocortical network is necessary for generating spike-and-wave discharges, their origin appears to lie outside the thalamus. Studies by Meeren and colleagues (2005) demonstrated unequivocally in a genetic rat model of absence epilepsy (WAG/Rij) that the site of origin is within the perioral region of the somatosensory cortex. These abnormal discharges spread rapidly across the cortex and initiate a corticothalamic cascade. Injection of a $GABA_B$ agonist into the ventrobasal thalamus or reticular nucleus of a rat with spontaneous absence seizures (Genetic Absence Epilepsy Rats from Strasbourg (GAERS)) exacerbates the seizures (Manning et al., 2003). By contrast, injection of a $GABA_B$ antagonist into the same regions suppresses the spike-and-wave discharges. The same results are obtained when a $GABA_B$ antagonist is administered systemically. This might indicate that interference with the GABAergic innervation from the reticular nucleus on the thalamocortical neurons disrupts the thalamocortical loop.

Whether or not $GABA_B$ antagonists prove to be of therapeutic benefit in absence epilepsy, these data clearly indicate that activation of $GABA_B$ receptors in such patients, such as by the administration of baclofen, should be avoided otherwise it may worsen the condition. Likewise, inhibitors of GABA metabolism or transport, such as vigabatrin and tiagabine, should be contraindicated in absence patients as well.

A major problem when attempting to use baclofen for the treatment of CNS-mediated disorders is its poor penetration into the brain and the need to administer high doses, resulting in intolerable side effects. While for patients being treated for spasticity where the drug target is in the spinal cord, the use of an intrathecal indwelling catheter for the administration of small doses has been a successful strategy (see above), this mode of administration is impractical for other indications. Thus, new approaches are needed for minimizing tolerance to agonists and their side effects. The recent introduction of an agonist prodrug is one possible approach. In this regard, arbaclofen placarbil is a an *R*-baclofen prodrug which has better absorption, distribution, metabolism, and elimination properties than the parent drug and, because of this, may prove to be more generally useful (Lal et al., 2009).

G. Applications Outside the Brain

Of course, the use of $GABA_B$ agonists for actions outside the brain would not necessarily have such constraints and, indeed, any reduction in brain penetration would be an advantage. One potential use of a peripherally selective $GABA_B$ agonist is as inhibitor of the gastroesophageal reflux (Lehmann et al., 1999). This possibility has led to the evaluation of lesogaberan, a $GABA_B$ receptor agonist, in the treatment of this condition (Bredenoord, 2009).

Another possible use for peripherally acting $GABA_B$ receptor agonists is in the treatment of pancreatic cancer. It has been shown that the growth of pancreatic ductal adenocarcinoma and pancreatic duct epithelial cells is regulated by β-adrenoreceptors. The activity of β-adrenoreceptors in these cells is counteracted by $GABA_B$ receptor-mediated inhibition of adenylyl

TABLE I Potential Therapeutic Application of $GABA_B$ Receptor Ligands

Agonist/positive allosteric effects	*Locus of action?*	*Therapeutic potential*
Smooth muscle relaxation	Lung (bladder, intestine)	Antiasthma
Antinociception	Spinal cord, thalamus	Analgesia
Food intake modification	Higher centers	Enhanced feeding
Fat intake reduction	Higher centers	Binge eating
Drug addiction suppression	CNS–mesolimbic system	Drug abuse
Muscle relaxation	Spinal cord	Spasticity
Antitussive action	Cough center in medulla	Cough suppressant
Insulin/glucagon release	Pancreas	Diabetes
Suppression of panic behavior	Dorsal periaqueductal gray	Antianxiety/panic disorder
Neutrophil chemotaxis enhanced	Leucocytes	Inflammation
Oesophageal sphincter relaxation	Intestine	Gastroesophageal reflux disease
Adenylyl cyclase inhibition	Pancreas	Antiadenocarcinoma
Antagonist effects		
Cognitive enhancement	Higher centers	Cognitive deficits
Antiabsenceseizures	Thalamocortical region	Absence epilepsy
Antidepressant	Higher centers	Affective disorders

cyclase resulting in a reduction in abnormal cell growth (Schuller et al., 2008). This effect appears to be contrary to that observed in CNS membrane systems in which $GABA_B$ receptor activation enhances β-adrenoreceptor activation of adenylyl cyclase even though basal activity is inhibited (Karbon & Enna, 1985; Xu & Wojcik, 1986).

Other potential applications for $GABA_B$ ligands outside of the CNS might be predicted as the receptor is present in many peripheral organs. This includes the lung, where $GABA_B$ site activation reduces airway constriction in immunosensitized animals (Chapman et al., 1993). This has led to the suggestion that $GABA_B$ receptor agonists may be useful for the treatment of asthma (Table I).

VII. Conclusion

Given the research conducted over the past three decades, there are high expectations for the clinical potential of $GABA_B$ receptor agonists, antagonists, and positive and negative allosteric modulators. This is amply demonstrated by the contents of the present volume and the enthusiasm displayed by these experts in the field. I am very proud to be part of this undertaking and wish to extend my gratitude to Tom Blackburn and Sam Enna for making it happen. I feel privileged to have worked on such an exciting topic that has provided me with so much enjoyment for decades, including the opportunity to establish close personal and professional bonds with so many friends and colleagues, not least of whom are Tom and Sam. Of course, many of the studies that I published were performed in conjunction with PhD and postdoctoral students too numerous to mention. Nonetheless, I want to take this opportunity to thank them all for providing their "part of the jigsaw." All of them have contributed greatly to the project, in terms of both expertise and enthusiasm. So it just remains for me to hope that you enjoy reading the following contributions as much as I.

References

Adams, P. R., & Brown, D. A. (1975). Actions of γ-aminobutyric acid on sympathetic ganglion cells. *Journal of Physiology (London)*, *250*, 85–120.

Aley, K. O., & Kulkarni, S. K. (1991). Baclofen analgesia in mice: A $GABA_B$-mediated response. *Methods and Findings in Experimental and Clinical Pharmacology*, *13*, 681–686.

Alstermark, C., Amin, K., Dinn, S. R., et al. (2008). Synthesis and pharmacological evaluation of novel γ-aminobutyric acid type B ($GABA_B$) receptor agonists as gastroesophageal reflux inhibitors. *Journal of Medicinal Chemistry*, *51*, 4315–4320.

Ashby, C. R., Rohatgi, R., Ngosuwan, J., et al. (1999). Implication of the GABA(B) receptor in gamma vinyl-GABA's inhibition of cocaine-induced increases in nucleus accumbens dopamine. *Synapse*, *31*, 151–153.

Barral, J., Toro, S., Galarraga, E., et al. (2000). GABAergic presynaptic inhibition of rat neostriatal afferents is mediated by Q-type Ca^{2+} channels. *Neuroscience Letters*, *283*, 33–36.

Bowery, N. G., & Brown, D. A. (1974). Depolarising actions of γ-aminobutyric acid and related compounds on rat superior cervical ganglia in vitro. *British Journal of Pharmacology*, *50*, 205–218.

Bowery, N. G., Bettler, B., Froestl, W., et al. (2000). Mammalian gamma-aminobutyric acid (B) receptors: structure and function, *Pharmac Revs*, *54*(2), 247–264.

Bowery, N. G., Doble, A., Hill, D. R., et al. (1981). Bicuculline-insensitive GABA receptors on peripheral autonomic nerve terminals. *European Journal of Pharmacology*, *71*, 53–70.

Bowery, N. G., Hill, D. R., & Hudson, A. L. (1983). Characteristics of $GABA_B$ receptor binding sites on rat whole brain synaptic membranes. *British Journal of Pharmacology*, *78*, 191–206.

Bowery, N. G., Hill, D. R., Hudson, A. L., et al. (1980). (-)Baclofen decreases neurotransmitter release in the mammalian CNS by an action at a novel GABA receptor. *Nature*, *283*, 92–94.

Bowery, N. G., Price, G. W., & Hudson, A. L. (1987). $GABA_A$ and $GABA_B$ receptor site distribution in rat central nervous system. *Neuroscience*, *20*, 365–385.

Bredenoord, A. J. (2009). Lesogaberan, a GABA(B), agonist for the potential treatment of gastroesophageal reflux disease. *IDrugs*, *12*, 576–84.

Brusberg, M., Ravnefjord, A., Martinsson, R., et al. (2009). The GABA(B) receptor agonist, baclofen, and the positive allosteric modulator, CGP7930, inhibit visceral pain-related responses to colorectal distension in rats. *Neuropharmacology*, *56*, 362–367.

Calver, A. R., Medhurst, A. D., Robbins, M. J., et al. (2000). The expression of GABA(B1) and GABA(B2) receptor subunits in the CNS differs from that in peripheral tissues. *Neuroscience*, *100*, 155–170.

Chapman, R. W., Hey, J. A., Rizzo, C. A., et al. (1993). $GABA_B$ receptors in the lung. *Trends in Pharmacological Sciences*, *14*, 26–29.

Charles, K. J., Evans, M. L., Robbins, M. J., et al. (2001). Comparative immunohistochemical localisation of $GABA_{B1a}$, $GABA_{B1b}$, $GABA_{B2}$ subunits in rat brain, spinal cord, and dorsal root ganglion. *Neuroscience*, *106*, 447–467.

Chen, G., & Van der Pol, A. N. (1998). Presynaptic $GABA_B$ autoreceptor modulation of P/Q-type calcium channels and GABA release in rat suprachiasmatic nucleus neurons. *Journal of Neuroscience*, *18*, 1913–1922.

Chu, D.C.M., Albin, R. L., Young, A. B., et al. (1990). Distribution and kinetics of $GABA_B$ binding sites in rat central nervous system: A quantitative autoradiographic study. *Neuroscience*, *34*, 341–357.

Colmers, W. F., & Pittman, Q. J. (1989). Presynaptic inhibition by neuropeptide Y and baclofen in hippocampus: Insensitivity to pertussis toxin treatment. *Brain Research*, *498*, 99–104.

Corrigall, W. A., Coen, K. M., Adamson, K. L., et al. (2000). Response of nicotine self-administration in the rat to manipulations of mu-opioid and gamma-aminobutyric acid receptors in the ventral tegmental area. *Psychopharmacology*, *149*, 107–114.

Cryan, J. F., & Kaupmann, K. (2005). Don't worry "B" happy!: A role for $GABA_B$ receptors in anxiety and depression. *Trends in Pharmacological Sciences*, *26*, 36–43.

Cunningham, M., & Enna, S. J. (1997). Cellular and biochemical responses to $GABA_B$ receptor activation. In S. J. Enna & N. G. Bowery (Eds.), *The GABA receptors* (pp. 237–258). Totowa, NJ: Humana Press.

Curtis, D. R. (1975). Gamma-aminobutyric and glutamic acids as mammalian central transmitters. In S. Berl, D. D. Clarke, & D. Schneider (Eds.), *Metabolic compartmentation and neurotransmission* (pp. 11–36). New York: Plenum Press.

Curtis, D. R. (1977). Pre- and non-synaptic activities of GABA and related amino acids in the mammalian nervous system. In F. Fonnum (Ed.), *Amino acids as chemical transmitters* (pp. 55–86). New York: Plenum Press.

Davies, C. H., Davies, S. N., & Collingridge, G. L. (1990). Paired-pulse depression of monosynaptic GABA-mediated inhibitory postsynaptic responses in rat hippocampus. *Journal of Physiology*, *424*, 513–531.

Deisz, R. A. (1997). Electrophysiology of $GABA_B$ receptors. In S. J. Enna & N. G. Bowery (Eds.), *The GABA receptors* (pp. 157–207). Totowa, NJ: Humana Press.

Desarmenien, M., Feltz, P., Occhipinti, G., et al. (1984). Coexistence of $GABA_A$ and $GABA_B$ receptors on A and C primary afferents. *British Journal of Pharmacology*, *81*, 327–333.

DeSousa, N. J., Beninger, R., Jhamandas, K., et al. (1994). Stimulation of $GABA_B$ receptors in the basal forebrain selectivity impairs working memory of rats in the double Y-maze. *Brain Research*, *641*, 29–38.

Doze, V. A., Cohen, G. A., & Madison, D. V. (1995). Calcium channel involvement in $GABA_B$ receptor-mediated inhibition of GABA release in area CA1 of the rat hippocampus. *Journal of Neurophysiology*, *74*, 43–53.

Dunlap, K. (1981). Two types of γ-aminobutyric acid receptor on embryonic sensory neurons. *British Journal of Pharmacology*, *74*, 59–585.

Durkin, M. M., Gunwaldsen, C. A., Borowsky, B., et al. (1999). An in situ hybridization study of the distribution of the GABA(B2) protein mRNA in the rat CNS. *Molecular Brain Research*, *71*, 185–200.

Dutar, P., & Nicoll, R. A. (1988a). A physiological role for $GABA_B$ receptors in the central nervous system. *Nature*, *332*, 156–158.

Dutar, P., & Nicoll, R. A. (1988b). Pre- and postsynaptic $GABA_B$ receptors in the hippocampus have different pharmacological properties. *Neuron*, *1*, 585–591.

Enna, S. J., & Bowery, N. G. (2004). $GABA_B$ receptor alterations as indicators of physiological and pharmacological function. *Biochemical Pharmacology*, *68*, 1541–1548.

Enna, S. J., & Snyder, S. (1975). Properties of γ-aminobutyric acid (GABA) receptor binding in rat brain synaptic membrane fractions. *Brain Research*, *100*, 81–97.

Enna, S. J., & Snyder, S. (1977). Influences of ions, enzymes, and detergents on γ-aminobutyric acid receptor binding in synaptic membranes of rat brain. *Molecular Pharmacology*, *13*, 442–453.

Erdo, S. L., & Bowery, N. G. (Eds.). (1986). *GABAergic mechanisms in the mammalian periphery*. New York: Raven.

Froestl, W., Gallagher, M., Jenkins, H., et al. (2004). SGS: 742: The first $GABA_B$ receptor antagonist in clinical trials. *Biochemical Pharmacology*, *68*, 1479–1487.

Gage, P. W. (1992). Activation and modulation of neuronal K^+ channels by GABA. *Trends in Neurosciences*, *15*, 46–51.

Genkova-Papazova, M. G., Petkova, B., Shishkova, N., et al. (2000). The GABA-B antagonist CGP 36742 prevent PTZ-kindling-provoked amnesia in rats. *European Neuropsychopharmacology*, *10*, 273–278.

Harrison, N. L., Lambert, N. A., & Lovinger, D. M. (1990). Presynaptic $GABA_B$ receptors on rat hippocampal neurons. In N. G. Bowery, H. Bittiger, & H.-R. Olpe (Eds.), *$GABA_B$ receptors in mammalian function* (pp. 208–221). Chichester: Wiley.

Hill, D. R. (1985). $GABA_B$ receptor modulation of adenylate cyclase activity in rat brain slices. *British Journal of Pharmacology*, *84*, 249–257.

Hill, D. R., & Bowery, N. G. (1981). ^{3}H-Baclofen and ^{3}H-GABA bind to bicuculline-insensitive GABAB sites in rat brain. *Nature*, *290*, 149–152.

Ipponi, A., Lamberti, C., Medica, A., et al. (1999). Tiagabine antinociception in rodents depends on $GABA_B$ receptor activation: Parallel antinociceptive testing and medial thalamus GABA microdialysis. *European Journal of Pharmacology*, *368*, 205–211.

Isaacson, J. S. (1998). $GABA_B$ receptor-mediated modulation of presynaptic currents and excitatory transmission at a fast central synapse. *Journal of Neurophysiology*, *80*, 1571–1576.

Isaacson, J. S., Solis, J. M., & Nicoll, R. A. (1993). Local and diffuse synaptic actions of GABA in the hippocampus. *Neuron*, *10*, 165–175.

John, J. P., Sunyer, B., Höger, H., et al. (2009). Hippocampal synapsin isoform levels are linked to spatial memory enhancement by SGS742. *Hippocampus*, *19*, 731–738.

Jones, K. A., Borowsky, B., Tamm, J. A., et al. (1998). $GABA_B$ receptors function as a heteromeric assembly of the subunits $GABA_BR1$ and $GABA_BR2$. *Nature*, *396*, 674–679.

Karbon, E. W., & Enna, S. J. (1985). Characterization of the relationship between gamma-aminobutyric acid B agonists and transmitter-coupled cyclic nucleotide-generating systems in rat brain. *Molecular Pharmacology*, *27*(1), 53–59.

Karbon, E. W., Duman, R. S., & Enna, S. J. (1984). $GABA_B$ receptors and norepinephrine-stimulated cAMP production in rat brain cortex. *Brain Research*, *306*, 327–332.

Kaupmann, K., Malitschek, B., Schuler, V., et al. (1998). $GABA_B$-receptor subtypes assemble into functional heteromeric complexes. *Nature*, *396*, 683–687.

Kerr, D.I.B., & Ong, J. (1995). $GABA_B$ receptors. *Pharmacology & Therapeutics*, *67*, 187–246.

Kerr, D.I.B., Ong, J., Prager, R. H., et al. (1987). Phaclofen: A peripheral and central baclofen antagonist. *Brain Research*, *405*, 150–154.

Kleinrok, A., & Kilbinger, H. (1983). γ-Aminobutyric acid and cholinergic transmission in the guinea-pig ileum. *Naunyn-Schmiedeberg's Archives of Pharmacology*, *322*, 216–220.

Knight, A. R., & Bowery, N. G. (1996). The pharmacology of adenylyl cyclase modulation by $GABA_B$ in rat brain slices. *Neuropharmacology*, *35*, 703–712.

Krnjevic, K. (1974). Chemical nature of synaptic transmission in vertebrates. *Physiological Reviews*, *54*, 418–540.

Krnjevic, K., & Schwartz, S. (1967). The action of γ-aminobutyric acid on cortical neurons. *Experimental Brain Research*, *3*, 320–336.

Kuner, R., Kohr, G., Grunewald, S., et al. (1999). Role of heteromer formation in $GABA_B$ receptor function. *Science*, *283*, 74–77.

Lal, R., Sukbuntherng, J., Tai, E. H., et al. (2009). Arbaclofen placarbil, a novel R-baclofen prodrug: Improved absorption, distribution, metabolism, and elimination properties compared with R-baclofen. *Journal of Pharmacology and Experimental Therapeutics*, *330*, 911–921.

Lambert, N. A., & Wilson, W. A. (1996). High-threshold Ca^{2+} currents in rat hippocampal interneurones and their selective inhibition by activation of $GABA_B$ receptors. *Journal of Physiology*, *492*, 115–127.

Lasarge, C. L., Bañuelos, C., Mayse, J. D., et al. (2009). Blockade of GABA(B) receptors completely reverses age-related learning impairment. *Neuroscience*, *164*, 941–947.

Lehmann, A., Antonsson, M., Bremner-Danielsen, M., et al. (1999). Activation of the $GABA_B$ receptor inhibits transient lower esophageal sphincter relaxations in dogs. *Gastroenterology*, *117*, 1147–1154.

Levitan, E. S., Scofield, P. R., Burt, D. R., et al. (1988). Structural and functional basis for $GABA_A$ receptor heterogeneity. *Nature*, *335*, 76–77.

Liebman, J. M., & Pastor, G. (1980). Antinociceptive effects of baclofen and muscimol upon intraventricular administration. *European Journal of Pharmacology*, *61*, 225–230.

Ling, W., Shoptaw, S., & Majewska, D. (1998). Baclofen as an anti-craving medication: A preliminary clinical study. *Neuropsychopharmacology*, *18*, 403–404.

Lloyd, K. G., Thuret, F., & Pilc, A. (1985). Upregulation of gamma-aminobutyric acid ($GABA_B$) binding sites in rat frontal cortex: A common action of repeated administration of different classes of antidepressant and electroshock. *Journal of Pharmacology and Experimental Therapeutics*, *235*, 191–199.

Luscher, C., Jan, L. Y., & Stoffel, M., et al. (1997). G-protein-coupled inwardly rectifying K^+ channels (GIRKs) mediate postsynaptic but not presynaptic transmitter actions in hippocampal neurons. *Neuron*, *19*, 687–695.

Maccioni, P., Carai, M. A., Kaupmann, K., et al. (2009). Reduction of alcohol's reinforcing and motivational properties by the positive allosteric modulator of the GABAB receptor, BHF177, in alcohol-preferring rats. *Alcoholism, Clinical and Experimental Research*, *33*, 1749–56.

Malcangio, M., & Bowery, N. G. (1996). GABA and its receptors in the spinal cord. *Trends in Pharmacological Sciences*, *17*, 457–462.

Manning, J.-P., Richards, D. A., & Bowery, N. G. (2003). Pharmacology of absence epilepsy. *Trends in Pharmacological Sciences*, *24*, 542–549.

Margeta-Mitrovic, M., Mitrovic, I., Riley, R. C., et al. (1999). Immunohistochemical localization of GABA(B) receptors in the rat central nervous system. *Journal of Comparative Neurology*, *405*, 299–321.

Meeren, H., van Luitelaar, G., Lopes da Silva, F., et al. (2005). Evolving concepts on the pathophysiology of absence seizures. *Archives of Neurology*, *62*, 371–376.

Mombereau, C., Kaupmann, K., Froestl, W., et al. (2004). Genetic and pharmacological of a role for GABAB receptors in the modulation of anxiety- and antidepressant-like behavior. *Neuropsychopharmacology*, *29*, 1050–1062.

Mott, D. D., & Lewis, D. L. (1994). The pharmacology and function of $GABA_B$ receptors. *International Review of Neurobiology*, *36*, 97–223.

Nakagawa, Y., Sasaki, A., & Takashima, T. (1999). The $GABA_B$ receptor antagonist CGP36742 improves learned helplessness in rats. *European Journal of Pharmacology*, *381*, 1–7.

Newberry, N. R., & Nicoll, R. A. (1984). A bicuculline-resistant inhibitory post-synaptic potential in rat hippocampal pyramidal cells in vitro. *Journal of Physiology*, *360*, 161–185.

Nicoll, R. A. (2004). My close encounter with $GABA_B$ receptors. *Biochemical Pharmacology*, *68*, 1667–1674.

Ochs, G., Struppler, A., Meyerson, B. A., et al. (1989). Intrathecal baclofen for long term treatment of spasticity: A multicentre study. *Journal of Neurology, Neurosurgery, and Psychiatry*, *52*, 933–939.

Odagaki, Y., & Koyama, T. (2001). Identification of G alpha subtype(s) involved in gamma-aminobutyric acid(B) receptor-mediated high-affinity guanosine triphosphatase activity in rat cerebral cortical membranes. *Neuroscience Letters*, *297*, 137–141.

Odagaki, Y., Nishi, N., & Koyama, T. (2000). Functional coupling of GABA(B) receptors with G proteins that are sensitive to *N*-ethylmaleimide treatment, suramin and benzalkonium chloride in rat cerebral cortical membranes. *Journal of Neural Transmission*, *107*, 1101–1116.

Olpe, H.-R., & Karlsson, G. (1990). The effects of baclofen and two $GABA_B$ receptor antagonists on long term potentiation. *Naunyn-Schmiedeberg's Archives of Pharmacology*, *342*, 194–197.

Ong, J., & Kerr, D. I. B. (1983), $GABA_A$- and $GABA_B$- receptor-mediated modification of intestinal motility. *European Journal of Pharmacology*, *86*, 9–17.

Ong, J., & Kerr, D. I. B. (1990). GABA receptors in peripheral tissues. *Life Sciences*, *46*, 489–1501.

Orsnes, G. B., Sorensen, P. S., Larsen, T. K., et al. (2000). Effect of baclofen on gait in spastic MS patients. *Acta Neurologica Scandinavica*, *101*, 244–248.

Penn, R. D., Savoy, S. M., Corcos, D., et al. (1989). Intrathecal baclofen for severe spinal spasticity. *New England Journal of Medicine*, *320*, 1517–1521.

Premkumar, L. S., & Gage, P. W. (1994). Potassium channels activated by $GABA_B$ agonists and serotonin in cultured hippocampal neurons. *Journal of Neurophysiology*, *71*, 2570–2575.

Price, G. W., Kelly, J. S., & Bowery, N. G. (1987). The location of GABAB receptor binding sites in mammalian spinal cord. *Synapse*, *1*, 530–538.

Price, G. W., Wilkin, G. P., Turnbull, M. J., et al. (1984). Are baclofen-sensitive $GABA_B$ receptors present on primary afferent terminals of the spinal cord? *Nature*, *307*, 71–74.

Riley, R. C., Trafton, J. A., Chi, S. I., et al. (2001). Presynaptic regulation of spinal cord tachykinin signaling via GABA(B) but not GABA(A) receptor activation. *Neuroscience*, *103*, 725–737.

Roberts, D.C.S., & Andrews, M. M. (1997). Baclofen suppression of cocaine self-administration: Demonstration using a discrete trials procedure. *Psychopharmacology*, *131*, 271–277.

Santos, A. E., Carvalho, C. M., Macedo, T. A., et al. (1995). Regulation of intracellular [Ca^{2+}] GABA release by presynaptic $GABA_B$ receptors in rat cerebrocortical synaptosomes. *Neurochemistry International*, *27*, 397–406.

Schuler, V., Luscher, C., Blanchet, C., et al. (2001). Epilepsy, hyperalgesia, impaired memory, and loss of pre- and post-synaptic $GABA_B$ responses in mice lacking $GABA_{B1}$. *Neuron*, *31*, 47–58.
Schuller, H. M., Al-Wadei, H. A., Majidi, M. (2008). $GABA_B$ receptor is a novel drug target for pancreatic cancer. *Cancer*, *112*, 767–78.
Schofield, P. R., Darlison, M. G., Fujita, N., et al. (1987). Sequence and functional expression of the $GABA_A$ receptor shows a ligand-gated receptor superfamily. *Nature*, *328*, 221–227.
Smith, M. A., Yancey, D. L., Morgan, D., et al. (2004). Effects of positive allosteric modulators of the $GABA_B$ receptor on cocaine self-administration in rats. *Psychopharmacology*, *173*, 105–111.
Stackman, R. J., & Walsh, T. J. (1994). Baclofen produces dose related working memory impairments after intraseptal injection. *Behavioral and Neural Biology*, *61*, 181–185.
Sunyer, B., Diao, W. F., & Kang, S. U. (2008). Cognitive enhancement by SGS742 in OF1 mice is linked to specific hippocampal protein expression. *Journal of Proteome Research*, *7*, 5237–5253.
Teichgräber, L. A., Lehmann, T. N., Meencke, H. J., et al. (2009). Impaired function of $GABA_B$ receptors in tissues from pharmacoresistant epilepsy patients. *Epilepsia*, doi: 10.1111/j.1528-1167.2009.
Upton, N., & Blackburn, T. (1997). Pharmacology of mammalian $GABA_A$ receptors. In S. J. Enna & N. G. Bowery (Eds.), *The GABA receptors* (pp. 83–120). Totowa, NJ: Humana Press.
Urwyler, S. (2006). Allosteric modulation of $GABA_B$ receptors. In N. G. Bowery (Ed.), *Allosteric receptor modulation in drug targeting* (pp. 235–257). New York: Taylor & Francis.
Urwyler, S., Mosbacher, J., Lingenhoehl, K. et al. (2001). Positive allosteric modulation of native and recombinant γ-aminobutyric $acid_B$ receptors by 2,6-di-tert-butyl-4-(3-hydroxy-2,2-dimethyl-propyl)-phenol(CGP7930) and its aldehyde analog CGP13501. *Molecular Pharmacology*, *60*, 963–971.
Urwyler, S., Pozza, M. F., Lingenhoehl, K., et al. (2003). *N,N′*-Dicyclopentyl-2-methlsulpfanyl-5-nitro-pyrimidine-4,6-diamine(GS39783) and structurally related compounds: Novel allosteric enhancers of γ-aminobutyric $acid_B$. *Journal of Pharmacology and Experimental Therapeutics*, *307*, 322–330.
Wagner, P. G., & Deakin, M. S. (1993). $GABA_B$ receptors are couple to a barium-insensitive outward rectifying potassium conductance in premotor respiratory neurons. *Journal of Neurophysiology*, *69*, 286–289.
White, J. H., Wise, A., Main, M., et al. (1998). Heterodimerization is required for the formation of a functional $GABA_B$ receptor. *Nature*, *396*, 679–682.
Wilkin, G. P., Hill, D. R., Hudson, A. L., et al. (1981). Autoradiographic localisation of $GABA_B$ receptors in rat cerebellum. *Nature*, *294*, 584–597.
Wilson, P. R., & Yaksh, T. L. (1978). Baclofen is anti-nociceptive in the intrathecal space of animals. *European Journal of Pharmacology*, *51*, 323–330.
Wu, L. G., & Saggau, P. (1995). $GABA_B$ receptor-mediated presynaptic inhibition in guinea-pig hippocampus is caused by reduction of presynaptic Ca^{2+} influx. *Journal of Physiology*, *485*, 649–657.
Xi, Z. X., & Stein, E. A. (1999). Baclofen inhibits heroin self-administration behavior and mesolimbic dopamine release. *Journal of Pharmacology and Experimental Therapeutics*, *290*, 1369–1374.
Xu, J., & Wojcik, W. J. (1986). Gamma aminobutyric acid B receptor-mediated inhibition of adenylate cyclase in cultured cerebellar granule cells: Blockade by islet-activating protein. *Journal of Pharmacology and Experimental Therapeutics*, *239*, 568–573.
Zukin, S., Young, A., Snyder, S., et al. (1974). Gamma-aminobutyric acid binding to receptor sites in the rat central nervous system. *Proceedings of the National Academy of Sciences of the United States of America*, *71*, 4802–4807.

Wolfgang Froestl
AC Immune SA, Lausanne, Switzerland

Chemistry and Pharmacology of GABA$_B$ Receptor Ligands

Abstract

This chapter presents new clinical applications of the prototypic GABA$_B$ receptor agonist **baclofen** for the treatment of addiction by drugs of abuse, such as alcohol, cocaine, nicotine, morphine, and heroin, a novel baclofen prodrug **Arbaclofen placarbil**, the GABA$_B$ receptor agonist **AZD3355** (**Lesogabaran**) currently in Phase 2 clinical trials for the treatment of gastro-esophageal reflux disease, and four positive allosteric modulators of GABA$_B$ receptors (**CGP7930, GS39783, NVP-BHF177,** and **BHFF**), which have less propensity for the development of tolerance due to receptor desensitization than classical GABA$_B$ receptor agonists. All four compounds showed anxiolytic affects. In the presence of positive allosteric modulators the

Advances in Pharmacology, Volume 58

1054-3589/10 $35.00
10.1016/S1054-3589(10)58002-5

"classical" $GABA_B$ receptor antagonists **CGP35348** and **2-hydroxy-saclofen** showed properties of partial $GABA_B$ receptor agonists. Seven micromolar affinity $GABA_B$ receptor antagonists, **phaclofen; 2-hydroxy-saclofen; CGP's 35348, 36742, 46381, 51176;** and **SCH50911,** are discussed. **CGP36742 (SGS742)** showed statistically significant improvements of working memory and attention in a Phase 2 clinical trial in mild, but not in moderate Alzheimer patients. Eight nanomolar affinity $GABA_B$ receptor antagonists are presented (**CGP's 52432, 54626, 55845, 56433, 56999, 61334, 62349,** and **63360**) that were used by pharmacologists for numerous *in vitro* and *in vivo* investigations. **CGP's 36742, 51176, 55845, and 56433** showed antidepressant effects. Several compounds are also available as **radioligands,** such as **$[^3H]$CGP27492, $[^3H]$CGP54626, $[^3H]$CGP5699,** and **$[^3H]$ CGP62349.** Three novel fluorescent and three $GABA_B$ receptor antagonists with very high specific radioactivity (>2,000 Ci/mmol) are presented. **$[^{125}I]$ CGP64213** and the photoaffinity ligand **$[^{125}I]$CGP71872** allowed the identification of $GABA_{B1a}$ and $GABA_{B1b}$ receptors in the expression cloning work.

I. Introduction

Three years have passed since the author's review "Chemistry of $GABA_B$ Modulators" appeared in the book *The GABA Receptors*, 3rd edition, edited by S. J. Enna and H. Möhler (Froestl et al., 2007). In this chapter I wish to outline new findings not covered in the 2007 paper and will add information which has not been covered in the 1997 and 2007 overviews (Froestl & Mickel, 1997).

Norman G. Bowery discovered a novel GABA receptor in January 1980, when he found that GABA at a concentration of 4 μM decreased the release of $[^3H]$-noradrenaline from rat atria and of $[^3H]$-acetylcholine from preganglionic terminals in the rat superior cervical ganglion *in vitro* (Bowery et al., 1980). These effects could not be antagonized by the established GABA antagonist bicuculline. Bowery showed that the GABA analogue baclofen was as active as GABA in reducing evoked transmitter output and that the effect was stereoselective with the (*R*)-(−)-enantiomer being >100-fold more active than the (*S*)-(+)-enantiomer. The term $GABA_B$ receptor was designated in March 1981 (Hill & Bowery, 1981; see also the review: Bowery, 1982).

II. $GABA_B$ Receptor Agonists

Baclofen, Ba-34647, was synthesized in September 1962 by Heinrich Keberle of Ciba, Basel, Switzerland, based on the idea to enhance the

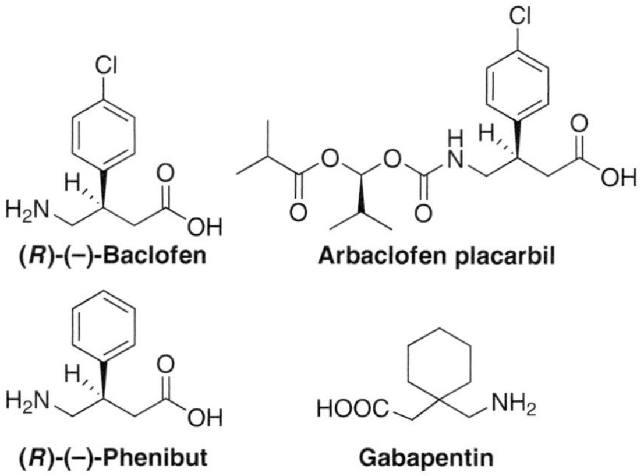

FIGURE 1 Structures of (*R*)-(−)-baclofen, its prodrug, (*R*)-(−)-Phenibut, and Gabapentin.

lipophilicity of GABA (calculated log P = −2.13; Cates, 1985) in order to achieve penetration of the blood–brain barrier (BBB; Keberle, Faigle, & Wilhelm, 1968). The lipophilicity of baclofen with a logD of −0.96 (Leisen et al., 2003) is still insufficient to bring the compound into the brain by passive diffusion. However, baclofen is transported into the brain by interaction with the large neutral amino acid transporter (Audus & Borchardt, 1986; van Bree et al., 1988, 1991). Conversely, the restricted distribution of baclofen in the brain interstitial fluid is due to an efficient efflux from the brain through the BBB regulated by a probenecid-sensitive organic anion transporter (Deguchi et al., 1995).

In 1975 William Bencze resolved baclofen into the two enantiomers, (*R*)-(−)-baclofen (CGP11973A; Fig. 1) and (*S*)-(+)-baclofen (CGP11974A). Racemic and (*R*)-(−)-baclofen depressed the patellar, flexor, linguomandibular, and the H-reflex in cats in a dose-dependent fashion (Olpe et al., 1978). Binding studies confirmed these findings: the IC_{50} values for (*R*)-(−)-baclofen, (*S*)-(+)-baclofen, and racemic baclofen for the inhibition of binding of $[^3H]$-baclofen to $GABA_B$ receptors of cat cerebellum are 15 nM, 1,770 nM, and 35 nM, respectively (Froestl et al., 1995a). The crystal structure of (*R*)-(−)-baclofen hydrochloride was published by Chang et al. (1982).

A. Clinical Use of Baclofen

Baclofen was marketed as **Lioresal** in 1972 and since then has been shown to exert beneficial clinical effects in many diverse indications:

1. As **muscle relaxant in spasticity** for patients with hemi- and tetraplegia (Brogden et al., 1974) including spasticity in **multiple sclerosis** (Brar et al., 1991; Giesser, 1985; Smith et al., 1991), after stroke (O'Brien

et al., 1996), associated with Lewy body dementia (Moutoussis & Orrell, 1996), and in **cerebral palsy** in children (Buonaguro et al., 2005; Scheinberg et al., 2006) and adults (Krach, 2009).

2. A very important breakthrough was achieved by administration of baclofen directly into the CSF via an implanted mini-pump, that is, **intrathecal baclofen.** Penn and coworkers achieved spectacular improvements of patients, some of whom were bedridden for up to 19 years, who regained their ability to walk (Kroin, 1992; Penn & Kroin, 1984, 1985, 1987). Many eminent neurologists took up this new method of treatment immediately (Becker et al., 1997; Becker et al., 1995; Dralle et al., 1985; Gianino et al., 1998; Ochs, 1993; Ochs et al., 1989; Rawlins, 1998). Quantitative measurements by Leisen et al. (2003) showed that intrathecal administration of doses up to 600 μg resulted in local plasma levels of 5–20 ng/g. In comparison, oral racemic baclofen at the very high dose of 100 mg gave concentrations of <3 ng/g in CSF. An additional benefit was a major reduction in the unwanted side effects with high doses required for systemic administration.
3. A welcome side effect of baclofen treatment is the marked **improvement of bladder function** in patients with spinal cord lesions (Frost et al., 1989; Nanninga et al., 1989; Rapidi et al., 2007; Taylor & Bates, 1979). Baclofen significantly reduced the K^+-evoked calcitonin gene-related peptide-like immunoreactivity (CGRP-LI) in guinea pig urinary bladder (Santicioli et al., 1991).
4. Oral and intrathecal baclofen is also the treatment of choice in **stiff-man syndrome** (Silbert et al., 1995; Stayer et al., 1997; Whelan, 1980). A patient bedridden for 3 years was able to walk on the 13th day of treatment with 90 mg baclofen daily (Miller & Korsvik, 1981).
5. And in **tetanus** (Boots et al., 2000; Demaziére et al., 1991; Engrand & Benhamou, 2001; Engrand et al., 1999; Müller et al., 1987; Santos et al., 2004).
6. Baclofen is an efficient medication in several manifestations of pain, such as in **trigeminal neuralgia.** Pioneered by Fromm et al., baclofen either alone or in combination with carbamazepine or phenytoin achieved substantial pain relief (Baker et al., 1985; Fromm, 1994; Fromm et al., 1980, 1984). Interestingly, Fromm observed that (*R*)-(−)-baclofen was 5 times more potent than racemic baclofen (Fromm & Terrence, 1987; Terrence et al., 1983). This may relate to papers by Sawynok and Dickson (1984, 1985) showing that (*S*)-(+)- = d-baclofen may act as an antagonist at baclofen receptors. This question has never been resolved.
7. Baclofen was effective in the treatment of **cluster headache** (Hering-Hanit & Gadoth, 2000, 2001) and in the prevention of **migraine** (Hering-Hanit, 1999). A very recent paper reveals functional

redundancy of $GABA_B$ receptors in peripheral nociceptors *in vivo*, which led to the conclusion that $GABA_B$ receptors in the peripheral nervous system play a less important role than those in the CNS in the regulation of pain (Gangadharan et al., 2009). Generally it is believed that baclofen exerts the analgesic properties by reduction of the release of L-glutamate, substance P, and calcitonin gene-related peptide (CGRP) from primary afferent terminals (Enna & McCarson, 2006; Kaupmann et al., 2001).

8. Baclofen is effective in reducing the craving for **alcohol.** Extensive clinical trials look promising even in patients suffering from *delirium tremens* (Addolorato et al., 2000, 2002a, 2002b, 2003, 2005, 2006a, 2006b, 2007, 2009; Evans & Bisaga, 2009; Flannery et al., 2004; Leggio, 2009).

   Many preclinical studies in experimental animals preceded the clinical trials (Colombo et al., 2002, 2004, 2006; Federici et al., 2009; Holstein et al., 2009; Knapp et al., 2007; Maccioni et al., 2005, 2008a, 2008b).

9. Baclofen effectively reduces the craving for **cocaine**. Clinical studies were started as early as 1998 (Haney et al., 2006; Kaplan et al., 2004; Lile et al., 2004; Ling et al., 1998; Shoptaw et al., 2003). Further studies are planned to evaluate baclofen's potential as a relapse prevention agent (Kahn et al., 2009). A drug combination of 30 mg t.i.d. baclofen with 100 mg t.i.d. amantadine has been evaluated by Rotheram-Fuller et al. (2007). The self-rated desire for cocaine in the drug-treated patients was significantly lower during cocaine administrations than in the placebo group.

   Many preclinical studies in experimental animals preceded these clinical trials (Cousins et al., 2002; Fadda et al., 2003; Filip & Frankowska, 2007, 2008; Hotsenpiller & Wolf, 2003; Weerts et al., 2005a; Weerts et al., 2007).

10. The results of the first clinical trial of baclofen to reduce craving for **nicotine** have been published in 2009 demonstrating a reduction of the number of cigarettes smoked per day and a significant lowering of craving (Franklin et al., 2009).

    Many preclinical studies in experimental animals preceded the clinical trial (Fattore et al., 2009; Le Foll et al., 2008; Levin et al., 2004; Markou, 2008; Markou et al., 2004; Paterson et al., 2004). The mechanism of action may be due to a reduction of release of dopamine by baclofen (Amantea & Bowery, 2004).

11. One clinical trial showed promising results that baclofen may be a novel therapeutic agent for **opiate withdrawal syndrome** (Akhondzadeh et al., 2000). Preclinical information is available for the attenuation of **morphine** withdrawal signs (Bartoletti et al., 2007; Bexis et al., 2001; Heinrichs et al., 2010; Leite-Morris et al., 2004;

Riahi et al., 2009; Sahraei et al., 2009; Zarrindast et al., 2008) and for the prevention of **heroin**-seeking behavior (Spano et al., 2007; Xi & Stein, 1999). Also symptoms of **benzodiazepine withdrawal** can be improved by baclofen (File et al., 1991).

12. Baclofen therapy for chronic **posttraumatic stress disorder** was effective in treating both the PTSD symptoms and the accompanying depression and anxiety (Drake et al., 2003). The number of **panic attacks** was significantly reduced by baclofen treatment (Breslow et al., 1989).
13. Peripheral $GABA_B$ receptors are involved in the actions of baclofen on **gastroesophageal reflux disease (GERD).** Preclinical findings in dogs (Lehmann, 2008, 2009; Lehmann et al., 2000; Lehmann et al., 1999) and ferrets (Blackshaw et al., 1999) paved the way for the first clinical trials (Cange et al., 2002; Koek et al., 2003; Lidums et al., 2000; Omari et al., 2006; van Herwaarden et al., 2002; Zhang et al., 2002). Currently AstraZeneca scientists are evaluating the novel $GABA_B$ receptor agonist **AZD3355** (Fig. 2) for this indication in Phase 2 clinical trials (*vide infra,* Chapter II. E).
14. Baclofen has been used for the treatment of **chronic hiccup.** In 37 patients with chronic hiccup (average duration 4.6 years) baclofen produced a long-term complete resolution in 18 cases and a considerable decrease in an additional 10 cases. Idiopathic chronic hiccup may result from gastroesophageal abnormalities (Guelaud et al., 1995; Turkyilmaz & Eroglu, 2008; Twycross, 2003).
15. Baclofen inhibits vagally induced **bronchoconstriction** (Chapman et al., 1993; Grimm et al., 1997; Prakash, 2009).
16. Baclofen has been used for the treatment of **cough** (Dicpinigaitis & Dobkin, 1997; Dicpinigaitis et al., 2000; Dicpinigaitis & Rauf, 1998).
17. First preclinical studies indicate that baclofen may inhibit audiogenic seizures in a mouse model of **fragile X syndrome** (Pacey et al., 2009; see also Zupan & Toth, 2008).

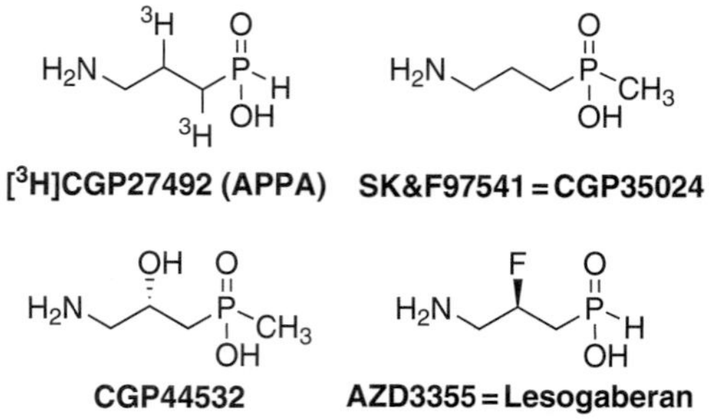

[^{3}H]CGP27492 (APPA) **SK&F97541 = CGP35024**

CGP44532 **AZD3355 = Lesogaberan**

FIGURE 2 Structures of $GABA_B$ receptor agonists.

B. Prodrugs of Baclofen

Scientists of XenoPort reported that the carbamate derivative of (*R*)-(−)-baclofen **Arbaclofen placarbil** (XP19986; Fig. 1) demonstrated enhanced colonic absorption, that is, 5-fold higher (*R*)-(−)-baclofen exposure in rats and 12-fold higher in monkeys compared with intracolonic administration of (*R*)-(−)-baclofen. Sustained release formulations of arbaclofen placarbil demonstrated sustained (*R*)-(−)-baclofen exposure in dogs with a bioavailability of up to 68% (Lal et al., 2009).

The methylester of baclofen readily penetrates into the brain, but its cleavage *in vivo* was smaller than expected (Leisen et al., 2003).

C. Phenibut

Phenibut, the des-chloro analogue of baclofen, was introduced as "a neuropsychotropic drug into clinical practice in Russia in the 1960s" under the trade name Citrocard (Lapin, 2001). Although it is structurally very close to baclofen, it has a significantly different spectrum of biological activities. It shows anxiolytic and nootropic properties and is used as a mood elevator and tranquilizer. Johnston and coworkers in Sydney evaluated the *in vitro* activities of (*R*)-(−) and (*S*)-(+)-β-phenyl-GABA. The IC_{50} values are 4.1 μM for the biologically active (*R*)-(−)-enantiomer (Fig. 1) and 699 μM for the (*S*)-(+)-enantiomer. Racemic baclofen showed an IC_{50} value of 0.06 μM (inhibition of binding of [^{3}H]-baclofen from rat cerebellar membranes; Allan et al., 1990). More recently Dambrova et al. (2008) disclosed K_i values for racemic baclofen, racemic phenibut, and (*R*)-(−)-phenibut as 6 μM, 177 μM, and 92 μM, respectively (inhibition of binding of [^{3}H]-CGP54626 on rat brain membranes). (*R*)-(−)-phenibut is a full agonist at $GABA_B$ receptors in brain slices (Ong et al., 1993). A dose of 100 mg/kg of (*R*)-phenibut significantly reduced immobility time in the forced swimming test, an effect which could be blocked by the $GABA_B$ receptor antagonist CGP35348. (*R*)-phenibut also showed analgesic activity in the tail-flick test and locomotor-depressing activity in the open field test (Dambrova et al., 2008).

D. Gabapentin

The results of Lacaille and coworkers in extensive electrophysiological experiments that the anticonvulsant and analgesic drug Gabapentin (Neurontin; Fig. 1) selectively interacts with the Sushi domain of the $GABA_{B1a}$/$GABA_{B2}$ heterodimer raised hopes to, finally, find subtype-selective $GABA_B$ receptor ligands (Bertrand et al., 2003a; Bertrand et al., 2001, 2003b; Ng et al., 2001). Similar results were published by Parker et al. (2004). These findings were already in conflict with the data by Stringer and Lorenzo

(1999), who state that the reduction of paired-pulse inhibition in the rat hippocampus by gabapentin is independent of $GABA_B$ receptor activation. Subsequent experiments by Lanneau et al. (2001), Jensen et al. (2002), Cheng et al. (2004), and Shimizu et al. (2004) led to the conclusion that gabapentin is not an agonist at $GABA_B$ receptors.

E. γ-Aminopropyl-phosphinic Acids as Novel $GABA_B$ Receptor Agonists

Starting from 1980 chemists of the Central Research Laboratories of Ciba–Geigy in Manchester prepared systematically phosphonous acid analogues of all natural α-, β-, and γ-amino acids and had these compounds tested in diverse pharmacological assays in Basel (Baylis et al., 1984; Dingwall et al., 1987). In fall 1984 it was discovered that the phosphonous acid analogue of GABA, that is, **CGP27492** (Fig. 2), showed an extraordinarily high affinity toward $GABA_B$ receptors ($IC_{50} = 2$ nM; inhibition of binding of [^{3}H]baclofen to $GABA_B$ receptors of cat cerebellum; Bittiger et al., 1988; Froestl et al., 1995a). Due to its 15 times higher potency, high specific binding, and the possibility to carry out filtration assays, **[^{3}H]CGP27492** has replaced [^{3}H]baclofen as a radioligand for $GABA_B$ receptor-binding assays (Hall et al., 1995). CGP27492, however, was inactive in various *in vivo* tests for potentially useful CNS effects.

The methyl-phosphinic acid derivative **CGP35024** (identical with **SK&F97541**; Fig. 2) was 7 times more potent against neuropathic hyperalgesia than (*R*)-(−)baclofen. It induced nociceptive responses at doses well below those that cause sedation (Patel et al., 2001).

Interestingly, CGP27492 and CGP35024 act as *antagonists* to $GABA_C$ receptors ($IC_{50} = 2.47\,\mu$M for CGP27492 at human $\rho 1$ $GABA_C$ receptors; $IC_{50} = 0.75\,\mu$M at human $\rho 1$ $GABA_C$ receptors; and $IC_{50} = 3.5\,\mu$M at human $\rho 2$ $GABA_C$ receptors for CGP35024, respectively) (Chebib et al., 1997, 1998).

CGP44532 showed muscle relaxant activity in the rotarod test in rats with ED_{50}'s of 0.4 mg/kg s.c. and 6.5 mg/kg p.o. and had a duration of action 3 times longer than that of baclofen. It showed a gastrointestinal and CNS side effect profile significantly superior to baclofen in *Rhesus* monkeys.

Repeated administration of CGP44532 at doses of 0.3 mg/kg s.c. for 5 days and of 3 mg/kg p.o. for 10 days produced significant antihyperalgesic effects in neuropathic rats with no evidence for tolerance (Enna et al., 1998). CGP44532 is not metabolized in liver S9 fractions of rat, dog, and man and has a bioavailability of 72% after oral administration to rats. A sophisticated method for this highly polar molecule (log P = −2.76) (*n*-octanol/water) was worked out by Blum et al. (2000).

CGP44532 was active in animal models of **psychosis**. It blocked the MK-801- and amphetamine-induced hyperactivity in a dose-dependent manner (Wieronska et al., 2010). For a review on $GABA_B$ receptors, schizophrenia and sleep dysfunction, see Kantrowitz et al. (2009).

CGP44532 was also tested in several animal models of suppression of craving for diverse drugs of abuse, such as **alcohol** (Colombo et al., 2002), **cocaine** in rats (Brebner et al., 1999; Dobrovitsky et al., 2002) and baboons (Weerts et al., 2005a, 2007), and **nicotine** (Paterson et al., 2004, 2005a, 2005b, 2008).

(*S*)-(−)-CGP44532 and its (*R*)-(+)-enantiomer CGP44533 act as *antagonists* to $GABA_C$ receptors ($IC_{50} = 17\,\mu M$ for CGP44532 and $IC_{50} = 5\,\mu M$ for CGP44533 at human $\rho 1$ $GABA_C$ receptors) (Hinton et al., 2008). A comparison of the actions of baclofen and the two enantiomers was published by Ong et al. (2001a) showing that CGP44532 was a far more potent agonist at $GABA_B$ autoreceptors.

The phosphonous acid derivative **AZD3355** (Fig. 2) exerts a highly valued peripheral mode of action by inhibiting transient lower esophageal sphincter relaxation (Alstermark et al., 2008; Lehmann et al., 2009). It is currently in Phase 2 clinical evaluation for the treatment of gastroesophageal reflux disease (GERD) under the generic name **Lesogaberan** (Bredenoord, 2009). At present Lesogaberan is the only drug interacting with $GABA_B$ receptors in clinical trials.

III. $GABA_B$ Receptor Partial Agonists

A. CGP47656

There is a rapid transition from γ-aminopropyl-methyl-phosphinic acids (CGP35024 or SF&F97541; Fig. 3) acting as a $GABA_B$ receptor agonist to the homologue γ-aminopropyl-ethyl-phosphinic acid CGP36216 (Fig. 3) acting as a $GABA_B$ receptor antagonist (Froestl et al., 1995a, 1995b; Ong

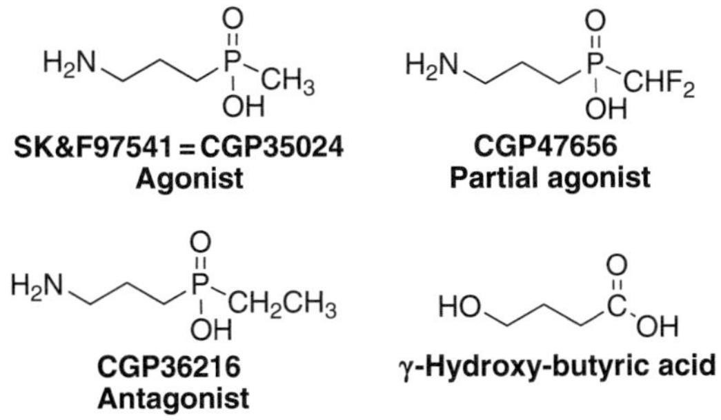

FIGURE 3 Structures of two $GABA_B$ receptor partial agonists.

et al., 2001b). The compound with a substituent in size between a methyl and an ethyl group, that is, the difluoromethyl derivative **CGP47656** (Fig. 3) with high affinity to $GABA_B$ receptors, is a partial $GABA_B$ receptor agonist ($IC_{50} = 89$ nM; inhibition of binding of [^{3}H]CGP27492 to $GABA_B$ receptors of rat cortex). It inhibited [^{3}H]GABA release in rat cortical slices, when the slices were stimulated at a frequency of 0.125 Hz ($IC_{50} = 10.6$ μM), whereas upon stimulation at a frequency of 2 Hz it increased the release of [^{3}H] GABA ($EC_{50} = 62$ μM; Froestl et al., 1995a). Raiteri and coworkers showed that CGP47656 acted as an antagonist on presynaptic $GABA_B$ autoreceptors in rat neocortex synaptosomes increasing the release of GABA but acted as a full agonist at presynaptic $GABA_B$ heteroreceptors inhibiting the release of somatostatin (Gemignani et al., 1994).

Urwyler and colleagues (2005) showed that CGP47656 displaced the $GABA_B$ receptor antagonist radioligand [^{3}H]CGP62349 (*vide infra*, Chapter V. B., Fig. 7) with a high- and a low-affinity component with K_i values of 71 nM and 1.07 μM, respectively. The affinities were increased by the addition of 30 μM of the positive allosteric modulator CGP7630 (to K_i values of 41 nM and 0.98 μM, respectively) and of 30 μM GS39783 (to K_i values of 35 nM and 0.87 μM, respectively). CGP47656 stimulated GTP(γ)[^{35}S] binding to recombinant $GABA_B$ receptors to maximally 25% of the maximal effect of GABA ($EC_{50} = 1.11$ μM). In the presence of the positive allosteric modulators CGP7930 or GS39783 (30 μM each), the maximal stimulation obtained with CGP47656 was increased by about fourfold. The potency of CGP47656 in this assay was increased by two- to threefold by the two modulators: 30 μM CGP7930: $EC_{50} = 0.39$ μM, maximal effect: 105% and 30 μM GS39783: $EC_{50} = 0.5$ μM, maximal effect: 109%.

B. γ-Hydroxy-Butyric Acid

γ-Hydroxy-butyric acid (GHB; Fig. 3) is a minor metabolite of GABA synthesized by GABA transaminase and succinic semialdehyde reductase. GHB is also a registered drug (Xyrem) for the treatment of excessive daytime sleepiness and cataplexy in patients with narcolepsy (Bernasconi et al., 1999). More recently GHB became infamous as a "date-rape drug" inducing euphoria, hallucinations, sedation, and relaxation (Wong et al., 2004). Several studies came to the conclusion that GHB acts as a weak $GABA_B$ receptor partial agonist ($K_i = 79$–126 μM; Mathivet et al., 1997; in recombinant $GABA_B$ R1/R2 receptors coexpressed with Kir3 channels in *Xenopus* oocytes: $EC_{50} = 5$ mM with a maximal stimulation of 69% when compared to (*R*)-(−)-baclofen; Lingenhoehl et al., 1999). Experiments in $GABA_{B1}$ receptor-deficient mice clearly showed that many of the *in vivo* effects of GHB were lost, such as hypolocomotion, hypothermia, and increase in striatal dopamine synthesis (Kaupmann et al., 2003), suggesting that the majority of exogenous GHB actions are mediated by $GABA_B$ receptors

(Crunelli et al., 2006). However, brains from GABA$_{B1}$ receptor-deficient mice still exhibit [^{3}H]GHB binding demonstrating that there are specific high-affinity GHB receptor-binding sites. Selective high-affinity GHB ligands with K_i's of up to 22 nM have been reported recently (Hog et al., 2008; Wellendorph et al., 2005). Interestingly, the widely used anti-inflammatory drug diclofenac binds to GHB receptors in rat brain (Wellendorph et al., 2009). Sustained efforts to clone the so far elusive GHB receptor(s) are currently ongoing (P. Wellendorph, personal communication). The paper by Andriamampandry et al. (2007) on the cloning of the putative human GHB receptor was rejected by many molecular biology experts.

IV. Positive Modulators of GABA$_B$ Receptors

Allosteric modulators are molecules that bind to a site on a neurotransmitter or hormone receptor which is topographically distinct from the orthosteric-binding pocket for agonists or competitive antagonists (Urwyler et al., 2005; Urwyler, 2006). Allosteric agents have little or no intrinsic agonistic activity of their own but induce conformational changes in the receptor protein, which affect its interaction with the endogenous neurotransmitter. These compounds have more pronounced *in vivo* responses and fewer side effects than full agonists (Marshall, 2005; Pin & Prézeau, 2007). A very important aspect is that positive allosteric modulators have less propensity for the development of tolerance due to receptor desensitization than classical GABA$_B$ receptor agonists (Gjoni & Urwyler, 2008, 2009).

In a high-throughput screen using a GTP(γ)[^{35}S] assay in membranes from CHO-K1 cells stably transfected with human GABA$_{B1b}$ and rat GABA$_{B2}$ cDNAs **CGP7930** (Fig. 4) was identified as positive modulator of GABA$_B$ receptor function (Urwyler et al., 2001). It potentiated GABA-stimulated GTP(γ)[^{35}S] binding at low micromolar concentrations and was inactive in the absence of GABA. It increased both agonist potency and maximal efficacy: EC$_{50}$ for CGP7930 in presence of 1 μM GABA in recombinant GABA$_B$ receptor heterodimer expressed in CHO cells was 4.6 μM, which increased to 1.87 μM in the presence of 20 μM GABA. The maximal effect was dose dependent: GABA at 20 μM stimulated basal activity to 301%, in presence of 1 μM CGP7930 to 328%, in presence of 3 μM CGP7930 to 377%, in presence of 10 μM CGP7930 to 394%, and in presence of 30 μM CGP7930 to 427%. CGP7930 enhanced efficacy and potency of GTP(γ)[^{35}S] binding to G$_{\alpha o}$ (Mannoury la Cour et al., 2008). CGP7930 potentiated baclofen-induced depression of dopamine neuron activity (Chen et al., 2005) and selectively enhanced the baclofen-induced modulation of synaptic inhibition in the rat hippocampal CA1 area (Chen et al., 2006). CGP7930 potentiated the stimulatory effects of (*R*)-(−)-baclofen and of GABA on basal and on corticotrophin-releasing

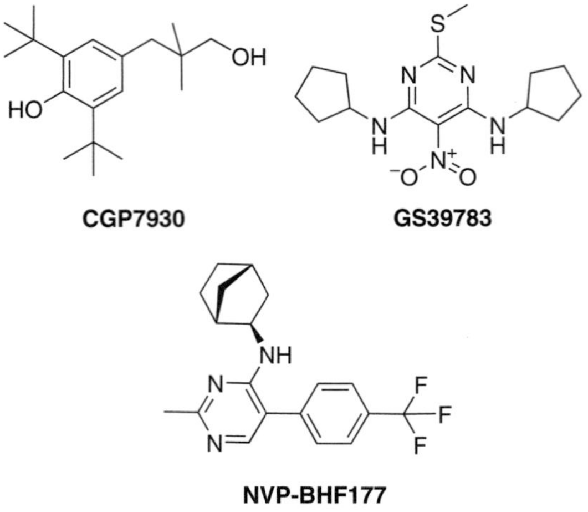

CGP7930 GS39783

NVP-BHF177

FIGURE 4 Structures of positive allosteric modulators of $GABA_B$ receptors.

hormone-stimulated adenylyl cyclase activities (Onali et al., 2003) and potentiated GTP(γ)[^{35}S] binding and inhibition of forskolin-stimulated adenylyl cyclase activity in *post mortem* human frontal cortex (Olianas et al., 2005). Activation of the $GABA_B$ receptor by GABA, baclofen, or by CGP7930 induced phosphorylation of the extracellular signal-regulated protein kinases (1/2) (ERK(1/2)) in cultured cerebellar granular cells, which in turn mediated cAMP(cyclic adenosine monophosphate)-responsive element binding protein (CREB) phosphorylation (Tu et al., 2007) and significantly reduced the number of apoptotic cerebellar granular cells in a cellular model of apoptosis (Tu et al., 2010).

CGP7930 efficiently potentiated the sedative/hypnotic effect of baclofen and γ-hydroxybutyric acid (Carai et al., 2004a). CGP7930 showed antidepressant-like effects in the modified forced swimming test significantly decreasing immobility time and anxiolytic properties in the elevated zero maze test significantly increasing time spent in the open area of the maze (Frankowska et al., 2007) and in the stress-induced hypothermia test (Jacobson & Cryan, 2008). AstraZeneca scientists found that baclofen and CGP7930 inhibit mechanically induced visceral pain (Brusberg et al., 2009).

CGP7930 reduced **alcohol intake** in alcohol-preferring rats (Liang et al., 2006; Orru et al., 2005). CGP7930 counteracted the **cocaine**-discontinuation-induced enhancement of immobility time in the forced swimming test (Filip et al., 2007; Frankowska et al., 2010).

CGP7930 interacts with the heptahelical domain of the $GABA_{B2}$ receptor subunit thus being the first-described ligand of the $GABA_{B2}$ receptor (Binet et al., 2004; Pin et al., 2004).

For an excellent review on the properties of CGP7930, see Adams and Lawrence (2007).

A close analogue of CGP7930 selectively potentiates the action of baclofen at $GABA_B$ autoreceptors, but not $GABA_B$ heteroreceptors (Parker et al., 2008).

A more potent compound was identified in **GS39783** (Fig. 4) increasing the potency of GABA about eightfold: EC_{50} of GABA in absence of GS39783 was 3.59 μM and in presence of 30 μM GS39783: $EC_{50} = 0.45$ μM. The maximal intrinsic efficacy increased from 100% to 217% (Urwyler et al., 2003, 2005). GS39783 enhances $GABA_B$ receptor-mediated inhibition of cAMP formation in the rat (Gjoni et al., 2006). GS39783 showed anxiolytic-like effects in the elevated plus maze in rats and the elevated zero maze in mice and rats (Cryan et al., 2004) and decreased anxiety in the light–dark box but did not show any effects in the forced swim test (Mombereau et al., 2004). GS39783 blocked the MK-801- and amphetamine-induced hyperactivity in a dose-dependent manner (Wieronska et al., 2010).

GS39783 reduced **alcohol** self-administration in alcohol-preferring rats (Maccioni et al., 2007, 2008a; Orru et al., 2005). GS39783 also attenuates the reward-facilitating effects of **cocaine** in rats (Lhuillier et al., 2007; Slattery et al., 2005a; Smith et al., 2004) and **nicotine** in rats (Mombereau et al., 2007; Paterson et al., 2008). The $GABA_B$ receptor-positive modulation-induced blockade of the rewarding properties of nicotine is associated with a reduction in nucleus accumbens delta fosB accumulation (Mombereau et al., 2007).

By the point mutations G706T and A708P Novartis molecular biologists were able to locate precisely the binding site of GS39783 in the sixth transmembrane domain of the $GABA_{B2}$ receptor (Dupuis et al., 2006).

Novartis chemists optimized the genotoxic lead structure GS39783, in collaboration with NIH, in particular NIDA and NIMH (Grant U01 MH069062) to obtain nontoxic positive modulators of $GABA_B$ receptors, such as **NVP-BHF177** (Fig. 4; Guery et al., 2007). The primary goal of the NIH collaboration was to find novel drugs for the treatment of craving after smoking cessation (Paterson et al., 2008), but NVP-BHF177 also reduced alcohol intake (Maccioni et al., 2009).

Roche scientists optimized both lead structures GS39783 and CGP7930. A 2005 patent disclosed structures based on GS39783 (Fig. 5; Malherbe et al., 2005). CGP7930 was further developed to the benzofuranon derivative (+)-BHFF (Fig. 5; Alker et al., 2008; Malherbe et al., 2008).

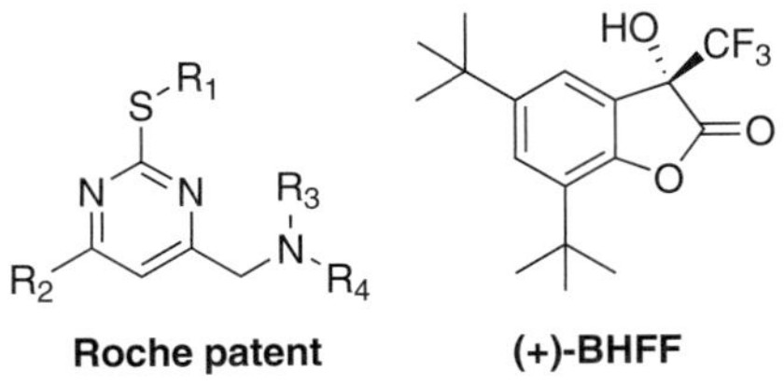

FIGURE 5 Structures of positive allosteric modulators of $GABA_B$ receptors of Roche.

In the GTP(γ)[^{35}S] binding assay in CHO cells stably expressing $G_{\alpha 16}$-$GABA_{B(1a,2a)}$ receptors the EC_{50} for GABA increased by a factor of 15.3-fold in presence of 0.3 μM of racemic BHFF reaching an E_{max} of 149%, for (+)-BHFF the EC_{50} for GABA increased by a factor of 87.3-fold in presence of 0.3 μM of (+)-BHFF reaching an E_{max} of 181%. Racemic BHFF reversed stress-induced hypothermia at doses of 3, 10, 30, and 100 mg/kg in mice. The mean absolute p.o. bioavailability was 100% (Malherbe et al., 2008).

V. $GABA_B$ Receptor Antagonists

A. First Generation $GABA_B$ Receptor Antagonists

In a simple classification we define first generation $GABA_B$ receptor antagonists as compounds displaying IC_{50} values of >1 μM.

Kerr et al. (1987) disclosed information on the first $GABA_B$ receptor antagonist **phaclofen** (IC_{50} = 130 μM; Fig. 6) in March 1987. Phaclofen blocked the slow inhibitory postsynaptic potential revealing an important physiological role for $GABA_B$ receptors (Dutar & Nicoll, 1988; see also the review by Nicoll, 2004). The active enantiomer is (*R*)-(−)-phaclofen (IC_{50} 76 μM), whereas (*S*)-(+)-phaclofen showed an IC_{50} > 1 mM (inhibition of

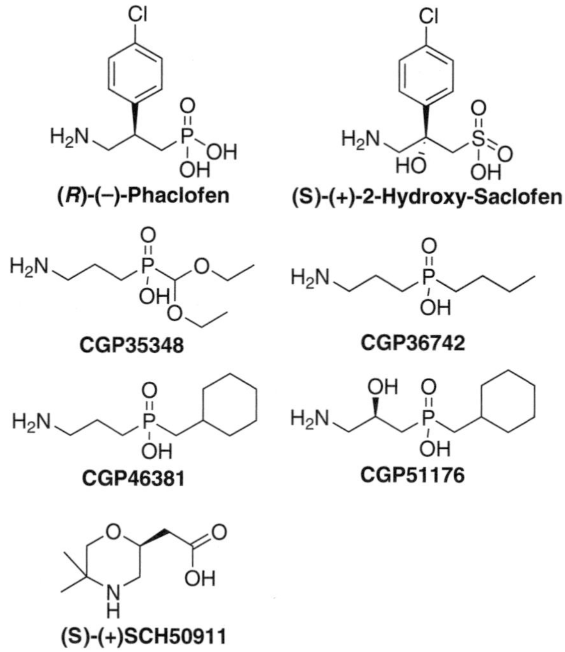

FIGURE 6 Structures of first generation $GABA_B$ receptor antagonists.

binding of [^{3}H]-(*R*)-(−)-baclofen from $GABA_B$ receptors of rat cerebellar membranes; Frydenvang et al., 1994).

The second $GABA_B$ receptor antagonist **2-hydroxy-saclofen** was published by Kerr et al. (1988; $IC_{50} = 11\,\mu M$; Fig. 6). The active enantiomer is (*S*)-(+)-2-hydroxy-saclofen (for the pharmacology, see Kerr et al., 1995; for the synthesis, see Prager et al., 1995).

The first $GABA_B$ receptor antagonists capable of penetrating the BBB were **CGP35348** (Olpe et al., 1990; $IC_{50} = 27\,\mu M$, active after i.p. administration only), **CGP36742, CGP46381, CGP51176** (IC_{50}'s $= 38\,\mu M$, $4\,\mu M$, and $6\,\mu M$, respectively; Froestl et al., 1995b), and **SCH50911**, a pure (*S*)-(+)-enantiomer ($IC_{50} = 1.1\,\mu M$; Bolser et al., 1995; Frydenvang, Enna, & Krogsgaard-Larsen, 1997; Fig. 6). The last four $GABA_B$ receptor antagonists are all active after p.o. administration.

Saegis Pharmaceuticals, Half Moon Bay near San Francisco, undertook a double-blind, placebo-controlled Phase 2 clinical trial in 110 mild cognitive impairment patients treated for 8 weeks with 600 mg t.i.d. of **CGP36742 (SGS742)** in 2002. The result showed **significant improvement in working memory, psychomotor speed, and attention** with SGS742 as compared with placebo (Froestl et al., 2004; Tomlinson et al., 2004).

A second double-blind, placebo-controlled Phase 2 trial was undertaken in 2005 in 271 mild to moderate Alzheimer disease patients (MMSE 16–26). Patients were treated for 12 weeks with either oral SGS742 600 mg t.i.d. ($n = 137$) or placebo ($n = 134$). Evaluation of the cognitive performances was carried out using the ADAS-Cog, CGIC, and cognitive tests from the CANTAB test battery. After 12 weeks treatment there was no statistically significant improvement of the ADAS-Cog score of patients treated with SGS742 versus placebo. Detailed analyses showed that the **mild Alzheimer disease** subpopulation in the patient group **did improve**, whereas the moderate Alzheimer disease patients did not.

Many **preclinical studies** were carried out recently **with CGP36742 (SGS742):** Helm et al. (2005) showed that SGS742 improves spatial memory and reduces protein binding to the cAMP response element (CRE) in hippocampus of rats. SGS742 may act by relieving CREB2-mediated suppression of transcription pathways for long-term memory storage. It is possible that by suppression of CREB2 the balance between CREB1 and CREB2 is restored. Gallagher and colleagues showed that in rats with age-impaired memory CREB1 is significantly reduced, while CREB2 is not (Brightwell et al., 2004). Sunyer et al. (2008a) investigated specific hippocampal protein expression of 11 signaling, chaperone and metabolic enzyme proteins causing cognitive enhancement by SGS742. Sunyer et al. (2009) found elevated hippocampal levels of phosphorylated protein kinase A. Also synapsin Ia levels were enhanced after treatment with SGS742 (John et al., 2009). SGS742 did not involve major known signaling cascades in mice (Sunyer et al., 2008b). These data complement earlier reports about enhanced release of somatostatin by

SGS742 (Bonanno et al., 1999; Nyitrai et al., 2003; Pittaluga et al., 2001; Raiteri, 2008) and of nerve growth factor (NGF) and brain-derived neurotrophic factor (BDNF) (Heese et al., 2000). The memory-enhancing effects of SGS742 depend on the mouse strains (Sunyer et al., 2007).

Electrophysiological studies using i.c.v. administration of CGP35348 revealed that the $GABA_B$ receptor antagonist enhances theta and gamma rhythms in the hippocampus of rats (Leung & Shen, 2007).

A recent report by Pilc and coworkers describes the **antidepressant actions of CGP36742** and **CGP51176** (Nowak et al., 2006) confirming earlier reports by Nakagawa and Takashima (1997) and by Nakagawa et al. (1999). The hypothesis of Pilc and Lloyd (1984), presented more than 25 years ago, that antidepressant effects are due to an upregulation of $GABA_B$ receptors is valid (see also Pilc & Novak, 2005). Pratt and Bowery (1993) could show that chronic administration of 100 mg/kg i.p. of CGP36742 for 21 days caused an upregulation of $GABA_B$ receptors in rat frontal cortex by 50% above control levels. In addition, the enhanced release of BDNF may also contribute to the antidepressant effects of CGP36742 (Heese et al., 2000). Also anxiolytic activity of CGP36742 was shown recently (Partyka et al., 2007). For the differential upregulation of $GABA_{B1a}$, $GABA_{B1b}$, and $GABA_{B2}$ receptor subtypes in spinal cord in response to treatment with antidepressants, such as amitriptyline and fluoxetine, see Enna and Bowery (2004), Sands et al. (2004), McCarson et al. (2005), and McCarson et al. (2006).

Marescaux and coworkers showed that **CGP36742** is a valuable drug to treat **absence (petit-mal) epilepsy** in GAERS rats (= genetic absence epilepsy rats of Strasbourg; Vergnes et al., 1997; see also Getova et al., 1997; in PTZ-kindled rats: Genkova-Papazova et al., 2000; and in WAG/Rij rats: Kaminski et al., 2001). Proconvulsive effects of CGP36742 were observed only at very high doses of 800–2,400 mg/kg (Vergnes et al., 1997). The pharmacology of absence epilepsy has been recently updated by Bowery (2006) and by Manning et al. (2003). The role of $GABA_B$ receptor antagonists in arresting cortical seizures has been reviewed by Mares and Kubova (2008). CGP35348 suppressed rhythmic metrazol activity (a model for human absences) but increased the incidence of clonic seizures (Mares & Slamberova, 2006).

The orally active drug CGP36742 may be the drug of choice for the treatment of patients suffering from a **deficiency of succinate semialdehyde dehydrogenase**. Experiments in the laboratories of Michael Gibson with the i.p. active compound **CGP35348** showed that mice deficient in succinate semialdehyde dehydrogenase can be rescued from lethal seizures (Gupta et al., 2002; Hogema et al., 2001).

γ-Hydroxybutyrate (GHB) dose dependently reduced the number of food pellets earned and impaired a motor task in baboons. Pretreatment with CGP36742 antagonized GHB-induced suppression of food-maintained behavior and performance on a fine-motor task. Signs of abdominal discomfort, ataxia, and muscle relaxation produced by GHB were also reduced (Goodwin et al., 2005, 2006; Weerts et al., 2005b).

Pretreatment with the $GABA_B$ receptor antagonists CGP35348 and SCH 50911 completely prevented γ-butyrolactone (a prodrug of GHB)-induced hypothermia, motor-incoordination, and sedation/hypnosis in DBA mice (Carai et al., 2008). These experiments confirm *in vitro* experiments of Waldmeier (1991), who showed that CGP35348 antagonized the effects of baclofen, γ-butyrolactone, and HA 966, a GHB agonist.

CGP36742 also blocks $GABA_C$ receptors ($IC_{50} = 62\,\mu M$; inhibition of 1 μM of GABA at $\rho 1$ $GABA_C$ receptors expressed in *Xenopus* oocytes; Chebib et al., 1997). Whether the **effects on slow wave sleep** are due to $GABA_B$ or $GABA_C$ receptor actions can only be decided when the now-available selective $GABA_C$ receptor antagonists (Chebib et al., 2009) are tested in the same paradigm (Deschaux et al., 2006).

The BBB penetration of CGP36742 in rat brain could be measured with high precision in a microdialysis/mass spectrometry assay (Andrén et al., 1998).

CGP46381 proved to be superior to CGP35348 in suppressing generalized absence seizures in lethargic, stargazer, and γ-hydroxybutyrate-treated model mice (Aizawa et al., 1997).

SCH50911 (Fig. 6) showed pronounced antiabsence effects in the lethargic mouse model (Hosford et al., 1995). It produced a marked protection against γ-hydroxybutyric acid mortality in mice (Carai et al., 2004b, 2005). Carai et al. (2002) observed proconvulsive effects in rats undergoing ethanol withdrawal, which Richards and Bowery (1996) did not observe in GAERS rats. Chronic administration of SCH50911 led to an upregulation of $GABA_B$ receptors (Pibiri et al., 2005). 30 mg/kg i.p. of either SCH50911, CGP46381 (both Fig. 6) or CGP52432 (Fig. 7) resulted in a significant stimulation of locomotor activity in rats, which could be blocked by haloperidol suggesting an enhanced release of dopamine by the $GABA_B$ receptor antagonists (Colombo et al., 2001). These findings corroborate previous electrophysiological studies by Erhardt et al. (1999).

I. Two $GABA_B$ Receptor Antagonists Act as Partial Agonists in the Presence of Positive Modulators

In the presence of either **CGP7930** or **GS39783** (30 μM; *vide supra* in Chapter IV) two ("classical") $GABA_B$ receptor antagonists (*vide supra* in Chapter V. A), **CGP35348** (Olpe et al., 1990; Fig. 6) and **2-hydroxy-saclofen** (Kerr et al., 1988; Fig. 6), stimulated GTP(γ)[^{35}S] binding to recombinant $GABA_B$ receptors to maximally 30–40% of the full GABA effect, that is, they act as partial agonists at the $GABA_B$ receptor: for CGP35348: in presence of 30 μM CGP7930: $EC_{50} = 2.1\,\mu M$, maximal effect: 36% and in presence of 30 μM GS39783: $EC_{50} = 3.0\,\mu M$, maximal effect: 37%; for 2-hydroxy-saclofen: in presence of 30 μM CGP7930: $EC_{50} = 7.76\,\mu M$,

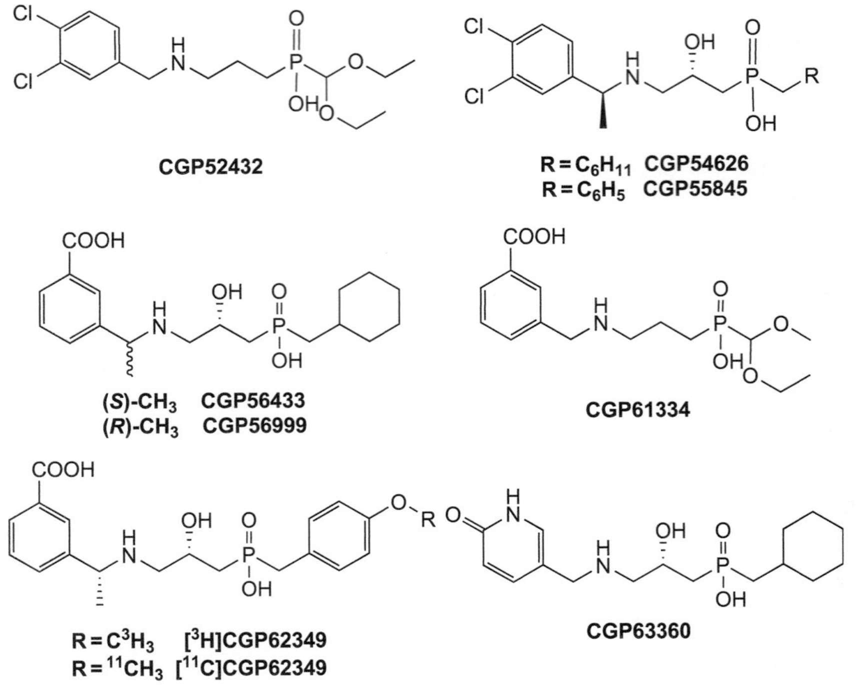

FIGURE 7 Structures of second generation $GABA_B$ receptor antagonists.

maximal effect: 31% and in presence of 30 μM GS39783: $EC_{50} = 11$ μM, maximal effect: 35% (Urwyler et al., 2005).

B. Second Generation $GABA_B$ Receptor Antagonists

In a simple classification we define second generation $GABA_B$ receptor antagonists as compounds displaying IC_{50} values in the nanomolar range.

The big breakthrough to nanomolar affinity $GABA_B$ receptor antagonists was achieved by Stuart J. Mickel in November 1990 by substituting the amino-group of γ-aminopropyl-phosphinic acids with selected benzyl substituents. **CGP52432** displays an IC_{50} of 55 nM, **CGP54626A** of 4 nM, **CGP55845A** of 6 nM, **CGP56433A** of 80 nM, **CGP56999A** of 2 nM, **CGP61334** of 36 nM, **CGP62349** of 2 nM (Froestl et al., 1996) and **CGP63360A** of 39 nM (all inhibition of binding of $[^3H]$CGP27492 from rat cerebral cortex membranes; Fig. 7).

In particular, **CGP55845A** and **CGP56433A** have been extensively tested in diverse *in vivo* paradigms. Interestingly, both drugs show about equal potencies at inhibitory synapses in the CA1 region of the rat hippocampus (Pozza et al., 1999).

CGP55845A completely reversed **age-related learning impairment** in aged Fisher 344 rats (Lasarge et al., 2009). It improved learning in rats in an active avoidance test (Getova & Bowery, 1998) and in rats with γ-hydroxy-butyrolactone-induced absence syndrome in active and passive avoidance tests (Getova & Bowery, 2001). CGP55845A allowed studying the impaired function of $GABA_B$ receptors in tissues from pharmacoresistant epilepsy patients (Teichgräber et al., 2009). Suppression of epileptiform activity by CGP55845A was investigated *in vitro* by Yanovsky and Misgeld (2003). Paired-pulse depression in rat cortex was fully antagonized by CGP55845A, but not by CGP35348 (Deisz, 1999).

CGP56433A improved learning and memory retention in a rat model of absence epilepsy (Getova et al., 1997).

CGP55845A and **CGP56433A** selectively increased swimming time in a modified rat forced swimming model of **depression** (Cryan & Kaupmann, 2005; Mombereau et al., 2004; Slattery & Cryan, 2006; Slattery et al., 2005b).

CGP56433A attenuated the effect of baclofen on cocaine but not heroin self-administration in the rat (Brebner et al., 2002). It significantly elevated brain stimulation reward thresholds at a dose of 10 mg/kg s.c. Even more pronounced effects were obtained with a combination of 7.5 mg/kg s.c. of CGP56433A plus 0.5 mg/kg s.c. of the $GABA_B$ receptor agonist CGP44532 (Macey et al., 2001).

CGP55845A proved to be beneficial in an animal model of **Down's syndrome** after administration of 0.5 mg/kg i.p. once a day during 3 weeks (A. Kleschevnikov & W. C. Mobley, personal communication).

CGP52432A was used recently for the elucidation of glycine release in mouse spinal cord and hippocampus (Romei et al., 2009) and for the functional mapping of $GABA_B$ receptor subtypes in the thalamus of mice (Ulrich et al., 2007). CGP52432 caused a significant stimulation of locomotor activity in rats via increased dopamine release (Colombo et al., 2001).

Administration of **CGP55845A** (25, 50, or 100 μg/kg) into the ventral tegmental area of rats elicited a concentration-dependent increase in dopamine levels but did not alter glutamate levels (Giorgetti et al., 2002). CGP55845 entirely abolished the reduction of the mean amplitude of eIPSC's by SNAP-5114, a specific GABA transporter 3 (GAT-3) blocker (Kirmse et al., 2009).

CGP56999A is probably the most potent $GABA_B$ receptor antagonist synthesized to date ($K_i = 0.25$ nM; Urwyler et al., 2005). Daily treatment with 1 mg/kg i.p. significantly attenuated the 6-hydroxydopamine-induced decline in nigrostriatal dopamine and increased the expression of BDNF in the ipsilateral striatum indicating an application for the treatment of Parkinson's disease (Enna et al., 2006, see also Heese et al., 2000 for enhanced NGF and BDNF release). CGP56999A at 3–6 mg/kg i.p. caused delayed clonic convulsions (Vergnes et al., 1997). The mechanism of the proconvulsive effect has been studied by Qu et al. (2010). Early-life seizures result in a long-lasting reduction in $GABA_B$ receptor-mediated transmission in dentate gyrus.

Because of its extraordinary potency **CGP56999A** is the preferred tool for electrophysiologists studying paired-pulse depression (Leung et al.,

2008) or spike backpropagation (Leung & Peloquin, 2006). γ-Hydroxybutyrate caused a rapid increase in the phosphorylation level of the CREB and of MAP (mitogen-activated protein) kinases in mouse hippocampus. These effects were blocked with CGP56999A (Ren & Mody, 2003, 2006).

CGP62349A has been used to study the physiological role of $GABA_B$ receptors in the **peripheral nervous system**, in particular on the synthesis of specific myelin proteins by Schwann cells (Magnaghi et al., 2004, 2008). It caused an increased release of acetylcholine in the small intestine of dogs (Kawakami et al., 2004).

CGP63360A ($IC_{50} = 39$ nM inhibition of the $GABA_B$ receptor *agonist ligand* [^{3}H]CGP27492 and $IC_{50} = 9$ nM inhibition of the $GABA_B$ receptor *antagonist ligand* [^{125}I]CGP64213, both from membranes of rat cortex) showed striking **learning and memory-improving effects** in active and passive avoidance tests in rats and mice at daily doses as low as 0.1 mg/kg p.o. It increased the number of avoidances on the third, fourth, and fifth day of the learning sessions and in the memory retention test on the twelfth day (Getova & Dimitrova, 2007). It suppressed seizures in a genetic model of absence epilepsy with ED_{50}'s of 2 mg/kg i.p. and 40 mg/kg p.o. without causing proconvulsive effects up to 300 mg/kg p.o. (W. Froestl, A. Boehrer, R. Bernasconi, & C. Marescaux, unpublished results).

Several $GABA_B$ receptor antagonists are also available as **radioligands**, such as [^{3}H]CGP54626 (Bittiger et al., 1992), [^{3}H]CGP5699, and [^{3}H] CGP62349 (Bittiger et al., 1996). In particular **[^{3}H]CGP62349** with a specific radioactivity of 85 Ci/mmol proved to be of great value for in depth investigations in rat, monkey, and human brain structures (Ambardekar et al., 1999; Keir et al., 1999; Princivalle et al., 2002, 2003; Sloviter et al., 1999). Saturation binding experiments with [^{3}H]CGP62349 in the presence of 30 μM of the positive allosteric modulators CGP7690 and GS39783 decreased the K_i values from 0.55 to 0.85 nM (with CGP7930) and 1.02 nM (with GS39783). Similar effects were measured for CGP52432, CGP54626, CGP56999, and SCH50911 indicating that these compounds act as **silent competitive $GABA_B$ receptor antagonists** (Urwyler et al., 2005). However, all five compounds behaved as **inverse $GABA_B$ receptor agonists** in the desensitized $GABA_B$ stable cell line potentiating 7β-forskolin-stimulated cAMP production. CGP62349 achieved a maximal increase of cAMP production of 279% over 7β-forskolin control with an EC_{50} of 3.9 nM (Gjoni & Urwyler, 2009).

As yet it has not been possible to produce a positron emission tomography ligand for $GABA_B$ receptors based on these compounds, because [^{11}C]CGP62349 did not reach sufficiently high levels in monkey brain (Todde et al., 1997, 2000).

Recently a specific photoaffinity and fluorescent $GABA_B$ receptor antagonist ligand has been published by Li et al. (2008; **Probe 1** in Fig. 8). It significantly inhibited $GABA_B$ receptor activation by GABA with an IC_{50}

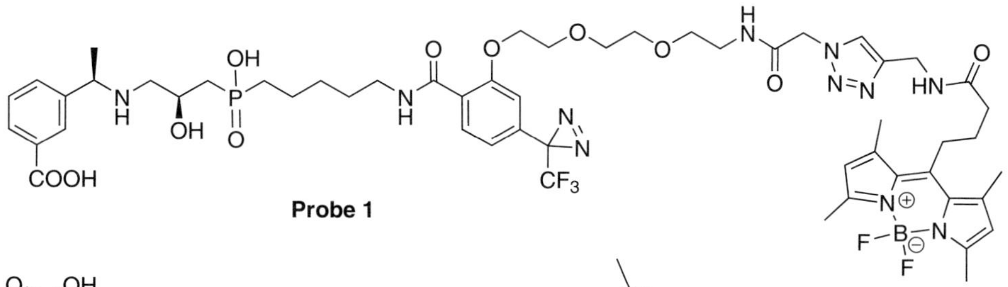

Probe 1

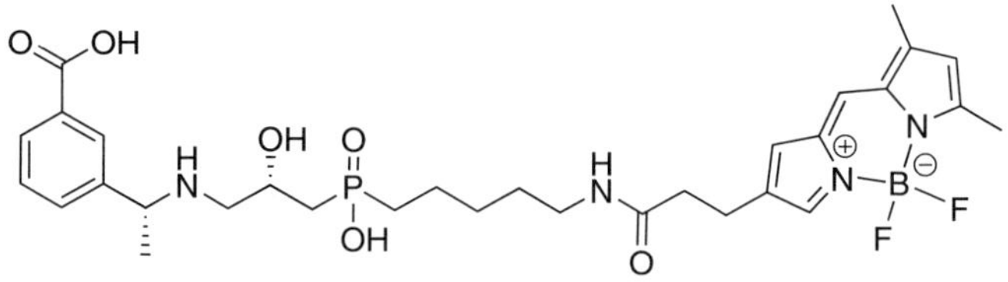

NVP-AAJ415

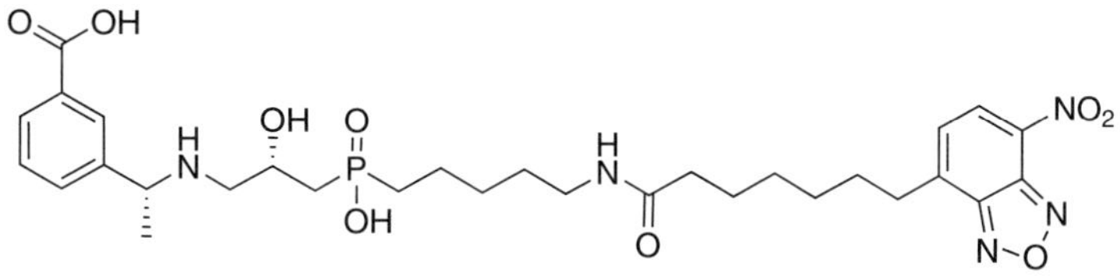

NVP-AAK515

FIGURE 8 Structures of fluorescent $GABA_B$ receptor antagonist ligands.

of 1.03 μM. Novartis chemists prepared two fluorescent $GABA_B$ receptor antagonist ligands, that is, NVP-AAJ415 ($IC_{50} = 22$ nM) and NVP-AAK515 ($IC_{50} = 8$ nM, both inhibition of binding of [^{3}H]CGP27492 on rat cerebral cortex membranes; Fig. 8; W. Froestl and H. Bittiger, unpublished results).

C. Third Generation $GABA_B$ Receptor Antagonists

The third generation $GABA_B$ receptor antagonists display IC_{50} values in the nanomolar range and are substituted with radioisotopes of very high specific radioactivity exceeding 2,000 Ci/mmol such as ^{125}I. This high specific radioactivity caused a rapid decomposition of our compounds, when we attempted to place the ^{125}I substituent on the benzylamino-group of γ-aminopropyl-phosphinic acids. Probably the strong radiation by ^{125}I caused a radical scission of the labile tertiary C–H bond at the branched benzyl position, which, however, is indispensable for high affinity. When we introduced the ^{125}I substituent in a second aromatic ring placed far away from the labile tertiary benzylic hydrogen, we finally obtained stable molecules. **[^{125}I] CGP64213** ($IC_{50} = 1.2$ nM, i.e., inhibition of binding of [^{125}I]CGP64213 to $GABA_B$ receptors on rat cerebral cortex membranes; Fig. 9) served as a means of identifying the clones of $GABA_{B1a}$ and $GABA_{B1b}$ receptor isoforms after transfection into COS cells (Kaupmann et al., 1997a). The photoaffinity ligand **[^{125}I]CGP71872** ($IC_{50} = 1$ nM; Fig. 9) allowed, for the first time, determination of the high molecular weights of $GABA_{B1a}$ and $GABA_{B1b}$ receptors of 130 and 100 kD. These two bands are conserved in brain membranes of human, rat, mouse, chicken, frog, and zebrafish but are not found in drosophila and nematodes (Bettler et al., 2004). The syntheses of both ligands are described with full experimental details in our patent (Kaupmann et al., 1997b). The synthesis of [^{125}I]CGP71872 has been described 2 years later also by chemists at Merck (Belley et al., 1999).

[^{125}I]CGP84963 ($IC_{50} = 6$ nM, i.e., inhibition of binding of [^{125}I] CGP64213 to $GABA_B$ receptors on rat cerebral cortex membranes; Fig. 9) combines, in one molecule, a $GABA_B$ receptor binding part, an azidosalicylic acid as photoaffinity moiety separated by a spacer of three GABA molecules from 2-iminobiotin, which binds to avidin in a reversible, pH-dependent fashion. This compound was specifically prepared to achieve isolation and purification of the extracellular *N*-terminal $GABA_{B1}$ receptor fragment for crystallization and X-ray studies of the $GABA_{B1}$ receptor-binding site (Froestl et al., 1999). The 31 step synthesis of this compound is described in detail in Froestl et al. (2003). A collaboration with a protein crystallographer yielded first interesting results. Deriu et al. (2005) succeeded to identify the minimal functional domain of the $GABA_{B1b}$ receptor, which still binds the competitive radioligand [^{125}I]CGP71872 and leads to a functional, GABA-responding receptor when coexpressed with $GABA_{B2}$. Mutant $GABA_{B1b}$ receptors, in

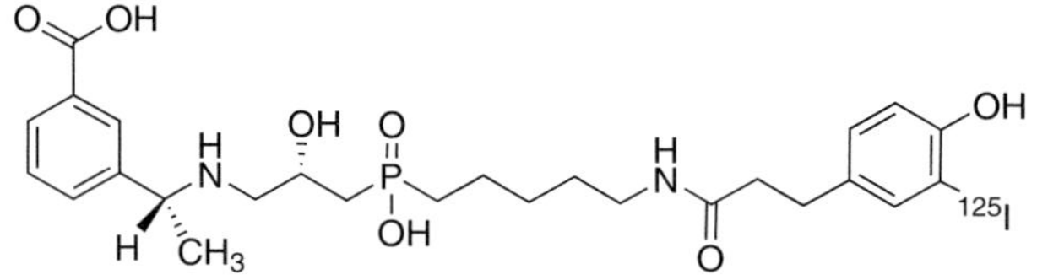

$[^{125}I]$CGP64213

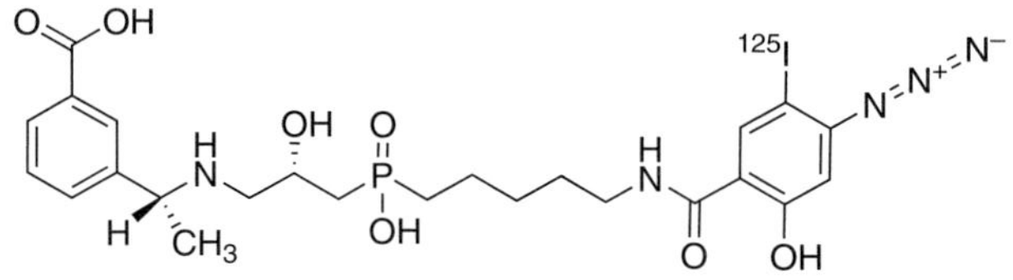

$[^{125}I]$CGP71872

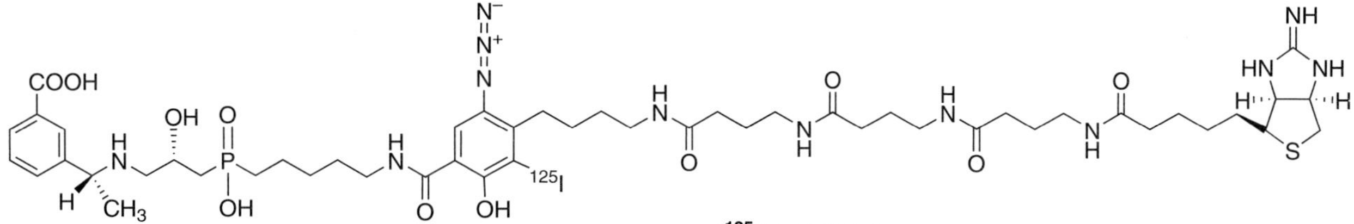

$[^{125}I]$CGP84963

FIGURE 9 Structures of third generation $GABA_B$ receptor antagonists.

which all potential N-glycosylation sites were removed, retained their receptor function. The work was terminated at this stage.

Therefore, the only X-ray structure of a class 3 G-protein-coupled receptor remains the extracellular ligand-binding region of the metabotropic glutamate receptor subtype 1 published by Kunishima et al. (2000) and reviewed by Jingami et al. (2003).

VI. Conclusion

Throughout the research efforts in the field of $GABA_B$ receptors a particularly close interaction between medicinal chemists and pharmacologists was key to success. With the synthesis of baclofen in 1962 Heinrich Keberle (Keberle et al., 1968) provided the ligand, which Norman Bowery, a coworker of the Research Division of Ciba Laboratories Ltd. in Horsham, Sussex, UK, for 7 years (Bowery, 1982) used to identify the baclofen-sensitive, bicuculline-insensitive $GABA_B$ receptor in 1980, 18 years after the synthesis of baclofen. Rolf Prager and his chemists synthesized phaclofen and 2-hydroxy-saclofen, which were extensively characterized pharmacologically and electrophysiologically by David Kerr, Jennifer Ong, David Curtis, and Graham Johnston (Kerr et al., 1987, 1988). In the late 1980 and early 1990s many of the CNS neurotransmitter receptors were cloned, mostly by expressing cDNAs in *Xenopus* oocytes and subsequent characterization by electrophysiological measurements. This procedure worked fine for the muscarinic acetylcholine receptor (Kubo et al., 1986), for the $GABA_A$ receptor (Schofield et al., 1987) and for the 5-HT_{1C} receptor measuring chloride currents (Lübbert et al., 1987), for the AMPA receptor measuring sodium currents (Hollmann et al., 1989), and for the NMDA receptor (Moriyoshi et al., 1991) and for the metabotropic glutamate receptor measuring calcium currents (Masu et al., 1991), but not for the $GABA_B$ receptors, because the postsynaptic outward potassium currents are too small (Sekiguchi et al., 1990; Taniyama et al., 1991; Woodward & Miledi, 1992). Four chemists and four technicians at Ciba–Geigy, Basel, and additional chemists at the Ciba–Geigy Central Research Laboratories in Manchester Trafford Park and in Macclesfield (Froestl et al., 2003) worked on phosphinic acid derivatives from 1985 to 1995 until they finally obtained high affinity and high specific radioactivity $GABA_B$ receptor antagonists, such as [^{125}I]CGP71872, which allowed the determination of the molecular weights of $GABA_{B1a}$ and $GABA_{B1b}$ receptors and [^{125}I]CGP64213, which was used for the identification of cDNA clones after transfection into COS cells after (unsuccessfully) screening of 720,000 clones of a commercial cDNA library (lacking the large clones of 4.4 kb) followed by the successful screening of 620,000 clones of a freshly prepared cDNA library, 17 years after Bowery's characterization of $GABA_B$ receptors by pharmacological means

(Kaupmann et al., 1997a). The $GABA_B$ receptor was the only receptor that was cloned using high affinity and high specific radioactivity ligands.

More and more data from sophisticated *in vivo* experiments are compiled on the potential use of $GABA_B$ receptor ligands in CNS diseases, such as depression, cognition deficits, and Down syndrome ($GABA_B$ receptor antagonists), anxiety and treatment of drug abuse ($GABA_B$ receptor positive modulators), and schizophrenia ($GABA_B$ receptor agonists). In terms of drug development, the most advanced $GABA_B$ receptor agonist Lesogaberan is currently in Phase 2 clinical trials for the treatment of gastroesophageal reflux disease.

Acknowledgments

There are more than 3,000 entries in Scifinder on publications on nonmarketed ligands for $GABA_B$ receptors, such as phaclofen, 2-hydroxy-saclofen, and compounds from AstraZeneca (AZD compound), Ciba–Geigy Pharma (CGP compounds), Geigy Agro (Geigy Saat = GS compound), GSK (SK&F compound), Novartis (NVP compounds), Roche (BHFF), and Schering (SCH compound). All papers would deserve to be presented in detail. However, the list of references in this chapter is restricted to 350 meaning that only 10% of the work of highly competent biologists and pharmacologists working *in vitro* and *in vivo* can be discussed. The selection is necessarily a subjective process. I apologize to all authors, whose work has not been discussed here. I wish to explicitly acknowledge the valuable contributions of all scientists working in this fascinating area of Neuroscience Research.

Conflict of Interest Statement: There is no conflict of interest.

List of Nonstandard Abbreviations

AMP	adenosine monophosphate
BDNF	brain-derived neurotrophic factor
cAMP	cyclic adenosine monophosphate
CREB	cAMP-responsive element binding protein
ERK(1/2)	extracellular signal-regulated kinases (1/2);
GAERS	genetic absence epilepsy rat of Strasbourg
GHB	γ-hydroxy-butyric acid;
MAP kinase	mitogen-activated protein kinase
NGF	nerve growth factor

References

Adams, C. L., & Lawrence, A. J. (2007). CGP7930: A positive allosteric modulator of the $GABA_B$ receptor. *CNS Drug Reviews*, *13*, 308–316.

Addolorato, G., Caputo, F., Capristo, E., et al. (2000). Ability of baclofen in reducing alcohol craving and intake: II—preliminary clinical evidence. *Alcoholism, Clinical and Experimental Research*, *24*, 67–71.

Addolorato, G., Caputo, F., Capristo, E., et al. (2002a). Rapid suppression of alcohol withdrawal syndrome by baclofen. *American Journal of Medicine*, *112*, 226–229.

Addolorato, G., Caputo, F., Capristo, E., et al. (2002b). Baclofen efficacy in reducing alcohol craving and intake: A preliminary double-blind randomized controlled study. *Alcohol and Alcoholism*, *37*, 504–508.

Addolorato, G., Leggio, L., Abenavoli, L., et al. (2003). Suppression of alcohol delirium tremens by baclofen administration: A case report. *Clinical Neuropharmacology*, *26*, 258–262.

Addolorato, G., Leggio, L., Abenavoli, L., et al. (2005). Baclofen for outpatient treatment of alcohol withdrawal syndrome. *Journal of Family Practice*, *54*, 24.

Addolorato, G., Leggio, L., Abenavoli, L., et al. (2006a). Baclofen in the treatment of alcohol withdrawal syndrome: A comparative study vs diazepam. *American Journal of Medicine*, *119*, 276–278.

Addolorato, G., Leggio, L., Agabio, R., et al. (2006b). Baclofen: A new drug for the treatment of alcohol dependence. *International Journal of Clinical Practice*, *60*, 1003–1008.

Addolorato, G., Leggio, L., Cardone, S., et al. (2009). Role of the $GABA_B$ receptor system in alcoholism and stress: Focus on clinical studies and treatment perspectives. *Alcohol*, *43*, 559–563.

Addolorato, G., Leggio, L., Ferrulli, A., et al. (2007). Effectiveness and safety of baclofen for maintenance of alcohol abstinence in alcohol-dependent patients with liver cirrhosis: A randomized, double-blind study. *Lancet*, *370*, 1915–1922.

Aizawa, M., Ito, Y., & Fukuda, H. (1997). Pharmacological profiles of generalized absence seizures in lethargic, stargazer and γ-hydroxybutyrate-treated model mice. *Neuroscience Research*, *29*, 17–25.

Akhondzadeh, S., Ahmadi-Abhari, S. A., Assadi, S. M., et al. (2000). Double-blind randomized controlled trial of baclofen vs. clonidine in the treatment of opiates withdrawal. *Journal of Clinical Pharmacy and Therapeutics*, *25*, 347–353.

Alker, A. M., Grillet, F., Malherbe, P., et al. (2008). Efficient one-pot synthesis of the $GABA_B$ positive allosteric modulator (R,S)-5,7-di-*tert*-butyl-3-hydroxy-3-trifluoromethyl-*3 H*-benzofuran-2-one. *Synthetic Communications*, *38*, 3398–3405.

Allan, R. D., Bates, M. C., Drew, C. A., et al. (1990). A new synthesis, resolution and *in vitro* activities of (*R*)- and (*S*)-β-phenyl-GABA. *Tetrahedron*, *46*, 2511–2524.

Alstermark, C., Amin, K., Dinn, S. R., et al. (2008). Synthesis and pharmacological evaluation of novel gamma-aminobutyric acid type B ($GABA_B$) receptor agonists as gastroesophageal reflux inhibitors. *Journal of Medicinal Chemistry*, *51*, 4315–4320.

Amantea, D., & Bowery, N. G. (2004). Reduced inhibitory action of a $GABA_B$ receptor agonist on [^{3}H]-dopamine release from rat ventral tegmental area *in vitro* after chronic nicotine administration. *BMC Pharmacology*, *4*, 24.

Ambardekar, A. V., Ilinsky, I. A., Froestl, W., et al. (1999). Distribution and properties of $GABA_B$ antagonist [^{3}H]CGP62349 binding in the *Rhesus* monkey thalamus and basal ganglia and the influence of lesions in the reticular thalamic nucleus. *Neuroscience*, *93*, 1339–1347.

Andrén, P. E., Emmett, M. R., DaGue, B. B., et al. (1998). Blood-brain barrier penetration of 3-aminopropyl-*n*-butylphosphinic acid (CGP 36742) in rat brain by microdialysis/mass spectrometry. *Journal of Mass Spectrometry*, *33*, 281–287.

Andriamampandry, C., Taleb, O., Kemmel, V., et al. (2007). Cloning and functional characterization of a γ-hydroxybutyrate receptor identified in the human brain. *FASEB Journal*, *21*, 885–895.

Audus, K. L., & Borchardt, R. T. (1986). Characteristics of the large neutral amino acid transport system of bovine brain microvessel endothelial cell monolayers. *Journal of Neurochemistry*, *47*, 484–488.

Baker, K. A., Taylor, J. W., & Lilly, G. E. (1985). Treatment of trigeminal neuralgia: Use of baclofen in combination with carbamazepine. *Clinical Pharmacy*, *4*, 93–96.

Bartoletti, M., Ricci, F., & Gaiardi, M. (2007). A $GABA_B$ agonist reverses the behavioral sensitization to morphine in rats. *Psychopharmacology*, *192*, 79–85.

Baylis, E. K., Campbell, C. D., & Dingwall, J. G. (1984). 1-Aminophosphonous acids. Part 1. Isosteres of the protein amino acids. *Journal of the Chemical Society. Perkin Transactions*, *1*, 2845–2853.

Becker, R., Alberti, O., & Bauer, B. L. (1997). Continuous intrathecal baclofen infusion in severe spasticity after traumatic or hypoxic brain injury. *Journal of Neurology*, *244*, 160–166.

Becker, W. J., Harris, C. J., Long, M. L., et al. (1995). Long-term intrathecal baclofen therapy in patients with intractable spasticity. *Canadian Journal of Neurological Sciences*, *22*(3), 208–217.

Belley, M., Sullivan, R., Reeves, A., et al. (1999). Synthesis of the nanomolar photoaffinity $GABA_B$ receptor ligand CGP71872 reveals diversity in the tissue distribution of $GABA_B$ receptor forms. *Bioorganic & Medicinal Chemistry*, *7*, 2697–2704.

Bernasconi, R., Mathivet, P., Bischoff, S., & Marescaux, C. (1999). γ-Hydroxy-butyric acid: An endogenous neuromodulator with abuse potential? *Trends in Pharmacological Sciences*, *20*, 135–141.

Bertrand, S., Morin, F., & Lacaille, J. C. (2003a). Different actions of gabapentin and baclofen in hippocampus from weaver mice. *Hippocampus*, *13*, 525–528.

Bertrand, S., Ng, G. Y., Purisai, M. G., et al. (2001). The anticonvulsant, antihyperalgesic agent gabapentin is an agonist at brain gamma-aminobutyric acid type B receptors negatively coupled to voltage-dependent calcium channels. *Journal of Pharmacology and Experimental Therapeutics*, *298*, 15–24.

Bertrand, S., Nouel, D., Morin, F., et al. (2003b). Gabapentin actions on Kir3 currents and *N*-type Ca^{2+} channels via $GABA_B$ receptors in hippocampal pyramidal cells. *Synapse*, *50*, 95–109.

Bettler, B., Kaupmann, K., Mosbacher, J., & Gassmann, M. (2004). Molecular structure and physiological functions of $GABA_B$ receptors. *Physiological Reviews*, *84*, 835–867.

Bexis, S., Ong, J., & White, J. (2001). Attenuation of morphine withdrawal signs by the $GABA_B$ receptor agonist baclofen. *Life Sciences*, *70*, 395–401.

Binet, V., Brajon, C., Le Corre, L., et al. (2004). The heptahelical domain of $GABA_{B2}$ is activated directly by CGP7930, a positive allosteric modulator of the $GABA_B$ receptor. *Journal of Biological Chemistry*, *279*, 29085–29091.

Bittiger, H., Reymann, N., Froestl, W., & Mickel, S. J. (1992). ^{3}H-CGP54626: A potent antagonist radioligand for $GABA_B$ receptors. *Pharmacology Communications* *2*, 23.

Bittiger, H., Reymann, N., Hall, R. G., et al. (1988). CGP27492, a new potent and selective radioligand for $GABA_B$ receptors (Abstract). *European Journal of Neuroscience*, *16* (Suppl.), 10.

Bittiger, H., Bellouin, C., Froestl, W., et al. (1996). [^{3}H]CGP62349: A new potent $GABA_B$ receptor antagonist radioligand. *Pharmacology Reviews and Communications*, *8*, 97–98.

Blackshaw, L. A., Staunton, E., Lehmann, A., et al. (1999). Inhibition of transient LES relaxations and reflux in ferrets by GABA receptor agonists. *American Journal of Physiology* *277*(4 Pt 1), G867–G874.

Blum, W., Aichholz, R., Ramstein, P., et al. (2000). Determination of the $GABA_B$ receptor agonist CGP44532 (3-amino-2-hydroxypropylmethylphosphinic acid) in rat plasma after pre-column derivatization by micro-high-performance liquid chromatography combined with negative tandem mass spectrometry. *Journal of Chromatography. B, Biomedical Sciences and Applications*, *748*, 349–359.

Bolser, D. C., Blythin, D. J., Chapman, R. W., et al. (1995). The pharmacology of SCH 50911: A novel, orally-active GABA-beta receptor antagonist. *Journal of Pharmacology and Experimental Therapeutics*, *274*, 1393–1398.

Bonanno, G., Carita, F., Cavazzani, P., et al. (1999). Selective block of rat and human neocortex $GABA_B$ receptors regulating somatostatin release by a $GABA_B$ antagonist endowed with cognition enhancing activity. *Neuropharmacology*, *38*, 1789–1795.

Boots, R. J., Lipman, J., O'Callaghan, J., et al. (2000). The treatment of tetanus with intrathecal baclofen. *Anaesthesia and Intensive Care*, *28*, 438–442.

Bowery, N. G. (1982). Baclofen: 10 years on. *Trends in Pharmacological Sciences*, *3*, 400–403.

Bowery, N. G. (2006). $GABA_B$ receptor: A site of therapeutic benefit. *Current Opinion in Pharmacology*, *6*, 37–43.

Bowery, N. G., Hill, D. R., Hudson, A. L., et al. (1980). (−)-Baclofen decreases neurotransmitter release in the mammalian CNS by an action at a novel GABA receptor. *Nature*, *283* (5742), 92–94.

Brar, S. P., Smith, M. B., Nelson, L. M., et al. (1991). Evaluation of treatment protocols on minimal to moderate spasticity in multiple sclerosis. *Archives of Physical Medicine and Rehabilitation*, *72*, 186–189.

Brebner, K., Froestl, W., Andrews, M., et al. (1999). The $GABA_B$ agonist CGP 44532 decreases cocaine self-administration in rats: Demonstration using a progressive ratio and a discrete trials procedure. *Neuropharmacology*, *38*, 1797–1804.

Brebner, K., Froestl, W., & Roberts, D. C. (2002). The $GABA_B$ antagonist CGP56433A attenuates the effect of baclofen on cocaine but not heroin self-administration in the rat. *Psychopharmacology*, *160*, 49–55.

Bredenoord, A. J. (2009). Lesogaberan, a $GABA_B$ agonist for the potential treatment of gastroesophageal reflux disease. *IDrugs*, *12*, 576–584.

Breslow, M. F., Fankhauser, M. P., Potter, R. L., et al. (1989). Role of γ-amino-butyric acid in antipanic drug efficacy. *American Journal of Psychiatry*, *146*, 353–356.

Brightwell, J. J., Gallagher, M., & Colombo, P. J. (2004). Hippocampal CREB1 but not CREB2 is decreased in aged rats with spatial memory impairment. *Neurobiology of Learning and Memory*, *81*, 19–26.

Brogden, R. N., Speight, T. M., & Avery, G. S. (1974). Baclofen: A preliminary report of its pharmacological properties and therapeutic efficacy in spasticity. *Drugs*, *8*, 1–14.

Brusberg, M., Ravnefjord, A., Martinsson, R., et al. (2009). The $GABA_B$ receptor agonist, baclofen, and the positive allosteric modulator, CGP7930, inhibit visceral pain-related responses to colorectal distension in rats. *Neuropharmacology*, *56*, 362–367.

Buonaguro, V., Scelsa, B., Curci, D., et al. (2005). Epilepsy and intrathecal baclofen therapy in children with cerebral palsy. *Pediatric Neurology*, *33*, 110–113.

Cange, L., Johnsson, E., Rydholm, H., et al. (2002). Baclofen-mediated gastro-oesophageal acid reflux control in patients with established reflux disease. *Alimentary Pharmacology & Therapeutics*, *16*, 869–873.

Carai, M. A., Brunetti, G., Lobina, C., et al. (2002). Proconvulsive effect of the $GABA_B$ receptor antagonist, SCH 50911, in rats undergoing ethanol withdrawal syndrome. *European Journal of Pharmacology*, *445*, 195–199.

Carai, M. A., Colombo, G., & Gessa, G. L. (2004b). Protection by the $GABA_B$ receptor antagonist, SCH 50911, of γ-hydroxybutyric acid-induced mortality in mice. *European Journal of Pharmacology*, *503*, 77–80.

Carai, M. A., Colombo, G., & Gessa, G. L. (2005). Resuscitative effect of a γ-aminobutyric acid B receptor antagonist on γ-hydroxybutyric acid mortality in mice. *Annals of Emergency Medicine*, *45*, 614–619.

Carai, M.A.M., Colombo, G., Froestl, W., et al. (2004a). *In vivo* effectiveness of CGP7930, a positive allosteric modulator of the $GABA_B$ receptor. *European Journal of Pharmacology, 504*, 213–216.
Carai, M. A., Lobina, C., Maccioni, P., et al. (2008). γ-Aminobutyric $acid_B$ ($GABA_B$)-receptor mediation of different *in vivo* effects of γ-butyrolactone. *Journal of Pharmacological Sciences, 106*, 199–207.
Cates, L. A. (1985). Anticonvulsant phosphorus analogs of GABA. *Drug Development Research, 5*, 395–399.
Chang, C.H.; Yang, D. S., Yoo, C. S., et al. (1982). Structure and absolute configuration of (*R*)-baclofen monochloride. *Acta Crystallographica, B38*, 2065–2067.
Chapman, R. W., Danko, G., del Prado, M., et al. (1993). Further evidence for prejunctional $GABA_B$ inhibition of cholinergic and peptidergic bronchoconstriction in guinea pigs: Studies with new agonists and antagonists. *Pharmacology, 46*, 315–323.
Chebib, M., Hinton, T., Schmid, K. L., et al. (2009). Novel, potent, and selective $GABA_C$ antagonists inhibit myopia development and facilitate learning and memory. *Journal of Pharmacology and Experimental Therapeutics, 328*, 448–457.
Chebib, M., Mewett, K. N., & Johnston, G.A.R. (1998). $GABA_C$ receptor antagonists differentiate between human rho1 and rho2 receptors expressed in *Xenopus* oocytes. *European Journal of Pharmacology, 357*, 227–234.
Chebib, M., Vandenberg, R. J., Froestl, W., & Johnston, G.A.R. (1997). Unsaturated phosphinic analogues of γ-aminobutyric acid as $GABA_C$ receptor antagonists. *European Journal of Pharmacology, 329*, 223–229.
Chen, Y., Menendez-Roche, N., & Sher, E. (2006). Differential modulation by the $GABA_B$ receptor allosteric potentiator 2,6-di-tert-butyl-4-(3-hydroxy-2,2-dimethylpropyl)-phenol (CGP7930) of synaptic transmission in the rat hippocampal CA1 area. *Journal of Pharmacology and Experimental Therapeutics, 317*, 1170–1177.
Chen, Y., Phillips, K., Minton, G., et al. (2005). $GABA_B$ receptor modulators potentiate baclofen-induced depression of dopamine neuron activity in the rat ventral tegmental area. *British Journal of Pharmacology, 144*, 926–932.
Cheng, J. K., Lee, S. Z., Yang, J. R., et al. (2004). Does gabapentin act as an agonist at native $GABA_B$ receptors? *Journal of Biomedical Science 11*(3), 346–355.
Colombo, G., Addolorato, G., Agabio, R., et al. (2004). Role of $GABA_B$ receptor in alcohol dependence: Reducing effect of baclofen on alcohol intake and alcohol motivational properties in rats and amelioration of alcohol withdrawal syndrome and alcohol craving in human alcoholics. *Neurotoxicity Research, 6*, 403–414.
Colombo, G., Melis, S., Brunetti, G., et al. (2001). $GABA_B$ receptor inhibition causes locomotor stimulation in mice. *European Journal of Pharmacology, 433*, 101–104.
Colombo, G., Serra, S., Brunetti, G., et al. (2002). The $GABA_B$ receptor agonists baclofen and CGP44532 prevent acquisition of alcohol drinking behaviour in alcohol-preferring rats. *Alcohol and Alcoholism, 37*, 499–503.
Colombo, G., Serra, S., Vacca, G., et al. (2006). Baclofen-induced suppression of alcohol deprivation effect in Sardinian alcohol-preferring (sP) rats exposed to different alcohol concentrations. *European Journal of Pharmacology, 550*, 123–126.
Cousins, M. S., Roberts, D. C., & de Wit, H. (2002). $GABA_B$ receptor agonists for the treatment of drug addiction: A review of recent findings. *Drug and Alcohol Dependence, 65*, 209–220.
Crunelli, V., Emri, Z., & Leresche, N. (2006). Unravelling the brain targets of γ-hydroxybutyric acid. *Current Opinion in Pharmacology, 6*, 44–52.
Cryan, J. F., & Kaupmann, K. (2005). Don't worry "B" happy!: A role for $GABA_B$ receptors in anxiety and depression. *Trends in Pharmacological Sciences, 26*, 36–43.
Cryan, J. F., Kelly, P. H., Chaperon, F., et al. (2004). Behavioral characterization of the novel $GABA_B$ receptor-positive modulator GS39783 (*N,N′*-dicyclopentyl-2-methylsulfanyl-5-

nitro-pyrimidine-4,6-diamine): Anxiolytic-like activity without side effects associated with baclofen or benzodiazepines. *Journal of Pharmacology and Experimental Therapeutics, 310*, 952–963.

Dambrova, M., Zvejniece, L., Liepinsh, E., et al. (2008). Comparative pharmacological activity of optical isomers of phenibut. *European Journal of Pharmacology, 583*, 128–134.

Deguchi, Y., Inabe, K., Tomiyasu, K., et al. (1995). Study on brain interstitial fluid distribution and blood-brain barrier transport of baclofen in rats by microdialysis. *Pharmaceutical Research, 12*, 1838–1844.

Deisz, R. A. (1999). $GABA_B$ receptor-mediated effects in human and rat neocortical neurones *in vitro*. *Neuropharmacology, 38*, 1755–1766.

Demazière, J., Saissy, J. M., Vitris, M., et al. (1991). Intermittent intrathecal baclofen for severe tetanus. *Lancet, 337*(8738), 427.

Deriu, D., Gassmann, M., Firbank, S., et al. (2005). Determination of the minimal functional ligand-binding domain of the $GABA_{B1b}$ receptor. *Biochemical Journal, 386*(Pt 3), 423–431.

Deschaux, O., Froestl, W., & Gottesmann, C. (2006). Influence of a $GABA_B$ and $GABA_C$ receptor antagonist on sleep-waking cycle in the rat. *European Journal of Pharmacology, 535*, 177–181.

Dicpinigaitis, P. V., & Dobkin, J. B. (1997). Antitussive effect of the GABA-agonist baclofen. *Chest, 111*, 996–999.

Dicpinigaitis, P. V., Grimm, D. R., & Lesser, M. (2000). Baclofen-induced cough suppression in cervical spinal cord injury. *Archives of Physical Medicine and Rehabilitation, 81*, 921–923.

Dicpinigaitis, P. V., & Rauf, K. (1998). Treatment of chronic, refractory cough with baclofen. *Respiration, 65*, 86–88.

Dingwall, J. G., Ehrenfreund, J., Hall, R. G., et al. (1987). Synthesis of γ-aminopropylphosphonous acids using hypophosphorous acid synthons. *Phosphorus, Sulfur, and Silicon and the Related Elements, 30*, 571–574.

Dobrovitsky, V., Pimentel, P., Duarte, A., et al. (2002). CGP44532, a $GABA_B$ receptor agonist, is hedonically neutral and reduces cocaine-induced enhancement of reward. *Neuropharmacology, 42*, 626–632.

Drake, R. G., Davis, L. L., Cates, M. E., et al. (2003). Baclofen treatment for chronic posttraumatic stress disorder. *Annals of Pharmacotherapy, 37*, 1177–1181.

Dralle, D., Müller, H., Zierski, J., et al. (1985). Intrathecal baclofen for spasticity. *Lancet, 326* (8462), 1003.

Dupuis, D. S., Relkovic, D., Lhuillier, L., et al. (2006). Point mutations in the transmembrane region of $GABA_{B2}$ facilitate activation by the positive modulator *N,N'*-dicyclopentyl-2-methylsulfanyl-5-nitro-pyrimidine-4,6-diamine (GS39783) in the absence of the $GABA_{B1}$ subunit. *Molecular Pharmacology, 70*, 2027–2036.

Dutar, P., & Nicoll, R. A. (1988). A physiological role for $GABA_B$ receptors in the central nervous system. *Nature, 332*(6160), 156–158.

Engrand, N., & Benhamou, D. (2001). About intrathecal baclofen in tetanus. *Anaesthesia and Intensive Care, 29*, 306–307.

Engrand, N., Guerot, E., Rouamba, A., et al. (1999). The efficacy of intrathecal baclofen in severe tetanus. *Anesthesiology, 90*, 1773–1776.

Enna, S. J., & Bowery, N. G. (2004). $GABA_B$ receptor alterations as indicators of physiological and pharmacological function. *Biochemical Pharmacology, 68*, 1541–1548.

Enna, S. J., Harstad, E. B., & McCarson, K. E. (1998). Regulation of neurokinin-1 receptor expression by $GABA_B$ receptor agonists. *Life Sciences, 62*, 1525–1530.

Enna, S. J., & McCarson, K. E. (2006). The role of GABA in the mediation and perception of pain. *Advances in Pharmacology, 54*, 1–27.

Enna, S. J., Reisman, S. A., & Stanford, J. A. (2006). CGP 56999A, a $GABA_B$ receptor antagonist, enhances expression of brain-derived neurotrophic factor and attenuates dopamine depletion in the rat corpus striatum following a 6-hydroxydopamine lesion of the nigrostriatal pathway. *Neuroscience Letters*, *406*, 102–106.

Erhardt, S., Nissbrandt, H., & Engberg, G. (1999). Activation of nigral dopamine neurons by the selective $GABA_B$-receptor antagonist SCH 50911. *Journal of Neural Transmission*, *106*, 383–394.

Evans, S. M., & Bisaga, A. (2009). Acute interaction of baclofen in combination with alcohol in heavy social drinkers. *Alcoholism, Clinical and Experimental Research*, *33*, 19–30.

Fadda, P., Scherma, M., Fresu, A., et al. (2003). Baclofen antagonizes nicotine-, cocaine-, and morphine-induced dopamine release in the nucleus accumbens of rat. *Synapse*, *50*, 1–6.

Fattore, L., Spano, M. S., Cossu, G., et al. (2009). Baclofen prevents drug-induced reinstatement of extinguished nicotine-seeking behaviour and nicotine place preference in rodents. *European Neuropsychopharmacology*, *19*, 487–498.

Federici, M., Nistico, R., Giustizieri, M., et al. (2009). Ethanol enhances $GABA_B$-mediated inhibitory postsynaptic transmission on rat midbrain dopaminergic neurons by facilitating GIRK currents. *European Journal of Neuroscience*, *29*, 1369–1377.

File, S. E., Mabbutt, P. S., & Andrews, N. (1991). Diazepam withdrawal responses measured in the social interaction test of anxiety and their reversal by baclofen. *Psychopharmacology*, *104*, 62–66.

Filip, M., & Frankowska, M. (2007). Effects of $GABA_B$ receptor agents on cocaine priming, discrete contextual cue and food induced relapses. *European Journal of Pharmacology*, *571*, 166–173.

Filip, M., & Frankowska, M. (2008). $GABA_B$ receptors in drug addiction. *Pharmacological Reports*, *60*, 755–770.

Filip, M., Frankowska, M., & Przegalinski, E. (2007). Effects of $GABA_B$ receptor antagonist, agonists and allosteric positive modulator on the cocaine-induced self-administration and drug discrimination. *European Journal of Pharmacology*, *574*, 148–157.

Flannery, B. A., Garbutt, J. C., Cody, M. W., et al. (2004). Baclofen for alcohol dependence: A preliminary open-label study. *Alcoholism, Clinical and Experimental Research*, *28*, 1517–1523.

Franklin, T. R., Harper, D., Kampman, K., et al. (2009). The $GABA_B$ agonist baclofen reduces cigarette consumption in a preliminary double-blind placebo-controlled smoking reduction study. *Drug and Alcohol Dependence*, *103*, 30–36.

Frankowska, M., Filip, M., & Przegalinski, E. (2007). Effects of $GABA_B$ receptor ligands in animal tests of depression and anxiety. *Pharmacological Reports*, *59*, 645–655.

Frankowska, M., Golda, A., Wydra, K., et al. (2010). Effects of imipramine or $GABA_B$ receptor ligands on the immobility, swimming and climbing in the forced swim test in rats following discontinuation of cocaine self-administration. *European Journal of Pharmacology*, *627*, 142–149.

Froestl, W., Bettler, B., Bittiger, H., et al. (1999). Ligands for the isolation of $GABA_B$ receptors. *Neuropharmacology*, *38*, 1641–1646.

Froestl, W., Bettler, B., Bittiger, H., et al. (2003). Ligands for expression cloning and isolation of $GABA_B$ receptors. *Farmaco*, *58*, 173–183.

Froestl, W., Cooke, N. G., & Mickel, S. J. (2007). Chemistry of $GABA_B$ modulators. In S. J. Enna & H. Möhler (Eds.), *The GABA receptors* (3rd ed., pp. 239–251). Totowa, NJ: Humana Press.

Froestl, W., Gallagher, M., Jenkins, H., et al. (2004). SGS742: The first $GABA_B$ receptor antagonist in clinical trials. *Biochemical Pharmacology*, *68*, 1479–1487.

Froestl, W., & Mickel, S. J. (1997). Chemistry of $GABA_B$ modulators. In S. J. Enna & N. G. Bowery (Eds.), *The GABA receptors* (2nd ed., pp. 271–296). Totowa, NJ: Humana Press.

Froestl, W., Mickel, S. J., Hall, R. G., et al. (1995a). Phosphinic acid analogues of GABA. 1. New potent and selective $GABA_B$ agonists. *Journal of Medicinal Chemistry*, *38*, 3297–3312.

Froestl, W., Mickel, S. J., Schmutz, M., & Bittiger, H. (1996). Potent, orally active $GABA_B$ receptor antagonists. *Pharmacology Reviews and Communications*, *8*, 127–133.

Froestl, W., Mickel, S. J., von Sprecher, G., et al. (1995b). Phosphinic acid analogues of GABA. 2. Selective, orally active $GABA_B$ antagonists. *Journal of Medicinal Chemistry*, *38*, 3313–3331.

Fromm, G. H. (1994). Baclofen as an adjuvant analgesic. *Journal of Pain and Symptom Management*, *9*, 500–509.

Fromm, G. H., & Terrence, C. F. (1987). Comparison of L-baclofen and racemic baclofen in trigeminal neuralgia. *Neurology*, *37*, 1725–1728.

Fromm, G. H., Terrence, C. F., & Chattha, A. S. (1984). Baclofen in the treatment of trigeminal neuralgia: Double-blind study and long-term follow-up. *Annals of neurology*, *15*, 240–244.

Fromm, G. H., Terrence, C. F., Chattha, A. S., et al. (1980). Baclofen in trigeminal neuralgia: Its effect on the spinal trigeminal nucleus: A pilot study. *Archives of Neurology*, *37*, 768–771.

Frost, F., Nanninga, J., Penn, R., et al. (1989). Intrathecal baclofen infusion. Effect on bladder management programs in patients with myelopathy. *American Journal of Physical Medicine & Rehabilitation*, *68*, 112–115.

Frydenvang, K., Enna, S. J., & Krogsgaard-Larsen, P. (1997). (−)-(*R*)-5,5-dimethylmorpholinyl-2-acetic acid ethyl ester hydrochloride. *Acta Crystallographica*, *C53*, 1088–1091.

Frydenvang, K., Hansen, J. J., Krogsgaard-Larsen, P., et al. (1994). $GABA_B$ antagonists: Resolution, absolute stereochemistry, and pharmacology of (*R*)- and (*S*)-phaclofen. *Chirality*, *6*, 583–589.

Gangadharan, V., Agarwal, N., Brugger, S., et al. (2009). Conditional gene deletion reveals functional redundancy of $GABA_B$ receptors in peripheral nociceptors *in vivo*. *Molecular Pain*, *5*, 68.

Gemignani, A., Paudice, P., Bonanno, G., & Raiteri, M. (1994). Pharmacological discrimination between γ-aminobutyric acid type B receptors regulating cholecystokinin and somatostatin release from rat neocortex synaptosomes. *Molecular Pharmacology*, *46*, 558–562.

Genkova-Papazova, M. G., Petkova, B., Shishkova, N., et al. (2000). The $GABA_B$ antagonist CGP 36742 prevents PTZ-kindling-provoked amnesia in rats. *European Neuropsychopharmacology*, *10*, 273–278.

Getova, D., & Bowery, N. G. (1998). The modulatory effects of high affinity $GABA_B$ receptor antagonists in an active avoidance learning paradigm in rats. *Psychopharmacology*, *137*, 369–373.

Getova, D., Bowery, N. G., & Spassov, N. (1997). Effects of $GABA_B$ receptor antagonists on learning and memory retention in a rat model of absence epilepsy. *European Journal of Pharmacology*, *320*, 9–13.

Getova, D. P., & Bowery, N. G. (2001). Effects of high-affinity $GABA_B$ receptor antagonists on active and passive avoidance responding in rodents with γ-hydroxybutyrolactone-induced absence syndrome. *Psychopharmacology*, *157*, 89–95.

Getova, D. P., & Dimitrova, D. D. (2007). Effects of $GABA_B$ receptor antagonists CGP63360, CGP76290A and CGP76291A on learning and memory processes in rodents. *Central European Journal of Medicine*, *2*, 280–293.

Gianino, J. M., York, M. M., Paice, J. A., et al. (1998). Quality of life: Effect of reduced spasticity from intrathecal baclofen. *Journal of Neuroscience Nursing*, *30*, 47–54.

Giesser, B. (1985). Multiple sclerosis. Current concepts in management. *Drugs*, *29*, 88–95.

Giorgetti, M., Hotsenpiller, G., Froestl, W., & Wolf, M. E. (2002). *In vivo* modulation of ventral tegmental area dopamine and glutamate efflux by local $GABA_B$ receptors is altered after repeated amphetamine treatment. *Neuroscience*, *109*, 585–595.

Gjoni, T., Desrayaud, S., Imobersteg, S., & Urwyler, S. (2006). The positive allosteric modulator GS39783 enhances $GABA_B$ receptor-mediated inhibition of cyclic AMP formation in rat striatum *in vivo*. *Journal of Neurochemistry*, *96*, 1416–1422.

Gjoni, T., & Urwyler, S. (2008). Receptor activation involving positive allosteric modulation, unlike full agonism, does not result in $GABA_B$ receptor desensitization. *Neuropharmacology, 55*, 1293–1299.
Gjoni, T., & Urwyler, S. (2009). Changes in the properties of allosteric and orthosteric $GABA_B$ receptor ligands after a continuous, desensitizing agonist pretreatment. *European Journal of Pharmacology, 603*, 37–41.
Goodwin, A. K., Froestl, W., & Weerts, E. M. (2005). Involvement of γ-hydroxybutyrate (GHB) and GABA-B receptors in the acute behavioral effects of GHB in baboons. *Psychopharmacology, 180*, 342–351.
Goodwin, A. K., Griffiths, R. R., Brown, P. R., et al. (2006). Chronic intragastric administration of γ-butyrolactone produces physical dependence in baboons. *Psychopharmacology, 189*, 71–82.
Grimm, D. R., DeLuca, R. V., Lesser, M., et al. (1997). Effects of $GABA_B$ agonist baclofen on bronchial hyperreactivity to inhaled histamine in subjects with cervical spinal cord injury. *Lung, 175*, 333–341.
Guelaud, C., Similowski, T., Bizec, J. L., et al. (1995). Baclofen therapy for chronic hiccup. *European Respiratory Journal, 8*, 235–237.
Guery, S., Floersheim, P., Kaupmann, K., & Froestl, W. (2007). Syntheses and optimization of new GS39783 analogues as positive allosteric modulators of $GABA_B$ receptors. *Bioorganic & Medicinal Chemistry Letters, 17*, 6206–6211.
Gupta, M., Greven, R., Jansen, E. E., et al. (2002). Therapeutic intervention in mice deficient for succinate semialdehyde dehydrogenase (γ-hydroxybutyric aciduria). *Journal of Pharmacology and Experimental Therapeutics, 302*, 180–187.
Hall, R. G., Kane, P. D., Bittiger, H., & Froestl, W. (1995). Phosphinic acid analogues of γ-aminobutyric acid (GABA). Synthesis of a new radioligand. *Journal of Labelled Compounds & Radiopharmaceuticals, 36*, 129–135.
Haney, M., Hart, C. L., & Foltin, R. W. (2006). Effects of baclofen on cocaine self-administration: Opioid- and nonopioid-dependent volunteers. *Neuropsychopharmacology, 31*, 1814–1821.
Heese, K., Otten, U., Mathivet, P., et al. (2000). $GABA_B$ receptor antagonists elevate both mRNA and protein levels of the neurotrophins nerve growth factor (NGF) and brain-derived neurotrophic factor (BDNF) but not neurotrophin-3 (NT-3) in brain and spinal cord of rats. *Neuropharmacology, 39*, 449–462.
Heinrichs, S. C., Leite-Morris, K. A., Carey, R. J., et al. (2010). Baclofen enhances extinction of opiate conditioned place preference. *Behavioural Brain Research, 207*, 353–359.
Helm, K. A., Haberman, R. P., Dean, S. L., et al. (2005). $GABA_B$ receptor antagonist SGS742 improves spatial memory and reduces protein binding to the cAMP response element (CRE) in the hippocampus. *Neuropharmacology, 48*, 956–964.
Hering-Hanit, R. (1999). Baclofen for prevention of migraine. *Cephalalgia, 19*, 589–591.
Hering-Hanit, R., & Gadoth, N. (2000). Baclofen in cluster headache. *Headache, 40*, 48–51.
Hering-Hanit, R., & Gadoth, N. (2001). The use of baclofen in cluster headache. *Current Pain and Headache Reports, 5*, 79–82.
Hill, D. R., & Bowery, N. G. (1981). ^{3}H-baclofen and ^{3}H-GABA bind to bicuculline-insensitive $GABA_B$ sites in rat brain. *Nature, 290*(5802), 149–152.
Hinton, T., Chebib, M., & Johnston, G.A.R. (2008). Enantioselective actions of 4-amino-3-hydroxybutanoic acid and (3-amino-2-hydroxypropyl)methylphosphinic acid at recombinant $GABA_C$ receptors. *Bioorganic & Medicinal Chemistry Letters, 18*, 402–404.
Hog, S., Wellendorph, P., Nielsen, B., et al. (2008). Novel high-affinity and selective biaromatic 4-substituted γ-hydroxybutyric acid (GHB) analogues as GHB ligands: Design, synthesis, and binding studies. *Journal of Medicinal Chemistry, 51*, 8088–8095.
Hogema, B. M., Gupta, M., Senephansiri, H., et al. (2001). Pharmacologic rescue of lethal seizures in mice deficient in succinate semialdehyde dehydrogenase. *Nature Genetics, 29*, 212–216.

Hollmann, M., O'Shea-Greenfield, A., Rogers, S. W., & Heinemann, S. (1989). Cloning by functional expression of a member of the glutamate receptor family. *Nature*, *342*, 643–648.

Holstein, S. E., Dobbs, L., & Phillips, T. J. (2009). Attenuation of the stimulant response to ethanol is associated with enhanced ataxia for a $GABA_A$, but not a $GABA_B$, receptor agonist. *Alcoholism, Clinical and Experimental Research*, *33*, 108–120.

Hosford, D. A., Wang, Y., Liu, C. C., & Snead, O. C., III. (1995). Characterization of the antiabsence effects of SCH 50911, a GABA-B receptor antagonist, in the lethargic mouse, γ-hydroxybutyrate, and pentylenetetrazole models. *Journal of Pharmacology and Experimental Therapeutics*, *274*, 1399–1403.

Hotsenpiller, G., & Wolf, M. E. (2003). Baclofen attenuates conditioned locomotion to cues associated with cocaine administration and stabilizes extracellular glutamate levels in rat nucleus accumbens. *Neuroscience*, *118*, 123–134.

Jacobson, L. H., & Cryan, J. F. (2008). Evaluation of the anxiolytic-like profile of the $GABA_B$ receptor positive modulator CGP7930 in rodents. *Neuropharmacology*, *54*, 854–862.

Jensen, A. A., Mosbacher, J., Elg, S., et al. (2002). The anticonvulsant gabapentin (neurontin) does not act through gamma-aminobutyric acid-B receptors. *Molecular Pharmacology*, *61*, 1377–1384.

Jingami, H., Nakanishi, S., & Morikawa, K. (2003). Structure of the metabotropic glutamate receptor. *Current Opinion in Neurobiology*, *13*, 271–278.

John, J. P., Sunyer, B., Höger, H., et al. (2009). Hippocampal synapsin isoform levels are linked to spatial memory enhancement by SGS742. *Hippocampus*, *19*, 731–738.

Kahn, R., Biswas, K., Childress, A. R., et al. (2009). Multi-center trial of baclofen for abstinence initiation in severe cocaine-dependent individuals. *Drug and Alcohol Dependence*, *103*, 59–64.

Kaminski, R. M., Van Rijn, C. M., Turski, W. A., et al. (2001). AMPA and $GABA_B$ receptor antagonists and their interaction in rats with a genetic form of absence epilepsy. *European Journal of Pharmacology*, *430*, 251–259.

Kantrowitz, J., Citrome, L., & Javitt, D. (2009). $GABA_B$ receptors, schizophrenia and sleep dysfunction: A review of the relationship and its potential clinical and therapeutic implications. *CNS Drugs*, *23*, 681–691.

Kaplan, G. B., McRoberts, R. L., III, & Smokler, H. J. (2004). Baclofen as adjunctive treatment for a patient with cocaine dependence and schizoaffective disorder. *Journal of Clinical Psychopharmacology*, *24*, 574–575.

Kaupmann, K., Bettler, B., Bittiger, H., Froestl, W., & Mickel, S. J. (1997b). Metabotropic $GABA_B$ receptors, receptor-specific ligands and their uses. WO 97/46675; Publ. date: 11 Dec 1997, Priority: 30 May 1996 for Novartis AG.

Kaupmann, K., Cryan, J. F., Wellendorph, P., et al. (2003). Specific γ-hydroxy-butyrate-binding sites but loss of pharmacological effects of γ-hydroxybutyrate in $GABA_{B1}$-deficient mice. *European Journal of Neuroscience*, *18*, 2722–2730.

Kaupmann, K., Froestl, W., & Bettler, B. (2001). $GABA_B$ receptors. In *Embryonic Encyclopedia of Life Sciences*, Nature Publishing Group, London, http://www.els.net.

Kaupmann, K., Huggel, K., Heid, J., et al. (1997a). Expression cloning of $GABA_B$ receptors uncovers similarity to metabotropic glutamate receptors. *Nature*, *386*, 239–246.

Kawakami, S., Uezono, Y., Makimoto, N., et al. (2004). Characterization of $GABA_B$ receptors involved in inhibition of motility associated with acetylcholine release in the dog small intestine: Possible existence of a heterodimer of $GABA_{B1}$ and $GABA_{B2}$ subunits. *Journal of Pharmacological Sciences*, *94*, 368–375.

Keberle, H., Faigle, J. W., & Wilhelm, M. (1968). Procedure for the preparation of new aminioacids. Swiss Patent 449046, Publ. date: 11 Apr 1968; Priority: 09 Jul 1963, *Chem Abstr. 69*, 106273f.

Keir, M. J., Barakat, M. J., Dev, K. K., et al. (1999). Characterisation and partial purification of the $GABA_B$ receptor from the rat cerebellum using the novel antagonist [^{3}H]CGP 62349. *Brain Research. Molecular Brain Research*, *71*, 279–289.

Kerr, D. I., Ong, J., Doolette, D. J., et al. (1995). The (*S*)-enantiomer of 2-hydroxysaclofen is the active $GABA_B$ receptor antagonist in central and peripheral preparations. *European Journal of Pharmacology*, *287*, 185–189.

Kerr, D. I., Ong, J., Johnston, G. A. R., et al. (1988). 2-Hydroxy-saclofen: An improved antagonist at central and peripheral $GABA_B$ receptors. *Neuroscience Letters*, *92*, 92–96.

Kerr, D. I., Ong, J., Prager, R. H., et al. (1987). Phaclofen: A peripheral and central baclofen antagonist. *Brain Research*, *405*, 150–154.

Kirmse, K., Kirischuk, S., & Grantyn, R. (2009). Role of GABA transporter 3 in GABAergic synaptic transmission at striatal output neurons. *Synapse*, *63*, 921–929.

Knapp, D. J., Overstreet, D. H., & Breese, G. R. (2007). Baclofen blocks expression and sensitization of anxiety-like behavior in an animal model of repeated stress and ethanol withdrawal. *Alcoholism, Clinical and Experimental Research*, *31*, 582–595.

Koek, G. H., Sifrim, D., Lerut, T., et al. (2003). Effect of the $GABA_B$ agonist baclofen in patients with symptoms and duodeno-gastro-oesophageal reflux refractory to proton pump inhibitors. *Gut*, *52*, 1397–1402.

Krach, L. E. (2009). Intrathecal baclofen use in adults with cerebral palsy. *Developmental Medicine and Child Neurology*, *51*(Suppl. 4), 106–112.

Kroin, J. S. (1992). Intrathecal drug administration. Present use and future trends. *Clinical Pharmacokinetics*, *22*, 319–326.

Kubo, T., Fukuda, K., Mikami, A., et al. (1986). Cloning, sequencing and expression of complementary DNA encoding the muscarinic acetylcholine receptor. *Nature*, *323*, 411–416.

Kunishima, N., Shimada, Y., Tsuji, Y., et al. (2000). Structural basis of glutamate recognition by a dimeric metabotropic glutamate receptor. *Nature*, *407*, 971–977.

Lal, R., Sukbuntherng, J., Tai, E. H., et al. (2009). Arbaclofen placarbil, a novel *R*-baclofen prodrug: Improved absorption, distribution, metabolism, and elimination properties compared with *R*-baclofen. *Journal of Pharmacology and Experimental Therapeutics*, *330*, 911–921.

Lanneau, C., Green, A., Hirst, W. D., et al. (2001). Gabapentin is not a $GABA_B$ receptor agonist. *Neuropharmacology*, *41*, 965–975.

Lapin, I. (2001). Phenibut (beta-phenyl-GABA): A tranquilizer and nootropic drug. *CNS Drug Reviews*, *7*, 471–481.

Lasarge, C. L., Banuelos, C., Mayse, J. D., & Bizon, J. L. (2009). Blockade of $GABA_B$ receptors completely reverses age-related learning impairment. *Neuroscience*, *164*, 941–947.

Le Foll, B., Wertheim, C. E., & Goldberg, S. R. (2008). Effects of baclofen on conditioned rewarding and discriminative stimulus effects of nicotine in rats. *Neuroscience Letters*, *443*, 236–240.

Leggio, L. (2009). Understanding and treating alcohol craving and dependence: Recent pharmacological and neuroendocrinological findings. *Alcohol and Alcoholism*, *44*, 341–352.

Lehmann, A. (2008). Novel treatments of GERD: Focus on the lower esophageal sphincter. *European Review for Medical and Pharmacological Sciences*, *12*(Suppl. 1), 103–110.

Lehmann, A. (2009). $GABA_B$ receptors as drug targets to treat gastroesophageal reflux disease. *Pharmacology & Therapeutics*, *122*, 239–245.

Lehmann, A., Antonsson, M., Bremner-Danielsen, M., et al. (1999). Activation of the $GABA_B$ receptor inhibits transient lower esophageal sphincter relaxations in dogs. *Gastroenterology*, *117*, 1147–1154.

Lehmann, A., Antonsson, M., Holmberg, A. A., et al. (2009). (*R*)-(3-amino-2-fluoropropyl) phosphinic acid (AZD3355), a novel $GABA_B$ receptor agonist, inhibits transient lower esophageal sphincter relaxation through a peripheral mode of action. *Journal of Pharmacology and Experimental Therapeutics*, *331*, 504–512.

Lehmann, A., Hansson-Branden, L., & Karrberg, L. (2000). Effects of repeated administration of baclofen on transient lower esophageal sphincter relaxation in the dog. *European Journal of Pharmacology*, *403*, 163–167.

Leisen, C., Langguth, P., Herbert, B., et al. (2003). Lipophilicities of baclofen ester prodrugs correlate with affinities to the ATP-dependent efflux pump P-glycoprotein: Relevance for their permeation across the blood-brain barrier? *Pharmaceutical Research*, *20*, 772–778.

Leite-Morris, K. A., Fukudome, E. Y., Shoeb, M. H., et al. (2004). $GABA_B$ receptor activation in the ventral tegmental area inhibits the acquisition and expression of opiate-induced motor sensitization. *Journal of Pharmacology and Experimental Therapeutics*, *308*, 667–678.

Leung, L. S., & Peloquin, P. (2006). $GABA_B$ receptors inhibit backpropagating dendritic spikes in hippocampal CA1 pyramidal cells *in vivo*. *Hippocampus*, *16*, 388–407.

Leung, L. S., Peloquin, P., & Canning, K. J. (2008). Paired-pulse depression of excitatory postsynaptic current sinks in hippocampal CA1 *in vivo*. *Hippocampus*, *18*, 1008–1020.

Leung, L. S., & Shen, B. (2007). $GABA_B$ receptor blockade enhances theta and gamma rhythms in the hippocampus of behaving rats. *Hippocampus*, *17*, 281–291.

Levin, E. D., Weber, E., & Icenogle, L. (2004). Baclofen interactions with nicotine in rats: Effects on memory. *Pharmacology, Biochemistry, and Behavior*, *79*, 343–348.

Lhuillier, L., Mombereau, C., Cryan, C. F., et al. (2007). $GABA_B$ receptor-positive modulation decreases selective molecular and behavioral effects of cocaine. *Neuropsychopharmacology*, *32*, 388–398.

Li, X., Cao, J. H., Li, Y., Rondard, P., et al. (2008). Activity-based probe for specific photoaffinity labeling γ-aminobutyric acid B ($GABA_B$) receptors on living cells: Design, synthesis, and biological evaluation. *Journal of Medicinal Chemistry*, *51*, 3057–3060.

Liang, J. H., Chen, F., Krstew, E., et al. (2006). The $GABA_B$ receptor allosteric modulator CGP7930, like baclofen, reduces operant self-administration of ethanol in alcohol-preferring rats. *Neuropharmacology*, *50*, 632–639.

Lidums, I., Lehmann, A., Checklin, H., et al. (2000). Control of transient lower esophageal sphincter relaxations and reflux by the $GABA_B$ agonist baclofen in normal subjects. *Gastroenterology*, *118*, 7–13.

Lile, J. A., Stoops, W. W., Allen, T. S., et al. (2004). Baclofen does not alter the reinforcing, subject-rated or cardiovascular effects of intranasal cocaine in humans. *Psychopharmacology*, *171*, 441–449.

Ling, W., Shoptaw, S., & Majewska, D. (1998). Baclofen as a cocaine anti-craving medication: A preliminary clinical study. *Neuropsychopharmacology*, *18*, 403–404.

Lingenhoehl, K., Brom, R., Heid, J., et al. (1999). γ-Hydroxybutyrate is a weak agonist at recombinant $GABA_B$ receptors. *Neuropharmacology*, *38*, 1667–1673.

Lübbert, H., Hoffman, B. J., Snutch, T. P., et al. (1987). cDNA cloning of a serotonin 5-HT_{1C} receptor by electrophysiological assays of mRNA-injected *Xenopus* oocytes. *Proceedings of the National Academy of Sciences of the United States of America*, *84*, 4332–4336.

Macey, D. J., Froestl, W., Koob, G. F., & Markou, A. (2001). Both $GABA_B$ receptor agonist and antagonists decreased brain stimulation reward in the rat. *Neuropharmacology*, *40*, 676–685.

Maccioni, P., Bienkowski, P., Carai, M. A., et al. (2008b). Baclofen attenuates cue-induced reinstatement of alcohol-seeking behavior in Sardinian alcohol-preferring (sP) rats. *Drug and Alcohol Dependence*, *95*, 284–287.

Maccioni, P., Carai, M. A., Kaupmann, K., et al. (2009). Reduction of alcohol's reinforcing and motivational properties by the positive allosteric modulator of the $GABA_B$ receptor, BHF177, in alcohol-preferring rats. *Alcoholism, Clinical and Experimental Research*, *33*, 1749–1756.

Maccioni, P., Fantini, N., Froestl, W., et al. (2008a). Specific reduction of alcohol's motivational properties by the positive allosteric modulator of the $GABA_B$ receptor, GS39783—comparison with the effect of the $GABA_B$ receptor direct agonist, baclofen. *Alcoholism, Clinical and Experimental Research*, *32*, 1558–1564.

Maccioni, P., Pes, D., Orru, A., et al. (2007). Reducing effect of the positive allosteric modulator of the $GABA_B$ receptor, GS39783, on alcohol self-administration in alcohol-preferring rats. *Psychopharmacology*, *193*, 171–178.

Maccioni, P., Serra, S., Vacca, G., et al. (2005). Baclofen-induced reduction of alcohol reinforcement in alcohol-preferring rats. *Alcohol*, *36*, 161–168.

Magnaghi, V., Ballabio, M., Camozzi, F., et al. (2008). Altered peripheral myelination in mice lacking $GABA_B$ receptors. *Molecular and Cellular Neurosciences*, *37*, 599–609.

Magnaghi, V., Ballabio, M., Cavarretta, I. T., et al. (2004). $GABA_B$ receptors in Schwann cells influence proliferation and myelin protein expression. *European Journal of Neuroscience*, *19*, 2641–2649.

Malherbe, P., Masciadri, R., Norcross, R. D., et al. (2008). Characterization of (*R*,*S*)-5,7-di-tert-butyl-3-hydroxy-3-trifluoromethyl-3H-benzofuran-2-one as a positive allosteric modulator of $GABA_B$ receptors. *British Journal of Pharmacology*, *154*, 797–811.

Malherbe, P., Masciadri, R., Prinssen, E., et al. (2005). Aminomethylpyrimidines as allosteric enhancers of the $GABA_B$ receptors. , US 2005/0197337 A1; Publ. date: 08 Sep 2005, Priority: EP2004-100830, 02 Mar 2004, for Hoffmann-La Roche Inc.

Manning, J. P., Richards, D. A., & Bowery, N. G. (2003). Pharmacology of absence epilepsy. *Trends in Pharmacological Sciences*, *24*, 542–549.

Mannoury la Cour, C., Herbelles, C., Pasteau, P., et al. (2008). Influence of positive allosteric modulators on $GABA_B$ receptor coupling in rat brain: A scintillation proximity assay characterisation of G protein subtypes. *Journal of Neurochemistry*, *105*, 308–323.

Mares, P., & Kubova, H. (2008). What is the role of neurotransmitter systems in cortical seizures? *Physiological Research*, *57*(Suppl. 3), S111–S120.

Mares, P., & Slamberova, R. (2006). Opposite effects of a $GABA_B$ antagonist in two models of epileptic seizures in developing rats. *Brain Research Bulletin*, *71*, 160–166.

Markou, A. (2008). Neurobiology of nicotine dependence. *Philosophical Transactions of the Royal Society of London. Series B, Biological Sciences*, *363*(1507), 3159–3168.

Markou, A., Paterson, N. E., & Semenova, S. (2004). Role of γ-aminobutyric acid (GABA) and metabotropic glutamate receptors in nicotine reinforcement: Potential pharmacotherapies for smoking cessation. *Annals of the New York Academy of Sciences*, *1025*, 491–503.

Marshall, F. H. (2005). Is the $GABA_B$ heterodimer a good drug target? *Journal of Molecular Neuroscience*, *26*, 169–176.

Masu, M., Tanabe, Y., Tsuchida, K., et al. (1991). Sequence and expression of a metabotropic glutamate receptor. *Nature*, *349*, 760–765.

Mathivet, P., Bernasconi, R., de Barry J., et al. (1997). Binding characteristics of γ-hydroxybutyric acid as a weak but selective $GABA_B$ receptor agonist. *European Journal of Pharmacology*, *321*, 67–75.

McCarson, K. E., Duric, V., Reisman, S. A., et al. (2006). $GABA_B$ receptor function and subunit expression in the rat spinal cord as indicators of stress and the antinociceptive response to antidepressants. *Brain Research*, *1068*, 109–117.

McCarson, K. E., Ralya, A., Reisman, S. A., & Enna, S. J. (2005). Amitriptyline prevents thermal hyperalgesia and modifications in rat spinal cord $GABA_B$ receptor expression and function in an animal model of neuropathic pain. *Biochemical Pharmacology*, *71*, 196–202.

Miller, F., & Korsvik, H. (1981). Baclofen in the treatment of stiff-man syndrome. *Annals of Neurology*, *9*(5), 511–512.

Mombereau, C., Kaupmann, K., Froestl, W., et al. (2004). Genetic and pharmacological evidence of a role for $GABA_B$ receptors in the modulation of anxiety- and antidepressant-like behavior. *Neuropsychopharmacology*, *29*, 1050–1062.

Mombereau, C., Lhuillier, L., Kaupmann, K., et al. (2007). $GABA_B$ receptor-positive modulation-induced blockade of the rewarding properties of nicotine is associated with a reduction in nucleus accumbens DeltaFosB accumulation. *Journal of Pharmacology and Experimental Therapeutics*, *321*, 172–177.

Moriyoshi, K., Masu, M., Ishii, T., et al. (1991). Molecular cloning and characterization of the rat NMDA receptor. *Nature*, *354*, 31–37.

Moutoussis, M., & Orrell, W. (1996). Baclofen therapy for rigidity associated with Lewy body dementia. *British Journal of Psychiatry*, *169*, 795.

Müller, H., Borner, U., Zierski, J., et al. (1987). Intrathecal baclofen for treatment of tetanus-induced spasticity. *Anesthesiology*, *66*, 76–79.

Nakagawa, Y., Sasaki, A., & Takashima, T. (1999). The $GABA_B$ receptor antagonist CGP36742 improves learned helplessness in rats. *European Journal of Pharmacology*, *381*, 1–7.

Nakagawa, Y., & Takashima, T. (1997). The $GABA_B$ receptor antagonist CGP36742 attenuates the baclofen- and scopolamine-induced deficit in Morris water maze task in rats. *Brain Research*, *766*, 101–106.

Nanninga, J. B., Frost, F., & Penn, R. (1989). Effect of intrathecal baclofen on bladder and sphincter function. *Journal of Urology*, *142*, 101–105.

Ng, G. Y., Bertrand, S., Sullivan, R., et al. (2001). Gamma-aminobutyric acid type B receptors with specific heterodimer composition and postsynaptic actions in hippocampal neurons are targets of anticonvulsant gabapentin action. *Molecular Pharmacology*, *59*, 144–152.

Nicoll, R. A. (2004). My close encounter with $GABA_B$ receptors. *Biochemical Pharmacology*, *68*, 1667–1674.

Nowak, G., Partyka, A., Palucha, A., et al. (2006). Antidepressant-like activity of CGP 36742 and CGP 51176, selective $GABA_B$ receptor antagonists, in rodents. *British Journal of Pharmacology*, *149*, 581–590.

Nyitrai, G., Kekesi, K. A., Emri, Z., et al. (2003). $GABA_B$ receptor antagonist CGP-36742 enhances somatostatin release in the rat hippocampus in vivo and in vitro. *European Journal of Pharmacology*, *478*(2–3), 111–119.

O'Brien, C. F., Seeberger, L. C., & Smith, D. B. (1996). Spasticity after stroke. Epidemiology and optimal treatment. *Drugs Aging*, *9*, 332–340.

Ochs, G., Struppler, A., Meyerson, B. A., et al. (1989). Intrathecal baclofen for long-term treatment of spasticity: A multi-centre study. *Journal of Neurology, Neurosurgery, and Psychiatry*, *52*, 933–939.

Ochs, G. A. (1993). Intrathecal baclofen. *Bailliére's Clinical Neurology*, 2, 73–86.

Olianas, M. C., Ambu, R., Garau, L., et al. (2005). Allosteric modulation of $GABA_B$ receptor function in human frontal cortex. *Neurochemistry International*, *46*, 149–158.

Olpe, H. R., Demieville, H., Baltzer, V., et al. (1978). The biological activity of D- and L-baclofen (Lioresal). *European Journal of Pharmacology*, *52*, 133–136.

Olpe, H. R., Karlsson, G., Pozza, M. F., et al. (1990). CGP 35348: A centrally active blocker of $GABA_B$ receptors. *European Journal of Pharmacology*, *187*, 27–38.

Omari, T. I., Benninga, M. A., Sansom, L., et al. (2006). Effect of baclofen on esophagogastric motility and gastroesophageal reflux in children with gastroesophageal reflux disease: A randomized controlled trial. *Journal of Pediatrics*, *149*, 468–474.

Onali, P., Mascia, F. M., & Olianas, M. C. (2003). Positive regulation of $GABA_B$ receptors dually coupled to cyclic AMP by the allosteric agent CGP7930. *European Journal of Pharmacology*, *471*, 77–84.

Ong, J., Bexis, S., Marino, V., et al. (2001a). Comparative activities of the enantiomeric $GABA_B$ receptor agonists CGP 44532 and 44533 in central and peripheral tissues. *European Journal of Pharmacology*, *412*, 27–37.

Ong, J., Bexis, S., Marino, V., et al. (2001b). CGP 36216 is a selective antagonist at $GABA_B$ presynaptic receptors in rat brain. *European Journal of Pharmacology*, *415*, 191–195.

Ong, J., Kerr, D. I., Doolette, D. J., et al. (1993). (*R*)-(−)-beta-phenyl-GABA is a full agonist at $GABA_B$ receptors in brain slices but a partial agonist in the ileum. *European Journal of Pharmacology*, *233*, 169–172.

Orru, A., Lai, A., Lobina, C., et al. (2005). Reducing effect of the positive allosteric modulators of the $GABA_B$ receptor, CGP7930 and GS39783, on alcohol intake in alcohol-preferring rats. *European Journal of Pharmacology*, *525*, 105–111.

Pacey, L. K., Heximer, S. P., & Hampson, D. R. (2009). Increased $GABA_B$ receptor-mediated signaling reduces the susceptibility of fragile X knockout mice to audiogenic seizures. *Molecular Pharmacology*, *76*, 18–24.

Parker, D. A., Marino, V., Ong, J., et al. (2008). The CGP7930 analogue 2,6-di-tert-butyl-4-(3-hydroxy-2-spiropentylpropyl)-phenol (BSPP) potentiates baclofen action at $GABA_B$ autoreceptors. *Clinical and Experimental Pharmacology & Physiology*, *35*, 1113–1115.

Parker, D. A., Ong, J., Marino, V., et al. (2004). Gabapentin activates presynaptic $GABA_B$ heteroreceptors in rat cortical slices. *European Journal of Pharmacology*, *495*, 137–143.

Partyka, A., Klodzinska, A., Szewczyk, B., et al. (2007). Effects of $GABA_B$ receptor ligands in rodent tests of anxiety-like behavior. *Pharmacological Reports*, *59*, 757–762.

Patel, S., Naeem, S., Kesingland, A., et al. (2001). The effects of $GABA_B$ agonists and gabapentin on mechanical hyperalgesia in models of neuropathic and inflammatory pain in rats. *Pain*, *90*, 217–226.

Paterson, N. E., Bruijnzeel, A. W., Kenny, P. J., et al. (2005a). Prolonged nicotine exposure does not alter $GABA_B$ receptor-mediated regulation of brain reward function. *Neuropharmacology*, *49*, 953–962.

Paterson, N. E., Froestl, W., & Markou, A. (2004). The $GABA_B$ receptor agonists baclofen and CGP44532 decreased nicotine self-administration in the rat. *Psychopharmacology*, *172*, 179–186.

Paterson, N. E., Froestl, W., & Markou, A. (2005b) Repeated administration of the $GABA_B$ receptor agonist CGP44532 decreased nicotine self-administration, and acute administration decreased cue-induced reinstatement of nicotine-seeking in rats. *Neuropsychopharmacology*, *30*, 119–128.

Paterson, N. E., Vlachou, S., Guery, S., et al. (2008). Positive modulation of $GABA_B$ receptors decreased nicotine self-administration and counteracted nicotine-induced enhancement of brain reward function in rats. *Journal of Pharmacology and Experimental Therapeutics*, *326*, 306–314.

Penn, R. D., & Kroin, J. S. (1984). Intrathecal baclofen alleviates spinal cord spasticity. *Lancet*, *323*(8385), 1078.

Penn, R. D., & Kroin, J. S. (1985). Continuous intrathecal baclofen for severe spasticity. *Lancet*, *326*(8447), 125–127.

Penn, R. D., & Kroin, J. S. (1987). Long-term intrathecal baclofen infusion for treatment of spasticity. *Journal of Neurosurgery*, *66*, 181–185.

Pibiri, F., Carboni, G., Carai, M. A., et al. (2005). Up-regulation of $GABA_B$ receptors by chronic administration of the $GABA_B$ receptor antagonist SCH 50,911. *European Journal of Pharmacology*, *515*, 94–98.

Pilc, A., & Lloyd, K. G. (1984). Chronic antidepressants and GABA "B" receptors: A GABA hypothesis of antidepressant drug action. *Life Sciences*, *35*, 2149–2154.

Pilc, A., & Nowak, G. (2005). GABAergic hypotheses of anxiety and depression: Focus on $GABA_B$ receptors. *Drugs Today*, *41*, 755–766.

Pin, J.-P., Kniazeff, J., Binet, V., et al. (2004). Activation mechanism of the heterodimeric $GABA_B$ receptor. *Biochemical Pharmacology*, *68*, 1565–1572.

Pin, J.-P., & Prézeau, L. (2007). Allosteric modulators of $GABA_B$ receptors: Mechanism of action and therapeutic perspective. *Current Neuropharmacology*, *5*, 195–201.

Pittaluga, A., Feligioni, M., Ghersi, C., et al. (2001). Potentiation of NMDA receptor function through somatostatin release: A possible mechanism for the cognition-enhancing activity of $GABA_B$ receptor antagonists. *Neuropharmacology*, *41*, 301–310.

Pozza, M. F., Manuel, N. A., Steinmann, M., et al. (1999). Comparison of antagonist potencies at pre- and post-synaptic $GABA_B$ receptors at inhibitory synapses in the CA1 region of the rat hippocampus. *British Journal of Pharmacology*, *127*, 211–219.

Prager, R. H., Schafer, K., Hamon, D. P., et al. (1995). The synthesis of (*R*)-(−) and (*S*)-(+)-hydroxy-saclofen. *Tetrahedron*, *51*, 11465–11472.

Prakash, Y. S. (2009). Gamma-aminobutyric acid: Something old, something new for bronchodilation. *Anesthesiology*, *110*, 696–697.

Pratt, G. D., & Bowery, N. G. (1993). Repeated administration of desipramine and a $GABA_B$ receptor antagonist, CGP 36742, discretely up-regulates $GABA_B$ receptor binding sites in rat frontal cortex. *British Journal of Pharmacology*, *110*, 724–735.

Princivalle, A. P., Duncan, J. S., Thom, M., & Bowery, N. G. (2002). Studies of $GABA_B$ receptors labelled with [^{3}H]-CGP62349 in hippocampus resected from patients with temporal lobe epilepsy. *British Journal of Pharmacology*, *136*, 1099–1106.

Princivalle, A. P., Richards, D. A., Duncan, J. S., et al. (2003). Modification of $GABA_{B1}$ and $GABA_{B2}$ receptor subunits in the somatosensory cerebral cortex and thalamus of rats with absence seizures (GAERS). *Epilepsy Research*, *55*, 39–51.

Qu, L., Boyce, R., & Leung, L. S. (2010). Seizures in the developing brain result in a long-lasting decrease in $GABA_B$ inhibitory postsynaptic currents in the rat hippocampus. *Neurobiology of Disease*, *37*, 704–710.

Raiteri, M. (2008). Presynaptic metabotropic glutamate and $GABA_B$ receptors. *Handbook of Experimental Pharmacology*, *184*, 373–407.

Rapidi, C. A., Panourias, I. G., Petropoulou, K., et al. (2007). Management and rehabilitation of neuropathic bladder in patients with spinal cord lesion. *Acta Neurochirurgica. Supplement*, *97*(Pt 1), 307–314.

Rawlins, P. (1998). Patient management of cerebral origin spasticity with intrathecal baclofen. *Journal of Neuroscience Nursing*, *30*(1), 32–35 and 40–46.

Ren, X., & Mody, I. (2003). γ-Hydroxybutyrate reduces mitogen-activated protein kinase phosphorylation via $GABA_B$ receptor activation in mouse frontal cortex and hippocampus. *Journal of Biological Chemistry*, *278*, 42006–42011.

Ren, X., & Mody, I. (2006). γ-Hydroxybutyrate induces cyclic AMP-responsive element-binding protein phosphorylation in mouse hippocampus: An involvement of $GABA_B$ receptors and cAMP-dependent protein kinase activation. *Neuroscience*, *141*, 269–275.

Riahi, E., Mirzaii-Dizgah, I., Karimian, S. M., et al. (2009). Attenuation of morphine withdrawal signs by a $GABA_B$ receptor agonist in the locus coeruleus of rats. *Behavioural Brain Research*, *196*, 11–14.

Richards, D. A., & Bowery, N. G. (1996). Anti-seizure effects of the $GABA_B$ antagonist, SCH50911, in the genetic absence epilepsy rat of Strasbourg (GAERS). *Pharmacology Reviews and Communications*, *8*, 227–230.

Romei, C., Luccini, E., Raiteri, M., & Raiteri, L. (2009). $GABA_B$ presynaptic receptors modulate glycine exocytosis from mouse spinal cord and hippocampus glycinergic nerve endings. *Pharmacological Research*, *59*, 154–159.

Rotheram-Fuller, E., De la Garza, R., II, Mahoney, J. J., III, et al.(2007). Subjective and cardiovascular effects of cocaine during treatment with amantadine and baclofen in combination. *Psychiatry Research*, *152*, 205–210.

Sahraei, H., Etemadi, L., Rostami, P., et al. (2009). $GABA_B$ receptors within the ventral tegmental area are involved in the expression and acquisition of morphine-induced place preference in morphine-sensitized rats. *Pharmacology, Biochemistry, and Behavior*, *91*, 409–416.

Sands, S. A., Reisman, S. A., & Enna, S. J. (2004). Effect of antidepressants on $GABA_B$ receptor function and subunit expression in rat hippocampus. *Biochemical Pharmacology*, *68*, 1489–1495.

Santicioli, P., Tramontana, M., Del Bianco, E., et al. (1991). $GABA_A$ and $GABA_B$ receptors modulate the K^+-evoked release of sensory CGRP from the guinea pig urinary bladder. *Life Sciences*, *48*, L69–L72.

Santos, M. L., Mota-Miranda, A., Alves-Pereira, A., et al. (2004). Intrathecal baclofen for the treatment of tetanus. *Clinical Infectious Diseases*, *38*(3), 321–328.

Sawynok, J., & Dickson, C. (1984). D-Baclofen: Is it an antagonist at baclofen receptors? *Progress in Neuro-psychopharmacology & Biological Psychiatry*, *8*, 729–731.

Sawynok, J., & Dickson, C. (1985). D-Baclofen is an antagonist at baclofen receptors mediating antinociception in the spinal cord. *Pharmacology*, *31*, 248–259.

Scheinberg, A., Hall, K., Lam, L. T., et al. (2006). Oral baclofen in children with cerebral palsy: A double-blind cross-over pilot study. *Journal of Paediatrics and Child Health*, *42*, 715–720.

Schofield, P. R., Darlison, M. G., Fujita, N., et al. (1987). Sequence and functional expression of the $GABA_A$ receptor shows a ligand-gated receptor super-family. *Nature*, *328*, 221–227.

Sekiguchi, M., Sakuta, H., Okamoto, K., & Sakai, Y. (1990). $GABA_B$ receptors expressed in *Xenopus* oocytes by guinea pig cerebral mRNA are functionally coupled with Ca^{2+}-dependent Cl^- channels and with K^+ channels, through GTP-binding proteins. *Brain Research. Molecular Brain Research*, *8*, 301–309.

Shimizu, S., Honda, M., Tanabe, M., et al. (2004). $GABA_B$ receptors do not mediate the inhibitory actions of gabapentin on spinal reflexes in rats. *Journal of Pharmacological Sciences*, *96*, 444–449.

Shoptaw, S., Yang, X., Rotheram-Fuller, E. J., et al. (2003). Randomized placebo-controlled trial of baclofen for cocaine dependence: Preliminary effects for individuals with chronic patterns of cocaine use. *Journal of Clinical Psychiatry*, *64*, 1440–1448.

Silbert, P. L., Matsumoto, J. Y., McManis, P. G., et al. (1995). Intrathecal baclofen therapy in stiff-man syndrome: A double-blind, placebo-controlled trial. *Neurology*, *45*, 1893–1897.

Slattery, D. A., & Cryan, J. F. (2006). The role of $GABA_B$ receptors in depression and antidepressant-related behavioural responses. *Drug Development Research*, *67*, 477–494.

Slattery, D. A., Desrayaud, S., & Cryan, J. F. (2005b). $GABA_B$ receptor antagonist-mediated antidepressant-like behavior is serotonin-dependent. *Journal of Pharmacology and Experimental Therapeutics*, *312*, 290–296.

Slattery, D. A., Markou, A., Froestl, W., & Cryan, J. F. (2005a). The $GABA_B$ receptor-positive modulator GS39783 and the $GABA_B$ receptor agonist baclofen attenuate the reward-facilitating effects of cocaine: Intercranial self-stimulation studies in the rat. *Neuropsychopharmacology*, *30*, 2065–2072.

Sloviter, R. S., Ali-Akbarian, L., Elliott, R. C., Bowery, B. J., & Bowery, N. G. (1999). Localization of $GABA_B$ (R1) receptors in the rat hippocampus by immunocytochemistry and high resolution autoradiography, with specific reference to its localization in identified hippocampal interneuron subpopulations. *Neuropharmacology*, *38*, 1707–1721.

Smith, C. R., LaRocca, N. G., Giesser, B. S., et al. (1991). High-dose oral baclofen: Experience in patients with multiple sclerosis. *Neurology*, *41*, 1829–1831.

Smith, M. A., Yancey, D. L., Morgan, D., et al. (2004). Effects of positive allosteric modulators of the $GABA_B$ receptor on cocaine self-administration in rats. *Psychopharmacology*, *173*, 105–111.

Spano, M. S., Fattore, L., Fratta, W., et al. (2007). The $GABA_B$ receptor agonist baclofen prevents heroin-induced reinstatement of heroin-seeking behavior in rats. *Neuropharmacology*, *52*, 1555–1562.

Stayer, C., Tronnier, V., Dressnandt, J., et al. (1997). Intrathecal baclofen therapy for stiff-man syndrome and progressive encephalomyelopathy with rigidity and myoclonus. *Neurology*, *49*, 1591–1597.

Stringer, J. L., & Lorenzo, N. (1999). The reduction in paired-pulse inhibition in the rat hippocampus by gabapentin is independent of $GABA_B$ receptor receptor activation. *Epilepsy Research*, *33*, 169–176.

Sunyer, B., Diao, W. F., Kang, S. U., et al. (2008a). Cognitive enhancement by SGS742 in OF1 mice is linked to specific hippocampal protein expression. *Journal of Proteome Research*, *7*, 5237–5253.

Sunyer, B., Patil, S., Frischer, C., et al. (2007). Strain-dependent effects of SGS742 in the mouse. *Behavioural Brain Research*, *181*, 64–75.

Sunyer, B., Shim, K. S., An, G., et al. (2009). Hippocampal levels of phosphorylated protein kinase A (phosphor-S96) are linked to spatial memory enhancement by SGS742. *Hippocampus*, *19*, 90–98.

Sunyer, B., Shim, K. S., Höger, H., & Lubec, G. (2008b). The cognitive enhancer SGS742 does not involve major known signaling cascades in OF1 mice. *Neurochemical Research*, *33*, 1384–1392.

Taniyama, K., Takeda, K., Ando, H., et al. (1991). Expression of the $GABA_B$ receptor in *Xenopus* oocytes and inhibition of the response by activation of protein kinase C. *FEBS Letters*, *278*, 222–224.

Taylor, M. C., & Bates, C. P. (1979). A double-blind crossover trial of baclofen—a new treatment for the unstable bladder syndrome. *British Journal of Urology*, *51*, 504–505.

Teichgräber, L. A., Lehmann, T. N., Meencke, H. J., et al. (2009). Impaired function of $GABA_B$ receptors in tissues from pharmacoresistant epilepsy patients. *Epilepsia*, *50*, 1697–1716.

Terrence, C. F., Sax, M., Fromm, G. H., et al. (1983). Effect of baclofen enantiomorphs on the spinal trigeminal nucleus and steric similarities of carbamazepine. *Pharmacology*, *27*, 85–94.

Todde, S., Froestl, W., Stampf P., et al. (1997). [^{11}C]CGP62349, a $GABA_B$ receptor antagonist ligand: Radiosynthesis and PET studies in monkeys. *Journal of Labelled Compounds & Radiopharmaceuticals*, *40*, 651–652.

Todde, S., Moresco, R. M., Froestl, W., et al. (2000). Synthesis and *in vivo* evaluation of [^{11}C] CGP62349, a new $GABA_B$ receptor antagonist. *Nuclear Medicine and Biology*, *27*, 565–569.

Tomlinson, J., Cummins, H., Wendt, J., et al. (2004). SGS742, a novel $GABA_B$ receptor antagonist improves cognition in patients with mild cognitive impairment. *Meeting of the American Association of Neurology*, , San Francisco, April 27, Poster Po 2.053.

Tu, H., Rondard, P., Xu, C., et al. (2007). Dominant role of $GABA_{B2}$ and Gβγ for $GABA_B$ receptor-mediated-ERK1/2/CREB pathway in cerebellar neurons. *Cellular Signalling*, *19*, 1996–2002.

Tu, H., Xu, C., Zhang, W., et al. (2010). $GABA_B$ receptor activation protects neurons from apoptosis via IGF-1 receptor transactivation. *Journal of Neuroscience*, *30*, 749–759.

Turkyilmaz, A., & Eroglu, A. (2008). Use of baclofen in the treatment of esophageal stent-related hiccups. *Annals of Thoracic Surgery*, *85*, 328–330.

Twycross, R. (2003). Baclofen for hiccups. *American Journal of Hospice & Palliative Care*, *20*, 262.

Ulrich, D., Besseyrias, V., & Bettler, B. (2007). Functional mapping of $GABA_B$ -receptor subtypes in the thalamus. *Journal of Neurophysiology*, *98*, 3791–3795.

Urwyler, S. (2006). Allosteric modulation of $GABA_B$ receptors. In N. G. Bowery (Ed.), *Allosteric Receptor Modulation in Drug Targeting* (pp. 235–258). New York, NY: Informa Healthcare (Taylor & Francis).

Urwyler, S., Gjoni, T., Koljatic, J., et al. (2005). Mechanisms of allosteric modulation at $GABA_B$ receptors by CGP7930 and GS39783: Effects on affinities and efficacies of orthosteric ligands with distinct intrinsic properties. *Neuropharmacology*, *48*, 343–353.

Urwyler, S., Mosbacher, J., Lingenhoehl, K., et al. (2001). Positive allosteric modulation of native and recombinant $GABA_B$ receptors by 2,6-di-tert.-butyl-4-(3-hydroxy-2,2-dimethyl-propyl)-phenol (CGP7930) and its aldehyde analogue CGP13501. *Molecular Pharmacology, 60*, 963–971.

Urwyler, S., Pozza, M. F., Lingenhoehl, K., et al. (2003). *N,N'*-dicyclopentyl-2-methylsulfanyl-5-nitro-pyrimidine-4,6-diamine (GS39783) and structurally related compounds: Novel allosteric enhancers of γ-aminobutyric $acid_B$ receptor function. *Journal of Pharmacology and Experimental Therapeutics, 307*, 322–330.

van Bree, J. B., Audus, K. L., & Borchardt, R. T. (1988). Carrier-mediated transport of baclofen across monolayers of bovine brain endothelial cells in primary culture. *Pharmaceutical Research, 5*, 369–371.

van Bree, J. B., Heijligers-Feijen, C. D., De Boer, A. G., et al. (1991). Stereoselective transport of baclofen across the blood-brain barrier in rats as determined by the unit impulse response methodology. *Pharmaceutical Research, 8*, 259–262.

van Herwaarden, M. A., Samsom, M., Rydholm, H., et al. (2002). The effect of baclofen on gastro-oesophageal reflux, lower oesophageal sphincter function and reflux symptoms in patients with reflux disease. *Alimentary Pharmacology & Therapeutics, 16*, 1655–1662.

Vergnes, M., Boehrer, A., Simler, S., et al. (1997). Opposite effects of $GABA_B$ receptor antagonists on absences and convulsive seizures. *European Journal of Pharmacology, 332*, 245–255.

Waldmeier, P. C. (1991). The $GABA_B$ antagonist, CGP 35348, antagonizes the effects of baclofen, γ-butyrolactone and HA 966 on rat striatal dopamine synthesis. *Naunyn-Schmiedeberg's Archives of Pharmacology, 343*, 173–178.

Weerts, E. M., Froestl, W., & Griffiths, R. R. (2005a). Effects of GABAergic modulators on food and cocaine self-administration in baboons. *Drug and Alcohol Dependence, 80*, 369–376.

Weerts, E. M., Goodwin, A. K., Griffiths, R. R., et al. (2005b). Spontaneous and precipitated withdrawal after chronic intragastric administration of γ-hydroxybutyrate (GHB) in baboons. *Psychopharmacology, 179*, 678–687.

Weerts, E. M., Froestl, W., Kaminski, B. J., et al. (2007). Attenuation of cocaine-seeking by $GABA_B$ receptor agonists baclofen and CGP44532 but not the GABA reuptake inhibitor tiagabine in baboons. *Drug and Alcohol Dependence, 89*, 206–213.

Wellendorph, P., Hog, S., Greenwood, J. R., et al. (2005). Novel cyclic γ-hydroxybutyrate (GHB) analogs with high affinity and stereoselectivity of binding to GHB sites in rat brain. *Journal of Pharmacology and Experimental Therapeutics, 315*, 346–351.

Wellendorph, P., Hog, S., Skonberg, C., & Bräuner-Osborne, H. (2009). Phenylacetic acids and the structurally related non-steroidal anti-inflammatory drug diclofenac bind to specific γ-hydroxybutyric acid sites in rat brain. *Fundamental & Clinical Pharmacology, 23*, 207–213.

Whelan, J. L. (1980). Baclofen in treatment of the "stiff-man" syndrome. *Archives of Neurology, 37*, 600–601.

Wieronska, J. M., Wabno, J., Froestl, W., Hess, G., & Pilc, A. (2010). The antipsychotic action of novel orthosteric and allosteric positive modulators of $GABA_B$ receptor in mice. *British Journal of Pharmacology*, 2010 (in press).

Wong, C. G., Gibson, K. M., & Snead, O. C., III. (2004). From the street to the brain: Neurobiology of the recreational drug γ-hydroxybutyric acid. *Trends in Pharmacological Sciences, 25*, 29–34.

Woodward, R. M., & Miledi, R. (1992). Sensitivity of *Xenopus* oocytes to changes in extracellular pH: Possible relevance to proposed expression of atypical mammalian $GABA_B$ receptors. *Brain Research. Molecular Brain Research, 16*, 204–210.

Xi, Z. X., & Stein, E. A. (1999). Baclofen inhibits heroin self-administration behavior and mesolimbic dopamine release. *Journal of Pharmacology and Experimental Therapeutics*, *290*, 369–1374.

Yanovsky, Y., & Misgeld, U. (2003). Suppression of epileptiform activity by $GABA_B$ receptors in wild type and weaver hippocampus "in vitro." *Epilepsy Research*, *52*, 263–273.

Zarrindast, M. R., Hoghooghi, V., & Rezayof, A. (2008). Inhibition of morphine-induced amnesia in morphine-sensitized mice: Involvement of dorsal hippocampal GABAergic receptors. *Neuropharmacology*, *54*, 569–576.

Zhang, Q., Lehmann, A., Rigda, R., et al. (2002). Control of transient lower oesophageal sphincter relaxations and reflux by the $GABA_B$ agonist baclofen in patients with gastro-oesophageal reflux disease. *Gut*, *50*, 19–24.

Zupan, B., & Toth, M. (2008). Inactivation of the maternal fragile X gene results in sensitization of $GABA_B$ receptor function in the offspring. *Journal of Pharmacology and Experimental Therapeutics*. *327*, 820–826.

Fiona H. Marshall and Steven M. Foord

Heptares Therapeutics, Biopark, Welwyn Garden City, Hertfordshire, UK

Heterodimerization of the $GABA_B$ Receptor—Implications for GPCR Signaling and Drug Discovery

Abstract

The identification of the molecular nature of the $GABA_B$ receptor and the demonstration of its heterodimeric structure has led to extensive studies investigating the mechanism of activation and signaling. Phylogenetic studies suggest that the formation of the heterodimer is a relatively recent event arising in conjunction with the evolution of the central

Advances in Pharmacology, Volume 58

1054-3589/10 $35.00
10.1016/S1054-3589(10)58003-7

nervous system. Heterodimerization has now been demonstrated for many other G-protein-coupled receptors (GPCRs) and plays a role in signaling and trafficking. This presents both challenges and opportunities for GPCR drug discovery. In the case of the GABA$_B$ receptor the best hope for the development of new drugs directed at this receptor is from allosteric modulators. This chapter summarizes our current understanding of the molecular function of the GABA$_B$ receptor and recent developments in the identification of allosteric modulators. The broader implication of heterodimerization on GPCR function and drug discovery is also discussed.

I. Introduction

The metabotropic GABA$_B$ receptor is a member of the superfamily of G-protein-coupled receptors (GPCRs). The GPCR superfamily consists of 800 receptors (Foord et al., 2005; Fredriksson et al., 2003) which share the common feature of having seven membrane-spanning helices which are located within the plasma membrane and an extracellular N-terminus which varies significantly in size and may contain a number of protein subdomains involved in ligand binding (Lagerstrom & Schioth, 2008). Within the superfamily there are three distinct subfamilies, related by sequence homology, known as Family A (rhodopsin family), B (secretin family), and C (metabotropic family). GPCRs are widely regarded as the most important class of receptors for drug discovery since currently around 30% of the marketed prescription drugs act at this class of receptors (Jacoby et al., 2006). Within this large family of conserved receptors GABA$_B$ receptors stand out as having the unique property of being an obligate heterodimer. This unusual aspect of GPCR structure resulted in a significant delay in the molecular identification of the receptor. Its discovery in 1998 by three different groups of researchers (Marshall et al., 1999) has had a dramatic impact on our understanding of the biology of the entire GPCR superfamily. The discovery led to the recognition that homodimerization or the formation of higher order oligomers is most likely the normal state for the majority of GPCRs (Milligan, 2007). Furthermore the ability of GPCRs to form heterodimers with related as well as distinct receptors in different tissues has been shown to have an impact on their trafficking, signaling, and pharmacology (Milligan, 2006). In this chapter we review the discovery of the GABA$_B$ receptor heterodimer and the evolution of this receptor across different species. We discuss the structure and function of this unusual GPCR, the heterodimerization of GPCRs, and the impact of this on GPCR drug discovery.

II. Identification of the GABA$_B$ Receptors

A. Expression Cloning of the GABA$_{B1}$ Receptor

The late 1980s and 1990s and the sequencing of the human genome heralded a rich age in GPCR molecular pharmacology. Many receptors which had previously been described pharmacologically were cloned using techniques such as partial sequencing of purified proteins isolated from native tissue (e.g., hamster β-adrenergic receptor (Lefkowitz, 2004)) or by expression cloning based on ligand binding (Kieffer et al., 1992), or function (Segre & Goldring, 1993). This led to the identification of many other well-described receptors based on sequence homology (Libert et al., 1989) as well as the identification of new subtypes not identified by pharmacology such as the histamine H_4 receptor (Nakamura et al., 2000). New families with completely novel pharmacology were discovered such as the receptors for orexins, sphingolipids, and carboxylic acids as well as the many so-called orphan receptors whose native ligands and functions remain undiscovered (Civelli, 2005). Despite all this frenetic activity for many years the GABA$_B$ receptor remained one of the few GPCRs that was well described pharmacologically but could not be cloned.

Many groups across the world attempted a variety of strategies to clone this important receptor. Two teams attempted protein purification from bovine (Nakayasu et al., 1993) and pig (Facklam & Bowery, 1993) brain. Purification was attempted using a baclofen affinity column as well as a monoclonal antibody. Several attempts were made at expression cloning in *Xenopus* oocytes including Barnard, who had previously cloned the GABA$_A$ receptor (Schofield et al., 1987), and Tanaka and Sekiguchi in Japan (Sekiguchi et al., 1990; Taniyama et al., 1991; Woodward & Miledi, 1992) These all proved unsuccessful.

The breakthrough in expression cloning in the end came from chemistry rather than molecular biology. Wolfgang Froestl team at Novartis (then Ciba–Geigy (Froestl et al., 2003)) designed the photo-affinity radioligand [^{125}I]-CGP71872 which upon irradiation bound irreversibly to native GABA$_B$ receptors in rat brain membranes. This indicated that there were two proteins with molecular weight 130 and 100 kD (unlike the 80 kD protein previously purified). Since commercial cDNA libraries were unlikely to contain clones of this size, a new cDNA library was prepared from rat brain membranes. Screening of this library by Kaupmann and Bettler resulted in the identification of the first GABA$_B$ receptors. Two splice variants were identified: GABA$_{B1A}$ which had 960 amino acids and GABA$_{B1B}$ which differed only in the N-terminal extracellular domain and had 844 amino acids. The 1A subunit includes a tandem pair of consensus sequences for the complement protein (CP) module (also known as Sushi domains) that are missing in 1B.

Although this work was published in *Nature* in March 1997 (Kaupmann et al., 1997) it was clear that this receptor did not fulfill all the characteristics of the native $GABA_B$ receptor. Firstly although the affinity of antagonists in competition studies was close to that of native receptor in rat brain membranes, agonist affinity was significantly lower at the newly cloned receptor. Of more concern was the fact that expression of the receptor in heterologous systems failed to couple to any known effectors of the receptor. Subsequently it was shown that in most systems $GABA_{B1}$ is not expressed at the cell surface (Couve et al., 1998).

B. Heterodimerization with $GABA_{B2}$ Subunit

Searches of sequence databases with the $GABA_{B1}$ sequences revealed the existence of a second closely related receptor which is now referred to as $GABA_{B2}$. This receptor was 35% homologous to $GABA_{B1}$; however, unlike $GABA_{B1}$, it was expressed at the cell surface when expressed alone but nevertheless failed to respond to GABA or other similar amino acids (Marshall, 2000). Further studies demonstrated that these receptors showed a remarkable overlap in expression throughout the brain. Furthermore in a yeast two-hybrid study designed to identify a molecular chaperone for $GABA_{B1}$ the strongest hit was the second receptor $GABA_{B2}$. It was soon clear that fully functional $GABA_B$ receptors consist of a heterodimer of both subunits $GABA_{B1}$ and $GABA_{B2}$ (Jones et al., 1998; Kaupmann et al., 1998; White et al., 1998). Following coexpression $GABA_{B2}$ traffics $GABA_{B1}$ to the cell surface generating a receptor capable of high agonist affinity binding, G-protein coupling, and effector activation (Bettler et al., 2004).

III. The Phylogeny of $GABA_B$ Receptors $GABA_B$ Receptors in Different Species

A. $GABA_B$ Receptors in Different Species

Both subunits of the $GABA_B$ receptor along with the metabotropic glutamate receptors (mGluRs), calcium sensing receptor (CaSR), taste receptors (T1R), pheromone receptors (V2R), and certain odor receptors have the same basic architecture. Each protein is made up of two distinct protein domains: the "Venus fly trap domain" or VFTD and the seven transmembrane (7TM) domain of the Family C GPCRs. The $GABA_B$ receptor subunits are slightly different from the others in that they do not have a cysteine-rich domain linking the VFTD and 7TM domains (Lagerstrom & Schioth, 2008; Pin et al., 2003). This may change the flexibility of the "hinge" between the two domains. The crystallization of the VFTD from the mGluR1 receptor including glutamate and ligand-free forms (Kunishima et al., 2000) shows that glutamate binding changes the conformation of the VFTD which must be communicated to the 7TM domain. This VFTD to 7TM mechanism is amongst the oldest employed to activate G proteins, and it is worth considering how it appears to have arisen during the course of evolution.

The evolution of Family C GPCRs has been driven by three successive gene duplication events. About 899 million years ago (mya) duplication resulted in two distinct groups: the GABA$_B$-like receptors and the other Family C receptors. A second duplication at about 638 mya resulted in the mGluR group forming within the "others." The third duplication at about 573–565 mya separated the sensing receptors (CaSR and T1R) and the two GABA$_B$ subunits. These figures have been derived from a phylogenetic analysis of the VFTD sequences from species including the slime mould *Dictyostelium discoideum*, the fly *Drosophila melanogaster*, and the worm *Ceanorhabditis elegans* (Cao et al., 2009). It is interesting that a similar analysis of the 7TM domains of the same receptors gives very similar results (Pin et al., 2004). From this we might infer that the relationship between the VFTD and 7TM domains is close.

These findings suggest that the heterodimeric nature of the GABA$_B$ receptor is actually amongst the most "recent" events in the evolution of Family C GPCRs. However, "recent" is a relative term. The GABA$_B$ receptor subunits from the fruit fly *D. melanogaster* are so conserved that they can complement those of the rat (Mezler et al., 2001). Given the divergence between arthropods and mammals occurred about 540 mya we are still describing an extremely old and extraordinarily well conserved mechanism. The sea squirt, *Ciona intestinalis*, has an evolutionary pedigree almost as old as the arthropods and contains about the same number of genes (roughly 16,000—half the number found in mice and humans). The barrel-shaped adult squirt attaches to rocks, piers, boats and the sea bottom and feeds by siphoning seawater through its body and using a basket-like internal filter to capture plankton and oxygen. This seems a very basic body plan; however, its egg develops into a small tadpole comprised of only about 2,500 cells with a notochord and a primitive nervous system and in this stage of its life *C. intestinalis* is highly mobile. Whilst there are differences between the GPCRs in *C. intestinalis* and the two insects analyzed so far (fruit fly and honey bee) (Fredriksson et al., 2005) both have Family C GPCRs much like a mammal. By this time, about 573–565 mya, the third Family C gene duplication had occurred and the GABA$_B$ receptor as we recognize it had appeared (Kamesh et al., 2008).

If the third gene duplication gave rise to the GABA$_B$ subunits then what was the nature of the GABA$_B$ receptor before that point? The slime mould *D. discoideum* responds to GABA via a GPCR with a 100-fold greater affinity for GABA than glutamate. However, this animal predates even the first gene duplication that gave rise to GABA$_B$ and other class C GPCRs. Confusingly, its genome contains 17 class C GPCRs all of which resemble

$GABA_B$ receptors from higher eukaryotes (Eichinger et al., 2005; Prabhu et al., 2007a, 2007b). The role of these receptors is starting to be understood thanks to the development of *D. discoideum* as a model organism and the application of molecular genetics. *D. discoideum* forms spores and this process is controlled via GPCRs by GABA, glutamate, and SDF-3 (see below). To summarize, *D. discoideum* expresses a combined GABA/glutamate receptor termed GrlE. GABA has a 100-fold higher affinity than glutamate but the levels of glutamate may be correspondingly higher. Glutamate activates a G protein $G_{\alpha 9}$ via GrlE and GABA activates $G_{\alpha 7}$. These G proteins are antagonistic with regard to spore formation (Anjard & Loomis, 2006). GrlE is the most "$GABA_B$-like" of the 17 Family C GPCRs in its genome and may represent a prototypical receptor. As for the remaining 16 Family C GPCRs, they appear to indicate how far this organism has evolved this particular receptor template. Another receptor, GrlA, controls the production of GABA via $G_{\alpha 4}$. However, its ligand, SDF-3, is neither glutamate nor GABA but a small and as yet unidentified hydrophobic molecule that shows some of the pharmacology of a hydrocortisone-like steroid. Pharmacological inhibition of steroidogenesis during the development of *Dictyostelium* blocked the production of SDF-3. Moreover, the response to SDF-3 could be blocked by the steroid antagonist mifepristone, whereas hydrocortisone and other steroids mimicked the effects of SDF-3 when added in the nanomolar range (Anjard et al., 2009). Another receptor, GrlJ, has already been identified as having a role in a different stage of the *D. discoideum* life cycle (Prabhu et al., 2007b). Other "simple" animals appear to have evolved the Family C template yet further. Placozoans are the simplest in structure of all nonparasitic multicellular animals and have genomes comparable in their protein-coding capacity to *D. discoideum*. They were first found growing on the walls of aquaria. The genome of the placozoan *Trichoplax adhaerens* has been reported to contain 85 class C GPCRs. XP_002109202.1 and XP_002113243.1 are 32% identical to $GABA_{B2}$ and XP_002113243.1 is 34% identical to $GABA_{B1}$ (Srivastava et al., 2008).

B. The Heptahelical (7TM) Domain

All of the Family C GPCRs that have been paired with ligands have contained the VFTD. The $GABA_B$ receptor-like receptor ($GABA_B$ RL) shows 30% sequence identity in its TM domains to the $GABA_B$ receptor subunits. It lacks a VFTD but contains cysteine residues in its truncated amino terminus (unlike the $GABA_B$ receptor subunits). Besides $GABA_B$ RL a further six "truncated" Family C orphans (e.g., RAIG1/GPRC5A) have been identified that more closely resemble mGluR receptors and again contain cysteine residues in their amino termini (Lagerstrom & Schioth, 2008). Whether the 7TM domain alone can be activated to the extent that

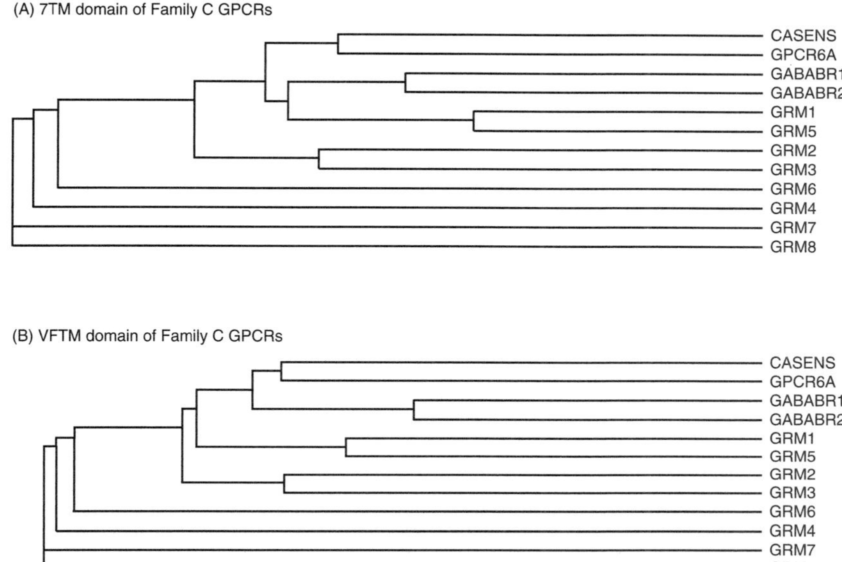

FIGURE 1 Family C GPCR sequence alignments. Human Family C GPCR sequences from (A) TMD and (B) VFTD were aligned using Clustal W and the relationships between sequences displayed using it Cladogram tool (http://www.ebi.ac.uk/Tools/clustalw2/index.html). The receptors were split at the extracellular boundary of transmembrane 1; no other sequence manipulation was performed. The figures show that the sequence changes in the two domains mirror each other such that the relationships between the receptors appear very similar irrespective of which domain is chosen for analysis.

it couples to G proteins has not been definitively shown as there are no ligands. However, in *Xenopus laevis* melanophores the overexpression of GABA$_B$ RL and RAIG1 leads to pigment dispersion characteristic of the activation of $G_{\alpha i/o}$. The GABA$_B$ RL was $G_{\alpha i}/G_{\alpha o}$ whilst mGluR1 and CaSR were $G_{\alpha s}/G_{\alpha q}$-coupled (C. Jayawickreme, personal communication). The 7TM domains of Family C GPCRs show little homology to those of other GPCR families; however, within Family C positions are conserved in helixes 3, 6, and 7. Some of these are also conserved in Family A receptors and this is a common ancestry and similar mechanism of activation (Lagerstrom & Schioth, 2008). Figure 1 shows the relationship of both the TMD and the VFTD across Family C receptors.

C. The Venus FlyTrap Domain

The VFTD is an ancient protein module that shares sequence homology with bacterial periplasmic amino acid-binding proteins (PBPs). Bacteria use these proteins to bind and transport a wide range of small molecules, such as amino acids and vitamins, that can help their growth and division (Felder et al., 1999; Pin et al., 2003). Mutations have often arisen in the "hinge" between the two lobes of the VFTD in Family C GPCRs. This region is associated with ligand binding in glutamate receptors (through direct observation of the mGluR1 VFTD structure (Kunishima et al., 2000) and in other Family C GPCRs by mutagenesis or subtle species differences (Pin et al., 2003)). Liu's group (Cao et al., 2009) performed a detailed sequence analysis and modeled 269 significantly constrained sites using the crystal structure of the mGluR1 VFTD domain (Kunishima et al., 2000). They showed that the sites that typify each receptor type within Family C cluster around the "binding pocket" of the VFTD. The changes in mGluRs are different than those in the T1Rs which differ again from those in the $GABA_{B1}$ subunit, for example. Independently, Kuang et al. (2006) predicted that the pharmacology of the mGluRs was distinguished from "sensory" Family C receptors by only a few residues. They tested their hypothesis on the broad specificity 5.24 amino acid-sensing receptor in the fish (which have an expanded group of these receptors) by mutating its VFTD in three places to "recapitulate the ancestral receptor." The pharmacology of the "ancestor" was more like a glutamate receptor and less like a basic amino acid receptor (it now responded to glutamate and less well to arginine). Candidates for the residues which determine GABA binding by the VFTD of $GABA_B$ have been defined by mutagenesis and modeling based on the mGluR1 VFTD structure (see below).

The changes in the $GABA_B$ subunits are interesting in that the R1 subunit shows strong selection pressure in the binding pocket whereas the $GABA_{B2}$ subunit shows almost none consistent with the role of the $GABA_{B2}$ subunit as a signaling rather than binding subunit. This finding was initially made experimentally when it was found that mutagenesis of the $GABA_{B2}$ subunit "binding domain" had essentially no effect on subsequent $GABA_B$ receptor function (Kniazeff et al., 2002).

IV. Structure of $GABA_B$ Receptors

A. The Heptahelical Transmembrane Domain

To date there are no X-ray crystal structures of any Family C transmembrane regions; however, recently there have been a number of X-ray structures of Family A GPCRs. Rhodopsin remains the prototypical GPCR for structure determination (Palczewski, 2006), and recently crystal structures have been obtained of the activated form of rhodopsin in complex with a peptide from the G protein transducin (Park et al., 2008; Scheerer et al., 2008). Family A structures are also available for the β-adrenergic receptors (Cherezov et al., 2007; Warne et al., 2008) and the adenosine A2a receptor

(Jaakola et al., 2008). Although these receptors have very little homology with Family C receptors they likely evolved from a common ancestor and therefore will share some structural features. It would be surprising if the mechanism involved in transmitting the signal-binding event through the TMD to bind and activate common G proteins is very different.

It is becoming clear from recent structural work on rhodopsin that upon agonist binding there are changes in the conformations of the side chains of highly conserved residues that contribute to both a hydrogen bond network within the TMD. These changes are transmitted through the receptor ultimately resulting in an outward rigid body movement of TM6 at the cytoplasmic face opening up a binding pocket for the G protein to bind. Interestingly many of these conserved residues which appear to function as molecular micro switches in Family A receptors (Nygaard et al., 2009) have equivalent highly conserved residues in Family C receptors.

Of the 19 highly conserved residues in the TMD of Family C, 7 are also found in the rhodopsin family (Pin et al., 2003). The NPxxY motif in TM7 is found in 92% of Family A GPCRs and includes an important tyrosine residue which undergoes a key role in receptor activation. In the inactive state of rhodopsin this Tyr interacts with helix 8; however, in the opsin complex the tyrosine is rotated toward TM6 stabilizing the interaction between TM6 and TM7. Many Family C receptors contain a similar xPKxY motif in TM7; however, in GABA$_{B1}$ this is PKxR and in GABA$_{B2}$ it is PKxI (Pin et al., 2003). It is possible that the arginine in GABA$_{B1}$ could form similar hydrogen-bonding interactions to tyrosine; however, the presence of isoleucine in GABA$_{B2}$ suggests that different interactions may occur upon activation and G-protein coupling of this TMD.

The most highly conserved micro switch in Family A is the Arg which forms part of the DRY motif in TM3 which in rhodopsin forms a salt bridge to glutamic residues in TM3 and TM5 known as the "ionic lock." This lock is broken during activation and instead the Arg rotates to make a hydrogen bond interaction with a Tyr in TM5 contributing to the formation of a G-protein-binding pocket. Interestingly GABA$_{B2}$ but not GABA$_{B1}$ has an Arg in this position in TM3.

Finally within TM6 there is a highly conserved Trp present in most Family A and C receptors. This Trp is located at the bottom of the small molecule-binding pocket and may play an early role in the receptor activation process. Surprisingly unlike most other Family C receptors including all the metabotropic receptors there is no Trp at this position in any of the GABA$_B$ receptors. This suggests that the triggering point for activation of the GABA$_B$ heterodimer may be quite different to other GPCRs.

Overall it seems that the structure of the GABA$_B$ receptor TMD and the conformational changes which occur during activation will show some similarities to Family A receptors; however, until we obtain at least a Family

C TMD structure and preferably a structure of the $GABA_B$ heterodimer we will not fully understand the molecular basis of $GABA_B$ function.

B. The Venus FlyTrap Domain

The crystal structures of several VFTD from mGluRs have now been obtained (Jingami et al., 2003; Kunishima et al., 2000; Muto et al., 2007; Tsuchiya et al., 2002). The first crystal structures were of mGluR1 VFT in glutamate-bound and two ligand-free forms. These proteins crystallize as dimers which interact via a large hydrophobic surface in lobe 1. The glutamate-bound and free-form crystals were nearly identical (Kunishima et al., 2000). Subsequently crystal structures of antagonist S-MCPG ((S)-(α) methyl-4-carboxyphenylglycine)-bound and gadolinium-bound forms were also obtained (Tsuchiya et al., 2002). In these structures the lobes are connected by three short loops and come together in a clamshell-like shape. The glutamate-bound structure included both closed and open protomers (considered to be the active A conformation) whilst the antagonist form consisted of two open protomers (considered to be the inactive R conformation). Closure of at least one of the VFTDs as a result of glutamate binding is suggested to stabilize the active conformation of the VFTDs which involves a direct association of lobes 2. This interaction may result in a movement of the VFTD relative to the TMD and thus an activation of the GPCR (Fig. 2).

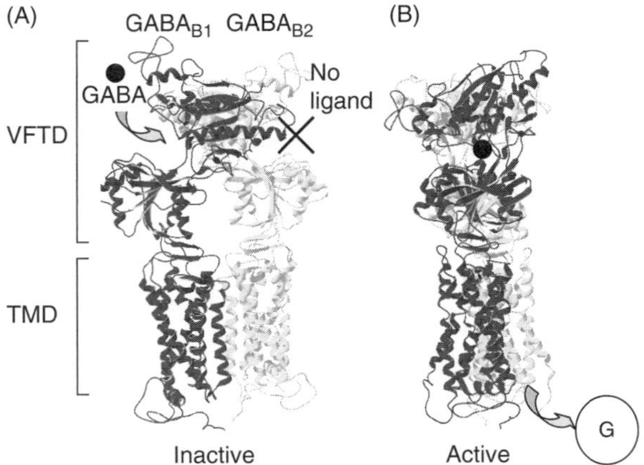

FIGURE 2 Proposed structure of the $GABA_B$ heterodimer in active and inactive states. The structure of the heterodimeric $GABA_B$ receptor in its inactive (A) and active (B) states with the extracellular domains modeled on the VFTD of the related mGluR1. The two related subunits, $GABA_{B1}$ (in the front, in black) and $GABA_{B2}$ (in the back, in light gray), are shown. Binding of the agonist GABA to the VFTD alters the conformation to a closed state which results in a subsequent activation of the 7-TM domain of the $GABA_{B2}$ subunit which is responsible for G-protein coupling. Figure is taken from Pin and Prezeau (2007) and is reproduced with kind permission from Bentham Science Publishers Ltd.

More recently the structures of the group II mGluR3 and group III mGluR7 VFTDs have been obtained (Muto et al., 2007). Surprisingly the mGluR3 structure complexed with agonist adopted the closed/closed R conformation suggesting other possibilities for an activation mechanism.

Although we do not yet have a structure of the GABA$_B$ VFTD it is likely to closely resemble that of the mGluRs. The exact mechanism of action which transmits agonist binding from the VFTD to the TMD remains to be confirmed; however, in the case of the GABA$_B$ receptor since only one pair within the dimer binds ligand, a model in which the VFTDs can form an active conformation with an open/closed combination remains likely. Mutagenesis studies supporting the role of lobe 2 interactions during receptor activation of the GABA$_B$ receptor are described below (Rondard et al., 2008).

V. Function of the GABA$_B$ Receptor Heterodimer

A. Binding of GABA to the GABA$_{B1}$ Subunit

Within the heterodimer it is clear that each subunit subserves a different function. The GABA$_{B1}$ subunit is primarily involved in ligand binding. The extracellular VFTD of GABA$_{B1}$ binds the agonists GABA and baclofen, as well as competitive antagonists such as CGP64213 (Galvez et al., 1999, 2000; Malitschek et al., 1999). Site-directed mutagenesis studies have been used to determine the key residues involved in ligand binding. Ser^{246} and Asp^{471} in lobe 1 are critical for agonist binding. Ser^{246} is also involved in the binding of the antagonist CGP64213. Mutation of this residue to alanine decreases agonist affinity determined in functional assays by 1,000-fold. This residue is thought to contact the carboxylic acid of GABA and baclofen via a hydrogen bond. Asp^{471} is thought to form an ionic interaction with the amino group of the agonists. Mutation of this residue reduces agonist activation as well as antagonist binding. This residue is highly conserved across GABA$_{B1}$ subunit from different species. Several residues in lobe 1 have been identified which decrease antagonist but not agonist binding. These include Ser^{265}, Tyr^{266}, Phe^{463}, and Tyr^{470}.

Although lobe 1 appears to play a pivotal role in ligand-binding mutagenesis of Tyr^{366} in lobe 2 also decreased the affinity of GABA and baclofen (Galvez et al., 2000). Presumably this residue must form its interaction when the VFTD is in its closed conformation. According to the model of Galvez et al., in this form the hydroxyl group Tyr^{366} could form a hydrogen bond from its oxygen to the carboxylic group of baclofen. Mutation of this residue prevents receptor activation by baclofen and converts it to a competitive antagonist.

Although GABA$_{B1}$ is able to bind to GABA it cannot signal in the absence of the GABA$_{B2}$ subunit. One role of GABA$_{B2}$ is to traffic GABA$_{B1}$

to the cell surface; mutation of the endoplasmic reticulum (ER) retention sequence on $GABA_{B1}$ allows the receptor to reach the cell surface alone; however, even at the cell surface $GABA_{B1}$ does not signal in the absence of $GABA_{B2}$ (Calver et al., 2001; Margeta-Mitrovic et al., 2000; Pagano et al., 2001). Based on the function of VFTDs in other receptors, in particular mGluR1 (Kunishima et al., 2000), it is thought that binding of agonists results in a stabilization of the closed state of the VFTD. This has been supported by mutagenesis studies where cysteine residues were introduced into both lobes 1 and 2 that form a disulfide bond in the closed form of the receptor (Kniazeff et al., 2004; Pin et al., 2004). When this mutant was coexpressed with the $GABA_{B2}$ subunit a significant constitutive activity was observed as seen by an elevated inositol phosphate formation through the activation of the chimeric G protein $G_{\alpha q19}$. This activity was abolished by the addition of the reducing agent dithiothreitol confirming the role of the mutant disulfide bond. When this receptor was locked in this active conformation it no longer bound antagonists (Kniazeff et al., 2004).

In summary, the role of the $GABA_{B1}$ subunit is to bind agonists resulting in closure of the $GABA_{B1}$ VFT—the first step in receptor activation.

B. Role of the $GABA_{B2}$ Subunit in Ligand Binding

The VFTD of the $GABA_{B2}$ subunit is similar in sequence to other VFT modules consistent with the possibility that it could be activated by an agonist ligand independent of $GABA_{B1}$. Indeed there are regions of the brain where $GABA_{B2}$ is expressed in the absence of $GABA_{B1}$ (Lopez-Bendito et al., 2002; Ng & Yung, 2001). Radioligand-binding studies with standard $GABA_B$ ligands have failed to show any specific binding to $GABA_{B2}$ (Kniazeff et al., 2002; Pin et al., 2004); however, in the case of agonists these have low affinity and do not make good radioligands. There have been a small number of reports published where GABA and baclofen have been shown to activate $GABA_{B2}$ alone (Martin et al., 1999); however, this is not a common finding. Mutagenesis studies which have defined key residues for binding of agonists to the VFTD of $GABA_{B1}$ indicate that none of these residues are present in $GABA_{B2}$ subunits despite a high degree of conservation in $GABA_{B1}$ from different species (Pin et al., 2004). It is possible that $GABA_{B2}$ may bind a different as yet undiscovered ligand which could function as a co-ligand of the receptor; however, the putative binding site within the $GABA_{B2}$ VFTD is poorly conserved across subunits from different species suggesting a lack of evolutionary pressure to conserve a ligand-binding site. Extensive mutagenesis of this region has no effect on the functioning of the $GABA_B$ receptor (Kniazeff et al., 2002).

Although $GABA_{B2}$ appears not to be directly involved in agonist binding it does make an important contribution. In the absence of $GABA_{B2}$, agonist binding to $GABA_{B1}$ is of low affinity. This is not due to misfolding or incomplete glycosylation of the receptor (Galvez et al., 2001). Coexpression of $GABA_{B2}$ results in a 10-fold increase in agonist affinity for $GABA_{B1}$. The interaction between the VFTD of the two subunits has been studied using chimeric receptor where the VFTDs have been swapped between the two subunits (Galvez et al., 2001) as well as using isolated VFTDs linked to single transmembrane domains (TMDs) or GPI anchor domains (Liu et al., 2004). These studies have demonstrated a cooperative interaction between the VFTDs of both subunits. In chimeric receptors an increase in agonist binding to $GABA_{B1}$ results from coexpression both with $GABA_{B2}$ as well as with a chimera consisting of the extracellular domain of $GABA_{B2}$ attached to the TM domain of $GABA_{B1}$ (Galvez et al., 2001). This increase in affinity is due to two effects. Firstly it opposes negative allosteric effects (NAMs) between the VFTD and the TMD of $GABA_{B1}$. In the absence of the TMD the VFTD of $GABA_{B1}$ binds agonists with high affinity. This negative effect of the transmembrane region on agonist binding is also found in the mGluRs (Peltekova et al., 2000). Interaction with the TMD favors the open (low agonist affinity) state of the VFTD and this is reduced in the presence of $GABA_{B2}$ (Liu et al., 2004). Secondly a direct interaction occurs between the two VFTDs. Fluorescence resonance energy transfer (FRET) studies have shown that there is a direct binding between the VFTDs of both subunits. Addition of the isolated VFTD of $GABA_{B2}$ increased agonist affinity for $GABA_{B1}$ threefold. This increase in affinity is most likely due to the ability of $GABA_{B2}$ VFTD to stabilize the closed state of the agonist-bound $GABA_{B1}$ receptor.

In line with proposed mechanisms for interactions between the VFTDs of mGLUR1 described above it was found that preventing association between lobes 2 of $GABA_{B1}$ and $GABA_{B2}$ by the introduction of an N-glycosylation site at the lobe 2 interface had no effect on receptor assembly or agonist binding but completely prevented receptor activation (Pin et al., 2009; Rondard et al., 2008).

C. Agonist-Induced Activation of G-protein Signaling

The chimeric, mutated, or truncated receptors have been very useful in elucidating the role of the subunits in G-protein coupling and signaling (Galvez et al., 2001; Margeta-Mitrovic et al., 2001). Firstly the chimeras where the VFTDs were swapped onto different TMDs have demonstrated the requirement of the TMD of $GABA_{B2}$ in G-protein coupling. In chimeras where both TMDs were derived from $GABA_{B1}$ there was no G-protein coupling (despite having the normal GB1/2 VFTD combination) whereas in chimeras where both TMDs were from $GABA_{B2}$ G-protein coupling was possible, although it was lower than the wild-type receptor (Galvez et al., 2001). In more detailed mutagenesis studies the loops and C-terminal tail of $GABA_{B2}$ were sequentially removed and replaced with random peptides (Margeta-Mitrovic et al., 2001). Modified receptors were coexpressed with the potassium channels GIRK1/GIRK2 (Kir3.1/3.2) in *Xenopus* oocytes. All intracellular regions of $GABA_{B2}$ were found to be required for coupling to GIRK channels whilst in contrast none of the intracellular regions of $GABA_{B1}$ were required.

The important regions of $GABA_{B2}$ involved in G-protein coupling have been further elucidated by single point mutations introduced into the intracellular loops (Duthey et al., 2002; Robbins et al., 2001). Mutation of Leu^{686} to proline or serine in the third intracellular loop of $GABA_{B2}$ significantly reduced the ability of the receptor to couple to the chimeric G-protein ($G_{\alpha qi9}$) in HEK293 cells or to native G-proteins inhibiting calcium channels in primary neurons. Similar residues have found to be important in the G-protein coupling of mGluR1 and the calcium-sensing receptor (Francesconi & Duvoisin, 1998). In contrast mutation of the equivalent residue in $GABA_{B1}$ had no effect on G-protein coupling. Mutations of residues Lys^{586}, Met^{587}, and Lys^{590} in the second intracellular loop of $GABA_{B2}$ also prevented agonist-induced signaling of the receptor (Robbins et al., 2001).

Although G-protein coupling is clearly mediated via $GABA_{B2}$, the TMD of $GABA_{B1}$ also plays a role (Galvez et al., 2001). Coexpression of $GABA_{B2}$ with a chimera consisting of $GABA_{B1}$ VFTD and $GABA_{B2}$ TMD is able to generate a functional receptor. However, the maximal effects of saturating agonist concentrations (measured by inositol phosphate production upon coexpression with G_{qi9}) is found to be half of that obtained with a wild-type receptor or chimeric combinations which include TMDs from both subunits. In addition the affinity of agonists measured in competition-binding studies was also lower compared to wild-type receptors. This suggests that cooperative interactions occur between the two subunit TMDs to increase G-protein coupling efficiency.

Taken together the studies carried out on engineered receptors together with information from other Family C receptors now give us a clear understanding of the mechanism of activation and signaling of the $GABA_B$ heterodimer. The binding of GABA to the VFTD of the $GABA_{B1}$ subunit results in its closure. This is stabilized by a direct interaction with the VFTD of $GABA_{B2}$. This results in a conformational change in the VFTD dimer relative to the TMD of the receptor. This conformational change is transmitted to the TMD, stabilizing an active conformational state which is capable of G-protein binding to the intracellular regions of $GABA_{B2}$. The G-protein is fully activated in a process that requires interaction between the TMDs of $GABA_{B1}$ and $GABA_{B2}$. Thus the $GABA_B$ receptor is a complex multidomain protein with allosteric interactions between the subunits necessary for correct receptor function.

It is clear that a heterodimer is the minimal unit required for $GABA_B$ receptor functioning; however, several studies suggest that $GABA_B$ receptors assemble into larger multimeric complexes. Using FRET studies with tagged receptors or with monoclonal antibodies to the N-termini of the receptor subunits it was found that a strong FRET signal could be obtained between GABA$_{B1}$ subunits but not through GABA$_{B2}$ subunits. This suggests that GABA$_B$ receptors assemble into oligomers which involve a close interaction between GABA$_{B1}$ subunits. Similar multimeric complexes could also be detected in brain membranes and cultured neurons (Maurel et al., 2008; Pin et al., 2009). Surprisingly disruption of the oligomers by overexpression of a truncated GABA$_{B1}$ subunit appeared to increase signaling measured by calcium mobilization though G$_{qi9}$ by twofold. Oligomerization may be a method to regulate G-protein signaling efficiency of the receptor perhaps to allow interaction with other proteins in the complex involved in G-protein independent signaling (Pin et al., 2009).

D. Mechanism of Action of Allosteric Modulators

The majority of drugs targeted at GPCRs bind to the orthosteric binding site and are competitive with the endogenous agonist. For Family C receptors targeting the orthosteric site which normally binds small charged amino acids (or cations in the case of CaSR) has proved problematical, and compounds which can bind to these sites tend to have poor drug properties such as pharmacokinetics. In the case of the mGluR family the ligand-binding site is highly conserved and orthosteric ligands tend to have problems with selectivity. Since glutamate and GABA represent the key excitatory and inhibitory neurotransmitters throughout the CNS, the use of drugs which turn on or off the receptor is prone to significant side effects.

A major breakthrough in the discovery of drugs for Family C receptors (and indeed many other GPCRs) has been the development of allosteric modulators (Conn et al., 2009; Wang et al., 2009). These ligands bind at sites distinct from the orthosteric site and in the case of Family C receptors the binding of allosteric modulators is principally within the TMD of the receptors. Allosteric modulators can behave in a positive (PAM) or negative (NAM) manner to modulate ligand affinity and/or efficacy. A major benefit of allosteric regulators is that their effect will depend on the local concentration of endogenous agonist ligand such that their action is "use-dependent" and therefore are less likely to produce side effects and tolerance.

In the case of GABA$_B$ receptors positive allosteric modulators (PAMs) have been identified (Pin & Prezeau, 2007). These agents either have no activity in their own right or are very weak partial agonists; however, they can increase both the potency and the efficacy of GABA. The two PAMs which are best characterized for GABA$_B$ receptors are CGP7930 and GS39783 (Fig. 3). These ligands can enhance both agonist potency by approximately 5 to 10 fold and efficacy by approximately 2-fold (Fig. 4). These effects can occur with a number of different GABA agonist including

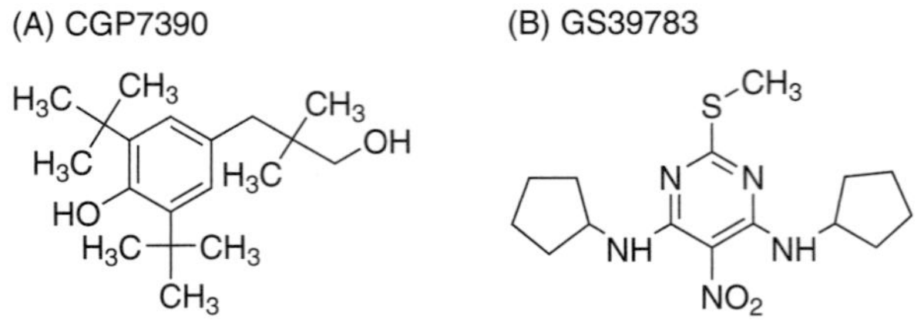

(A) CGP7390

(B) GS39783

FIGURE 3 Chemical structures of the $GABA_B$ receptor positive allosteric modulators (A) CGP7390 and (B) GS39783 (see Pin & Prezeau, 2007).

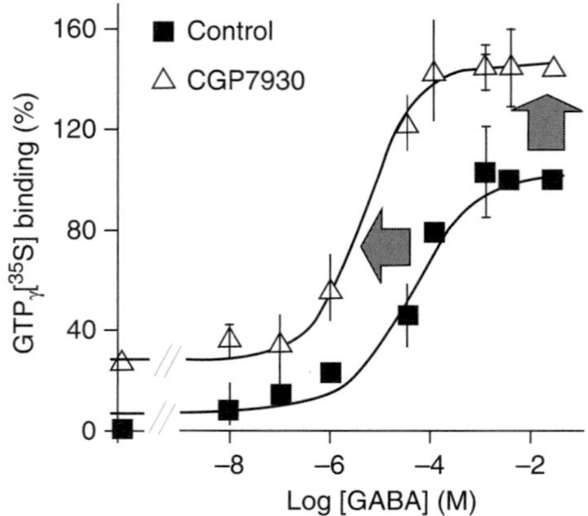

FIGURE 4 The effect of the positive allosteric modulator CGP7390 on $GABA_B$ receptor signaling. CGP7390 (100 μM) increases the potency and efficacy of GABA in a GTPγS-binding assay to recombinant $GABA_B$ receptors expressed in HEK 293 cells. Figure is taken from Pin and Prezeau (2007) and is reproduced with kind permission from Bentham Science Publishers Ltd.

GABA, baclofen, and 3-aminopropylphosphonic acid (APPA) as well as partial agonists such as CGP47656.

The mode of action of CGP7930 has been studied again using chimeric receptors (Binet et al., 2004). This compound does have some weak agonist activity in the absence of orthosteric agonist. Using chimeric subunits it was found that CGP7930 could activate the $GABA_{B2}$ when this was expressed alone in the absence of $GABA_{B1}$ and that this effect was mediated by the TMD (Binet et al., 2004). GS39783 is inactive at the Drosophila $GABA_B$ receptor and so Kaupmann's group made chimeric receptor-swapping regions and residues between *Drosophila* and rat receptors (Dupuis et al., 2006). Only chimeric receptors which included the TMD of rat $GABA_{B2}$ were modulated by GS39783. Further mutagenesis studies led to the identification of a mutant receptor (G706T, A708P in TMVI of rGB2) which could be activated by GS39783 in the absence of orthosteric agonist.

The identification of allosteric modulators opens up new opportunities for drug discovery targeted at the GABA$_B$ receptor. Activation of GABA$_B$ receptors has therapeutic potential in a number of disease states including pain, anxiety, and drug addiction (Marshall, 2005; Pin & Prezeau, 2007); however, the use of directly acting agents such as baclofen is limited by side effects such as sedation, muscle relaxation, cognitive impairment, and hypothermia. A number of studies have been carried out with GABA$_B$ PAMs which suggest an improved therapeutic index. GS39783 has no effect on locomotion, cognition, temperature, or narcosis but had anxiolytic-like effects in a range of anxiety models in both rats and mice (Cryan et al., 2004). Similar anxiolytic effects in the absence of side effects have been observed with CGP7930 (Jacobson & Cryan, 2008). PAMs for GABA$_B$ receptors also have utility in the treatment of drug addiction (Smith et al., 2004). Both CGP7390 and GS3973 inhibit cocaine self-administration (Smith et al., 2004) and alcohol consumption in inbred alcohol-preferring rats (Liang et al., 2006; Orru et al., 2005). See also Tyacke et al. (2010), Chapter 15. GABA$_B$ PAMs are currently being evaluated in clinical trials for gastroesophageal reflux (see Lehmann et al., 2010, Chapter 13).

The discovery of allosteric modulators and an understanding of their site and mechanism of action show great potential for the development of new therapeutic agents targeted at the GABA$_B$ receptor. Perhaps nearly 30 years after Professor Bowery discovered the mechanism of action of baclofen, another agent targeting this receptor, will eventually reach the market.

VI. GABA$_B$ as a Model System for GPCR Dimerization

Although there were several early lines of evidence that GPCRs could form dimers or higher order oligomers (Salahpour et al., 2000) it was not until the discovery of the GABA$_B$ receptor heterodimer that the role of dimerization in normal GPCR function was recognized. Since then there has been an explosion in the number of groups studying GPCR dimerization to the extent that probably too many aspects of GPCR biology now seem to be explained through dimerization. A wide range of biochemical, biophysical, and pharmacological techniques have been applied to demonstrate the formation of GPCR dimers including co-immunoprecipitation, FRET, complementation of inactivated mutant receptors, and pharmacological cross talk; however, many of these are prone to artifacts or are feature of overexpressed systems (Milligan & Bouvier, 2005). Nevertheless there is now a large body of data supporting the idea that the majority of GPCRs function as dimers. Furthermore there is increasing evidence that many GPCRs can form heterodimers, and these may have important physiological consequences as well as representing potentially novel drug targets (Milligan, 2009; Rozenfeld & Devi, 2010).

As described in this chapter as well as others in this book $GABA_B$ receptor heterodimerization plays a role in both receptor trafficking and signaling. It is now clear that these functions of dimerization for the $GABA_B$ receptor also apply to other GPCR receptor systems.

A. Role of Dimerization on GPCR Trafficking

The C-terminal tail of $GABA_{B1}$ includes an ER retention motif such that the subunit is retained within the ER in the absence of $GABA_{B2}$. Coexpression of $GABA_{B2}$ results in a masking of this motif and a trafficking of the receptor to the cell surface. The ER retention motif in $GABA_{B1}$ has been used to study the trafficking of other GPCRs. When the C-terminal tail of the β_2-adrenergic receptor was replaced with that of $GABA_B$ it resulted in the retention of the β_2AR in the ER (Salahpour et al., 2004). Interestingly this construct also acted as a dominant negative to prevent cell surface expression of the wild-type receptor β_2AR when coexpressed. This suggests that homodimerization of β_2AR occurs in the ER which was confirmed by bioluminescent resonance energy transfer (BRET) studies showing that BRET occurs between dimers both in the plasma membrane and ER. In addition mutations with the receptor which prevent dimerization impaired cell surface trafficking by 50%.

Heterodimerization within the ER has also been shown to be a prerequisite for the correct cell-surface expression of some GPCRs (Milligan, 2010). The adrenergic receptor α_{1D}-AR is poorly expressed at the cell surface and accumulates in the ER. When this receptor is coexpressed with α_{1B}-AR but not α_{1A}-AR the receptor pair forms a heterodimer resulting in translocation of α_{1D}-AR to the cell surface enabling α_{1D}-AR mediated signaling. Truncation of the N and C terminal domains showed that these were not required for the interaction (Hague et al., 2004). Similar effects were found with the β2AR in that it could also heterodimerize with and translocate α_{1D}-AR to the cell surface (Uberti et al., 2005). Other examples of heterodimerization playing a role in cell surface trafficking include the chemokine receptor pair CXCR1/CXCR2 (Wilson et al., 2005) and the taste receptors TIR2/TIR3 or TIR1/TIR3 (Nelson et al., 2001; Nelson et al., 2002).

There are a number of naturally occurring mutations in GPCRs which lead to disease, and in some cases these mutations result in a loss of function as a result of receptors failing to reach the cell surface (Milligan, 2010; Ulloa-Aguirre & Conn, 2009). Such mutants can often behave as dominant negatives reducing the surface expression of coexpressed wild-type receptors, and this has an implication for heterozygotes who have a loss of function of greater than the expected 50%. Receptors which fail to traffic to the surface may have mutations which prevent dimerization. In the case of CaSR which similar to $GABA_B$ is a member of Family C mutations has been identified which led to an accumulation of dimeric receptors in the ER and a lack of functional cell surface receptors. This results in familial hypocalciuric hypercalcemia (Pidasheva et al., 2006).

Another example of obligate heterodimerization being required for correct cell surface trafficking of a GPCR is that of the calcitonin gene-related peptide (CGRP) receptor (Foord & Marshall, 1999). In this case the GPCR, calcitonin receptor-like receptor (CRLR) heterodimerizes with a 1-TM protein called RAMP1. This occurs in the ER and the complex moves to the cell surface thereby generating a functional CGRP receptor. In this case the RAMP protein also contributes to defining the pharmacology of the receptor. Heterodimerization of CRLR with a different RAMP protein changes the ligand specificity of the receptor from CGRP to adrenomedullin. RAMPs can also dimerize with other GPCRs to alter their pharmacology and signaling properties (Sexton et al., 2006).

It is now clear that the role of heterodimerization in trafficking the $GABA_B$ receptor to the cell surface is not unique to that receptor but actually plays a key role in trafficking many GPCRs to the cell surface. The identification of the $GABA_B$ receptor dimerization has stimulated research in this field and this may lead to novel treatments for diseases caused by mutations which alter dimerization and cell-surface expression of GPCRs.

B. Effect of Dimerization on GPCR Signaling

The $GABA_B$ receptor has proved to be an excellent model system to study how a GPCR heterodimer couples to G proteins. The surface area of a heterotrimeric G protein is approximately twice the size of the TMD of a GPCR (Hamm, 2001) suggesting that a GPCR dimer would only bind a single G protein. In the case of the $GABA_B$ receptor this makes sense since only one receptor subunit ($GABA_{B2}$ receptor) is responsible for interacting with and activating the $G\alpha$ subunit. In the case of heterodimerization between two functional 7-TM units this opens the possibility for one receptor to regulate signaling of the partner receptor and also may allow a receptor to signal through a different G protein than usual via transactivation of a heterodimeric partner.

A well-studied example is that of opioid receptor heterodimerization (Rozenfeld & Devi, 2010). When coexpressed μ and δ opioid receptors form dimers and higher order oligomers which have both different pharmacology and signaling properties to the receptors expressed individually (George et al., 2000). The μ and δ oligomeric complex has a 10-fold lower affinity for the μ- and δ-selective agonists DAMGO([D-Ala2, N-MePhe4, Gly-ol]-enkephalin) and DPDPE ([d-Pen2,d-Pen5]-enkephalin), respectively, compared to the individual receptors and in addition the rank order of agonist affinities for a series of selected agonists was different to either single receptor. This suggests that either a new binding pocket is formed in the dimer or allosteric interactions between the receptors alter the shape of the ligand-binding pockets of the receptors. Furthermore the heterodimer, unlike the single receptors, was insensitive to pertussis toxin and resistant to internalization suggesting that the heterodimer could couple to a different G protein and interact with a different set of regulating proteins. More recently it has been shown that heterodimerization of μ and δ receptors results in a recruitment of β-arrestin 2 which shifts the pattern of ERK1/2 phosphorylation. Addition of agonist alters the equilibrium between hetero- and homodimers switching signaling pathways from β-arrestin mediated to G-protein coupled (Rozenfeld & Devi, 2007).

Another interesting example occurs in the dopaminergic system. Dopamine D1 and D2 receptors that normally couple to stimulate and inhibit adenylyl cyclase, respectively, when coexpressed can form heterodimers. The heterodimers, unlike either of the single receptors, are able to couple dopamine activation to the Gq pathway resulting in the activation of phospholipase C and mobilization of intracellular Ca^{2+} (Lee et al., 2004).

Until recently most of the evidence for GPCR dimerization impacting on signaling came from recombinant systems; however, very recently *in vivo* evidence demonstrating the physiological significance of receptor dimers has emerged both from vertebrates and from invertebrates. To demonstrate that dimerization could occur *in vivo* Rivero-Muller et al. (2010) made transgenic mice in which the native luteinizing hormone (LH) receptor was knocked out. Modified receptors which were deficient either in ligand binding or in signaling (through a deletion between TM6 and 7) were introduced into the null background and were able to reconstitute completely normal LH signaling through intermolecular functional complementation. This also suggests that receptors are capable of transactivation *in vivo*. Another recent example demonstrating the physiological significance of heterodimerization is taken from the sea squirt, *C. intestinalis* (see above; Sakai, Aoyama, Kusakabe, Tsuda, & Satake, 2009). This organism has four Gonadotropin releasing hormone receptors (GnRHRs) including Ci-GnRHR4 which was considered to be an orphan receptor since it has no affinity for any of the *Ciona* GnrH ligands. Heterodimers of GnRHR4 with GnRHR1 can clearly be demonstrated in *Ciona* ovaries. Furthermore coexpression of these receptors in HEK293 cells resulted in a potentiation of calcium mobilization and ERK phosphorylation relative to R1 alone. Given that there remain over 100 orphan GPCRs it is possible that some of these may just be partners or regulators of other receptors—in the same way as $GABA_{B2}$ without having a ligand in their own right.

C. Relevance to GPCR Drug Discovery

The discovery of the $GABA_B$ heterodimer was a wakeup call to the pharmaceutical industry that GPCR signaling may be more complex than was previously thought. The simplistic idea that one compound would bind to a single receptor and activate or inactive its G-protein-mediated signaling pathway has now been replaced with the knowledge that GPCRs can function as multiprotein complexes both at the level of the receptor in the membrane as well as the diverse set of signaling pathways and proteins that they interact with in the cytoplasm.

With the growing evidence demonstrating the physiological relevance of heterodimerization it may be possible to develop heterodimer-specific drugs. Changes in the pharmacology of ligands when measured at the heterodimer suggest that there are changes in the binding pocket as a result of receptor dimerization. Furthermore new binding sites may be generated, for example, at the dimer interface which may be exploited therapeutically. So far only a small number of heterodimer-biased ligands have been described at the opioid receptors (Waldhoer et al., 2005) and dopamine receptors (Rashid et al., 2007). This number is set to increase as a number of companies have developed heterodimer-specific assays to specifically screen for heterodimer-selective ligands (Cara Therapeutics http://www.caratherapeutics.com/; Dimerex http://www.dimerix.com).

Allosteric modulators which bind to one receptor and cause an increased or decreased response in the dimer partner is currently the most interesting strategy to target heterodimers. The GABA$_B$ heterodimer has proved a useful model in this regard. The allosteric ligands which have been identified for this receptor demonstrate the possibilities for receptor cross talk to modulate the potency and efficacy of agonist responses. In the case of Family C receptors allosteric regulators which bind within the TMD of the receptor offer the best opportunity for drugging this receptor class. PAMs and NAMs, particularly for the mGluRs, are in clinical trials for diseases including Parkinson's disease, schizophrenia, and depression (Niswender & Conn, 2010). Allosteric modulators of the calcium-sensing receptor are on the market for the treatment of (Trivedi et al., 2008) hyperparathyroidism. Heterodimerization between adenosine A2a receptors and dopamine D2 receptors in Parkinson's disease (Ferre et al., 2008) may increase the efficacy of A2a receptor antagonists in clinical development. Finally PAMs at the GABA$_B$ receptor are being investigated for diseases including pain, urinary incontinence, and gastroesophageal reflux (e.g., Addex Pharmaceuticals http://www.addexpharma.com).

VII. Conclusions

Since the cloning of the GABA$_B$ receptor and the surprising discovery of its heterodimeric nature there has been considerable work carried out to understand the molecular nature of the heterodimer and the mechanism of activation of the receptor. Much of this work has been carried out in the laboratories of Jean-Phillipe Pin (Universites Montpellier) whose group has undertaken extensive studies in making receptor chimeras and multiple mutant receptors. This work has also involved the study of the mechanism of action of the closely related mGluRs. In the case of $GABA_B$ receptors it is clear that the $GABA_{B1}$ receptor subunit is involved in binding the agonist ligand; this most likely results in closure of the extracellular VFTD of the receptor and a change in conformation of a dimer between the VFTD of the two receptor subunits. Conformational changes in the extracellular domain are transmitted to the transmembrane-spanning domain ultimately resulting in coupling to and activation of the G protein. It is the $GABA_{B2}$ receptor that is responsible for coupling to the G protein.

Although the $GABA_B$ receptor remains the only obligate GPCR heterodimer there are now many other examples of heterodimerization within this family (Milligan, 2009). Heterodimerization plays a role in receptor trafficking and signaling and importantly for drug discovery may change the pharmacology. The possibility of designing drugs targeting heterodimers is now being investigated.

Allosteric modulators which target the 7-TM domain offer the best opportunities for drugging the Family C subclass of GPCRs. In the case of the $GABA_B$ receptor a number of PAMs have been identified and these are being progressed to clinical trials in a number of disease areas.

Thirty years after the identification of the $GABA_B$ receptor by Professor Bowery and colleagues and close to 90 years after the synthesis of baclofen we may finally be close to having new drugs directed at the $GABA_B$ receptor. Even if this is not the case the discovery of the receptor and its subsequent heterodimerization has certainly led to a sea change in our understanding of GPCR function and will pave the way to new classes of therapeutic agents at this important family of proteins.

Conflict of Interest Statement: Dr Fiona Marshall is a cofounder of Heptares Therapeutics a GPCR drug discovery specializing in structure-based approaches to GPCRs.

Abbreviations

APPA	3-aminopropylphosphonic acid
CaSR	calcium sensing receptor
CGRP	calcitonin gene-related peptide
CP	complement protein
DAMGO	[D-Ala2, N-MePhe4, Gly-ol]-enkephalin
DPDPE	[d-Pen2,d-Pen5]-enkephalin
FRET	fluorescence resonance energy transfer
GABA	γ-amino butyric acid
GnRHR	gonadotropin releasing hormone receptor

GPCR	G-protein-coupled receptor
LH	luteinizing hormone
mGluR	metabotropic glutamate receptor
Mya	million years ago
NAM	negative allosteric modulator
PAM	positive allosteric modulator
PBP	periplasmic amino acid-binding proteins
TMD	transmembrane domain
T1R	taste receptor T1R
VFTD	Venus flytrap domain
7TM	7-transmembrane

References

Anjard, C., & Loomis, W. F. (2006). GABA induces terminal differentiation of dictyostelium through a GABA$_B$ receptor. *Development*, *133*(11), 2253–2261.

Anjard, C., Su, Y., & Loomis, W. F. (2009). Steroids initiate a signaling cascade that triggers rapid sporulation in dictyostelium. *Development*, *136*(5), 803–812.

Bettler, B., Kaupmann, K., Mosbacher, J., & Gassmann, M. (2004). Molecular structure and physiological functions of GABA(B) receptors. *Physiological Reviews*, *84*(3), 835–867.

Binet, V., Brajon, C., Le Corre, L., Acher, F., Pin, J. P., & Prezeau, L. (2004). The heptahelical domain of GABA(B2) is activated directly by CGP7930, a positive allosteric modulator of the GABA(B) receptor. *Journal of Biological Chemistry*, *279*(28), 29085–29091.

Binet, V., Goudet, C., Brajon, C., Le Corre, L., Acher, F., Pin, J. P., et al. (2004). Molecular mechanisms of GABA(B) receptor activation: New insights from the mechanism of action of CGP7930, a positive allosteric modulator. *Biochemical Society Transactions*, *32*(Pt 5), 871–872.

Calver, A. R., Robbins, M. J., Cosio, C., Rice, S. Q., Babbs, A. J., Hirst, W. D., et al. (2001). The C-terminal domains of the GABA(b) receptor subunits mediate intracellular trafficking but are not required for receptor signaling. *Journal of Neuroscience*, *21*(4), 1203–1210.

Cao, J., Huang, S., Qian, J., Huang, J., Jin, L., Su, Z., et al. (2009). Evolution of the class C GPCR venus flytrap modules involved positive selected functional divergence. *BMC Evolutionary Biology*, *9*, 67.

Cherezov, V., Rosenbaum, D. M., Hanson, M. A., Rasmussen, S. G., Thian, F. S., Kobilka, T. S., et al. (2007). High-resolution crystal structure of an engineered human beta2-adrenergic G protein-coupled receptor. *Science*, *318*(5854), 1258–1265.

Civelli, O. (2005). GPCR deorphanizations: The novel, the known and the unexpected transmitters. *Trends in Pharmacological Sciences*, *26*(1), 15–19.

Conn, P. J., Christopoulos, A., & Lindsley, C. W. (2009). Allosteric modulators of GPCRs: A novel approach for the treatment of CNS disorders. *Nature Reviews. Drug Discovery*, *8*(1), 41–54.

Couve, A., Filippov, A. K., Connolly, C. N., Bettler, B., Brown, D. A., & Moss, S. J. (1998). Intracellular retention of recombinant GABAB receptors. *Journal of Biological Chemistry*, *273*(41), 26361–26367.

Cryan, J. F., Kelly, P. H., Chaperon, F., Gentsch, C., Mombereau, C., Lingenhoehl, K., et al. (2004). Behavioral characterization of the novel GABAB receptor-positive modulator GS39783 (*N,N′*-dicyclopentyl-2-methylsulfanyl-5-nitro-pyrimidine-4,6-diamine): Anxiolytic-like activity without side effects associated with baclofen or benzodiazepines. *Journal of Pharmacology and Experimental Therapeutics*, *310*(3), 952–963.

Dupuis, D. S., Relkovic, D., Lhuillier, L., Mosbacher, J., & Kaupmann, K. (2006). Point mutations in the transmembrane region of GABAB2 facilitate activation by the positive modulator *N, N′*-dicyclopentyl-2-methylsulfanyl-5-nitro-pyrimidine-4,6-diamine (GS39783) in the absence of the GABAB1 subunit. *Molecular Pharmacology*, *70*(6), 2027–2036.

Duthey, B., Caudron, S., Perroy, J., Bettler, B., Fagni, L., Pin, J. P., et al. (2002). A single subunit (GB2) is required for G-protein activation by the heterodimeric GABA(B) receptor. *Journal of Biological Chemistry*, *277*(5), 3236–3241.

Eichinger, L., Pachebat, J. A., Glockner, G., Rajandream, M. A., Sucgang, R., Berriman, M., et al. (2005). The genome of the social amoeba *Dictyostelium discoideum*. *Nature*, *435* (7038), 43–57.

Facklam, M., & Bowery, N. G. (1993). Solubilization and characterization of GABAB receptor binding sites from porcine brain synaptic membranes. *British Journal of Pharmacology*, *110*(4), 1291–1296.

Felder, C. B., Graul, R. C., Lee, A. Y., Merkle, H. P., & Sadee, W. (1999). The venus flytrap of periplasmic binding proteins: An ancient protein module present in multiple drug receptors. *AAPS PharmSci*, *1*(2), E2.

Ferre, S., Quiroz, C., Woods, A. S., Cunha, R., Popoli, P., Ciruela, F., et al. (2008). An update on adenosine A2A-dopamine D2 receptor interactions: Implications for the function of G protein-coupled receptors. *Current Pharmaceutical Design*, *14*(15), 1468–1474.

Foord, S. M., & Marshall, F. H. (1999). RAMPs: Accessory proteins for seven transmembrane domain receptors. *Trends in Pharmacological Sciences*, *20*(5), 184–187.

Foord, S. M., Bonner, T. I., Neubig, R. R., Rosser, E. M., Pin, J.-P., Davenport, A. P., et al. (2005). International Union of Pharmacology. XLVI. G protein-coupled receptor list. *Pharmacological Reviews*, *57*(2), 279–288.

Francesconi, A., & Duvoisin, R. M. (1998). Role of the second and third intracellular loops of metabotropic glutamate receptors in mediating dual signal transduction activation. *Journal of Biological Chemistry*, *273*(10), 5615–5624.

Fredriksson, R., Lagerstrom, M. C., Lundin, L. G., & Schioth, H. B. (2003). The G-protein-coupled receptors in the human genome form five main families. Phylogenetic analysis, paralogon groups, and fingerprints. *Molecular Pharmacology*, *63*(6), 1256–1272.

Fredriksson, R., Lagerstrom, M. C., & Schioth, H. B. (2005). Expansion of the superfamily of G-protein-coupled receptors in chordates. *Annals of the New York Academy of Sciences*, *1040*, 89–94.

Froestl, W., Bettler, B., Bittiger, H., Heid, J., Kaupmann, K., Mickel, S. J., et al. (2003). Ligands for expression cloning and isolation of GABA(B) receptors. *Farmaco*, *58*(3), 173–183.

Galvez, T., Duthey, B., Kniazeff, J., Blahos, J., Rovelli, G., Bettler, B., et al. (2001). Allosteric interactions between GB1 and GB2 subunits are required for optimal GABA(B) receptor function. *EMBO journal*, *20*(9), 2152–2159.

Galvez, T., Parmentier, M. L., Joly, C., Malitschek, B., Kaupmann, K., Kuhn, R., et al. (1999). Mutagenesis and modeling of the GABAB receptor extracellular domain support a venus flytrap mechanism for ligand binding. *Journal of Biological Chemistry*, *274*(19), 13362–13369.

Galvez, T., Urwyler, S., Prezeau, L., Mosbacher, J., Joly, C., Malitschek, B., et al. (2000). Ca(2+) requirement for high-affinity gamma-aminobutyric acid (GABA) binding at GABA(B) receptors: Involvement of serine 269 of the GABA(B)R1 subunit. *Molecular Pharmacology*, *57*(3), 419–426.

George, S. R., Fan, T., Xie, Z., Tse, R., Tam, V., Varghese, G., et al. (2000). Oligomerization of mu- and delta-opioid receptors. Generation of novel functional properties. *Journal of Biological Chemistry*, *275*(34), 26128–26135.

Hague, C., Uberti, M. A., Chen, Z., Hall, R. A., & Minneman, K. P. (2004). Cell surface expression of alpha1D-adrenergic receptors is controlled by heterodimerization with alpha1B-adrenergic receptors. *Journal of Biological Chemistry*, *279*(15), 15541–15549.

Hamm, H. E. (2001). How activated receptors couple to G proteins. *Proceedings of the National Academy of Sciences of the United States of America*, *98*(9), 4819–4821.

Jaakola, V. P., Griffith, M. T., Hanson, M. A., Cherezov, V., Chien, E. Y., Lane, J. R., et al. (2008). The 2.6 Angstrom crystal structure of a human A2A adenosine receptor bound to an antagonist. *Science*, *322*(5905), 1211–1217.

Jacobson, L. H., & Cryan, J. F. (2008). Evaluation of the anxiolytic-like profile of the GABAB receptor positive modulator CGP7930 in rodents. *Neuropharmacology*, *54*(5), 854–862.

Jacoby, E., Bouhelal, R., Gerspacher, M., & Seuwen, K. (2006). The 7 TM G-protein-coupled receptor target family. *ChemMedChem*, *1*(8), 760–782.

Jingami, H., Nakanishi, S., & Morikawa, K. (2003). Structure of the metabotropic glutamate receptor. *Current Opinion in Neurobiology*, *13*(3), 271–278.

Jones, K. A., Borowsky, B., Tamm, J. A., Craig, D. A., Durkin, M. M., Dai, M., et al. (1998). GABA(B) receptors function as a heteromeric assembly of the subunits GABA(B)R1 and GABA(B)R2. *Nature*, *396*(6712), 674–679.

Kamesh, N., Aradhyam, G. K., & Manoj, N. (2008). The repertoire of G protein-coupled receptors in the sea squirt *Ciona intestinalis*. *BMC Evolutionary Biology*, *8*, 129.

Kaupmann, K., Huggel, K., Heid, J., Flor, P. J., Bischoff, S., Mickel, S. J., et al. (1997). Expression cloning of GABA(B) receptors uncovers similarity to metabotropic glutamate receptors. *Nature*, *386*(6622), 239–246.

Kaupmann, K., Malitschek, B., Schuler, V., Heid, J., Froestl, W., Beck, P., et al. (1998). GABA (B)-receptor subtypes assemble into functional heteromeric complexes. *Nature*, *396* (6712), 683–687.

Kieffer, B. L., Befort, K., Gaveriaux-Ruff, C., & Hirth, C. G. (1992). The delta-opioid receptor: Isolation of a cDNA by expression cloning and pharmacological characterization. *Proceedings of the National Academy of Sciences of the United States of America*, *89*(24), 12048–12052.

Kniazeff, J., Bessis, A. S., Maurel, D., Ansanay, H., Prezeau, L., & Pin, J. P. (2004). Closed state of both binding domains of homodimeric mGlu receptors is required for full activity. *Nature Structural & Molecular Biology*, *11*(8), 706–713.

Kniazeff, J., Galvez, T., Labesse, G., & Pin, J. P. (2002). No ligand binding in the GB2 subunit of the GABA(B) receptor is required for activation and allosteric interaction between the subunits. *Journal of Neuroscience*, *22*(17), 7352–7361.

Kniazeff, J., Saintot, P. P., Goudet, C., Liu, J., Charnet, A., Guillon, G., et al. (2004). Locking the dimeric GABA(B) G-protein-coupled receptor in its active state. *Journal of Neuroscience*, *24*(2), 370–377.

Kuang, D., Yao, Y., MacLean, D., Wang, M., Hampson, D. R., & Chang, B. S. W. (2006). Ancestral reconstruction of the ligand-binding pocket of family C G protein-coupled receptors. *PNAS*, *103*(38), 14050–14055.

Kunishima, N., Shimada, Y., Tsuji, Y., Sato, T., Yamamoto, M., Kumasaka, T., et al. (2000). Structural basis of glutamate recognition by a dimeric metabotropic glutamate receptor. *Nature*, *407*(6807), 971–977.

Lagerstrom, M. C., & Schioth, H. B. (2008). Structural diversity of G protein-coupled receptors and significance for drug discovery. *Nature Reviews. Drug Discovery*, *7*(4), 339–357.

Lee, S. P., So, C. H., Rashid, A. J., Varghese, G., Cheng, R., Lanca, A. J., et al. (2004). Dopamine D1 and D2 receptor Co-activation generates a novel phospholipase C-mediated calcium signal. *Journal of biological chemistry*, *279*(34), 35671–35678.

Lefkowitz, R. J. (2004). Historical review: A brief history and personal retrospective of seven-transmembrane receptors. *Trends in Pharmacological Sciences*, *25*(8), 413–422.

Lehmann, A., Jörgen, M. Jensen, J. M., & Boeckxstaens, G. E. (2010). GABAB receptor agonism as a novel therapeutic modality in the treatment of gastroesophageal reflux disease. In *GABAB receptor pharmacology*, Advances in Pharmacology Vol. 58, pp. 288 ed., Blackburn and Enna (eds.): Elsevier Inc.

Liang, J. H., Chen, F., Krstew, E., Cowen, M. S., Carroll, F. Y., Crawford, D., et al. (2006). The GABA(B) receptor allosteric modulator CGP7930, like baclofen, reduces operant self-administration of ethanol in alcohol-preferring rats. *Neuropharmacology*, *50*(5), 632–639.

Libert, F., Parmentier, M., Lefort, A., Dinsart, C., Van Sande, J., Maenhaut, C., et al. (1989). Selective amplification and cloning of four new members of the G protein-coupled receptor family. *Science*, *244*(4904), 569–572.

Liu, J., Maurel, D., Etzol, S., Brabet, I., Ansanay, H., Pin, J. P., et al. (2004). Molecular determinants involved in the allosteric control of agonist affinity in the GABAB receptor by the GABAB2 subunit. *Journal of Biological Chemistry*, *279*(16), 15824–15830.

Lopez-Bendito, G., Shigemoto, R., Kulik, A., Paulsen, O., Fairen, A., & Lujan, R. (2002). Expression and distribution of metabotropic GABA receptor subtypes GABABR1 and GABABR2 during rat neocortical development. *European Journal of Neuroscience*, *15* (11), 1766–1778.

Malitschek, B., Schweizer, C., Keir, M., Heid, J., Froestl, W., Mosbacher, J., et al. (1999). The N-terminal domain of gamma-aminobutyric acid(B) receptors is sufficient to specify agonist and antagonist binding. *Molecular Pharmacology*, *56*(2), 448–454.

Margeta-Mitrovic, M., Jan, Y. N., & Jan, L. Y. (2000). A trafficking checkpoint controls GABA (B) receptor heterodimerization. *Neuron*, *27*(1), 97–106.

Margeta-Mitrovic, M., Jan, Y. N., & Jan, L. Y. (2001). Function of GB1 and GB2 subunits in G protein coupling of GABAB receptors. *PNAS*, *98*(25), 14649–14654.

Marshall, F. H. (2000). Molecular insight develops our understanding of the GABA(B) receptor. *Current Opinion in Drug Discovery & Development*, *3*(5), 597–604.

Marshall, F. H. (2005). Is the GABA B heterodimer a good drug target?. *Journal of Molecular Neuroscience*, *26*(2–3), 169–176.

Marshall, F. H., Jones, K. A., Kaupmann, K., & Bettler, B. (1999). GABAB receptors—the first 7TM heterodimers. *Trends in Pharmacological Sciences*, *20*(10), 396–399.

Martin, S. C., Russek, S. J., & Farb, D. H. (1999). Molecular identification of the human GABABR2: Cell surface expression and coupling to adenylyl cyclase in the absence of GABABR1. *Molecular and Cellular Neuroscience*, *13*(3), 180–191.

Maurel, D., Comps-Agrar, L., Brock, C., Rives, M. L., Bourrier, E., Ayoub, M. A., et al. (2008). Cell-surface protein-protein interaction analysis with time-resolved FRET and snap-tag technologies: Application to GPCR oligomerization. *Nature Methods*, *5*(6), 561–567.

Mezler, M., Muller, T., & Raming, K. (2001). Cloning and functional expression of GABA(B) receptors from *Drosophila*. *European Journal of Neuroscience*, *13*(3), 477–486.

Milligan, G. (2006). G-protein-coupled receptor heterodimers: Pharmacology, function and relevance to drug discovery. *Drug Discovery Today*, *11*(11–12), 541–549.

Milligan, G. (2007). G protein-coupled receptor dimerisation: Molecular basis and relevance to function. *Biochimica et Biophysica Acta*, *1768*(4), 825–835.

Milligan, G. (2009). G protein-coupled receptor hetero-dimerization: Contribution to pharmacology and function. *British Journal of Pharmacology*, *158*(1), 5–14.

Milligan, G. (2010). The role of dimerisation in the cellular trafficking of G-protein-coupled receptors. *Current Opinion in Pharmacology*, *10*(1), 23–29.

Milligan, G., & Bouvier, M. (2005). Methods to monitor the quaternary structure of G protein-coupled receptors. *FEBS Journal*, *272*(12), 2914–2925.

Muto, T., Tsuchiya, D., Morikawa, K., & Jingami, H. (2007). Structures of the extracellular regions of the group II/III metabotropic glutamate receptors. *Proceedings of the National Academy of Sciences of the United States of America*, *104*(10), 3759–3764.

Nakamura, T., Itadani, H., Hidaka, Y., Ohta, M., & Tanaka, K. (2000). Molecular cloning and characterization of a new human histamine receptor, HH4R. *Biochemical and Biophysical Research Communications*, *279*(2), 615–620.

Nakayasu, H., Nishikawa, M., Mizutani, H., Kimura, H., & Kuriyama, K. (1993). Immunoaffinity purification and characterization of gamma-aminobutyric acid (GABA)B receptor from bovine cerebral cortex. *Journal of Biological Chemistry*, *268*(12), 8658–8664.

Nelson, G., Chandrashekar, J., Hoon, M. A., Feng, L., Zhao, G., Ryba, N. J., et al. (2002). An amino-acid taste receptor. *Nature*, *416*(6877), 199–202.

Nelson, G., Hoon, M. A., Chandrashekar, J., Zhang, Y., Ryba, N. J., & Zuker, C. S. (2001). Mammalian sweet taste receptors. *Cell*, *106*(3), 381–390.

Ng, T. K., & Yung, K. K. (2001). Differential expression of GABA(B)R1 and GABA(B)R2 receptor immunoreactivity in neurochemically identified neurons of the rat neostriatum. *Journal of Comparative Neurology*, *433*(4), 458–470.

Niswender, C. M., & Conn, P. J. (2010). Metabotropic glutamate receptors: Physiology, pharmacology, and disease. *Annual Review of Pharmacology and Toxicology*, *50*, 295–322.

Nygaard, R., Frimurer, T. M., Holst, B., Rosenkilde, M. M., & Schwartz, T. W. (2009). Ligand binding and micro-switches in 7TM receptor structures. *Trends in Pharmacological Sciences*, *30*(5), 249–259.

Orru, A., Lai, P., Lobina, C., Maccioni, P., Piras, P., Scanu, L., et al. (2005). Reducing effect of the positive allosteric modulators of the GABA(B) receptor, CGP7930 and GS39783, on alcohol intake in alcohol-preferring rats. *European Journal of Pharmacology*, *525*(1–3), 105–111.

Pagano, A., Rovelli, G., Mosbacher, J., Lohmann, T., Duthey, B., Stauffer, D., et al. (2001). C-terminal interaction is essential for surface trafficking but not for heteromeric assembly of GABA(b) receptors. *Journal of Neuroscience*, *21*(4), 1189–1202.

Palczewski, K. (2006). G protein-coupled receptor rhodopsin. *Annual Review of Biochemistry*, *75*, 743–767.

Park, J. H., Scheerer, P., Hofmann, K. P., Choe, H. W., & Ernst, O. P. (2008). Crystal structure of the ligand-free G-protein-coupled receptor opsin. *Nature*, *454*(7201), 183–187.

Peltekova, V., Han, G., Soleymanlou, N., & Hampson, D. R. (2000). Constraints on proper folding of the amino terminal domains of group III metabotropic glutamate receptors. *Brain Research. Molecular Brain Research*, *76*(1), 180–190.

Pidasheva, S., Grant, M., Canaff, L., Ercan, O., Kumar, U., & Hendy, G. N. (2006). Calcium-sensing receptor dimerizes in the endoplasmic reticulum: Biochemical and biophysical characterization of CASR mutants retained intracellularly. *Human Molecular Genetics*, *15*(14), 2200–2209.

Pin, J. P., Comps-Agrar, L., Maurel, D., Monnier, C., Rives, M. L., Trinquet, E., et al. (2009). G-protein-coupled receptor oligomers: Two or more for what? Lessons from mGlu and GABA(B) receptors. *Journal of Physiology*, *587*(Pt 22), 5337–5344.

Pin, J. P., Galvez, T., & Prezeau, L. (2003). Evolution, structure, and activation mechanism of family 3/C G-protein-coupled receptors. *Pharmacology & Therapeutics*, *98*(3), 325–354.

Pin, J. P., Kniazeff, J., Binet, V., Liu, J., Maurel, D., Galvez, T., et al. (2004). Activation mechanism of the heterodimeric GABA(B) receptor. *Biochemical Pharmacology*, *68*(8), 1565–1572.

Pin, J. P., & Prezeau, L. (2007). Allosteric modulators of GABA(B) receptors: Mechanism of action and therapeutic perspective. *Current Neuropharmacology*, *5*(3), 195–201.

Prabhu, Y., Mondal, S., Eichinger, L., & Noegel, A. A. (2007a). A GPCR involved in post aggregation events in *Dictyostelium discoideum*. *Developmental Biology*, *312*(1), 29–43.

Prabhu, Y., Muller, R., Anjard, C., & Noegel, A. A. (2007b). GrlJ, a dictyostelium GABAB-like receptor with roles in post-aggregation development. *BMC Developmental Biology*, *7*, 44.

Rashid, A. J., So, C. H., Kong, M. M., Furtak, T., El-Ghundi, M., Cheng, R., et al. (2007). D1-D2 dopamine receptor heterooligomers with unique pharmacology are coupled to rapid activation of Gq/11 in the striatum. *Proceedings of the National Academy of Sciences of the United States of America*, *104*(2), 654–659.

Rivero-Muller, A., Chou, Y. Y., Ji, I., Lajic, S., Hanyaloglu, A. C., Jonas, K., et al. (2010). Rescue of defective G protein-coupled receptor function in vivo by intermolecular cooperation. *Proceedings of the National Academy of Sciences of the United States of America*, *107*(5), 2319–2324.

Robbins, M. J., Calver, A. R., Filippov, A. K., Hirst, W. D., Russell, R. B., Wood, M. D., et al. (2001). GABA(B2) is essential for G-protein coupling of the GABA(B) receptor heterodimer. *Journal of Neuroscience*, *21*(20), 8043–8052.

Rondard, P., Huang, S., Monnier, C., Tu, H., Blanchard, B., Oueslati, N., et al. (2008). Functioning of the dimeric GABA(B) receptor extracellular domain revealed by glycan wedge scanning. *EMBO Journal*, *27*(9), 1321–1332.

Rozenfeld, R., & Devi, L. A. (2007). Receptor heterodimerization leads to a switch in signaling: Beta-arrestin2-mediated ERK activation by mu-delta opioid receptor heterodimers. *FASEB Journal*, *21*(10), 2455–2465.

Rozenfeld, R., & Devi, L. A. (2010). Receptor heteromerization and drug discovery. *Trends in Pharmacological Sciences*, *31*(3), 124–130.

Sakai, T., Aoyama, M., Kusakabe, T., Tsuda, M., & Satake, H. (2009). Functional diversity of signaling pathways through G protein-coupled receptor heterodimerization with a species-specific orphan receptor subtype. *Molecular Biology and Evolution*, *27*, 1097–1106.

Salahpour, A., Angers, S., & Bouvier, M. (2000). Functional significance of oligomerization of G-protein-coupled receptors. *Trends in Endocrinology and Metabolism*, *11*(5), 163–168.

Salahpour, A., Angers, S., Mercier, J. F., Lagace, M., Marullo, S., & Bouvier, M. (2004). Homodimerization of the beta2-adrenergic receptor as a prerequisite for cell surface targeting. *Journal of Biological Chemistry*, *279*(32), 33390–33397.

Scheerer, P., Park, J. H., Hildebrand, P. W., Kim, Y. J., Krauss, N., Choe, H. W., et al. (2008). Crystal structure of opsin in its G-protein-interacting conformation. *Nature*, *455*(7212), 497–502.

Schofield, P. R., Darlison, M. G., Fujita, N., Burt, D. R., Stephenson, F. A., Rodriguez, H., et al. (1987). Sequence and functional expression of the GABA A receptor shows a ligand-gated receptor super-family. *Nature*, *328*(6127), 221–227.

Segre, G. V., & Goldring, S. R. (1993). Receptors for secretin, calcitonin, parathyroid hormone (PTH)/PTH-related peptide, vasoactive intestinal peptide, glucagonlike peptide 1, growth hormone-releasing hormone, and glucagon belong to a newly discovered G-protein-linked receptor family. *Trends in Endocrinology and Metabolism*, *4*(10), 309–314.

Sekiguchi, M., Sakuta, H., Okamoto, K., & Sakai, Y. (1990). GABAB receptors expressed in *Xenopus* oocytes by guinea pig cerebral mRNA are functionally coupled with Ca2(+)-dependent Cl– channels and with K+ channels, through GTP-binding proteins. *Brain Research. Molecular Brain Research*, *8*(4), 301–309.

Sexton, P. M., Morfis, M., Tilakaratne, N., Hay, D. L., Udawela, M., Christopoulos, G., et al. (2006). Complexing receptor pharmacology: Modulation of family B G protein-coupled receptor function by RAMPs. *Annals of the New York Academy of Sciences*, *1070*, 90–104.

Smith, M. A., Yancey, D. L., Morgan, D., Liu, Y., Froestl, W., & Roberts, D. C. (2004). Effects of positive allosteric modulators of the GABAB receptor on cocaine self-administration in rats. *Psychopharmacology (Berl)*, *173*(1–2), 105–111.

Srivastava, M., Begovic, E., Chapman, J., Putnam, N. H., Hellsten, U., Kawashima, T., et al. (2008). The *Trichoplax* genome and the nature of placozoans. *Nature*, *454*(7207), 955–960.

Taniyama, K., Takeda, K., Ando, H., & Tanaka, C. (1991). Expression of the GABAB receptor in *Xenopus* oocytes and desensitization by activation of protein kinase C. *Advances in Experimental Medicine & Biology*, *287*, 413–420.

Trivedi, R., Mithal, A., & Chattopadhyay, N. (2008). Recent updates on the calcium-sensing receptor as a drug target. *Current Medicinal Chemistry*, *15*(2), 178–186.

Tsuchiya, D., Kunishima, N., Kamiya, N., Jingami, H., & Morikawa, K. (2002). Structural views of the ligand-binding cores of a metabotropic glutamate receptor complexed with an antagonist and both glutamate and Gd3+. *Proceedings of the National Academy of Sciences of the United States of America*, *99*(5), 2660–2665.

Tyacke, R., Lingford-Hughes, A., Reed, L. J., & Nutt, D. J. (2010). GABAB receptors in addiction and its treatment. In *GABAB receptor pharmacology*, Advances in Pharmacology Vol. 58, 288 ed. Blackburn and Enna: Elsevier Inc.

Uberti, M. A., Hague, C., Oller, H., Minneman, K. P., & Hall, R. A. (2005). Heterodimerization with beta2-adrenergic receptors promotes surface expression and functional activity of alpha1D-adrenergic receptors. *Journal of Pharmacology and Experimental Therapeutics*, *313*(1), 16–23.

Ulloa-Aguirre, A., & Conn, P. M. (2009). Targeting of G protein-coupled receptors to the plasma membrane in health and disease. *Frontiers in Bioscience*, *14*, 973–994.

Waldhoer, M., Fong, J., Jones, R. M., Lunzer, M. M., Sharma, S. K., Kostenis, E., et al. (2005). A heterodimer-selective agonist shows in vivo relevance of G protein-coupled receptor dimers. *Proceedings of the National Academy of Sciences of the United States of America*, *102*(25), 9050–9055.

Wang, L., Martin, B., Brenneman, R., Luttrell, L. M., & Maudsley, S. (2009). Allosteric modulators of g protein-coupled receptors: Future therapeutics for complex physiological disorders. *Journal of Pharmacology and Experimental Therapeutics*, *331*(2), 340–348.

Warne, T., Serrano-Vega, M. J., Baker, J. G., Moukhametzianov, R., Edwards, P. C., Henderson, R., et al. (2008). Structure of a beta1-adrenergic G-protein-coupled receptor. *Nature*, *454*(7203), 486–491.

White, J. H., Wise, A., Main, M. J., Green, A., Fraser, N. J., Disney, G. H., et al. (1998). Heterodimerization is required for the formation of a functional GABA(B) receptor. *Nature*, *396*(6712), 679–682.

Wilson, S., Wilkinson, G., & Milligan, G. (2005). The CXCR1 and CXCR2 receptors form constitutive homo- and heterodimers selectively and with equal apparent affinities. *Journal of Biological Chemistry*, *280*(31), 28663–28674.

Woodward, R. M., & Miledi, R. (1992). Sensitivity of *Xenopus* oocytes to changes in extracellular pH: Possible relevance to proposed expression of atypical mammalian GABAB receptors. *Brain Research. Molecular Brain Research*, *16*(3–4), 204–210.

Dietmar Benke
Institute of Pharmacology and Toxicology, University of Zurich, Zurich, Switzerland

Mechanisms of $GABA_B$ Receptor Exocytosis, Endocytosis, and Degradation

Abstract

$GABA_B$ receptors belong to the family of G-protein-coupled receptors, which mediate slow inhibitory neurotransmission in the central nervous system. They are promising drug targets for a variety of neurological disorders and play important functions in regulating synaptic plasticity. Signaling strength is critically dependent on the availability of the receptors at the cell surface. Several distinct highly regulated trafficking mechanisms ensure the presence of adequate receptor numbers in the plasma membrane. The rate of exocytosis of newly synthesized receptors

Advances in Pharmacology, Volume 58
1054-3589/10 $35.00
10.1016/S1054-3589(10)58004-9

from the endoplasmic reticulum via the Golgi apparatus to the cell surface as well as the rates of their endocytosis and degradation determines the retention time of receptors at the cell surface. This chapter focuses on the recently emerged mechanisms of $GABA_B$ receptor exocytosis, endocytosis, recycling, and degradation.

I. Introduction

$GABA_B$ receptors are G-protein-coupled receptors composed of the two subunits $GABA_{B1}$ and $GABA_{B2}$. Each subunit comprises a large extracellular N-terminal domain, seven transmembrane spanning regions, and a large intracellular C-terminal domain. The C-terminal domains of both subunits contain a coiled-coil structure, which is involved in heterodimerization of the receptor and constitutes a binding domain for several interacting proteins (Bettler & Tiao, 2006). $GABA_B$ receptors mediate slow inhibitory neurotransmission in the central nervous system and play a crucial role in neuronal plasticity underlying learning and memory. The regulation of receptor signaling is one fundamental process contributing to neuronal plasticity. Accordingly, numerous mechanisms have evolved that modulate receptor activity and availability. In particular, mechanisms regulating cell-surface expression of receptors are predestinated to adjust signaling strength to changing physiological conditions. There are several distinct and highly regulated processes that affect the presence and lifetime of a receptor at the cell surface (Fig. 1): (i) The rate of synthesis, maturation, and assembly of receptor proteins, (ii) transport and targeting of receptors to a particular site at the neuronal surface, (iii) retention of the receptor at that site, (iv) receptor endocytosis, (v) receptor recycling, and eventually (vi) degradation of the receptors. Extensive investigations during the past years provided first insights into some of these mechanisms for $GABA_B$ receptors. This chapter reviews recent progress that has been achieved analyzing exocytosis, endocytosis, recycling, and degradation of $GABA_B$ receptors.

II. Cell-Surface Trafficking of $GABA_B$ Receptors

Newly synthesized cell-surface receptors are processed and passed through distinct membrane compartments before reaching the plasma membrane, including the endoplasmic reticulum (ER), *cis*-Golgi network, and *trans*-Golgi network (Fig. 2). While being translated, receptor proteins are inserted into the membrane of the ER. Folding into the proper three-dimensional structure is assisted by chaperone proteins and first

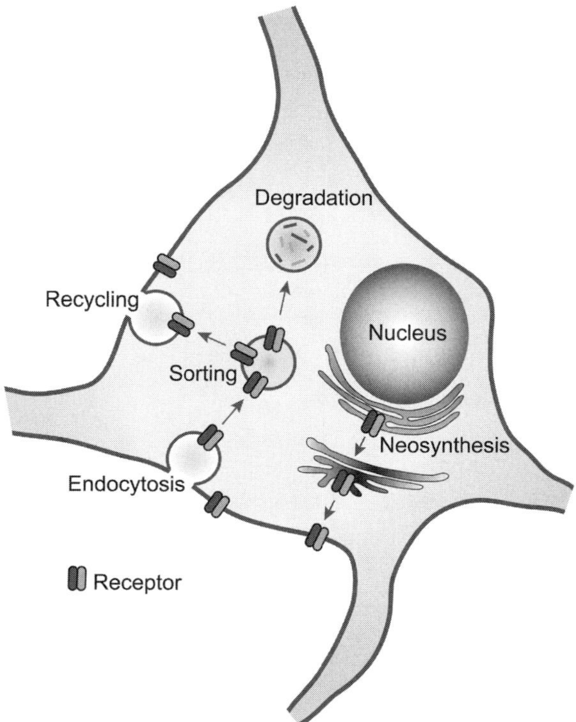

FIGURE 1 Schematic overview of trafficking pathways determining cell-surface availability of plasma membrane receptors. After neosynthesis and assembly of subunits in the ER the receptors are transported to the Golgi apparatus where they are further processed and finally targeted via the *trans*-Golgi network to the plasma membrane. Cell-surface receptors are removed from the plasma membrane by endocytosis and first enter early endosomes from which they are either recycled back to the plasma membrane or sorted to lysosomes for degradation. The rates of exocytosis, internalization, recycling, and degradation determine the size and half-life of the receptor population at the cell surface. Therefore, altering any of those rates, for example, by external stimuli or under pathophysiological conditions, will affect cell-surface expression of receptors.

posttranslational modifications such as core glycosylation and disulfide bond formation take place. Correctly folded and assembled proteins exit the ER at specific sites (transitional ER or ER exit sites) and are transported via coat protein complex II (COPII)-coated vesicles to the Golgi apparatus where the proteins are further processed and finally transported via the *trans*-Golgi network to the plasma membrane. Trafficking of cell-surface receptors from the ER to the plasma membrane is a tightly controlled process which assures that only correctly folded, assembled, and fully functional receptors reach the cell surface.

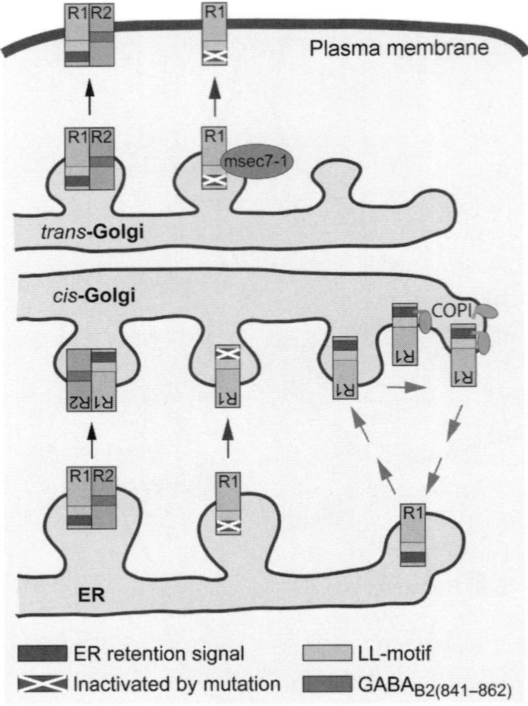

FIGURE 2 Cell-surface trafficking of $GABA_B$ receptors. Newly synthesized $GABA_B$ receptor subunits are assembled within the ER. Association of $GABA_{B1}$ and $GABA_{B2}$ inactivates an ER retention signal in the C-terminal domain of $GABA_{B1}$ which permits ER exit (left hand side). Unassembled $GABA_{B1}$ subunits may leave the ER to enter the *cis*-Golgi where they are recognized by COPI via direct interaction with the ER retention signal of $GABA_{B1}$. Association with COPI recruits $GABA_{B1}$ to retrograde transport back to the ER (right hand side). This quality control system ensures that only $GABA_{B1}$ assembled with $GABA_{B2}$ are transported to the cell surface. If the ER retention signal is mutated inactive $GABA_{B1}$ is able to reach the plasma membrane (middle). This process is, however, inefficient due to a LL-motif within the coiled-coil domain of $GABA_{B1}$, which causes accumulation of $GABA_{B1}$ in the *trans*-Golgi. Interaction of the GEF msec7-1 presumably with the coiled-coil domain of $GABA_{B1}$ containing the LL-motif facilitates cell-surface expression of $GABA_{B1}$ subunits with mutational inactivated ER retention signal. A 20 amino acid amino sequence in $GABA_{B2}$ contains additional trafficking signals ($GABA_{B2(841–862)}$). Deletion of this sequence results in an impaired cell-surface expression and reduced axonal targeting of $GABA_B$ receptors.

A. Regulation of ER Exit by an ER Retention Signal in $GABA_{B1}$

At present, only limited information is available about folding, processing, assembly, and cell-surface targeting of $GABA_B$ receptors. Soon after cloning of the $GABA_B$ receptor subunits it became clear that cell-surface expression of functional receptors requires the heterodimerization of the $GABA_{B1}$ and $GABA_{B2}$ subunits. An arginine-rich ER retention signal

(RSRR) present in the intracellular C-terminal tail at the distal end of the coiled-coil domain of $GABA_{B1}$ efficiently prevents forward trafficking of unassembled $GABA_{B1}$ (Calver et al., 2001; Margeta-Mitrovic et al., 2000; Pagano et al., 2001). Analysis of transgenic mice finally proved that the ER retention signal provides a control element that determines the expression of functional $GABA_B$ receptors at the cell surface *in vivo*. In mice expressing a C-terminally truncated $GABA_{B2}$, which leaves the ER retention signal of $GABA_{B1}$ exposed after heterodimerization, no $GABA_{B1}$ reached the cell surface and accordingly no $GABA_B$ receptor function was observed (Thuault et al., 2004). ER retention of $GABA_{B1}$ appears to involve the coat protein complex I (COPI), which is a central component of the retrograde transport of proteins from the *cis*-Golgi back to the ER (Brock et al., 2005). COPI directly interacts with the ER retention signal of $GABA_{B1}$ and is thought to retrieve unassembled $GABA_{B1}$ from the Golgi apparatus (Fig. 2). The presence of unassembled $GABA_{B1}$ in the *cis*-Golgi compartment supports this view (Brock et al., 2005). Mutational analysis indicates that the correct spacing of the ER retention signal to the membrane is important for its function (Gassmann et al., 2005). According to this data heterodimerization of $GABA_B$ receptors induces conformational changes that render the ER retention signal inactive. A direct masking or shielding of the ER retention signal of $GABA_{B1}$ by the $GABA_{B2}$ coiled-coil domain does not appear to be required for its inactivation.

B. Regulation of *trans*-Golgi to Plasma Membrane Targeting by a Dileucine Motif in $GABA_{B1}$

In addition to the ER retention signal, a dileucine (LL) motif located proximal to the ER retention signal within the coiled-coil domain of $GABA_{B1}$ appears to regulate intracellular transport of $GABA_{B1}$ (Restituito et al., 2005). LL motifs are known to be involved in the regulation of exocytosis, endocytosis, and targeting of proteins to distinct cellular compartments (Marchese et al., 2008; Pandey, 2009; Tan et al., 2004). When the ER retention signal in $GABA_{B1}$ is inactivated by mutation, $GABA_{B1}$ exits from the ER but accumulates in the *trans*-Golgi network and the transport to the cell surface is inefficient. Additional mutational inactivation of the LL-motif was required for efficient cell-surface targeting of $GABA_{B1}$ (Restituito et al., 2005). Interestingly, direct interaction of the guanine-nucleotide-exchange factor (GEF) msec7-1, most likely with the coiled-coil domain of $GABA_{B1}$, facilitated cell-surface expression of $GABA_{B1}$ (Restituito et al., 2005). msec7-1 is a GEF for ADP-ribosylation factors which regulate vesicular membrane transport (Casanova, 2007).

These findings support the concept of a sequential regulation of $GABA_B$ receptor cell-surface transport by at least two distinct amino acid motifs in

$GABA_{B1}$ (Fig. 2). The ER-retention signal interacts with COPI and prevents forward trafficking of unassembled $GABA_{B1}$ subunits by shuttling it back from the *cis*-Golgi to the ER. The LL-motif seems to regulate $GABA_{B1}$ transport at a later stage at the level of the *trans*-Golgi network by a direct interaction with the GEF msec7-1. However, at present the function of the LL-motif and its interaction with msec7-1 in assembled heterodimeric $GABA_B$ receptors is unclear. In line with its location, the LL-motif is likely to be masked upon heterodimerization with $GABA_{B2}$, and it may not directly be involved in regulating the trafficking of assembled $GABA_B$ receptors.

C. Regulation of Forward Trafficking by a C-terminal $GABA_{B2}$ Sequence

Apart from the ER retention signal and LL-motif in $GABA_{B1}$, an amino acid sequence in $GABA_{B2}$ has been identified containing information for trafficking (Pooler et al., 2009). Mutational analysis revealed that the amino acid sequence 841–862 in the C-terminal domain of $GABA_{B2}$ is important for cell-surface trafficking. When the sequence was absent, $GABA_{B2}$—expressed alone or in combination with $GABA_{B1}$—exhibited a reduced cell-surface expression. Partial sensitivity to endoglycosidase H indicates an impaired ER exit of mutant $GABA_{B2}$. Since none of the so far identified $GABA_B$ receptor interacting proteins bind to $GABA_{B2(841–862)}$ as yet unknown protein(s) may promote ER exit of $GABA_{B2}$. The $GABA_{B2(841–862)}$ sequence therefore represents a highly promising bait for yeast two-hybrid screens that might yield novel $GABA_B$ receptor-interacting proteins involved in regulating forward trafficking of the receptors.

D. $GABA_{B1}$ Transport along Dendritic ER

Neurons are highly polarized cells that extend the ER and Golgi network far into their dendrites and possibly axons (Kennedy & Ehlers, 2006). As protein synthesis for the majority of membrane proteins occurs at the somatic ER, the question arises whether the assembled $GABA_B$ receptor complex or unassembled individual subunits are transported into neurites. Interestingly, a recent study observed only a low level of colocalization of $GABA_{B1}$ and $GABA_{B2}$ in intracellular compartments of the soma and dendrites of hippocampal neurons as compared to the plasma membrane (Ramirez et al., 2009). The data suggest that $GABA_B$ receptor subunits are synthesized at the somatic ER and then are individually transported along the ER into dendrites where they assemble into heterodimeric receptor complexes and leave the ER for trafficking to the plasma membrane. This model requires a mechanism that efficiently regulates assembly of $GABA_B$ receptors within the ER and thus would be another control system to determine the cell-surface availability of functional $GABA_B$ receptors. The model further assumes specific transport mechanisms for $GABA_B$ receptor subunits along the ER within neurites. Support for a specific $GABA_B$ receptor subunit transport within the ER of dendrites is provided by the finding that the multiple coiled-coil domain-containing protein Marlin-1 selectively interacts with $GABA_{B1}$ and the molecular motor kinesin-I (Vidal et al., 2007). Kinesin-I is a molecular motor for the directional microtubule-dependent transport in axons and dendrites (Hirokawa et al., 2009). Overexpression of a kinesin-I mutant that lacked cargo-binding domains inhibited $GABA_{B1}$ transport into dendrites. This observation suggests that newly synthesized $GABA_{B1}$ proteins are transported within the ER into dendrites where they assemble with $GABA_{B2}$ and then exit the ER. This conclusion is supported by the observation that $GABA_{B1a}$ can be transported into axons in the absence of $GABA_{B2}$ (Biermann et al., 2010). However, $GABA_{B2}$ requires the presence of $GABA_{B1a}$ for entry into axons (Biermann et al., 2010), indicating that at least for presynaptic localization of functional $GABA_B$ receptors the assembled heterodimeric complex is transported along the axon. On the other hand, there is data indicating that $GABA_{B2}$ may be required for dendritic and axonal targeting of $GABA_{B1}$ in cultured hippocampal neurons (Pooler et al., 2009), contradicting the intriguing model of individual transport of $GABA_B$ receptor subunits in dendrites. Unfortunately, there is not enough detailed electron microscopic data on the intracellular colocalization of $GABA_{B1}$ and $GABA_{B2}$ in brain sections to unambiguously identify the presence of a large pool of unassembled $GABA_{B1}$ and $GABA_{B2}$ in the ER of dendrites. Depending on the brain area and type of neuron, the proportion of intracellular versus cell-surface staining of $GABA_{B1}$ and $GABA_{B2}$ varies considerably. For instance, in neurons of the globus pallidus (Chen et al., 2004) and substantia nigra (Boyes & Bolam, 2003) the majority of $GABA_{B1}$ was found to be associated with intracellular sites, whereas most of $GABA_{B2}$ was located in the plasma membrane. This observation may favor independent dendritic transport of $GABA_{B1}$ and $GABA_{B2}$. On the other hand, in dendrites of neurons in thalamic nuclei (Villalba et al., 2006), neocortex (Lopez-Bendito et al., 2002), and hippocampus (Kulik et al., 2003) both subunits were predominantly present in the plasma membrane and displayed colocalization at intracellular sites. Hence, depending on the neuronal population there might be distinct trafficking and targeting mechanisms.

III. Endocytosis of $GABA_B$ Receptors

At the plasma membrane endocytosis is a determinant of the availability of the receptors for signal transduction. There are two principal modes of internalization: constitutive and agonist-induced endocytosis. Particularly agonist-induced endocytosis provides a mechanism that ensures reduction of receptor signaling and represents an important part of cellular plasticity. In combination with recycling it provides a mechanism that recovers the receptors from desensitization (Gainetdinov et al., 2004; Luttrell & Lefkowitz, 2002). Initial reports indicated that desensitization of $GABA_B$ receptors is distinct from the classical mechanism of G-protein-coupled receptors (Perroy et al., 2003; Pontier et al., 2006), which involves phosphorylation of the receptors upon activation, recruitment of arrestin followed by internalization, dephosphorylation of the receptor, and recycling it back to the plasma membrane. Instead of promoting internalization of $GABA_B$ receptors, phosphorylation appears to stabilize the receptors in the plasma membrane (Couve et al., 2002; Fairfax et al., 2004; Kuramoto et al., 2007). Based on these results it was concluded that $GABA_B$ receptors are very stable at the cell surface and display no agonist-induced internalization. However, subsequent studies concentrating on the mechanisms of $GABA_B$ receptor endocytosis using more refined experimental tools provided evidence that $GABA_B$ receptors undergo rapid constitutive endocytosis and recycling as well as agonist-induced internalization, as described below.

A. Constitutive Internalization of $GABA_B$ Receptors

First evidence for constitutive endocytosis of $GABA_B$ receptors came from studies of the internalization of recombinant $GABA_{B1b,2}$ receptors expressed in human embryonic kidney (HEK) 293 cells (Grampp et al., 2007). In this study, cell-surface receptors were either tagged with an antibody recognizing an extracellular epitope on $GABA_{B1b}$ or by cell-surface biotinylation. With both approaches robust intracellular accumulation of internalized $GABA_B$ receptors was observed. Within 10 min small clusters of internalized receptors were detected in the vicinity of the plasma membrane, indicating that $GABA_B$ receptors undergo rapid constitutive internalization. This observation was confirmed by a study using $GABA_{B1a}$ tagged with the minimal binding site motif for α-bungarotoxin (BBS), which binds fluorescently labeled α-bungarotoxin almost irreversibly (Wilkins et al., 2008), and by a study on human influenza hemagglutinin (HA)-tagged receptors expressed in COS-7 cells (Pooler et al., 2009). Further work on epitope-tagged $GABA_B$ receptors transiently expressed in neurons (Vargas et al., 2008; Wilkins et al., 2008) and native neuronal $GABA_B$ receptors (Grampp et al., 2008; Vargas et al., 2008) confirmed that fast constitutive endocytosis is not due to overexpression of the receptors and is not restricted to non-neuronal cells. The data accumulated so far indicate that the intact dimeric receptor complex is internalized but not dissociated monomeric $GABA_B$ receptor subunits (Grampp et al., 2008; Laffray et al., 2007; Vargas et al., 2008). Interestingly, although the level of internalization is similar for ectopically expressed and neuronal receptors (~40–60%), the time constant for internalization of ectopically expressed receptors (Pooler et al., 2009; Wilkins et al., 2008) is considerably slower ($\tau \sim 40$ min) than for BBS-tagged $GABA_B$ receptors transiently expressed in hippocampal neurons ($\tau \sim 7.5$–14 min; Wilkins et al., 2008). It is conceivable that a fast internalization rate in neurons provides the basis for rapid changes in the number of cell-surface receptors by increasing or decreasing the rate of internalization.

B. Pathway of $GABA_B$ Receptor Endocytosis and Endosomal Sorting

G-protein-coupled receptors principally can internalize via dynamin-dependent or dynamin-independent mechanisms. Both mechanisms may use either clathrin-coated or caveolin-coated vesicles for internalization of the receptors. So far, there is significant evidence that $GABA_B$ receptors internalize via the classical dynamin and clathrin-dependent pathway (Fig. 3). Dynamin is a GTPase that mediates membrane fission and is essential for clathrin-dependent endocytosis (Mettlen et al., 2009). Overexpression of a nonfunctional mutant of dynamin inhibited endocytosis of $GABA_B$ receptors expressed in HEK 293 cells (Grampp et al., 2007) and neurons (Vargas et al., 2008), indicating that functional dynamin is required for internalization of the receptors. $GABA_B$ receptors appear to

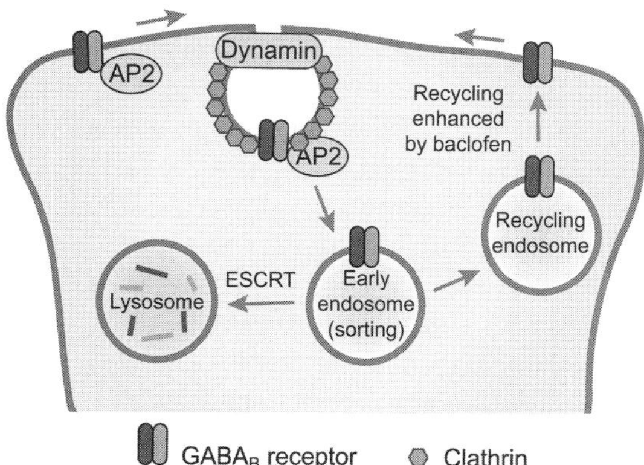

FIGURE 3 Endocytic pathways of $GABA_B$ receptors. $GABA_B$ receptors are most likely recruited by the AP2 complex to clathrin-coated pits and are constitutively endocytosed in a dynamin-dependent manner. From early endosomes, $GABA_B$ receptors are either sorted via the ESCRT machinery to lysosomes for degradation or recycled back to the plasma membrane. Recycling and degradation of $GABA_B$ receptors appear to be highly regulated and tightly balanced: activation by baclofen accelerates recycling of receptors whereas blocking recycling redirects the receptors to lysosomes.

predominantly internalize via clathrin-coated vesicles since treatments that interfere with this pathway, such as hypertonic sucrose, low K^+, and chlorpromazine, efficiently inhibited internalization of neuronal and ectopically expressed $GABA_B$ receptors (Grampp et al., 2007; Laffray et al., 2007; Vargas et al., 2008). Further support for the endocytosis via the clathrin-dependent pathway is provided by the observation that the adaptor protein complex 2 (AP2) complex, which recruits cargo to clathrin-coated pits (Schmid, 1997), colocalizes and co-immunoprecipitates with $GABA_B$ receptors (Grampp et al., 2007, 2008; but see Vargas et al., 2008).

Although $GABA_B$ receptors expressed in CHO cells and neurons have been shown to be present in lipid rafts (Becher et al., 2001, 2004) no evidence for raft/caveolin-dependent endocytosis have been found so far. Agents known to inhibit caveolin-dependent endocytosis (nystatin, filipin) did not affect internalization of $GABA_B$ receptors expressed in HEK 293 cells (Grampp et al., 2007) or neurons (Laffray et al., 2007). In addition, neither cell surface nor internalized $GABA_B$ receptors expressed in HEK 293 cells display significant colocalization with caveolin (Grampp et al., 2007). However, these experiments do not exclude that a minor fraction of $GABA_B$ receptor uses this endocytic pathway.

Endocytosed membrane proteins, irrespective of whether they are internalized via clathrin-dependent or independent mechanisms, first enter the compartment of early endosomes, which represents the vesicular organelle for sorting of cargo. From early endosomes proteins can be targeted to specific routes: to fast recycling endosomes, to a compartment for slow recycling or to multivesicular bodies (late endosomes), and finally lysosomes for degradation (Fig. 3). Colocalization studies using antibodies directed against marker proteins for these compartments suggest that $GABA_B$ receptors may use all these endosomal trafficking routes. In cultured neurons $GABA_B$ receptors have been found to colocalize with markers for early endosomes (Rab5, EEA1; Grampp et al., 2008), fast recycling endosomes (Rab4; Grampp et al., 2008), slow recycling endosomes (Rab11, Grampp et al., 2008; Laffray et al., 2007; Vargas et al., 2008), and late endosomes/lysosomes (Lamp1, cathepsin; Grampp et al., 2008; Laffray et al., 2007).

Together, these findings indicate that $GABA_B$ receptors are recruited via the AP2 complex to clathrin-coated pits and internalized in a dynamin-dependent manner. Endocytosed $GABA_B$ receptors enter first the compartment of early endosomes from which they are sorted either to recycling endosomes for reinsertion into the plasma membrane or to lysosomes for degradation (Fig. 3).

C. Constitutive Recycling of $GABA_B$ Receptors

There is solid functional evidence from cortical and hippocampal neurons that internalized $GABA_B$ receptors constitutively recycle back to the cell surface, which is in line with their colocalization with the marker proteins

Rab4 and Rab11, respectively, for fast and slow recycling endosomes (see above). Using an immunofluorescence assay (Vargas et al., 2008) or cell-surface biotinylation experiments (Grampp et al., 2008) a significant fraction of internalized $GABA_B$ receptors was found to reappear at the plasma membrane. About 40–50% of endocytosed $GABA_B$ receptors were estimated to recycle back to the cell surface within 15 min (Grampp et al., 2008). Constitutive recycling of $GABA_B$ receptors seems to be in a tightly regulated balance with their degradation in lysosomes. Inhibition of recycling with monensin—which blocks the fusion of vesicles with the plasma membrane and thus prevents the reinsertion of the receptors—resulted in fast lysosomal degradation of the receptors in cortical neurons. About 50% of the entire receptor population was degraded within 30 min (Grampp et al., 2008). Thus, one important function of recycling might be to prevent entry of internalized receptors into degenerative pathways. The mechanism that is involved in sorting $GABA_B$ receptors from early endosomes to the recycling compartment remains to be established. In analogy to the well-characterized recycling mechanism of the transferrin receptor, sorting nexin 4 (SNX4) might be one potential protein involved. Ablation of SNX4 resulted in redirecting transferrin receptors to lysosomes for degradation (Traer et al., 2007).

The physiological function of constitutive $GABA_B$ receptor recycling is as yet unknown. However, endocytic recycling appears to play an important role in a variety of cellular processes including learning and memory (Grant & Donaldson, 2009). During the early phase of LTP, recycling endosomes are recruited in a myosin Vb-dependent manner into dendritic spines (Wang et al., 2008) and constitutively recycling AMPA receptors are inserted with enhanced rate into the plasma membrane (Park et al., 2004). These events result in spine growth, increased level of synaptic AMPA receptors, and synaptic efficacy. Accordingly, constitutive recycling of $GABA_B$ receptors may provide a mechanism for the neuron to be prepared for rapid adjustment of the cell-surface number of receptors available for signaling to changing physiological conditions.

D. Agonist-Induced Endocytosis of $GABA_B$ Receptors

Endocytosis of many G-protein-coupled receptors is triggered or stimulated by ligand-induced activation. Agonist-induced endocytosis of receptors may direct receptors to degradation, which is an efficient way to downregulate signaling by reducing the receptor number at the cell surface. Alternatively, for many G-protein-coupled receptors agonist-induced internalization in combination with recycling provides a mechanism for reactivating desensitized receptors. However, activation of $GABA_B$ receptors with baclofen did neither result in a reduction of cell-surface receptors nor in their intracellular accumulation, irrespective whether the receptors were expressed ectopically (Fairfax et al., 2004; Grampp et al.,

2007; Mutneja et al., 2005; Perroy et al., 2003; but see Gonzalez-Maeso et al., 2003) or in neurons (Fairfax et al., 2004; Grampp et al., 2008; Vargas et al., 2008). However, the experimental settings of those studies did not take into account that rapid recycling of the receptors might have obscured the effects of receptor activation. Indeed, Laffray et al. (2007) showed that blocking recycling resulted in an increased intracellular accumulation of $GABA_B$ receptors upon stimulating spinal neurons with baclofen. Subsequent studies showed that baclofen accelerates internalization of $GABA_B$ receptor expressed in HEK cells without changing significantly the amount of internalized receptors (Wilkins et al., 2008) and enhanced recycling of receptors in cortical neurons (Grampp et al., 2008).

Further support for agonist-induced endocytosis of $GABA_B$ receptors is provided by a study that analyzed $GABA_B$ receptor-mediated internalization of calcium channels in chick sensory neurons (Puckerin et al., 2006). Activation of $GABA_B$ receptors with baclofen resulted in a transient intracellular colocalization of $GABA_B$ receptors with calcium channels and arrestin. So far, there is no evidence that activated $GABA_B$ receptors directly bind arrestin, which is involved in agonist-induced endocytosis of many G-protein-coupled receptors. Coexpression of arrestin 2-EGFP or arrestin 3-EGFP with $GABA_B$ receptors in COS-7 cells (Fairfax et al., 2004) did not result in a recruitment of arrestin to $GABA_B$ receptors upon activation. There is, however, evidence that $GABA_B$ receptors are recruited to a preformed complex of calcium channels and arrestin which is then rapidly internalized (Puckerin et al., 2006).

Together, the current data indicate that activation of $GABA_B$ receptors accelerates the rates of their endocytosis and recycling. This may be a relevant mechanism for recovering desensitized $GABA_B$ receptors. One mechanism proposed for desensitization of $GABA_B$ receptor involves the interaction of *N*-ethylmaleimide-sensitive fusion (NSF) protein followed by phosphorylation of the receptor by protein kinase C (Pontier et al., 2006). An enhanced rate of recycling upon agonist activation is expected to result in a fast dephosphorylation of receptors in the endosomal compartment and accelerated reinsertion of resensitized receptors into the plasma membrane. This mechanism would ensure the fast recovery of desensitized $GABA_B$ receptors into a functional state.

IV. Degradation of $GABA_B$ Receptors

Degradation of most plasma membrane proteins takes place in lysosomes. The degenerative pathway of such proteins is initiated by their internalization and sorting from early endosomes via late endosomes to lysosomes. Lysosomes are heterogeneously shaped vacuoles that contain a variety of soluble hydrolases for degradation of proteins (Saftig & Klumperman, 2009).

First evidence for lysosomal degradation came from analyzing the fate of endocytosed $GABA_B$ receptors transiently expressed in HEK 293 cells (Grampp et al., 2007). Internalized $GABA_B$ receptors were found to accumulate in an endosomal sorting compartment from which they disappear over time (Fig. 3). Agents that interfere with lysosomal function (chloroquine) or directly inhibit lysosomal proteases (leupeptin, pepstatin A) prevented the loss of receptors. In cultured neurons blocking lysosomal proteases with leupeptin resulted likewise in the intracellular accumulation of endocytosed $GABA_B$ receptors (Grampp et al., 2008). Finally, inhibition of recycling, which leads to a rapid loss of $GABA_B$ receptors, was effectively prevented by leupeptin (Grampp et al., 2008). Thus, there is now ample data documenting the lysosomal degradation of endocytosed $GABA_B$ receptors.

The discovery of GISP (G-protein-coupled receptor interacting scaffolding protein) as $GABA_{B1}$-interacting protein (Kantamneni et al., 2007) provided first hints for the potential lysosomal sorting mechanism of $GABA_B$ receptors. GISP is derived from the A-kinase anchoring protein (AKAP)-9 gene, which also encodes AKAP450, AKAP350, and AKAP9. AKAPs are scaffold proteins that interact with multiple proteins to assemble signaling complexes (Carnegie et al., 2009). GISP specifically binds to $GABA_{B1}$ via coiled-coil domain interaction. Coexpression of $GABA_B$ receptors with GISP enhanced cell-surface expression of the receptors. This effect of GISP appears to be a result of its binding to and inhibition of TSG101 (tumor susceptibility gene 101; Kantamneni et al., 2008). TSG101 is an integral protein of the ESCRT (endosomal sorting complex required for transport) machinery that targets mono and Lys63-linked polyubiquitinated proteins to lysosomes for degradation (Raiborg & Stenmark, 2009). TSG101 appears to play a critical role in the degradation of $GABA_B$ receptors since its siRNA-mediated knockdown considerably increased $GABA_B$ receptor expression in HEK 293 cells (Kantamneni et al., 2008). So far, it is not yet established that $GABA_B$ receptors targeted to lysosomes for degradation are ubiquitinated. However, the observation that prolonged application of the proteasome inhibitor MG132 retards degradation of $GABA_B$ receptors (Kantamneni et al., 2008) supports the view that $GABA_B$ receptors are ubiquitinated and sorted via the ESCRT machinery to lysosomes for degradation (Fig. 3). MG132 is known to rapidly deplete cells from free ubiquitin thereby preventing also proteasome-unrelated functions of ubiquitin such as sorting proteins via the ESCRT machinery to lysosomes.

V. Conclusions

Trafficking events critically determine the signaling of plasma membrane receptors by controlling their temporal and spatial availability. The current experimental data accumulated during the past decade provide first insights into trafficking mechanisms of $GABA_B$ receptors. Forward trafficking of newly synthesized receptors is one process that contributes to the cell-surface availability of $GABA_B$ receptors (Fig. 2). Although it is now well established that exocytosis of $GABA_B$ receptors from the ER via the Golgi apparatus to the plasma membrane critically depends on the heterodimerization of $GABA_{B1}$ with $GABA_{B2}$ and is regulated by at least three distinct amino acid motifs it is currently not clear in which form the receptors are transported into dendrites and axons. First findings suggest that $GABA_{B1}$ might be transported independent of $GABA_{B2}$ within the ER into dendrites (Vidal et al., 2007). If true, this would imply that assembly of $GABA_{B1}$ and $GABA_{B2}$ is a spatially and temporally controlled mechanism that might be distinct for somatic, dendritic, and axonal receptors. Such mechanism would require that newly synthesized receptor subunits bound for dendrites or axons are protected from assembly, most probably by chaperons, until they reach their destination where they assemble and exit the ER. This would add a new dimension for the regulation of cell-surface trafficking of $GABA_B$ receptors.

In addition, it is currently unknown if and how cell-surface trafficking of $GABA_B$ receptors is regulated by physiological stimuli such as neuronal activity. In analogy to recent findings on forward trafficking of $GABA_A$ receptors (Saliba et al., 2007, 2009) neuronal activity-dependent ubiquitination and subsequent degradation of newly synthesized receptors by the ER-associated degradation pathway may be one mechanism to regulate the amount of receptors for cell-surface delivery by external stimuli.

At the cell surface, $GABA_B$ receptors undergo fast constitutive endocytosis via the classical dynamin- and clathrin-dependent pathway and are rapidly recycled back to the cell surface (Fig. 3). On the one hand, constitutive recycling may prevent $GABA_B$ receptors from sorting via the ESCRT machinery to lysosomes for degradation. On the other hand, constitutive recycling may provide a mechanism that permits the neuron to rapidly up- or downregulate cell-surface receptors in response to changing physiological conditions. External stimuli that might regulate cell-surface expression of $GABA_B$ receptors remain to be uncovered but first indications suggest that glutamate may be one candidate (Vargas et al., 2008; Wilkins et al., 2008). This would perfectly make sense given the fact that $GABA_B$ receptors are abundantly expressed at glutamatergic synapses.

Major future efforts are expected to focus on the analysis of $GABA_B$ receptor trafficking under disease states. It is conceivable that under pathophysiological conditions the rates of exocytosis, endocytosis, or degradation are altered, which would result in up- or downregulation of $GABA_B$ receptors. For instance, in a mouse model of temporal lobe epilepsy and an *in vitro* model of ischemia the levels of hippocampal $GABA_B$ receptors were found to be decreased (Cimarosti et al., 2009; Straessle et al., 2003). The mechanisms that lead to the downregulation of $GABA_B$ receptors are as yet unknown. However, one potential mechanism implicated in the decreased GABA$_B$ receptor levels under ischemic conditions may involve the GABA$_B$ receptor interacting protein CHOP (CCAAT/enhancer-binding protein homologous protein). CHOP is a transcription factor that is induced upon ER stress (Ramji & Foka, 2002) and is strongly upregulated following transient cerebral ischemia (Oida et al., 2008; Tajiri et al., 2004). Interaction of GABA$_B$ receptors with CHOP has been shown to downregulate cell-surface expression of the receptors probably by retaining the receptors in the ER and thus preventing their forward trafficking (Sauter et al., 2005). It may well be that induction of CHOP and its interaction with GABA$_B$ receptors under ischemic conditions lead to a retention of the receptors in the ER and finally in their degradation.

Clearly, this is only the very beginning of our understanding how trafficking events ensure the adequate number of GABA$_B$ receptors at the cell surface under a given physiological or pathophysiological state. Both, exocytosis and endocytosis of GABA$_B$ receptors appear to be highly regulated processes that depend on a variety of complex protein–protein interactions. It will be a major future challenge to identify these proteins to gain new insights into the mechanisms and regulation of GABA$_B$ receptor trafficking.

Acknowledgments

The work of the author is supported by the Swiss Science Foundation (Grant 31003A_121963). The author thanks Hanns Möhler for critical comments on the chapter.

Conflict of Interest: The author is employee of the University of Zurich and has no conflict of interest.

Abbreviations

AKAP	A-kinase anchoring protein
AP2	adaptor protein complex 2
CHOP	CCAAT/enhancer-binding protein homologous protein
COPI	coat protein complex I
COPII	coat protein complex II
ER	endoplasmic reticulum
ESCRT	endosomal sorting complex required for transport
GISP	G-protein-coupled receptor interacting scaffolding protein
HA	human influenza hemagglutinin
HEK	human embryonic kidney
LL motif	dileucine motif

NSF	*N*-ethylmaleimide-sensitive fusion protein
SNX4	sorting nexin 4
TSG101	tumor susceptibility gene 101

References

Becher, A., Green, A., Ige, A. O., Wise, A., White, J. H., & McIlhinney, R. A. (2004). Ectopically expressed gamma-aminobutyric acid receptor B is functionally down-regulated in isolated lipid raft-enriched membranes. *Biochemical and Biophysical Research Communications, 321*, 981–987.

Becher, A., White, J. H., & McIlhinney, R. A. (2001). The gamma-aminobutyric acid receptor B, but not the metabotropic glutamate receptor type-1, associates with lipid rafts in the rat cerebellum. *Journal of Neurochemistry, 79*, 787–795.

Bettler, B., & Tiao, J. Y. (2006). Molecular diversity, trafficking and subcellular localization of $GABA_B$ receptors. *Pharmacology & Therapeutics, 110*, 533–543.

Biermann, B., Ivankova-Susankova, K., Bradaia, A., Abdel Aziz, S., Besseyrias, V., Kapfhammer, J. P., et al. (2010). The sushi domains of $GABA_B$ receptors function as axonal targeting signals. *Journal of Neuroscience, 30*, 1385–1394.

Boyes, J., & Bolam, J. P. (2003). The subcellular localization of $GABA_B$ receptor subunits in the rat substantia nigra. *European Journal of Neuroscience, 18*, 3279–3293.

Brock, C., Boudier, L., Maurel, D., Blahos, J., & Pin, J. P. (2005). Assembly-dependent surface targeting of the heterodimeric $GABA_B$ receptor is controlled by COPI but not 14-3-3. *Molecular Biology of the Cell, 16*, 5572–5578.

Calver, A. R., Robbins, M. J., Cosio, C., Rice, S. Q. J., Babbs, A. J., Hirst, W. D., et al. (2001). The C-terminal domains of the $GABA_B$ receptor subunits mediate intracellular trafficking but are not required for receptor signaling. *Journal of Neuroscience, 21*, 1203–1210.

Carnegie, G. K., Means, C. K., & Scott, J. D. (2009). A-kinase anchoring proteins: From protein complexes to physiology and disease. *IUBMB Life, 61*, 394–406.

Casanova, J. E. (2007). Regulation of arf activation: The sec7 family of guanine nucleotide exchange factors. *Traffic, 8*, 1476–1485.

Chen, L., Boyes, J., Yung, W. H., & Bolam, J. P. (2004). Subcellular localization of $GABA_B$ receptor subunits in rat globus pallidus. *Journal of Comparative Neurology, 474*, 340–352.

Cimarosti, H., Kantamneni, S., & Henley, J. M. (2009). Ischaemia differentially regulates $GABA_B$ receptor subunits in organotypic hippocampal slice cultures. *Neuropharmacology, 56*, 1088–1096.

Couve, A., Thomas, P., Calver, A. R., Hirst, W. D., Pangalos, M. N., Walsh, F. S., et al. (2002). Cyclic AMP-dependent protein kinase phosphorylation facilitates $GABA_B$ receptor-effector coupling. *Nature Neuroscience, 5*, 415–424.

Fairfax, B. P., Pitcher, J. A., Scott, M. G., Calver, A. R., Pangalos, M. N., Moss, S. J., et al. (2004). Phosphorylation and chronic agonist treatment atypically modulate $GABA_B$ receptor cell surface stability. *Journal of Biological Chemistry, 279*, 12565–12573.

Gainetdinov, R. R., Premont, R. T., Bohn, L. M., Lefkowitz, R. J., & Caron, M. G. (2004). Desensitization of G protein-coupled receptors and neuronal functions. *Annual Review of Neuroscience, 27*, 107–144.

Gassmann, M., Haller, C., Stoll, Y., Aziz, S. A., Biermann, B., Mosbacher, J., et al. (2005). The RXR-type endoplasmic reticulum-retention/retrieval signal of $GABA_{B1}$ requires distant spacing from the membrane to function. *Molecular Pharmacology, 68*, 137–144.

Gonzalez-Maeso, J., Wise, A., Green, A., & Koenig, J. A. (2003). Agonist-induced desensitization and endocytosis of heterodimeric GABA$_B$ receptors in CHO-K1 cells. *European Journal of Pharmacology*, *481*, 15–23.

Grampp, T., Notz, V., Broll, I., Fischer, N., & Benke, D. (2008). Constitutive, agonist-accelerated, recycling and lysosomal degradation of GABA$_B$ receptors in cortical neurons. *Molecular and Cellular Neuroscience*, *39*, 628–637.

Grampp, T., Sauter, K., Markovic, B., & Benke, D. (2007). γ-Aminobutyric acid type B receptors are constitutively internalized via the clathrin-dependent pathway and targeted to lysosomes for degradation. *Journal of Biological Chemistry*, *282*, 24157–24165.

Grant, B. D., & Donaldson, J. G. (2009). Pathways and mechanisms of endocytic recycling. *Nature Reviews Molecular Cell Biology*, *10*, 597–608.

Hirokawa, N., Noda, Y., Tanaka, Y., & Niwa, S. (2009). Kinesin superfamily motor proteins and intracellular transport. *Nature Reviews Molecular Cell Biology*, *10*, 682–696.

Kantamneni, S., Correa, S. A., Hodgkinson, G. K., Meyer, G., Vinh, N. N., Henley, J. M., et al. (2007). GISP: A novel brain-specific protein that promotes surface expression and function of GABA$_B$ receptors. *Journal of Neurochemistry*, *100*, 1003–1017.

Kantamneni, S., Holman, D., Wilkinson, K. A., Correa, S. A., Feligioni, M., Ogden, S., et al. (2008). GISP binding to TSG101 increases GABA$_B$ receptor stability by down-regulating ESCRT-mediated lysosomal degradation. *Journal of Neurochemistry*, *107*, 86–95.

Kennedy, M. J., & Ehlers, M. D. (2006). Organelles and trafficking machinery for postsynaptic plasticity. *Annual Review of Neuroscience*, *29*, 325–362.

Kulik, A., Vida, I., Lujan, R., Haas, C. A., Lopez-Bendito, G., Shigemoto, R., et al. (2003). Subcellular localization of metabotropic GABA$_B$ receptor subunits GABA$_{B1a/b}$ and GABA$_{B2}$ in the rat hippocampus. *Journal of Neuroscience*, *23*, 11026–11035.

Kuramoto, N., Wilkins, M. E., Fairfax, B. P., Revilla-Sanchez, R., Terunuma, M., Tamaki, K., et al. (2007). Phospho-dependent functional modulation of GABA$_B$ receptors by the metabolic sensor AMP-dependent protein kinase. *Neuron*, *53*, 233–247.

Laffray, S., Tan, K., Dulluc, J., Bouali-Benazzouz, R., Calver, A. R., Nagy, F., et al. (2007). Dissociation and trafficking of rat GABA$_B$ receptor heterodimer upon chronic capsaicin stimulation. *European Journal of Neuroscience*, *25*, 1402–1416.

Lopez-Bendito, G., Shigemoto, R., Kulik, A., Paulsen, O., Fairen, A., & Lujan, R. (2002). Expression and distribution of metabotropic GABA receptor subtypes GABA$_B$R1 and GABA$_B$R2 during rat neocortical development. *European Journal of Neuroscience*, *15*, 1766–1778.

Luttrell, L. M., & Lefkowitz, R. J. (2002). The role of β-arrestins in the termination and transduction of G-protein-coupled receptor signals. *Journal of Cell Science*, *115*, 455–465.

Marchese, A., Paing, M. M., Temple, B. R., & Trejo, J. (2008). G protein-coupled receptor sorting to endosomes and lysosomes. *Annual Review of Pharmacology and Toxicology*, *48*, 601–629.

Margeta-Mitrovic, M., Jan, Y. N., & Jan, L. Y. (2000). A trafficking checkpoint controls GABA$_B$ receptor heterodimerization. *Neuron*, *27*, 97–106.

Mettlen, M., Pucadyil, T., Ramachandran, R., & Schmid, S. L. (2009). Dissecting dynamin's role in clathrin-mediated endocytosis. *Biochemical Society Transactions*, *37*, 1022–1026.

Mutneja, M., Berton, F., Suen, K. F., Lüscher, C., & Slesinger, P. A. (2005). Endogenous RGS proteins enhance acute desensitization of GABA$_B$ receptor-activated GIRK currents in HEK-293t cells. *Pflugers Archiv*, *450*, 61–73.

Oida, Y., Shimazawa, M., Imaizumi, K., & Hara, H. (2008). Involvement of endoplasmic reticulum stress in the neuronal death induced by transient forebrain ischemia in gerbil. *Neuroscience*, *151*, 111–119.

Pagano, A., Rovelli, G., Mosbacher, J., Lohmann, T., Duthey, B., Stauffer, D., et al. (2001). C-terminal interaction is essential for surface trafficking but not for heteromeric assembly of GABA$_B$ receptors. *Journal of Neuroscience*, *21*, 1189–1202.

Pandey, K. N. (2009). Functional roles of short sequence motifs in the endocytosis of membrane receptors. *Frontiers in Bioscience*, *14*, 5339–5360.

Park, M., Penick, E. C., Edwards, J. G., Kauer, J. A., & Ehlers, M. D. (2004). Recycling endosomes supply AMPA receptors for LTP. *Science*, *305*, 1972–1975.

Perroy, J., Adam, L., Qanbar, R., Chénier, S., & Bouvier, M. (2003). Phosphorylation-independent desensitization of $GABA_B$ receptor by GRK4. *EMBO Journal*, *22*, 3816–3824.

Pontier, S. M., Lahaie, N., Ginham, R., St-Gelais, F., Bonin, H., Bell, D. J., et al. (2006). Coordinated action of NSF and PKC regulates $GABA_B$ receptor signaling efficacy. *EMBO Journal*, *25*, 2698–2709.

Pooler, A. M., Gray, A. G., & McIlhinney, R. A. (2009). Identification of a novel region of the $GABA_{B2}$ C-terminus that regulates surface expression and neuronal targeting of the $GABA_B$ receptor. *European Journal of Neuroscience*, *29*, 869–878.

Puckerin, A., Liu, L., Permaul, N., Carman, P., Lee, J., & Diverse-Pierluissi, M. A. (2006). Arrestin is required for agonist-induced trafficking of voltage-dependent calcium channels. *Journal of Biological Chemistry*, *281*, 31131–31141.

Raiborg, C., & Stenmark, H. (2009). The ESCRT machinery in endosomal sorting of ubiquitylated membrane proteins. *Nature*, *458*, 445–452.

Ramirez, O. A., Vidal, R. L., Tello, J. A., Vargas, K. J., Kindler, S., Härtel, S., et al. (2009). Dendritic assembly of heteromeric γ-aminobutyric acid type B receptor subunits in hippocampal neurons. *Journal of Biological Chemistry*, *284*, 13077–13085.

Ramji, D. P., & Foka, P. (2002). CCAAT/enhancer-binding proteins: Structure, function and regulation. *Biochemical Journal*, *365*, 561–575.

Restituito, S., Couve, A., Bawagan, H., Jourdain, S., Pangalos, M. N., Calver, A. R., et al. (2005). Multiple motifs regulate the trafficking of $GABA_B$ receptors at distinct checkpoints within the secretory pathway. *Molecular and Cellular Neuroscience*, *28*, 747–756.

Saftig, P., & Klumperman, J. (2009). Lysosome biogenesis and lysosomal membrane proteins: Trafficking meets function. *Nature Reviews Molecular Cell Biology*, *10*, 623–635.

Saliba, R. S., Gu, Z., Yan, Z., & Moss, S. J. (2009). Blocking L-type voltage-gated Ca^{2+} channels with dihydropyridines reduces γ-aminobutyric acid type A receptor expression and synaptic inhibition. *Journal of Biological Chemistry*, *284*, 32544–32550.

Saliba, R. S., Michels, G., Jacob, T. C., Pangalos, M. N., & Moss, S. J. (2007). Activity-dependent ubiquitination of $GABA_A$ receptors regulates their accumulation at synaptic sites. *Journal of Neuroscience*, *27*, 13341–13351.

Sauter, K., Grampp, T., Fritschy, J. M., Kaupmann, K., Bettler, B., Mohler, H., et al. (2005). Subtype-selective interaction with the transcription factor CCAAT/enhancer-binding protein (C/EBP) homologous protein (CHOP) regulates cell surface expression of $GABA_B$ receptors. *Journal of Biological Chemistry*, *280*, 33566–33572.

Schmid, S. L. (1997). Clathrin-coated vesicle formation and protein sorting: An integrated process. *Annual Review of Biochemistry*, *66*, 511–548.

Straessle, A., Loup, F., Arabadzisz, D., Ohning, G. V., & Fritschy, J. M. (2003). Rapid and long-term alterations of hippocampal $GABA_B$ receptors in a mouse model of temporal lobe epilepsy. *European Journal of Neuroscience*, *18*, 2213–2226.

Tajiri, S., Oyadomari, S., Yano, S., Morioka, M., Gotoh, T., Hamada, J.-I., et al. (2004). Ischemia-induced neuronal cell death is mediated by the endoplasmic reticulum stress pathway involving CHOP. *Cell Death & Differentiation*, *11*, 403–415.

Tan, C. M., Brady, A. E., Nickols, H. H., Wang, Q., & Limbird, L. E. (2004). Membrane trafficking of G protein-coupled receptors. *Annual Review of Pharmacology and Toxicology*, *44*, 559–609.

Thuault, S. J., Brown, J. T., Sheardown, S. A., Jourdain, S., Fairfax, B., Spencer, J. P., et al. (2004). The $GABA_{B2}$ subunit is critical for the trafficking and function of native $GABA_B$ receptors. *Biochemical Pharmacology*, *68*, 1655–1666.

Traer, C. J., Rutherford, A. C., Palmer, K. J., Wassmer, T., Oakley, J., Attar, N., et al. (2007). SNX4 coordinates endosomal sorting of TfnR with dynein-mediated transport into the endocytic recycling compartment. *Nature Cell Biology*, *9*, 1370–1380.

Vargas, K. J., Terunuma, M., Tello, J. A., Pangalos, M. N., Moss, S. J., & Couve, A. (2008). The availability of surface GABA$_B$ receptors is independent of γ-aminobutyric acid but controlled by glutamate in central neurons. *Journal of Biological Chemistry*, *283*, 24641–24648.

Vidal, R. L., Ramirez, O. A., Sandoval, L., Koenig-Robert, R., Hartel, S., & Couve, A. (2007). Marlin-1 and conventional kinesin link GABA$_B$ receptors to the cytoskeleton and regulate receptor transport. *Molecular and Cellular Neuroscience*, *35*, 501–512.

Villalba, R. M., Raju, D. V., Hall, R. A., & Smith, Y. (2006). GABA$_B$ receptors in the centromedian/parafascicular thalamic nuclear complex: An ultrastructural analysis of GABA$_{BR1}$ and GABA$_{BR2}$ in the monkey thalamus. *Journal of Comparative Neurology*, *496*, 269–287.

Wang, Z., Edwards, J. G., Riley, N., Provance, D. W. Jr., Karcher, R., Li, X. D., et al. (2008). Myosin vb mobilizes recycling endosomes and AMPA receptors for postsynaptic plasticity. *Cell*, *135*, 535–548.

Wilkins, M. E., Li, X., & Smart, T. G. (2008). Tracking cell surface GABA$_B$ receptors using an α-bungarotoxin tag. *Journal of Biological Chemistry*, *283*, 34745–34752.

Miho Terunuma[*], **Menelas N. Pangalos**[†], **and Stephen J. Moss**[*,‡]

[*]Department of Neuroscience, Tufts University School of Medicine, Boston, MA, USA

[†]AstraZeneca, Alderley Park, UK

[‡]Department of Neuroscience, Physiology and Pharmacology, University College London, London, UK

Functional Modulation of $GABA_B$ Receptors by Protein Kinases and Receptor Trafficking

Abstract

$GABA_B$ receptors ($GABA_BR$) are heterodimeric G protein-coupled receptors (GPCRs) that mediate slow and prolonged inhibitory signals in the central nervous system. The signaling of GPCRs is under stringent control and is subject to regulation by multiple posttranslational mechanisms. The β-adrenergic receptor is a prototypic GPCR. Like most GPCRs, prolonged exposure of this receptor to agonist induces phosphorylation of multiple intracellular residues that is largely dependent upon the activity of

1054-3589/10 $35.00
10.1016/S1054-3589(10)58005-0

G protein-coupled receptor kinases (GRKs). Phosphorylation terminates receptor–effector coupling and promotes both interaction with β-arrestins and removal from the plasma membrane via clathrin-dependent endocytosis. Emerging evidence for $GABA_BRs$ suggests that these GPCRs do not conform to this mode of regulation. Studies using both native and recombinant receptor preparations have demonstrated that $GABA_BRs$ do not undergo agonist-induced internalization and are not GRK substrates. Moreover, whilst $GABA_BRs$ undergo clathrin-dependent constitutive endocytosis, it is generally accepted that their rates of internalization are not modified by prolonged agonist exposure. Biochemical studies have revealed that $GABA_BRs$ are phosphorylated on multiple residues within the cytoplasmic domains of both the R1 and R2 subunits by cAMP-dependent protein kinase and 5′AMP-dependent protein kinase (AMPK). Here we discuss the role that this phosphorylation plays in determining $GABA_BR$ effector coupling and their trafficking within the endocytic pathway and go on to evaluate the significance of $GABA_BR$ phosphorylation in controlling neuronal excitability under normal and pathological conditions.

I. Introduction

$GABA_BRs$ are GPCRs that mediate the slow and prolonged inhibitory action of GABA, the principal inhibitory neurotransmitter in the brain, via activation of Gαi- and Gαo-type heteromeric type G proteins. $GABA_BRs$ mediate their inhibitory action via multiple effectors: post-synaptically they activate inwardly rectifying K^+ channels (GIRKs), leading to hyperpolarization, whilst pre-synaptically they inhibit voltage-gated N/P/Q type Ca^{2+} channels, leading to reduced neurotransmitter release (Bettler et al., 2004; Bettler & Tiao, 2006; Couve, Moss, & Pangalos, 2000, Couve et al., 2004). Through their activation of Gαi they also inhibit the activity of adenylate cyclase to reduce PKA signaling pathways.

Given the central roles of $GABA_BRs$ in mediating neuronal inhibition there is considerable interest in understanding the cellular mechanisms neurons use to modulate their activity. This is determined in part by the efficacy of the effector coupling of $GABA_BRs$ and their residence time on the plasma membrane. Classically the signaling of monomeric GPCRs, as exemplified by the prototypic β-adrenergic receptor, is tightly controlled by agonist desensitization. Prolonged exposure to agonist leads to decreased effector coupling due to enhanced phosphorylation by G protein-coupled receptor kinases (GRKs). This decreased effector coupling ultimately leads to endocytosis followed by recycling or degradation (Ferguson, 2001; Gainetdinov et al., 2003; Marchese et al., 2008; von Zastrow, 2003). In this review we discuss the molecular mechanisms that neurons use to regulate the functional activity of $GABA_BRs$ and the roles that these processes play as determinants of the efficacy of neuronal inhibition.

II. Phosphorylation of $GABA_BR$ and Its Functional Modulation

It is well documented that the activation and inactivation of GPCRs is modulated by the phosphorylation within the carboxyl (C)-terminus and the third intracellular loop by various protein kinases (Ferguson, 2001; Marchese et al., 2008; von Zastrow, 2003). In general, phosphorylation of GPCRs by intracellular kinases induces the desensitization of the receptor followed by interaction with cytosolic cofactor proteins called arrestins, which uncouple the receptor from G proteins. GPCRs are internalized from the plasma membrane principally via clathrin-dependent endocytosis. They are then recycled back to the plasma membrane for re-insertion or targeted for lysozomal degradation.

To evaluate the significance of this mode of regulation for $GABA_BRs$, measurements of phosphorylation have been performed in both expression systems and in neurons using labeling with ^{32}P-orthophosphate followed by immunoprecipitation with or without agonist exposure. Whilst $GABA_BRs$ exhibit significant levels of basal phosphorylation, this is not subject to agonist-induced modulation. Consistent with this, neither R1 or R2 subunits are substrates of GRKs 1–4 when co-expressed in HEK-293 cells. The major intracellular domains of these proteins are not phosphorylated by purified GRKs *in vitro* (Fairfax et al., 2004; Perroy et al., 2003). In contrast to other GPCRs, $GABA_BRs$ are not GRK substrates and do not appear to undergo agonist-induced internalization. We discuss which kinases mediate $GABA_BR$ phosphorylation and consider the significance of this covalent modification in determining the efficacy of receptor signaling and trafficking itineraries. We then go on to assess the physiological significance of phosphorylation and its impact on $GABA_BR$ signaling under normal and pathological conditions.

A. PKA

The phosphorylation of GPCRs by PKA has been shown to regulate receptor activity (Bouvier et al., 1988; Bunemann & Hosey, 1999; Moffett et al., 1996). For instance, for the β-adrenergic receptor and rhodopsin receptors, PKA phosphorylation enhances receptor desensitization by inducing internalization. Consistent with other GPCRs, a role for PKA in mediating $GABA_BR$ receptor desensitization had been postulated prior to their structural characterization (Yoshimura et al., 1995). In agreement with this study a strong consensus site for PKA phosphorylation is found at serine 892 (S892) within the R2 subunit (Q-R-R-L-S-L). Studies using *in vitro* phosphorylation with purified PKA coupled with metabolic labeling of recombinant receptors expressed in fibroblast has confirmed that S892 is the sole site of phosphorylation for PKA-mediated phosphorylation within $GABA_BRs$. Using phospho-specific antibodies, significant levels of basal phosphorylation of S892 are evident within the brain, a phenomenon that is dependent principally upon the activity of PKA (Couve et al., 2001).

The role of PKA-mediated phosphorylation in mediating $GABA_BR$ effector coupling has been examined by measuring their ability to activate GIRKs. Upon brief exposure to agonist phosphorylation of S892 via PKA decreases the time-dependent decrease (rundown) in the efficacy of $GABA_BR$-dependent activation of GIRKs in both HEK-293 cells and hippocampal neurons (Couve et al., 2001). This effect likely results from stabilization of $GABA_BRs$ on the neuronal plasma membrane. In contrast to the majority of GPCRs, PKA-mediated phosphorylation of S892 in the R2 subunit thus appears to promote $GABA_BR$ signaling. PKA regulation of $GABA_BR$ might not occur as a result of agonist stimulation but as a result of cAMP accumulation by neurotransmitters that activate Gαs-coupled receptors. It remains to be determined which signaling pathways mediate S892 phosphorylation and thus $GABA_BR$ effector coupling; however, it is well accepted that prolonged activation of $GABA_BRs$ via the activation of Gαi leads to decreased PKA activity. Consistent with this, prolonged activation of $GABA_BRs$ leads to dephosphorylation of S892 and receptor degradation (Fairfax et al., 2004). Thus the phosphorylation status of S892, which is in part dependent upon agonist activation, is likely to play critical roles in maintaining $GABA_BR$ cell surface stability and the strength of synaptic inhibition.

B. PKC

PKC activity had long been known to attenuate $GABA_BR$-mediated inhibition of neurotransmitter release and receptor–effector coupling (Dutar & Nicoll, 1988; Thompson & Gahwiler, 1992; Taniyama et al., 1991, 1992). More recently, the phosphorylation of R1 subunit by PKC has been reported by Bouvier and colleagues (Pontier et al., 2006). The activation of PKC induces the phosphorylation of R1 subunit followed by *N*-ethylmaleimide-sensitive fusion (NSF) protein dissociation from $GABA_BR$ and promotes desensitization. Like PKA, PKC phosphorylation does not induce $GABA_BR$ internalization. NSF is a molecular chaperone that is known to be a crucial element in membrane fusion events. It has also been identified as a regulator of the α-amino-3-hydroxy-5-methyl-4-isoxazole-propionic acid receptor (AMPAR) (Nishimune et al., 1998; Noel et al., 1999; Osten et al., 1998), the γ-aminobutyric acid (GABA) type A receptor ($GABA_AR$) (Goto et al., 2005), and β2 adrenergic receptor cell surface expression (Cong et al., 2001). In Chinese hamster ovary (CHO) cells $GABA_BR$ activity promotes PKC recruitment to the plasma membrane, inducing R1 subunit phosphorylation. Phosphorylation of R1 subunit is disrupted by preventing the binding of NSF to GABA$_B$Rs using selective peptides; the association of NSF to GABA$_B$Rs is clearly crucial for GABA$_B$R phosphorylation. As measured in CHO cells, recruitment of PKC dependent upon NSF blocked baclofen-induced G protein activation by GABA$_B$Rs but did not modify basal activity as measured using ^{35}S-GTP-γS binding. It remains to be determined which residues within GABA$_B$Rs are PKC substrates and to ascertain the relevance of this novel regulatory mechanism for neuronal GABA$_B$Rs.

C. AMPK

It is evident that, in addition to PKA, GABA$_B$Rs are substrates for AMPK (Kuramoto et al., 2007). AMPK is a serine/threonine protein kinase that exists as a heterotrimer consisting of the catalytic α-subunit and regulatory β- and γ-subunits (Spasic et al., 2009). This kinase acts as an energy sensor that is rapidly activated when the cellular levels of AMP increase due to high metabolic activity or the pathological states of anoxia and ischemia (Carling, 2005; Kahn et al., 2005). To date the majority of AMPK substrates are metabolic enzymes; however, AMPK also binds with high affinity to the C-terminus of R1 but not R2 subunit. Interestingly, AMPK phosphorylates the C-tails of both R1 and R2 subunits *in vitro*. Using mass spectrometry and Edman degradation analysis, two phosphorylation sites in R1 subunit (S917/923) and one site in R2 subunit (S783) have been identified. The physiological relevance of this phosphorylation has been examined by measuring the effects of AMPK activity on the activation of GIRKs in both expression systems and neurons. This revealed that the phosphorylation of R2 subunit on S783 decreases the desensitization of GABA$_B$Rs by stabilizing the receptor complex at the plasma membrane. These results suggest that, similar to PKA, the AMPK-mediated phosphorylation of GABA$_B$Rs either inhibits receptor endocytosis or increases receptor exocytosis to maintain synaptic inhibition.

III. Phosphorylation-Independent Desensitization of GABA$_B$Rs by Protein Kinases

Desensitization to prevent overstimulation is a common feature of GPCRs. For most GPCRs, a time-dependent desensitization of the receptor appears to be mediated by direct phosphorylation of the GPCRs by GRK followed by arrestin- and dynamin-dependent receptor internalization via clathrin-coated vesicle and recycling/degradation signaling pathways (Marchese et al., 2008; von Zastrow, 2003). Whilst GABA$_B$Rs do not appear to be directly phosphorylated by GRKs, GRK4 and GRK5 have been reported to play a central role in agonist-induced desensitization (Kanaide et al., 2007; Perroy et al., 2003). For instance, the suppression of

GRK4 expression in cerebellar granule cells strongly inhibits $GABA_BR$ desensitization (Perroy et al., 2003). Similarly, in *Xenopus* oocytes and baby hamster kidney (BHK) cells, expression of $GABA_BR$ and GIRKs does not result in desensitization unless co-expressed with GRK4 or GRK5 (Kanaide et al., 2007). Interestingly, the association of GRKs to R2 subunits is also observed but the $GABA_BR$ phosphorylation is independent of GRK association (Fairfax et al., 2004; Kanaide et al., 2007; Perroy et al., 2003). Unlike most GPCRs, GRKs may thus function as anchoring proteins to regulate $GABA_BR$ activity but not phosphorylation.

IV. $GABA_BR$ Endocytic Sorting and the Control of Receptor Cell Surface Stability

It is generally agreed that $GABA_BRs$ do not show agonist-induced endocytosis; however, they do exhibit significant rates of constitutive endocytosis (Couve et al., 2001; Fairfax et al., 2004; Perroy et al., 2003; Vargas et al., 2008). Under basal conditions, $GABA_BRs$ endocytose as dimers via clathrin- and dynamin-dependent mechanisms and localize to Rab11-positive recycling endosomes. After constitutive endocytosis, large numbers of $GABA_BRs$ recycle back to the plasma membrane to maintain steady-state cell surface numbers, presumably reflecting the long cell surface half-lives for these proteins as measured via biotinylation (Fairfax et al., 2004; Vargas et al., 2008). Interestingly endocytosis is detected only in dendrites, not in axons. The mechanisms underlying this compartmentalization of $GABA_BR$ endocytosis remain obscure; however, R1 subunit has many splice variants, which likely results in multiple distinct modes of phosphorylation. Consistent with steady-state measurement $GABA_BR$ endocytosis is agonist-independent; however, cell surface accumulation is dramatically reduced by exposure to glutamate via a mechanism dependent upon the activity of the proteasome (Vargas et al., 2008). In this study glutamate treatment did not result in enhanced intracellular accumulation of $GABA_BRs$, suggesting that this excitatory neurotransmitter may regulate lysozomal targeting of $GABA_BRs$. Further studies are needed to determine how glutamate treatment of neurons leads to degradation of $GABA_BRs$.

V. $GABA_BR$ Phosphorylation and Diseases

A. Addiction

Recently $GABA_BR$ agonist and allosteric modulators have been successfully used to treat the symptoms associated with withdrawal from cocaine, opiates, nicotine, and ethanol (Addolorato et al., 2009; Filip & Frankowska,

2008; Frankowska et al., 2009). This may be due to the phosphorylation of $GABA_B$Rs. Kalivas and colleagues reported that repeated cocaine administration leads to an increase in basal extracellular GABA in the nucleus accumbens and to dephosphorylation of R2 subunits (Xi et al., 2003). Because prolonged activation of $GABA_B$Rs decreases cAMP production and PKA activation through Giα, elevated extracellular GABA by repeated cocaine injection may regulate S892 phosphorylation in R2 subunit. In this study, the authors performed an immunoprecipitation assay followed by immunoblotting using anti-phospho-serine antibody. Additional experiments using phospho-specific antibody against S892 and S783 is clearly essential. Repeated cocaine administration can regulate the release of various neurotransmitters, and thus would also be able to regulate many of the second messenger kinases that phosphorylate $GABA_B$Rs.

B. Ischemia

Our group recently demonstrated that the phosphorylation of $GABA_B$Rs by AMPK indicates a possible neuroprotective role for $GABA_B$Rs (Kuramoto et al., 2007). AMPK is expressed in neurons throughout the brain. Rapid AMPK activation has been observed following ischemia, hypoxia, and glucose deprivation (Carling, 2005; Culmsee et al., 2001; Gadalla et al., 2004; Kahn et al., 2005; Li et al., 2007; Spasic et al., 2009). Phosphorylation of S783, but not S892, in R2 subunit has been documented in the CA3 and dentate gyrus of hippocampus after ischemic injury induced by MCAO (middle cerebral artery occlusion). The overexpression of R2 subunits in cultured hippocampal neurons also shows phosphorylation of S783 and enhanced neuronal survival in experiments using a model of anoxia (oxygen–glucose deprivation). Since mutation of S783 to alanine significantly decreases neuronal survival compared to wild-type in this model, phosphorylation of S783 in R2 subunit is likely to be a critical mechanism for neuronal survival.

VI. Conclusion

$GABA_B$Rs are unique GPCRs that function as heterodimers composed of R1 and R2 subunits. They mediate slow and prolonged inhibitory signals. Deficits in $GABA_B$R function have been reported in many neurological and psychiatric disorders. We have discussed here current findings on the involvement of protein kinases in $GABA_B$R modulation; however, the existence of phosphorylation-dependent $GABA_B$R endocytosis is still unclear. Identifying the molecular machinery that regulates $GABA_B$R trafficking is thus key in developing novel therapeutics to treat neurological and psychiatric disorders.

Acknowledgments

MT is recipient of a National Scientist Development Award from the American Heart Association. SJM is supported by National Institute of Neurological Disorders and Stroke Grants NS047478, NS048045, NS051195, NS056359, NS065725, and NS054900.

Conflict of Interest statement: The authors declare no conflict of interest.

Abbreviations

AMPA	α-amino-3-hydroxy-5-methyl-4-isoxazolepropionic acid;
AMPK	AMP-activated protein kinase;
BHK	baby hamster kidney;
cAMP	cyclic adenosine monophosphate;
CHO	Chinese hamster ovary;
GABA	γ-aminobutyric acid;
$GABA_BR$	$GABA_B$ receptor;
GIRKs	G protein-activated inwardly rectifying K^+ channels;
GPCRs	G protein-coupled receptors;
GRKs	G protein-coupled receptor kinases;
HEK	human embryonic kidney;
MCAO	middle cerebral artery occlusion;
NSF	*N*-ethylmaleimide-sensitive fusion;
PKA	cAMP-dependent protein kinase;
PKC	protein kinase C

References

Addolorato, G., Leggio, L., Cardone, S., et al. (2009). Role of the $GABA_B$ receptor system in alcoholism and stress; focus on clinical studies and treatment perspectives. *Alcohol*, *43*, 559–563.

Bettler, B., Kaupmann, K., Mosbacher, J., et al. (2004). Molecular structure and physiological functions of $GABA_B$ receptors. *Physiological Reviews*, *84*, 835–867.

Bettler, B., & Tiao, J. Y. (2006). Molecular diversity, trafficking and subcellular localization of $GABA_B$ receptors. *Pharmacology & Therapeutics*, *110*, 533–543.

Bouvier, M., Hausdorff, W. P., De Blasi, A., et al. (1988). Removal of phosphorylation sites from the β2-adrenergic receptor delays the onset of agonist-promoted desensitization. *Nature*, *333*, 370–373.

Bunemann, M., & Hosey, M. M. (1999). G-protein coupled receptor kinases as modulators of G-protein signaling. *Journal of Physiology (Paris)*, *517*, 5–23.

Carling, D. (2005). AMP-activated protein kinase; balancing the scales. *Biochimie*, *87*, 87–91.

Cong, M., Perry, S. J., Hu, L. A., et al. (2001). Binding of the beta2 adrenergic receptor to *N*-ethylmaleimide-sensitive factor regulates receptor recycling. *Journal of Biological Chemistry*, *276*, 45145–45152.

Couve, A., Calver, A. R., Fairfax, B. P., et al. (2004). Unravelling the unusual signaling properties of the GABA$_B$ receptor. *Biochemical Pharmacology*, *68*, 1527–1536.

Couve, A., Moss, S. J., & Pangalos, M. N. (2000). GABAB rceptors: A new paradigm in G protein signaling. *Molecular and Cellular Neuroscience*, *16*, 296–312.

Couve, A., Thomas, P., Calver, A. R., et al. (2001). Cyclic AMP-dependent protein kinase phosphorylation facilitates GABA$_B$ receptor–effector coupling. *Nature Neuroscience*, *5*, 415–424.

Culmsee, C., Monnig, J., Kemp, B. E., et al. (2001). AMP-activated protein kinase is highly expressed in neurons in the developing rat brain and promotes neuronal survival following glucose deprivation. *Journal of Molecular Neuroscience*, *17*, 45–58.

Dutar, P., & Nicoll, R. A. (1988). Pre- and postsynaptic GABAB receptors in the hippocampus have different pharmacological properties. *Neuron*, *1*, 585–591.

Fairfax, B. P., Pitcher, J. A., Scott, M. G., et al. (2004). Phosphorylation and chronic agonist treatment atypically modulate GABA$_B$ receptor cell surface stability. *Journal of Biological Chemistry*, *279*, 12565–12573.

Ferguson, S. S. G. (2001). Evolving concepts in G protein-coupled receptor endocytosis: The role in receptor desensitization and signaling. *Pharmacological Reviews*, *53*, 1–24.

Filip, M., & Frankowska, M. (2008). GABA$_B$ receptors in drug addiction. *Pharmacological Reports*, *60*, 755–770.

Frankowska, M., Nowak, E., & Fillip, M. (2009). Effects of GABA$_B$ receptor agonists on cocaine hyperlocomotor and sensitizing effects in rats. *Pharmacological Reports*, *61*, 1042–1049.

Gadalla, A. E., Pearson, T., Currie, A. J., et al. (2004). AICA ribose both activates AMP-activated protein kinase and competes with adenosine for the nucleotide transporter in the CA1 region of the rat hippocampus. *Journal of Neurochemistry*, *88*, 1272–1282.

Gainetdinov, R. R., Bohn, L. M., Sotnikova, T. D., et al. (2003). Dopaminergic supersensitivity in G protein-coupled receptor kinase 6-deficient mice. *Neuron*, *24*, 291–303.

Goto, H., Terunuma, M., Kanematsu, T., et al. (2005). Direct interaction of *N*-ethylmaleimide-sensitive factor with GABA(A) receptor beta subunits. *Molecular and Cellular Neuroscience*, *30*, 197–206.

Kahn, B. B., Alquier, T., Carling, D., et al. (2005). AMP-activated protein kinase; ancient energy gauge provides clues to modern understanding of metabolism. *Cell Metabolism*, *1*, 15–25.

Kanaide, M., Uezono, Y., Matsumoto, M., et al. (2007). Desensitization of GABA$_B$ receptor signaling by formation of protein complexes of GABA$_{B2}$ subunit with GRK4 or GRK5. *Journal of Cellular Physiology*, *210*, 237–245.

Kuramoto, N., Wilkins, M. E., Fairfax, B. P., et al. (2007). Phospho-dependent functional modulation of GABA$_B$ receptors by the metabolic sensor AMP-dependent protein kinase. *Neuron*, *53*, 233–247.

Li, J., Zeng, Z., Viollet, B., et al. (2007). Neuroprotective effects of adenosine monophosphate-activated protein kinase inhibition and gene deletion in stroke. *Stroke*, *38*, 2992–2999.

Marchese, A., Paing, M. M., Temple, B. R., et al. (2008). G protein-coupled receptor sorting to endosomes and lysosomes. *Annual Review of Pharmacology and Toxicology*, *48*, 601–629.

Moffett, S., Adam, L., Bonin, H., et al. (1996). Palmitoylated cysteine 341 modulates phosphorylation of the β2-adrenergic receptor by the cAMP-dependent protein kinase. *Journal of Biological Chemistry*, *271*, 21490–21497.

Nishimune, A., Isaac, J. T., Molnar, E., et al. (1998). NSF binding to GluR2 regulates synaptic transmission. *Neuron*, *21*, 87–97.

Noel, J., Ralph, G. S., Pickard, L., et al. (1999). Surface expression of AMPA receptors in hippocampal neurons is regulated by an NSF-dependent mechanism. *Neuron*, *23*, 365–376.

Osten, P., Srivastava, S., Inman, G. J., et al. (1998). The AMPA receptor GluR2 C terminus can mediate a reversible, ATP-dependent interaction with NSF and alpha- and beta-SNAPs. *Neuron*, *21*, 99–110.

Perroy, J., Adam, L., Qanbar, R., et al. (2003). Phosphorylation-independent desensitization of $GABA_B$ receptor by GRK4. *EMBO Journal*, *22*, 3816–3824.

Pontier, S. M., Lahaie, N., Ginham, R., et al. (2006). Coordinated action of NSF and PKC regulates $GABA_B$ receptor signaling efficacy. *EMBO Journal*, *25*, 2698–2709.

Spasic, M. R., Callaerts, P., & Norga, K. K. (2009). AMP-activated protein kinase (AMPK) molecular crossroad for metabolic control and survival of neurons. *Neuroscientist*, *15*, 309–316.

Taniyama, K., Niwa, M., Kataoka, Y., et al. (1992). Activation of protein kinase C suppresses the gamma-aminobutyric acid b receptor-mediated inhibition of the vesicular release of noradrenaline and acetylcholine. *Journal of Neurochemistry*, *58*, 1239–1245.

Taniyama, K., Takeda, K., Ando, H., et al. (1991). Expression of the GABAB receptor in *Xenopus* oocytes and desensitization by activation of protein kinase C. *Advances in Experimental Medicine & Biology*, *287*, 413–420.

Thompson, S. M., & Gahwiler, B. H. (1992). Comparison of the actions of baclofen at pre- and postsynaptic receptors in the rat hippocampus in vitro. *Journal of Physiology*, *451*, 329–345.

Vargas, K. J., Terunuma, M., Tello, J. A., et al. (2008). The availability of surface $GABA_B$ receptors is independent of γ-aminobutyric acid but controlled by glutamate in central neurons. *Journal of Biological Chemistry*, *283*, 24641–24648.

Von Zastrow, M. (2003). Mechanisms regulating membrane trafficking of G protein-coupled receptors in the endocytic pathway. *Life Sciences*, *74*, 217–224.

Xi, Z.-X., Ramamoorthy, S., Shen, H., et al. (2003). GABA transmission in the nucleus accumbens is altered after withdrawal from repeated cocaine. *Journal of Neuroscience*, *23*(8), 3498–3505.

Yoshimura, M., Yoshida, S., & Taniyama, K. (1995). Desensitization by cyclic AMP-dependent protein kinase of $GABA_B$ receptor expressed in *Xenopus* oocytes. *Life Sciences*, *57*, 2397–2401.

Claire L. Padgett and Paul A. Slesinger
Peptide Biology Laboratory, The Salk Institute for Biological Studies, La Jolla, CA, USA

GABA$_B$ Receptor Coupling to G-proteins and Ion Channels

Abstract

GABA$_B$ receptors have been found to play a key role in regulating membrane excitability and synaptic transmission in the brain. The GABA$_B$ receptor is a G-protein coupled receptor (GPCR) that associates with a subset of G-proteins (pertussis toxin sensitive Gi/o family), that in turn regulate specific ion channels and trigger cAMP cascades. In this review, we describe the relationships between the GABA$_B$ receptor, its effectors and associated proteins that mediate GABA$_B$ receptor function within the brain. We discuss a unique feature of the GABA$_B$ receptor, the requirement for heterodimerization to produce functional receptors, as well as an increasing body of evidence that suggests GABA$_B$ receptors comprise a macromolecular signaling

Advances in Pharmacology, Volume 58

1054-3589/10 $35.00
10.1016/S1054-3589(10)58006-2

heterocomplex, critical for efficient targeting and function of the receptors. Within this complex, $GABA_B$ receptors associate specifically with Gi/o G-proteins that regulate voltage-gated Ca^{2+} (Ca_V) channels, G-protein activated inwardly rectifying K^+ (GIRK) channels, and adenylyl cyclase. Numerous studies have revealed that lipid rafts, scaffold proteins, targeting motifs in the receptor, and regulators of G-protein signaling (RGS) proteins also contribute to the function of $GABA_B$ receptors and affect cellular processes such as receptor trafficking and activity-dependent desensitization. This complex regulation of $GABA_B$ receptors in the brain may provide opportunities for new ways to regulate GABA-dependent inhibition in normal and diseased states of the nervous system.

I. Introduction

GABA is the major inhibitory neurotransmitter in the brain that activates two types of receptors; ligand-gated chloride-selective receptors (e.g., $GABA_A$ and $GABA_\rho$) and metabotropic G-protein coupled $GABA_B$ receptors. Activation of both receptors triggers inhibitory postsynaptic currents (IPSCs) that are fast ($GABA_A$-mediated) and slow ($GABA_B$-mediated), resulting in reduced neuronal excitability. Over the past 30 years, the role of the $GABA_B$ receptor has been explored in detail and this chapter focuses on the relationship of the $GABA_B$ receptor with G-proteins, ion channels, and enzymes — the functional effectors that mediate the $GABA_B$ response within the cell. This review also describes the significance of these relationships in the context of putative $GABA_B$-containing signaling complexes and functional regulation within the brain.

First identified by Bowery & Hudson (1979) as an anomalous, bicuculline-insensitive GABAergic current, the $GABA_B$ receptor is now better understood due to the development of a pharmacologically distinct set of agonists (e.g., baclofen) (Bowery et al., 1979) and antagonists (Bowery, 1993). The $GABA_B$ receptor is a G-protein coupled receptor (GPCR) that associates with a subset of pertussis toxin sensitive G-proteins (Gi/o family). Activation of the receptor triggers GTP-dependent release of G-protein heterotrimers ($G\alpha$-GTP and $G\beta\gamma$) which, in turn, regulate specific ion channels and trigger other secondary messenger cascades that affect neuronal excitability (Bormann, 1988). The $GABA_B$ receptor was first identified in the periphery (Bowery & Hudson, 1979) but has since been shown to also function throughout the CNS. $GABA_B$ currents have been identified in the dorsal root ganglion (DRG) (Diverse-Pierluissi et al., 1999; Dunlap & Fischbach, 1981), hippocampus (Andrade et al., 1986; Charles et al., 2001; Chen & Lambert, 2000; Davies et al., 1991; Fairfax et al., 2004; Gassmann et al., 2004; Kulik et al., 2003; Leaney, 2003; Otis et al., 1993; Otmakhova & Lisman, 2004; Premkumar & Gage, 1994; Ramirez et al.,

2009; Schuler et al., 2001; Sohn et al., 2007; Tiao et al., 2008; Toselli & Taglietti, 1993; Tosetti et al., 2003, 2004; Vidal et al., 2007; Vigot et al., 2006), midbrain (Cardozo & Bean, 1995; Federici et al., 2009; O'Callaghan et al., 1996; Rohrbacher et al., 1997), cerebellum (Becher et al., 2001; Billinton et al., 1999; Fernandez-Alacid et al., 2009; Huston et al., 1995; Slesinger et al., 1997; Turgeon & Albin, 1993), supraoptic nucleus (SON) (Li & Stern, 2004), laterodorsal tegmental area (Chieng & Christie, 1995), ventral tegmental area (VTA) (Cruz et al., 2004), suprachiasmatic nucleus (SCN) (Chen & van den Pol, 1998), and thalamocortical cells (Crunelli & Leresche, 1991), suggesting that their activation and regulation may underlie a range of actions in normal and diseased states of the nervous system.

II. $GABA_B$ Receptor Structure

In the mid-1990s, cloning of the $GABA_B$ receptor led to the identification of two $GABA_B$ gene products, $GABA_{B1}$ and $GABA_{B2}$, also referred to as GABA R1 (Kaupmann et al., 1997) and R2 (Jones et al., 1998; Kaupmann et al., 1998; Kuner et al., 1999; White et al., 1998), that together form a functional $GABA_B$ receptor. Interestingly, there was much speculation prior to the cloning on the number of $GABA_B$ receptor subunits due to the pharmacological and functional profile of the receptors (see below). At first the two subunits seemed inadequate to address the diverse roles of the receptor. However, two splice variants of the $GABA_{B1}$ subunit (1a and 1b) were identified that differ in the N-terminal domain (Kaupmann et al., 1997), the $GABA_{B1a}$ subunit containing two protein interacting "sushi" domains (Fig. 1). It has since become apparent that $GABA_B$ functional diversity arises from a complex control of expression and function (see heterodimerization below) and not from a plethora of subunit isoforms.

The $GABA_{B1}$ and $GABA_{B2}$ subunits share 35% sequence homology and both resemble a classical GPCR subunit containing an extracellular N-terminal domain, seven-transmembrane domains, and a C-terminal intracellular domain. Functional characterization revealed a remarkable property of $GABA_B$ receptors—the $GABA_{B1}$/$GABA_{B2}$ subunits must be coexpressed to form a functional $GABA_B$ receptor; when expressed individually, the subunits failed to form physiologically normal receptors (Couve et al., 1998; Jones et al., 1998; Kaupmann et al., 1998; Kuner et al., 1999). It has since been discovered that the $GABA_{B1}$ subunit contains an ER retention signal, (RXR(R)), which prevents forward trafficking of the receptor (Margeta-Mitrovic et al., 2000). Dimerization with the $GABA_{B2}$ subunit shields the ER retention signal and permits surface expression of both $GABA_{B1}$ and $GABA_{B2}$ (Calver et al., 2001; Margeta-Mitrovic et al., 2000). Yeast two-hybrid analysis also revealed that the C-terminus of both the $GABA_{B1}$ and $GABA_{B2}$ subunits were critical for heterodimerization

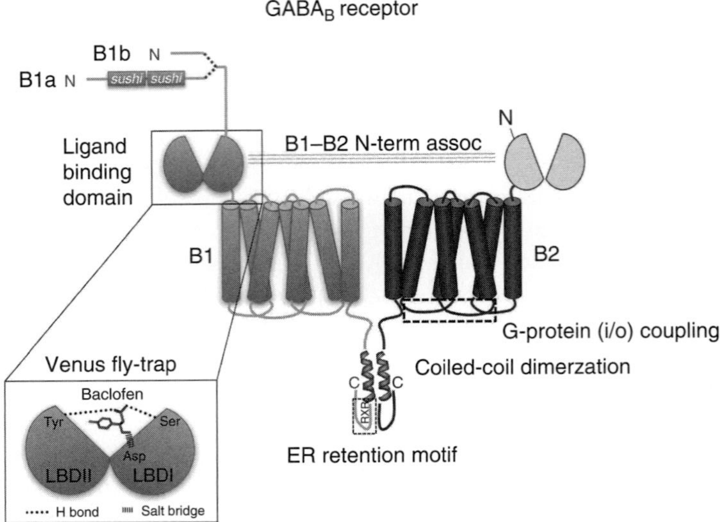

FIGURE 1 GABA$_B$ receptor structure. The functional GABA$_B$ receptor is a heterodimer formed by a GABA B1 and B2 subunit. Each of the subunits possesses extracellular N-termini, seven-transmembrane domains, and intracellular C-termini. The functional receptors heterodimerize via a C-terminal coiled-coil domain that shields an ER retention motif on B1, promoting cell surface expression. Two splice variants of the B1 subunit (1A and 1B) exist, differing by the presence of two "sushi" axonal targeting domains in the 1a subunit. Activation of the receptor occurs when ligand (GABA or baclofen) binds to the N-termini of the B1 subunit in a venus flytrap mode of ligand binding (zoom panel). Essential ligand-binding amino acids have been highlighted. The B2 subunit then confers functional activity coupling to Gi/o G-proteins via its intracellular loops.

(White et al., 1998) and CD spectroscopic analysis of a 30 amino acid sequence in these C-termini revealed a coiled-coil domain between the GABA$_{B1}$ and GABA$_{B2}$ subunits required for the subunit-specific formation of the receptor (Kammerer et al., 1999) (Fig. 1). Interestingly, the GABA$_{B2}$ subunit C-terminus also regulates lateral diffusion in hippocampal neurons suggesting that it helps control receptor expression levels at the plasma membrane (Pooler & McIlhinney, 2007).

The extracellular N-terminal domain of the GABA$_{B1}$ subunit is important for ligand binding. Creating a series of GABA$_B$ receptor chimeras and truncations revealed that the GABA$_{B1}$ N-terminus was sufficient to confer wildtype-like ligand binding pharmacology (Malitschek et al., 1999). Later, attempts to determine the minimal functional ligand binding region found that, in fact, most of the N-terminal domain contributes to ligand binding (Deriu et al., 2005). Based on sequence homology and site-directed mutagenesis, it has been proposed that GABA$_B$ receptors contain a "venus flytrap" model for ligand binding (Galvez et al., 1999). In the venus flytrap model, the initial step in receptor activation by agonist induces the closure of two ligand binding lobes (lobes I and II; Fig. 1). In the $GABA_{B1}$ subunit, hydrogen bonds and salt bridges form between lobe I and GABA and a critical tyrosine in lobe II (Galvez et al., 2000). The tyrosine residue is so critical that baclofen can be converted from an agonist to an antagonist by mutation of the tyrosine to an alanine (Galvez et al., 2000); this finding also supports the conclusion that both agonists and antagonists bind to this region of the receptor.

Though the $GABA_{B1}$ subunit binds agonist, both $GABA_{B1}$ and $GABA_{B2}$ N-terminal domains are required for effective receptor function. As expected, heterodimeric receptors that are mutated to contain two $GABA_{B2}$ N-termini are unable to bind GABA (Margeta-Mitrovic et al., 2001b). However, heterodimeric receptors containing two $GABA_{B1}$ N-termini do not display wild-type signaling; agonists inhibit instead of activating downstream effectors (Margeta-Mitrovic et al., 2001b). Interestingly, swapping the N-termini ($GABA_{B1}$ N-termini attached to $GABA_{B2}$ transmembrane/C-termini and vice versa) resulted in receptors indistinguishable from wildtype. The presence of the $GABA_{B2}$ N-terminus has also been shown to increase ligand affinity of the $GABA_{B1}$ subunit (Galvez et al., 2001). Thus, it is becoming evident that both $GABA_{B1}$ and $GABA_{B2}$ N-termini are critical for effective ligand activation of $GABA_B$ receptors.

Ligand binding to the extracellular N-terminal domain leads to activation of G-proteins on the cytoplasmic side of the membrane. Previous studies with other GPCRs had implicated the intracellular loops in coupling to G-proteins (Arora et al., 1995; Cypess et al., 1999; Moro et al., 1993). The molecular determinants required for G-protein signaling in the $GABA_B$ receptor were first identified by a series of chimeric studies. The heptahelical domain (seven-transmembrane domains, intracellular loops, and C-terminus) of the $GABA_{B2}$ subunit conferred the receptors G-protein sensitivity (Galvez et al., 2001). More refined replacements indicated that exchange of any of the three intracellular loops of $GABA_{B2}$ with their $GABA_{B1}$ counterparts resulted in nonfunctional receptors narrowing the G-protein associations to the intracellular loops of $GABA_{B2}$ (Margeta-Mitrovic et al., 2001a). Further, point mutations within the second and third intracellular loops of $GABA_{B2}$ decreased or completely disrupted G-protein activation (Duthey et al., 2002; Havlickova et al., 2002). Thus, a working model for $GABA_B$ receptor activation is that the $GABA_{B1}$ subunit responds to ligands while the $GABA_{B2}$ subunit couples to G-proteins via the intracellular loops (Fig. 1).

The development of subunit-specific $GABA_B$ receptor knockout mice has revealed a complex role for $GABA_B$ receptor subunits in the brain. $GABA_{B1}$ knockout mice do not exhibit detectable $GABA_B$ receptor agonist biochemistry or electrophysiology (Prosser et al., 2001; Schuler et al., 2001;

Quéva et al., 2003), consistent with the loss of $GABA_{B1}$ ligand binding site. Interestingly, atypical $GABA_B$ electrophysiological responses can be detected in the hippocampus of $GABA_{B2}$ knockout mouse (Gassmann et al., 2004). This study suggests the $GABA_{B1}$ subunit either functions independently in these neurons (unlikely due to the ER retention sequence) or is capable of forming functional relationships with other $GABA_{B2}$-like subunits. This is supported by data measuring $GABA_{B1}$ RNA and protein in peripheral tissue in the absence of the $GABA_{B2}$ subunit despite measurable $GABA_B$ function within these tissues (Calver et al., 2000). Similarly, it has been shown that the $GABA_{B2}$ subunits are capable of forming functional relationships with other GPCR subunits such as the M_2 muscarinic receptor (Boyer et al., 2009), suggesting that alternate $GABA_B$–GPCR heterodimer relationships are possible. Moreover, both $GABA_{B1}$ and $GABA_{B2}$ were shown to independently interact with other receptors such as the extracellular calcium sensing receptor (ECaR), leading to increased surface expression of the receptor (Chang et al., 2007). Future research may prove that $GABA_{B1}$ and $GABA_{B2}$ promiscuity increases signaling opportunities and potential pathway crosstalk.

III. $GABA_B$ Receptor Protein Expression

$GABA_B$ receptors have been identified within both the peripheral (Lehmann, 2009; Meza, 1998; Ong et al., 1990) and central nervous systems. Some of the highest regions of $GABA_{B1}$ and $GABA_{B2}$ expression within the CNS include the neocortex, hippocampus, thalamus, and cerebellum. Notably, the $GABA_{B2}$ subunit is found in brain regions lacking the $GABA_{B1}$ subunit (Charles et al., 2001). These findings suggest that functional $GABA_B$ receptors are found in many regions of the brain.

At the light microscopy level, $GABA_{B1}$ and $GABA_{B2}$ subunits were found to overlap in the dendrites of the hippocampus (Kulik et al., 2003) and high-resolution immunocytochemistry revealed a predominance of $GABA_B$ receptors within spines and shafts of dendrites in hippocampal pyramidal neurons, shown to correspond to the location of excitatory (glutamatergic) synapses (Kulik et al., 2006) and inhibitory (GABAergic) terminals (Kulik et al., 2003). Similarly, electron microscopic studies have localized $GABA_B$ receptors with their effectors to glutamatergic synapses in Purkinje cells of the cerebellum (Fernandez-Alacid et al., 2009). Recent studies have combined live fluorescence microscopy, biochemistry, and quantitative colocalization to show that $GABA_{B1}$ and $GABA_{B2}$ subunits have very different mobility properties within hippocampal neurons; $GABA_{B1}$ and $GABA_{B2}$ subunits heterodimerize primarily at the plasma membrane rather than pre-assembling in the Golgi for vesicular transport (Ramirez et al., 2009). Excitingly, this finding raises the possibility that assembly of functional

$GABA_B$ receptors at the plasma membrane is a highly dynamic process that may be regulated by intracellular and extracellular cues. However, the $GABA_{B1}$ subunit has been shown to redistribute from the distal axons to the soma of hippocampal neurons in the $GABA_{B2}$ knockout mouse suggesting an earlier relationship between $GABA_{B2}$ and $GABA_{B1}$ that targets/maintains $GABA_{B1}$ subunit presence in the distal terminals (Gassmann et al., 2004). Further research is needed to fully understand the requirements of heterodimerization and subcellular targeting.

Electron microscopy studies have found the $GABA_{B1}$ and $GABA_{B2}$ subunits expressed pre- and postsynaptically (Kulik et al., 2003), and electrophysiological studies have corroborated their functional presence at presynaptic (Chen & van den Pol, 1998; Iyadomi et al., 2000; Li et al., 2002; Santos et al., 1995; Takahashi et al., 1998; Wu & Saggau, 1995) and postsynaptic sites (Deisz et al., 1997; Harayama et al., 1998; Li & Stern, 2004). Interestingly the two isoforms of the $GABA_{B1}$ subunit (1a and 1b) have unique expression patterns; *in situ* hybridization studies revealed $GABA_{B1a}$ mRNA expression predominantly in the granule cell parallel fiber terminals of the cerebellum (considered presynaptic) while $GABA_{B1b}$ was found predominantly in Purkinje cells (considered postsynaptic) (Billinton et al., 1999). Using a combined genetic, physiological, and morphological approach the $GABA_{B1a}$ subunits have since been shown to assemble with $GABA_{B2}$ presynaptically to inhibit glutamate release in the hippocampus. Conversely, the $GABA_{B1b/B2}$ subunits predominantly mediate postsynaptic inhibition (Perez-Garci et al., 2006; Vigot et al., 2006). Further, transfected $GABA_{B1a}$ subunits localize to the distal axons of the CA3 neurons suggesting that the protein interacting "sushi" motifs that are unique to the N-terminus of the $GABA_{B1a}$ subunit promote subcellular localization of the receptor, perhaps by association with unknown partner proteins (see heterocomplex formation section).

IV. $GABA_B$ Receptor Effectors

A. G-proteins

Ligand binding to the $GABA_B$ receptor triggers GDP/GTP exchange at the $G\alpha$ subunit of the heterotrimeric (α, β, and γ) G-proteins, resulting in dissociation of the $G\alpha$-GTP from the $G\beta\gamma$ dimer. In turn, these G-proteins trigger activation of downstream effectors, including ion channels and enzymes (see below; Fig. 2). The downstream activity is terminated by the intrinsic GTPase activity of the $G\alpha$ subunit, which hydrolyzes GTP to GDP, and promotes reassociation of the G-protein with the $GABA_B$ receptor. This GTPase activity has been quantified in postmortem brain tissue, where pharmacological activation of $GABA_B$ receptors is accompanied by GTPase

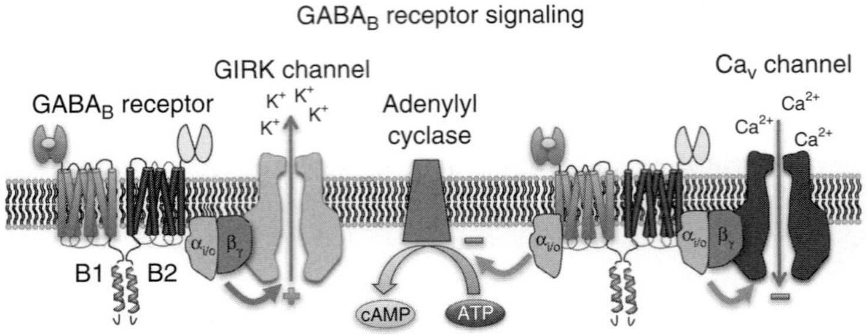

FIGURE 2 The primary $GABA_B$ receptor effectors. Activation of the G-protein coupled $GABA_B$ receptor stimulates GTP-dependent G-protein (Gi/o) dissociation of the Gα and Gβγ dimer. The Gαi/o subunit has been shown to inhibit adenylyl cyclase while the Gβγ dimer is capable of modulating voltage-gated Ca^{2+} (Ca_V) or G protein-gated inwardly rectifying K^+ (GIRK) channels, resulting in potent neuronal inhibition. Effector specificity may be regulated by heterocomplex formation, guided by targeting protein partners and subcellular localization.

activity (Odagaki etal., 1998), and in studies that use hydrolysis-resistant analogs of GDP and GTP, GDPβS and GTPγS, respectively, to irreversibly promote $GABA_B$ receptor effects (Dolphin & Scott, 1987). Using G-protein anti-sera and toxins, it has been shown that the $GABA_B$ receptor selectively activates a subset of heterotrimeric G-proteins that are pertussis toxin (PTX) sensitive, the Gαi and Gαo family (Menon-Johansson et al., 1993). Interestingly Gαo and Gαi1, but not Gαi2, increased GABA binding to NEM-treated brain membranes suggesting these G-proteins' specific involvement in $GABA_B$ function (Morishita et al., 1990). However, functional coupling of $GABA_B$ receptors with K^+ channels appears to involve a different subset of G-proteins, Gαo and Gαi2 (Leaney & Tinker, 2000). Interestingly, only the Gi subtype has been isolated from the bovine cerebral cortex and couples the $GABA_B$ receptor to adenylyl cyclase (Nishikawa et al., 1997). Increasingly, evidence suggests that the specificity of the G-protein subtype is region- and/or function-specific.

B. Voltage-Gated Calcium Channels

One of the first confirmed ion channel effectors of the $GABA_B$ receptor was the voltage-gated Ca^{2+} (Ca_V) channel in the early 1980s; calcium-dependent action potentials of the DRG were impaired by GABA treatment (Dunlap & Fischbach, 1981). This was later confirmed by voltage clamp analysis that revealed two calcium currents: transient and sustained (Robertson & Taylor, 1986). Baclofen reduced the peak amplitude of the sustained current. These findings suggested the $GABA_B$ receptor might mediate inhibitory effects within neurons via the Ca_V channel. Six types of Ca_V channel, made up of multiple and various subunits, have been classified by pharmacology and electrophysiology (Catterall, 2000). The Ca_V channels are termed L-, T-, N-, P-, Q-, and R-type. The T-type Ca_V channels are low-voltage-activated, predominantly performing a pacemaker role; the other subtypes are high-voltage-activated. The $GABA_B$ receptor has been shown to couple specifically to the high-voltage-activated P-, Q-, and N- but not L- or R-type Ca_V channels (Guyon & Leresche, 1995; Harayama et al., 1998; Huston et al., 1995; Li & Stern, 2004; Mintz & Bean, 1993; Vigot et al., 2006).

The electrophysiological relationship of $GABA_B$-Ca_V has been extensively studied throughout the brain including the hippocampus (Doze et al., 1995; Wu & Saggau, 1995), SON (Li & Stern, 2004), DRG (Menon-Johansson et al., 1993), and midbrain (Cardozo & Bean, 1995). Agonist activation of $GABA_B$ receptors stimulates Gi/o proteins (Menon-Johansson et al., 1993; Obrietan & van den Pol, 1999) releasing Gβγ to bind to, and inhibit, the high voltage-activated Ca_V channel (Dolphin, 2003) by 10–50 % (Guyon & Leresche, 1995; Huston et al., 1990; Menon-Johansson et al., 1993). This current inhibition is voltage-independent (Deisz & Lux, 1985) and is driven by a slowing of the Ca_V channel activation kinetics; as a result calcium influx is reduced, thereby decreasing various calcium-dependent, excitatory neuronal activities. For example, presynaptically, $GABA_B$ receptor-dependent inhibition of Ca_V currents leads to reduced neurotransmitter release. In cerebellar granule neurons, application of baclofen inhibits the amplitude of Ca_V currents by 30–50%, inhibiting glutamate release by 25–30% (Huston et al., 1990). Presynaptic $GABA_B$-inhibited Ca_V currents have also been found at GABA synapses (Chen & van den Pol, 1998; Doze et al., 1995) where the receptor acts as an autoreceptor. In the SCN, for example, evoked IPSCs can be completely ablated by baclofen administration, resulting from presynaptic inhibition of GABA release through inhibition of Ca_V (Chen & van den Pol, 1998). Similarly depolarization-induced GABA release in CA1 of the hippocampus is reduced by presynaptic $GABA_B$ autoreceptors that inhibit Ca_V currents (Doze et al., 1995). High-frequency stimulation promotes long-term potentiation (LTP) in CA1 by depressing GABA release (Davies et al., 1991) and therefore implicates the $GABA_B$ receptor in synaptic plasticity.

Postsynaptically, the $GABA_B$ receptor modulation of Ca_V currents can also regulate neuronal excitability. In the SON, $GABA_B$ receptor-dependent inhibition of Ca_V currents (10–25%) (Harayama et al., 1998) reduced firing capabilities due to diminished depolarizations (Li & Stern, 2004). Interestingly, this neuronal regulation was more predominant in vasopressin neurons (as opposed to the oxytocin neurons) suggesting cell-type-specific regulation.

Taken together these studies demonstrate the integral and diverse role $GABA_B$ modulation of Ca_V currents plays both pre- and postsynaptically to affect neuronal excitability and synaptic plasticity.

C. Potassium Channels

$GABA_B$-induced K^+ currents were identified in 1985 in hippocampal slices (Gähwiler & Brown, 1985; Inoue et al., 1985a, 1985b). The currents responded to changes in extracellular K^+ and were blocked by either barium or caesium, suggesting inwardly rectifying K^+-selective channels. Unlike other regions of the brain in which Ca^{2+} was shown to be the primary ionic conductance controlled by baclofen, the $GABA_B$ receptor was shown to also be K^+-linked in the hippocampus. It has since been shown that the $GABA_B$ receptor couples to a specific type of K^+ channel—the G-protein-gated inwardly rectifying K^+ (GIRK or Kir3) channel (Misgeld et al., 1995). GIRK channels assemble in a cell-specific, homo- or heterotetrameric manner (made up of four different subunits, GIRK1-4) and despite some controversy over the Gα or G$\beta\gamma$ subunits in the 1980s, it is now accepted that the G$\beta\gamma$ dimer binds directly to and opens GIRK channels (Reuveny et al., 1994; Wickman et al., 1994). GIRK channel activation by a Gi/o-coupled GPCR results in hyperpolarization of the neuron, inhibiting neuronal activity (Dascal, 1997). In this way, similar to the $GABA_B$-Ca_V currents, the $GABA_B$-GIRK currents are considered inhibitory.

Similar to $GABA_B$ expression patterns, the $GABA_B$-GIRK currents have now been identified in many brain regions, including the hippocampus (Gähwiler & Brown, 1985; Premkumar & Gage, 1994; Kulik et al., 2006), locus coeruleus (Osmanovic & Shefner, 1988), substantia nigra (Lacey et al., 1988; Watts et al., 1996), VTA (Cruz et al., 2004; Labouèbe et al., 2007), cerebellum (Slesinger et al., 1997), thalamus (Wallenstein, 1994), and paraventricular nucleus and SON (Slugg et al., 2003). Activation of the $GABA_B$ receptor increases membrane conductance and reduces neuronal excitability by direct activation of the GIRK channel (Misgeld et al., 1995; Watts et al., 1996). Critically, loss of GIRK channels leads to uncontrollable neuronal activity; GIRK2 knockout mice display increased susceptibility to seizures and are hyperactive (Signorini et al., 1997; Blednov et al., 2001). The significance of the inhibitory role of the $GABA_B$-GIRKs is further highlighted by recent research within the VTA. Blocking the $GABA_B$ receptor or the GIRK channel within the VTA prevents $GABA_A$-mediated IPSC auto-inhibition of DA neurons (Michaeli & Yaka, 2010). These results suggest that presynaptically localized $GABA_B$-GIRK currents modulate DAergic output from the VTA. The sensitivity of postsynaptic $GABA_B$-GIRK currents within the VTA has been described further (Cruz et al., 2004; Labouèbe et al., 2007); in the VTA, the $GABA_B$ receptor coupling efficiency (EC_{50}) with GIRK channels is an order of magnitude higher in DA neurons, compared to GABA neurons (Cruz et al., 2004). Due to the arrangement of GABA neurons impinging on DA neurons in the VTA, and the difference in coupling efficiency, low concentrations of baclofen (or endogenous GABA) will inhibit GABA neurons via $GABA_B$-GIRK, thereby increasing VTA activity, while higher concentrations will directly activate GABA$_B$-GIRKs on the DA neurons to inhibit VTA output (Cruz et al., 2004). Differing expression patterns of GIRK channel subunits in each of the cell types account for this difference in coupling efficiency, enabling bi-directional effects of baclofen/GABA underlying the critical relationship between GABA$_B$ receptor and effector (Cruz et al., 2004). Recently, the expression of a regulator of G-protein signaling (RGS) 2 protein was shown to further influence the efficacy of GABA$_B$ receptor signaling in DA neurons (Labouèbe et al., 2007) (see heterocomplex formation and functional regulation sections below). These studies begin to suggest that the role of the GABA$_B$ receptor is far more complex and subtle than just inhibiting action potential propagation.

D. Enzymes

GABA$_B$ receptor activation of enzymes offers another pathway for modulation of neuronal excitability. GABA$_B$ receptors regulate adenylyl cyclase, an enzyme that converts ATP to the second messenger cAMP, through activated Gαi G-proteins (Nishikawa et al., 1997). The Gαi coupled to the GABA$_B$ receptor inhibits adenylyl cyclase (Federman et al., 1992), reducing cAMP levels and kinase activity within the cell. Paradoxically, however, evidence suggests that under some circumstances GABA$_B$ receptors are also capable of assisting in the activation of adenylyl cyclase via the Gβγ dimer (Robichon et al., 2004). Adenylyl cyclase is stimulated by Gαs proteins to increase cAMP levels and promote downstream kinase activity. Research suggests that Gβγ from the GABA$_B$ receptor is capable of increasing adenylyl cyclase affinity for Gαs from the β-adrenergic receptor (while the Gαi from GABA$_B$ is sequestered by RGS proteins) resulting in a synergistic acceleration of adenylyl cyclase activity (Robichon et al., 2004). Propagation of adenylyl cyclase activity by the GABA$_B$ receptor results in cAMP surges that activate protein kinase A (PKA). PKA is capable of altering cellular processes through phosphorylation and has been shown to modify both ion channel function directly and further downstream enzymes and proteins. cAMP surges also regulate cAMP response element binding (CREB) protein, a transcription factor known to regulate CRE-element-containing genes such as tyrosine hydroxylase and neuropeptides (Mayr & Montminy, 2001)—key elements in synaptic signaling. Moreover, in cerebellar neurons, GABA$_B$ activation promotes Ras activity to induce ERK(1/2) phosphorylation, activation of which triggers a kinase cascade leading to CREB activation (Tu et al., 2007).

Taken together, it appears that GABA$_B$ receptor is capable of mediating long-term effects within the central nervous system beyond "slow" adaptations to neuronal excitability via the slow IPSC. GABA$_B$ receptors can modify protein expression and function by phosphorylation and even alter transcription patterns via CREB. Increasingly, there is also evidence that the

$GABA_B$ receptor is capable of promoting both inhibitory and, in partnership with Gs-coupled GPCRs, stimulatory intracellular pathways. This is an interesting aspect to $GABA_B$ function within the brain that will need to be explored in more detail over the coming years.

E. G-Protein Independent Activation

The β-arrestins are small proteins that sterically disrupt GPCR coupling to G-proteins but more recently have also been shown to regulate phosphorylated GPCR expression (Shenoy & Lefkowitz, 2005). They are critical to cellular function contributing to responses such as chemotaxis and cell survival. In sensory chick neurons, β-arrestin has been shown to associate with Ca_V channels and, upon agonist activation, the $GABA_B$ receptor is recruited to the β-arrestin-channel and internalized as a complex (Puckerin et al., 2006). In this way fast, β-arrestin-mediated internalization of $GABA_B$ may contribute to neuronal excitability in a G-protein-independent manner. However, others have found that $GABA_B$ receptors are stable at the neuronal cell surface, showing little endocytic internalization correlating with a lack of agonist-induced receptor phosphorylation and arrestin recruitment in heterologous expression systems (Fairfax et al., 2004). Interestingly, it appears that cAMP-dependent kinases may phosphorylate the $GABA_B$ receptor to promote cell surface stability. The apparent paradoxical stabilizing/destabilizing effects of $GABA_B$ phosphorylation may prove yet another complexity of $GABA_B$ receptor regulation and signaling.

Recent evidence also suggests that $GABA_B$ receptors may be capable of G-protein-independent activation by receptor crosstalk. In cerebellar Purkinje cells, for example, the $GABA_B$ receptor acts as a Ca^{2+} sensor and modulates metabotropic glutamatergic activity by proposed association and co-factor-like activation of mGluR1 (Tabata et al., 2004). However, whether this is truly attributed to G-protein-independent signaling or $GABA_{B1}$ subunit heterodimerization with mGluR1 remains to be determined (see heterodimerization above).

V. Formation of a Macromolecular Signaling Heterocomplex

Recent studies suggest that small heterocomplexes can be generated to control the trafficking of the receptor to subcellular targets. For example, a pair of "sushi" repeats that distinguish the N-terminus of the $GABA_{B1a}$ and the $GABA_{B1b}$ subunit appear to promote subcellular localization (Vigot et al., 2006). The $GABA_{B1a}$ subunits form heterodimers with $GABA_{B2}$ to inhibit glutamate release in the hippocampus while the $GABA_{B1b}$ subunits form heteroreceptors postsynaptically to mediate inhibition. In fact, the sushi motifs are capable of redirecting the GABA$_{B1a}$-containing receptors away from the default dendrite targeting to the axons (Biermann et al., 2010; Vigot et al., 2006). Similarly, when the sushi domains are fused to membrane-associated protein CD8α, its normally uniform expression is redirected almost exclusively to the axons suggesting the motifs function by interacting with axonally bound proteins along intracellular sorting pathways (Biermann et al., 2010; Tiao et al., 2008). The identity of these axonal proteins have yet to be determined but their recruitment to a putative receptor complex may aid targeting.

An interaction with specific cytoskeletal proteins or lipid domains could explain the crossover or separation of signaling pathways. Recent studies have identified some of these scaffold proteins that associate with GABA$_B$ receptors. Both Marlin 1 (a microtubule binding G-protein) and kinesin-I (a molecular motor) interact with the GABA$_B$ receptor and are proposed to link the receptor with the tubulin cytoskeleton of neurons, presumably to promote localization and cellular movement (Couve et al., 2004; Vidal et al., 2007). The GABA$_B$ receptor has also been reported to associate with 14-3-3 proteins (specifically eta and zeta) (Couve et al., 2001). The 14-3-3 proteins often mediate signal transduction suggesting their association with the GABA$_B$ receptor may contribute to signaling by association with additional binding partners. Similarly the GABA$_B$ receptor can localize within the plasma membrane by associating with lipid rafts (Becher et al., 2001). Enriched in cholesterol and sphingolipids, the lipid rafts contain specific populations of membrane proteins and are capable of changing their size and composition in response to cellular cues (Lingwood et al., 2009). Association of the GABA$_B$ receptor with the lipid raft microdomain could be one mechanism assisting the formation of a heterocomplex on the plasma membrane.

Recent studies suggest the GABA$_B$ receptors may form higher order oligomers (Maurel et al., 2008); using spectroscopic techniques, multiple GABA$_{B1}$ subunits have been shown to be proximal to one another in both heterologous COS-7 cells and the brain suggesting the formation of GABA$_B$ oligomers. Further, these oligomers have been shown to exhibit decreased G-protein coupling efficiency (Maurel et al., 2008). Taken together the GABA$_B$ receptor complexes appear capable of functionally regulating receptor efficacy making their formation potentially critical to cellular function. Further investigation is warranted to fully understand the role of these higher order oligomers *in vivo*.

There is also an increasing body of evidence that suggests the heteromeric GABA$_B$ receptors form a larger macromolecular signaling complex with G-proteins, scaffold proteins, and effectors to facilitate region- and subcellular-specific regulation of GABA$_B$ activity (Fig. 2). The time-course/kinetics of GIRK channel activation would suggest the receptor, G-proteins, and channels reside within the immediate vicinity of each other (Hille, 2001). In support of this, microscopic and spectroscopic techniques have indicated that some effectors are in close proximity to the GABA$_B$ receptor.

For example, high-resolution immuno-electronmicroscopy shows $GABA_B$ receptors and GIRK channels colocalize on the same dendritic spines of pyramidal neurons of the hippocampus (Kulik et al., 2006). Interestingly, the GIRK channels and $GABA_B$ receptors are largely segregated in the dendritic shafts, suggesting at least two types of targeting in these neurons. The effectors may also be involved in recruiting stabilizing protein elements into a complex. For example, various PDZ proteins, known to promote subcellular localization, have been shown to interact with GIRK channels (e.g., SNX27 (Lunn et al., 2007) and PSD-95 (Inanobe et al., 1999)), and association of signaling components (e.g., Src kinase, Ras pathway members) has been reported for the Ca_V channel (Richman et al., 2004). Finally, RGS proteins have been shown to functionally regulate $GABA_B$-effector efficacy in a cue-dependent manner suggesting their local, cell-specific recruitment (Brown & Sihra, 2008; Fowler et al., 2007; Labouèbe et al., 2007; Lomazzi et al., 2008; Mutneja et al., 2005).

Direct evidence for the formation of a quaternary complex made up from $GABA_B$ receptor, G-protein, effector and RGS proteins has come from fluorescence resonance energy transfer (FRET) techniques. FRET occurs when two proteins are in close proximity and has been measured between the $GABA_B$ receptor and RGS4, G-proteins or GIRK channels, but not between GIRK and $G\alpha o$ (Fowler et al., 2007), directly between $G\beta\gamma$-proteins and GIRK channels (Raveh et al., 2008) and between GIRK3 and RGS2 (Labouèbe et al., 2007). Further, the formation of a stable physical interaction between G-protein and RGS (independent of G-protein functional state) that is capable of directly contributing to signaling kinetics has also been described (Benians et al., 2005). Taken together, these data suggest that the formation of the quaternary heterocomplex may be a central feature of $GABA_B$ receptor signaling providing increased potential therapeutic targets for manipulating $GABA_B$ signaling to treat disease.

VI. $GABA_B$ Receptor-Dependent Desensitization

Persistent stimulation of $GABA_B$ receptors that couple to GIRK channels can generate two different kinds of GIRK current responses. In some cases, stimulation of the receptor leads to a GIRK current that faithfully reports the time course of receptor activation (i.e., no decrement with time). Alternatively, persistent stimulation of the receptor can elicit a current that decreases with time, which is referred to as desensitization. In fact, longer stimulation of the receptor can lead to a persistent reduction in $GABA_B$ receptor-activated current. The extent of $GABA_B$ receptor-dependent desensitization can be variable, depending on the type of neuron. For example, in peri-aquaductal gray neurons, stimulation of $GABA_B$ receptors produces little desensitization of GIRK currents while stimulation of those in laterodorsal tegmental area results in >25% desensitization (Chieng & Christie, 1995). Likewise in the VTA, baclofen stimulation produces a desensitizing GIRK current in DA neurons but not in GABAergic neurons (Cruz et al., 2004). In hippocampus, $GABA_B$ receptor-mediated control of GABA release exhibits acute desensitization while both $GABA_B$ receptor-mediated inhibition of glutamate release and postsynaptic $GABA_B$ receptor-mediated inhibition show little desensitization (Tosetti et al., 2004). In general, the desensitizing response is an important adaptive response of the cell to prevent excessive signaling for particular signaling pathways.

There are several possible ways to generate a desensitizing current. Changes in the signaling of the GPCR, the G-proteins, and/or effector could all lead to smaller responses with time. A well-known mechanism of receptor-dependent desensitization involves phosphorylation of the receptor (Kovoor et al., 1995; Shui et al., 2001). For example, stimulation of β2-adrenergic receptors leads to phosphorylation of the GPCR by receptor kinases (GRKs), and subsequent uncoupling of the G-protein with the receptor (Kovoor et al., 1995; Shui et al., 2001). Prolonged stimulation (tens of minutes) then triggers receptor internalization, which by rapid recycling has been implicated in resensitization of the GPCR (Ferguson, 2001). Other pathways for producing a receptor-dependent desensitizing current have been described that involve changes in G-protein turnover (Chuang et al., 1998; Jeong & Ikeda, 2001; Mutneja et al., 2005; Saitoh et al., 2001), in the membrane phospholipid PIP_2 (Kobrinsky et al., 2000), and/or inhibitory G-proteins (Blanchet & Luscher, 2002).

To study the mechanism of $GABA_B$ receptor-dependent desensitization, several labs have heterologously expressed $GABA_B$ receptors and GIRK channels in mammalian cells (e.g., HEK293, COS, CHO) or *Xenopus* oocytes. Similar to hippocampal pyramidal neurons, brief (2 s) but repetitive stimulation of $GABA_{B1/B2}$ receptors decreases the amplitude of GIRK current by ~60% over a period of 10–15 min in HEK293 cells (Couve et al., 2002). PKA phosphorylation of the C-terminal tail attenuated this decrease, suggesting that the stability of $GABA_B$ receptors in the membrane is enhanced with PKA-dependent phosphorylation (Couve et al., 2002). Prolonged stimulation (1 h) of $GABA_B$ receptors produces desensitization when G-protein receptor kinase 4 (GRK4) is coexpressed with the receptor in HEK-293 cells (Perroy et al., 2003). Surprisingly, the kinase domain of GRK4 was not required for internalization, suggesting that phosphorylation is not involved (Perroy et al., 2003). In CHO cells, long-term (>2 h) stimulation of $GABA_B$ receptors also induces internalization (Gonzalez-Maeso et al., 2003), suggesting these cells express an endogenous GRK. In HEK293 cells, on the other hand, a 60 s application of $GABA_B$ agonist induced 60% GIRK current desensitization but did not appear to induce $GABA_B$ receptor internalization (Mutneja et al., 2005). Instead, the turnover of G-proteins appear to be involved in generating a desensitizing GIRK current (Mutneja et al., 2005). Heterologous desensitization was observed with mu-opioid receptor and $GABA_B$ receptors, suggesting that both GPCRs share a common pool of G-proteins and channels (Mutneja et al., 2005). Desensitization was attenuated with coexpression of $G\alpha_q$-Q209L-ΔN, a putative RGS binding protein, as well as a RGS-insensitive Gαo (Mutneja et al., 2005). How do RGS proteins induce receptor-dependent desensitization? One function of RGS proteins is to accelerate the intrinsic GTPase activity of Gα subunits, which promotes the formation of the inactive Gαβγ heterotrimer (Fig. 3). Modeling this effect of RGS on the G-protein turnover cycle shows that acute desensitization of receptor-induced GIRK currents is more extensive if a GTPase-activating protein, such as RGS4, accelerates the GTPase activity of Gα subunits (Chuang et al., 1998).

An increasing number of studies have established a link between RGS protein activity and GIRK current desensitization. Ectopic expression of RGS-insensitive Gαo G-proteins in sympathetic neurons dramatically reduced α2 adrenergic receptor-dependent desensitization of GIRK currents (Jeong & Ikeda, 2001). Similarly, coexpression of RGS proteins in oocytes enhances GIRK current desensitization following stimulation of opioid (Chuang et al., 1998) and muscarinic receptors (Saitoh et al., 2001). In a separate study, expression of two Gα subunits with different activation kinetics modulates the extent of $GABA_B$ receptor-dependent desensitization (Tosetti et al., 2004). In addition to GIRK channels, RGS proteins appear to modulate desensitization of GPCR-mediated inhibition of Ca_V channels (Chen & Lambert, 2000; Diverse-Pierluissi et al., 1999; Tosetti et al., 2003).

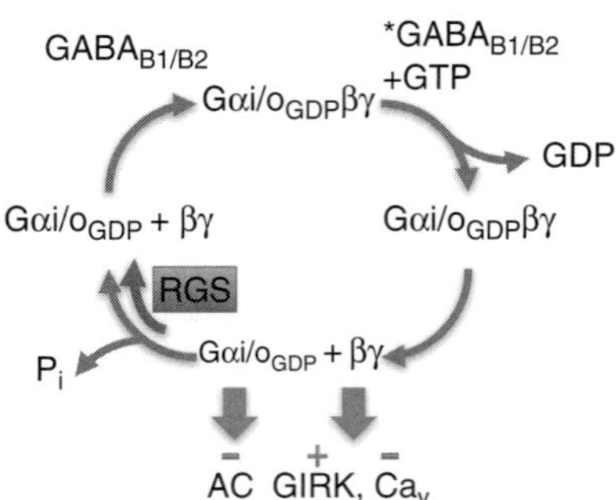

FIGURE 3 G-protein turnover affects $GABA_B$ receptor function. Agonist activation of $GABA_B$ receptor ($^*GABA_{B1/B2}$) leads to GTP exchange for GDP on Gαi/o. The activated heterotrimeric complex ($G\alpha i/o_{GTP}\beta\gamma$) then signals to different effectors (e.g., adenylyl cyclase, GIRK, and Ca_V). The intrinsic GTPase activity of Gαi/o hydrolyzes GTP to GDP, allowing the inactive heterotrimer ($G\alpha i/o_{GDP}\beta\gamma$) to reform. The presence of RGS accelerates the GTPase activity of Gαi/o, leading to less $G\alpha i/o_{GTP}$ and βγ, leading to desensitization of GIRK and Ca_V currents, and shift in the $GABA_B$ coupling efficiency to higher concentrations. Adapted from Chuang et al. (1998).

VII. Conclusion

The inhibitory $GABA_B$ receptor system is found throughout the brain and is increasingly implicated in many critical neuronal pathways and diseases including hormone secretion (Li & Stern, 2004), body temperature regulation (Quéva et al., 2003), learning and memory (Scanziani, 2000; Schuler et al., 2001), addiction (Cruz et al., 2004; Jayaram & Steketee, 2004, 2005), absence seizures (Hosford et al., 1992; Liu et al., 1992), epilepsy (Straessle et al., 2003), hyperalgesia (Schuler et al., 2001), gastro-esophageal reflux disease (Lehmann, 2009), and anxiety (Knapp et al., 2007). Understanding the basic molecular mechanisms of $GABA_B$ receptor activity will be crucial in directing targeted pharmaceutical development.

Conflict of Interest Statement: The authors have no conflicts of interest to declare.

Abbreviations

Ca_V	Voltage-gated Ca^{2+}
DRG	Dorsal root ganglion
FRET	Fluorescence resonance energy transfer
GABA	γ-aminobutyric acid
GIRK	G protein-gated inwardly rectifying K^+
RGS	Regulators of G-protein signaling
SON	Supraoptic nucleus
VTA	Ventral tegmental area

References

Andrade, R., Malenka, R. C., & Nicoll, R. A. (1986). A G protein couples serotonin and $GABA_B$ receptors to the same channels in hippocampus. *Science*, *234*, 1261–1265.

Arora, K. K., Sakai, A., & Catt, K. J. (1995). Effects of second intracellular loop mutations on signal transduction and internalization of the gonadotropin-releasing hormone receptor. *Journal of Biological Chemistry*, *270*, 22820–22826.

Becher, A., White, J. H., & McIlhinney, R. A. (2001). The gamma-aminobutyric acid receptor B, but not the metabotropic glutamate receptor type-1, associates with lipid rafts in the rat cerebellum. *Journal of Neurochemistry*, *79*, 787–795.

Benians, A., Nobles, M., Hosny, S., et al. (2005). Regulators of G-protein signaling form a quaternary complex with the agonist, receptor, and G-protein: A novel explanation for the acceleration of signaling activation kinetics. *Journal of Biological Chemistry*, *280*, 13383–13394.

Biermann, B., Ivankova-Susankova, K., Bradaia, A., et al. (2010). The sushi domains of $GABA_B$ receptors function as axonal targeting signals. *Journal of Neuroscience, 30*, 1385–1394.

Billinton, A., Upton, N., & Bowery, N. G. (1999). $GABA_B$ receptor isoforms, GBR1a and GBR1b, appear to be associated with pre- and post-synaptic elements, respectively, in rat and human cerebellum. *British Journal of Pharmacology, 126*, 1387–1392.

Blanchet, C., & Luscher, C. (2002). Desensitization of mu-opioid receptor-evoked potassium currents: Initiation at the receptor, expression at the effector. *Proceedings of the National Academy of Sciences of the United States of America, 99*, 4674–4679.

Blednov, Y. A., Stoffel, M., Chang, S. R., et al. (2001). GIRK2 deficient mice. Evidence for hyperactivity and reduced anxiety. *Physiology & Behavior, 74*, 109–117.

Bormann, J. (1988). Electrophysiology of $GABA_A$ and $GABA_B$ receptor subtypes. *Trends in Neuroscience, 11*, 112–116.

Bowery, N. G. (1993). $GABA_B$ receptor pharmacology. *Annual Review of Pharmacology and Toxicology, 33*, 109–147.

Bowery, N. G., Doble, A., Hill, D. R., et al. (1979). Baclofen: A selective agonist for a novel type of GABA receptor[proceedings]. *British Journal of Pharmacology, 67*, 444–445.

Bowery, N. G., & Hudson, A. L. (1979). Gamma-aminobutyric acid reduces the evoked release of [3h]-noradrenaline from sympathetic nerve terminals [proceedings]. *British Journal of Pharmacology, 66*, 108.

Boyer, S., Clancy, S., Terunuma, M., et al. (2009). Direct interaction of $GABA_B$ receptors with M_2 muscarinic receptors enhances muscarinic signaling. *Journal of Neuroscience, 29*, 15796–15809.

Brown, D. A., & Sihra, T. S. (2008). Presynaptic signaling by heterotrimeric G-proteins. *Handbook of Experimental Pharmacology*, 207–260.

Calver, A. R., Medhurst, A. D., Robbins, M. J., et al. (2000). The expression of GABA(B1) and GABA(B2) receptor subunits in the CNS differs from that in peripheral tissues. *Neuroscience, 100*, 155–170.

Calver, A. R., Robbins, M. J., Cosio, C., et al. (2001). The C-terminal domains of the GABA(b) receptor subunits mediate intracellular trafficking but are not required for receptor signaling. *Journal of Neuroscience, 21*, 1203–1210.

Cardozo, D. L., & Bean, B. P. (1995). Voltage-dependent calcium channels in rat midbrain dopamine neurons: Modulation by dopamine and $GABA_B$ receptors. *Journal Neurophysiology, 74*, 1137–1148.

Catterall, W. A. (2000). Structure and regulation of voltage-gated Ca^{2+} channels. *Annual Review of Cell and Developmental Biology, 16*, 521–555.

Chang, W., Tu, C., Cheng, Z., et al. (2007). Complex formation with the type B gamma-aminobutyric acid receptor affects the expression and signal transduction of the extracellular calcium-sensing receptor. Studies with HEK-293 cells and neurons. *Journal of Biological Chemistry, 282*, 25030–25040.

Charles, K. J., Evans, M. L., Robbins, M. J., et al. (2001). Comparative immunohistochemical localisation of GABA(B1a), GABA(B1b) and GABA(B2) subunits in rat brain, spinal cord and dorsal root ganglion. *Neuroscience, 106*, 447–467.

Chen, H., & Lambert, N. A. (2000). Endogenous regulators of G protein signaling proteins regulate presynaptic inhibition at rat hippocampal synapses. *Proceedings of the National Academy of Sciences of the United States of America, 97*, 12810–12815.

Chen, G., & van den Pol, A. N. (1998). Presynaptic $GABA_B$ autoreceptor modulation of P/Q-type calcium channels and GABA release in rat suprachiasmatic nucleus neurons. *Journal of Neuroscience, 18*, 1913–1922.

Chieng, B., & Christie, M. J. (1995). Hyperpolarization by $GABA_B$ receptor agonists in mid-brain periaqueductal gray neurones in vitro. *British Journal of Pharmacology, 116*, 1583–1588.

Chuang, H. H., Yu, M., Jan, Y. N., et al. (1998). Evidence that the nucleotide exchange and hydrolysis cycle of G proteins causes acute desensitization of G-protein gated inward rectifier K+ channels. *Proceedings of the National Academy of Sciences of the United States of America*, *95*, 11727–11732.

Couve, A., Filippov, A. K., Connolly, C. N., et al. (1998). Intracellular retention of recombinant GABA$_B$ receptors. *Journal of Biological Chemistry*, *273*, 26361–26367.

Couve, A., Kittler, J. T., Uren, J. M., et al. (2001). Association of GABA(B) receptors and members of the 14-3-3 family of signaling proteins. *Molecular and Cellular Neuroscience*, *17*, 317–328.

Couve, A., Restituito, S., Brandon, J. M., et al. (2004). Marlin-1, a novel RNA-binding protein associates with GABA receptors. *Journal of Biological Chemistry*, *279*, 13934–13943.

Couve, A., Thomas, P., Calver, A. R., et al. (2002). Cyclic AMP-dependent protein kinase phosphorylation facilitates GABA$_B$ receptor–effector coupling. *Nature Neuroscience*, *5*, 415–424.

Crunelli, V., & Leresche, N. (1991). A role for GABA$_B$ receptors in excitation and inhibition of thalamocortical cells. *Technical Innovations N Solution*, *14*, 16–21.

Cruz, H. G., Ivanova, T., Lunn, M. L., et al. (2004). Bi-directional effects of GABA$_B$ receptor agonists on the mesolimbic dopamine system. *Nature Neuroscience*, *7*, 153–159.

Cypess, A. M., Unson, C. G., Wu, C. R., et al. (1999). Two cytoplasmic loops of the glucagon receptor are required to elevate cAMP or intracellular calcium. *Journal of Biological Chemistry*, *274*, 19455–19464.

Dascal, N. (1997). Signalling via the G protein-activated K+ channels. *Cellular Signalling*, *9*, 551–573.

Davies, C. H., Starkey, S. J., Pozza, M. F., et al. (1991). GABA autoreceptors regulate the induction of LTP. *Nature*, *349*, 609–611.

Deisz, R. A., Billard, J.-M., & Zieglgansberger, W. (1997). Presynaptic amd postsynaptic GABA$_B$ receptors of neocortical neurons of the rat in vitro: Differences in pharmacology and ionic mechanisms. *Synapse*, *25*, 62–72.

Deisz, R. A., & Lux, H. D. (1985). Gamma-aminobutyric acid-induced depression of calcium currents of chick sensory neurons. *Neuroscience Letters*, *56*, 205–210.

Deriu, D., Gassmann, M., Firbank, S., et al. (2005). Determination of the minimal functional ligand-binding domain of the GABA$_{B1b}$ receptor. *Biochemical Journal*, *386*, 423–431.

Diverse-Pierluissi, M. A., Fischer, T., Jordan, J. D., et al. (1999). Regulators of G protein signaling proteins as determinants of the rate of desensitization of presynaptic calcium channels. *Journal of Biological Chemistry*, *274*, 14490–14494.

Dolphin, A. C. (2003). G protein modulation of voltage-gated calcium channels. *Pharmacological Reviews*, *55*, 607–627.

Dolphin, A. C., & Scott, R. H. (1987). Calcium channel currents and their inhibition by (-)-baclofen in rat sensory neurones: Modulation by guanine nucleotides. *Journal of Physiology (Paris)*, *386*, 1–17.

Doze, V. A., Cohen, G. A., & Madison, D. V. (1995). Calcium channel involvement in GABA$_B$ receptor-mediated inhibition of GABA release in area CA1 of the rat hippocampus. *Journal Neurophysiology*, *74*, 43–53.

Dunlap, K., & Fischbach, G. D. (1981). Neurotransmitters decrease the calcium conductance activated by depolarization of embryonic chick sensory neurones. *Journal of Physiology (Paris)*, *317*, 519–535.

Duthey, B., Caudron, S., Perroy, J., et al. (2002). A single subunit (GB2) is required for G-protein activation by the heterodimeric GABA(B) receptor. *Journal of Biological Chemistry*, *277*, 3236–3241.

Fairfax, B. P., Pitcher, J. A., Scott, M. G., et al. (2004). Phosphorylation and chronic agonist treatment atypically modulate GABA$_B$ receptor cell surface stability. *Journal of Biological Chemistry*, *279*, 12565–12573.

Federici, M., Nistico, R., Giustizieri, M., et al. (2009). Ethanol enhances $GABA_B$-mediated inhibitory postsynaptic transmission on rat midbrain dopaminergic neurons by facilitating GIRK currents. *European Journal of Neuroscience*, *29*, 1369–1377.

Federman, A. D., Conklin, B. R., Schrader, K. A., et al. (1992). Hormonal stimulation of adenylyl cyclase through gi-protein βγ subunits. *Nature*, *356*, 159–161.

Ferguson, S. S. (2001). Evolving concepts in G protein-coupled receptor endocytosis: The role in receptor desensitization and signaling. *Pharmacological Reviews*, *53*, 1–24.

Fernandez-Alacid, L., Aguado, C., Ciruela, F., et al. (2009). Subcellular compartment-specific molecular diversity of pre- and post-synaptic GABA-activated GIRK channels in Purkinje cells. *Journal Neurochemistry*, *110*, 1363–1376.

Fowler, C. E., Aryal, P., Suen, K. F., et al. (2007). Evidence for association of $GABA_B$ receptors with kir3 channels and RGS4 proteins. *Journal of Physiology (Paris)*, *580*, 51–65.

Gähwiler, B. H., & Brown, D. A. (1985). $GABA_B$-receptor-activated K^+ current in voltage-clamped CA3 pyramidal cells in hippocampal cultures. *Proceedings of the National Academy of Sciences of the United States of America*, *82*, 1558–1562.

Galvez, T., Duthey, B., Kniazeff, J., et al. (2001). Allosteric interactions between GB1 and GB2 subunits are required for optimal $GABA_B$ receptor function. *EMBO Journal*, *20*, 2152–2159.

Galvez, T., Parmentier, M. L., Joly, C., et al. (1999). Mutagenesis and modeling of the $GABA_B$ receptor extracellular domain support a venus flytrap mechanism for ligand binding. *Journal of Biological Chemistry*, *274*, 13362–13369.

Galvez, T., Prezeau, L., Milioti, G., et al. (2000). Mapping the agonist-binding site of $GABA_B$ type 1 subunit sheds light on the activation process of $GABA_B$ receptors. *Journal of Biological Chemistry*, *275*, 41166–41174.

Gassmann, M., Shaban, H., Vigot, R., et al. (2004). Redistribution of $GABA_{B1}$ protein and atypical $GABA_B$ responses in $GABA_{B2}$-deficient mice. *Journal of Neuroscience*, *24*, 6086–6097.

Gonzalez-Maeso, J., Wise, A., Green, A., et al. (2003). Agonist-induced desensitization and endocytosis of heterodimeric $GABA_B$ receptors in CHO-K1 cells. *European Journal of Pharmacology*, *481*, 15–23.

Guyon, A., & Leresche, N. (1995). Modulation by different $GABA_B$ receptor types of voltage-activated calcium currents in rat thalamocortical neurones. *Journal of Physiology (Paris)*, *485*(Pt 1), 29–42.

Harayama, N., Shibuya, I., Tanaka, K., et al. (1998). Inhibition of N- and P/Q-type calcium channels by postsynaptic $GABA_B$ receptor activation in rat supraoptic neurones. *Journal of Physiology (Paris)*, *509*(Pt 2), 371–383.

Havlickova, M., Prezeau, L., Duthey, B., et al. (2002). The intracellular loops of the GB2 subunit are crucial for G-protein coupling of the heteromeric gamma-aminobutyrate B receptor. *Molecular Pharmacology*, *62*, 343–350.

Hille, B. (2001). *Ion Channels of Excitable Membranes* (3rd ed.). Sunderland, Mass: Sinauer.

Hosford, D. A., Clark, S., Cao, Z., et al. (1992). The role of $GABA_B$ receptor activation in absence seizures of lethargic (lh/lh) mice. *Science*, *257*, 398–401.

Huston, E., Cullen, G. P., Burley, J. R., et al. (1995). The involvement of multiple calcium channel sub-types in glutamate release from cerebellar granule cells and its modulation by $GABA_B$ receptor activation. *Neuroscience*, *68*, 465–478.

Huston, E., Scott, R. H., & Dolphin, A. C. (1990). A comparison of the effect of calcium channel ligands and $GABA_B$ agonists and antagonists on transmitter release and somatic calcium channel currents in cultured neurons. *Neuroscience*, *38*, 721–729.

Inanobe, A., Yoshimoto, Y., Horio, Y., et al. (1999). Characterization of G-protein-gated K^+ channels composed of kir3.2 subunits in dopaminergic neurons of the substantia nigra. *Journal of Neuroscience*, *19*, 1006–1017.

Inoue, M., Matsuo, T., & Ogata, N. (1985a). Baclofen activates voltage-dependent and 4-aminopyridine sensitive K^+ conductance in guinea-pig hippocampal pyramidal cells maintained in vitro. *British Journal of Pharmacology*, *84*, 833–841.

Inoue, M., Matsuo, T., & Ogata, N. (1985b). Characterization of pre- and postsynaptic actions of (-)-baclofen in the guinea-pig hippocampus in vitro. *British Journal of Pharmacology*, *84*, 843–851.

Iyadomi, M., Iyadomi, I., Kumamoto, E., et al. (2000). Presynaptic inhibition by baclofen of miniature EPSCs and IPSCs in substantia gelatinosa neurons of the adult rat spinal dorsal horn. *Pain*, *85*, 385–393.

Jayaram, P., & Steketee, J. D. (2004). Effects of repeated cocaine on medial prefrontal cortical GABA$_B$ receptor modulation of neurotransmission in the mesocorticolimbic dopamine system. *Journal of Neurochemistry*, *90*, 839–847.

Jayaram, P., & Steketee, J. D. (2005). Effects of cocaine-induced behavioural sensitization on GABA transmission within rat medial prefrontal cortex. *European Journal of Neuroscience*, *21*, 2035–2039.

Jeong, S. W., & Ikeda, S. R. (2001). Differential regulation of G protein-gated inwardly rectifying K^+ channel kinetics by distinct domains of RGS8. *Journal of Physiology (Paris)*, *535*, 335–347.

Jones, K. A., Borowsky, B., Tamm, J. A., et al. (1998). GABA$_B$ receptors function as a heteromeric assembly of the subunits GABA$_B$ R1 and GABA$_B$ R2. *Nature*, *396*, 674–678.

Kammerer, R. A., Frank, S., Schulthess, T., et al. (1999). Heterodimerization of a functional GABA$_B$ receptor is mediated by parallel coiled-coil alpha-helices. *Biochemistry*, *38*, 13263–13269.

Kaupmann, K., Huggel, K., Heid, J., et al. (1997). Expression cloning of GABA$_B$ receptors uncovers similarity to metabotropic glutamate receptors. *Nature*, *386*, 239–246.

Kaupmann, K., Malitschek, B., Schuler, V., et al. (1998). GABA$_B$-receptor subtypes assemble into functional heteromeric complexes. *Nature*, *396*, 683–687.

Knapp, D. J., Overstreet, D. H., & Breese, G. R. (2007). Baclofen blocks expression and sensitization of anxiety-like behavior in an animal model of repeated stress and ethanol withdrawal. *Alcoholism Clinical and Experimental Research*, *31*, 582–595.

Kobrinsky, E., Mirshahi, T., Zhang, H., et al. (2000). Receptor-mediated hydrolysis of plasma membrane messenger PIP$_2$ leads to K^+-current desensitization. *Nature Cell Biology*, *2*, 507–514.

Kovoor, A., Henry, D. J., & Chavkin, C. (1995). Agonist-induced desensitization of the mu opioid receptor-coupled potassium channel (GIRK1). *Journal of Biological Chemistry*, *270*, 589–595.

Kulik, A., Vida, I., Fukazawa, Y., et al. (2006). Compartment-dependent colocalization of kir3.2-containing K^+ channels and GABA$_B$ receptors in hippocampal pyramidal cells. *Journal of Neuroscience*, *26*, 4289–4297.

Kulik, A., Vida, I., Lujan, R., et al. (2003). Subcellular localization of metabotropic GABA$_B$ receptor subunits GABA$_{B1a/b}$ and GABA$_{B2}$ in the rat hippocampus. *Journal of Neuroscience*, *23*, 11026–11035.

Kuner, R., Köner, G., Grünewald, S., et al. (1999). Role of heteromer formation in GABA$_B$ receptor function. *Science*, *283*, 74–77.

Labouèbe, G., Lomazzi, M., Cruz, H., et al. (2007). RGS2 modulates coupling between GABA$_B$ receptors and GIRK channels in dopamine neurons of the ventral tegmental area. *Nature Neuroscience*, *12*, 1559–1568.

Lacey, M. G., Mercuri, N. B., & North, R. A. (1988). On the potassium conductance increase activated by GABA$_B$ and dopamine D$_2$ receptor in rat substantia nigra neurones. *Journal of Physiology (Paris)*, *401*, 437–453.

Leaney, J. L. (2003). Contribution of kir3.1, kir3.2a and kir3.2c subunits to native G protein-gated inwardly rectifying potassium currents in cultured hippocampal neurons. *European Journal of Neuroscience*, *18*, 2110–2118.

Leaney, J. L., & Tinker, A. (2000). The role of members of the pertussis toxin-sensitive family of G proteins in coupling receptors to the activation of the G protein-gated inwardly rectifying potassium channel. *Proceedings of the National Academy of Sciences of the United States of America*, *97*, 5651–5656.

Lehmann, A. (2009). $GABA_B$ receptors as drug targets to treat gastroesophageal reflux disease. *Pharmacology & Therapeutics*, *122*, 239–245.

Li, D. P., Chen, S. R., Pan, Y. Z., et al. (2002). Role of presynaptic muscarinic and GABA(B) receptors in spinal glutamate release and cholinergic analgesia in rats. *Journal of Physiology (Paris)*, *543*, 807–818.

Li, Y., & Stern, J. E. (2004). Activation of postsynaptic $GABA_B$ receptors modulate the firing activity of supraoptic oxytocin and vasopressin neurones: Role of calcium channels. *Journal of Neuroendocrinology*, *16*, 119–130.

Lingwood, D., Kaiser, H. J., Levental, I., et al. (2009). Lipid rafts as functional heterogeneity in cell membranes. *Biochemical Society Transactions*, *37*, 955–960.

Liu, Z., Vergnes, M., Depaulis, A., et al. (1992). Involvement of intrathalamic $GABA_B$ neurotransmission in the control of absence seizures in the rat. *Neuroscience*, *48*, 87–93.

Lomazzi, M., Slesinger, P. A., & Luscher, C. (2008). Addictive drugs modulate GIRK-channel signaling by regulating RGS proteins. *Trends in Pharmacological Sciences*, *29*, 544–549.

Lunn, M. L., Nassirpour, R., Arrabit, C., et al. (2007). A unique sorting nexin regulates trafficking of potassium channels via a PDZ domain interaction. *Nature Neuroscience*, *10*, 1249–1259.

Malitschek, B., Schweizer, C., Keir, M., et al. (1999). The N-terminal domain of gamma-aminobutyric acid (B) receptors is sufficient to specify agonist and antagonist binding. *Molecular Pharmacology*, *56*, 448–454.

Margeta-Mitrovic, M., Jan, Y. N., & Jan, L. Y. (2000). A trafficking checkpoint controls GABA (B) receptor heterodimerization. *Neuron*, *27*, 97–106.

Margeta-Mitrovic, M., Jan, Y. N., & Jan, L. Y. (2001a). Function of GB1 and GB2 subunits in G protein coupling of GABA(B) receptors. *Proceedings of the National Academy of Sciences of the United States of America*, *98*, 14649–14654.

Margeta-Mitrovic, M., Jan, Y. N., & Jan, L. Y. (2001b). Ligand-induced signal transduction within heterodimeric GABA(B) receptor. *Proceedings of the National Academy of Sciences of the United States of America*, *98*, 14643–14648.

Maurel, D., Comps-Agrar, L., Brock, C., et al. (2008). Cell-surface protein-protein interaction analysis with time-resolved FRET and snap-tag technologies: Application to GPCR oligomerization. *Nature Methods*, *5*, 561–567.

Mayr, B., & Montminy, M. (2001). Transcriptional regulation by the phosphorylation-dependent factor CREB. *Nature Reviews Molcular Cell Biology*, *2*, 599–609.

Menon-Johansson, A. S., Berrow, N., & Dolphin, A. C. (1993). G(o) transduces $GABA_B$-receptor modulation of N-type calcium channels in cultured dorsal root ganglion neurons. *Pflugers Archiv*, *425*, 335–343.

Meza, G. (1998). GABA as an afferent neurotransmitter in the vestibular sensory periphery of vertebrates. *Neurobiology (Budapest, Hungary)*, *6*, 109–125.

Michaeli, A., & Yaka, R. (2010). Dopamine inhibits $GABA_A$ currents in ventral tegmental area dopamine neurons via activation of presynaptic G-protein coupled inwardly-rectifying potassium channels. *Neuroscience*, *165*, 1159–1169.

Mintz, I. M., & Bean, B. P. (1993). $GABA_B$ receptor inhibition of P-type Ca2+ channels in central neurons. *Neuron*, *10*, 889–898.

Misgeld, U., Bijak, M., & Jarolimek, W. (1995). A physiological role for $GABA_B$ receptors and the effects of baclofen in the mammalian central nervous system. *Progress in Neurobiology*, *46*, 423–462.

Morishita, R., Kato, K., & Asano, T. (1990). $GABA_B$ receptors couple to G proteins go, go* and gi1 but not to gi2. *FEBS Letters*, *271*, 231–235.

Moro, O., Lameh, J., Hogger, P., et al. (1993). Hydrophobic amino acid in the i2 loop plays a key role in receptor–G protein coupling. *Journal of Biological Chemistry*, *268*, 22273–22276.

Mutneja, M., Berton, F., Suen, K. -F., et al. (2005). Endogenous RGS proteins enhance acute desensitization of GABA$_B$ receptor-activated GIRK currents in HEK-293T cells. *Pflugers Archiv*, *450*, 61–73.

Nishikawa, M., Hirouchi, M., & Kuriyama, K. (1997). Functional coupling of gi subtype with GABA$_B$ receptor/adenylyl cyclase system: Analysis using a reconstituted system with purified GTP-binding protein from bovine cerebral cortex. *Neurochemistry International*, *31*, 21–25.

Obrietan, K., & van den Pol, A. N. (1999). GABA$_B$ receptor-mediated regulation of glutamate-activated calcium transients in hypothalamic and cortical neuron development. *Journal Neurophysiology*, *82*, 94–102.

O'Callaghan, J. F., Jarolimek, W., Lewen, A., et al. (1996). (-)-Baclofen-induced and constitutively active inwardly rectifying potassium conductances in cultured rat midbrain neurons. *Pflugers Archiv*, *433*, 49–57.

Odagaki, Y., Nishi, N., Ozawa, H., et al. (1998). Measurement of receptor-mediated functional activation of G proteins in postmortem human brain membranes. *Brain Research*, *789*, 84–91.

Ong, J., Harrison, N. L., Hall, R. G., et al. (1990). 3-Aminopropanephosphinic acid is a potent agonist at peripheral and central presynaptic GABA$_B$ receptors. *Brain Research*, *526*, 138–142.

Osmanovic, S. S., & Shefner, S. A. (1988). Baclofen increases the potassium conductance of rat locus coeruleus neurons recorded in brain slices. *Brain Research*, *438*, 124–136.

Otis, T. S., De Koninck, Y., & Mody, I. (1993). Characterization of synaptically elicited GABA$_B$ responses using patch-clamp recordings in rat hippocampal slices. *Journal of Physiology (Paris)*, *463*, 391–407.

Otmakhova, N. A., & Lisman, J. E. (2004). Contribution of lh and GABA$_B$ to synaptically induced afterhyperpolarizations in CA1: A brake on the NMDA response. *Journal Neurophysiology*, *92*, 2027–2039.

Perez-Garci, E., Gassmann, M., Bettler, B., et al. (2006). The GABA$_B$ 1b isoform mediates long-lasting inhibition of dendritic Ca^{2+} spikes in layer 5 somatosensory pyramidal neurons. *Neuron*, *50*, 603–616.

Perroy, J., Adam, L., Qanbar, R., et al. (2003). Phosphorylation-independent desensitization of GABA(B) receptor by GRK4. *EMBO Journal*, *22*, 3816–3824.

Pooler, A. M., & McIlhinney, R. A. (2007). Lateral diffusion of the GABA$_B$ receptor is regulated by the GABA$_{B2}$ C terminus. *Journal of Biological Chemistry*, *282*, 25349–25356.

Premkumar, L. S., & Gage, P. W. (1994). Potassium channels activated by GABA$_B$ agonists and serotonin in cultured hippocampal neurons. *Journal Neurophysiology*, *71*, 2570–2575.

Prosser, H. M., Gill, C. H., Hirst, W. D., et al. (2001). Epileptogenesis and enhanced prepulse inhibition in GABA(B1)-deficient mice. *Molecular and Cellular Neuroscience*, *17*, 1059–1070.

Puckerin, A., Liu, L., Permaul, N., et al. (2006). Arrestin is required for agonist-induced trafficking of voltage-dependent calcium channels. *Journal of Biological Chemistry*, *281*, 31131–31141.

Quéva, C., Bremmer-Danielsen, M., Edlund, A., et al. (2003). Effects of GABA agonists on body temperature regulation in GABA$_{B1}$-/-mice. *British Journal of Pharmacology*, *140*, 315–322.

Ramirez, O. A., Vidal, R. L., Tello, J. A., et al. (2009). Dendritic assembly of heteromeric gamma-aminobutyric acid type B receptor subunits in hippocampal neurons. *Journal of Biological Chemistry*, *284*, 13077–13085.

Raveh, A., Riven, I., & Reuveny, E. (2008). The use of FRET microscopy to elucidate steady state channel conformational rearrangements and G protein interaction with the GIRK channels. *Methods in Molecular Biology*, *491*, 199–212.

Reuveny, E., Slesinger, P. A., Inglese, J., et al. (1994). Activation of the cloned muscarinic potassium channel by G protein βγ subunits. *Nature*, *370*, 143–146.

Richman, R. W., Tombler, E., Lau, K. K., et al. (2004). N-type Ca^{2+} channels as scaffold proteins in the assembly of signaling molecules for $GABA_B$ receptor effects. *Journal of Biological Chemistry*, *279*, 24649–24658.

Robertson, B., & Taylor, W. R. (1986). Effects of gamma-aminobutyric acid and (-)-baclofen on calcium and potassium currents in cat dorsal root ganglion neurones in vitro. *British Journal of Pharmacology*, *89*, 661–672.

Robichon, A., Tinette, S., Courtial, C., et al. (2004). Simultaneous stimulation of GABA and beta adrenergic receptors stabilizes isotypes of activated adenylyl cyclase heterocomplex. *BMC Cell Biology*, *5*, 25.

Rohrbacher, J., Jarolimek, W., Lewen, A., et al. (1997). $GABA_B$ receptor-mediated inhibition of spontaneous inhibitory synaptic currents in rat midbrain culture. *Journal of Physiology (Paris)*, *500*(Pt 3), 739–749.

Saitoh, O., Masuho, I., Terakawa, I., et al. (2001). Regulator of G protein signaling 8 (RGS8) requires its NH2 terminus for subcellular localization and acute desensitization of G protein-gated K+ channels. *Journal of Biological Chemistry*, *276*, 5052–5058.

Santos, A. E., Carvalho, C. M., Macedo, T. A., et al. (1995). Regulation of intracellular [ca2+] and GABA release by presynaptic $GABA_B$ receptors in rat cerebrocortical synaptosomes. *Neurochemistry International*, *27*, 397–406.

Scanziani, M. (2000). GABA spillover activates postsynaptic $GABA_B$ receptors to control rhythmic hippocampal activity. *Neuron*, *25*, 673–681.

Schuler, V., Luscher, C., Blanchet, C., et al. (2001). Epilepsy, hyperalgesia, impaired memory, and loss of pre- and postsynaptic GABA(B) responses in mice lacking GABA(B(1)). *Neuron*, *31*, 47–58.

Shenoy, S. K., & Lefkowitz, R. J. (2005). Seven-transmembrane receptor signaling through beta-arrestin. *Science's STKE: Signal Transduction Knowledge Environment*, *2005*, cm10.

Shui, Z., Yamanushi, T. T., & Boyett, M. R. (2001). Evidence of involvement of GIRK1/GIRK4 in long-term desensitization of cardiac muscarinic K^+ channels. *American Journal of Physiology. Heart and Circulatory Physiology*, *280*, H2554–H2562.

Signorini, S., Liao, Y. J., Duncan, S. A., et al. (1997). Normal cerebellar development but susceptibility to seizures in mice lacking G protein-coupled, inwardly rectifying K^+ channel GIRK2. *Proceedings of the National Academy of Sciences of the United States of America*, *94*, 923–927.

Slesinger, P. A., Stoffel, M., Jan, Y. N., et al. (1997). Defective gamma-aminobutyric acid type B receptor-activated inwardly rectifying K^+ currents in cerebellar granule cells isolated from weaver and *girk2* null mutant mice. *Proceedings of the National Academy of Sciences of the United States of America*, *94*, 12210–12217.

Slugg, R. M., Zheng, S. X., Fang, Y., et al. (2003). Baclofen inhibits guinea pig magnocellular neurones via activation of an inwardly rectifying K+ conductance. *Journal of Physiology (Paris)*, *551*, 295–308.

Sohn, J. W., Lee, D., Cho, H., et al. (2007). Receptor-specific inhibition of $GABA_B$-activated K+ currents by muscarinic and metabotropic glutamate receptors in immature rat hippocampus. *Journal of Physiology (Paris)*, *580*, 411–422.

Straessle, A., Loup, F., Arabadzisz, D., et al. (2003). Rapid and long-term alterations of hippocampal $GABA_B$ receptors in a mouse model of temporal lobe epilepsy. *European Journal of Neuroscience*, *18*, 2213–2226.

Tabata, T., Araishi, K., Hashimoto, K., et al. (2004). Ca2+ activity at $GABA_B$ receptors constitutively promotes metabotropic glutamate signaling in the absence of GABA.

Proceedings of the National Academy of Sciences of the United States of America, *101*, 16952–16957.

Takahashi, T., Kajikawa, Y., & Tsujimoto, T. (1998). G-protein-coupled modulation of presynaptic calcium currents and transmitter release by a GABA$_B$ receptor. *Journal of Neuroscience*, *18*, 3138–3146.

Tiao, J. Y., Bradaia, A., Biermann, B., et al. (2008). The sushi domains of secreted GABA(B1) isoforms selectively impair GABA(B) heteroreceptor function. *Journal of Biological Chemistry*, *283*, 31005–31011.

Toselli, M., & Taglietti, V. (1993). Baclofen inhibits high-threshold calcium currents with two distinct modes in rat hippocampal neurons. *Neuroscience Letters*, *164*, 134–136.

Tosetti, P., Bakels, R., Colin-Le Brun, I., et al. (2004). Acute desensitization of presynaptic GABA$_B$-mediated inhibition and induction of epileptiform discharges in the neonatal rat hippocampus. *European Journal of Neuroscience*, *19*, 3227–3234.

Tosetti, P., Pathak, N., Jacob, M. H., et al. (2003). RGS3 mediates a calcium-dependent termination of G protein signaling in sensory neurons. *Proceedings of the National Academy of Sciences of the United States of America*, *100*, 7337–7342.

Tu, H., Rondard, P., Xu, C., et al. (2007). Dominant role of GABA$_{B2}$ and G$\beta\gamma$ for GABA$_B$ receptor-mediated ERK1/2/CREB pathway in cerebellar neurons. *Cellular Signalling*, *19*, 1996–2002.

Turgeon, S. M., & Albin, R. L. (1993). Pharmacology, distribution, cellular localization, and development of GABA$_B$ binding in rodent cerebellum. *Neuroscience*, *55*, 311–323.

Vidal, R. L., Ramirez, O. A., Sandoval, L., et al. (2007). Marlin-1 and conventional kinesin link GABA$_B$ receptors to the cytoskeleton and regulate receptor transport. *Molecular and Cellular Neuroscience*, *35*, 501–512.

Vigot, R., Barbieri, S., Brauner-Osborne, H., et al. (2006). Differential compartmentalization and distinct functions of GABA$_B$ receptor variants. *Neuron*, *50*, 589–601.

Wallenstein, G. V. (1994). Simulation of GABA$_B$-receptor-mediated K^+ current in thalamocortical relay neurons: Tonic firing, bursting, and oscillations. *Biological Cybernetics*, *71*, 271–280.

Watts, A. E., Williams, J. T., & Henderson, G. (1996). Baclofen inhibition of the hyperpolarization-activated cation current, lh, in rat substantia nigra zona compacta neurons may be secondary to potassium current activation. *Journal Neurophysiology*, *76*, 2262–2270.

White, J. H., Wise, A., Main, M. J., et al. (1998). Heterodimerization is required for the formation of a functional GABA$_B$ receptor. *Nature*, *396*, 679–682.

Wickman, K. D., Iniguez-Lluhl, J. A., Davenport, P. A., et al. (1994). Recombinant G-protein $\beta\gamma$-subunits activate the muscarinic-gated atrial potassium channel. *Nature*, *368*, 255–257.

Wu, L. G., & Saggau, P. (1995). GABA$_B$ receptor-mediated presynaptic inhibition in guinea-pig hippocampus is caused by reduction of presynaptic Ca2+ influx. *Journal of Physiology (Paris)*, *485*(Pt 3), 649–657.

Toshihide Tabata[*] and Masanobu Kano[†]

[*]Laboratory for Neural Information Technology, Graduate School of Sciences and Engineering, University of Toyama, Toyama, Japan

[†]Department of Neurophysiology, Graduate School of Medicine, University of Tokyo, Bunkyo-ku, Tokyo, Japan

$GABA_B$ Receptor-Mediated Modulation of Metabotropic Glutamate Signaling and Synaptic Plasticity in Central Neurons

Abstract

In mammalian brains, γ-amino butyric acid (GABA) is the most ubiquitous inhibitory neurotransmitter and neuromodulator. The $G_{i/o}$ protein-coupled GABA receptor termed B-type GABA receptor ($GABA_BR$) has been recognized as one of the major mediators of the inhibitory effects of GABA. Several years ago, Hirono et al. and our group independently found that $GABA_BR$ mediates non-inhibitory effects in cerebellar Purkinje cells. In

Advances in Pharmacology, Volume 58

1054-3589/10 $35.00
10.1016/S1054-3589(10)58007-4

this cell type, $GABA_BR$ co-localizes with type-1 metabotropic glutamate receptor (mGluR1), a $G_{q/11}$ protein-coupled receptor around the postsynaptic membrane of the excitatory synapses. At that site, $GABA_BR$ is not exposed to the direct bombardment of GABA released from the terminals of inhibitory neurons. Instead, the receptor may sense a low concentration of GABA and Ca^{2+} usually contained in the extracellular fluid and a relatively high concentration of GABA spilt over from the neighboring active inhibitory synapses. In response to these ambient ligands, $GABA_BR$ increases the ligand affinity of mGluR1 independently of $G_{i/o}$ protein and augments mGluR1-coupled intracellular signaling via $G_{i/o}$ protein. These $GABA_BR$-mediated modulations may facilitate mGluR1-mediated neuronal responses including cerebellar long-term depression, a form of synaptic plasticity crucial for cerebellar motor learning. In this article, we present current knowledge on a new role of $GABA_BR$ as an ambience-dependent regulator of synaptic signaling.

I. Introduction

γ-amino butyric acid (GABA) is the most ubiquitous inhibitory neurotransmitter and neuromodulator in mammalian brains. B-type GABA receptor ($GABA_BR$), a $G_{i/o}$ protein-coupled receptor has been recognized as a mediator of the inhibitory effects of GABA (for review, see Bettler et al., 2004; Bowery et al., 2002). For example, in the presynaptic terminals of many types of neurons, $GABA_BR$ activation leads to inhibition of voltage-gated Ca^{2+} channels. The resultant reduction of Ca^{2+} influx attenuates action potential-triggered neurotransmitter release and decreases the efficacy of synaptic transmission. In the dendrites and/or soma of many types of neurons, $GABA_BR$ activation leads to opening of G protein-coupled inwardly rectifying K^+ (GIRK) channels (Bichet et al., 2003; Jan & Jan, 1997). The resultant K^+ conductance counteracts spontaneous or synaptically driven depolarization and thereby impedes action potential discharge.

Several years ago, Hirono et al. (Hirono et al., 2001) and our group (Tabata et al., 2004) independently discovered novel non-inhibitory actions of $GABA_BR$ in cerebellar Purkinje cells. Purkinje cells are known as one of the "hot spots" of $GABA_BR$ throughout the brain (Jones et al., 1998; Kaupmann et al., 1998; Kuner et al., 1999). Nevertheless, the function of $GABA_BR$ in this cell type has not been fully elucidated. What deepens the mystery is that a majority of the $GABA_BR$ proteins reside in the dendritic spines innervated by *excitatory* (*glutamatergic*) but not inhibitory (GABAergic) presynaptic neurons (Fig. 1). In those spines, $GABA_BR$ co-localizes with type-1 metabotropic glutamate receptor (mGluR1), a $G_{q/11}$ protein-coupled receptor for the excitatory neurotransmitter [for review, see (Kano et al., 2008; Nakanishi, 1994)]. The two groups revealed that ligand binding to

GABA$_B$R leads to enhancement of mGluR1-mediated neuronal responses in two ways. One is a $G_{i/o}$ protein-dependent augmentation of mGluR1-coupled intracellular signaling. The other is a $G_{i/o}$ protein-independent increase in the ligand affinity of mGluR1. Although these two phenomena produce apparently similar effects on mGluR1-mediated responses, they are absolutely distinct in respect of the underlying mechanisms. To distinguish these phenomena in the following sections, we term them *mGluR1 signaling augmentation* and *mGluR1 sensitization*, respectively. mGluR1 signaling in Purkinje cells is an essential factor to induce a form of synaptic plasticity crucial for cerebellar motor learning [for review, see (Kano et al., 2008; Tabata & Kano, 2009)]. Thus, GABA$_B$R-mediated mGluR1 signaling modulation could have an important role in regulation of synaptic plasticity and learning.

In this article, we present current knowledge on the functional interplay between GABA$_B$R, mGluR1, and their immediately related molecules in cerebellar Purkinje cells and discuss the possible physiological significance of GABA$_B$R-mediated mGluR1 signaling modulation.

II. Synaptic Organization Around Cerebellar Purkinje Cells

To provide a background for understanding the cellular environment where GABA$_B$R exerts its actions, we give a brief explanation of the synaptic organization around cerebellar Purkinje cells in mammalian brains [for review, see (Ito, 2006; Llinas et al., 2003)]. This cell type is the sole output neuron of the cerebellar cortex. Purkinje cells spread their large dendritic arbors in the molecular layer of the cerebellar cortex and receive synaptic inputs from excitatory fibers and inhibitory interneurons.

Each Purkinje cell receives excitatory synaptic inputs from more than 100,000 parallel fibers (PFs) and a single climbing fiber (CF). PFs are the axons of granule cells and convey motor commands from the upper centers and sensory information from various parts of the body (Ito, 2006). Glutamate released from the presynaptic terminals of PFs binds to amino-3-hydroxy-5-methyl-4-isoxazolepropionic acid-type ionotropic glutamate receptor (AMPAR) and mGluR1 residing at the tip and annuli of the dendritic spines of the Purkinje cell, respectively (for details, see below). GABA$_B$R co-localizes with mGluR1. In response to glutamate, AMPAR opens its built-in cation pore and generates a fast PF-Purkinje cell excitatory postsynaptic potential (PF-PC EPSP). mGluR1 opens transient receptor potential C3 (TRPC3) cation channels via G_q protein, and this generates a slow PF-PC EPSP (Hartmann et al., 2004, 2008; Hartmann & Konnerth, 2008). Purkinje cells integrate these EPSPs and based on the result of this computation, control the motor system (Ito, 2006). This promotes motor accuracy and coordination (Kawato et al., 2003; Llinas & Welsh, 1993; Ohyama et al., 2003; Thach et al., 1992).

A CF is the axon of a neuron whose soma exists in the inferior olive and informs errors between the planned and executed motion (Ito, 2006). Each branch of a CF forms many synaptic contacts widely over the dendrites of a Purkinje cell. Glutamatergic transmission at these synapses generates a large fast CF-Purkinje cell EPSP (CF-PC EPSP), which often elicits a burst of a spike and spikelets (complex spikes) (Llinas et al., 2003). mGluR1 signaling driven by a certain group of PFs and CF-driven complex spikes together trigger cerebellar long-term depression (LTD) which is a long-lasting attenuation of fast EPSPs at the synapses between the group of PFs and the Purkinje cell (see below).

GABAergic transmission from interneurons to a Purkinje cell evokes an inhibitory postsynaptic potential (IPSP) mediated largely by A-type ionotropic GABA receptor ($GABA_AR$). PFs innervate inhibitory interneurons. Thus, PF excitation evokes sequentially a PF-PC EPSP and an IPSP driven by PF-innervated interneurons in the same Purkinje cell. Such an IPSP delimits a time-window for the preceding EPSP to influence the whole-cell activity, and this may promote temporal precision of computation in Purkinje cells.

III. Possible Ligands for $GABA_BR$ in Cerebellar Purkinje Cells

A. Basal "Ambient" GABA

Some studies employing immuno-electron microscopy (Fritschy et al., 1999; Ige et al., 2000; Kulik et al., 2002) show that in rodent cerebellar Purkinje cells, $GABA_BR$ is concentrated at the annuli of the dendritic spines innervated by *excitatory* PFs but not those innervated by inhibitory interneurons (Fig. 1). Could $GABA_BR$ have any role without being exposed to the direct bombardment of GABA released from inhibitory neurons?

One possibility lies in "ambient" GABA. A study employing microdialysis (Bohlen et al., 1979) reports that mammalian cerebrospinal fluids contain ~65 nM of free GABA under normal (unstimulated) conditions. This GABA is thought to originate largely from the brain because its concentration increases when the whole-brain GABA level is pharmacologically elevated. Some recent microdialysis studies (Bist & Bhatt, 2009; Paredes et al., 2009) detected extracellular concentrations of ~120 nM in rodent cerebella. At these concentrations, GABA is thought to bind to a considerable fraction of the $GABA_BR$ proteins based on the GABA-$GABA_BR$ affinity. The GABA dose for the half-maximal blockade against the binding of ^{3}H-baclofen and [^{125}I]CGP64213 to $GABA_BR$ (IC_{50}) measured *in vitro* are 40 nM and 270 nM, respectively (Galvez et al., 2000; Hill & Bowery, 1981). Thus, ambient GABA might constitutively induce weak $GABA_BR$ signaling in Purkinje cells under physiological conditions.

B. GABA Spilt Over from Synapses

The concentration of ambient GABA may rise locally near an active neural circuitry. In rodent cerebellar slices, high-frequency electrical stimuli given to the molecular layer induce release of a large amount of GABA from the presynaptic terminals of interneurons. A part of this GABA appears to be spilt over from the synaptic cleft (Dittman & Regehr, 1997) (Fig. 1). As a result, ~10 μM of GABA appears to act on the neighboring PF-Purkinje cell synapses. This level of GABA is thought to activate a majority of the $GABA_BR$ proteins. In cultured mouse Purkinje cells, 3 μM of baclofen, a $GABA_BR$-selective agonist with a similar potency to GABA[(1)] induces a GIRK current (Tabata et al., 2005).

The ambient GABA concentration indeed rises in a neuronal activity-dependent manner *in vivo*. One of the microdialysis studies (Paredes et al., 2009) collected samples through the rat cerebellar cortex and interpositus nuclei and detected a 30–50% increase in the GABA concentration during a training of delayed eye-blink conditioning. This behavioral task reinforces association of unconditioned eye-blink responses to an aversive stimulus with the preceding conditioning stimulus (in the case of the above study, an air puff and a pure tone, respectively). A study using mGluR1-knockout and mGluR1-rescue mice (Kishimoto et al., 2002) showed that learning in this task depends on the neuronal activity involving mGluR1 of Purkinje cells.

C. Extracellular Ca^{2+}

Some studies using heterologous expression cells (Galvez et al., 2000) indicate another possibility that $GABA_BR$ senses Ca^{2+} usually contained in the extracellular fluid ($Ca^{2+}{}_o$). $GABA_BR$ belongs to family C G protein-coupled receptors (GPCRs) and has a structural similarity to Ca^{2+}-sensing receptor (CaR), another family C member that serves as a blood Ca^{2+} sensor in the prarathyroid [for review, see (Bettler et al., 2004)]. Ca^{2+} itself does not induce $G_{i/o}$ protein activation by $GABA_BR$. However, studies performing [^{35}S]GTPγS binding assays on membrane from heterologous expression cells and rat brains (Galvez et al., 2000; Wise et al., 1999) showed that the continuous presence of a physiological level (1 mM) of $Ca^{2+}{}_o$ lowers the GABA dose to induce the half-maximal level of $GABA_BR$-mediated G protein activation (EC_{50}, 7.7 or 2.4 μM) by a factor of 10–80 from that in the absence of $Ca^{2+}{}_o$ (72 or 200 μM). A point mutation study (Galvez et al., 2000) identified a Ca^{2+}-interacting site near the GABA-binding site of $GABA_BR$ subunit 1. These observations indicate that $Ca^{2+}{}_o$ induces a conformational change which allosterically modulates the GABA binding property of $GABA_BR$.

[1] For example, a study employing [^{35}S]GTPγS binding assay (Galvez et al., 2000) reports that the EC_{50} for $GABA_BR$-mediated G protein activation are 2.4 μM for GABA and 3.2 μM for baclofen in the presence of Ca^{2+} (1 mM) in the bath.

In heterologous expression cells, the extent of $Ca^{2+}{}_o$-induced enhancement of $GABA_BR$-mediated GIRK current reaches the maximum with 1–10 mM of $Ca^{2+}{}_o$ (Wise et al., 1999). As well, the extent of $Ca^{2+}{}_o$-induced enhancement of $GABA_BR$-mediated G protein activation reaches the maximum with ~1 mM of $Ca^{2+}{}_o$ (Galvez et al., 2000). Thus, a majority of the $GABA_BR$ proteins in Purkinje cells might interact with $Ca^{2+}{}_o$ under physiological conditions.

IV. Functions of $GABA_BR$ Near the Excitatory Synapses

A. Function Exerted Through GIRK Channels

Given that ambient GABA and/or $Ca^{2+}{}_o$ can stimulate $GABA_BR$ in cerebellar Purkinje cells, new questions arise: in response to such a ligand, what does $GABA_BR$ do near the excitatory synapses? What is the biological merit of such "ectopic" localization? First, we focus on the functional relationship to GIRK channels because a study employing immuno-electron microscopy and co-immunoprecipitation (Fernandez-Alacid et al., 2009) showed that mouse Purkinje cells express some types of GIRK channel subunits in the dendritic spines and shafts and that GIRK channel subunits form complexes with $GABA_BR$ in mouse cerebellum. If $GABA_BR$ in a PF-innervated dendritic spine could open the closely associated GIRK channels, the resultant K^+ conductance could efficiently shunt AMPAR-mediated fast PF-PC EPSPs to be generated in the same spine (effects on mGluR1-mediated responses will be discussed later).

We assessed this possibility in cultured mouse Purkinje cells (Tabata et al., 2005). We compared voltage ramp-evoked whole-cell currents recorded before and during an application of 3 μM of baclofen. Because baclofen has a similar potency to GABA (see footnote 1), baclofen at this dose mimics the effect of spilt GABA (~10 μM, see above). Purkinje cells usually display a large inwardly rectifying current carried mostly by K^+ passing through IRK channels (Falk et al., 1995; Miyashita & Kubo, 1997). Application of baclofen further adds an inwardly rectifying K^+ component. This additional component is susceptible to pertussis toxin (PTX), a $G_{i/o}$ protein inhibitor and several GIRK channel blockers including tertiapin-Q, Ba^{2+}, and Cs^{2+}. These results show that $GABA_BR$ is indeed coupled to GIRK channels in Purkinje cells.

However, baclofen-induced GIRK channel conductance does not greatly increase the total membrane conductance (Tabata et al., 2005). At a membrane potential of −130 mV, the amplitude of the baclofen-induced component is as small as 1/10–1/20 of that of the basal current. Current-clamp recordings from the somata of cultured mouse Purkinje cells showed that a whole-cell application of baclofen does not reduce the amplitude of an excitatory potential evoked by AMPA applied iontophoretically to the dendrites. These results indicate that the GIRK channel conductance little affects the conduction of AMPAR-mediated PF-PC EPSPs from the dendrites to the soma.

Although the direct effect on the dendritic excitatory potential is minor, the GABA$_B$R-mediated GIRK conductance appears to reduce the excitability of Purkinje cells by changing the membrane potential near the soma. Some studies employing light microscopic immunohistology (Iizuka et al., 1997; Murer et al., 1997) detected a GIRK channel subunit in and near the somata of rat Purkinje cells. In cultured mouse Purkinje cells, 3 μM of baclofen continuously hyperpolarizes the resting potential by $\sim$3.5 mV (Tabata et al., 2005). This hyperpolarization negatively offsets the *peak level* of AMPA-evoked excitatory potentials measured at the soma without changing the peak amplitude (Tabata et al., 2005). Such a shift might hamper EPSP-driven and/or spontaneous excitatory conductances to elevate the membrane potential towards the threshold of action potential discharge. The GABA$_B$R-mediated GIRK conductance indeed influences the neuronal excitability *in situ*. In Purkinje cells in mouse cerebellar slices, baclofen (10 μM) decreases the rate of spontaneous action potential discharges, and this effect is susceptible to Ba^{2+} and tertiapin-Q (Tabata et al., 2005).

These findings demonstrate that GABA$_B$R exerts a physiological function through opening GIRK channels whereas this does not require the ectopic localization of GABA$_B$R.

B. GABA$_B$R-Mediated mGluR1 Signaling Augmentation

In cultured cerebellar Purkinje cells, activation of A1 adenosine receptor (A1R), another $G_{i/o}$ protein-coupled receptor little affects the amplitude of excitatory currents evoked by dendritic application of AMPA (Tabata et al., 2007). This result together with the findings described in the previous section suggests that the signaling cascade involving GABA$_B$R and $G_{i/o}$ protein does not modulate AMPAR signaling itself. Thus, our concern shifts to the functional relationship of GABA$_B$R to mGluR1, the other excitatory receptor co-localizing in the dendritic spines (Fig. 1).

Several years ago, Hirono et al. (2001) found that application of baclofen augments mGluR1 signaling in Purkinje cells in mouse cerebellar slices (we call this effect GABA$_B$R-mediated mGluR1 signaling augmentation). They evaluated the intensity of mGluR1 signaling by slow PF-PC EPSP and excitatory postsynatptic current (EPSC, an inward current response observed under voltage clamp which is mediated by the same machinery as that of slow EPSP) evoked by a high-frequency burst of electrical stimuli to the PFs. These responses reflect the opening of TRPC3 channel which is gated by the mGluR1-G_q protein signaling cascade (Hartmann & Konnerth, 2008) (Fig. 1). CGP62349, a GABA$_B$R-selective antagonist, abolishes the baclofen-induced augmentation of the slow EPSC, confirming the involvement of GABA$_B$R (Hirono et al., 2001).

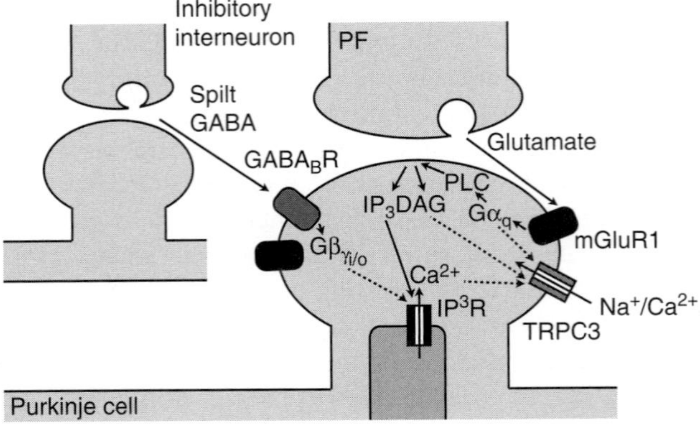

FIGURE 1 GABA$_B$R-mediated mGluR1 signaling augmentation. As shown schematically, GABA$_B$R and mGluR1 co-localize at the annuli of the PF-innervated dendritic spines of cerebellar Purkinje cells. Spillover from the neighboring inhibitory synapses elevates the local concentration of ambient GABA around a Purkinje cell. In response to a relatively high concentration of GABA, GABA$_B$R activates an enough large amount of $G_{i/o}$ protein to augment mGluR1-coupled intracellular signaling (mGluR1 signaling augmentation). This augmentation can be monitored as an increase in the amplitude of the slow PF-PC EPSP or EPSC under current or voltage clamp, respectively. These responses reflect cation flow through TRPC3 channel, which is coupled to mGluR1 via the α subunit of G_q protein ($G\alpha_q$) and are evoked when mGluR1 binds glutamate diffusing out of the cleft of the PF-PC synapse. It has been suggested in heterologous expression cells that the $\beta\gamma$ subunit complex, which is cleaved from activated $G_{i/o}$ protein ($G\beta\gamma_{i/o}$), may potentiate (but not necessarily trigger for itself) PLC, which produces IP_3 and DAG from a plasma membrane component. Potentiation of PLC may promote IP_3 and DAG production. IP_3 augments mGluR1-mediated Ca^{2+} release from the intracellular store to the cytoplasm. DAG and cytoplasmic Ca^{2+} may facilitate the function of TRPC3 channel, and this increases the amplitude of slow EPSPs or EPSCs. Dotted lines indicate signaling involving unknown messengers.

N-ethylmaleimide, a wide-spectrum inhibitor against intracellular signaling molecules including $G_{i/o}$ protein also abolishes the augmentation (Hirono et al., 2001). Our study in cultured mouse Purkinje cells (Tabata et al., 2007) showed that a high dose of an A1R agonist produces a similar augmentation of a R,S-3,5-dihydroxyphenylglycine (DHPG, a mGluR1/5-selective agonist)-evoked, mGluR1-mediated inward current and that this augmentation is susceptible to PTX. These observations together suggest the $G_{i/o}$ protein-dependence of the GABA$_B$R-mediated augmentation (Fig. 1).

Baclofen-induced augmentation is observed for the inward current evoked by 500 μM of DHPG in cultured Purkinje cells (Tabata et al., 2007). This effect cannot be ascribed to an increase in the DHPG-bound fraction of the mGluR1 proteins because this dose of DHPG is shown to saturate most of the mGluR1 proteins (Tabata et al., 2002). Thus, $G_{i/o}$ protein may augment mGluR1-coupled intracellular signaling rather than increase the ligand affinity of mGluR1 (Fig. 1).

Although it is unclear how $G_{i/o}$ protein acts on the intracellular signaling cascade, some insights (Fig. 1) can be drawn from studies using heterologous expression cells. In cells co-expressing mGluR1 and GABA$_B$R, GABA$_B$R-mediated augmentation of mGluR1-mediated responses is decreased by additional expression of β-adrenergic receptor kinase C-terminus which traps G protein βγ subunit complex and is occluded by additional expression of an excessive amount of the βγ subunits (Rives et al., 2009). These observations suggest that among the subunits detached from activated $G_{i/o}$ protein, the βγ subunit complex is thought to be important for the subsequent reaction. The βγ subunit complex may potentiate (but not necessarily trigger for itself) phospholipase C (PLC) (Quitterer & Lohse, 1999), which produces diacylglycerol (DAG) and inositol trisphosphate (IP_3) from the plasma membrane component. PLC is a major target of $G_{q/11}$ protein. Thus, mGluR1 activation triggers massive production of DAG and IP_3 by PLC. PLC potentiation by the βγ subunit complex further promotes DAG and IP_3 production. IP_3 elevates the cytoplasmic Ca^{2+} concentration by opening IP_3 receptor (IP_3R) on the surface of the intracellular Ca^{2+} store. DAG and cytoplasmic Ca^{2+} may facilitate the function of TRPC3 channel [for review, see (Hartmann & Konnerth, 2008)]. In support of the possibility that $G_{i/o}$ protein links GABA$_B$R to the downstream cascade of mGluR1 signaling, baclofen augments DHPG-induced, mGluR1-mediated Ca^{2+} release from the intracellular store in cultured mouse Purkinje cells (Kamikubo et al., 2007). This effect is indeed mediated by $G_{i/o}$ protein because it is not observed in PTX-pretreated cells. Moreover, baclofen does not elevate the resting cytoplasmic Ca^{2+} level (Kamikubo et al., 2007), suggesting that GABA$_B$R signaling itself does not trigger Ca^{2+} release.

One problem with the above scheme is that $G_{i/o}$ protein-dependent potentiation has been clearly shown for PLCβ3 but not PLCβ4 (Park et al., 1993). The cerebellar cortex consists of complementary stripes each of which contains Purkinje cells enriched with either PLC subtype (Sarna et al., 2006). Does the GABA$_B$R-mediated mGluR1 signaling augmentation take place in the PLCβ3-poor Purkinje cells? If so, is there any other cross-talk point for the GABA$_B$R and mGluR1 signaling cascades? To answer these questions, detailed analyses should be performed on native molecules in Purkinje cells.

C. GABA$_B$R-Mediated mGluR1 Sensitization

1. mGluR1 Sensitization

Kubo's group (Kubo et al., 1998; Miyashita & Kubo, 2000) reported that $Ca^{2+}{}_o$ interacts with mGluR1, and this increases the glutamate affinity of heterologously expressed mGluR1. As inspired by their work, we performed physiological analyses in cultured mouse cerebellar Purkinje cells and observed an apparently similar effect of Ca^{2+}_{o} (Tabata et al., 2002). Later, this effect turned out to be mediated by a different mechanism than that of the effect observed in the heterologous cells; in Purkinje cells, Ca^{2+}_{o} modulates mGluR1 not directly but indirectly via $GABA_BR$ (Tabata et al., 2004)(Fig. 2A).

mGluR1 in Purkinje cells bathed in Ca^{2+}-free saline or in heterologous expression cells displays a dose–response curve with a Hill coefficient (steepness of the curve over the working dose range) of ~1 (Miyashita & Kubo, 2000; Tabata et al., 2002, 2004)(Fig. 2B). This value of the coefficient indicates that each functional receptor possesses only one ligand-binding site or multiple non-cooperative ligand-binding sites. Most of the functional

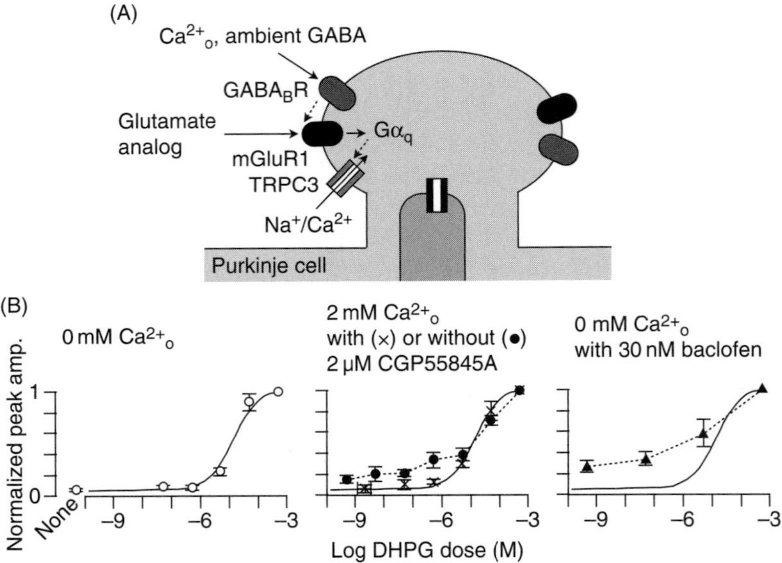

FIGURE 2 $GABA_BR$-mediated mGluR1 sensitization. (A) In cerebellar Purkinje cells, $GABA_BR$ increases the ligand affinity of mGluR1 in a $G_{i/o}$ protein-independent fashion (mGluR1 sensitization) in response to Ca^{2+}_{o} and ambient GABA. Mechanisms underlying the functional interplay between $GABA_BR$ and mGluR1 remain unknown (dotted arrow tying these receptors). This sensitization can be monitored as a change in the dose–response relation of a glutamate analog (e.g., DHPG)-evoked inward current recorded from cultured Purkinje cells under voltage clamp. This current is thought to be mediated by a signaling cascade involving G_q protein α subunit and TRPC3 channel like the slow PF-PC EPSP and EPSC (cf. Fig. 1). (B) Dose–response relations of the DHPG-evoked inward currents recorded from cultured mouse Purkinje cells in the absence or presence of the labeled agents. Note that Ca^{2+}_{o} or baclofen, a $GABA_BR$-selective agonist, enables mGluR1 to respond to lower doses of DHPG. CGP55845A, a $GABA_BR$-selective blocker, suppresses this effect of Ca^{2+}_{o}, suggesting that $GABA_BR$ mediates it. Sigmoid curve in each plot indicates a Hill function with a Hill coefficient of 1 fitted to the data obtained in the absence of Ca^{2+}_{o}. [Adapted from (Tabata et al., 2004), copyright (2004) National Academy of Sciences, U.S.A.]

receptors on cell surface consist of homo-dimerized mGluR1 subunits, each of which has a glutamate-binding site in its large extracellular domain [for review see (Jingami et al., 2003; Kubo & Tateyama, 2005)]. Thus, a functional mGluR1 as a whole possesses two glutamate-binding sites. Although there could be a cooperativity between the two glutamate-binding sites, it is not obvious in the dose–response curve presumably because glutamate binding to one of the dimerized mGluR1 subunits may be enough to initiate signal transduction as suggested by molecular and atomic studies [for review, see (Jingami et al., 2003; Kubo & Tateyama, 2005)][(2)]. Such binding leads to a change of the relative position of the extracellular domains of the dimerized subunits. As a result, the transmembrane and intracellular domains of the two subunits approach to each other. This promotes $G_{q/11}$ protein activation by the receptor.

We measured the peak amplitudes of the mGluR1-mediated inward currents, applying various doses of DHPG to cultured mouse Purkinje cells (Tabata et al., 2002, 2004)(Fig. 2A, B). In the absence of $Ca^{2+}{}_o$, the dose–response curve of native mGluR1 is similar to that of heterologously expressed mGluR1; the curve is well fitted by a Hill function with a Hill coefficient of 1 (sigmoid curve in Fig. 2B). Only 0.5 μM or higher doses of DHPG evokes significant mGluR1-mediated currents. Addition of $Ca^{2+}{}_o$ to the bath solution alters the dose-response relation in a complicated manner. In the presence of a physiological concentration (2 mM) of $Ca^{2+}{}_o$, even nanomolar doses of DHPG evokes inward currents (we term this effect mGluR1 sensitization) while the saturating dose (500 μM) unchanged; as a result, the working dose range in which the peak amplitude changes quasi-linearly with the dose becomes broader.

2. *Possible Mechanisms Underlying mGluR1 Sensitization*

Multivalent cations including $Ca^{2+}{}_o$ can alone activate mGluR1-mediated responses in cultured Purkinje cells (Tabata et al., 2002). Thus, we first postulated that $Ca^{2+}{}_o$-induced mGluR1 sensitization is due to direct interaction of mGluR1 with $Ca^{2+}{}_o$. However, another possibility is raised by a report suggesting complex formation between CaR and mGluR1 in cerebellar neurons (Gama et al., 2001). To date, various GPCRs have been reported to form hetero-oligomeric complexes in which the constituents functionally modulate one another [for review, see (Bouvier, 2001; Kubo & Tateyama, 2005)].

We examined whether $Ca^{2+}{}_o$-sensitive GPCRs (in the case of Purkinje cells, CaR and GABA$_B$R) mediate $Ca^{2+}{}_o$ action onto mGluR1 in cultured

[2] A molecular study in heterologous expression cells (Kniazeff et al., 2004) shows that glutamate binding to both of the dimerized mGuR1 subunits is required to induce the full activity of mGluR1. However, partial activation by glutamate binding to only one of the dimerized subunits is thought to occur more frequently (Kubo & Tateyama, 2005).

Purkinje cells. A β-amyloid fragment which is known to activate CaR does not enhance the DHPG-induced, mGluR1-mediated inward current (Tabata et al., 2004). An immunohistological study failed to detect CaR in mouse cerebellar Purkinje cell (M. Watanabe, personal communication). These observations exclude the involvement of CaR. By contrast, CGP55845A, a $GABA_BR$-selective antagonist abolishes the $Ca^{2+}{}_o$-induced mGluR1 sensitization (Tabata et al., 2004)(Fig. 2B). In the continuous presence of a dose (2 μM) of CPG55845A, which is thought to block most of the $GABA_BR$ proteins[(3)], the dose–response curve is well fitted by a Hill function with a Hill coefficient of 1 irrespective of the $Ca^{2+}{}_o$ concentration. GABA and baclofen have a similar effect to $Ca^{2+}{}_o$, increasing the amplitudes of the inward currents to lower doses of DHPG (Fig. 2B). Purkinje cells derived from $GABA_BR$ subunit 1-knockout mice do not display significant mGluR1-mediated inward currents to nanomolar doses of DHPG even in the presence of $Ca^{2+}{}_o$ (2 mM). These results demonstrate that the $Ca^{2+}{}_o$-induced mGluR1 sensitization is due to indirect action of $Ca^{2+}{}_o$ mediated by $GABA_BR$.

Both $Ca^{2+}{}_o$- and baclofen-induced mGluR1 sensitization is not susceptible to a pretreatment with PTX, which is shown to completely suppress a baclofen-induced GIRK current in cultured mouse Purkinje cells (Tabata et al., 2004). Without the aid of its own messenger $G_{i/o}$ protein, how could $GABA_BR$ influence the target? We performed co-immunoprecipitation for the synapse-rich P2 membrane fraction of samples obtained from adult mouse cerebella and detected complex formation between $GABA_BR$ and mGluR1 (Tabata et al., 2004). A study employing freeze-fracture immunoelectron microscopy (Rives et al., 2009) detected co-localization of $GABA_BR$ and mGluR1 at a nanometric scale in the dendritic spines of mouse Purkinje cells. These observations suggest that these receptors reside very close to each other in Purkinje cells. This might allow direct communication or indirect communication via a non-G protein local intermediate molecule(s) between $GABA_BR$ and mGluR1. Direct communication has been suggested for various GPCRs forming hetero-oligomeric complexes [for review, see (Bouvier, 2001)]. As to local intermediate molecules, there is increasing evidence that GPCRs form complexes with various non-G protein transmembrane and intracellular proteins (Bockaert et al., 2004). As exemplified by D5 dopamine receptor and $GABA_AR$, a GPCR and its interacting protein may modulate each other [see (Bockaert et al., 2004)]. Moreover, as exemplified by non-receptor tyrosine kinase janus kinase 2 which mediates type 1 angiotensin II receptor signaling, certain receptor-interacting proteins can act as mediators of agonist-induced GPCR signaling independently of G protein pathways [for review, see (Ritter & Hall, 2009)].

[3] In rat cerebral cortical tissues, CGP55458A reverses a baclofen effect on cAMP production with an IC_{50} of 130 nM (Cunningham & Enna, 1996) and facilitates GABA release with an EC_{50} of 8.08 nM presumably due to relief of presynaptic inhibition (Waldmeier et al., 1994).

The exact mechanism underlying complex formation between GABA$_B$R and mGluR1 is unknown. A recent study (Rives et al., 2009) reports that GABA$_B$R and mGluR1 do not form complexes in heterologous expression cells. Thus, a molecule(s) peculiar to Purkinje cells might be necessary to help complex formation.

The mechanism underlying the co-localization of GABA$_B$R and mGluR1 at the annuli of the dendritic spines is also unknown. Co-localization in the dendritic spines is observed for Purkinje cells in the dissociated culture system which lacks the spatial organization of neurons and glial cells seen in the molecular layer of the cerebellum (Tabata et al., 2000). Thus, the subcellular distribution of these receptors might be regulated by an intrinsic mechanism(s) of the Purkinje cells. The immuno-electron microscopic studies (Kulik et al., 2002; Rives et al., 2009) showed that mGluR1 is more concentrated on the annuli of dendritic spines than GABA$_B$R; a low density of GABA$_B$R is scattered to the dendritic shafts and spine tips. This indicates that mGluR1 might play a leading role in co-localization. mGluR1 itself might be stabilized in the dendritic spines by Homer proteins which may anchor the receptor to IP$_3$R on the endoplasmic reticulum [for review, see (Mikoshiba, 2007)]. Consistent with this notion, in rat cerebellar Purkinje cells, the full-length variant of mGluR1 (mGluR1a) which can interact with Homers is more concentrated at the annuli than a splice variant with short C-termini (mGluR1b) which cannot interact with Homers (Mateos et al., 2000)[for review on mGluR1-homer interaction, see (Fagni et al., 2000)].

3. *Another Example of G$_{i/o}$ Protein-Independent Action of GPCR in Cerebellar Purkinje Cells*

The above findings suggest a possibility that a G$_{i/o}$ protein-coupled receptor exerts its effect in a G$_{i/o}$ protein-independent manner in cerebellar Purkinje cells. Another line of our studies supports that this is the case for native preparations. Mouse Purkinje cells display an immunoreactivity against A1R in the dendritic structures including the spines (Kamikubo et al., Soc. Neurosci. Annual Meeting, 320.11, 2009). In cultured mouse Purkinje cells, [R]-N^6-(1-methyl-2-phenylethyl)adenosine (R-PIA), an A1R-selective agonist induces an inwardly rectifying K^+ current susceptible to PTX (Tabata et al., 2007). Thus, A1R is coupled to G$_{i/o}$ protein in Purkinje cells. We found that several A1R agonists induce the continuous reduction of the DHPG-evoked, mGluR1-mediated inward current and Ca^{2+} release from the intracellular store (Tabata et al., 2007). A1R indeed mediates this inhibition because it is not observed in cells derived from A1R-knockout mice. When an extremely high dose of an A1R agonist is applied, the reduction is preceded by a transient augmentation of the mGluR1-mediated inward current. A pretreatment with PTX abolishes selectively the transient augmentation but not the continuous reduction. The fact that $G_{i/o}$ protein-dependent and -independent A1R signaling produce the opposing effects on the same target (i.e., the mGluR1-mediated response) clearly shows the existence of distinct $G_{i/o}$ protein-dependent and -independent signaling pathways in Purkinje cells.

V. Possible Physiological Significance of $GABA_BR$-Mediated Effects on mGluR1 Signaling

A. Expansion of the Dynamic Range of Slow PF-PC EPSP

mGluR1 in the dendritic spines of cerebellar Purkinje cells is known to be involved in several neuronal events with much different time-spans. For example, mGluR1-mediated PF-PC EPSP may influence information integration in the Purkinje cell until it decays, taking several seconds (Batchelor & Garthwaite, 1997; Tempia et al., 1998). Various forms of mGluR1-mediated synaptic plasticity switch the mode of information integration to a new state which will last for seconds to days (Tabata & Kano, 2009). mGluR1 is required for the late phase of redundant CF synapse elimination undergoing for three weeks after the birth (Kano et al., 2008). In principle, $GABA_BR$-mediated modulatory effects could possibly influence all of these mGluR1-related events. However, we focus on the relatively short events because our current knowledge on functional interaction of $GABA_BR$ and mGluR1 is based on the acute *ex-vivo* measurements.

The slow PF-PC EPSP is generated immediately after mGluR1 receives glutamate from the PFs (see above). One of the features of the slow EPSP is that its amplitude gradually increases with the frequency and number of electrical stimuli given to PFs (Batchelor & Garthwaite, 1997; Tempia et al., 1998). mGluR1 at the annuli of the dendritic spines is thought to experience a relatively low concentration of glutamate because glutamate diffusing out of the synaptic cleft is partly taken up by perisynaptic transporters before reaching the vicinity of mGluR1 (Brasnjo & Otis, 2001). The concentration of glutamate around mGluR1 may rise with the amount of glutamate released from the PF terminal which reflects the frequency and number of action potentials discharged by the PF. $Ca^{2+}{}_{o}$- or ambient GABA-induced, $GABA_BR$-mediated mGluR1 sensitization broadens the working dose range of mGluR1 (see above), and this might emphasize the above feature of the slow EPSP. The graded amplitudes of slow PF-PC EPSP may encode motor commands and sensory information carried by the PFs with a high resolution, and this might contribute to precise motor control by Purkinje cells. Consistent with this notion, the TRPC3 channel-knockout mice, in which the slow EPSP is not generated, displays discoordinated gait (Hartmann et al., 2008).

B. Facilitation of Cerebellar LTD

1. Molecular Basis for Cerebellar LTD

Next, we discuss how $GABA_BR$ influences cerebellar LTD. Cerebellar LTD is a reduction of PF-PC EPSPs developing gradually and lasting over hours following repetitive, synchronized discharges (e.g., 100–600 sets at 1–4 Hz) of a certain group of PFs and a CF innervating the same Purkinje cell. Cerebellar LTD changes the manner of integration of motor commands and sensory information conveyed by PFs in the PC and may thereby serve as a physiological basis for cerebellar motor learning.

The molecular mechanisms underlying cerebellar LTD have been analyzed extensively [for review, see (Ito, 2002; Tabata & Kano, 2009)]. Pharmacological inhibition or genetic knock-out of mGluR1 abolishes cerebellar LTD (Aiba et al., 1994; Shigemoto et al., 1994). Purkinje cell-specific genetic rescue of mGluR1 in the mGluR1-KO mice restores cerebellar LTD (Ichise et al., 2000). These observations show that mGluR1 in Purkinje cells is essential for inducing cerebellar LTD. In response to glutamate released from the PFs, mGluR1 activates $G_{q/11}$ protein. $G_{q/11}$ protein in turn activates PLC. PLC produces DAG and IP_3. IP_3 opens IP_3R on the intracellular Ca^{2+} store. On the other hand, strong synaptic depolarization driven by the CF (see above) opens voltage-gated Ca^{2+} channels on the plasma membrane. The opening of these channels elevates the cytoplasmic Ca^{2+} concentration. This Ca^{2+} together with DAG activates protein kinase C (PKC). PKC and some other activity-dependent signaling molecules cooperatively facilitate endocytosis of AMPAR from the surface of the PF-innervated dendritic spines. The resultant decrease of AMPAR proteins reduces the efficacy of synaptic transmission at the PF-Purkinje cell synapses.

2. $GABA_BR$-Mediated mGluR1 Sensitization May Secure LTD Induction

$GABA_BR$ resides not only on the postsynaptic side but also on presynaptic side of PF-Purkinje cell synapses (Lujan & Shigemoto, 2006). Activation of presynaptic $GABA_BR$ is shown to reduce glutamate release from the PF terminals [see (Hirono et al., 2001)]. Thus, in cerebellar slice preparations, pharmacological manipulation of $GABA_BR$ may affect both the presynaptic inhibition and postsynaptic mGluR1 modulation. To focus on the effect mediated by postsynaptic $GABA_BR$, we performed experiments in cultured mouse Purkinje cells (Fig. 3A). In this preparation, a long-lasting reduction of Purkinje cell's responsiveness to glutamate can be induced by repetition of conjunctive dendritic application of glutamate and somatic depolarization (we hereafter call this phenomenon *in-vitro* LTD). These stimuli mimic synaptic inputs from the PFs and CF, respectively. The stimuli themselves are insensitive to any pharmacological agents. It is shown that glutamate-evoked intracellular signaling from mGluR1 to PKC and

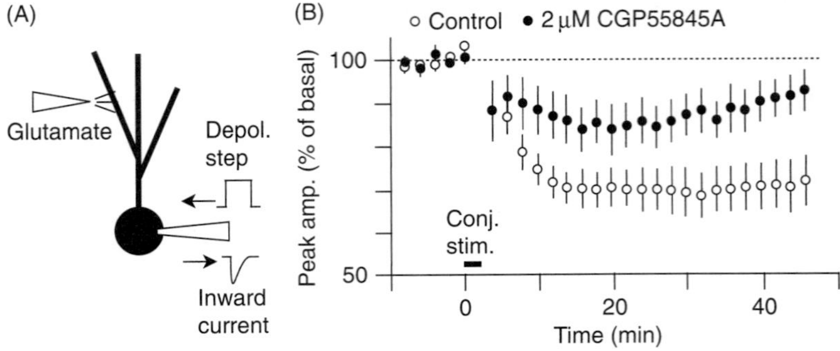

FIGURE 3 $GABA_BR$-mediated mGluR1 sensitization may secure LTD induction. (A) and (B) $Ca^{2+}{}_o$-induced $GABA_BR$ action may be required for induction of *in-vitro* LTD. As schematically shown in (A), glutamate was delivered locally to the dendrites of a cultured mouse cerebellar Purkinje cell, using iontophoresis. The glutamate-responsiveness of the Purkinje cell was evaluated by the peak amplitude of a glutamate-evoked inward current recorded from the soma under voltage clamp. To induce LTD, glutamate was applied for six times in conjunction with depolarizing voltage steps given to the soma. These two types of stimuli mimic synaptic inputs from the PFs and CF, respectively. CGP55845A, which is shown to block $Ca^{2+}{}_o$ binding to $GABA_BR$, abolishes *in-vitro* LTD (B). Plots indicate the peak amplitudes of the glutamate-evoked currents measured in the normal and CGP55845A-containing saline. Thick bar indicates the period of the conjunctive stimuli. The bath solution always contained $Ca^{2+}{}_o$ (2 mM). [Adapted from (Tabata et al., 2004), copyright (2004) National Academy of Sciences, U.S.A.]

depolarization-evoked Ca^{2+} influx through voltage-gated channels are sufficient factors for inducing *in-vitro* LTD (Linden et al., 1991).

In the presence of a physiological concentration (2 mM) of $Ca^{2+}{}_o$, several sets of conjunctive stimuli induce *in-vitro* LTD. The amplitude of glutamate-evoked inward currents measured under voltage clamp is reduced by ~30%, and this state is maintained over 45 min (Tabata et al., 2004) (Fig. 3B). When 2 μM of CGP55845A is added to the bath solution, the amplitude first decreases by ~15% and then gradually increases towards the basal level. This dose of CGP55845A is shown to completely abolish $Ca^{2+}{}_o$-induced, $GABA_BR$-mediated mGluR1 sensitization (see above, Fig. 2B)(Tabata et al., 2004). Our experiment employing simultaneous voltage-clamp recording and fluorometry (Tabata et al., 2004) showed that CGP55845A little affects the depolarization-evoked Ca^{2+} influx in cultured Purkinje cells. These results together suggest that induction of *in-vitro* LTD observed in the presence of $Ca^{2+}{}_o$ may be secured by $Ca^{2+}{}_o$-induced, $GABA_BR$-mediated mGluR1 sensitization.

3. *$GABA_BR$-Mediated mGluR1 Signaling Augmentation May Facilitate Cerebellar LTD*

We examined how $GABA_BR$-mediated, $G_{i/o}$ protein-dependent mGluR1 signal augmentation influences on *in-vitro* LTD in cultured mouse Purkinje cells (Kamikubo et al., 2007). A relatively high dose (3 μM) of GABA or baclofen increases the depth of *in-vitro* LTD from the control level (24%) to 33 or 45%, respectively (Fig. 4A). This dose of baclofen is shown to induce a GIRK current in a $G_{i/o}$ protein-dependent fashion (Tabata et al., 2005). A similar dose (1 μM) of baclofen is shown to augment mGluR1-mediated inward current evoked by a saturating dose of DHPG (see above). Moreover, GABA has a similar potency for GABA$_B$R to baclofen (see footnote 1). Thus, both agonists, at the dose used in the LTD experiments, are thought to induce $G_{i/o}$ protein-dependent mGluR1 signaling augmentation. GABA$_B$R indeed exerts the effect on *in-vitro* LTD via $G_{i/o}$ protein because a pretreatment with PTX completely abolishes it and because mastoparan, a $G_{i/o}$ protein agonist, mimics it.

GABA$_B$R facilitates *in-vitro* LTD by enhancing the mechanisms underlying the induction but not maintenance of *in-vitro* LTD (Kamikubo et al., 2007). The facilitation of *in-vitro* LTD is observed even when application of baclofen is restricted to the period of the conjunctive stimuli. By contrast, the facilitation is not observed when baclofen is applied after the conjunctive stimuli.

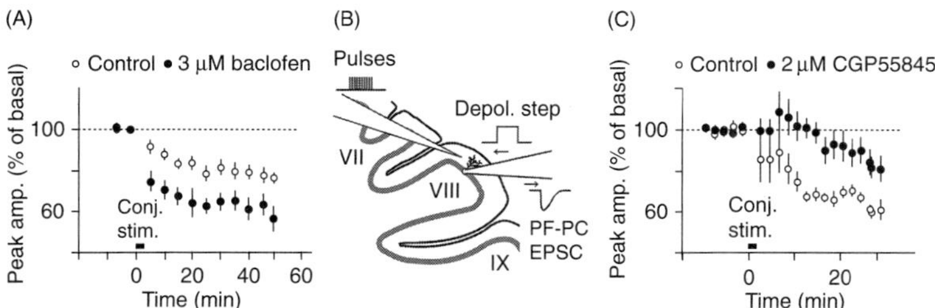

FIGURE 4 GABA$_B$R-mediated mGluR1 signaling augmentation may facilitate cerebellar LTD. (A) A relatively high dose of baclofen, which is shown to induce GABA$_B$R-mediated mGluR1 signaling augmentation, facilitates *in-vitro* LTD. Plots indicate the peak amplitudes of the glutamate-evoked currents as functions of time measured from cultured mouse cerebellar Purkinje cells in the normal and baclofen-containing saline. *In-vitro* LTD was induced by six sets of the conjunctive glutamate iontophoresis and depolarizing steps (see Fig. 3A). Thick bar indicates the period of the conjunctive stimuli. The bath solution always contained $Ca^{2+}{}_o$ (2 mM). [Adapted from (Kamikubo et al., 2007)]. (B) and (C) GABA$_B$R also mediates a facilitative effect on LTD *in situ*. As schematically shown in (B), PF-PC EPSCs were recorded from the soma of a Purkinje cell in the cortical lamina VII or VIII of a cerebellar slice under voltage clamp. To induce LTD, 30 trains of high-frequency electrical pulses (100 Hz for 100 ms) were given extracellularly to the PFs in conjunction with somatic depolarizing voltage steps which mimic CF synaptic inputs. Such high-frequency stimulation of PFs is shown to induce GABA spillover from the terminals of the PF-innervated inhibitory interneurons. Spilt GABA is thought to facilitate LTD because CGP55845 (2 μM), a GABA$_B$R-selective antagonist, attenuates LTD (C). Plots indicate the peak amplitudes of PF-PC EPSCs recorded with or without CGP55845 in the bath as functions of time. The thick bar indicates the period of the conjunctive stimuli. [Adapted from (Kamikubo et al., 2007)].

Among the sufficient factors to induce *in-vitro* LTD (see above), mGluR1-coupled intracellular signaling appears to be important for the facilitation (Kamikubo et al., 2007). Our fluorometry showed that baclofen augments DHPG-evoked, mGluR1-coupled Ca^{2+} release from the intracellular store. However, baclofen did not augment the amplitude of depolarization-evoked Ca^{2+} influx, the other factor for *in-vitro* LTD. These results suggest that $GABA_BR$-mediated mGluR1 signaling augmentation facilitates *in-vitro* LTD.

Facilitation by a high dose of GABA may occur for cerebellar LTD under physiological conditions. In mouse cerebellar slices, we applied bursts of high-frequency electrical stimuli to PFs in conjunction with depolarizing steps to the soma, which mimic CF inputs (Kamikubo et al., 2007)(Fig. 4B). Such a burst is shown to induce GABA spillover from the terminals of the PF-innervated inhibitory interneurons and $G_{i/o}$ protein-dependent mGluR1 signaling augmentation in Purkinje cells (Dittman & Regehr, 1997; Hirono et al., 2001). In the absence of test drugs, the amplitude of PF-PC EPSC is depressed by ~40% at ~30 min of the conjunctive stimuli (Fig. 4C). In the presence of 2 μM of CGP55845, which is thought to block GABA binding at most of the $GABA_BR$ proteins (see footnote 3), the amplitude is depressed by only ~20%. This result indicates that $G_{i/o}$ protein-dependent mGluR1 signaling augmentation may regulate cerebellar LTD under physiological conditions.

C. Balance Between the Effects Mediated by Pre- and Postsynaptic $GABA_BR$

Under physiological conditions *in vivo*, ambient GABA may bind to both pre- and postsynaptic $GABA_BR$. Thus, for assessment of the physiological role of $GABA_BR$ macroscopically at the level of whole cerebellum, it is important to consider synergy or interference between the effects mediated by pre- and postsynaptic $GABA_BR$.

In mouse cerebellar slices, baclofen attenuates AMPAR-mediated fast PF-PC EPSCs over a wide dose range of 0.3–10 μM (Hirono et al., 2001). Because AMPAR itself is not sensitive to baclofen in cultured mouse Purkinje cells (Tabata et al., 2005), this attenuation can be ascribed to a decrease of glutamate release from the presynaptic terminals of the PFs. The microdialysis studies (Bist & Bhatt, 2009; Paredes et al., 2009) suggest that ~120 nM of ambient GABA exists in unstimulated rodent cerebellum and that the concentration increases by 30–50% during eye-blink conditioning. A study in rat cerebellar slices (Dittman & Regehr, 1997) suggests that spillover from the neighboring inhibitory synapses may elevate the local concentration of ambient GABA around Purkinje cells to ~10 μM. Thus, $GABA_BR$ in Purkinje cells may experience 0.1–10 μM of GABA *in vivo*. This dose range of GABA may exert a similar action to that of the above dose range of baclofen because potency for $GABA_BR$ is similar between GABA and baclofen (see footnote 1). Therefore, presynaptic $GABA_BR$ is thought to mediate a decrease of glutamate release from the PF terminals under physiological conditions.

By contrast, baclofen induces different effects on mGluR1-mediated slow PF-PC EPSCs at lower and higher doses; that is, augmentation at doses of 0.1–0.3 μM, while attenuation at doses of 1–30 μM (Hirono et al., 2001). This indicates that the postsynaptic $GABA_BR$-mediated mGluR1 signaling augmentation may be saturated with a relatively low concentration of ambient GABA; the augmentation may be overwhelmed by the influence of the presynaptic $GABA_BR$-mediated decrease of glutamate with higher concentrations of ambient GABA.

The relationship between the frequency or number of action potentials discharged by the PFs and the amount of GABA spilt over from the terminal of the PF-innerved inhibitory interneuron has not been extensively investigated. However, some studies in mouse cerebellar slices suggest at least that under certain conditions, the postsynaptic $GABA_BR$-mediated action overwhelms the presynaptic $GABA_BR$-mediated action and as a net, produces a facilitative effect on mGluR1-mediated responses. In one of the studies (Hirono et al., 2001), a burst of electrical pulses to PFs (100 Hz, 100 ms) was used to evoke fast and slow PF-PC EPSCs. Under this condition, an application of CGP34629, a $GABA_BR$-selective antagonist, augments fast EPSCs while attenuating slow EPSCs. This change of fast EPSCs suggests that a certain level of GABA is spilt over from the interneuron terminals and somewhat decreases glutamate release from the PF terminals before the drug application. However, this level of GABA may also induce a strong postsynaptic $GABA_BR$-mediated augmentation of mGluR1-mediated slow EPSCs, and this is thought to mask the influence of the decrease of glutamate release. In the other study (Kamikubo et al., 2007), similar bursts were given to PFs in conjunction with somatic depolarizing steps to induce LTD (Fig. 4B, C). Addition of CGP55845, a $GABA_BR$-selective antagonist to the bath solution decreases the depth of LTD (Fig. 4C). This result suggests that in the absence of the drug, spilt GABA induces $GABA_BR$-mediated mGluR1 signaling augmentation which facilitates LTD induction.

VI. Conclusion

In cerebellar Purkinje cells, $GABA_BR$ resides around the postsynaptic membrane of the excitatory but not inhibitory synapses. $GABA_BR$ at such a site may respond to $Ca^{2+}{}_o$ and ambient GABA. In response to lower concentrations of these ligands, $GABA_BR$ increases the ligand affinity of mGluR1 via $G_{i/o}$ protein-independent pathway ($GABA_BR$-mediated mGluR1 sensitization). This effect might enable slow PF-PC EPSPs to efficiently encode the motor commands and sensory information carried by the PFs and might contribute to precise motor control by Purkinje cells. $GABA_BR$-mediated mGluR1 sensitization might also contribute to securing LTD induction in Purkinje cells. On the other hand, in response to higher concentrations of GABA, $GABA_BR$ augments mGluR1-coupled signaling via $G_{i/o}$ protein ($GABA_BR$-mediated mGluR1 signaling augmentation). This effect may facilitate cerebellar LTD, a form of synaptic plasticity crucial for cerebellar motor learning. These findings provide a new insight into the physiological roles of $GABA_BR$ in mammalian brains.

Issues to be addressed in future studies include explorations of (i) how $GABA_BR$ is closely associated with mGluR1, (ii) how $G_{i/o}$ protein activated by $GABA_BR$ modulates mGluR1-coupled intracellular signaling, and (iii) how $GABA_BR$ modulates the ligand affinity of mGluR1 without the aid of $G_{i/o}$ protein in cerebellar Purkinje cells. These explorations require detailed molecular dissection in native preparations such as cerebellar slices or cultures. The behavioral-level contributions of these $GABA_BR$-mediated effects also remain to be explored. There are several behavioral tasks that can assess motor learning that depends on mGluR1 activity in cerebellar Purkinje cells. These include rota-rod test, delayed eye-blink conditioning, and optokinetic reflex adaptation. Combination of such a task and pharmacological or genetic manipulation of functional interaction between $GABA_BR$, mGluR1, and their intermediate molecules would yield fruitful results.

Acknowledgments

Some of our studies described in this article were supported by the Grants-in-Aid for Scientific Research (21026011, 20500284, 20022025, 19045019 to T.T.; 17023021 and 21220006 to M.K.; Ministry of Education, Culture, Sports, Science and Technology, Japan) and the University of Toyama Presidential Grant (to T.T.).

Conflict of Interest Statement: T.T. and M.K. are employees of the University of Toyama and the University of Tokyo, respectively. Both universities are independent national corporations.

Non-standard abbreviations

A1R	adenosine A_1 receptor
AMPAR	amino-3-hydroxy-5-methyl-4-isoxazolepropionic acid-type glutamate receptor
CaR	Ca^{2+}-sensing receptor
CF	climbing fiber
DAG	diacylglycerol

DHPG	R,S-3,5-dihydroxyphenylglycine
EPSC	excitatory postsynaptic current
EPSP	excitatory postsynaptic potential
GABA	γ-amino butyric acid
GABA$_A$R	A-type GABA receptor
GABA$_B$R	B-type GABA receptor
GIRK channel	G protein-coupled inwardly rectifying K^+ channel
GPCR	G protein-coupled receptor
IP$_3$	inositol trisphosphate
IP$_3$R	IP$_3$ receptor
IPSP	inhibitory postsynaptic potential
LTD	long-term depression
mGluR1	type-1 metabotropic glutamate receptor
PF	parallel fiber
PF-PC EPSC	parallel fiber-Purkinje cell EPSC
PF-PC EPSP	parallel fiber-Purkinje cell EPSP
PKC	protein kinase C
PLC	phospholipase C
R-PIA	[R]-N^6-(1-methyl-2-phenylethyl)adenosine
TRPC3 channel	transient receptor potential C3 channel

References

Aiba, A., Kano, M., Chen, C., et al. (1994). Deficient cerebellar long-term depression and impaired motor learning in mGluR1 mutant mice. *Cell*, *79*, 377–388.

Batchelor, A. M., & Garthwaite, J. (1997). Frequency detection and temporally dispersed synaptic signal association through a metabotropic glutamate receptor pathway. *Nature*, *385*, 74–77.

Bettler, B., Kaupmann, K., Mosbacher, J., et al. (2004). Molecular structure and physiological functions of GABA$_B$ receptors. *Physiological Reviews*, *84*, 835–867.

Bichet, D., Haass, F. A., & Jan, L. Y. (2003). Merging functional studies with structures of inward-rectifier K^+ channels. *Nature Reviews. Neuroscience*, *4*, 957–967.

Bist, R., & Bhatt, D. K. (2009). The evaluation of effect of alpha-lipoic acid and vitamin E on the lipid peroxidation, gamma-amino butyric acid and serotonin level in the brain of mice (mus musculus) acutely intoxicated with lindane. *Journal of the Neurological Sciences*, *276*, 99–102.

Bockaert, J., Roussignol, G., Becamel, C., et al. (2004). GPCR-interacting proteins (GIPs): nature and functions. *Biochemical Society Transactions*, *32*, 851–855.

Bohlen, P., Huot, S., & Palfreyman, M. G. (1979). The relationship between GABA concentrations in brain and cerebrospinal fluid. *Brain Research*, *167*, 297–305.

Bouvier, M. (2001). Oligomerization of G-protein-coupled transmitter receptors. *Nature Reviews. Neuroscience*, *2*, 274–286.

Bowery, N. G., Bettler, B., Froestl, W., et al. (2002). International Union of Pharmacology. XXXIII. Mammalian gamma-aminobutyric $acid_B$ receptors: Structure and function. *Pharmacological Reviews*, *54*, 247–264.

Brasnjo, G., & Otis, T. S. (2001). Neuronal glutamate transporters control activation of postsynaptic metabotropic glutamate receptors and influence cerebellar long-term depression. *Neuron*, *31*, 607–616.

Cunningham, M. D., & Enna, S. J. (1996). Evidence for pharmacologically distinct GABAB receptors associated with cAMP production in rat brain. *Brain Research*, *720*, 220–224.

Dittman, J. S., & Regehr, W. G. (1997). Mechanism and kinetics of heterosynaptic depression at a cerebellar synapse. *Journal of Neuroscience*, *17*, 9048–9059.

Fagni, L., Chavis, P., Ango, F., et al. (2000). Complex interactions between mGluRs, intracellular Ca^{2+} stores and ion channels in neurons. *Trends in Neurosciences*, *23*, 80–88.

Falk, T., Meyerhof, W., Corrette, B. J., et al. (1995). Cloning, functional expression and mRNA distribution of an inwardly rectifying potassium channel protein. *FEBS Letters*, *367*, 127–131.

Fernandez-Alacid, L., Aguado, C., Ciruela, F., et al. (2009). Subcellular compartment-specific molecular diversity of pre- and post-synaptic GABA-activated GIRK channels in Purkinje cells. *Journal of Neurochemistry*, *110*, 1363–1376.

Fritschy, J.-M., Meskenaite, V., Weinmann, O., et al. (1999). $GABA_B$-receptor splice variants GB1a and GB1b in rat brain: Developmental regulation, cellular distribution and extrasynaptic localization. *European Journal of Neuroscience*, *11*, 761–768.

Galvez, T., Urwyler, S., Prezeau, L., et al. (2000). Ca^{2+} requirement for high-affinity gamma-aminobutyric acid (GABA) binding at $GABA_B$receptors: Involvement of serine 269 of the $GABA_BR1$ subunit. *Molecular Pharmacology*, *57*, 419–426.

Gama, L., Wilt, S. G., & Breitwieser, G. E. (2001). Heterodimerization of calcium sensing receptors with metabotropic glutamate receptors in neurons. *Journal of Biological Chemistry*, *276*, 39053–39059.

Hartmann, J., Blum, R., Kovalchuk, Y., et al. (2004). Distinct roles of galpha(q) and galpha11 for Purkinje cell signaling and motor behavior. *Journal of Neuroscience*, *24*, 5119–5130.

Hartmann, J., Dragicevic, E., Adelsberger, H., et al. (2008). TRPC3 channels are required for synaptic transmission and motor coordination. *Neuron*, *59*, 392–398.

Hartmann, J., & Konnerth, A. (2008). Mechanisms of metabotropic glutamate receptor-mediated synaptic signaling in cerebellar Purkinje cells. *Acta Physiologica (Oxford)*.

Hill, D. R., & Bowery, N. G. (1981). 3H-baclofen and 3h-GABA bind to bicuculline-insensitive GABA B sites in rat brain. *Nature*, *290*, 149–152.

Hirono, M., Yoshioka, T., & Konishi, S. (2001). $GABA_B$ receptor activation enhances mGluR-mediated responses at cerebellar excitatory synapses. *Nature Neuroscience*, *4*, 1207–1216.

Ichise, T., Kano, M., Hashimoto, K., et al. (2000). mGluR1 in cerebellar Purkinje cells essential for long-term depression, synapse elimination, and motor coordination. *Science*, *288*, 1832–1835.

Ige, A. O., Bolam, J. P., Billinton, A., et al. (2000). Cellular and sub-cellular localisation of $GABA_{B1}$ and $GABA_{B2}$ receptor proteins in the rat cerebellum. *Brain research. Molecular Brain Research*, *83*, 72–80.

Iizuka, M., Tsunenari, I., Momota, Y., et al. (1997). Localization of a G-protein-coupled inwardly rectifying K^+ channel, CIR, in the rat brain. *Neuroscience*, *77*, 1–13.

Ito, M. (2002). The molecular organization of cerebellar long-term depression. *Nature Reviews. Neuroscience*, *3*, 896–902.

Ito, M. (2006). Cerebellar circuitry as a neuronal machine. *Progress in Neurobiology*, *78*, 272–303.

Jan, L. Y., & Jan, Y. N. (1997). Voltage-gated and inwardly rectifying potassium channels. *Journal of Physiology*, *505*, 267–282.

Jingami, H., Nakanishi, S., & Morikawa, K. (2003). Structure of the metabotropic glutamate receptor. *Current Opinion in Neurobiology*, *13*, 271–278.

Jones, K. A., Borowsky, B., Tamm, J. A., et al. (1998). GABA$_B$ receptors function as a heteromeric assembly of the subunits GABA$_B$R1 and GABA$_B$R2. *Nature*, *396*, 674–679.

Kamikubo, Y., Tabata, T., Kakizawa, S., et al. (2007). Postsynaptic GABA$_B$ receptor signalling enhances LTD in mouse cerebellar Purkinje cells. *Journal of Physiology*, *585*, 549–563.

Kano, M., Hashimoto, K., & Tabata, T. (2008). Type-1 metabotropic glutamate receptor in cerebellar Purkinje cells: A key molecule responsible for long-term depression, endocannabinoid signalling and synapse elimination. *Philosophical Transactions of the Royal Society of London. Series B, Biological sciences*, *363*, 2173–2186.

Kaupmann, K., Malitschek, B., Schuler, V., et al. (1998). GABA$_B$-receptor subtypes assemble into functional heteromeric complexes. *Nature*, *396*, 683–687.

Kawato, M., Kuroda, T., Imamizu, H., et al. (2003). Internal forward models in the cerebellum: FMRI study on grip force and load force coupling. *Progress in Brain Research*, *142*, 171–188.

Kishimoto, Y., Fujimichi, R., Araishi, K., et al. (2002). mGluR1 in cerebellar Purkinje cells is required for normal association of temporally contiguous stimuli in classical conditioning. *European Journal of Neuroscience*, *16*, 2416–2424.

Kniazeff, J., Bessis, A. S., Maurel, D., et al. (2004). Closed state of both binding domains of homodimeric mGlu receptors is required for full activity. *Nature Structural & Molecular Biology*, *11*, 706–713.

Kubo, Y., Miyashita, T., & Murata, Y. (1998). Structural basis for a Ca^{2+}-sensing function of the metabotropic glutamate receptors. *Science*, *279*, 1722–1725.

Kubo, Y., & Tateyama, M. (2005). Towards a view of functioning dimeric metabotropic receptors. *Current Opinion in Neurobiology*, *15*, 289–295.

Kulik, A., Nakadate, K., Nyiri, G., et al. (2002). Distinct localization of GABA$_B$ receptors relative to synaptic sites in the rat cerebellum and ventrobasal thalamus. *European Journal of Neuroscience*, *15*, 291–307.

Kuner, R., Kohr, G., Grunewald, S., et al. (1999). Role of heteromer formation in GABA$_B$ receptor function. *Science*, *283*, 74–77.

Linden, D. J., Dickinson, M. H., Smeyne, M., et al. (1991). A long-term depression of AMPA currents in cultured cerebellar Purkinje neurons. *Neuron*, *7*, 81–89.

Llinas, R. R., Walton, K. D., & Lang, E. J. (2003). Cerebellum. In G. M. Shepherd (Ed.), *The synaptic organization of the brain* (pp. 271–309). New York: Oxford University Press.

Llinas, R., & Welsh, J. P. (1993). On the cerebellum and motor learning. *Current Opinion in Neurobiology*, *3*, 958–965.

Lujan, R., & Shigemoto, R. (2006). Localization of metabotropic GABA receptor subunits GABA$_B$1 and GABA$_B$2 relative to synaptic sites in the rat developing cerebellum. *European Journal of Neuroscience*, *23*, 1479–1490.

Mateos, J. M., Benitez, R., Elezgarai, I., et al. (2000). Immunolocalization of the mGluR1b splice variant of the metabotropic glutamate receptor 1 at parallel fiber-Purkinje cell synapses in the rat cerebellar cortex. *Journal of Neurochemistry*, *74*, 1301–1309.

Mikoshiba, K. (2007). IP_3 receptor/Ca^{2+} channel: From discovery to new signaling concepts. *Journal of Neurochemistry*, *102*, 1426–1446.

Miyashita, T., & Kubo, Y. (1997). Localization and developmental changes of the expression of two inward rectifying K^+-channel proteins in the rat brain. *Brain Research*, *750*, 251–263.

Miyashita, T., & Kubo, Y. (2000). Extracellular Ca^{2+} sensitivity of mGluR1alpha associated with persistent glutamate response in transfected CHO cells. *Receptors Channels*, *7*, 25–40.

Murer, G., Adelbrecht, C., Lauritzen, I., et al. (1997). An immunocytochemical study on the distribution of two G-protein-gated inward rectifier potassium channels (GIRK2 and GIRK4) in the adult rat brain. *Neuroscience*, *80*, 345–357.

Nakanishi, S. (1994). Metabotropic glutamate receptors: Synaptic transmission, modulation, and plasticity. *Neuron*, *13*, 1031–1037.

Ohyama, T., Nores, W. L., Murphy, M., et al. (2003). What the cerebellum computes. *Trends in Neurosciences*, *26*, 222–227.

Paredes, D. A., Cartford, M. C., Catlow, B. J., et al. (2009). Neurotransmitter release during delay eyeblink classical conditioning: Role of norepinephrine in consolidation and effect of age. *Neurobiology of Learning and Memory*, *92*, 267–282.

Park, D., Jhon, D. Y., Lee, C. W., et al. (1993). Activation of phospholipase C isozymes by G protein beta gamma subunits. *Journal of Biological Chemistry*, *268*, 4573–4576.

Quitterer, U., & Lohse, M. J. (1999). Crosstalk between Galpha$_i$- and Galpha$_q$-coupled receptors is mediated by Gbetagamma exchange. *Proceedings of the National Academy of Sciences of the United States of America*, *96*, 10626–10631.

Ritter, S. L., & Hall, R. A. (2009). Fine-tuning of GPCR activity by receptor-interacting proteins. *Nature Reviews Molecular Cell Biology*, *10*, 819–830.

Rives, M. L., Vol, C., Fukazawa, Y., et al. (2009). Crosstalk between GABA(B) and mGlu1a receptors reveals new insight into GPCR signal integration. *EMBO Journal*, *28*, 2195–2208.

Sarna, J. R., Marzban, H., Watanabe, M., et al. (2006). Complementary stripes of phospholipase cbeta3 and cbeta4 expression by Purkinje cell subsets in the mouse cerebellum. *Journal of Comparative Neurology*, *496*, 303–313.

Shigemoto, R., Abe, T., Nomura, S., et al. (1994). Antibodies inactivating mGluR1 metabotropic glutamate receptor block long-term depression in cultured Purkinje cells. *Neuron*, *12*, 1245–1255.

Tabata, T., Aiba, A., & Kano, M. (2002). Extracellular calcium controls the dynamic range of neuronal metabotropic glutamate receptor responses. *Molecular and Cellular Neuroscience*, *20*, 56–68.

Tabata, T., Araishi, K., Hashimoto, K., et al. (2004). Ca^{2+} activity at $GABA_B$ receptor constitutively promotes metabotropic glutamate signaling in the absence of GABA. *Proceedings of the National Academy of Sciences of the United States of America*, *101*, 16952–16957.

Tabata, T., Haruki, S., Nakayama, H., et al. (2005). GABAergic activation of an inwardly rectifying K+ current in mouse cerebellar Purkinje cells. *Journal of Physiology*, *563*, 443–457.

Tabata, T., & Kano, M. (2009). Synaptic plasticity in the cerebellum. In A. Lajtha, & K. Mikoshiba (Eds.). *Handbook of neurochemistry and molecular neurobiology* (3rd ed., pp. 63–86). USA: Springer IS.

Tabata, T., Kawakami, D., Hashimoto, K., et al. (2007). G protein-independent neuromodulatory action of adenosine on metabotropic glutamate signalling in mouse cerebellar Purkinje cells. *Journal of Physiology*, *581*, 693–708.

Tabata, T., Sawada, S., Araki, K., et al. (2000). A reliable method for culture of dissociated mouse cerebellar cells enriched for Purkinje neurons. *Journal of Neuroscience Methods*, *104*, 45–53.

Tempia, F., Miniaci, M. C., Anchisi, D., et al. (1998). Postsynaptic current mediated by metabotropic glutamate receptors in cerebellar Purkinje cells. *Journal of Neurophysiology*, *80*, 520–528.

Thach, W. T., Goodkin, H. P., & Keating, J. G. (1992). The cerebellum and the adaptive coordination of movement. *Annual Review of Neuroscience*, *15*, 403–442.

Waldmeier, P. C., Wicki, P., Feldtrauer, J. J., et al. (1994). GABA and glutamate release affected by GABA$_B$ receptor antagonists with similar potency: No evidence for pharmacologically different presynaptic receptors. *British Journal of Pharmacology*, *113*, 1515–1521.

Wise, A., Green, A., Main, M. J., et al. (1999). Calcium sensing properties of the GABA$_B$ receptor. *Neuropharmacology*, *38*, 1647–1656.

Guillermo Gonzalez-Burgos

Department of Psychiatry, Translational Neuroscience Program,
University of Pittsburgh School of Medicine, Pittsburgh, Pennsylvania, USA

GABA Transporter GAT1: A Crucial Determinant of $GABA_B$ Receptor Activation in Cortical Circuits?

Abstract

The GABA transporter 1 (GAT1), the main plasma membrane GABA transporter in brain tissue, mediates translocation of GABA from the extracellular to the intracellular space. Whereas GAT1-mediated uptake could generally terminate the synaptic effects of GABA, recent studies suggest a more complex physiological role. This chapter reviews evidence suggesting that in hippocampal and neocortical circuits, GAT1-mediated GABA

Advances in Pharmacology, Volume 58

1054-3589/10 $35.00
10.1016/S1054-3589(10)58008-6

transport regulates the electrophysiological effects of $GABA_B$ receptor ($GABA_BR$) activation by synaptically-released GABA. Contrasting with synaptic $GABA_A$ receptors, $GABA_BRs$ display high GABA binding affinity, slow G protein-coupled mediated signaling, and a predominantly extrasynaptic localization. Such $GABA_BR$ properties determine production of slow inhibitory postsynaptic potentials (IPSPs) and slow presynaptic effects. Such effects possibly require diffusion of GABA far away from the release sites, and consequently both $GABA_BR$-mediated IPSPs and presynaptic effects are strongly enhanced when GAT1-mediated uptake is blocked. Studies are reviewed here which indicate that $GABA_BR$-mediated IPSPs seem to be produced by dendrite-targeting GABA neurons including specifically, although perhaps not exclusively, the neurogliaform cell class. In contrast, the GABA interneuron subtypes that synapse onto the perisomatic membrane of pyramidal cells mostly signal via synaptic $GABA_ARs$. This chapter reviews data suggesting that neurogliaform cells produce electrophysiological effects onto other neurons in the cortical cell network via $GABA_BR$-mediated volume transmission that is highly regulated by GAT1 activity. Therefore, the role of GAT1 in controlling $GABA_BR$-mediated signaling is markedly different from its regulation of $GABA_AR$-mediated fast synaptic transmission.

I. Introduction

$GABA_B$ receptors ($GABA_BRs$) are prominent in hippocampal and neocortical circuits. Compared with receptors of the $GABA_A$ family ($GABA_ARs$), the distinctive properties of $GABA_BRs$ suggest that these receptors play different roles in regulating the flow of neural activity in cortical circuits. The physiological role of $GABA_BRs$ at the cellular level was identified in early studies [summarized in (Nicoll, 2004)] following the demonstration that GABA acts through receptors different from the bicuculline-sensitive sites (Bowery et al., 1980). Whereas substantial progress has been made in further defining the role of $GABA_BRs$ (Ulrich & Bettler, 2007), the regulation of $GABA_BR$ activation in physiological conditions is still poorly understood. Here, I review findings from experimental studies suggesting that in cortical microcircuits the activity of the GABA transporter 1 is an important determinant of $GABA_BR$ activation. The review focuses on the interplay between GAT1-mediated GABA uptake and $GABA_BRs$ which may play a role in determining the electrophysiological effects of synaptically-released GABA.

II. The Plasma Membrane GABA Transporter 1

The GABA transporter 1 (GAT1) is the predominant plasma membrane GABA transporter in cortical tissue (Guastella et al., 1990), where it is localized in neuronal and glial membranes near synapses (see Section III).

Below I review some of the properties of the GAT1 protein, its role in translocation of GABA across the plasma membrane, and some evidence on the mechanisms regulating GAT1 activity.

A. Properties of the GAT1 Protein and Translocation of GABA by GAT1

The gene encoding the amino acid sequence of GAT1 (*slc6a1*) belongs to the *slc6* family of genes for plasma membrane transporters, together with transporters for dopamine, serotonin, norepinephrine, and glycine, as well as the GABA transporters GAT2, GAT3, and GAT4 (Gether et al., 2006). The GAT1 *slc6a1* gene encodes a 599 amino acid protein with 12 transmembrane domains. To form functional transporter molecules, these proteins are probably assembled as dimers (Gether et al., 2006; Moss et al., 2009).

As the other *slc6* transporters, GAT1 translocates its substrate by co-transport with Na^+ in a process that requires Cl^- (Fig. 1A), although it is not clear whether Cl^- itself is translocated or not (Bicho & Grewer, 2005; Chen et al., 2004). GABA translocation occurs with a relatively slow kinetics, in the order of tens of GABA molecules per second (Bicho & Grewer, 2005; Chen et al., 2004). Such kinetics is significantly slower than the rapid kinetics of synaptic GABA-receptor binding and receptor channel gating (Farrant & Kaila, 2007). Interestingly, basal levels of GAT1 expression in the plasma membrane suggest a density of 300–500 transporters/μm^2 (Chiu et al., 2002; Wang & Quick, 2005). Therefore, depending on their exact membrane localization (see Section III.B. below), GAT1 molecules could provide ample high-affinity (IC50 ~6 nM) binding sites (Borden, 1996) to rapidly buffer synaptically-released GABA independently of its slow translocation rate (Diamond & Jahr, 1997).

GAT1 activity can be inhibited with various pharmacological compounds, some of which are highly selective to block GAT1 as opposed to other GABA transporters (Borden, 1996). Among the GAT1-selective commercially available inhibitors are NO711, also named NNC 711 (1,2,5,6-Tetrahydro-1-[2-[[(diphenylmethylene)amino]oxy] ethyl]-3-pyridinecarboxylic acid hydrochloride), and tiagabine ((*R*)-1-[4,4-bis(3-methylthiophen-2-yl)but-3-enyl] piperidine-3-carboxylic acid). Pharmacological inhibition of GAT1 activity is typically considered to decrease GABA uptake (Borden, 1996)—that is, the translocation of extracellular GABA into the intracellular compartment (Fig. 1A). GAT1 in GABAergic nerve terminals may therefore provide cytosolic GABA to rapidly refill synaptic vesicles without need of GABA synthesis de novo. Glial GAT1-mediated uptake (Fig. 1B) may produce a pool of GABA that is transformed into glutamine and thus is not readily available for neuronal release (Schousboe & Waagepetersen, 2006).

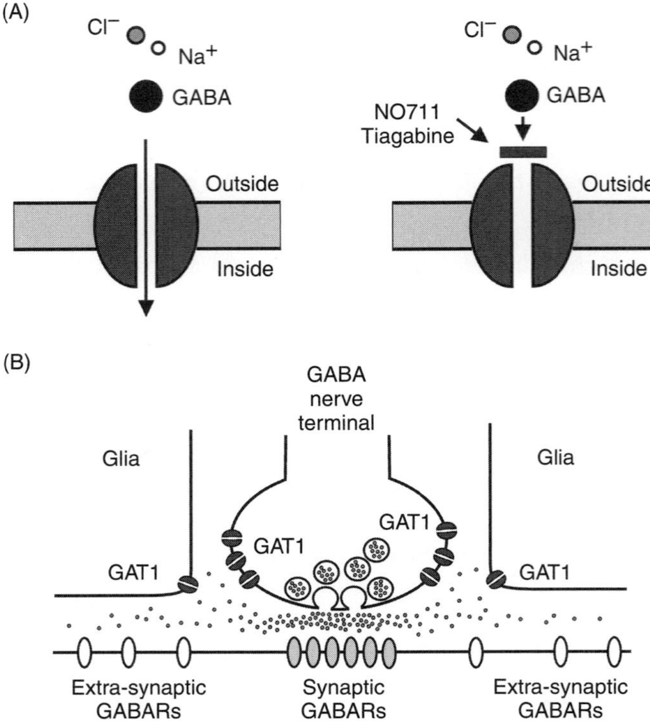

FIGURE 1 (A) Left, Scheme showing that GAT1 translocates GABA through the membrane via a process that requires Na^+ and Cl^-. Na^+ is co-transported by GAT1, contributing to an ionic current typically associated with GABA transport across the membrane. Whether Cl^- is transported as well, is not entirely clear. (A) Right, Several pharmacological compounds inhibit GABA transport in brain tissue, including NO711 and tiagabine, which are selective inhibitors of GAT1. GAT1 inhibitors are thought to mainly inhibit GABA uptake (that is, translocation through the plasma membrane from outside to inside), but can inhibit GABA translocation in the opposite direction as well, in conditions in which GAT1 mediates GABA release (see text). (B) Diagram showing a model, based on electron microscopy studies, for the localization of GAT1 at the ultrastructural level. The model suggests that GAT1 is expressed mainly in neurons, but some expression is observed in the glia as well. In neurons, GAT1 is apparently found mostly in the membrane of presynaptic GABA nerve terminals. The data regarding subcellular localization of GAT1 suggest that the transporter is found in extrasynaptic regions of the presynaptic nerve terminal membrane, relatively distant from the GABA release sites and postsynaptic receptors. This model assumes that neither GAT1-mediated uptake nor the binding of GABA to the transporters (before uptake begins) can effectively control the amount of GABA available to activate synaptic $GABA_A$ receptors after rapid release into the synaptic cleft.

However, glial glutamine can be released and taken up by neurons to be converted into GABA, possibly entering the releasable pool (Cherubini & Conti, 2001). In any event, a clear role of GAT1-mediated GABA uptake is to reduce extracellular GABA levels. An outstanding question is whether removal of extracellular GABA by GAT1-mediated uptake can influence the activation of GABA receptors by synaptically-released GABA, a topic that will be considered in sections below.

Although GAT1 is thought to mainly mediate GABA uptake, some evidence suggests that GAT1 may translocate GABA in the opposite direction, thus producing non-vesicular GABA release (Attwell et al., 1993). For instance, GAT1 may mediate GABA release when the intracellular Na^+ concentration is increased (Wu et al., 2007). Moreover, the GABA translocation direction can be reversed quickly enough to produce GAT1-mediated GABA release in response to action potentials (Wu et al., 2007). GAT1-mediated release is Ca^{2+}-independent and is insensitive to inhibitors of vesicular GABA release, but is inhibited by the pharmacological GAT1 inhibitor NO711 (Wu et al., 2007).

Under what conditions does GAT1-mediated uptake or release predominate? Whereas answering such question requires additional research, it is interesting to note that in humans, the GAT1 inhibitor tiagabine has antiepileptic effects (Borden, 1996; Laroche, 2007). GAT1 blockade with tiagabine also increases the tonic currents produced by either exogenous or endogenous GABA (Frahm et al., 2001). In addition, electrophysiological experiments in tissue from GAT1 knock-out mice are consistent with an increase in extracellular GABA concentration (Chiu et al., 2005; Jensen et al., 2003). These general observations suggest that in cortical networks GAT1 predominantly operates in the uptake direction (Fig. 1A).

B. Regulation of GAT1-Mediated GABA Transport

GAT1-mediated transport is regulated by controlling its membrane surface expression. Importantly, phosphorylation of several residues in its intracellular loops regulates trafficking of the GAT1 protein (Chen et al., 2004; Quick et al., 2004). A crucial factor regulating phosphorylation-dependent GAT1 trafficking is the extracellular concentration of GABA. A short-term (30 min) raise in extracellular GABA increases GAT1 surface expression by slowing the transporter internalization in a manner dependent on GAT1 phosphorylation on tyrosine (Bernstein & Quick, 1999). Conversely, long-term elevation of extracellular GABA (24 h) reduces GAT1 surface expression by a process involving $GABA_A$ receptor activation and GAT1 phosphorylation on serine (Hu & Quick, 2008). One possibility is that the chronic (hours-long) elevation of extracellular GABA is sensed as a sustained increase in the levels of GABA-mediated inhibition in the circuit. An increased demand for inhibition may then be transduced, by some homeostatic mechanism, into decreased GAT1 levels that potentiate tonic inhibition by increasing the levels of so-called ambient GABA (Ortinski et al., 2006). Regulation of GAT1 levels in the plasma membrane may be an important pathophysiological or compensatory mechanism in several psychiatric and neurological disorders including schizophrenia (Lewis et al., 2005), epilepsy (Fueta et al., 2003; Lee et al., 2006), cerebral ischemia (Frahm et al., 2004), and alcohol abuse (Hu et al., 2004). Not surprisingly, genetically-engineered GAT1 deficiency in mice produces several behavioral alterations (Cai et al., 2006; Chiu et al., 2005; Gong et al., 2009; Liu et al., 2007).

III. Cellular and Subcellular Localization of $GABA_B$Rs and GAT1

$GABA_B$Rs and GAT1 are relatively ubiquitous and co-localize in multiple cortical and subcortical brain regions. However, a critical issue for understanding the potential functional interactions between $GABA_B$Rs and GAT1 in cortical circuits is their localization at the ultrastructural level. Especially important is the localization of $GABA_B$Rs and GAT1 relative to the GABA release sites. Whereas some ultrastructural studies, reviewed below, examined the localization of $GABA_B$Rs or GAT1, no studies thus far examined whether these two proteins are co-localized, or found within short distance, at the ultrastructural level.

A. Localization of $GABA_B$Rs

Functional $GABA_B$Rs most likely are heteromers containing GABAB1 and GABAB2 subunits (Mohler & Fritschy, 1999; Perez-Garci et al., 2006). Immunogold labeling electron microscopy studies show that GABAB1 and GABAB2 subunit proteins co-localize in the plasma membrane of neurons at both pre- and post-synaptic sites (Perez-Garci et al., 2006).

$GABA_B$R subunits are found in the plasma membrane of nerve terminals at GABAergic/symmetric and glutamatergic/asymmetric synapses (Gonchar et al., 2001; Kulik et al., 2003; Vigot et al., 2006), suggesting that $GABA_B$Rs modulate GABA and glutamate release. Importantly, the plasma membrane of presynaptic nerve terminals in the central nervous system generally lacks excitatory or inhibitory synaptic contacts, raising questions about the source of GABA for presynaptic $GABA_B$R activation. In GABAergic terminals, some $GABA_B$Rs are found close to the GABA release sites (Gonchar et al., 2001), suggesting those $GABA_B$Rs act as autoreceptors. The $GABA_B$Rs localized in the plasma membrane of glutamatergic nerve terminals are by definition extrasynaptic, and must be activated by GABA diffusing from nearby GABAergic synapses.

In addition to nerve terminals, some $GABA_B$Rs are found in the postsynaptic membrane of GABAergic/symmetric synapses (Gonchar et al., 2001), where they may be directly activated by synaptically-released GABA. However, $GABA_B$Rs are undetectable in the postsynaptic membrane of many other GABA/symmetric synapses (Gonchar et al., 2001). $GABA_BR$ subunits can also be found in the plasma membrane of dendritic shafts and spines in the vicinity of glutamate/asymmetric synapses (Gonchar et al., 2001; Kulik et al., 2003; Kulik et al., 2006; Lopez-Bendito et al., 2002; Vigot et al., 2006). Although the dendritic membrane of pyramidal neurons (including the spines) may receive some GABA synapses (Andrasfalvy & Mody, 2006; Megias et al., 2001; Papp et al., 2001), GABAB1 subunits in dendritic spines and shafts are commonly found in regions of membrane lacking an inhibitory presynaptic partner (Kulik et al., 2003). This suggests that many of the dendritic $GABA_BRs$, which presumably regulate the excitatory effects of glutamate inputs, are extrasynaptic. The extrasynaptic $GABA_BR$ subunits found in dendrites typically co-cluster with Kir3-type K^+ channel proteins (Kulik et al., 2006). Importantly (see Section IV), $GABA_BR$ activation is commonly coupled, via G protein activation, to Kir K^+ channel gating, producing a hyperpolarizing K^+ current (Luscher et al., 1997).

B. Localization of GAT1

Electron microscopy studies of cortical tissue demonstrated that the GAT1 protein is expressed in the plasma membrane of neurons and glial cells. More specifically, GAT1 is frequently found in presynaptic nerve terminals that are part of an inhibitory/symmetric synaptic contact, as well as in glial processes near inhibitory synapses (Fig. 1B) (Conti et al., 1998; Conti et al., 2004; Mahendrasingam et al., 2003; Minelli et al., 1995; Minelli et al., 1996; Ong et al., 1998; Vitellaro-Zuccarello et al., 2003; Yan et al., 1997).

In contrast to the ultrastructural studies of $GABA_BR$ localization reviewed in Section III.A. (which employed immunogold particle methods), most electron microscopy studies of GAT1 localization employed immunocytochemistry based on diaminobenzidine peroxidation for detection of GAT1-positive structures. Using this method, the oxidation reaction product is relatively well confined inside the immunoreactive structure, but does not reveal the specific membrane compartments where the protein of interest is localized. Therefore, unlike the case of the $GABA_BR$ studies, the immunoperoxidase studies of GAT1 localization do not accurately describe if GAT1 proteins are preferentially localized in synaptic or extrasynaptic compartments of the neuronal plasma membrane. The few studies employing immunogold techniques to localize GAT1 in central nervous system tissue found label in the plasma membrane of glial process close to GABA/symmetric synapses as well as in the plasma membrane of presynaptic terminals of GABA/symmetric synapses, where the GAT1 labels are found extrasynaptically, relatively far from the GABA release sites (Conti et al., 1998; Mahendrasingam et al., 2003).

In summary, GAT1 proteins appear to be localized in GABAergic terminals as well as in glial processes near GABA synapses (Conti et al., 2004). The limited data available from immunogold labeling studies show that GAT1 proteins do not localize in very close proximity to GABA release sites or postsynaptic GABA receptors, suggesting that GAT1 activity mainly regulates the GABA concentration in the extracellular space outside the synaptic cleft (Fig. 1B).

IV. GABA$_B$R Activation by GABA Released from Endogenous Sources

In hippocampal and neocortical circuits GABA is synthesized and released by GABAergic interneurons. While a minority (20–30% of all neurons), interneurons are nevertheless a highly heterogeneous cell type, probably constituted by multiple subtypes (Markram et al., 2004). A crucial question to understand GABA$_B$R function in cortical circuits is therefore whether GABA$_B$Rs are typically activated by GABA released by all or only by specific subtypes of GABA neurons. Given their highly complex features, classification of GABA neurons is complicated and it is indeed still unclear how many different interneuron subtypes actually exist (Ascoli et al., 2008; Zaitsev et al., 2009). Before considering the importance of GAT1 activity for GABA$_B$R activation, in this Section I review the different types of electrophysiological responses produced by GABA$_B$Rs, when possible identifying the interneuronal source of GABA producing the response.

A. GABA$_B$R-Mediated Inhibitory Postsynaptic Potentials

Inhibitory postsynaptic potentials (IPSPs) can be elicited using extracellular electrical stimulation. Using this approach, pioneer electrophysiological studies demonstrated that synaptic GABA release can elicit IPSPs by activation of GABA$_B$Rs coupled to the gating of a K^+ current (Connors et al., 1988; Dutar & Nicoll, 1988). These GABA$_B$R-mediated IPSPs (GABA$_B$R-IPSPs) thus differ in their ionic basis from IPSPs produced by GABA$_A$R-gated Cl^- currents. GABA$_B$R-IPSPs are significantly slower in rise and decay times than GABA$_A$R-IPSPs (Fig. 2A) (Connors et al., 1988; Deisz, 1999; Dutar & Nicoll, 1988; McCormick, 1989). Such slower kinetics is expected, since GABA$_B$Rs belong to the G protein-coupled receptor family, gating K^+ channels through the various molecular steps involved in signaling via G protein stimulation.

Eliciting GABA$_B$R-IPSPs typically requires stronger extracellular stimulation compared with producing GABA$_A$R-IPSPs (Bertrand & Lacaille, 2001; Dutar & Nicoll, 1988; Scanziani, 2000). Stronger stimulus intensities increase the number of stimulated axons, and therefore the number of active synapses. Stronger stimuli may be required to elicit $GABA_B$R-IPSPs if at most GABA synapses transmission is purely $GABA_A$R-mediated and stimulating the uncommon $GABA_B$R-containing synapses is a low probability event. A second possibility is that many of the $GABA_B$Rs activated by synaptically-released GABA are extrasynaptic, their activation requiring pooling of extracellular GABA released by multiple synapses. In this case, stronger stimuli may be needed because $GABA_B$R activation requires recruiting synapses above a threshold or minimal number. These two scenarios are not mutually exclusive and actually are both supported by the electron microscopy data reviewed in Section III.A (Gonchar et al., 2001; Kulik et al., 2006; Vigot et al., 2006).

Using extracellular electrical stimulation, the identity of the stimulated axons (that is, to what GABAergic neuron subtype such axons belong) cannot be determined. Importantly, interneuron subtypes are broadly divided into those synapsing at or near the pyramidal cell soma (perisomatic-targeting) versus those making synapses in more distal dendrites (dendritic-targeting). Do $GABA_B$Rs mediate responses at specific compartments of the postsynaptic cell membrane? Applying focal extracellular stimulation, it is feasible to elicit in pyramidal cells IPSPs by selective activation of perisomatic-targeting versus dendritic-targeting GABAergic axons (Bertrand & Lacaille, 2001; Gonzalez-Burgos et al., 2009; Gulledge & Stuart, 2003). Using focal stimulation, we found recently that $GABA_B$R-IPSPs are preferentially evoked by stimulation of dendritic-targeting synapses (Gonzalez-Burgos et al., 2009). Importantly, such dendritic $GABA_B$R-IPSPs co-exist with dendritic $GABA_A$R-IPSPs and are produced with the same stimulus strength (Gonzalez-Burgos et al., 2009). These results suggest that $GABA_B$Rs predominantly mediate signaling by interneurons that target pyramidal cell dendrites.

B. Postsynaptic $GABA_B$R Activation During Unitary Synaptic Transmission

Although synaptic stimulation with extracellular electrodes may in cases be targeted to specific membrane compartments, this approach still has multiple limitations. For instance, it cannot determine whether the $GABA_A$R-IPSPs and $GABA_B$R-IPSPs originate at the same or separate synapse populations (Nurse & Lacaille, 1997). A state-of-the-art method to stimulate well-identified GABAergic synaptic connections is simultaneous recording from synaptically-connected pairs of neurons (Debanne et al., 2008). In these experiments, current is injected into a neuron to elicit an action potential while recording membrane potential or current simultaneously in another cell (Fig. 2B). Action potentials produced in a GABAergic interneuron synaptically-connected onto a postsynaptic cell produce responses called "single-axon" or "unitary" IPSCs/IPSPs. Unitary IPSCs/

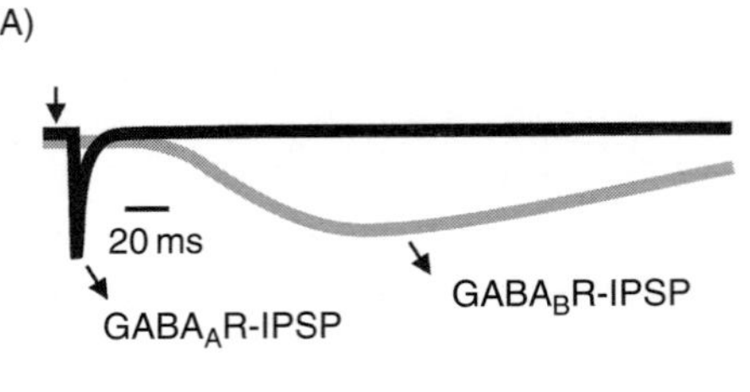
(A)
20 ms
GABA$_A$R-IPSP
GABA$_B$R-IPSP

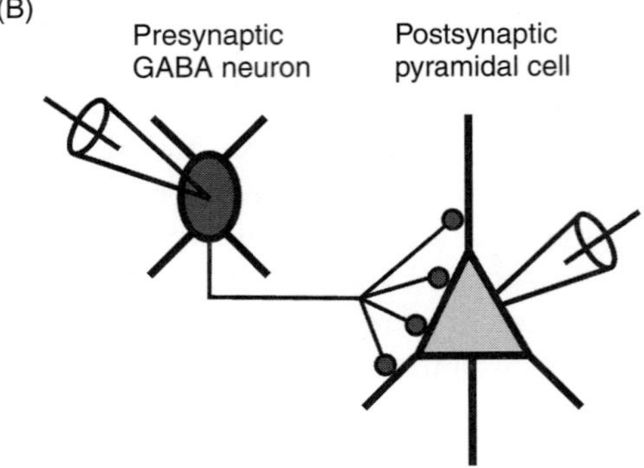
(B)
Presynaptic
GABA neuron
Postsynaptic
pyramidal cell

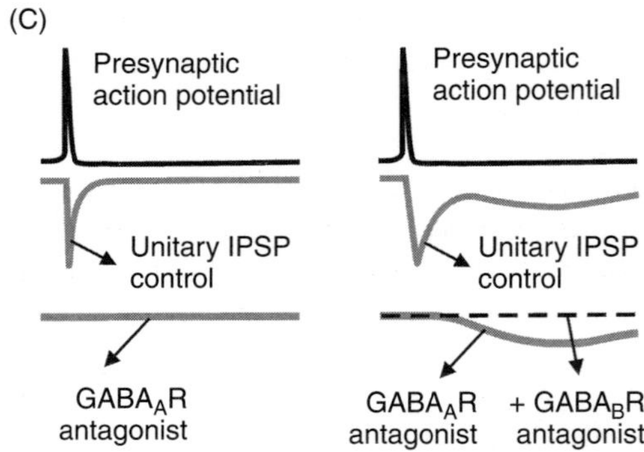
(C)
Presynaptic
action potential
Unitary IPSP
control
GABA$_A$R
antagonist
Presynaptic
action potential
Unitary IPSP
control
GABA$_A$R
antagonist
+ GABA$_B$R
antagonist

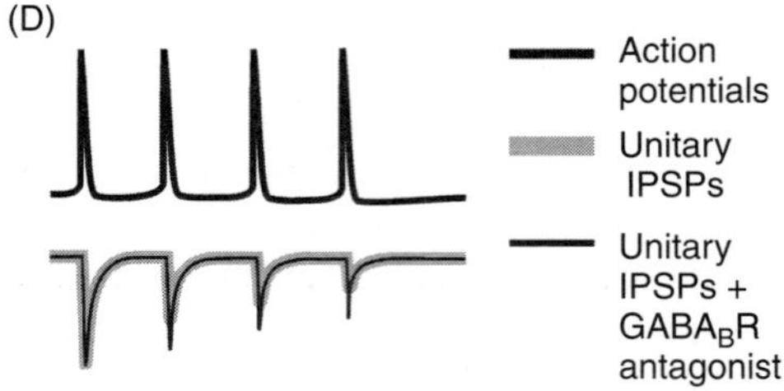
(D)
Action
potentials
Unitary
IPSPs
Unitary
IPSPs +
GABA$_B$R
antagonist

IPSPs therefore result from activation of a single GABA neuron, although they typically involve multiple synaptic contacts between the presynaptic axon and the postsynaptic cell (Fig. 2B). In paired recordings, the recorded neurons can be identified by their morphology, biochemical and electrical properties. Moreover, using GABA receptor antagonists, whether $GABA_BRs$ or $GABA_ARs$ mediate the unitary IPSCs/IPSPs can be determined (Fig. 2C).

In recordings from synaptically-connected pairs, $GABA_AR$ antagonists completely abolish the unitary responses evoked by a variety of interneuron subtypes (Fig. 2C, left), including the fast-spiking basket cells and chandelier neurons in the hippocampus (Ali et al., 1999; Bartos et al., 2002; Buhl et al., 1994; Buhl et al., 1995; Buhl et al., 1995; Doischer et al., 2008; Thomson & Destexhe, 1999; Thomson et al., 2000) and neocortex (Gonzalez-Burgos et al., 2005; Szabadics et al., 2006; Szabadics et al., 2007; Tamas et al., 1997; Tamas et al., 1998; Tamas et al., 2003; Thomson & Destexhe, 1999; Thomson et al., 1996). Non-fast-spiking basket cells similarly produce fast unitary IPSCs/IPSPs consistent with being mediated by $GABA_ARs$ (Galarreta et al., 2008; Glickfeld & Scanziani, 2006; Glickfeld et al., 2008; Hefft & Jonas, 2005). Basket cells (fast-spiking and non-fast spiking) and chandelier neurons innervate pyramidal

---

FIGURE 2 (A) Scheme comparing the fast time-course of $GABA_AR$-IPSPs (black trace) with the significantly slower time-course of $GABA_BR$-IPSPs (gray trace). Note that relative to the stimulus (arrow), the onset of the $GABA_BR$-IPSPs is substantially delayed compared to the $GABA_AR$-IPSP. In addition, the rising phase of the $GABA_BR$-IPSP is much slower and its decay time is at least ten-times slower compared with the $GABA_AR$-IPSP. (B) A drawing illustrating the typical experimental arrangement used to obtain electrophysiological recordings from synaptically-connected pairs of neurons. In this example, a presynaptic GABA neuron makes a synaptic connection, composed of four synaptic contacts, onto a postsynaptic pyramidal cell. The glass electrode pipettes can be used to inject stimuli or to record membrane potential or current from either neuron. The IPSPs produced in this kind of study are called single-axon or unitary IPSPs, even though they involve multiple synaptic contacts. (C) Examples of $GABA_AR$- and $GABA_BR$-mediated unitary IPSPs observed in recordings from synaptically-connected pairs. The example on the left shows a case in which the presynaptic GABA neuron produces in the postsynaptic cell a unitary IPSP purely mediated by $GABA_ARs$. The great majority of GABA neuron subtypes identified thus far produce unitary IPSPs of this kind, including basket cells that target the perisomatic region of the pyramidal cell membrane, as well as some dendrite-targeting GABA neuron subtypes. The example on the right illustrates a unitary IPSP produced by the combined activation of $GABA_A$ and $GABA_B$ receptors. Only interneurons of the neurogliaform cell subtype have so far been clearly identified to produce unitary IPSPs with these characteristics. Note that the time-course of the $GABA_AR$-IPSP produced by neurogliaform cells is substantially slower than the time-course of $GABA_AR$-IPSPs produced by other interneuron subtypes. (D) Illustration of the autoreceptor effects produced on unitary $GABA_AR$-IPSPs by presynaptic $GABA_BR$ in some types of GABA synapses. Note that, in control conditions, the strength of subsequent $GABA_AR$-IPSPs produced by stimulus trains shows progressive depression. After application of a $GABA_BR$ antagonist, the strength of subsequent $GABA_AR$-IPSPs increases, showing that $GABA_BRs$ contribute to the IPSP depression. The initial unitary IPSP in the train is, however, unaltered by the $GABA_BR$ antagonist, showing that autoreceptor effects are produced by GABA released by a recently preceding stimulus.

cells onto the perisomatic membrane compartment (Somogyi et al., 1998). Thus, perisomatic-targeting GABA neurons in general transmit via $GABA_A$Rs without significant contribution of postsynaptic $GABA_B$Rs (Freund & Katona, 2007). In addition, the unitary IPSCs/IPSPs produced by various dendrite-targeting interneuron subtypes in hippocampus and neocortex are similarly mediated by postsynaptic $GABA_A$R activation (Ali & Thomson, 2008; Bertrand & Lacaille, 2001; Kapfer et al., 2007; Murayama et al., 2009; Vida et al., 1998).

Whereas most GABA neuron subtypes produce purely $GABA_A$R-mediated unitary IPSCs/IPSPs (Fig. 2C, left), recent studies identified an interneuron subtype producing unitary $GABA_B$R-IPSPs. In paired recordings, interneurons of the neurogliaform cell class produce in pyramidal neurons slow unitary IPSPs that are never completely abolished by $GABA_A$R antagonists (Fig. 2C, right) (Tamas et al., 2003). Moreover, unitary IPSPs produced by neurogliaform cells after $GABA_A$R blockade rise and decay slowly, have long onset latency and are blocked by $GABA_B$R antagonists (Tamas et al., 2003). Therefore, neurogliaform cells produce unitary IPSPs simultaneously mediated by $GABA_A$Rs and $GABA_B$Rs (Fig. 2C, right). Interestingly, when stimulated repetitively, GABA release by neurogliaform cells exhausts quickly, eventually depressing almost completely the unitary IPSP (Tamas et al., 2003).

Are neurogliaform cells the only interneuron subtype producing $GABA_B$R-IPSPs by synaptic GABA release? The quick and strong depression of GABA release observed during repetitive stimulation of neurogliaform cells differs from the finding that repetitive extracellular stimulation produces $GABA_B$R-IPSPs more efficiently than single stimuli (Gonzalez-Burgos et al., 2009; Isaacson et al., 1993; Scanziani, 2000). These findings suggest that in addition to neurogliaform cells, other interneuron subtypes produce unitary $GABA_B$R-IPSPs. In one study, interneurons different from the neurogliaform cell class produced, exclusively during repetitive stimulation, slow unitary IPSPs insensitive to $GABA_A$R antagonists (Thomson & Destexhe, 1999). Moreover, in some cases, the slow $GABA_B$R-IPSPs were observed in the absence of a $GABA_A$R-mediated IPSP (Thomson & Destexhe, 1999), contrasting with the neurogliaform cell unitary IPSPs mediated by both $GABA_B$R and $GABA_A$Rs (Tamas et al., 2003). Such results confirm that some non-neurogliaform interneurons produce $GABA_B$R-IPSPs, although the particular cell class remains to be identified (Thomson & Destexhe, 1999). Recently, it was reported that fast-spiking basket neurons, thought to produce exclusively $GABA_A$R-mediated unitary IPSPs, can elicit $GABA_B$R-IPSPs (Oswald et al., 2009), although only if the interneuron is stimulated at high frequency (80 Hz) (Oswald et al., 2009).

C. Presynaptic $GABA_B$R Effects

As described in Section III.A., $GABA_B$Rs are localized in presynaptic nerve terminals at both GABA and glutamate synapses. At inhibitory synapses, presynaptic $GABA_BRs$ may downregulate GABA release when activated by GABA released from the same nerve terminal—in other words, acting as autoreceptors (Fig. 2D). Alternatively, presynaptic $GABA_BRs$ can be activated by GABA released from adjacent GABA synapses or by ambient GABA. The activation of presynaptic $GABA_BRs$ at glutamate synapses must depend on heterosynaptic effects, given that glutamate synapses do not release GABA. An exception is the case of the mossy fiber synapses in the immature hippocampus, which are glutamatergic and also release GABA. Presynaptic effects of GABA at such synapses are mostly $GABA_AR$-mediated (Alle & Geiger, 2007), but $GABA_BRs$ may contribute as well (Safiulina & Cherubini, 2009).

Presynaptic and postsynaptic effects of $GABA_BR$ activation appear to be mediated by different mechanisms (Deisz et al., 1997). For instance, presynaptic $GABA_BR$ effects are intact in synapses of Kir K^+ channel knock-out mice that lack postsynaptic $GABA_BR$ effects (Luscher et al., 1997). Several studies have actually shown that $GABA_BRs$ inhibit voltage-dependent Ca^{2+} channels, an effect that may mediate presynaptic regulation of transmitter release by $GABA_BRs$ (Mintz & Bean, 1993; Takahashi et al., 1998; Thompson et al., 1993). At some GABA synapses, however, presynaptic $GABA_B$ autoreceptors inhibit transmitter release but do not regulate action potential-evoked presynaptic Ca^{2+} transients (Price et al., 2008).

Autoreceptor effects at GABA synapses are demonstrated by stimulating the synapses repetitively so that GABA released by a preceding stimulus regulates subsequent release (Fig. 2D). If presynaptic $GABA_BRs$ inhibit transmitter release, then the size of subsequent responses relative to preceding ones should increase when $GABA_BRs$ are blocked (Fig. 2D). Such an effect was demonstrated using extracellular stimulation of GABAergic axons eliciting $GABA_AR$-mediated IPSCs in hippocampal pyramidal cells (Cobb et al., 1999; Davies & Collingridge, 1993; Davies et al., 1990). $GABA_BR$ antagonists increase the amplitude of subsequent unitary IPSPs/IPSCs produced by repetitive stimulation of neurogliaform cells, consistent with downregulation of release via autoreceptors (Olah et al., 2009; Price et al., 2005, 2008). In contrast, $GABA_BR$ autoreceptor effects are absent at connections made by fast-spiking basket neurons (Olah et al., 2009).

Presynaptic $GABA_BRs$ may also regulate GABA release when activated tonically—that is, independent of the effects of GABA released by a recently preceding stimulus. The sources of GABA for tonic effects are difficult to evaluate, but may include GABA released by other GABA synapses, ambient GABA, or eventually GABA released homosynaptically at low frequency (tonic autoreceptor effects). In the last case, the effects of GABA released by a previous stimulus must persist until the next stimulus arrives, even if the inter-stimulus interval is relatively long during low-frequency stimulation. Some studies showed that $GABA_BR$ antagonists enhance $GABA_AR$-mediated

IPSCs evoked with low-frequency stimulation (Buhl et al., 1994; Buhl et al., 1995; Gonzalez-Burgos et al., 2009; Jensen et al., 2003; Lei & McBain, 2003; Price et al., 2008). However, in other cases $GABA_BR$ antagonists failed to modulate $GABA_AR$-IPSCs evoked at low frequency, even if the $GABA_AR$-IPSCs are strongly depressed by presynaptic effects of the $GABA_BR$ agonist baclofen (Neu et al., 2007). These results suggest that tonic activation of presynaptic $GABA_BRs$ by endogenous GABA is synapse type-specific and is not observed in some synapses at which presynaptic $GABA_BRs$ are nevertheless readily activated by exogenous agonists.

As described in Section III.A., $GABA_BRs$ are also localized in glutamate nerve terminals, where $GABA_BRs$ may produce downregulation of glutamate release (Davies & Collingridge, 1996; Isaacson et al., 1993; Lei & McBain, 2003; Pan et al., 2009; Porter & Nieves, 2004). Glutamate release may be actually modulated by presynaptic $GABA_BRs$ activated by endogenous GABA. For instance, burst stimulation of inhibitory inputs just prior to stimulation of glutamate inputs produces a depression of glutamate transmission that is reversed by a $GABA_BR$ antagonist (Davies & Collingridge, 1996; Isaacson et al., 1993). In the absence of prior stimulation of inhibitory inputs, $GABA_BR$ antagonists do not affect EPSCs, showing absence of tonic activation of presynaptic $GABA_BRs$ at glutamate terminals (Lei & McBain, 2003). Whereas various sources may provide endogenous GABA to activate presynaptic $GABA_BRs$ at glutamate synapses, a recent study showed that GABA released by neurogliaform cells is a prominent source of GABA for such presynaptic effects (Olah et al., 2009). More specifically, properly timed stimulation of GABA release by neurogliaform cells depressed, via presynaptic $GABA_BRs$, excitatory transmission between pyramidal neurons or between pyramidal cells and interneurons (Fig. 3B) (Olah et al., 2009).

V. GAT1 Activity and $GABA_BR$ Activation

The functional relevance of GAT1-mediated GABA transport has been assessed by either pharmacological blockade of GAT1 activity or in genetically-engineered GAT1-deficient mice. Such mice have significant behavioral alterations (not reviewed here), which suggest that in cortical circuits GAT1 plays a crucial role that cannot be compensated by other GABA transporters (Cai et al., 2006; Chiu et al., 2005; Gong et al., 2009). Below, I review the evidence from studies addressing the functional significance of GAT1 activity for GABA-mediated inhibition in cortical circuits. The effects of GAT1-mediated transport on $GABA_AR$-IPSCs are reviewed first, since they are crucial to understand the role of GAT1 for $GABA_BR$ activation.

A. GAT1-Mediated Uptake Prevents GABA Spillover onto Synaptic $GABA_A$Rs

A functional role typically assigned to GABA transporter-mediated uptake is the termination of the postsynaptic effect of GABA following presynaptic transmitter release. However, $GABA_A$R-IPSCs produced at single synapses are not altered in time-course or amplitude after GAT1 blockade with NO711 or in GAT1-deficient mice (Bragina et al., 2008; Gonzalez-Burgos et al., 2009; Jensen et al., 2003; Overstreet & Westbrook, 2003). These results are opposed to the prediction that if GAT1 normally controls the time-course and magnitude of $GABA_A$R activation then GAT1 blockade would enhance and prolong IPSCs.

Unitary IPSCs produced by individual fast-spiking GABA neurons typically are not altered by GAT1 blockade (Overstreet & Westbrook, 2003; Szabadics et al., 2007). However, if a fast-spiking GABA neuron unitary IPSC is specifically mediated by adjacent synapses, it is significantly prolonged by GAT1 blockade (Fig. 3A), suggesting that proximity of the synapses involved is a critical determinant of GAT1 effects (Overstreet & Westbrook, 2003). Interestingly, $GABA_A$R-IPSCs/IPSPs evoked by extracellular stimulation, which activates multiple axons (and therefore multiple synapses) together, are prolonged by GAT1 blockade or in GAT1-deficient mice (Bragina et al., 2008; Caillard et al., 1998; Gonzalez-Burgos et al., 2009; Keros & Hablitz, 2005; Overstreet & Westbrook, 2003; Thompson & Gahwiler, 1992). Importantly, GAT1 blockade or knock-out increase the duration of $GABA_A$R-IPSCs evoked by multiple-synapse stimulation, but do not increase their amplitude (Fig. 3A) (Bragina et al., 2008; Gonzalez-Burgos et al., 2009; Overstreet & Westbrook, 2003).

The absence of GAT1 effects on single-synapse responses is consistent with the idea that GAT1 does not regulate the time-course of synaptic $GABA_A$R activation following GABA release into an individual synaptic cleft. Possibly, within-synapse GABA diffusion and $GABA_A$R channel gating are significantly faster than the relatively slow speed of GABA uptake by GAT1 (see Section II). Via high-affinity binding, GAT1 potentially could rapidly buffer GABA molecules released into the synaptic cleft independent of a slow translocation rate. However, ultrastructural studies suggest an extra- or peri-synaptic localization of GAT1 (Figs. 1B and 3A) which may preclude effective GABA buffering by binding.

The prolongation by GAT1 blockade of $GABA_A$R-IPSC duration when multiple adjacent synapses are stimulated together may be the consequence of GAT1-controlled GABA diffusion between synapses, or spillover (Fig. 3A). Possibly, GABA spillover prolongs the $GABA_A$R-IPSCs but does not increase their peak amplitude because IPSC amplitude is determined by rapid within-synapse GABA diffusion (Farrant & Kaila, 2007; Mozrzymas, 2004). In contrast, diffusion of GABA between adjacent synapses when

GAT1 is blocked, probably is not rapid enough to contribute to the IPSC peak (Barbour, 2001), thus only increasing the IPSC decay time.

Overall, GAT1 may help preventing the effects of GABA spillover, thus preserving synapse independence—that is, the temporal and spatial specificity of fast $GABA_A$R-mediated transmission. In addition, GAT1 may prevent the densensitization of synaptic $GABA_A$Rs that in the absence of GAT1 activity may be exposed to ambient GABA. For groups of GABA synapses that are sufficiently distant, spillover is unlikely and $GABA_A$R-IPSCs

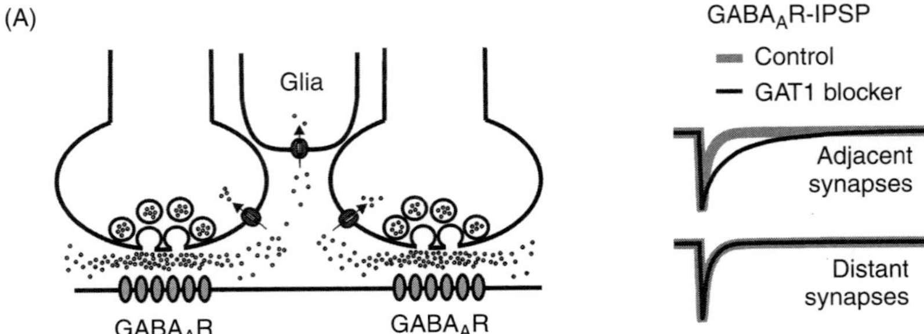

FIGURE 3 (A) Left, A scheme showing the effects of GAT1-mediated uptake on $GABA_A$R activation when GABA is released by two adjacent synapses. Note that, under this model, at each synapse $GABA_A$Rs are readily activated by GABA released into the synaptic cleft by release sites directly opposed to the synaptic receptors. GAT1 localized in the nerve terminals, and possibly in the glia as well, uptakes most of the GABA that otherwise may diffuse and reach the adjacent synapse. This model suggests that the main function of GAT1 is controlling the effects on synaptic $GABA_A$Rs of GABA diffusing between synapses. (A) Right, Example traces showing the effect of GAT1 blockade on IPSPs produced by stimulation of adjacent synapses. Note that GAT1 blockade prolongs the decay time of the IPSPs when the synapses are adjacent. However, if the stimulated synapses are distant, the likelihood of spillover effects is very low and the IPSPs are insensitive to GAT1 blockade. Note that the duration, measured in control conditions, of IPSPs produced by adjacent synapses (thus prolonged by GAT1 blockade) is identical to the duration of IPSPs produced by distant synapses (that are insensitive to GAT1 blockade). (B) Left, A model for $GABA_B$R activation by synaptically-released GABA based on data suggesting that $GABA_B$R activation is mainly produced by GABA released by dendrite-targeting neurogliaform cells. These cells have been actually suggested to activate $GABA_B$Rs and $GABA_A$Rs via GABA spillover given that most of their terminals do not form actual synaptic junctions (see text). Such absence of synaptic junctions is represented in the scheme as a wider space separating release sites and the receptors activated by GABA. Note that GAT1 localized in the neurogliaform cell terminals, in terminals of other GABA neurons as well (not shown in the scheme) and in glia, significantly regulates activation of GABA receptors by GABA released by neurogliaform cells. Such GABA may also reach, under GAT1 control, presynaptic $GABA_B$Rs localized in excitatory glutamatergic synapses. (B) Right, Example traces showing that, under GAT1 control, neurogliaform cell-released GABA produces slow IPSPs by activation of $GABA_B$Rs, as well as downregulation of EPSPs via presynaptic $GABA_B$R activation. Note that, as suggested by recent data, when a neurogliaform cell is activated shortly before EPSPs are elicited, the release of GABA mostly affects, via presynaptic $GABA_B$R activation, subsequent EPSPs but not the initial EPSP. This suggests very specific mechanisms of action that have not been yet investigated.

(Continued)

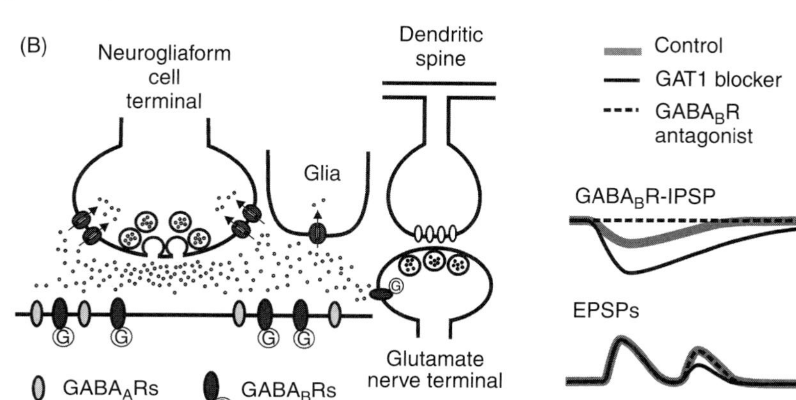

FIGURE 3 (*Continued*).

produced by activation of such synapses are not affected by GAT1 blockade (Fig. 3A). Conversely, GAT1-mediated uptake may be critical to prevent the effects of GABA spillover and $GABA_A$R-IPSCs mediated by adjacent synapses are thus prolonged by GAT1 blockade (Fig. 3A). Interestingly, we have recently reported that GAT1-mediated uptake may completely prevent the effects of spillover on $GABA_A$R-IPSPs (Gonzalez-Burgos et al., 2009). For instance, we found that IPSPs that are sensitive to GAT1 blockade (thus probably mediated by adjacent synapses susceptible of spillover), when measured with GAT1 activity intact, have similar duration compared with IPSPs that are GAT1 blockade-insensitive (thus probably mediated by distant synapses not susceptible to spillover).

B. GAT1 Activity and $GABA_B$R-Mediated IPSPs

The data reviewed in Section V.A suggest that GAT1 activity prevents the activation of synaptic $GABA_A$Rs by GABA released at heterologous nearby synapses. Although GAT1 may completely prevent the heterosynaptic effects of GABA diffusion (Gonzalez-Burgos et al., 2009), it is possible that GAT1 does not completely decrease GABA spillover, but reduces it to levels below those necessary to activate synaptic $GABA_A$Rs, which have low affinity for GABA (Mozrzymas, 2004). If so, a basal level of GABA diffusion out of the synaptic cleft may normally occur with GAT1 activity intact, and could activate high-affinity receptors such as the extrasynaptic $GABA_A$Rs (Glykys & Mody, 2007) and the $GABA_B$Rs (Bowery & Enna, 2000).

Consistent with such possibility is the finding that tonic $GABA_AR$-mediated currents, typically mediated by extrasynaptic $GABA_ARs$, are measurable with GAT1 activity intact and, moreover, are increased by GAT1 blockade or in GAT1-deficient mice (Bragina et al., 2008; Glykys & Mody, 2007; Jensen et al., 2003). Surprisingly, $GABA_BRs$ do not seem to produce tonic currents either with GAT1 activity intact or when this activity is reduced in GAT1-deficient mice (Jensen et al., 2003).

Considering the role of GAT1 in regulating $GABA_AR$ activation, a crucial question is whether similar mechanisms regulate $GABA_BR$ activation. As reviewed in Section IV, $GABA_BR$-IPSPs have much slower decay time-course (decay time >>100 ms) than $GABA_AR$-IPSPs (decay time < 20 ms). Therefore, $GABA_BR$-IPSPs are less likely to contribute to temporally precise fast synaptic transmission and GAT1-mediated control of $GABA_BR$-IPSP decay may be less critical for the rapid flow of neural activity in cortical circuits.

Production of fast $GABA_AR$-IPSCs requires rapid gating of the $GABA_ARs$ which in synapses is achieved by exposure of the receptors to a brief and high concentration transient of GABA in the synaptic cleft (Farrant & Kaila, 2007; Mozrzymas, 2004). However, the high affinity of $GABA_BRs$ and the slow kinetics of their postsynaptic effects suggest that production of $GABA_BR$-IPSPs does not require exposure of $GABA_BRs$ to GABA released into the synaptic cleft. Indeed, whereas an ultrastructural study found some $GABA_BRs$ concentrated postsynaptically at symmetric synapses (Gonchar et al., 2001), other studies using three-dimensional quantitative immunogold particle electron microscopy suggest that $GABA_BR$ subunits are localized mainly extra- or peri-synaptically in dendritic shafts and extrasynaptically near glutamate synapses in dendritic spines (Kulik et al., 2003, 2006).

As reviewed in Section IV, neurogliaform cells are the only interneuron subtype thus far identified to produce unitary $GABA_BR$-IPSPs (Price et al., 2005; Tamas et al., 2003). Neurogliaform cells mainly target the dendritic spines and shafts of pyramidal cells (Olah et al., 2009; Szabadics et al., 2007; Tamas et al., 2003), and in addition to producing slow unitary $GABA_BR$-IPSCs, produce unusually slow $GABA_AR$-IPSCs (Olah et al., 2009; Szabadics et al., 2007; Tamas et al., 2003). Such properties of the IPSCs lead others to hypothesize that the GABA receptors mediating neurogliaform cell responses are extra- or peri-synaptic (Fig. 3B) (Szabadics et al., 2007). For instance, neurogliaform cell-mediated unitary IPSCs have pharmacological properties similar to those of extrasynaptic $GABA_ARs$ (Szabadics et al., 2007). In addition, the effects of low-affinity $GABA_AR$ antagonists on neurogliaform cell-mediated unitary IPSCs are consistent with receptor activation by spillover (Szabadics et al., 2007). Notably, both $GABA_AR$- and $GABA_BR$-mediated unitary IPSCs produced by neurogliaform cells are extremely sensitive to GAT1 blockade by NO711 (Szabadics et al., 2007). NO711 application substantially prolongs the

$GABA_BR$-IPSPs and, in addition, significantly increases their amplitude (Gonzalez-Burgos et al., 2009; Isaacson et al., 1993; Szabadics et al., 2007). These finding suggests that, contrasting with $GABA_AR$-IPSPs, GAT1 regulates the strength of the $GABA_BR$-IPSPs, represented by their peak amplitude, possibly by regulating the number of $GABA_BRs$ bound to GABA (Fig. 3B).

Electrophysiological studies therefore suggest that vesicular GABA release from neurogliaform cell terminals produces, under GAT1 control, slow activation of extra- or peri-synaptic $GABA_ARs$ and $GABA_BRs$ by spillover. Only one study thus far examined whether $GABA_BRs$ are postsynaptic to neurogliaform cell release sites, but did not provide detailed quantitative analysis of the receptor localization (Price et al., 2005). Importantly, an ultrastructural analysis of neurogliaform cell nerve terminals showed that most (78%) of the synaptic vesicle-containing varicosities in the neurogliaform cell axon do not form actual synaptic junctions (Olah et al., 2009). Moreover, the nerve terminals apparently connecting individual neurogliaform cells onto postsynaptic neurons to produce slow unitary IPSPs lack clear synaptic junctions (Olah et al., 2009). This observation confirms that neurogliaform cells produce responses in their target cells without involving synaptic junctions and possibly by activation of extra- or peri-synaptic receptors by spillover (Fig. 3B).

An important morphological property of neurogliaform cells described by Santiago Ramon y Cajal more than 100 years ago is that their axonal arborization branches profusely near the cell body (Kawaguchi & Kubota, 1997; Olah et al., 2007, 2009; Povysheva et al., 2007; Zaitsev et al., 2009). Such profuse branching and a high density of boutons per unit of axon length (Olah et al., 2009) increase the probability of GABA spillover between neurogliaform cell terminals, potentially overcoming the GABA uptake system. The fact that GAT1 blockade with NO711 significantly enhances the neurogliaform cell-mediated $GABA_BR$- and $GABA_AR$-IPSPs (Szabadics et al., 2007) demonstrates that GAT1-mediated GABA uptake critically controls the postsynaptic response to GABA released by these neurons.

Whereas neurogliaform cell-mediated GABA release shows rapid exhaustion during repetitive stimulation (Tamas et al., 2003), several studies (see Section IV) showed that repetitive stimulation applied extracellularly or to non-neurogliaform cells is more efficient than single stimuli to produce $GABA_BR$-IPSPs (Gonzalez-Burgos et al., 2009; Isaacson et al., 1993; Oswald et al., 2009; Scanziani, 2000; Thomson & Destexhe, 1999). Moreover, the $GABA_BR$-IPSPs produced by repetitive stimulation are significantly enhanced by GAT1 blockade (Gonzalez-Burgos et al., 2009; Isaacson et al., 1993; Scanziani, 2000). These results support the conclusion that GAT1-mediated uptake strongly regulates $GABA_BR$-IPSPs produced by interneurons of other subtypes in addition to neurogliaform cells. Such conclusion raises the question of whether GAT1 controls the activation, by other interneuron subtypes, of $GABA_BRs$ that are normally targeted by neurogliaform cell-released GABA (Fig. 3B). For example, some paired recordings studies have shown that single-pulse or repetitive stimulation of an interneuron may not produce $GABA_AR$-IPSPs or $GABA_BR$-IPSPs in a simultaneously recorded neuron, but that subsequent inhibition of GAT1 revealed a substantial $GABA_BR$-IPSP (Scanziani, 2000). Whether such $GABA_BR$ effects observed only after GAT1 is blocked reflect activation of $GABA_BRs$ normally activated by neurogliaform cells remains to be determined.

C. GAT1 Activity and Presynaptic $GABA_BR$ Activation

As reviewed in previous sections, some $GABA_BRs$ have been localized to presynaptic nerve terminals at GABA and glutamate synapses, and electrophysiological experiments have demonstrated that presynaptic $GABA_BRs$ regulate transmitter release. Does GAT1-mediated uptake control activation of presynaptic $GABA_BRs$? Some studies have shown that $GABA_BR$ antagonists fail to produce presynaptic effects at GABA synapses, suggesting that the amount of GABA normally accessing presynaptic sites is too small to produce significant autoreceptor activation (Caillard et al., 1998; Gonzalez-Burgos et al., 2009; Neu et al., 2007; Olah et al., 2009). However, GAT1 blockade reveals presynaptic effects of $GABA_BRs$, possibly by increasing the exposure of such presynaptic receptors to GABA (Caillard et al., 1998; Gonzalez-Burgos et al., 2009; Lei & McBain, 2003). Why presynaptic $GABA_BRs$ at some GABA synapses do not produce autoreceptor effects (Caillard et al., 1998; Gonzalez-Burgos et al., 2009; Neu et al., 2007; Olah et al., 2009) but can be activated by exogenous agonists or by endogenous GABA following GAT1 blockade is a puzzling issue that remains to be resolved.

Activation of presynaptic $GABA_BRs$ localized in glutamate synapses requires GABA diffusion for some distance in the extracellular space (Fig. 3B). It is therefore reasonable that GAT1 blockade enhances inhibition of glutamate release by such presynaptic receptors (Isaacson et al., 1993; Lei & McBain, 2003; Olah et al., 2009). An identified source of GABA for activation of presynaptic $GABA_BRs$ localized in glutamate nerve terminals are the neurogliaform cells (Olah et al., 2009). Interestingly, neurogliaform cells elicit presynaptic effects within a volume of tissue closely matching the distribution of their axon, which gives rise to a high density of branches around their soma (Olah et al., 2009). The presynaptic $GABA_BR$-mediated modulation of glutamate release by neurogliaform cells is enhanced by GAT1 blockade with NO711 (Olah et al., 2009). In contrast, NO711 does not reveal presynaptic $GABA_BR$ effects at glutamate synapses by GABA released by fast-spiking basket neurons (Olah et al., 2009).

VI. Conclusions

This chapter reviews studies suggesting that in cortical circuits GAT1-mediated GABA transport is important for the cellular electrophysiological responses produced by $GABA_BR$ activation by endogenous GABA. $GABA_BRs$ produce slow hyperpolarizing IPSPs in pyramidal cells and also regulate GABA and glutamate release via presynaptic effects (Figs. 2 and 3B). These physiological effects of $GABA_BR$ activation are enhanced by pharmacological inhibition of GAT1 activity, showing that GAT1 tightly regulates the amount of GABA available for $GABA_BR$ activation (Fig. 3).

GAT1 critically controls the activation of $GABA_ARs$, playing a "protective" role by preventing the stimulation of $GABA_ARs$ at a given synapse by GABA released at other synapses (Fig. 3A). Does GAT1 also protect $GABA_BRs$ from inter-synaptic cross-talk by GABA diffusion between synapses? If, as indicated by several studies, a majority of the $GABA_BRs$ are actually localized extra- or peri-synaptically, $GABA_BRs$ are therefore normally activated by GABA diffusing far away from the release sites (Fig. 3B). Therefore, instead of playing such a protective role, GAT1-mediated uptake may be a physiological mechanism regulating $GABA_BR$ activation.

The experimental observations reviewed here are consistent with the idea that $GABA_BRs$ signal via volume transmission. Such mechanism operates when a neurotransmitter is released into extracellular space compartments that are not directly opposed to membranes containing postsynaptic receptors (Fuxe et al., 2007). The properties of neurogliaform cell-mediated responses reviewed in this chapter, including their control by GAT1 activity, are actually consistent with $GABA_BR$-mediated volume transmission. Whether additional interneuron subtypes produce $GABA_BR$-mediated responses remains to be established. If so, then GAT1 may play two different functions: first, regulating the strength of physiological $GABA_BR$-mediated responses by controlling the amount of GABA reaching those receptors and, second, preventing the activation of $GABA_BRs$ normally targeted by a particular interneuron subtype, for instance neurogliaform cells, by GABA released by other interneurons (Fig. 3B).

If $GABA_BRs$ exclusively mediate signaling by neurogliaform interneurons, then their general role in cortical circuits is tightly linked to the neurogliaform cell function. Neurogliaform cells are present in various cortical areas and their physiological properties are being characterized in the cortex and hippocampus of rodents (Kawaguchi & Kubota, 1997; Price et al., 2005; Tamas et al., 2003), monkeys (Krimer et al., 2005; Povysheva et al., 2007; Zaitsev et al., 2009), and humans (Olah et al., 2007). Importantly, a main function of interneurons is the production of synchronized oscillations in hippocampal and cortical circuits (Klausberger & Somogyi,

2008). Interneurons involved in the mechanisms of network oscillations may fire at frequencies within the 30–80 Hz range (gamma oscillations) or 4–10 Hz (theta oscillations). Because GABA release from neurogliaform cells strongly depresses during repetitive activity, if these interneurons participate in network oscillations, then their repetitive firing would produce severe depression of GABA release. Since neurogliaform cells actually fire coupled to theta oscillations (Fuentealba et al., 2010), an interesting question is therefore whether during high-frequency firing that depresses GABA release, these interneurons release neuromodulators such as neuropeptide Y (Karagiannis et al., 2009) that could complement the $GABA_BR$-mediated signaling. Importantly, neurogliaform cell activity is decreased by neurosteroids (Olah et al., 2009) and enhanced by norephinephrine (Kawaguchi & Shindou, 1998), and thus can be modulated by the slow paracrine effects of these endogenous substances.

As highlighted elsewhere (Olah et al., 2009), solitary spikes in neurogliaform cells would fill, with a "cloud" of GABA, a volume of tissue closely matching the geometry of their axonal arbor. Following such spike, $GABA_BR$ activation via volume transmission would produce long-lasting $GABA_BR$-IPSPs in a relatively large group of pyramidal neurons and also would decrease excitation between pairs of pyramidal cells connected via synapses located in the same volume of tissue. Interestingly, the time-courses of presynaptic effects of $GABA_BRs$ on excitatory transmission and of $GABA_BR$-IPSPs are very similar (Isaacson et al., 1993). If so, then neurogliaform cell spikes would create a time-window of several hundred milliseconds during which pyramidal cell excitability is reduced by the $GABA_BR$-IPSP and pyramidal cell excitation is reduced by the inhibition of glutamate release by presynaptic $GABA_BRs$. Such an effect may contribute to terminating persistent activity in cortical networks (Mann et al., 2009), an activity pattern that may be essential for cognitive function dependent on working memory. The data reviewed in this chapter show that up- or down-regulating GAT1-mediated uptake could potentiate or weaken the strength and duration of such $GABA_BR$-mediated effects. Furthermore, regulation of GAT1-mediated uptake could potentially expand or contract the effective volume of tissue in which neurogliaform cells produce $GABA_BR$-mediated volume transmission.

Acknowledgments

The author is grateful to Dr David A Lewis and the members of the Translational Neuroscience Program for continuous support. Work by the author and colleagues which is cited in this manuscript was funded by NIH grants MH-45156 and MH-51234.

Conflict of interest: The author declares no conflict of interest.

Non-standard abbreviations

Ca^{2+}	calcium ion
Cl^-	chloride ion
$GABA_AR$	GABA A receptor
$GABA_BR$	GABA B receptor
GAT1	plasma membrane GABA transporter 1
IPSC	inhibitory postsynaptic current
IPSP	inhibitory postsynaptic potential
K^+	potassium ion
Kir	inward-rectifying potassium channel
Na^+	sodium ion

References

Ali, A. B., Bannister, A. P., & Thomson, A. M. (1999). IPSPs elicited in CA1 pyramidal cells by putative basket cells in slices of adult rat hippocampus. *European Journal of Neuroscience*, *11*, 1741–1753.

Ali, A. B., & Thomson, A. M. (2008). Synaptic alpha 5 subunit-containing GABAA receptors mediate IPSPs elicited by dendrite-preferring cells in rat neocortex. *Cerebral Cortex*, *18*, 1260–1271.

Alle, H., & Geiger, J. R. (2007). GABAergic spill-over transmission onto hippocampal mossy fiber boutons. *Journal of Neuroscience*, *27*, 942–950.

Andrasfalvy, B. K., & Mody, I. (2006). Differences between the scaling of miniature IPSCs and EPSCs recorded in the dendrites of CA1 mouse pyramidal neurons. *Journal of Physiology (Paris)*, *576*, 191–196.

Ascoli, G. A., Alonso-Nanclares, L., Anderson, S. A., Barrionuevo, G., Benavides-Piccione, R., Burkhalter, A., et al. (2008). Petilla terminology: Nomenclature of features of GABAergic interneurons of the cerebral cortex. *Nature Reviews. Neuroscience*, *9*, 557–568.

Attwell, D., Barbour, B., & Szatkowski, M. (1993). Nonvesicular release of neurotransmitter. *Neuron*, *11*, 401–407.

Barbour, B. (2001). An evaluation of synapse independence. *Journal of Neuroscience*, *21*, 7969–7984.

Bartos, M., Vida, I., Frotscher, M., Meyer, A., Monyer, H., Geiger, J. R., et al. (2002). Fast synaptic inhibition promotes synchronized gamma oscillations in hippocampal interneuron networks. *Proceedings of the National Academy of Sciences of the United States of America*, *99*, 13222–13227.

Bernstein, E. M., & Quick, M. W. (1999). Regulation of gamma-aminobutyric acid (GABA) transporters by extracellular GABA. *Journal of Biological Chemistry*, *274*, 889–895.

Bertrand, S., & Lacaille, J. C. (2001). Unitary synaptic currents between lacunosum-moleculare interneurones and pyramidal cells in rat hippocampus. *Journal of Physiology (Paris)*, *532*, 369–384.

Bicho, A., & Grewer, C. (2005). Rapid substrate-induced charge movements of the GABA transporter GAT1. *Biophysical Journal*, *89*, 211–231.

Borden, L. A. (1996). GABA transporter heterogeneity: Pharmacology and cellular localization. *Neurochemistry International*, *29*, 335–356.

Bowery, N. G., & Enna, S. J. (2000). Gamma-aminobutyric acid(B) receptors: First of the functional metabotropic heterodimers. *Journal of Pharmacology and Experimental Therapeutics*, *292*, 2–7.

Bowery, N. G., Hill, D. R., Hudson, A. L., Doble, A., Middlemiss, D. N., Shaw, J., et al. (1980). (-)Baclofen decreases neurotransmitter release in the mammalian CNS by an action at a novel GABA receptor. *Nature*, *283*, 92–94.

Bragina, L., Marchionni, I., Omrani, A., Cozzi, A., Pellegrini-Giampietro, D. E., Cherubini, E., et al. (2008). GAT-1 regulates both tonic and phasic GABA(A) receptor-mediated inhibition in the cerebral cortex. *Journal of Neurochemistry*, *105*, 1781–1793.

Buhl, E. H., Cobb, S. R., Halasy, K., & Somogyi, P. (1995). Properties of unitary IPSPs evoked by anatomically identified basket cells in the rat hippocampus. *European Journal of Neuroscience*, *7*, 1989–2004.

Buhl, E. H., Halasy, K., & Somogyi, P. (1994). Diverse sources of hippocampal unitary inhibitory postsynaptic potentials and the number of synaptic release sites. *Nature*, *368*, 823–828.

Cai, Y. Q., Cai, G. Q., Liu, G. X., Cai, Q., Shi, J. H., Shi, J., et al. (2006). Mice with genetically altered GABA transporter subtype I (GAT1) expression show altered behavioral responses to ethanol. *Journal of Neuroscience Research*, *84*, 255–267.

Caillard, O., McLean, H. A., Ben-Ari, Y., & Gaiarsa, J. L. (1998). Ontogenesis of presynaptic GABAB receptor-mediated inhibition in the CA3 region of the rat hippocampus. *Journal of Neurophysiology*, *79*, 1341–1348.

Chen, N. H., Reith, M. E., & Quick, M. W. (2004). Synaptic uptake and beyond: The sodium- and chloride-dependent neurotransmitter transporter family SLC6. *Pflugers Archiv*, *447*, 519–531.

Cherubini, E., & Conti, F. (2001). Generating diversity at GABAergic synapses. *Trends in Neurosciences*, *24*, 155–162.

Chiu, C. S., Brickley, S., Jensen, K., Southwell, A., Mckinney, S., Cull-Candy, S., et al. (2005). GABA transporter deficiency causes tremor, ataxia, nervousness, and increased GABA-induced tonic conductance in cerebellum. *Journal of Neuroscience*, *25*, 3234–3245.

Chiu, C. S., Jensen, K., Sokolova, I., Wang, D., Li, M., Deshpande, P., et al. (2002). Number, density, and surface/cytoplasmic distribution of GABA transporters at presynaptic structures of knock-in mice carrying GABA transporter subtype 1-green fluorescent protein fusions. *Journal of Neuroscience*, *22*, 10251–10266.

Cobb, S. R., Manuel, N. A., Morton, R. A., Gill, C. H., Collingridge, G. L., & Davies, C. H. (1999). Regulation of depolarizing GABA(A) receptor-mediated synaptic potentials by synaptic activation of GABA(B) autoreceptors in the rat hippocampus. *Neuropharmacology*, *38*, 1723–1732.

Connors, B. W., Malenka, R. C., & Silva, L. R. (1988). Two inhibitory postsynaptic potentials, and GABAA and GABAB receptor-mediated responses in neocortex of rat and cat. *Journal of Physiology*, *406*, 443–468.

Conti, F., Melone, M., De Biasi, S., Minelli, A., Brecha, N. C., & Ducati, A. (1998). Neuronal and glial localization of GAT-1, a high-affinity gamma-aminobutyric acid plasma membrane transporter, in human cerebral cortex: With a note on its distribution in monkey cortex. *Journal of Comparative Neurology*, *396*, 51–63.

Conti, F., Minelli, A., & Melone, M. (2004). GABA transporters in the mammalian cerebral cortex: Localization, development and pathological implications. *Brain Research. Molecular Brain Research*, *45*, 196–212.

Davies, C. H., & Collingridge, G. L. (1993). The physiological regulation of synaptic inhibition by GABAB autoreceptors in rat hippocampus. *Journal of Physiology (Paris)*, *472*, 245–265.

Davies, C. H., & Collingridge, G. L. (1996). Regulation of EPSPs by the synaptic activation of GABAB autoreceptors in rat hippocampus. *Journal of Physiology (Paris)*, *496*(Pt 2), 451–470.
Davies, C. H., Davies, S. N., & Collingridge, G. L. (1990). Paired-pulse depression of monosynaptic GABA-mediated inhibitory postsynaptic responses in rat hippocampus. *Journal of Physiology (Paris)*, *424*, 513–531.
Debanne, D., Boudkkazi, S., Campanac, E., Cudmore, R. H., Giraud, P., Fronzaroli-Molinieres, L., et al. (2008). Paired-recordings from synaptically coupled cortical and hippocampal neurons in acute and cultured brain slices. *Nature Protocols*, *3*, 1559–1568.
Deisz, R. A. (1999). The GABA(B) receptor antagonist CGP 55845a reduces presynaptic GABA(B) actions in neocortical neurons of the rat in vitro. *Neuroscience*, *93*, 1241–1249.
Deisz, R. A., Billard, J. M., & Zieglgansberger, W. (1997). Presynaptic and postsynaptic GABAB receptors of neocortical neurons of the rat in vitro: Differences in pharmacology and ionic mechanisms. *Synapse*, *25*, 62–72.
Diamond, J. S., & Jahr, C. E. (1997). Transporters buffer synaptically released glutamate on a submillisecond time scale. *Journal of Neuroscience*, *17*, 4672–4687.
Doischer, D., Hosp, J. A., Yanagawa, Y., Obata, K., Jonas, P., Vida, I., et al. (2008). Postnatal differentiation of basket cells from slow to fast signaling devices. *Journal of Neuroscience*, *28*, 12956–12968.
Dutar, P., & Nicoll, R. A. (1988). A physiological role for GABAB receptors in the central nervous system. *Nature*, *332*, 156–158.
Farrant, M., & Kaila, K. (2007). The cellular, molecular and ionic basis of GABAA receptor signalling. *Progress in Brain Research*, *160*, 59–87.
Frahm, C., Engel, D., & Draguhn, A. (2001). Efficacy of background GABA uptake in rat hippocampal slices. *NeuroReport*, *12*, 1593–1596.
Frahm, C., Haupt, C., Weinandy, F., Siegel, G., Bruehl, C., & Witte, O. W. (2004). Regulation of GABA transporter mRNA and protein after photothrombotic infarct in rat brain. *Journal of Comparative Neurology*, *478*, 176–188.
Freund, T. F., & Katona, I. (2007). Perisomatic inhibition. *Neuron*, *56*, 33–42.
Fuentealba, P., Klausberger, T., Karayannis, T., Suen, W. Y., Huck, J., Tomioka, R., et al. (2010). Expression of COUP-TFII nuclear receptor in restricted GABAergic neuronal populations in the adult rat hippocampus. *Journal of Neuroscience*, *30*, 1595–1609.
Fueta, Y., Vasilets, L. A., Takeda, K., Kawamura, M., & Schwarz, W. (2003). Down-regulation of GABA-transporter function by hippocampal translation products: Its possible role in epilepsy. *Neuroscience*, *118*, 371–378.
Fuxe, K., Dahlstrom, A., Hoistad, M., Marcellino, D., Jansson, A., Rivera, A., et al. (2007). From the golgi-cajal mapping to the transmitter-based characterization of the neuronal networks leading to two modes of brain communication: Wiring and volume transmission. *Brain Research Reviews*, *55*, 17–54.
Galarreta, M., Erdelyi, F., Szabo, G., & Hestrin, S. (2008). Cannabinoid sensitivity and synaptic properties of 2 GABAergic networks in the neocortex. *Cerebral Cortex*, *18*, 2296–2305.
Gether, U., Andersen, P. H., Larsson, O. M., & Schousboe, A. (2006). Neurotransmitter transporters: Molecular function of important drug targets. *Trends in Pharmacological Sciences*, *27*, 375–383.
Glickfeld, L. L., Atallah, B. V., & Scanziani, M. (2008). Complementary modulation of somatic inhibition by opioids and cannabinoids. *Journal of Neuroscience*, *28*, 1824–1832.
Glickfeld, L. L., & Scanziani, M. (2006). Distinct timing in the activity of cannabinoid-sensitive and cannabinoid-insensitive basket cells. *Nature Neuroscience*, *9*, 807–815.
Glykys, J., & Mody, I. (2007). Activation of GABAA receptors: Views from outside the synaptic cleft. *Neuron*, *56*, 763–770.

Gonchar, Y., Pang, L., Malitschek, B., Bettler, B., & Burkhalter, A. (2001). Subcellular localization of GABA(B) receptor subunits in rat visual cortex. *Journal of Comparative Neurology*, *431*, 182–197.

Gong, N., Li, Y., Cai, G. Q., Niu, R. F., Fang, Q., Wu, K., et al. (2009). GABA transporter-1 activity modulates hippocampal theta oscillation and theta burst stimulation-induced long-term potentiation. *Journal of Neuroscience*, *29*, 15836–15845.

Gonzalez-Burgos, G., Krimer, L. S., Povysheva, N. V., Barrionuevo, G., & Lewis, D. A. (2005). Functional properties of fast spiking interneurons and their synaptic connections with pyramidal cells in primate dorsolateral prefrontal cortex. *Journal of Neurophysiology*, *93*, 942–953.

Gonzalez-Burgos, G., Rotaru, D. C., Zaitsev, A. V., Povysheva, N. V., & Lewis, D. A. (2009). GABA transporter GAT1 prevents spillover at proximal and distal GABA synapses onto primate prefrontal cortex neurons. *Journal of Neurophysiology*, *101*, 533–547.

Guastella, J., Nelson, N., Nelson, H., Czyzyk, L., Keynan, S., Miedel, M. C., et al. (1990). Cloning and expression of a rat brain GABA transporter. *Science*, *249*, 1303–1306.

Gulledge, A. T., & Stuart, G. J. (2003). Excitatory actions of GABA in the cortex. *Neuron*, *37*, 299–309.

Hefft, S., & Jonas, P. (2005). Asynchronous GABA release generates long-lasting inhibition at a hippocampal interneuron-principal neuron synapse. *Nature Neuroscience*, *8*, 1319–1328.

Hu, J. H., Ma, Y. H., Jiang, J., Yang, N., Duan, S. H., Jiang, Z. H., et al. (2004). Cognitive impairment in mice over-expressing gamma-aminobutyric acid transporter 1 (GAT1. *NeuroReport*, *15*, 9–12.

Hu, J., & Quick, M. W. (2008). Substrate-mediated regulation of gamma-aminobutyric acid transporter 1 in rat brain. *Neuropharmacology*, *54*, 309–318.

Isaacson, J. S., Solis, J. M., & Nicoll, R. A. (1993). Local and diffuse synaptic actions of GABA in the hippocampus. *Neuron*, *10*, 165–175.

Jensen, K., Chiu, C. S., Sokolova, I., Lester, H. A., & Mody, I. (2003). GABA transporter-1 (GAT1)-deficient mice: Differential tonic activation of GABAA versus GABAB receptors in the hippocampus. *Journal of Neurophysiology*, *90*, 2690–2701.

Kapfer, C., Glickfeld, L. L., Atallah, B. V., & Scanziani, M. (2007). Supralinear increase of recurrent inhibition during sparse activity in the somatosensory cortex. *Nature Neuroscience*, *10*, 743–753.

Karagiannis, A., Gallopin, T., David, C., Battaglia, D., Geoffroy, H., Rossier, J., et al. (2009). Classification of NPY-expressing neocortical interneurons. *Journal of Neuroscience*, *29*, 3642–3659.

Kawaguchi, Y., & Kubota, Y. (1997). GABAergic cell subtypes and their synaptic connections in rat frontal cortex. *Cerebral Cortex*, *7*, 476–486.

Kawaguchi, Y., & Shindou, T. (1998). Noradrenergic excitation and inhibition of GABAergic cell types in rat frontal cortex. *Journal of Neuroscience*, *18*, 6963–6976.

Keros, S., & Hablitz, J. J. (2005). Subtype-specific GABA transporter antagonists synergistically modulate phasic and tonic GABAA conductances in rat neocortex. *Journal of Neurophysiology*, *94*, 2073–2085.

Klausberger, T., & Somogyi, P. (2008). Neuronal diversity and temporal dynamics: The unity of hippocampal circuit operations. *Science*, *321*, 53–57.

Krimer, L. S., Zaitsev, A. V., Czanner, G., Kroner, S., Gonzalez-Burgos, G., Povysheva, N. V., et al. (2005). Cluster analysis-based physiological classification and morphological properties of inhibitory neurons in layers 2-3 of monkey dorsolateral prefrontal cortex. *Journal of Neurophysiology*, *94*, 3009–3022.

Kulik, A., Vida, I., Fukazawa, Y., Guetg, N., Kasugai, Y., Marker, C. L., et al. (2006). Compartment-dependent colocalization of kir3.2-Containing K+ channels and GABAB receptors in hippocampal pyramidal cells. *Journal of Neuroscience*, *26*, 4289–4297.

Kulik, A., Vida, I., Lujan, R., Haas, C. A., Lopez-Bendito, G., Shigemoto, R., et al. (2003). Subcellular localization of metabotropic GABA(B) receptor subunits GABA(B1a/b) and GABA(B2) in the rat hippocampus. *Journal of Neuroscience*, *23*, 11026–11035.

Laroche, S. M. (2007). A new look at the second-generation antiepileptic drugs: A decade of experience. *Neurologist*, *13*, 133–139.

Lee, T. S., Bjornsen, L. P., Paz, C., Kim, J. H., Spencer, S. S., Spencer, D. D., et al. (2006). GAT1 and GAT3 expression are differently localized in the human epileptogenic hippocampus. *Acta Neuropathologica*, *111*, 351–363.

Lei, S., & McBain, C. J. (2003). GABA B receptor modulation of excitatory and inhibitory synaptic transmission onto rat CA3 hippocampal interneurons. *Journal of Physiology (Paris)*, *546*, 439–453.

Lewis, D. A., Hashimoto, T., & Volk, D. W. (2005). Cortical inhibitory neurons and schizophrenia. *Nature Reviews. Neuroscience*, *6*, 312–324.

Liu, G. X., Cai, G. Q., Cai, Y. Q., Sheng, Z. J., Jiang, J., Mei, Z., et al. (2007). Reduced anxiety and depression-like behaviors in mice lacking GABA transporter subtype 1. *Neuropsychopharmacology*, *32*, 1531–1539.

Lopez-Bendito, G., Shigemoto, R., Kulik, A., Paulsen, O., Fairen, A., & Lujan, R. (2002). Expression and distribution of metabotropic GABA receptor subtypes GABABR1 and GABABR2 during rat neocortical development. *European Journal of Neuroscience*, *15*, 1766–1778.

Luscher, C., Jan, L. Y., Stoffel, M., Malenka, R. C., & Nicoll, R. A. (1997). G protein-coupled inwardly rectifying K+ channels (GIRKs) mediate postsynaptic but not presynaptic transmitter actions in hippocampal neurons. *Neuron*, *19*, 687–695.

Mahendrasingam, S., Wallam, C. A., & Hackney, C. M. (2003). Two approaches to double post-embedding immunogold labeling of freeze-substituted tissue embedded in low temperature lowicryl HM20 resin. *Brain Research. Brain Research Protocols*, *11*, 134–141.

Mann, E. O., Kohl, M. M., & Paulsen, O. (2009). Distinct roles of GABA(A) and GABA(B) receptors in balancing and terminating persistent cortical activity. *Journal of Neuroscience*, *29*, 7513–7518.

Markram, H., Toledo-Rodriguez, M., Wang, Y., Gupta, A., Silberberg, G., & Wu, C. (2004). Interneurons of the neocortical inhibitory system. *Nature Reviews. Neuroscience*, *5*, 793–807.

McCormick, D. A. (1989). GABA as an inhibitory neurotransmitter in human cerebral cortex. *Journal of Neurophysiology*, *62*, 1018–1027.

Megias, M., Emri, Z., Freund, T. F., & Gulyas, A. I. (2001). Total number and distribution of inhibitory and excitatory synapses on hippocampal CA1 pyramidal cells. *Neuroscience*, *102*, 527–540.

Minelli, A., Brecha, N. C., Karschin, C., DeBiasi, S., & Conti, F. (1995). GAT-1, a high-affinity GABA plasma membrane transporter, is localized to neurons and astroglia in the cerebral cortex. *Journal of Neuroscience*, *15*, 7734–7746.

Minelli, A., DeBiasi, S., Brecha, N. C., Zuccarello, L. V., & Conti, F. (1996). GAT-3, a high-affinity GABA plasma membrane transporter, is localized to astrocytic processes, and it is not confined to the vicinity of GABAergic synapses in the cerebral cortex. *Journal of Neuroscience*, *16*, 6255–6264.

Mintz, I. M., & Bean, B. P. (1993). GABAB receptor inhibition of P-type ca2+ channels in central neurons. *Neuron*, *10*, 889–898.

Mohler, H., & Fritschy, J. M. (1999). GABAB receptors make it to the top–as dimers. *Trends in Pharmacological Sciences*, *20*, 87–89.

Moss, F. J., Imoukhuede, P. I., Scott, K., Hu, J., Jankowsky, J. L., Quick, M. W., et al. (2009). GABA transporter function, oligomerization state, and anchoring: Correlates with subcellularly resolved FRET. *Journal of General Physiology*, *134*, 489–521.

Mozrzymas, J. W. (2004). Dynamism of GABAA receptor activation shapes the "personality" of inhibitory synapses. *Neuropharmacology*, *47*, 945–960.

Murayama, M., Perez-Garci, E., Nevian, T., Bock, T., Senn, W., & Larkum, M. E. (2009). Dendritic encoding of sensory stimuli controlled by deep cortical interneurons. *Nature*, *457*, 1137–1141.

Neu, A., Foldy, C., & Soltesz, I. (2007). Postsynaptic origin of CB1-dependent tonic inhibition of GABA release at cholecystokinin-positive basket cell to pyramidal cell synapses in the CA1 region of the rat hippocampus. *Journal of Physiology (Paris)*, *578*, 233–247.

Nicoll, R. A. (2004). My close encounter with GABA(B) receptors. *Biochemical Pharmacology*, *68*, 1667–1674.

Nurse, S., & Lacaille, J. C. (1997). Do GABAA and GABAB inhibitory postsynaptic responses originate from distinct interneurons in the hippocampus? *Canadian Journal of Physiology and Pharmacology*, *75*, 520–525.

Olah, S., Fule, M., Komlosi, G., Varga, C., Baldi, R., Barzo, P., et al. (2009). Regulation of cortical microcircuits by unitary GABA-mediated volume transmission. *Nature*, *461*, 1278–1281.

Olah, S., Komlosi, G., Szabadics, J., Varga, C., Toth, E., Barzo, P., et al. (2007). Output of neurogliaform cells to various neuron types in the human and rat cerebral cortex. *Frontiers in Neural Circuits*, *1*, 4.

Ong, W. Y., Yeo, T. T., Balcar, V. J., & Garey, L. J. (1998). A light and electron microscopic study of GAT-1-positive cells in the cerebral cortex of man and monkey. *Journal of Neurocytology*, *27*, 719–730.

Ortinski, P. I., Turner, J. R., Barberis, A., Motamedi, G., Yasuda, R. P., Wolfe, B. B., et al. (2006). Deletion of the GABA(A) receptor alpha1 subunit increases tonic GABA(A) receptor current: A role for GABA uptake transporters. *Journal of Neuroscience*, *26*, 9323–9331.

Oswald, A. M., Doiron, B., Rinzel, J., & Reyes, A. D. (2009). Spatial profile and differential recruitment of GABAB modulate oscillatory activity in auditory cortex. *Journal of Neuroscience*, *29*, 10321–10334.

Overstreet, L. S., & Westbrook, G. L. (2003). Synapse density regulates independence at unitary inhibitory synapses. *Journal of Neuroscience*, *23*, 2618–2626.

Pan, B. X., Dong, Y., Ito, W., Yanagawa, Y., Shigemoto, R., & Morozov, A. (2009). Selective gating of glutamatergic inputs to excitatory neurons of amygdala by presynaptic GABAb receptor. *Neuron*, *61*, 917–929.

Papp, E., Leinekugel, X., Henze, D. A., Lee, J., & Buzsaki, G. (2001). The apical shaft of CA1 pyramidal cells is under GABAergic interneuronal control. *Neuroscience*, *102*, 715–721.

Perez-Garci, E., Gassmann, M., Bettler, B., & Larkum, M. E. (2006). The GABAB1b isoform mediates long-lasting inhibition of dendritic Ca2+ spikes in layer 5 somatosensory pyramidal neurons. *Neuron*, *50*, 603–616.

Porter, J. T., & Nieves, D. (2004). Presynaptic GABAB receptors modulate thalamic excitation of inhibitory and excitatory neurons in the mouse barrel cortex. *Journal of Neurophysiology*, *92*, 2762–2770.

Povysheva, N. V., Zaitsev, A. V., Kroener, S., Krimer, O. A., Rotaru, D. C., Gonzalez-Burgos, G., et al. (2007). Electrophysiological differences between neurogliaform cells from monkey and rat prefrontal cortex. *Journal of Neurophysiology*, *97*(2), 1030–1039.

Price, C. J., Cauli, B., Kovacs, E. R., Kulik, A., Lambolez, B., Shigemoto, R., et al. (2005). Neurogliaform neurons form a novel inhibitory network in the hippocampal CA1 area. *Journal of Neuroscience*, *25*, 6775–6786.

Price, C. J., Scott, R., Rusakov, D. A., & Capogna, M. (2008). GABA(B) receptor modulation of feedforward inhibition through hippocampal neurogliaform cells. *Journal of Neuroscience*, *28*, 6974–6982.

Quick, M. W., Hu, J., Wang, D., & Zhang, H. Y. (2004). Regulation of a gamma-aminobutyric acid transporter by reciprocal tyrosine and serine phosphorylation. *Journal of Biological Chemistry*, *279*, 15961–15967.

Safiulina, V. F., & Cherubini, E. (2009). At immature mossy fibers-CA3 connections, activation of presynaptic GABA(B) receptors by endogenously released GABA contributes to synapses silencing. *Frontiers in Cellular Neuroscience*, *3*, 1.

Scanziani, M. (2000). GABA spillover activates postsynaptic GABA(B) receptors to control rhythmic hippocampal activity. *Neuron*, *25*, 673–681.

Schousboe, A., & Waagepetersen, H. S. (2006). Glial modulation of GABAergic and glutamatergic neurotransmission. *Current Topics in Medicinal Chemistry*, *6*, 929–934.

Somogyi, P., Tamas, G., Lujan, R., & Buhl, E. H. (1998). Salient features of synaptic organisation in the cerebral cortex. *Brain Research Reviews*, *26*, 113–135.

Szabadics, J., Tamas, G., & Soltesz, I. (2007). Different transmitter transients underlie presynaptic cell type specificity of GABAA,slow and GABAA,fast. *Proceedings of the National Academy of Sciences of the United States of America*, *104*, 14831–14836.

Szabadics, J., Varga, C., Molnar, G., Olah, S., Barzo, P., & Tamas, G. (2006). Excitatory effect of GABAergic axo-axonic cells in cortical microcircuits. *Science*, *311*, 233–235.

Takahashi, T., Kajikawa, Y., & Tsujimoto, T. (1998). G-protein-coupled modulation of presynaptic calcium currents and transmitter release by a GABAB receptor. *Journal of Neuroscience*, *18*, 3138–3146.

Tamas, G., Buhl, E. H., & Somogyi, P. (1997). Fast IPSPs elicited via multiple synaptic release sites by different types of GABAergic neurone in the cat visual cortex. *Journal of Physiology (Paris)*, *500*(Pt 3), 715–738.

Tamas, G., Lorincz, A., Simon, A., & Szabadics, J. (2003). Identified sources and targets of slow inhibition in the neocortex. *Science*, *299*, 1902–1905.

Tamas, G., Somogyi, P., & Buhl, E. H. (1998). Differentially interconnected networks of GABAergic interneurons in the visual cortex of the cat. *Journal of Neuroscience*, *18*, 4255–4270.

Thompson, S. M., Capogna, M., & Scanziani, M. (1993). Presynaptic inhibition in the hippocampus. *Trends in Neurosciences*, *16*, 222–227.

Thompson, S. M., & Gahwiler, B. H. (1992). Effects of the GABA uptake inhibitor tiagabine on inhibitory synaptic potentials in rat hippocampal slice cultures. *Journal of Neurophysiology*, *67*, 1698–1701.

Thomson, A. M., Bannister, A. P., Hughes, D. I., & Pawelzik, H. (2000). Differential sensitivity to zolpidem of IPSPs activated by morphologically identified CA1 interneurons in slices of rat hippocampus. *European Journal of Neuroscience*, *12*, 425–436.

Thomson, A. M., & Destexhe, A. (1999). Dual intracellular recordings and computational models of slow inhibitory postsynaptic potentials in rat neocortical and hippocampal slices. *Neuroscience*, *92*, 1193–1215.

Thomson, A. M., West, D. C., Hahn, J., & Deuchars, J. (1996). Single axon IPSPs elicited in pyramidal cells by three classes of interneurones in slices of rat neocortex. *Journal of Physiology*, *496*, 81–102.

Ulrich, D., & Bettler, B. (2007). GABA(B) receptors: Synaptic functions and mechanisms of diversity. *Current Opinion in Neurobiology*, *17*, 293–303.

Vida, I., Halasy, K., Szinyei, C., Somogyi, P., & Buhl, E. H. (1998). Unitary IPSPs evoked by interneurons at the stratum radiatum-stratum lacunosum-moleculare border in the CA1 area of the rat hippocampus in vitro. *Journal of Physiology (Paris)*, *506*(Pt 3), 755–773.

Vigot, R., Barbieri, S., Brauner-Osborne, H., Turecek, R., Shigemoto, R., Zhang, Y. P., et al. (2006). Differential compartmentalization and distinct functions of GABAB receptor variants. *Neuron*, *50*, 589–601.

Vitellaro-Zuccarello, L., Calvaresi, N., & De Biasi, S. (2003). Expression of GABA transporters, GAT-1 and GAT-3, in the cerebral cortex and thalamus of the rat during postnatal development. *Cell and Tissue Research*, *313*, 245–257.

Wang, D., & Quick, M. W. (2005). Trafficking of the plasma membrane gamma-aminobutyric acid transporter GAT1. Size and rates of an acutely recycling pool. *Journal of Biological Chemistry*, *280*, 18703–18709.

Wu, Y., Wang, W., Diez-Sampedro, A., & Richerson, G. B. (2007). Nonvesicular inhibitory neurotransmission via reversal of the GABA transporter GAT-1. *Neuron*, *56*, 851–865.

Yan, X. X., Cariaga, W. A., & Ribak, C. E. (1997). Immunoreactivity for GABA plasma membrane transporter, GAT-1, in the developing rat cerebral cortex: Transient presence in the somata of neocortical and hippocampal neurons. *Brain Research. Developmental Brain Research*, *99*, 1–19.

Zaitsev, A. V., Povysheva, N. V., Gonzalez-Burgos, G., Rotaru, D., Fish, K. N., Krimer, L. S., et al. (2009). Interneuron diversity in layers 2–3 of monkey prefrontal cortex. *Cerebral Cortex*, *19*, 1597–1615.

Michael M. Kohl* **and Ole Paulsen***

*Department of Physiology, Development and Neuroscience, University of Cambridge, Cambridge, UK

The Roles of $GABA_B$ Receptors in Cortical Network Activity

Abstract

Temporally-structured cortical activity in the form of synchronized network oscillations and persistent activity is fundamental for cognitive processes such as sensory processing, motor control, working memory, and consolidation of long-term memory. The roles of fast glutamatergic excitation via AMPA, kainate, and NMDA receptors, as well as fast GABAergic inhibition via $GABA_A$ receptors, in such network activity have been studied in great detail. In contrast, we have only recently begun

Advances in Pharmacology, Volume 58

1054-3589/10 $35.00
10.1016/S1054-3589(10)58009-8

to appreciate the roles of slow inhibition via $GABA_B$ receptors in the control of cortical network activity. Here, we provide a framework for understanding the contributions of $GABA_B$ receptors in helping mediate, modulate, and moderate different types of physiological and pathological cortical network activity. We demonstrate how the slow time course of $GABA_B$ receptor-mediated inhibition is well suited to help *mediate* the slow oscillation, to *modulate* the power and spatial profile of gamma oscillations, and to *moderate* the relative spike timing of individual neurons during theta oscillations. We further suggest that $GABA_B$ receptors are interesting therapeutic targets in pathological conditions where cortical network activity is disturbed, such as epilepsy and schizophrenia.

I. Introduction

The many billions of neurons in the neocortex of the mammalian brain are heavily interconnected through chemical synapses, thereby creating a vast neuronal network. This network is arranged into many individual processing units known as microcircuits, allowing massive parallel processing of information. Microcircuits comprise sets of precisely interconnected excitatory and inhibitory neurons, and it is a fundamental challenge to understand how these microcircuits operate.

Cortical microcircuits form feedforward and feedback loops that generate spontaneous oscillatory activity, such as the slow oscillation during slow-wave sleep and the fast gamma oscillation during attention. It is likely that these oscillatory modes of network activity are fundamental to the normal function of these local microcircuits. Conversely, when these network activities are disturbed, pathology arises. Interference with the slow oscillation, for example, can lead to memory deficits and generalized epilepsy, and schizophrenia is associated with changes in gamma oscillations.

Thus, it is imperative that we study the normal synaptic function of microcircuits. To date, most studies have focused on the roles of fast excitation mediated by AMPA, kainate, and NMDA receptors and fast inhibition mediated by $GABA_A$ receptors in the control of network activity. Much less is understood about the roles of $GABA_B$ receptors. In this chapter, we will discuss recent findings that attribute a significant role of slow inhibition via $GABA_B$ receptors to the control of cortical network activity. We will first briefly review the functional properties of $GABA_B$ receptor-mediated inhibition, and then discuss its involvement in *mediating*, *modulating*, and *moderating* cortical network activity, before relating these functions to pathological network events.

II. GABA$_B$ Receptors in Cortical Microcircuits

A. Localization of GABA$_B$ Receptors

In the 1980s, a series of elegant studies by Norman Bowery (Bowery, 1993; Bowery et al., 1980; 1981a; 1981b; Bowery et al., 1982, 1983; Wilkin et al., 1981) suggested the existence of a second GABA receptor with a distinct pharmacology from the classical GABA$_A$ receptor. This bicuculline-insensitive GABA receptor was called the GABA$_B$ receptor and was later shown to consist of a heterodimer made up of two G-protein-coupled receptor subunits, GABA$_{B1a/b}$ and GABA$_{B2}$ (Jones et al., 1998; Kaupmann et al., 1998; White et al., 1998; for review see Bowery et al., 2002). The GABA$_{B1a/b}$ subunit is thought to be the site of agonist binding, whilst the GABA$_{B2}$ subunit activates the G-protein signaling system (Galvez et al., 2001; Robbins et al., 2001), and neither subunit is functional in isolation (Agnati et al., 2003; Bowery et al., 2002).

Immunohistochemical studies have localized GABA$_B$ receptors to both pre- and postsynaptic elements at both excitatory and inhibitory synapses and mainly in the perisynaptic and extrasynaptic plasma membrane (Gonchar & Burkhalter, 1999; Lopez-Bendito et al., 2002, 2004). The requirement of strong or repetitive extracellular stimulation to induce a GABA$_B$ receptor-mediated inhibitory postsynaptic potential (IPSP; Connors et al., 1988; Dutar & Nicoll, 1988; Thomson & Destexhe, 1999), and the absence of a GABA$_B$ receptor-mediated IPSP in paired recordings between synaptically-connected GABAergic interneurons and postsynaptic pyramidal cells (Buhl et al., 1994; Thomson et al., 1996), led to the suggestion that GABA$_B$ receptors are activated by spillover of synaptically-released GABA from neighboring synapses (for an early review, see Mody et al., 1994). However, it was later shown that activation of individual cells of a specific subclass of interneuron, the neurogliaform cell, could indeed elicit unitary GABA$_B$ receptor-mediated IPSPs (Tamás et al., 2003), although it was recently suggested that these cells still provide mostly non-synaptic, spatially non-specific input to the target cells (Oláh et al., 2009).

GABA$_B$ receptors exert their inhibitory effect by various means. Postsynaptically, GABA$_B$ receptors can open G-protein-coupled inwardly-rectifying potassium (GIRK) channels controlling a slow hyperpolarizing potassium conductance (Fig. 1A; Dutar & Nicoll, 1988; Lüscher et al., 1997). They can also directly inhibit postsynaptic calcium channels (Mintz & Bean, 1993), thereby controlling the generation of dendritic calcium spikes in cortical pyramidal cells (Fig. 1B; Pérez-Garci et al., 2006). On the presynaptic side, GABA$_B$ receptors have been found to reduce neurotransmitter release, as originally shown by Bowery and colleagues (Bowery et al., 1980). This was seen both at inhibitory synapses, mediated by synaptically-released GABA acting on presynaptic GABA$_B$ autoreceptors (Fig. 1C;

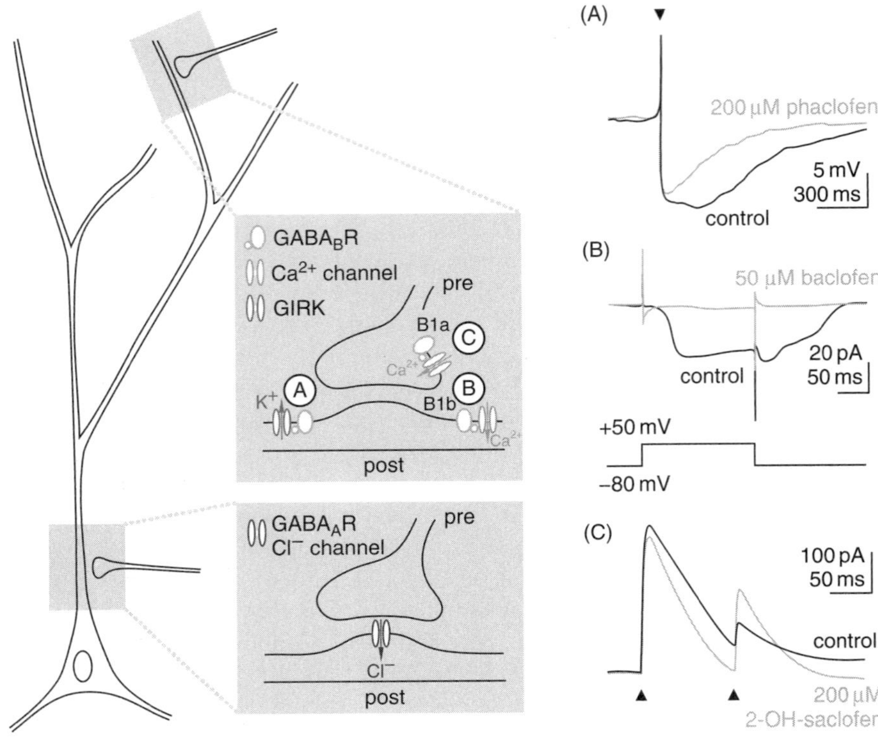

FIGURE 1 GABA$_B$ receptor effects through three different mechanisms. (A) Postsynaptically located GABA$_B$ receptors can open slow potassium conductances by acting on GIRK channels. The GABA$_B$ receptor antagonist phaclofen abolishes the associated slow IPSP. (B) Postsynaptic GABA$_B$ receptors can also inhibit postsynaptic dendritic calcium channels. A depolarizing voltage step applied in dendritic voltage-clamp during pharmacological Na^+ and K^+ channel block reveals calcium currents that are inhibited by the GABA$_B$ receptor agonist baclofen. (C) Presynaptically located GABA$_B$ autoreceptors can close presynaptic calcium channels and directly inhibit the vesicle release machinery. The GABA$_B$ receptor antagonist 2-hydroxy-saclofen prevents GABA$_B$ receptor-mediated paired-pulse depression. Traces in (A), (B), and (C) are modified, with permission, from Dutar & Nicoll (1988), Pérez-Garci et al. (2006), and Davies et al. (1990), respectively.

Davies et al., 1990; Davies et al., 1993), and at excitatory synapses, at which GABA "spillover" from neighboring inhibitory synapses was suggested to mediate heterosynaptic depression of excitatory synaptic transmission (Isaacson et al., 1993; Lei & McBain, 2003). Presynaptic inhibition of transmitter release upon activation of GABA$_B$ receptors appears to be mediated primarily through the inhibition of calcium channels (Takahashi et al., 1998), though an additional effect directly on the release apparatus is also possible (Dittman & Regehr, 1996; Sakaba & Neher, 2003). Presynaptic and postsynaptic inhibition appears to be mediated via $GABA_B$ receptors with distinct $GABA_{B1}$ subunits, $GABA_{B1a}$ and $GABA_{B1b}$, respectively (Pérez-Garci et al., 2006; Ulrich & Bettler, 2007; Vigot et al., 2006).

B. Functional Role of $GABA_B$ Receptors

To understand the functional roles of $GABA_B$ receptor-mediated inhibition in network function, it is useful to appreciate the conditions under which $GABA_B$ receptors are activated.

1. Conditions That Lead to Presynaptic $GABA_B$ Receptor Activation

Presynaptic $GABA_B$ receptors are found on many feedforward inhibitory neurons where they act as autoreceptors (Thompson et al., 1993; Vigot et al., 2006). They have also been found on glutamatergic terminals where they are thought to exert heterosynaptic effects.

Paired pulse activation of inhibitory synapses in the rat hippocampus at 10 Hz was shown to result in a depression of the second IPSP that was partially reversed by $GABA_B$ receptor antagonists (Fig. 1B; Davies et al., 1990). This effect was dependent on the stimulus strength, suggesting that $GABA_B$ autoreceptor activation requires a sufficient amount of synaptically-released GABA to accumulate (Lambert & Wilson, 1994). Given the higher affinity of $GABA_B$ receptors for GABA when compared to $GABA_A$ receptors, this suggests an extrasynaptic location of $GABA_B$ autoreceptors. Subsequently, various other cortical interneurons and hippocampal neurogliaform cells were found to modulate their own axon terminals through the activation of $GABA_B$ autoreceptors (Oláh et al., 2009; Price et al., 2005, 2008).

In contrast, presynaptic receptors on glutamatergic terminals were shown to be activated by GABA spillover from neighboring inhibitory synapses. Isaacson and colleagues were the first to show that repetitive, but not single, stimulation of inhibitory synapses resulted in a slow presynaptic inhibition of excitatory synaptic transmission that was blocked by $GABA_B$ receptor antagonists (Isaacson et al., 1993; Oláh et al., 2009).

2. Conditions That Lead to Postsynaptic $GABA_B$ Receptor Activation

Early studies in the hippocampus suggested that only strong stimulation could evoke $GABA_B$ receptor-mediated slow IPSPs (Dutar & Nicoll, 1988). In paired recordings from interneurons and pyramidal cells, only $GABA_A$ receptor-mediated events could be evoked (Buhl et al., 1994; Thomson et al., 1996), even with repetitive stimuli (Scanziani, 2000). Yet, addition of a GABA uptake blocker revealed a slow IPSC (Scanziani, 2000). Given the higher affinity of $GABA_B$ receptors for GABA, it is conceivable that postsynaptic $GABA_B$ receptors are also located extrasynaptically.

These findings suggest that in order to activate $GABA_B$ receptors, cooperativity among release sites is required. This, however, means that postsynaptic $GABA_B$ receptors are only activated during periods of synchronous activity under physiological conditions. Such synchronous activity occurs during network oscillations and studies have shown that the synaptic activation of $GABA_B$ receptors during cholinergically-induced theta rhythm in hippocampal slices modulates this oscillation (Scanziani, 2000). Similarly, prolonged burst firing in ferret perigeniculate neurons produced $GABA_B$ receptor-mediated slow IPSPs in neurons of the lateral geniculate nucleus (Kim et al., 1997).

However, whereas only fast $GABA_A$ receptor-mediated events were observed following unitary activation of perisomatic-targeting interneurons (Buhl et al., 1994; Thomson et al., 1996), dendritic-targeting neurogliaform cells have been shown to elicit $GABA_B$ receptor-mediated IPSPs following single action potentials in both neocortex (Tamás et al., 2003) and hippocampus (Price et al., 2008), and it has recently been demonstrated that this effect is possibly mediated by volume transmission through the synchronous release of GABA from many boutons of neurogliaform cells (Oláh et al., 2009).

In conclusion, it is becoming clear that activation and action of $GABA_B$ receptors are very distinct from those of $GABA_A$ receptors. $GABA_B$ receptors usually require sustained activity for activation, and $GABA_B$ receptor-mediated effects, whether mediated by enhancing potassium conductances or inhibiting calcium channels, are typically long-lasting.

III. Control of Network Activity by $GABA_B$ Receptors

The sustained synaptic activity required for $GABA_B$ receptor activation is usually met during rhythmic network activity or burst firing. This, in combination with $GABA_B$ receptor-mediated strong and long-lasting inhibitory effects, implicates $GABA_B$ receptors in the regulation of cortical network activity.

Depending on the temporal properties of network activity, $GABA_B$ receptors can (A) directly *mediate* slow network activity, (B) *modulate* the strength of fast network activity, and (C) *moderate* the relative spike timing of individual cells during network oscillations.

A. $GABA_B$ Receptor-Mediated Direct Control of Slow Network Activity

1. Cortex

In the early 1990s, Steriade and colleagues identified a slow oscillation (<1 Hz) in the cortex *in vivo* during anaesthesia and sleep (Steriade et al.,

1993a; Steriade et al., 1993c, 1993d). These slow oscillations show alternating persistently active "Up states" and quiescent "Down states," each about 1–2 s in duration. Slow oscillations survived extensive thalamic lesions as well as the transection of the corpus callosum (Steriade et al., 1993a, 1993c, 1993d). Furthermore, fully deafferented cortical slabs also exhibited slow oscillations (Timofeev et al., 2000), suggesting that such cortical activity can be generated independently of subcortical input.

The finding that slow oscillations are generated intrinsically in the cortex and can persist without thalamic input (Steriade et al., 1993d; Timofeev et al., 2000) enabled the introduction of reduced cortical preparations *in vitro* that allowed the study of the mechanisms underlying such persistent activity (Sanchez-Vives & McCormick, 2000; Shu et al., 2003).

McCormick and colleagues were the first to develop an acute slice model of the slow oscillation using ferret prefrontal and visual cortex exhibiting spontaneous Up and Down states (Sanchez-Vives & McCormick, 2000; Shu et al., 2003). Subsequently, spontaneous slow oscillations were also observed in thalamocortical slices (MacLean et al., 2005; Rigas & Castro-Alamancos, 2007) and in slices from the rodent entorhinal cortex (Cunningham et al., 2006; Mann et al., 2009). Using these preparations, it was found that the recurrent connectivity in the cortex is sufficient to drive persistent activity (Up states). Fast excitation (AMPA/kainate receptors) and inhibition ($GABA_A$ receptors) were found to scale proportionally, preventing run-away excitation that could otherwise lead to epileptiform activity (Mann et al., 2009; Shu et al., 2003).

However, the mechanisms underlying the termination of Up states are less clear. Cell-intrinsic properties and currents have been suggested to form the basis for Up-to-Down-state transitions (Bazhenov et al., 2002; Compte et al., 2003; Hill & Tononi, 2005; Milojkovic et al., 2005). However, the Up-and-Down-state transitions were found to be highly synchronous across the whole cortex, suggesting that a long-range synaptic mechanism might be at play (Volgushev et al., 2006). Using a model of Up and Down states in the rat entorhinal cortex, it was found that $GABA_B$ receptors are likely to be part of this mechanism. $GABA_B$ receptor block was found to increase Up-state duration and prevent stimulus-induced Down-state transitions (Mann et al., 2009; Fig. 2A). Over the course of the Up state, the membrane potential steadily dropped and input conductance increased. Additionally, following the Up state, Mann and colleagues observed a pronounced hyperpolarization, which was associated with a refractory period of up to 10 s during which new Up states could not be initiated. These time courses fit well with a $GABA_B$ receptor-mediated IPSP. The $GABA_B$ receptor-mediated Up-state termination might be due to presynaptic inhibition, postsynaptic hyperpolarizing currents, and/or the direct inhibition of postsynaptic calcium influx (Pérez-Garci et al., 2006), which could be necessary for the maintenance of Up states. The slow kinetics of $GABA_B$ receptors would

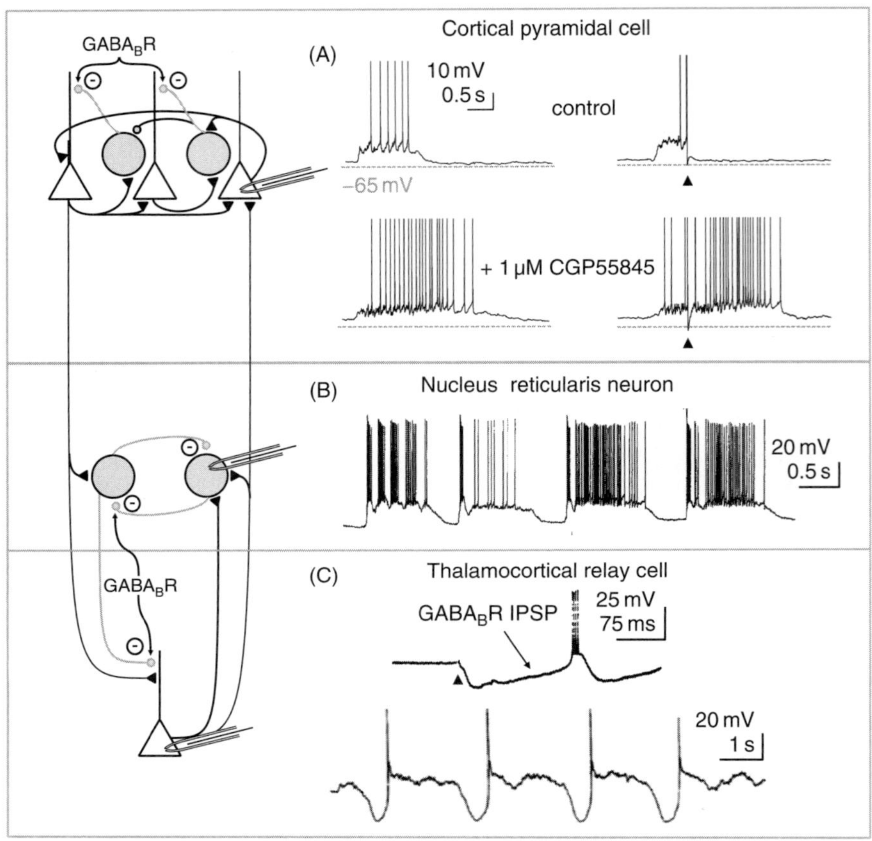

FIGURE 2 GABA$_B$ receptor mediation of slow oscillations. (A) GABA$_B$ receptors contribute to the termination of Up states. GABA$_B$ receptor block with CGP55845 reversibly increases the duration of Up states (left panel) and prevents Up-state termination by extracellular stimulation (arrowhead) in the superficial layers. (B) Slow oscillatory input from the cortex activates thalamic reticular neurons, whose reciprocal inhibition via GABA$_B$ receptor-mediated slow IPSPs facilitates rebound bursting. (C) Strong burst activity in thalamic reticular neurons activates GABA$_B$ receptors, which produce slow IPSPs in thalamocortical relay neurons (upper trace). The resultant hyperpolarization de-inactivates T-type calcium channels leading to calcium spikes (lower trace) that integrate with the ongoing slow oscillation in the cortex to produce sleep spindles and delta waves. Traces are modified, with permission, from Mann et al. (2009) (A), Steriade et al. (1996) (B), Roy et al. (1984) (C, upper trace) and Vincenzo Crunelli & Hughes (2010) (C, lower trace).

seem to match the slow time course of Up and Down states, allowing GABA$_B$ receptors to exert a strong synaptic effect and thereby control the frequency and synchronization of the slow oscillation.

2. Thalamus

The slow cortical oscillation provides rhythmic input to thalamic nuclei through corticothalamic projections. Such slow rhythmic input is thought to drive synchronized network activity through reciprocally-connected excitatory and inhibitory cells in the thalamus. These oscillations can in turn instruct cortical physiological network activity such as delta waves and sleep spindles, as well as pathological spike-and-wave discharges during generalized absence epilepsy.

Thalamocortical neurons and interneurons in the reticular formation both express T-type calcium channels, which mediate a low-threshold Ca^{2+} current. This Ca^{2+} current promotes the generation of low-threshold Ca^{2+} spike-mediated bursts of action potentials. Strong hyperpolarization of these thalamic neurons facilitates the de-inactivation of T-type calcium channels, thereby enhancing the generation of rebound action potential bursts. These bursts are characteristic for thalamic neurons and critical to the generation of normal thalamocortical rhythms, such as spindle waves during slow-wave sleep (Fig. 2B and C).

The rebound Ca^{2+} spike burst activity in thalamic reticular neurons has been shown to depend on recurrent inhibition amongst reticular neurons. This inhibition is mediated by $GABA_B$ receptors, which provide sufficient hyperpolarization to drive rebound Ca^{2+} spike burst activity (Ulrich & Huguenard, 1996). In turn, this burst activity coordinates the generation of rhythmic activity in thalamocortical neurons.

Using paired recordings from interneurons in the perigeniculate nucleus (PGN) and thalamocortical neurons in the dorsal lateral geniculate nucleus in acute slices from ferrets, Kim and colleagues dissociated the roles of $GABA_A$ and $GABA_B$ receptors in the generation of rebound burst activity (Kim et al., 1997). They showed that low-frequency (<100 Hz) discharge of PGN neurons recruits mainly $GABA_A$ receptor-mediated fast inhibition controlling the dendritic processing and probability of action potential generation in thalamocortical neurons. In contrast, high-frequency burst activity in PGN neurons resulted in large $GABA_A$ receptor-mediated compound IPSPs that were effective in eliciting rebound Ca^{2+} spike burst activity. These bursts of action potentials in thalamocortical neurons are essential for the generation of normal thalamocortical rhythms, such as spindle waves (for review, see Steriade et al., 1993b), and, through reciprocal connections, can excite more PGN and thalamic reticular neurons. Finally, prolonged and high-frequency burst discharge of PGN neurons, or perhaps the simultaneous activation of a number of PGN cells, was shown to result in the strong activation of $GABA_B$ receptors on thalamocortical neurons (Bal et al., 1995a, 1995b; Huguenard & Prince, 1994). The slow kinetics and prolonged time course of the $GABA_B$ receptor-mediated IPSPs (Fig. 2C, lower trace) resulted in the entrainment of slow synchronized oscillations, similar to those associated with the generation of spike-and-wave discharges during absence seizures (Crunelli & Leresche, 1991).

In summary, activity patterns in thalamocortical cells depend on the amplitude and duration of IPSPs received from thalamic reticular and perigeniculate interneurons. Slow-frequency tonic discharge results in $GABA_A$ receptor-mediated IPSPs of small amplitude. Burst activity triggers larger $GABA_A$ receptor-mediated IPSPs that enable the generation of rebound Ca^{2+} spike burst activity in thalamocortical neurons, which is thought to be essential for the generation of normal thalamocortical rhythms. Very strong burst activity in thalamic reticular and perigeniculate GABAergic neurons additionally recruits longer-lasting IPSPs in thalamocortical neurons through the activation of $GABA_B$ receptors. This strong inhibition results in the generation of slow synchronized oscillations that may underlie pathological network activity such as spike-and-wave discharges during generalized absence epilepsy (Crunelli & Leresche, 1991).

B. $GABA_B$ Receptors Modulate Fast Oscillations

Fast network activity occurs at too rapid a time scale for $GABA_B$ receptors to mediate this activity directly. However, through their long-lasting effects, $GABA_B$ receptors are capable of modulating the amplitude of fast oscillations such as gamma oscillations.

Gamma oscillations (30–70 Hz) have been associated with a wide range of cognitive processes, ranging from sensory processing (Gray, 1994; Singer, 1993) to memory formation (Fell et al., 2001; Sederberg et al., 2007) and selective attention (Fries et al., 2001), and even consciousness (Llinás et al., 1998). At the cellular level, gamma oscillations appear to be important for the temporal regulation of synaptic activity in the cortex and hippocampus and have thus been associated with a spike timing-dependent form of synaptic plasticity (Traub et al., 1998; Wespatat et al., 2004). Spontaneous gamma oscillations rely on recurrently connected excitatory networks, which can be found in the neocortex, amygdala, and CA3 region of the hippocampus.

Gamma oscillations can be induced *in vitro* by high-frequency stimulation (Sohal et al., 2009; Song et al., 2005; Whittington et al., 1995) and by pharmacological means (Brown et al., 2006; Buhl et al., 1998; Cunningham et al., 2003; Fisahn et al., 2004) in cortical (Cunningham et al., 2003; Sohal et al., 2009; Traub et al., 2005) and hippocampal slices (Brown et al., 2006; Buhl et al., 1998; Fisahn et al., 1998; Hormuzdi et al., 2001; Traub et al., 2000), allowing experimenters to investigate the underlying pharmacological mechanisms.

1. *Hippocampus*

In the hippocampus, gamma oscillations can be recorded *in vivo* nested within the theta rhythm during exploratory behavior (Bragin et al., 1995; Chrobak & Buzsáki, 1998; Csicsvari et al., 2003; Fell et al., 2001). They have been suggested to contribute to the encoding and retrieval of memory (Bauer et al., 2007; Hasselmo et al., 1996; Montgomery & Buzsáki, 2007; Varela et al., 2001).

The mechanisms of hippocampal gamma oscillations have been studied in detail using a number of *in vitro* models (Fisahn et al., 1998; Fischer et al., 2002; Hájos et al., 2000; Lebeau et al., 2002; Palhalmi et al., 2004; Whittington et al., 1995). It is becoming apparent that rhythmic inhibition provided by fast-spiking, perisomatic-targeting interneurons is crucial for entraining the hippocampal network in the gamma frequency band. Perisomatic-targeting interneurons are synchronized primarily by collaterals from their targets, CA3 pyramidal cells. It is this recurrent feedback loop that allows local gamma oscillations to control the firing of CA3 pyramidal cells, both spatially and temporally (Mann & Paulsen, 2005).

Fast, phasic inhibition through $GABA_A$ receptors seems to be an essential ingredient for the precise control of gamma oscillations. But recent data also implicate $GABA_B$ receptors in the control of generating such network activity. $GABA_B$ receptor block was shown to prolong stimulus-induced gamma oscillations, suggesting that they could serve as a synaptic control mechanism for fast inhibition-driven network oscillations (Whittington et al., 1995). This idea was expanded by studies using a kainate-based gamma oscillation model (Brown et al., 2007). In this model, transient stimulation was found to suppress re-emergent gamma oscillations for several hundred milliseconds, reminiscent of the time course of the $GABA_B$ receptor-mediated slow IPSP (Davies et al., 1990; Dutar & Nicoll, 1988; Isaacson et al., 1993). When applying paired pulses (1–3.3 Hz), this suppressive effect was smaller for the second pulse, suggesting a further role of $GABA_B$ autoreceptors in shaping the postsynaptic $GABA_B$ receptor-mediated inhibition. It has been suggested that the pro-cognitive effects of $GABA_B$ receptor antagonists (Getova & Bowery, 1998; Stäubli et al., 1999) could, at least in part, be mediated by the inhibition of $GABA_B$ receptor-mediated suppression of gamma oscillations.

2. *Neocortex*

Similar mechanisms control the emergence of gamma oscillations in the neocortex. Experimental (Atencio & Schreiner, 2008; Buhl et al., 1998; Cardin et al., 2009; Sohal et al., 2009) and computational studies (Börgers et al., 2005; Brunel & Wang, 2003; Cunningham et al., 2004) have shown that coupled networks consisting of interneurons and pyramidal cells can support synchronous network activity. Fast-spiking interneurons have a very high probability of connecting to pyramidal cells and thus present a major source of intracortical inhibition (Beierlein et al., 2003; Gonchar & Burkhalter, 1999; Holmgren et al., 2003; Thomson et al., 1996). Paired recordings from fast-spiking interneurons and pyramidal cells in the auditory cortex have shown that pairs with intersomatic distances of less than 50 μm have a very high probability of being reciprocally connected (Oswald et al., 2009). In contrast, at intersomatic distances of 50–100 μm, only about half the pairs were reciprocally connected (Fig. 3A). Selective stimulation of the inhibitory neurons showed that $GABA_B$ receptor-mediated inhibition was stronger for the non-reciprocally connected pairs that were further

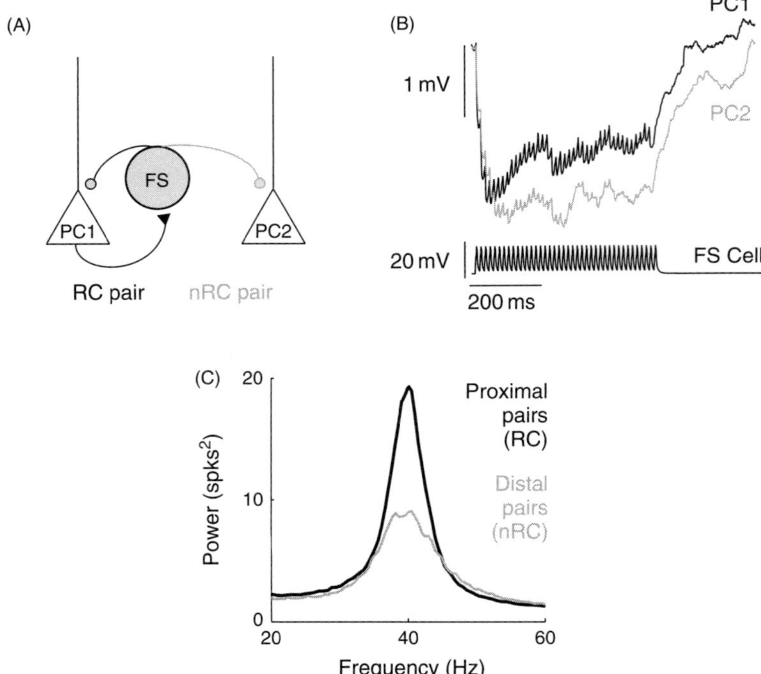

FIGURE 3 $GABA_B$ receptor modulation of gamma oscillations. (A) In the auditory cortex, simultaneous whole-cell recordings from fast-spiking interneurons (FS) and pyramidal cells (PC) reveal that for intersomatic distances <50 μm, most inhibitory connections occur in reciprocally-connected (RC) pairs whilst at greater distances, inhibitory connections are equally likely in RC and non-reciprocally connected (nRC) pairs. (B) 80 Hz stimulation of presynaptic FS cells produces a slow IPSP in nRC pyramidal cells, which is mediated by $GABA_B$ receptors. (C) A biophysical model shows that FS-induced feedforward gamma oscillations are strongest when the network inputs are confined to a small area and attenuated in distal pyramidal cells through the activation of $GABA_B$ receptors. (B) and (C) are modified, with permission, from Oswald et al. (2009).

apart (Fig. 3B). Consistent with other experimental data (Cardin et al., 2009), a computer model based on these data showed that local inputs elicited strong gamma oscillations (Oswald et al., 2009). In contrast, spatially more distributed activity recruited more $GABA_B$ receptor-mediated inhibition, resulting in a suppression of the self-emergent gamma activity (Fig. 3C; Oswald et al., 2009).

These data suggest that $GABA_B$ receptors allow local network activity to be modulated depending on the spatial distribution of afferent sensory input. This is of significance given the topographic mapping of afferent inputs in the sensory cortices. The precise mechanisms by which $GABA_B$ receptor activation is spatially segregated is not clear, but different activity patterns in reciprocally connected and non-connected pairs could result in different levels of activity-dependent internalization (Balasubramanian et al., 2004), (de)phosphorylation (Fairfax et al., 2004), or a decrease in expression levels (Vargas et al., 2008).

From these data, it becomes clear that $GABA_B$ receptor-mediated responses can have a strong modulatory effect on gamma activity in hippocampal and neocortical networks.

C. $GABA_B$ Receptors Moderate Spike Timing During Hippocampal Theta Activity

Theta oscillations (4–10 Hz) have been recorded in the hippocampus and associated cortical regions; they are most consistently present during REM sleep (Jouvet, 1969), as well as during various types of locomotor activities described by the subjective terms "voluntary," "preparatory," "orienting," or "exploratory" (Vanderwolf, 1969).

Theta oscillations are thought to provide a temporal metric, enabling the sequential activation of hippocampal and cortical cell assemblies, which is necessary for processes such as memory acquisition, recollection, and spatial orientation (Hasselmo, 2005; Lisman & Buzsáki, 2008). Whilst the circuit elements required to generate theta oscillations are already present in the hippocampus and cortex (Goutagny et al., 2009), theta oscillations are thought to be orchestrated by inhibitory input from subcortical structures. It has been shown that lesions and inactivation of the medial septum and diagonal band of Broca (MS-DBB), structures that provide cholinergic and GABAergic input to the hippocampus and entorhinal cortex, both resulted in the abolition of theta oscillations in all cortical targets, suggesting a major pacemaker role for the MS-DBB (Petsche et al., 1962). Similarly, stimulation of other subcortical nuclei was capable of inducing theta (for review, see Bland, 1986; Buzsáki, 2002).

The studies outlined above suggest that $GABA_A$ receptor-mediated inhibition is involved in the generation of theta oscillations. This leaves

$GABA_B$ receptors to moderate other tasks during theta activity. However, the precise roles of $GABA_B$ receptors in theta oscillations is not clear. Since at least 60% of inhibitory basket cells discharge synchronously during the theta cycle (Csicsvari et al., 1999), it is conceivable that the amount of GABA released could be sufficient to activate $GABA_B$ receptors (Dvorak-Carbone & Schuman, 1999; Scanziani, 2000). Indeed, $GABA_B$ receptor block during theta oscillations generated in hippocampal slices by muscarinic activation was found to increase the power of the theta oscillation (Konopacki et al., 1997). Similarly, *in vivo* administration of the $GABA_B$ receptor blocker CGP35348 was shown to increase the power of theta oscillations during walking, and septal $GABA_B$ receptor block resulted in an increase in theta harmonics during walking (Leung & Shen, 2007). Since *in vivo* recordings from hippocampal pyramidal cells have shown that theta oscillations have an amplitude minimum and phase reversal between −75 and −60 mV (Soltesz & Deschênes, 1993; Ylinen et al., 1995), it has been suggested that the hyperpolarization during theta is mediated by chloride rather than potassium currents. This finding makes it likely that, at least in part, the $GABA_B$ receptor-mediated effect on theta oscillations is due to presynaptic inhibition.

$GABA_B$ receptors might also help hippocampal pyramidal cells to keep intrinsic and extrinsic afferent inputs separate. During ongoing theta oscillations synaptic inputs can advance or delay action potentials depending on whether the input arrives at the ascending or descending phase of the theta cycle (Kwag & Paulsen, 2009). Hippocampal CA1 pyramidal cells receive two major inputs: intrinsic input from the CA3 area (Schaffer collaterals) and extrinsic input from the entorhinal cortex (tempero-ammonic input) (Fig. 4A). The tempero-ammonic pathway was found to recruit more $GABA_B$ receptor-mediated inhibition resulting in a more pronounced spike delay than for the Schaffer collateral input (Kwag & Paulsen, 2009; Fig. 4B and C). In this way, $GABA_B$ receptors might enable pyramidal cells to integrate intrinsic and extrinsic input differently during ongoing theta network activity (Otmakhova & Lisman, 2004).

Moreover, various *in silico* models have emphasized a role for slow inhibition mediated by $GABA_B$ receptors during network activity. For instance, a detailed computational model of the CA3 region by Wallenstein and Hasselmo (1997) showed that learning and recall performance is poor in the absence of $GABA_B$ receptor-mediated inhibition. They found that $GABA_B$ receptors predominantly suppress intrinsic, but not afferent, excitatory and inhibitory transmission during the early phases of the theta cycle, allowing for afferent input carrying information about location to be dominant before the waning of this suppression towards the end of the cycle. This would favor recall of the location sequence (Wallenstein & Hasselmo, 1997).

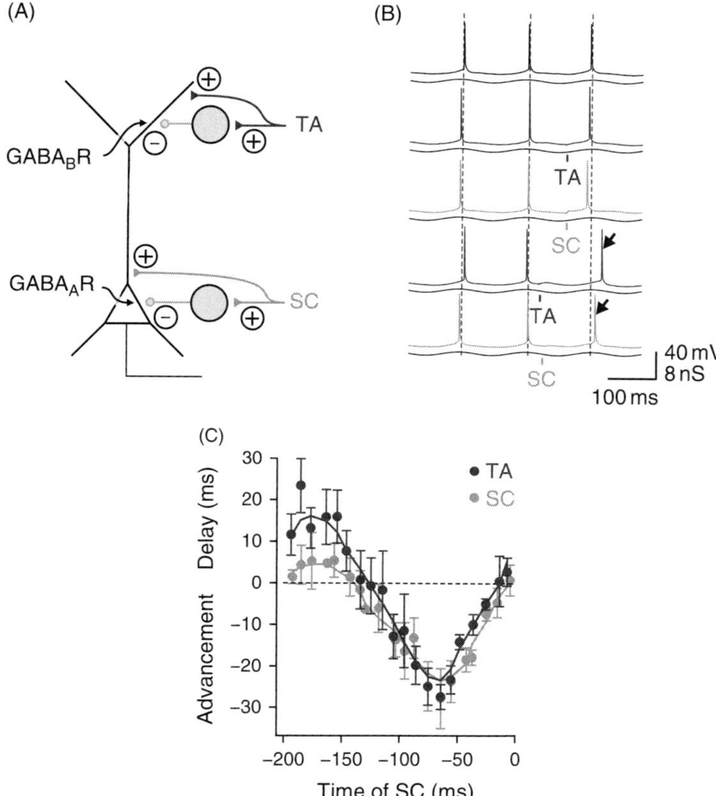

FIGURE 4 GABA$_B$ receptor moderation of spike timing during theta oscillations. (A) Hippocampal CA1 pyramidal cells receive two major excitatory input pathways: Schaffer collaterals (SC) from CA3 pyramidal cells and tempero-ammonic input (TA) from the entorhinal cortex. Feedforward inhibition in SC input is mediated by GABA$_A$ receptors whilst feedforward inhibition in TA input is additionally mediated by GABA$_B$ receptors. (B) During ongoing theta oscillations, CA1 cells spike reliably at the peak of the oscillation. Synaptic input during the ascending phase of the oscillation advances postsynaptic spikes, whilst input during the descending phase delays the spike (arrow). (C) The slower, but stronger feedforward inhibition mediated by GABA$_B$ receptors in TA input results in a larger spike delay than that in SC input. (B) and (C) are modified, with permission, from Kwag & Paulsen (2009).

D. GABA$_B$ Receptors in Pathological Network Activity

It is becoming apparent that GABA$_B$ receptors are essential for the stability of cortical network activity by *mediating*, *modulating*, and *moderating* such activity.

This notion is emphasized by the fact that high doses of GABA$_B$ receptor antagonists disrupt normal hippocampal and cortical oscillations and eventually lead to epiletiform activity (Leung et al., 2005; Vergnes et al., 1997). Similarly, $GABA_B$ receptor knock-out mice are prone to developing spontaneous seizures (Prosser et al., 2001; Schuler et al., 2001).

Because $GABA_B$ receptor activation can directly or indirectly exert a wide range of stimulating and dampening effects on excitatory and inhibitory synaptic transmission, it is not surprising that $GABA_B$ receptor antagonism can also have beneficial effects. For instance, in genetically seizure-prone animals, $GABA_B$ antagonists can block absence seizures (Hosford et al., 1992; Liu et al., 1992).

To understand the role of $GABA_B$ receptors in pathology and identify possible therapeutic targets, we will first have to elucidate the precise mechanisms of action. We have seen that $GABA_B$ receptors are instrumental in mediating the control of slow cortical and thalamic oscillations, including delta waves and sleep spindles, which are thought to coordinate the reactivation and redistribution of hippocampus-dependent memories to neocortical sites (Diekelmann & Born, 2010). Disturbances to these slow oscillations are known to produce sleeps disorders and associated memory deficits (for review, see Diekelmann & Born, 2010).

Severe disruption of the thalamic oscillator can even result in epileptiform activity and is thought to underlie juvenile absence seizures. The defining electroencephalogram activity during absence seizures is a large 3 Hz spike-and-wave complex. The circuitry underlying this spike-and-wave discharge involves the activation of $GABA_B$ receptors and is relatively well understood. Recent work using a genetic rat model of absence epilepsy has traced the origin of these seizures to increased excitation in the sensory cortices (Meeren et al., 2005). This activity spreads through the cortex and, via corticothalamic neurons, triggers bursts of action potentials in a large number of inhibitory reticular thalamic neurons. This burst results in a pronounced hyperpolarization of thalamocortical neurons in the ventrobasal nucleus via $GABA_B$ receptors, which, through the deinactivation of T-type calcium currents, results in a slower, more synchronized output that triggers the spike-and-wave discharges in the cortex. These discharges further excite reticular cells, rapidly resulting in the generalization of paroxysmal activity in the whole cortex (for review, see Crunelli & Leresche, 2002).

Using biophysical models of the thalamic and cortical neurons during spike-and-wave discharges, it has been shown that the cortical spike part coincides with thalamic burst and the wave component is generated by the $GABA_B$ receptor-mediated slow potassium current (Destexhe, 1998). The unique properties of $GABA_B$ receptor-mediated inhibition are also thought to underlie the characteristic 3 Hz spike-and-wave discharge. Consistent with these simulations, $GABA_B$ receptor agonist application in the ventrobasal and reticular thalamic nuclei of a rat with spontaneous absence seizures was shown to intensify the cortical discharges (Marescaux et al., 1992). Furthermore, blocking $GABA_B$ receptors in the ventrobasal thalamus, but also in the reticular nucleus (Hosford et al., 1992; Liu et al., 1992) and even in the cortex (Prosser et al., 2001), prevents spike-and-wave discharges in experimental models and thus presents a possible pharmacological target. Indeed, it is thought that the anti-absence drug clonazepam acts by diminishing $GABA_B$ receptor-mediated IPSPs in thalamocortical neurons, thereby reducing their tendency to burst in synchrony (Gibbs et al., 1996; Huguenard & Prince, 1994).

These data all suggest a significant role for $GABA_B$ receptors in the generation and control of epileptiform activity. This identifies the $GABA_B$ receptor as a promising therapeutic target for the treatment of seizures.

Furthermore, we have seen that $GABA_B$ receptors are also involved in modulating fast network activity, such as gamma oscillations, which in turn are implicated in cognitive processes. Psychiatric disorders such as schizophrenia are linked to changes in gamma activity (Uhlhaas & Singer, 2010), making the modulatory effect of $GABA_B$ receptor activation a possible therapeutic target here as well (Kantrowitz et al., 2009). Based on the beneficial effects of weak $GABA_B$ receptor antagonism on gamma and theta oscillations, a couple of behavioral studies have reported enhancement of cognitive performance in several different animal species following blockade of $GABA_B$ receptors (Carletti et al., 1993; Mondadori et al., 1993). Further studies are required to elucidate a possible link between the control of fast oscillations through $GABA_B$ receptors and psychiatric disorders associated with disturbed fast oscillations.

IV. Conclusion

Since Norman Bowery's seminal experiments in the 1980s, we have learned a great deal about the effect of $GABA_B$ receptor activation at the cellular level. However, the functional role of $GABA_B$ receptors at the network level is only just starting to emerge. Here, we have outlined a framework for understanding the contributions of $GABA_B$ receptors in helping to mediate, modulate, and moderate different types of physiological and pathological cortical network activity. We have reviewed how $GABA_B$ receptor activation in cortical and subcortical structures can help *mediate* the slow oscillation. Since the disturbance of the slow oscillation is associated with sleep disorders, memory deficits, and the generation of epileptiform activity, $GABA_B$ receptors present an attractive therapeutic target for the treatment of these conditions. Moreover, we have reviewed recent literature that implicates $GABA_B$ receptors in the *modulation* of the power and spatial profile of fast network oscillations. These fast oscillations have been associated with cognitive functions, implicating the $GABA_B$ receptor in potential treatments of cognitive disorders. Finally, we also present evidence for a role of $GABA_B$ receptors in the *moderation* of the relative spike timing of individual neurons during theta oscillations, thereby attributing $GABA_B$ receptors a significant role in the synaptic integration during large-scale network activity. The diffuse nature of many $GABA_B$ receptor-mediated effects presents challenges to the experimental study of the physiological role of these receptors in cortical microcircuits. Recent developments of optogenetic tools promise to deliver new powerful experimental techniques that will enable more precisely targeted studies of $GABA_B$ receptor function in both physiological and pathological cortical network activity.

Conflict of interest: The authors declare no conflict of interest.

Abbreviations

AMPA	α-amino-3-hydroxy-5-methyl-4-isoxazolepropionic acid
GABA	γ-aminobutyric acid
IPSP	inhibitory postsynaptic potential
NMDA	*N*-methyl-D-aspartate

References

Agnati, L. F., Ferré, S., Lluis, C., Franco, R., & Fuxe, K. (2003). Molecular mechanisms and therapeutical implications of intramembrane receptor/receptor interactions among heptahelical receptors with examples from the striatopallidal gaba neurons. *Pharmacological Reviews*, *55*, 509–550.

Atencio, C. A., & Schreiner, C. E. (2008). Spectrotemporal processing differences between auditory cortical fast-spiking and regular-spiking neurons. *Journal of Neuroscience*, *28*, 3897–3910.

Bal, T., von Krosigk, M., & McCormick, D. A. (1995a). Role of the ferret perigeniculate nucleus in the generation of synchronized oscillations in vitro. *Journal of Physiology*, *483*, 665–685.

Bal, T., von Krosigk, M., & McCormick, D. A. (1995b). Synaptic and membrane mechanisms underlying synchronized oscillations in the ferret lateral geniculate nucleus in vitro. *Journal of Physiology*, *483*, 641–663.

Balasubramanian, S., Teissére, J. A., Raju, D. V., & Hall, R. A. (2004). Hetero-oligomerization between $GABA_A$ and $GABA_B$ receptors regulates $GABA_B$ receptor trafficking. *Journal of Biological Chemistry*, *279*, 18840–18850.

Bauer, E. P., Paz, R., & Paré, D. (2007). Gamma oscillations coordinate amygdalo–rhinal interactions during learning. *Journal of Neuroscience*, *27*, 9369–9379.

Bazhenov, M., Timofeev, I., Steriade, M., & Sejnowski, T. J. (2002). Model of thalamocortical slow-wave sleep oscillations and transitions to activated states. *Journal of Neuroscience*, *22*, 8691–8704.

Beierlein, M., Gibson, J. R., & Connors, B. W. (2003). Two dynamically distinct inhibitory networks in layer 4 of the neocortex. *Journal of Neurophysiology*, *90*, 2987–3000.

Bland, B. H. (1986). The physiology and pharmacology of hippocampal formation theta rhythms. *Progress in Neurobiology*, *26*, 1–54.

Börgers, C., Epstein, S., & Kopell, N. J. (2005). Background gamma rhythmicity and attention in cortical local circuits: A computational study. *Proceedings of the National Academy of Sciences of the United States of America*, *102*, 7002–7007.

Bowery, N. G. (1993). $GABA_B$ receptor pharmacology. *Annual Review of Pharmacology and Toxicology*, *33*, 109–147.

Bowery, N. G., Bettler, B., Froestl, W., Gallagher, J. P., Marshall, F., Raiteri, M., et al. (2002). International Union of Pharmacology. XXXIII. Mammalian γ-aminobutyric acid(B) receptors: structure and function. *Pharmacological Reviews*, *54*, 247–264.

Bowery, N. G., Doble, A., Hill, D. R., Hudson, A. L., Shaw, J. S., Turnbull, M. J., et al. (1981a). Bicuculline-insensitive GABA receptors on peripheral autonomic nerve terminals. *European Journal of Pharmacology*, *71*, 53–70.

Bowery, N. G., Doble, A., Hill, D. R., Hudson, A. L., Turnbull, M. J., & Warrington, R. (1981b). Structure/activity studies at a baclofen-sensitive, bicuculline-insensitive GABA receptor. *Advances in Biochemical Psychopharmacology*, *29*, 333–341.

Bowery, N. G., Hill, D. R., & Hudson, A. L. (1982). Evidence that sl75102 is an agonist at $GABA_B$ as well as $GABA_A$ receptors. *Neuropharmacology*, *21*, 391–395.

Bowery, N. G., Hill, D. R., & Hudson, A. L. (1983). Characteristics of $GABA_B$ receptor binding sites on rat whole brain synaptic membranes. *British Journal of Pharmacology*, *78*, 191–206.

Bowery, N. G., Hill, D. R., Hudson, A. L., Doble, A., Middlemiss, D. N., Shaw, J., et al. (1980). (–)Baclofen decreases neurotransmitter release in the mammalian CNS by an action at a novel GABA receptor. *Nature*, *283*, 92–94.

Bragin, A., Jandó, G., Nádasdy, Z., Hetke, J., Wise, K., & Buzsáki, G. (1995). Gamma (40–100 hz) oscillation in the hippocampus of the behaving rat. *Journal of Neuroscience*, *15*, 47–60.

Brown, J. T., Davies, C. H., & Randall, A. D. (2007). Synaptic activation of GABA(b) receptors regulates neuronal network activity and entrainment. *European Journal of Neuroscience*, *25*, 2982–2990.

Brown, J. T., Teriakidis, A., & Randall, A. D. (2006). A pharmacological investigation of the role of GLU_{K5}-containing receptors in kainate-driven hippocampal gamma band oscillations. *Neuropharmacology*, *50*, 47–56.

Brunel, N., & Wang, X.-J. (2003). What determines the frequency of fast network oscillations with irregular neural discharges? I. Synaptic dynamics and excitation-inhibition balance. *Journal of Neurophysiology*, *90*, 415–430.

Buhl, E. H., Halasy, K., & Somogyi, P. (1994). Diverse sources of hippocampal unitary inhibitory postsynaptic potentials and the number of synaptic release sites. *Nature*, *368*, 823–828.

Buhl, E. H., Tamás, G., & Fisahn, A. (1998). Cholinergic activation and tonic excitation induce persistent gamma oscillations in mouse somatosensory cortex in vitro. *Journal of Physiology*, *513*, 117–126.

Buzsáki, G. (2002). Theta oscillations in the hippocampus. *Neuron*, *33*, 325–340.

Cardin, J., Carlén, M., Meletis, K., Knoblich, U., Zhang, F., Deisseroth, K., et al. (2009). Driving fast-spiking cells induces gamma rhythm and controls sensory responses. *Nature*, *459*, 663–667.

Carletti, R., Libri, V., & Bowery, N. G. (1993). The $GABA_B$ antagonist CGP36742 enhances spatial learning performance and antagonises baclofen-induced amnesia in mice. *British Journal of Pharmacology Supplementum*, *109*, P74.

Chrobak, J. J., & Buzsáki, G. (1998). Gamma oscillations in the entorhinal cortex of the freely behaving rat. *Journal of Neuroscience*, *18*, 388–398.

Compte, A., Sanchez-Vives, M. V., McCormick, D. A., & Wang, X. -J. (2003). Cellular and network mechanisms of slow oscillatory activity (<1 hz) and wave propagations in a cortical network model. *Journal of Neurophysiology*, *89*, 2707–2725.

Connors, B. W., Malenka, R. C., & Silva, L. R. (1988). Two inhibitory postsynaptic potentials, and $GABA_A$ and $GABA_B$ receptor-mediated responses in neocortex of rat and cat. *Journal of Physiology (Paris)*, *406*, 443–468.

Crunelli, V., & Hughes, S. W. (2010). The slow (<1 hz) rhythm of non-rem sleep: A dialogue between three cardinal oscillators. *Nature Neuroscience*, *13*, 9–17.

Crunelli, V., & Leresche, N. (1991). A role for $GABA_B$ receptors in excitation and inhibition of thalamocortical cells. *Trends in Neurosciences*, *14*, 16–21.

Crunelli, V., & Leresche, N. (2002). Childhood absence epilepsy: genes, channels, neurons and networks. *Nature Reviews Neuroscience*, *3*, 371–382.

Csicsvari, J., Hirase, H., Czurkó, A., Mamiya, A., & Buzsáki, G. (1999). Oscillatory coupling of hippocampal pyramidal cells and interneurons in the behaving rat. *Journal of Neuroscience*, *19*, 274–287.

Csicsvari, J., Jamieson, B., Wise, K. D., & Buzsáki, G. (2003). Mechanisms of gamma oscillations in the hippocampus of the behaving rat. *Neuron*, *37*, 311–322.

Cunningham, M. O., Davies, C. H., Buhl, E. H., Kopell, N., & Whittington, M. A. (2003). Gamma oscillations induced by kainate receptor activation in the entorhinal cortex in vitro. *Journal of Neuroscience*, *23*, 9761–9769.

Cunningham, M. O., Halliday, D. M., Davies, C. H., Traub, R. D., Buhl, E. H., & Whittington, M. A. (2004). Coexistence of gamma and high-frequency oscillations in rat medial entorhinal cortex in vitro. *Journal of Physiology*, *559*, 347–353.

Cunningham, M. O., Pervouchine, D. D., Racca, C., Kopell, N. J., Davies, C. H., Jones, R.S.G., et al. (2006). Neuronal metabolism governs cortical network response state. *Proceedings of the National Academy of Sciences of the United States of America*, *103*, 5597–5601.

Davies, C. H., Davies, S. N., & Collingridge, G. L. (1990). Paired-pulse depression of monosynaptic GABA-mediated inhibitory postsynaptic responses in rat hippocampus. *Journal of Physiology*, *424*, 513–531.

Davies, C. H., Pozza, M. F., & Collingridge, G. L. (1993). CGP 55845a: A potent antagonist of $GABA_B$ receptors in the ca1 region of rat hippocampus. *Neuropharmacology*, *32*, 1071–1073.

Destexhe, A. (1998). Spike-and-wave oscillations based on the properties of $GABA_B$ receptors. *Journal of Neuroscience*, *18*, 9099–9111.

Diekelmann, S., & Born, J. (2010). The memory function of sleep. *Nature Reviews Neuroscience*, *11*, 114–126.

Dittman, J. S., & Regehr, W. G. (1996). Contributions of calcium-dependent and calcium-independent mechanisms to presynaptic inhibition at a cerebellar synapse. *Journal of Neuroscience*, *16*, 1623–1633.

Dutar, P., & Nicoll, R. A. (1988). A physiological role for $GABA_B$ receptors in the central nervous system. *Nature*, *332*, 156–158.

Dvorak-Carbone, H., & Schuman, E. M. (1999). Long-term depression of temporoammonic-ca1 hippocampal synaptic transmission. *Journal of Neurophysiology*, *81*, 1036–1044.

Fairfax, B. P., Pitcher, J. A., Scott, M.G.H., Calver, A. R., Pangalos, M. N., Moss, S. J., et al. (2004). Phosphorylation and chronic agonist treatment atypically modulate $GABA_B$ receptor cell surface stability. *Journal of Biological Chemistry*, *279*, 12565–12573.

Fell, J., Klaver, P., Lehnertz, K., Grunwald, T., Schaller, C., Elger, C. E., et al. (2001). Human memory formation is accompanied by rhinal–hippocampal coupling and decoupling. *Nature Neuroscience*, *4*, 1259–1264.

Fisahn, A., Contractor, A., Traub, R. D., Buhl, E. H., Heinemann, S. F., & McBain, C. J. (2004). Distinct roles for the kainate receptor subunits GluR5 and GluR6 in kainate-induced hippocampal gamma oscillations. *Journal of Neuroscience*, *24*, 9658–9668.

Fisahn, A., Pike, F. G., Buhl, E. H., & Paulsen, O. (1998). Cholinergic induction of network oscillations at 40 hz in the hippocampus in vitro. *Nature*, *394*, 186–189.

Fischer, Y., Wittner, L., Freund, T. F., & Gähwiler, B. H. (2002). Simultaneous activation of gamma and theta network oscillations in rat hippocampal slice cultures. *Journal of Physiology*, *539*, 857–868.

Fries, P., Reynolds, J. H., Rorie, A. E., & Desimone, R. (2001). Modulation of oscillatory neuronal synchronization by selective visual attention. *Science*, *291*, 1560–1563.

Galvez, T., Duthey, B., Kniazeff, J., Blahos, J., Rovelli, G., Bettler, B., et al. (2001). Allosteric interactions between GB1 and GB2 subunits are required for optimal GABA$_B$ receptor function. *EMBO Journal*, *20*, 2152–2159.

Getova, D., & Bowery, N. G. (1998). The modulatory effects of high affinity GABA$_B$ receptor antagonists in an active avoidance learning paradigm in rats. *Psychopharmacology*, *137*, 369–373.

Gibbs, J. W., Schroder, G. B., & Coulter, D. A. (1996). GABA$_A$ receptor function in developing rat thalamic reticular neurons: whole cell recordings of GABA-mediated currents and modulation by clonazepam. *Journal of Neurophysiology*, *76*, 2568–2579.

Gonchar, Y., & Burkhalter, A. (1999). Differential subcellular localization of forward and feedback interareal inputs to parvalbumin expressing GABAergic neurons in rat visual cortex. *Journal of Comparative Neurology*, *406*, 346–360.

Goutagny, R., Jackson, J., & Williams, S. (2009). Self-generated theta oscillations in the hippocampus. *Nature Neuroscience*, *12*, 1491–1493.

Gray, J. A. (1994). A general model of the limbic system and basal ganglia: applications to schizophrenia and compulsive behavior of the obsessive type. *Revue Neurologique*, *150*, 605–613.

Hájos, N., Katona, I., Naiem, S. S., MacKie, K., Ledent, C., Mody, I., et al. (2000). Cannabinoids inhibit hippocampal GABAergic transmission and network oscillations. *European Journal of Neuroscience*, *12*, 3239–3249.

Hasselmo, M. E. (2005). What is the function of hippocampal theta rhythm?—Linking behavioral data to phasic properties of field potential and unit recording data. *Hippocampus*, *15*, 936–949.

Hasselmo, M. E., Wyble, B. P., & Wallenstein, G. V. (1996). Encoding and retrieval of episodic memories: role of cholinergic and GABAergic modulation in the hippocampus. *Hippocampus*, *6*, 693–708.

Hill, S., & Tononi, G. (2005). Modeling sleep and wakefulness in the thalamocortical system. *Journal of Neurophysiology*, *93*, 1671–1698.

Holmgren, C., Harkany, T., Svennenfors, B., & Zilberter, Y. (2003). Pyramidal cell communication within local networks in layer 2/3 of rat neocortex. *Journal of Physiology*, *551*, 139–153.

Hormuzdi, S. G., Pais, I., LeBeau, F. E., Towers, S. K., Rozov, A., Buhl, E. H., et al. (2001). Impaired electrical signaling disrupts gamma frequency oscillations in connexin 36-deficient mice. *Neuron*, *31*, 487–495.

Hosford, D. A., Clark, S., Cao, Z., Wilson, W. A., Lin, F. H., Morrisett, R. A., et al. (1992). The role of GABA$_B$ receptor activation in absence seizures of lethargic (lh/lh) mice. *Science*, *257*, 398–401.

Huguenard, J. R., & Prince, D. A. (1994). Intrathalamic rhythmicity studied in vitro: nominal t-current modulation causes robust antioscillatory effects. *Journal of Neuroscience*, *14*, 5485–5502.

Isaacson, J. S., Solís, J. M., & Nicoll, R. A. (1993). Local and diffuse synaptic actions of GABA in the hippocampus. *Neuron*, *10*, 165–175.

Jones, K. A., Borowsky, B., Tamm, J. A., Craig, D. A., Durkin, M. M., Dai, M., et al. (1998). GABA$_B$ receptors function as a heteromeric assembly of the subunits GABA$_B$R1 and GABA$_B$R2. *Nature*, *396*, 674–679.

Jouvet, M. (1969). Biogenic amines and the states of sleep. *Science*, *163*, 32–41.

Kantrowitz, J., Citrome, L., & Javitt, D. (2009). GABA$_B$ receptors, schizophrenia and sleep dysfunction: A review of the relationship and its potential clinical and therapeutic implications. *CNS Drugs*, *23*, 681–691.

Kaupmann, K., Malitschek, B., Schuler, V., Heid, J., Froestl, W., Beck, P., et al. (1998). $GABA_B$-receptor subtypes assemble into functional heteromeric complexes. *Nature*, *396*, 683–687.

Kim, U., Sanchez-Vives, M. V., & McCormick, D. A. (1997). Functional dynamics of gabaergic inhibition in the thalamus. *Science*, *278*, 130–134.

Konopacki, J., Golebiewski, H., Eckersdorf, B., Blaszczyk, M., & Grabowski, R. (1997). Theta-like activity in hippocampal formation slices: the effect of strong disinhibition of $GABA_A$ and $GABA_B$ receptors. *Brain Research*, *775*, 91–98.

Kwag, J., & Paulsen, O. (2009). The timing of external input controls the sign of plasticity at local synapses. *Nature Neuroscience*, *12*, 1219–1221.

Lambert, N. A., & Wilson, W. A. (1994). Temporally distinct mechanisms of use-dependent depression at inhibitory synapses in the rat hippocampus in vitro. *Journal of Neurophysiology*, *72*, 121–130.

Lebeau, F.E.N., Towers, S. K., Traub, R. D., Whittington, M. A., & Buhl, E. H. (2002). Fast network oscillations induced by potassium transients in the rat hippocampus in vitro. *Journal of Physiology*, *542*, 167–179.

Lei, S., & McBain, C. J. (2003). $GABA_B$ receptor modulation of excitatory and inhibitory synaptic transmission onto rat CA3 hippocampal interneurons. *Journal of Physiology (Paris)*, *546*, 439–453.

Leung, L. S., Canning, K. J., & Shen, B. (2005). Hippocampal afterdischarges after $GABA_B$-receptor blockade in the freely moving rat. *Epilepsia*, *46*, 203–216.

Leung, L. S., & Shen, B. (2007). $GABA_B$ receptor blockade enhances theta and gamma rhythms in the hippocampus of behaving rats. *Hippocampus*, *17*, 281–291.

Lisman, J., & Buzsáki, G. (2008). A neural coding scheme formed by the combined function of gamma and theta oscillations. *Schizophrenia Bulletin*, *34*, 974–980.

Liu, Z., Vergnes, M., Depaulis, A., & Marescaux, C. (1992). Involvement of intrathalamic $GABA_B$ neurotransmission in the control of absence seizures in the rat. *Neuroscience*, *48*, 87–93.

Llinás, R., Ribary, U., Contreras, D., & Pedroarena, C. (1998). The neuronal basis for consciousness. *Philosophical Transactions of the Royal Society of London. Series B, Biological Sciences*, *353*, 1841–1849.

Lopez-Bendito, G., Shigemoto, R., Kulik, A., Paulsen, O., Fairen, A., & Lujan, R. (2002). Expression and distribution of metabotropic GABA receptor subtypes $GABA_BR1$ and $GABA_BR2$ during rat neocortical development. *European Journal of Neuroscience*, *15*, 1766–1778.

López-Bendito, G., Shigemoto, R., Kulik, A., Vida, I., Fairén, A., & Luján, R. (2004). Distribution of metabotropic GABA receptor subunits $GABA_{B1a/b}$ and $GABA_{B2}$ in the rat hippocampus during prenatal and postnatal development. *Hippocampus*, *14*, 836–848.

Lüscher, C., Jan, L. Y., Stoffel, M., Malenka, R. C., & Nicoll, R. A. (1997). G protein-coupled inwardly rectifying K^+ channels (GIRKS) mediate postsynaptic but not presynaptic transmitter actions in hippocampal neurons. *Neuron*, *19*, 687–695.

MacLean, J. N., Watson, B. O., Aaron, G. B., & Yuste, R. (2005). Internal dynamics determine the cortical response to thalamic stimulation. *Neuron*, *48*, 811–823.

Mann, E. O., Kohl, M. M., & Paulsen, O. (2009). Distinct roles of $GABA_A$ and $GABA_B$ receptors in balancing and terminating persistent cortical activity. *Journal of Neuroscience*, *29*, 7513–7518.

Mann, E. O., & Paulsen, O. (2005). Mechanisms underlying gamma ("40 hz") network oscillations in the hippocampus — a mini-review. *Progress in Biophysics and Molecular Biology*, *87*, 67–76.

Marescaux, C., Vergnes, M., & Bernasconi, R. (1992). $GABA_B$ receptor antagonists: potential new anti-absence drugs. *Journal of Neural Transmission. Supplementum*, *35*, 179–188.

Meeren, H., van Luijtelaar, G., Lopes da Silva, F., & Coenen, A. (2005). Evolving concepts on the pathophysiology of absence seizures: the cortical focus theory. *Archives of Neurology*, *62*, 371–376.

Milojkovic, B. A., Radojicic, M. S., & Antic, S. D. (2005). A strict correlation between dendritic and somatic plateau depolarizations in the rat prefrontal cortex pyramidal neurons. *Journal of Neuroscience*, *25*, 3940–3951.

Mintz, I. M., & Bean, B. P. (1993). GABA$_B$ receptor inhibition of P-type Ca^{2+} channels in central neurons. *Neuron*, *10*, 889–898.

Mody, I., De Koninck, Y., Otis, T. S., & Soltesz, I. (1994). Bridging the cleft at GABA synapses in the brain. *Trends in Neurosciences*, *17*, 517–525.

Mondadori, C., Jaekel, J., & Preiswerk, G. (1993). CGP36742: the first orally active GABA$_B$ blocker improves the cognitive performance of mice, rats, and rhesus monkeys. *Behavioral and Neural Biology*, *60*, 62–68.

Montgomery, S. M., & Buzsáki, G. (2007). Gamma oscillations dynamically couple hippocampal CA3 and CA1 regions during memory task performance. *Proceedings of the National Academy of Sciences of the United States of America*, *104*, 14495–14500.

Oláh, S., Füle, M., Komlósi, G., Varga, C., Báldi, R., Barzó, P., et al. (2009). Regulation of cortical microcircuits by unitary GABA-mediated volume transmission. *Nature*, *461*, 1278–1281.

Oswald, A.-M.M., Doiron, B., Rinzel, J., & Reyes, A. D. (2009). Spatial profile and differential recruitment of GABA$_B$ modulate oscillatory activity in auditory cortex. *Journal of Neuroscience*, *29*, 10321–10334.

Otmakhova, N. A., & Lisman, J. E. (2004). Contribution of Ih and GABA$_B$ to synaptically induced afterhyperpolarizations in CA1: A brake on the NMDA response. *Journal of Neurophysiology*, *92*, 2027–2039.

Palhalmi, J., Paulsen, O., Freund, T. F., & Hajos, N. (2004). Distinct properties of carbachol- and DHPG-induced network oscillations in hippocampal slices. *Neuropharmacology*, *47*, 381–389.

Pérez-Garci, E., Gassmann, M., Bettler, B., & Larkum, M. E. (2006). The GABA$_{B1b}$ isoform mediates long-lasting inhibition of dendritic Ca^{2+} spikes in layer 5 somatosensory pyramidal neurons. *Neuron*, *50*, 603–616.

Petsche, H., Stumpf, C., & Gogolak, G. (1962). The significance of the rabbit's septum as a relay station between the midbrain and the hippocampus. I. The control of hippocampus arousal activity by the septum cells. *Electroencephalography and Clinical Neurophysiology*, *14*, 202–211.

Price, C. J., Cauli, B., Kovacs, E. R., Kulik, A., Lambolez, B., Shigemoto, R., et al. (2005). Neurogliaform neurons form a novel inhibitory network in the hippocampal CA1 area. *Journal of Neuroscience*, *25*, 6775–6786.

Price, C. J., Scott, R., Rusakov, D. A., & Capogna, M. (2008). GABA$_B$ receptor modulation of feedforward inhibition through hippocampal neurogliaform cells. *Journal of Neuroscience*, *28*, 6974–6982.

Prosser, H. M., Gill, C. H., Hirst, W. D., Grau, E., Robbins, M., Calver, A., et al. (2001). Epileptogenesis and enhanced prepulse inhibition in GABA$_{B1}$-deficient mice. *Molecular and Cellular Neuroscience*, *17*, 1059–1070.

Rigas, P., & Castro-Alamancos, M. A. (2007). Thalamocortical Up states: differential effects of intrinsic and extrinsic cortical inputs on persistent activity. *Journal of Neuroscience*, *27*, 4261–4272.

Robbins, M. J., Calver, A. R., Filippov, A. K., Hirst, W. D., Russell, R. B., Wood, M. D., et al. (2001). GABA$_{B2}$ is essential for G-protein coupling of the GABA$_B$ receptor heterodimer. *Journal of Neuroscience*, *21*, 8043–8052.

Roy, J. P., Clercq, M., Steriade, M., & Deschênes, M. (1984). Electrophysiology of neurons of lateral thalamic nuclei in cat: mechanisms of long-lasting hyperpolarizations. *Journal of Neurophysiology*, *51*, 1220–1235.

Sakaba, T., & Neher, E. (2003). Direct modulation of synaptic vesicle priming by GABA$_B$ receptor activation at a glutamatergic synapse. *Nature*, *424*, 775–778.

Sanchez-Vives, M. V., & McCormick, D. A. (2000). Cellular and network mechanisms of rhythmic recurrent activity in neocortex. *Nature Neuroscience*, *3*, 1027–1034.

Scanziani, M. (2000). GABA spillover activates postsynaptic $GABA_B$ receptors to control rhythmic hippocampal activity. *Neuron*, *25*, 673–681.

Schuler, V., Lüscher, C., Blanchet, C., Klix, N., Sansig, G., Klebs, K., et al. (2001). Epilepsy, hyperalgesia, impaired memory, and loss of pre- and postsynaptic $GABA_B$ responses in mice lacking $GABA_{B(1)}$. *Neuron*, *31*, 47–58.

Sederberg, P. B., Schulze-Bonhage, A., Madsen, J. R., Bromfield, E. B., McCarthy, D. C., Brandt, A., et al. (2007). Hippocampal and neocortical gamma oscillations predict memory formation in humans. *Cerebral Cortex*, *17*, 1190–1196.

Shu, Y., Hasenstaub, A., & McCormick, D. A. (2003). Turning on and off recurrent balanced cortical activity. *Nature*, *423*, 288–293.

Singer, W. (1993). Synchronization of cortical activity and its putative role in information processing and learning. *Annual Review of Physiology*, *55*, 349–374.

Sohal, V., Zhang, F., Yizhar, O., & Deisseroth, K. (2009). Parvalbumin neurons and gamma rhythms enhance cortical circuit performance. *Nature*, *459*, 698–702.

Soltesz, I., & Deschênes, M. (1993). Low- and high-frequency membrane potential oscillations during theta activity in CA1 and CA3 pyramidal neurons of the rat hippocampus under ketamine-xylazine anesthesia. *Journal of Neurophysiology*, *70*, 97–116.

Song, C., Murray, T. A., Kimura, R., Wakui, M., Ellsworth, K., Javedan, S. P., et al. (2005). Role of α7–nicotinic acetylcholine receptors in tetanic stimulation-induced γ oscillations in rat hippocampal slices. *Neuropharmacology*, *48*, 869–880.

Stäubli, U., Scafidi, J., & Chun, D. (1999). $GABA_B$ receptor antagonism: facilitatory effects on memory parallel those on LTP induced by TBS but not HFS. *Journal of Neuroscience*, *19*, 4609–4615.

Steriade, M., Contreras, D., Amzica, F., & Timofeev, I. (1996). Synchronization of fast (30–40 hz) spontaneous oscillations in intrathalamic and thalamocortical networks. *Journal of Neuroscience*, *16*, 2788–2808.

Steriade, M., Contreras, D., Curró Dossi, R., & Nuñez, A. (1993a). The slow (<1 hz) oscillation in reticular thalamic and thalamocortical neurons: Scenario of sleep rhythm generation in interacting thalamic and neocortical networks. *Journal of Neuroscience*, *13*, 3284–3299.

Steriade, M., McCormick, D. A., & Sejnowski, T. J. (1993b). Thalamocortical oscillations in the sleeping and aroused brain. *Science*, *262*, 679–685.

Steriade, M., Nuñez, A., & Amzica, F. (1993c). Intracellular analysis of relations between the slow (<1 hz) neocortical oscillation and other sleep rhythms of the electroencephalogram. *Journal of Neuroscience*, *13*, 3266–3283.

Steriade, M., Nuñez, A., & Amzica, F. (1993d). A novel slow (<1 hz) oscillation of neocortical neurons in vivo: Depolarizing and hyperpolarizing components. *Journal of Neuroscience*, *13*, 3252–3265.

Takahashi, T., Kajikawa, Y., & Tsujimoto, T. (1998). G-protein-coupled modulation of presynaptic calcium currents and transmitter release by a $GABA_B$ receptor. *Journal of Neuroscience*, *18*, 3138–3146.

Tamás, G., Lorincz, A., Simon, A., & Szabadics, J. (2003). Identified sources and targets of slow inhibition in the neocortex. *Science*, *299*, 1902–1905.

Thompson, S. M., Capogna, M., & Scanziani, M. (1993). Presynaptic inhibition in the hippocampus. *Trends in Neurosciences*, *16*, 222–227.

Thomson, A. M., & Destexhe, A. (1999). Dual intracellular recordings and computational models of slow inhibitory postsynaptic potentials in rat neocortical and hippocampal slices. *Neuroscience*, *92*, 1193–1215.

Thomson, A. M., West, D. C., Hahn, J., & Deuchars, J. (1996). Single axon IPSPs elicited in pyramidal cells by three classes of interneurones in slices of rat neocortex. *Journal of Physiology*, *496*, 81–102.

Timofeev, I., Grenier, F., Bazhenov, M., Sejnowski, T. J., & Steriade, M. (2000). Origin of slow cortical oscillations in deafferented cortical slabs. *Cerebral Cortex*, *10*, 1185–1199.

Traub, R. D., Bibbig, A., Fisahn, A., LeBeau, F. E., Whittington, M. A., & Buhl, E. H. (2000). A model of gamma-frequency network oscillations induced in the rat CA3 region by carbachol in vitro. *European Journal of Neuroscience*, *12*, 4093–4106.

Traub, R. D., Bibbig, A., Lebeau, F.E.N., Cunningham, M. O., & Whittington, M. A. (2005). Persistent gamma oscillations in superficial layers of rat auditory neocortex: experiment and model. *Journal of Physiology*, *562*, 3–8.

Traub, R. D., Spruston, N., Soltesz, I., Konnerth, A., Whittington, M. A., & Jefferys, G. R. (1998). Gamma-frequency oscillations: A neuronal population phenomenon, regulated by synaptic and intrinsic cellular processes, and inducing synaptic plasticity. *Progress in Neurobiology*, *55*, 563–575.

Uhlhaas, P. J., & Singer, W. (2010). Abnormal neural oscillations and synchrony in schizophrenia. *Nature Reviews Neuroscience*, *11*, 100–113.

Ulrich, D., & Bettler, B. (2007). GABA(b) receptors: synaptic functions and mechanisms of diversity. *Current Opinion in Neurobiology*, *17*, 298–303.

Ulrich, D., & Huguenard, J. R. (1996). γ-aminobutyric acid type B receptor-dependent burst-firing in thalamic neurons: A dynamic clamp study. *Proceedings of the National Academy of Sciences of the United States of America*, *93*, 13245–13249.

Vanderwolf, C. H. (1969). Hippocampal electrical activity and voluntary movement in the rat. *Electroencephalography and Clinical Neurophysiology*, *26*, 407–418.

Varela, F., Lachaux, J. P., Rodriguez, E., & Martinerie, J. (2001). The brainweb: phase synchronization and large-scale integration. *Nature Reviews Neuroscience*, *2*, 229–239.

Vargas, K. J., Terunuma, M., Tello, J. A., Pangalos, M. N., Moss, S. J., & Couve, A. (2008). The availability of surface GABA$_B$ receptors is independent of γ-aminobutyric acid but controlled by glutamate in central neurons. *Journal of Biological Chemistry*, *283*, 24641–24648.

Vergnes, M., Boehrer, A., Simler, S., Bernasconi, R., & Marescaux, C. (1997). Opposite effects of GABA$_B$ receptor antagonists on absences and convulsive seizures. *European Journal of Pharmacology*, *332*, 245–255.

Vigot, R., Barbieri, S., Bräuner-Osborne, H., Turecek, R., Shigemoto, R., Zhang, Y., et al. (2006). Differential compartmentalization and distinct functions of GABA$_B$ receptor variants. *Neuron*, *50*, 589–601.

Volgushev, M., Chauvette, S., Mukovski, M., & Timofeev, I. (2006). Precise long-range synchronization of activity and silence in neocortical neurons during slow-wave oscillations. *Journal of Neuroscience*, *26*, 5665–5672.

Wallenstein, G. V., & Hasselmo, M. E. (1997). GABAergic modulation of hippocampal population activity: sequence learning, place field development, and the phase precession effect. *Journal of Neurophysiology*, *78*, 393–408.

Wespatat, V., Tennigkeit, F., & Singer, W. (2004). Phase sensitivity of synaptic modifications in oscillating cells of rat visual cortex. *Journal of Neuroscience*, *24*, 9067–9075.

White, J. H., Wise, A., Main, M. J., Green, A., Fraser, N. J., Disney, G. H., et al. (1998). Heterodimerization is required for the formation of a functional GABA$_B$ receptor. *Nature*, *396*, 679–682.

Whittington, M. A., Traub, R. D., & Jefferys, J. G. (1995). Synchronized oscillations in interneuron networks driven by metabotropic glutamate receptor activation. *Nature*, *373*, 612–615.

Wilkin, G. P., Hudson, A. L., Hill, D. R., & Bowery, N. G. (1981). Autoradiographic localization of GABA$_B$ receptors in rat cerebellum. *Nature*, *294*, 584–587.

Ylinen, A., Soltész, I., Bragin, A., Penttonen, M., Sik, A., & Buzsáki, G. (1995). Intracellular correlates of hippocampal theta rhythm in identified pyramidal cells, granule cells, and basket cells. *Hippocampus*, *5*, 78–90.

Audrée Pinard, Riad Seddik, and Bernhard Bettler
Department of Biomedicine, Institute of Physiology, Pharmazentrum, University of Basel, Basel, Switzerland

$GABA_B$ Receptors: Physiological Functions and Mechanisms of Diversity

Abstract

$GABA_B$ receptors are the G-protein-coupled receptors (GPCRs) for γ-aminobutyric acid (GABA), the main inhibitory neurotransmitter in the central nervous system. $GABA_B$ receptors are implicated in the etiology of a variety of psychiatric disorders and are considered attractive drug targets. With the cloning of $GABA_B$ receptor subunits 13 years ago, substantial progress was made in the understanding of the molecular structure, physiology, and pharmacology of these receptors. However, it remained puzzling that native studies demonstrated a heterogeneity of $GABA_B$ responses

Advances in Pharmacology, Volume 58

1054-3589/10 $35.00
10.1016/S1054-3589(10)58010-4

that contrasted with a very limited diversity of cloned $GABA_B$ receptor subunits. Until recently, the only firmly established molecular diversity consisted of two $GABA_{B1}$ subunit isoforms, $GABA_{B1a}$ and $GABA_{B1b}$, which assemble with $GABA_{B2}$ subunits to generate heterodimeric $GABA_{B(1a,2)}$ and $GABA_{B(1b,2)}$ receptors. Using genetic, ultrastructural, biochemical, and electrophysiological approaches, it has been possible to identify functional properties that segregate with these two receptors. Moreover, receptor modifications and factors that can alter the receptor response have been identified. Most importantly, recent data reveal the existence of a family of auxiliary $GABA_B$ receptor subunits that assemble as tetramers with the C-terminal domain of $GABA_{B2}$ subunits and drastically alter pharmacology and kinetics of the receptor response. The data are most consistent with native $GABA_B$ receptors minimally forming dimeric assemblies of units composed of $GABA_{B1}$, $GABA_{B2}$, and a tetramer of auxiliary subunits. This represents a substantial departure from current structural concepts for GPCRs.

I. Introduction

GABA is the main inhibitory transmitter in the vertebrate central nervous system. GABAergic neurotransmission relies on two classes of receptors. Ionotropic $GABA_A$ receptors mediate fast GABA responses by triggering chloride channel openings. Metabotropic $GABA_B$ receptors are members of the family 3 (or family C) G-protein-coupled receptors (GPCRs) and mediate slower GABA responses by activating G-proteins and influencing second messenger systems (Bettler & Tiao, 2006; Bowery, 2006; Couve et al., 2000; Kornau, 2006; Ulrich & Bettler, 2007). These two classes of receptors were originally discriminated based on pharmacological differences (Bowery et al., 1980; Hill & Bowery, 1981). It was observed that the evoked release of adrenaline, dopamine, and serotonin was decreased by GABA acting on a receptor different from the known $GABA_A$ receptors. This decrease of evoked neurotransmitter release was surprisingly independent of the $GABA_A$ antagonist bicuculline but sensitive to baclofen. The terms $GABA_A$ and $GABA_B$ were introduced to distinguish bicuculline- from baclofen-sensitive receptors (Hill & Bowery, 1981). Since then, kinetic and pharmacological differences between native $GABA_B$ responses further suggested the existence of molecularly distinct $GABA_B$ receptor subtypes. However, cloning of $GABA_B$ receptors revealed an unexpectedly low structural diversity, unlike what is observed with the sequence-related metabotropic glutamate receptors (mGluRs). Only two heteromeric $GABA_B$ receptors consisting of $GABA_{B1a}$ or $GABA_{B1b}$ subunits in combination with $GABA_{B2}$ subunits were identified. It was therefore important to address how such a limited structural diversity leads to a variety of pharmacologically and kinetically distinct native $GABA_B$ responses. Here, we review evidence for heterogeneous native $GABA_B$ responses and discuss mechanisms that generate functional and pharmacological diversity in the $GABA_B$ receptor system.

II. Physiological Functions of $GABA_B$ Receptors

$GABA_B$ receptors influence their effectors via the $G\alpha$ and $G\beta\gamma$ subunits of the activated G protein. The first $GABA_B$ effector characterized was adenylate cyclase whose activity was shown to be inhibited via the $G\alpha_i/G\alpha_o$ subunits (Xu & Wojcik, 1986). Unfortunately, the physiological consequences of this modulation are poorly understood but include effects on transcription factors, kinases and intracellular Ca^{2+} signaling (Couve et al., 2002; New et al., 2006; Ren & Mody, 2003; Steiger et al., 2004). In comparison to $G\alpha$-mediated signaling, $G\beta\gamma$-mediated signaling is much better understood. The main $G\beta\gamma$-dependent effectors of presynaptic $GABA_B$ receptors are P/Q- and N-type voltage-dependent Ca^{2+} channels (Barral et al., 2000; Bussieres & El Manira, 1999; Chen & van den Pol, 1998). $GABA_B$ receptors inhibit these Ca^{2+} channels at excitatory and inhibitory terminals, thereby restricting neurotransmitter release. By definition, $GABA_B$ autoreceptors inhibit GABA release while $GABA_B$ heteroreceptors decrease the release of other neurotransmitters, including, for example, glutamate, dopamine, adrenaline, or serotonin. Depending on whether the terminal releases an inhibitory or excitatory neurotransmitter, presynaptic $GABA_B$ receptors increase or decrease the excitability of the postsynaptic neuron. Presynaptic $GABA_B$ receptors restrict neurotransmitter release not only by inhibiting Ca^{2+} channels but also by retarding the recruitment of synaptic vesicles (Sakaba & Neher, 2003). Recent evidence also suggests that presynaptic $GABA_B$ receptors couple to inwardly rectifying Kir3-type K^+ channels (also designated GIRK channels) to inhibit glutamate release (Fernandez-Alacid et al., 2009; Ladera et al., 2008). However, Kir3 channels are generally considered the main effectors of postsynaptic $GABA_B$ receptors (Luscher et al., 1997; Wagner & Dekin, 1993). $GABA_B$-mediated activation of Kir3 channels produces slow inhibitory postsynaptic potentials (IPSPs) by inducing K^+ efflux, which hyperpolarizes the membrane and shunts excitatory currents. Postsynaptic $GABA_B$ receptors also down-regulate Ca^{2+} channels, which inhibits dendritic Ca^{2+}-spike propagation (Perez-Garci et al., 2006).

Interestingly, $GABA_B$ receptor activation not only modulates neuronal firing but also triggers long-term changes in synaptic strength (Davies et al., 1991; Huang et al., 2005; Patenaude et al., 2003). Postsynaptic $GABA_B$ receptors limit the induction of long-term potentiation (LTP) by promoting blockade of *N*-methyl-D-aspartate (NMDA) receptors by Mg^{2+} (Otmakhova & Lisman, 2004). In contrast, autoreceptors facilitate LTP by restricting postsynaptic inhibition (Davies & Collingridge, 1996; Davies et al., 1991). $GABA_B$ heteroreceptors mediate heterosynaptic depression of glutamate release at hippocampal synapses (Guetg et al., 2009; Vogt & Nicoll, 1999). This was shown to modulate the expression of LTP at synapses in the hippocampus and the lateral amygdala (LA) (Shaban et al., 2006; Vigot et al., 2006). $GABA_B$ receptors were also identified on glial cells, including astrocytes in the hippocampus, where they contribute to heterosynaptic depression (Oka et al., 2006; Serrano et al., 2006) and to activity-dependent modulation of synaptic transmission (Kang et al., 1998).

Electrophysiological detection of $GABA_B$ receptor activity often necessitates strong stimulation intensities (Scanziani, 2000). This suggests that $GABA_B$ receptors are located at a certain distance from release sites, thus necessitating pooling of synaptically released GABA for receptor activation. Accordingly, ultrastructural studies detect presynaptic $GABA_B$ receptors mostly at extrasynaptic sites. $GABA_B$ heteroreceptors are enriched on the rim of glutamate release sites (Guetg et al., 2009; Kulik et al., 2003, 2006; Lacey et al., 2005) where they are activated under conditions of high GABA release (Chandler et al., 2003; Guetg et al., 2009; Vogt & Nicoll, 1999). Postsynaptic $GABA_B$ receptors are present on dendritic shafts and on the extrasynaptic membrane of spines (Guetg et al., 2009; Kulik et al., 2003). Presumably, $GABA_B$-mediated activation of Kir3 channels in the spines balances excitatory inputs by hyperpolarizing the membrane and inducing shunting inhibition (Isaacson, Solis, & Nicoll, 1993). Dendritic $GABA_B$ receptors also play a role in delaying and inhibiting back-propagating action potentials and Ca^{2+} spikes (Perez-Garci et al., 2006; Zilberter et al., 1999).

III. Heterogeneity of Native $GABA_B$ Responses

As highlighted in the previous section, $GABA_B$ receptors are involved in a variety of neuronal functions. During the past 20 years, heterogeneous native $GABA_B$ responses suggested the existence of multiple receptor subtypes (Bonanno & Raiteri, 1993b; Raiteri, 2006, 2008). Below, we review some of the pharmacological and functional evidence for differences between (A) pre- and postsynaptic $GABA_B$ receptors, (B) presynaptic $GABA_B$ auto- and heteroreceptors, and (C) postsynaptic $GABA_B$ receptors on different cell types.

A. Heterogeneity between Pre- and Postsynaptic $GABA_B$ Responses

Differences in native $GABA_B$ responses were first observed between pre- and postsynaptic receptors (Bowery, 1993; Nicoll, 2004). Electrophysiological studies in the neocortex and hippocampus showed different sensitivities of pre- and postsynaptic $GABA_B$ receptors to the antagonist phaclofen and to pertussis toxin, which suggested the existence of distinct receptors (Colmers & Williams, 1988; Deisz et al., 1993; Dutar & Nicoll, 1988; Harrison, 1990). However, studies in other brain regions or hippocampal cultures using more potent $GABA_B$ receptor antagonists did not detect any pharmacological difference between pre- and postsynaptic receptors (Seabrook et al., 1990; Thompson & Gahwiler, 1992; Yoon & Rothman, 1991). Still, several studies showed divergence in the potency of $GABA_B$ receptor agonists and antagonists in a variety of preparations (Chan et al., 1998; Colmers & Williams, 1988; Cruz et al., 2004; Deisz et al., 1993; Pozza et al., 1999; Yamada et al., 1999). For instance, an electrophysiological study determined the half-maximal inhibitory concentration (IC_{50}) of various $GABA_B$ antagonists needed to inhibit the $GABA_B$-mediated inhibitory postsynaptic potential ($IPSP_B$) and the paired-pulse widening of α-amino-3-hydroxy-5-methylisoazol-4-proprionate-mediated (AMPA) excitatory postsynaptic potentials (EPSPs) in the CA1 region of hippocampus (Pozza et al., 1999). The authors showed that 5- to 10-fold higher concentrations of antagonists were required to block presynaptic versus postsynaptic $GABA_B$ receptors (Table I). In addition, other studies showed that the selective $GABA_B$ receptor agonist baclofen inhibits inhibitory and excitatory postsynaptic currents (IPSCs and EPSCs, respectively) with a lower half-maximal effective concentration (EC_{50}) than the one required to activate postsynaptic K^+ currents in the ventral tegmental area (VTA) and in the CA3 region of the hippocampus (Cruz et al., 2004; R. Seddik unpublished observation). This indicates that baclofen more efficiently activates presynaptic $GABA_B$ receptors (Table 1). Importantly, differences in agonist potency can be due to differences in the expression level of receptor or effector protein, the coupling efficiency as well as the threshold for effector activation. Differences in agonist potency therefore do not necessarily reflect molecularly distinct receptor subtypes. Only differences in the rank order of drug potencies would clearly argue for receptor subtypes but no such differences were reported.

B. Heterogeneity between Presynaptic $GABA_B$ Responses

Considerable evidence has accumulated over the years to support differences between $GABA_B$ auto- and heteroreceptors (Bowery et al., 2002; Raiteri, 2006, 2008). This is predominantly supported by studies investigating the effects of various $GABA_B$ receptor antagonists on neurotransmitter release in different preparations (Bonanno et al., 1996, 1997, 1999; Fassio et al., 1994). As shown in Table II, different antagonist potencies were reported for auto- and heteroreceptors but also for heteroreceptors from rat and human neocortex. Additional studies showed pharmacological differences between autoreceptors in cerebral cortex and spinal cord suggesting extensive heterogeneity in

TABLE I Potency of Antagonists and Baclofen at Pre- and Postsynaptic $GABA_B$ Receptors

Region	*Drug*	*Read-out*	*IC_{50} or EC_{50} (μM)*	*Reference*
CA1 hippocampus	CGP36742	$IPSP_B$[a]	23	Pozza et al., 1999
	CGP52432	PPW of EPSP[b]	239	
	CGP55845A	$IPSP_B$	0.12	
		PPW of EPSP	0.68	
		$IPSP_B$	0.11	
		PPW of EPSP	0.74	
Dopaminergic neurons, VTA	Baclofen	K^+ current[a]	14.8	Cruz et al., 2004
		GABA release[b] ($IPSC_A$ inhibition)	0.5	
CA3 hippocampus	Baclofen	K^+ current[a]	55	R. Seddik, unpublished
		Glutamate release[b] (EPSC inhibition)	0.4	

VTA, ventral tegmental area; $IPSP_B$, $GABA_B$-mediated inhibitory postsynaptic potential; $IPSC_A$, $GABA_A$-mediated inhibitory postsynaptic current; PPW, paired-pulse widening; EPSP, excitatory postsynaptic potential; EPSC, excitatory postsynaptic current.

[a] Postsynaptic $GABA_B$ receptor read-out. IC_{50} values of $GABA_B$ antagonist tested against synaptically released GABA.

[b] Presynaptic $GABA_B$ receptor read-out.

TABLE II Potencies of Antagonists at Presynaptic $GABA_B$ Receptors in Neurotransmitter Release Studies with Rat and Human Neocortical Tissue

	Drug	*Neurotransmitter released*			
		GABA	*Glutamate*	*Somatostatin*	*Cholecystokinin*
Rat neocortex	Phaclofen	79.2	>300	62.6	66.1
	CGP35348	>300	4.2	3.6	3.5
	CGP52432	0.08	9.3	3.4	0.11
	CGP47656	3.1	Partial agonist	Agonist	>300
	CGP36742	>100	>100	0.14	>100
Human neocortex	Phaclofen	≈100	>300	N.D.	N.D.
	CGP35348	>100	≈10	24.4	13.9
	CGP52432	<<1	>1; <30	0.06	0.08
	CGP47656	<10	N.D.	Agonist	Agonist
	CGP36742	>100	>100	≈5	>100

Values are IC_{50} (μM) of $GABA_B$ receptor antagonists tested against (–)baclofen. N.D., not determined. Reproduced from Bowery et al., 2002.

presynaptic $GABA_B$ receptors (Bonanno & Raiteri, 1993a; Bonanno et al., 1998). However, the notion of distinct presynaptic $GABA_B$ receptors has been open to dispute since other release studies detected no pharmacological difference between $GABA_B$ auto- and heteroreceptors (Waldmeier et al., 1994). Waldmeier and colleagues showed that the antagonists CGP35348 and CGP52432, which previously were reported to have distinct potencies at auto- and heteroreceptors, regulate GABA and glutamate release with similar potencies. Still, pharmacologically distinct presynaptic $GABA_B$ receptors were also suggested by electrophysiological experiments. In the dorsolateral septal nucleus, the agonist CGP44533 inhibited glutamate-mediated postsynaptic currents with higher efficacy and potency than $GABA_A$-mediated currents. This suggests a pharmacological difference between $GABA_B$ receptors at glutamatergic and GABAergic terminals (Yu et al., 1999). Thus, despite some conflicting data, it appears that presynaptic $GABA_B$ receptors exhibit pharmacological differences. Based on such differences, some investigators proposed the existence of molecularly distinct $GABA_B$ receptor subtypes (Bonanno & Raiteri, 1993b; Raiteri, 2008).

In addition to differences in drug potencies, presynaptic $GABA_B$ receptors also differ in their distribution and desensitization properties. A striking example of a differential distribution of the $GABA_B$ receptors is observed in the glutamatergic afferents projecting to principal neurons and to interneurons in the LA (Pan et al., 2009). Preembedding immunogold labeling and electron microscopy showed that $GABA_B$ receptors in terminals contacting interneurons are located further away from active zones than those in terminals contacting principal neurons. Consequently, baclofen more efficiently inhibits glutamate release at terminals projecting to principal neurons than at terminals projecting to interneurons, suggesting that $GABA_B$-mediated control of release in the LA is differentially regulated. Distinct kinetic properties between presynaptic $GABA_B$ responses were also reported, in line with $GABA_B$ receptor heterogeneity. In CA3 neurons, $GABA_B$-mediated inhibition of GABA but not glutamate release desensitizes during continuous agonist application (Tosetti et al., 2004). It was suggested that the specific loss of $GABA_B$-mediated inhibition of GABA release, induced by receptor desensitization, triggers persistent epileptiform discharges in the neonatal hippocampus (Tosetti et al., 2004). Although far from being conclusive, a significant body of data suggests the existence of pharmacologically and functionally distinct $GABA_B$ receptors at glutamatergic and GABAergic terminals.

C. Heterogeneity of Postsynaptic $GABA_B$ Responses

A remarkable example of heterogeneous postsynaptic $GABA_B$ responses is found in the VTA, a brain structure involved in the rewarding effects of drugs of abuse (Luscher, 2009). The VTA contains GABAergic interneurons as well as dopaminergic (DA) neurons implicated in the reward circuitry. Using patch-clamp electrophysiology, Cruz and colleagues found that GABAergic and DA neurons produce $GABA_B$-mediated Kir3 currents with distinct pharmacological and kinetic properties (Cruz et al., 2004). At GABAergic neurons, baclofen induced a high-potency nondesensitizing current while it evoked a low-potency desensitizing current in DA neurons. Accordingly, $GABA_B$ agonists have bidirectional effects on the excitability of DA neurons. Low agonist concentrations selectively inhibit GABAergic neurons and thereby increase the firing frequency of DA neurons. High agonist concentrations directly inhibit DA neurons and hence reduce their firing frequency. Low-affinity $GABA_B$ agonists like the recreational drug γ-hydroxybutyrate (GHB) therefore carry an abuse liability while high-affinity agonists, such as baclofen, are of potential therapeutic benefit in the treatment of addiction. $GABA_B$ responses with distinct susceptibilities to antagonists were also observed by studying the inhibition of cAMP production in cerebral cortical slices (Cunningham & Enna, 1996). It was therefore proposed that distinct $GABA_B$ receptor subtypes regulate adenylyl cyclase activity.

IV. Functions of the Cloned $GABA_B$ Receptor Subtypes

Given the heterogeneity observed with native $GABA_B$ responses, many researchers in the field expected the existence of multiple $GABA_B$ receptor subtypes. However, cloning efforts only produced two molecular subtypes of $GABA_B$ receptors (Marshall et al., 1999). Molecular diversity in the $GABA_B$ system is based on the subunit isoforms $GABA_{B1a}$ and $GABA_{B1b}$, both of which combine with $GABA_{B2}$ to form heteromeric $GABA_{B(1a,2)}$ and $GABA_{B(1b,2)}$ receptors. Recombinant experiments showed that heteromerization is mandatory for cell surface expression and G-protein coupling of the receptor. Specifically, it was shown that $GABA_{B2}$ is critical for the transport of $GABA_{B1}$ to the cell surface and activation of the G-protein while $GABA_{B1}$ contains the agonist binding site (Fig. 1; Calver et al., 2001; Couve et al., 1998; Duthey et al., 2002; Galvez et al., 2001; Havlickova et al., 2002; Margeta-Mitrovic et al., 2001; Pagano et al., 2001; Robbins et al., 2001). Structurally, $GABA_{B1a}$ and $GABA_{B1b}$ differ in their N-terminal ectodomain by a pair of sushi domains (SDs) that are unique to $GABA_{B1a}$ (Fig. 1; Blein et al., 2004). SDs, also known as short consensus repeats, were originally observed in proteins of the complement cascade and mediate protein interactions in a wide variety of adhesion proteins (Lehtinen et al., 2004).

Since only two structurally distinct $GABA_B$ receptors with essentially identical functional and pharmacological properties in recombinant

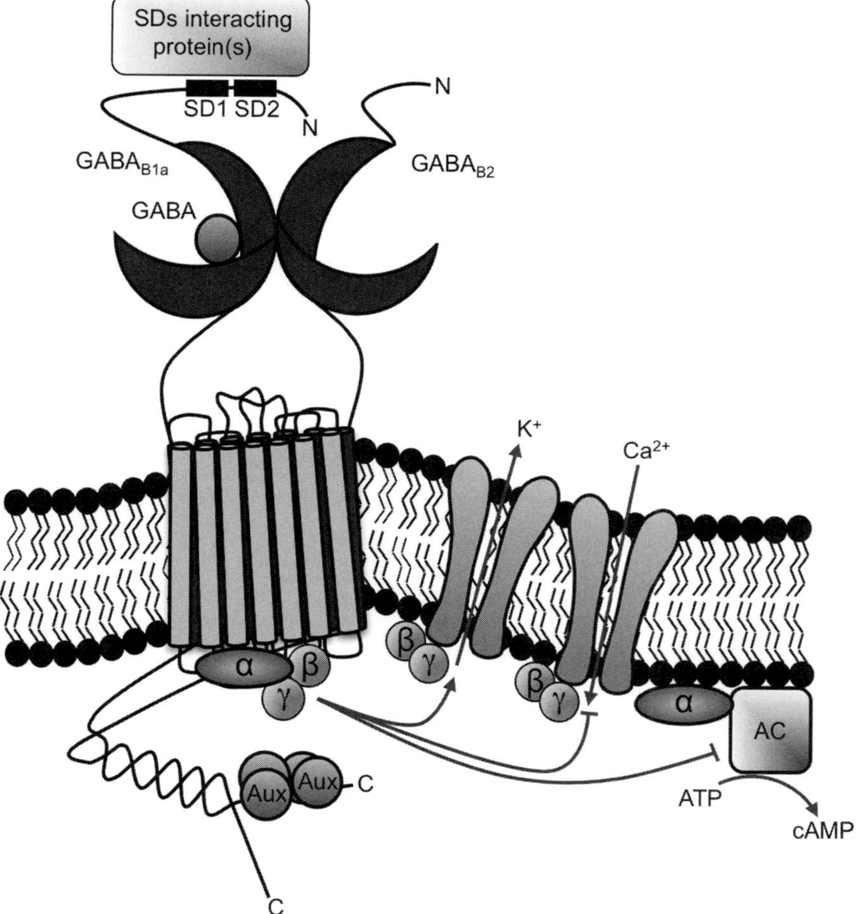

FIGURE 1 Schematic representation of the GABA$_{B(1a,2)}$ heterodimer and its downstream effectors. Native GABA$_B$ receptors are formed by the heteromerization of GABA$_{B1}$ and GABA$_{B2}$ subunits, which interact via C-terminal coiled-coil domains. Two GABA$_{B1}$ subunit isoforms, GABA$_{B1a}$ and the GABA$_{B1b}$, are generated by differential promoter usage from the GABA$_{B1}$ gene. GABA$_{B1a}$ (represented in the figure) and GABA$_{B1b}$ differ in their N-terminal ectodomain by a pair of SDs (SD1 and SD2) that are unique to GABA$_{B1a}$. Auxiliary subunits (Aux) associate as tetramers with the C-terminus of GABA$_{B2}$ subunit and determine the pharmacology and kinetics of the receptor response. GABA$_{B1a}$-containing heteroreceptors are retained in proximity of the presynaptic release machinery via interaction of their SDs with a protein that may additionally be involved in the selective delivery of GABA$_{B1a}$-containing receptor to glutamatergic terminals. The GABA$_{B1}$ subunit is essential for GABA binding whereas the GABA$_{B2}$ subunit is critical for G-protein coupling and receptor surface trafficking. The Gβγ subunits of the activated G-protein trigger inwardly rectifying Kir3 channels and inhibit voltage-gated Ca^{2+} channels. The Gα subunit inhibits adenylyl cyclase (AC).

expression systems were described, it became important to clarify whether the cloned subunits account for all classical $GABA_B$ functions.

A. $GABA_{B1}$ and $GABA_{B2}$ Subunits Are Essential for Classical $GABA_B$ Responses

Although largely overlapping, some differences in the distribution of $GABA_{B1}$ and $GABA_{B2}$ subunits were reported in various brain regions. Specifically, $GABA_{B2}$ mRNA was shown to be barely detectable in the caudate putamen, the medial basal hypothalamus, septum, and brainstem while $GABA_{B1}$ mRNA was abundant there (Clark et al., 2000). Additionally, immunohistochemical studies demonstrated that the expression of $GABA_{B1}$ and $GABA_{B2}$ protein differs in the rat striatum and in large interneurons of the human hippocampus (Billinton et al., 2000; Ng & Yung, 2001; Regard, Sato, & Coughlin, 2008). It is thus conceivable that in a neuronal context, $GABA_{B1}$ and $GABA_{B2}$ participate in distinct receptors, either alone or in association with additional proteins. Accordingly, the functions of $GABA_B$ subunits were studied in the absence of their partner subunit in $GABA_{B1}$- and $GABA_{B2}$-deficient mice ($GABA_{B1}^{-/-}$ and $GABA_{B2}^{-/-}$ mice). The analysis of these mice showed a significant down-regulation of $GABA_{B2}$ and $GABA_{B1}$ protein in $GABA_{B1}^{-/-}$ and $GABA_{B2}^{-/-}$ mice, respectively (Gassmann et al., 2004; Schuler et al., 2001). This suggests that $GABA_{B1}$ and $GABA_{B2}$ subunits cross-stabilize each other. In agreement with biochemical studies (Benke et al., 1999), this corroborates the predominantly heteromeric nature of native $GABA_B$ receptors. In addition, electrophysiological experiments showed that $GABA_B$-mediated inhibition of glutamate and GABA release is lost in $GABA_{B1}^{-/-}$ and $GABA_{B2}^{-/-}$ mice (Gassmann et al., 2004; Schuler et al., 2001). Postsynaptic $GABA_B$-mediated Kir3 currents were absent in $GABA_{B1}^{-/-}$ mice but atypical $GABA_{B1}$-mediated responses persisted in $GABA_{B2}^{-/-}$ mice. In $GABA_{B2}^{-/-}$ mice, $GABA_{B1}$ mediates a G-protein-dependent inhibition of constitutively active K^+ channels as opposed to the normal activation of Kir3 channels (Gassmann et al., 2004). In the absence of $GABA_{B2}$, $GABA_{B1}$ may therefore couple to different G-proteins than heteromeric $GABA_{B(1,2)}$ receptors. However, the existence of homomeric $GABA_{B1}$ receptors remains controversial since numerous *in vitro* experiments showed that $GABA_{B2}$ is necessary for activating the G-protein and for receptor trafficking to the cell surface (Bettler & Tiao, 2006; Couve et al., 1998; Duthey et al., 2002; Galvez et al., 2001; Havlickova et al., 2002; Margeta-Mitrovic et al., 2000, 2001; Pagano et al., 2001; Robbins et al., 2001). Therefore, it remains unclear whether atypical $GABA_{B1}$ responses in $GABA_{B2}^{-/-}$ mice are of physiological relevance or a consequence of the knockout situation. Overall, the studies with $GABA_{B1}^{-/-}$ and $GABA_{B2}^{-/-}$ mice clearly demonstrate that classical $GABA_B$ responses in the brain depend on the presence of both $GABA_{B1}$ and $GABA_{B2}$ subunits.

B. Localization and Functions of $GABA_{B1a}$ and $GABA_{B1b}$ Subunits

As outlined above, the only established molecular diversity in the $GABA_B$ receptor system is based on the $GABA_{B1a}$ and $GABA_{B1b}$ subunit isoforms that differ by a pair of SDs (Blein et al., 2004; Kaupmann et al., 1997). The two isoforms do not produce pharmacological or physiological differences in heterologous expression systems (Brauner-Osborne & Krogsgaard-Larsen, 1999; Ulrich & Bettler, 2007). The absence of isoform-specific $GABA_{B1}$ antibodies has prevented studying the synaptic localizations of $GABA_{B1a}$ and $GABA_{B1b}$. Moreover, most neurons express both $GABA_{B1}$ isoforms, which further complicates the identification of potential physiological or pharmacological specificities. The demonstration that $GABA_{B1a}$ and $GABA_{B1b}$ contribute to distinct native $GABA_B$ functions was made possible by the generation of mice deficient in $GABA_{B1a}$ ($GABA_{B1a}^{-/-}$) or $GABA_{B1b}$ ($GABA_{B1b}^{-/-}$) subunits (Vigot et al., 2006). $GABA_{B1a}$ and $GABA_{B1b}$ transcripts are generated from a single $GABA_{B1}$ gene by differential promoter usage (Bischoff et al., 1999; Steiger et al., 2004). By individually converting the $GABA_{B1a}$ or $GABA_{B1b}$ initiation codons into stop codons using a knock-in approach, one $GABA_{B1}$ isoform at the time was inactivated. The resulting $GABA_{B1a}^{-/-}$ and $GABA_{B1b}^{-/-}$ mice were analyzed using pharmacological or physiological activation of $GABA_B$ receptors. Either approach led to similar conclusions regarding the localization and functions of $GABA_{B1}$ subunit isoforms.

1. $GABA_{B1}$ Isoform-Specific Functions in Response to Pharmacological Activation

Using $GABA_{B1a}^{-/-}$ and $GABA_{B1b}^{-/-}$ mice, Vigot and colleagues showed that $GABA_{B1a}$ and $GABA_{B1b}$ mediate distinct functions as a result of their distinct localization within neurons (Vigot et al., 2006). At the Schaffer collateral to CA1 pyramidal neuron synapse, $GABA_{B1a}$ was predominantly localized at glutamatergic terminals. In contrast, $GABA_{B1b}$ was mostly localized in dendritic spines juxtaposed to glutamate release sites. The axonal versus dendritic distribution of $GABA_{B1a}$ and $GABA_{B1b}$ subunits was also studied in transfected organotypic hippocampal slice cultures and cultured hippocampal neurons (Guetg et al., 2009; Vigot et al., 2006). This confirmed the predominant localization of $GABA_{B1a}$ at glutamatergic terminals. In addition, these experiments revealed that both $GABA_{B1a}$ and $GABA_{B1b}$ are expressed in the dendrites but that only $GABA_{B1b}$ efficiently enters spine heads. These observations were further supported by whole-cell patch-clamp recordings from CA1 pyramidal neurons. Stimulation of the Schaffer collateral-commissural fibers generates EPSCs in the pyramidal cells. The amplitude of these EPSCs is reduced after activation of $GABA_B$ heteroreceptors due to the ensuing inhibition of glutamate release (Schuler et al.,

2001). Vigot and colleagues (2006) showed that this EPSC amplitude reduction is impaired in $GABA_{B1a}^{-/-}$ mice, implicating $GABA_{B1a}$-containing receptors in heteroreceptor function. Vigot and colleagues also recorded IPSCs in pyramidal cells to address which isoform(s) serve(s) as autoreceptors. Baclofen reduced IPSC amplitude to a similar extent in $GABA_{B1a}^{-/-}$ and $GABA_{B1b}^{-/-}$ mice, which supports that $GABA_{B1a}$ as well as $GABA_{B1b}$ subunits assemble into autoreceptors. Finally, the contribution of $GABA_{B1}$ subunit isoforms to postsynaptic inhibition was addressed by recording Kir3 currents. The amplitudes of K^+ currents in response to baclofen application were normal in wild-type and $GABA_{B1a}^{-/-}$ mice. However, baclofen-induced currents were reduced in $GABA_{B1b}^{-/-}$ mice suggesting that the principal isoform mediating postsynaptic inhibition in CA1 neurons is $GABA_{B1b}$.

Postsynaptic $GABA_B$ receptors and $GABA_B$ autoreceptors were shown to restrict and promote LTP, respectively (Davies & Collingridge, 1996; Davies et al., 1991). Vigot and colleagues (2006) studied LTP at the CA3–CA1 synapse of $GABA_{B1a}^{-/-}$ and $GABA_{B1b}^{-/-}$ mice. While LTP was normal in $GABA_{B1b}^{-/-}$ mice, LTP was clearly impaired in $GABA_{B1a}^{-/-}$ mice. This LTP deficit was also reflected in hippocampus-dependent memory tasks where $GABA_{B1a}^{-/-}$ mice exhibited a decreased performance in the recognition of known versus novel objects. Of note, $GABA_{B1}^{-/-}$ mice with no $GABA_{B1}$ subunits at all did not exhibit any LTP either. However, acute blockade of $GABA_B$ receptors with an antagonist was not sufficient to prevent the induction of LTP, suggesting that the constitutive absence of $GABA_{B1a}$ receptors in $GABA_{B1a}^{-/-}$ mice triggers adaptive changes that prevent LTP. To clarify this issue, the paired-pulse ratio of evoked EPSCs and the miniature excitatory postsynaptic current (mEPSC) amplitude and frequency were monitored in $GABA_{B1a}^{-/-}$ mice. While no differences were observed in the paired-pulse ratio and mEPSC amplitude, there was a significant increase in the mEPSC frequency. The increase in mEPSC frequency was likely caused by an increase in the number of functional synapses. In support of this, $GABA_{B1a}^{-/-}$ mice exhibited an increased coefficient of variation of the AMPA receptor component of EPSCs (CV_{AMPA}) while the CV_{NMDA} remained normal, suggesting an unmasking of silent synapses by insertion of AMPA receptors (Kullmann, 1994). This not only reveals a compensatory mechanism by which the absence of the $GABA_{B1a}$ isoform leads to altered synaptic functions but also demonstrates a role for $GABA_B$ heteroreceptors in synaptic plasticity processes. In conclusion, it appears that heteroreceptor activation by spillover of GABA is important to prevent loss of silent synapses and saturation of LTP (Vigot et al., 2006). As discussed below, the lack of $GABA_{B1a}$ at glutamatergic terminals also strongly influences LTP processes in the amygdala (Shaban et al., 2006).

In addition to data from CA3 to CA1 pyramidal neuron synapses, a prevalence of the $GABA_{B1a}$ isoform at glutamatergic terminals was also observed at other synapses. Ulrich and colleagues showed that at the corticothalamic fibers to thalamocortical relay synapse, glutamate release is exclusively controlled by GABA$_{B1a}$-containing receptors (Ulrich et al., 2007). They also showed that both GABA$_{B1a}$ and GABA$_{B1b}$ are expressed at postsynaptic sites and at GABAergic terminals. Likewise, baclofen reduced glutamate release in GABA$_{B1b}^{-/-}$ but not in GABA$_{B1a}^{-/-}$ mice at mossy fiber (MF)–CA3 pyramidal neuron synapses and at cortical afferents projecting to the amygdale (Guetg et al., 2009; Shaban et al., 2006). This supports that the selective association of GABA$_{B1a}$ with glutamatergic terminals is a feature that is conserved at most synapses.

2. *GABA$_{B1}$ Isoform-Specific Functions in Response to Physiological Activation*

The aforementioned studies on GABA$_{B1}$ isoform-specific functions all used pharmacological activation of GABA$_B$ receptors. Isoform-specific functions were also observed under conditions of physiological receptor activation. Pérez-Garci and colleagues (2006) studied GABA$_B$ receptors at synaptic inputs on layer 5 neocortical pyramidal cells using a physiological paradigm. L5 pyramidal cells extend their dendritic tufts to layer L1 where they receive excitatory and inhibitory synaptic inputs from converging L1 fibers and from several classes of interneurons (Hestrin & Armstrong, 1996; Perez-Garci et al., 2006). Excitatory synaptic inputs often occur hundreds of microns away from the site of action potential (AP) generation in the axon initial segment (Radnikow et al., 2002; Somogyi et al., 1998). Therefore, the excitatory signals transmitted by these remote inputs contribute little to the generation of APs (Stuart et al., 1997). It was suggested that these inputs instead modulate the generation of Na^+/Ca^{2+} APs or Ca^{2+} spikes in the dendritic tree (Larkum et al., 1999). Inhibitory inputs to the apical tufts produce a long-lasting inhibition of dendritic Ca^{2+} spikes (Larkum et al., 1999). Perez-Garci et al. (2006) studied the contribution of GABA$_{B1}$ subunit isoforms to the inhibition of Ca^{2+} spikes induced by extracellular stimulation in L1. Using the mice developed by Vigot et al. (2006), they showed that selectively GABA$_{B1b}$ is involved in Ca^{2+}-spike inhibition. Moreover, they showed that this inhibition was caused by a GABA$_B$-mediated down-regulation of dendritic Ca^{2+} channel activity and not by hyperpolarization through Kir3 channels. Interestingly, Pérez-Garci et al. (2006) also showed that physiological activation of GABA$_{B1a}$ autoreceptors inhibits GABA release. This clearly demonstrates that synaptically released GABA controls a number of physiological functions by activating GABA$_B$ receptors assembled from specific GABA$_{B1}$ isoforms.

Physiological activation of GABA$_B$ heteroreceptors at glutamatergic terminals was studied at hippocampal MF–CA3 pyramidal neuron synapses. Heterosynaptic depression at this synapse relies on the activation of heteroreceptors by synaptically released GABA (Chandler et al., 2003; Isaacson et al., 1993; Vogt & Nicoll, 1999). Using $GABA_{B1a}^{-/-}$ and $GABA_{B1b}^{-/-}$ mice, Guetg and colleagues (2009) showed that $GABA_{B1a}$ is more abundant than $GABA_{B1b}$ at MF terminals, consistent with findings at other glutamatergic terminals (Ulrich & Bettler, 2007). Heterosynaptic depression in response to endogenously released GABA was absent in $GABA_{B1a}^{-/-}$ mice, demonstrating that under physiological conditions only $GABA_{B1a}$-containing receptors assume heteroreceptor function.

Yet, another study reinforces the notion that selectively the $GABA_{B1a}$ isoform is involved in the physiological control of glutamate release. Projection neurons of the LA, which are important for classical fear conditioning, receive convergent inputs from thalamic and cortical sensory afferents (LeDoux, 2000; Rumpel et al., 2005). Coincident stimulation of both thalamic and cortical afferents induces LTP at cortical afferents through a presynaptic NMDA receptor-dependent mechanism (Humeau et al., 2003). Conversely, a tetanic stimulation of the cortical afferents alone only induces LTP at cortical afferents when $GABA_B$ receptors are antagonized (Shaban et al., 2006). This form of LTP is expressed presynaptically and is associated with a decrease in the paired-pulse ratio (Shaban et al., 2006). Considering the strong modulation of LA synaptic activity by local GABAergic inhibition (Bissiere et al., 2003; Li et al., 1996; Sugita et al., 1993), this suggests an involvement of presynaptic $GABA_B$ heteroreceptors. Activation of $GABA_B$ heteroreceptors on cortical afferents may explain the failure to induce LTP by the unique stimulation of cortical afferents. Indeed, induction of LTP was possible in $GABA_{B1a}^{-/-}$ mice lacking heteroreceptors (Shaban et al., 2006). In contrast, the $GABA_{B1b}$ subunit was responsible for postsynaptic inhibition. At thalamic afferents to projection neurons LTP is induced and expressed postsynaptically and therefore facilitated in $GABA_{B1b}^{-/-}$ mice. This study reinforces that $GABA_{B1a}$ and $GABA_{B1b}$ regulate pre- and postsynaptic inhibition, respectively. In further support of this, ultrastructural studies show that $GABA_{B1a}$ is predominantly localized at glutamatergic terminals while $GABA_{B1b}$ is most abundant in the dendritic spines of projection neurons. The behavioral consequence of a selective loss of $GABA_{B1a}$ heteroreceptors in the LA is a generalization of conditioned fear to unconditioned stimuli. Presumably, the lack of $GABA_B$ heteroreceptors in $GABA_{B1a}^{-/-}$ mice allows LTP induction at the cortical afferent in the absence of coincident thalamic afferent activity, thus inappropriately potentiating synaptic inputs and leading to a generalization of fear (Shaban et al., 2006).

The studies mentioned above shed light on specific functions of $GABA_{B1}$ isoforms. They also reveal that these specific functions are a consequence of a differential distribution of $GABA_{B1}$ isoforms to axonal and dendritic sites. Most interestingly, the available data also highlight similarities in the distribution of $GABA_{B1}$ isoforms at different synapses. In Fig. 2, we summarize the distribution of $GABA_B$ isoforms at glutamatergic and GABAergic synapses.

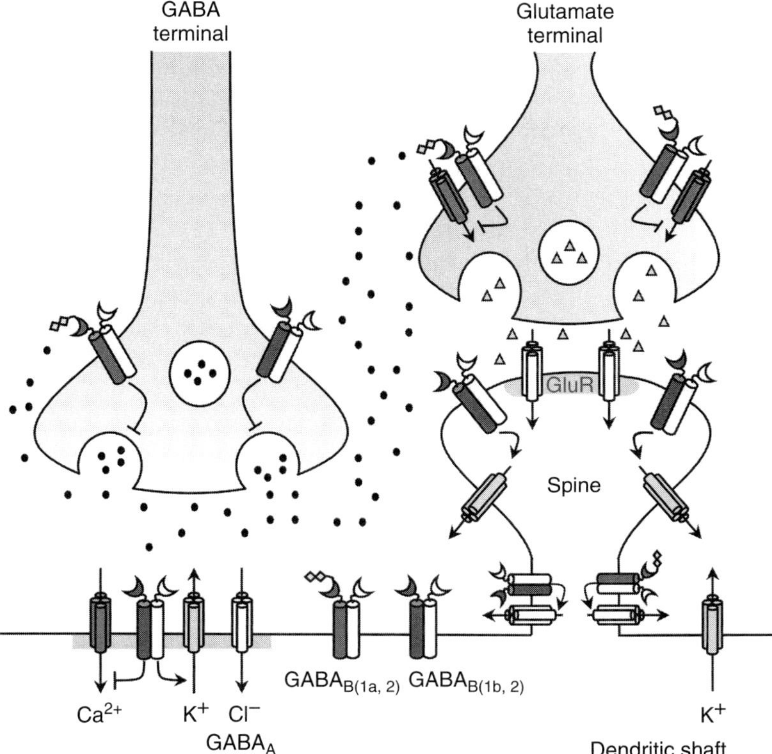

FIGURE 2 Distribution of $GABA_{B1}$ subunit isoforms to neuronal compartments. Schematic of a glutamatergic terminal contacting a dendritic spine and a GABAergic terminal contacting the dendritic shaft. $GABA_B$ receptors are localized presynaptically on glutamate and GABA terminals and postsynaptically on dendritic shafts and spines. Presynaptic $GABA_B$ receptors prevent neurotransmitter release by inhibiting voltage-sensitive Ca^{2+} channels. Specifically, $GABA_B$ autoreceptors inhibit the release of GABA whereas $GABA_B$ heteroreceptors inhibit the release of other neurotransmitters such as glutamate. Postsynaptic $GABA_B$ receptors mediate slow IPSPs by activating Kir3 channels and inhibit voltage-sensitive Ca^{2+} channels. $GABA_{B1a}$ assumes heteroreceptor function at glutamatergic terminals. These heteroreceptors are activated by GABA spillover from neighboring inhibitory terminals. Both $GABA_{B1a}$ and $GABA_{B1b}$ assume autoreceptor function at inhibitory terminals impinging on CA1 pyramidal and amygdala neurons. However, only $GABA_{B1a}$ assumes autoreceptor function at inhibitory terminals contacting the apical tufts of L5 cortical neurons. Dendritic $GABA_B$ receptors coupled to Kir3 channels of CA1 neurons mostly contain $GABA_{B1b}$. Dendritic $GABA_B$ receptors that inhibit dendritic Ca^{2+} channels at the apical tufts of L5 cortical neurons exclusively contain $GABA_{B1b}$. $GABA_{B1a}$ and $GABA_{B1b}$ are present in the spine neck and in dendritic shafts but only $GABA_{B1b}$ enters spine heads.

3. *Mechanisms Involved in the Selective Axonal Targeting of $GABA_{B1a}$*

The prevalence of $GABA_{B1a}$ at glutamatergic terminals is interesting from a trafficking point of view. $GABA_{B1a}$ and $GABA_{B1b}$ differ by a pair of SDs that is selectively present in $GABA_{B1a}$ (Blein et al., 2004). It is therefore reasonable to speculate that these SDs are involved in protein interactions that are required for trafficking and/or retention of $GABA_{B1a}$ at glutamatergic terminals. The role of the SDs in axonal targeting was studied using cultured hippocampal neurons as an experimental system (Biermann et al., 2010). Mutation of cysteines into serines in the SDs precludes axonal localization of $GABA_{B1a}$, presumably by impairing binding of the SDs to an axonally destined protein. Fusion of SDs to uniformly (CD8) or somatodendritically (mGluR1a) localized proteins redirects these proteins to the axon. Interestingly, surface expression of $GABA_{B1a}$ is not mandatory for axonal delivery. This suggests that $GABA_{B1a}$ reaches the terminal via intracellular vesicles rather than via lateral diffusion at the cell surface. A model explaining axonal targeting of $GABA_{B1a}$ was proposed. In this model both the $GABA_{B1a}$ and $GABA_{B1b}$ isoforms are targeted to the dendrites via C-terminal targeting signal(s) present in $GABA_{B1}$ and/or $GABA_{B2}$. A subpopulation of $GABA_{B1a}$ is then redistributed to the axonal compartment via the association of the SDs with axonally destined proteins. In support of this, the SDs of $GABA_{B1a}$ recognize binding sites in neuronal membrane preparations with low nanomolar affinity (Tiao et al., 2008). However, the specific binding partners remain to be identified. The expression level of these SD-binding partners potentially modulates axonal expression of $GABA_{B1a}$ and thus presynaptic $GABA_B$-mediated inhibition. Interestingly, the $GABA_{B1a}$ and $GABA_{B1b}$ subunits are themselves independently regulated at the transcriptional level (Steiger et al., 2004). This allows for an isoform-specific adjustment of $GABA_{B1}$ levels in axons and dendrites. The differential distribution of $GABA_{B1}$ isoforms to axonal and dendritic compartments exposes them to effector systems with distinct thresholds of activation, which may underlie some of the differences in agonist potencies that were observed with native $GABA_B$ responses.

V. Additional Mechanisms of Diversity

Pharmacological and kinetic differences between native $GABA_B$ responses may additionally arise from proteins influencing receptor properties, the G-protein cycle, or the effector system. GPCRs are commonly modulated via phosphorylation of their intracellular domains resulting in a decrease in receptor to effector coupling and/or a reduction of surface expression (Tobin et al., 2008; Tsao & von Zastrow, 2000). Desensitization of GPCRs is primarily mediated by G-protein kinases (GRKs) and arrestins. According to the prevailing model for desensitization, GRK phosphorylation of the agonist-bound receptor recruits arrestins, which precludes coupling to the G-protein and leads to receptor internalization (Ferguson et al., 1996). Little is known about modulation of $GABA_B$ receptors by phosphorylation. However, the available data tend to disagree with the generally accepted model predicting that phosphorylation increases receptor desensitization. While it was proposed that GRK_4 promotes agonist-dependent desensitization of $GABA_B$ receptors the mechanism surprisingly does not involve phosphorylation (Perroy et al., 2003). Other kinases were also proposed to modulate $GABA_B$ responses. Couve and collaborators showed that protein kinase A (PKA)-dependent phosphorylation of serine 862 in the cytoplasmic tail of $GABA_{B2}$ alleviates $GABA_B$ receptor desensitization, in clear contrast to the generally accepted view that phosphorylation leads to internalization of GPCRs (Couve et al., 2002). The mechanisms responsible for reducing $GABA_B$ receptor desensitization upon serine 862 phosphorylation involve increased receptor stability at the cell surface (Couve et al., 2002; Fairfax et al., 2004). Similarly, it was shown that AMP kinase directly binds to $GABA_{B1}$ and phosphorylates serine 783 in the cytoplasmic tail of $GABA_{B2}$ to enhance Kir3 channel coupling (Kuramoto et al., 2007). $GABA_B$ receptor phosphorylation therefore appears to influence desensitization and effector coupling, which additionally may explain some of the kinetic and pharmacological differences observed with native $GABA_B$ responses.

Regulators of G-protein signaling (RGS) are members of a multifunctional family of GTPase-accelerating proteins (GAPs) promoting GTP hydrolysis and hence accelerating termination of $G\beta\gamma$ signaling at effectors. This influences the EC_{50} of agonist-mediated responses (Shea & Linderman, 1997). RGS proteins comprise a large family of proteins, including RGS2, a modulator of Kir3 channel coupling (Doupnik et al., 2004). Accordingly, RGS proteins were shown to increase $GABA_B$-mediated desensitization of the Kir3.2 channel response in transfected HEK293 cells (Mutneja et al., 2005). $GABA_B$ receptors on DA and GABAergic neurons in the VTA couple to Kir3 channels with low and high efficacy, respectively (Cruz et al., 2004). According to Labouebe and colleagues, a selective expression of RGS2 in DA neurons is partly responsible for the low coupling efficiency (Labouebe et al., 2007). They showed that RGS2 inactivation increases $GABA_B$ receptor to Kir3 channel coupling in DA neurons, monitored as a decrease in the EC_{50} for baclofen activation. Hence, the presence of RGS2 in DA neurons restrains G-protein activation of the Kir3 channels. Interestingly, they further showed that the drug of abuse GHB increases the coupling efficiency in DA neurons by inhibiting RGS2 transcription (Labouebe et al., 2007). Yet, the low coupling efficiency observed in DA neurons was also explained by the subunit composition of Kir3 channels. Four different Kir3 (Kir3.1–3.4) subunits assemble to form functional homo- or heteromeric channels. Using single-cell RT PCR, it was shown that DA neurons exclusively express heteromeric Kir3.2/3.3 channels while GABAergic neurons also express Kir3.1, thus enabling the formation of Kir3.1/3.2 and Kir3.1/3.3 channels. Apparently, Kir3.1/3.2 and Kir3.1/3.3 channels are responsible for the higher coupling efficiency of $GABA_B$ receptors in GABAergic neurons (Cruz et al., 2004). Noteworthy, modulation of the coupling efficiency by

RGS2 is unique to Kir3.2/3.3 channels (Labouebe et al., 2007) further explaining the lower coupling efficiency in the DA neurons. Contribution of RGS proteins to $GABA_B$ signaling was also investigated using fluorescence resonance energy transfer (FRET), which revealed interactions between RGS4 and $GABA_{B1}$ or $GABA_{B2}$ (Fowler et al., 2007). In addition to $GABA_B$ receptors coupling to distinct channel subtypes, the extent of receptor–effector colocalization may also influence agonist potency (Koyrakh et al., 2005; Ulrich & Bettler, 2007). Pharmacological and functional diversity of native $GABA_B$ responses may therefore arise from direct receptor modifications, differences in effector coupling, as well as effector diversity.

A proteomic analysis was performed to test the hypothesis that functional diversity of native $GABA_B$ responses results from yet unknown subunits (Schwenk et al., 2010). Antibodies against $GABA_{B1}$ and $GABA_{B2}$ were used to affinity-purify receptor complexes from native tissue. The constituents of these complexes were then identified by high-resolution nanoflow liquid-chromatography tandem mass spectrometry. This revealed the existence of four sequence-related cytosolic proteins that bind as tetramers to the C-terminal domain of the $GABA_{B2}$ subunit (Fig. 1). These proteins exhibit distinct but overlapping distribution patterns in the brain. Expression of these proteins in heterologous cells or hippocampal neurons revealed that they determine both the pharmacology and the kinetics of the receptor response. Specifically, these proteins alter agonist potency, onset, and desensitization of the $GABA_B$ response. Most if not all $GABA_B$ receptors in the brain are associated with these proteins. Since these proteins tightly bind to the receptor and alter its kinetic and pharmacological properties, they qualify as auxiliary subunits of $GABA_B$ receptors. The data are most consistent with native $GABA_B$ receptors minimally forming dimeric assemblies of units composed of $GABA_{B1}$, $GABA_{B2}$, and a tetramer of auxiliary subunits. This represents a substantial departure from current structural concepts for GPCRs. Much of the heterogeneity observed with native $GABA_B$ responses may relate to these auxiliary subunits. Importantly, the association of the core $GABA_{B(1,2)}$ subunits with distinct auxiliary subunit combinations is expected to further expand the functional and pharmacological repertoire of $GABA_B$ receptors.

VI. Conclusions

Cloning of $GABA_B$ receptors in 1997/98 identified two $GABA_{B(1a,2)}$ and $GABA_{B(1b,2)}$ receptors with essentially indistinguishable properties in recombinant expression systems. However, a significant body of literature supports that native $GABA_B$ responses differ in their pharmacological and kinetic properties. It is emerging that some of these differences relate to a differential distribution of $GABA_{B(1a,2)}$ and $GABA_{B(1b,2)}$ receptors to axonal and dendritic compartments, which will expose them to effectors with distinct coupling efficiencies. This coupling may further be influenced by receptor phosphorylation, proteins that influence the G-protein cycle and effector diversity. An exciting new development in the field is the discovery of a family of auxiliary subunits that tightly associate with the core $GABA_{B(1a,2)}$ and $GABA_{B(1b,2)}$ receptors and markedly alter their G-protein signaling. It is expected that these auxiliary subunits are responsible for much of the observed heterogeneity of native $GABA_B$ responses. The existence of auxiliary $GABA_B$ subunits imparting distinct pharmacological and kinetic properties to the receptor will hopefully spark drug discovery efforts aiming at a more selective therapeutic interference with the $GABA_B$ receptor system.

Acknowledgments

We would like to thank M. Gassmann, K. Ivankova, and M. Rajalu for comments on the manuscript. Our work is supported by the Swiss Science Foundation (Grant 3100A0-117816), the Fridericus Stiftung, and the European Community's 7th Framework Programme (FP7/2007–2013) under Grant Agreement 201714.

Conflict of Interest Statement: The authors have no conflict of interest.

References

Barral, J., Toro, S., Galarraga, E., et al. (2000). GABAergic presynaptic inhibition of rat neostriatal afferents is mediated by Q-type Ca(2+) channels. *Neuroscience Letters*, *283*, 33–36.

Benke, D., Honer, M., Michel, C., et al. (1999). Gamma-aminobutyric acid type B receptor splice variant proteins GBR1a and GBR1b are both associated with GBR2 in situ and display differential regional and subcellular distribution. *Journal of Biological Chemistry*, *274*, 27323–27330.

Bettler, B., & Tiao, J. Y. (2006). Molecular diversity, trafficking and subcellular localization of $GABA_B$ receptors. *Pharmacology & Therapeutics*, *110*, 533–543.

Biermann, B., Ivankova-Susankova, K., Bradaia, A., et al. (2010). The sushi domains of $GABA_B$ receptors function as axonal targeting signals. *Journal of Neuroscience*, *30*, 1385–1394.

Billinton, A., Ige, A. O., Wise, A., et al. (2000). GABA(B) receptor heterodimer-component localisation in human brain. *Brain Research. Molecular Brain Research*, *77*, 111–124.

Bischoff, S., Leonhard, S., Reymann, N., et al. (1999). Spatial distribution of GABA(B)R1 receptor mRNA and binding sites in the rat brain. *Journal of Comparative Neurology*, *412*, 1–16.

Bissiere, S., Humeau, Y., & Luthi, A. (2003). Dopamine gates LTP induction in lateral amygdala by suppressing feedforward inhibition. *Nature Neuroscience*, *6*, 587–592.

Blein, S., Ginham, R., Uhrin, D., et al. (2004). Structural analysis of the complement control protein (CCP) modules of GABA(B) receptor 1a: Only one of the two CCP modules is compactly folded. *Journal of Biological Chemistry*, *279*, 48292–48306.

Bonanno, G., Carita, F., Cavazzani, P., et al. (1999). Selective block of rat and human neocortex GABA(B) receptors regulating somatostatin release by a GABA(B) antagonist endowed with cognition enhancing activity. *Neuropharmacology*, *38*, 1789–1795.

Bonanno, G., Fassio, A., Sala, R., et al. (1998). GABA(B) receptors as potential targets for drugs able to prevent excessive excitatory amino acid transmission in the spinal cord. *European Journal of Pharmacology*, *362*, 143–148.

Bonanno, G., Fassio, A., Schmid, G., et al. (1997). Pharmacologically distinct $GABA_B$ receptors that mediate inhibition of GABA and glutamate release in human neocortex. *British Journal of Pharmacology*, *120*, 60–64.

Bonanno, G., Gemignani, A., Schmid, G., et al. (1996). Human brain somatostatin release from isolated cortical nerve endings and its modulation through $GABA_B$ receptors. *British Journal of Pharmacology*, *118*, 1441–1446.

Bonanno, G., & Raiteri, M. (1993). Gamma-aminobutyric acid (GABA) autoreceptors in rat cerebral cortex and spinal cord represent pharmacologically distinct subtypes of the $GABA_B$ receptor. *Journal of Pharmacology and Experimental Therapeutics*, *265*, 765–770.

Bonanno, G., & Raiteri, M. (1993). Multiple $GABA_B$ receptors. *Trends in Pharmacological Sciences*, *14*, 259–261.

Bowery, N. G. (1993). $GABA_B$ receptor pharmacology. *Annual Review of Pharmacology and Toxicology*, *33*, 109–147.

Bowery, N. G. (2006). $GABA_B$ receptor: A site of therapeutic benefit. *Current Opinion in Pharmacology*, *6*, 37–43.

Bowery, N. G., Bettler, B., Froestl, W., et al. (2002). International Union of Pharmacology. XXXIII. Mammalian gamma-aminobutyric acid(B) receptors: Structure and function. *Pharmacological Reviews*, *54*, 247–264.

Bowery, N. G., Hill, D. R., Hudson, A. L., et al. (1980). (–)Baclofen decreases neurotransmitter release in the mammalian CNS by an action at a novel GABA receptor. *Nature*, *283*, 92–94.

Brauner-Osborne, H., & Krogsgaard-Larsen, P. (1999). Functional pharmacology of cloned heterodimeric $GABA_B$ receptors expressed in mammalian cells. *British Journal of Pharmacology*, *128*, 1370–1374.

Bussieres, N., & El Manira, A. (1999). GABA(B) receptor activation inhibits N- and P/Q-type calcium channels in cultured lamprey sensory neurons. *Brain Research*, *847*, 175–185.

Calver, A. R., Robbins, M. J., Cosio, C., et al. (2001). The C-terminal domains of the GABA(b) receptor subunits mediate intracellular trafficking but are not required for receptor signaling. *Journal of Neuroscience*, *21*, 1203–1210.

Chan, P. K., Leung, C. K., & Yung, W. H. (1998). Differential expression of pre- and postsynaptic GABA(B) receptors in rat substantia nigra pars reticulata neurones. *European Journal of Pharmacology*, *349*, 187–197.

Chandler, K. E., Princivalle, A. P., Fabian-Fine, R., et al. (2003). Plasticity of GABA(B) receptor-mediated heterosynaptic interactions at mossy fibers after status epilepticus. *Journal of Neuroscience*, *23*, 11382–11391.

Chen, G., & van den Pol, A. N. (1998). Presynaptic $GABA_B$ autoreceptor modulation of P/Q-type calcium channels and GABA release in rat suprachiasmatic nucleus neurons. *Journal of Neuroscience*, *18*, 1913–1922.

Clark, J. A., Mezey, E., Lam, A. S., et al. (2000). Distribution of the GABA(B) receptor subunit gb2 in rat CNS. *Brain Research*, *860*, 41–52.

Colmers, W. F., & Williams, J. T. (1988). Pertussis toxin pretreatment discriminates between pre- and postsynaptic actions of baclofen in rat dorsal raphe nucleus in vitro. *Neuroscience Letters*, *93*, 300–306.

Couve, A., Filippov, A. K., Connolly, C. N., et al. (1998). Intracellular retention of recombinant $GABA_B$ receptors. *Journal of Biological Chemistry*, *273*, 26361–26367.

Couve, A., Moss, S. J., & Pangalos, M. N. (2000). $GABA_B$ receptors: A new paradigm in G protein signaling. *Molecular and Cellular Neuroscience*, *16*, 296–312.

Couve, A., Thomas, P., Calver, A. R., et al. (2002). Cyclic AMP-dependent protein kinase phosphorylation facilitates GABA(B) receptor-effector coupling. *Nature Neuroscience*, *5*, 415–424.

Cruz, H. G., Ivanova, T., Lunn, M. L., et al. (2004). Bi-directional effects of GABA(B) receptor agonists on the mesolimbic dopamine system. *Nature Neuroscience*, *7*, 153–159.

Cunningham, M. D., & Enna, S. J. (1996). Evidence for pharmacologically distinct $GABA_B$ receptors associated with cAMP production in rat brain. *Brain Research*, *720*, 220–224.

Davies, C. H., & Collingridge, G. L. (1996). Regulation of EPSPs by the synaptic activation of $GABA_B$ autoreceptors in rat hippocampus. *Journal of Physiology*, *496*(Pt 2), 451–470.

Davies, C. H., Starkey, S. J., Pozza, M. F., et al. (1991). GABA autoreceptors regulate the induction of LTP. *Nature*, *349*, 609–611.

Deisz, R. A., Billard, J. M., & Zieglgansberger, W. (1993). Pre- and postsynaptic $GABA_B$ receptors of rat neocortical neurons differ in their pharmacological properties. *Neuroscience Letters*, *154*, 209–212.

Doupnik, C. A., Jaen, C., & Zhang, Q. (2004). Measuring the modulatory effects of RGS proteins on GIRK channels. *Methods in Enzymology*, *389*, 131–154.

Dutar, P., & Nicoll, R. A. (1988). Pre- and postsynaptic $GABA_B$ receptors in the hippocampus have different pharmacological properties. *Neuron*, *1*, 585–591.

Duthey, B., Caudron, S., Perroy, J., et al. (2002). A single subunit (GB2) is required for G-protein activation by the heterodimeric GABA(B) receptor. *Journal of Biological Chemistry*, *277*, 3236–3241.

Fairfax, B. P., Pitcher, J. A., Scott, M. G., et al. (2004). Phosphorylation and chronic agonist treatment atypically modulate $GABA_B$ receptor cell surface stability. *Journal of Biological Chemistry*, *279*, 12565–12573.

Fassio, A., Bonanno, G., Cavazzani, P., et al. (1994). Characterization of the GABA autoreceptor in human neocortex as a pharmacological subtype of the $GABA_B$ receptor. *European Journal of Pharmacology*, *263*, 311–314.

Ferguson, S. S., Barak, L. S., Zhang, J., et al. (1996). G-protein-coupled receptor regulation: Role of G-protein-coupled receptor kinases and arrestins. *Canadian Journal of Physiology and Pharmacology*, *74*, 1095–1110.

Fernandez-Alacid, L., Aguado, C., Ciruela, F., et al. (2009). Subcellular compartment-specific molecular diversity of pre- and post-synaptic GABA-activated GIRK channels in Purkinje cells. *Journal of Neurochemistry*, *110*, 1363–1376.

Fowler, C. E., Aryal, P., Suen, K. F., et al. (2007). Evidence for association of GABA(B) receptors with Kir3 channels and regulators of G protein signalling (RGS4) proteins. *Journal of Physiology*, *580*, 51–65.

Galvez, T., Duthey, B., Kniazeff, J., et al. (2001). Allosteric interactions between GB1 and GB2 subunits are required for optimal GABA(B) receptor function. *EMBO Journal*, *20*, 2152–2159.

Gassmann, M., Shaban, H., Vigot, R., et al. (2004). Redistribution of $GABA_{B(1)}$ protein and atypical $GABA_B$ responses in $GABA_{B(2)}$-deficient mice. *Journal of Neuroscience*, *24*, 6086–6097.

Guetg, N., Seddik, R., Vigot, R., et al. (2009). The $GABA_{B1a}$ isoform mediates heterosynaptic depression at hippocampal mossy fiber synapses. *Journal of Neuroscience*, *29*, 1414–1423.

Harrison, N. L. (1990). On the presynaptic action of baclofen at inhibitory synapses between cultured rat hippocampal neurones. *Journal of Physiology*, *422*, 433–446.

Havlickova, M., Prezeau, L., Duthey, B., et al. (2002). The intracellular loops of the GB2 subunit are crucial for G-protein coupling of the heteromeric gamma-aminobutyrate B receptor. *Molecular Pharmacology*, *62*, 343–350.

Hestrin, S., & Armstrong, W. E. (1996). Morphology and physiology of cortical neurons in layer I. *Journal of Neuroscience*, *16*, 5290–5300.

Hill, D. R., & Bowery, N. G. (1981). 3H-baclofen and 3h-GABA bind to bicuculline-insensitive GABA B sites in rat brain. *Nature*, *290*, 149–152.
Huang, C. S., Shi, S. H., Ule, J., et al. (2005). Common molecular pathways mediate long-term potentiation of synaptic excitation and slow synaptic inhibition. *Cell*, *123*, 105–118.
Humeau, Y., Shaban, H., Bissiere, S., et al. (2003). Presynaptic induction of heterosynaptic associative plasticity in the mammalian brain. *Nature*, *426*, 841–845.
Isaacson, J. S., Solis, J. M., & Nicoll, R. A. (1993). Local and diffuse synaptic actions of GABA in the hippocampus. *Neuron*, *10*, 165–175.
Kang, J., Jiang, L., Goldman, S. A., et al. (1998). Astrocyte-mediated potentiation of inhibitory synaptic transmission. *Nature Neuroscience*, *1*, 683–692.
Kaupmann, K., Huggel, K., Heid, J., et al. (1997). Expression cloning of GABA(B) receptors uncovers similarity to metabotropic glutamate receptors. *Nature*, *386*, 239–246.
Kornau, H. C. (2006). GABA(B) receptors and synaptic modulation. *Cell and Tissue Research*, *326*, 517–533.
Koyrakh, L., Lujan, R., Colon, J., et al. (2005). Molecular and cellular diversity of neuronal G-protein-gated potassium channels. *Journal of Neuroscience*, *25*, 11468–11478.
Kulik, A., Vida, I., Fukazawa, Y., et al. (2006). Compartment-dependent colocalization of Kir3.2-Containing K^+ channels and $GABA_B$ receptors in hippocampal pyramidal cells. *Journal of Neuroscience*, *26*, 4289–4297.
Kulik, A., Vida, I., Lujan, R., et al. (2003). Subcellular localization of metabotropic GABA(B) receptor subunits GABA(B1a/b) and GABA(B2) in the rat hippocampus. *Journal of Neuroscience*, *23*, 11026–11035.
Kullmann, D. M. (1994). Amplitude fluctuations of dual-component EPSCs in hippocampal pyramidal cells: Implications for long-term potentiation. *Neuron*, *12*, 1111–1120.
Kuramoto, N., Wilkins, M. E., Fairfax, B. P., et al. (2007). Phospho-dependent functional modulation of GABA(B) receptors by the metabolic sensor AMP-dependent protein kinase. *Neuron*, *53*, 233–247.
Labouebe, G., Lomazzi, M., Cruz, H. G., et al. (2007). RGS2 modulates coupling between $GABA_B$ receptors and GIRK channels in dopamine neurons of the ventral tegmental area. *Nature Neuroscience*, *10*, 1559–1568.
Lacey, C. J., Boyes, J., Gerlach, O., et al. (2005). GABA(B) receptors at glutamatergic synapses in the rat striatum. *Neuroscience*, *136*, 1083–1095.
Ladera, C., del Carmen Godino, M., Jose Cabanero, M., et al. (2008). Pre-synaptic GABA receptors inhibit glutamate release through GIRK channels in rat cerebral cortex. *Journal of Neurochemistry*, *107*, 1506–1517.
Larkum, M. E., Zhu, J. J., & Sakmann, B. (1999). A new cellular mechanism for coupling inputs arriving at different cortical layers. *Nature*, *398*, 338–341.
LeDoux, J. E. (2000). Emotion circuits in the brain. *Annual Review of Neuroscience*, *23*, 155–184.
Lehtinen, M. J., Meri, S., & Jokiranta, T. S. (2004). Interdomain contact regions and angles between adjacent short consensus repeat domains. *Journal of Molecular Biology*, *344*, 1385–1396.
Li, X. F., Armony, J. L., & LeDoux, J. E. (1996). $GABA_A$ and $GABA_B$ receptors differentially regulate synaptic transmission in the auditory thalamo-amygdala pathway: An in vivo microiontophoretic study and a model. *Synapse*, *24*, 115–124.
Luscher, C. (2009). Against addiction: Light at the end of the tunnel? *Journal of Physiology*, *587*, 3757.
Luscher, C., Jan, L. Y., Stoffel, M., et al. (1997). G protein-coupled inwardly rectifying K^+ channels (GIRKs) mediate postsynaptic but not presynaptic transmitter actions in hippocampal neurons. *Neuron*, *19*, 687–695.
Margeta-Mitrovic, M., Jan, Y. N., & Jan, L. Y. (2000). A trafficking checkpoint controls GABA (B) receptor heterodimerization. *Neuron*, *27*, 97–106.

Margeta-Mitrovic, M., Jan, Y. N., & Jan, L. Y. (2001). Function of GB1 and GB2 subunits in G protein coupling of GABA(B) receptors. *Proceedings of the National Academy of Sciences of the United States of America*, *98*, 14649–14654.

Marshall, F. H., Jones, K. A., Kaupmann, K., et al. (1999). $GABA_B$ receptors—the first 7TM heterodimers. *Trends in Pharmacological Sciences*, *20*, 396–399.

Mutneja, M., Berton, F., Suen, K. F., et al. (2005). Endogenous RGS proteins enhance acute desensitization of GABA(B) receptor-activated GIRK currents in HEK-293t cells. *Pflugers Archiv*, *450*, 61–73.

New, D. C., An, H., Ip, N. Y., et al. (2006). $GABA_B$ heterodimeric receptors promote Ca^{2+} influx via store-operated channels in rat cortical neurons and transfected Chinese hamster ovary cells. *Neuroscience*, *137*, 1347–1358.

Ng, T. K., & Yung, K. K. (2001). Differential expression of GABA(B)R1 and GABA(B)R2 receptor immunoreactivity in neurochemically identified neurons of the rat neostriatum. *Journal of Comparative Neurology*, *433*, 458–470.

Nicoll, R. A. (2004). My close encounter with GABA(B) receptors. *Biochemical Pharmacology*, *68*, 1667–1674.

Oka, M., Wada, M., Wu, Q., et al. (2006). Functional expression of metabotropic $GABA_B$ receptors in primary cultures of astrocytes from rat cerebral cortex. *Biochemical and Biophysical Research Communications*, *341*, 874–881.

Otmakhova, N. A., & Lisman, J. E. (2004). Contribution of Ih and $GABA_B$ to synaptically induced afterhyperpolarizations in CA1: A brake on the NMDA response. *Journal of Neurophysiology*, *92*, 2027–2039.

Pagano, A., Rovelli, G., Mosbacher, J., et al. (2001). C-terminal interaction is essential for surface trafficking but not for heteromeric assembly of GABA(b) receptors. *Journal of Neuroscience*, *21*, 1189–1202.

Pan, B. X., Dong, Y., Ito, W., et al. (2009). Selective gating of glutamatergic inputs to excitatory neurons of amygdala by presynaptic $GABA_b$ receptor. *Neuron*, *61*, 917–929.

Patenaude, C., Chapman, C. A., Bertrand, S., et al. (2003). $GABA_B$ receptor- and metabotropic glutamate receptor-dependent cooperative long-term potentiation of rat hippocampal $GABA_A$ synaptic transmission. *Journal of Physiology*, *553*, 155–167.

Pérez-Garci, E., Gassmann, M., Bettler, B., et al. (2006). The $GABA_{B1b}$ isoform mediates long-lasting inhibition of dendritic Ca^{2+} spikes in layer 5 somatosensory pyramidal neurons. *Neuron*, *50*, 603–616.

Perroy, J., Adam, L., Qanbar, R., et al. (2003). Phosphorylation-independent desensitization of GABA(B) receptor by GRK4. *EMBO Journal*, *22*, 3816–3824.

Pozza, M. F., Manuel, N. A., Steinmann, M., et al. (1999). Comparison of antagonist potencies at pre- and post-synaptic GABA(B) receptors at inhibitory synapses in the CA1 region of the rat hippocampus. *British Journal of Pharmacology*, *127*, 211–219.

Radnikow, G., Feldmeyer, D., & Lubke, J. (2002). Axonal projection, input and output synapses, and synaptic physiology of Cajal-Retzius cells in the developing rat neocortex. *Journal of Neuroscience*, *22*, 6908–6919.

Raiteri, M. (2006). Functional pharmacology in human brain. *Pharmacological Reviews*, *58*, 162–193.

Raiteri, M. (2008). Presynaptic metabotropic glutamate and $GABA_B$ receptors. *Handbook of Experimental Pharmacology*, *184*, 373–407.

Regard, J. B., Sato, I. T., & Coughlin, S. R. (2008). Anatomical profiling of G protein-coupled receptor expression. *Cell*, *135*, 561–571.

Ren, X., & Mody, I. (2003). Gamma-hydroxybutyrate reduces mitogen-activated protein kinase phosphorylation via GABA B receptor activation in mouse frontal cortex and hippocampus. *Journal of Biological Chemistry*, *278*, 42006–42011.

Robbins, M. J., Calver, A. R., Filippov, A. K., et al. (2001). GABA(B2) is essential for G-protein coupling of the GABA(B) receptor heterodimer. *Journal of Neuroscience*, *21*, 8043–8052.

Rumpel, S., LeDoux, J., Zador, A., et al. (2005). Postsynaptic receptor trafficking underlying a form of associative learning. *Science*, *308*, 83–88.

Sakaba, T., & Neher, E. (2003). Direct modulation of synaptic vesicle priming by GABA(B) receptor activation at a glutamatergic synapse. *Nature*, *424*, 775–778.

Scanziani, M. (2000). GABA spillover activates postsynaptic GABA(B) receptors to control rhythmic hippocampal activity. *Neuron*, *25*, 673–681.

Schuler, V., Luscher, C., Blanchet, C., et al. (2001). Epilepsy, hyperalgesia, impaired memory, and loss of pre- and postsynaptic GABA(B) responses in mice lacking GABA(B(1)). *Neuron*, *31*, 47–58.

Schwenk, J., Metz, M., Zolles, G., et al. (2010). Native GABA(B) receptors are heteromultimers with a family of auxiliary subunits. *Nature*, *465*, 231–235.

Seabrook, G. R., Howson, W., & Lacey, M. G. (1990). Electrophysiological characterization of potent agonists and antagonists at pre- and postsynaptic $GABA_B$ receptors on neurones in rat brain slices. *British Journal of Pharmacology*, *101*, 949–957.

Serrano, A., Haddjeri, N., Lacaille, J. C., et al. (2006). GABAergic network activation of glial cells underlies hippocampal heterosynaptic depression. *Journal of Neuroscience*, *26*, 5370–5382.

Shaban, H., Humeau, Y., Herry, C., et al. (2006). Generalization of amygdala LTP and conditioned fear in the absence of presynaptic inhibition. *Nature Neuroscience*, *9*, 1028–1035.

Shea, L., & Linderman, J. J. (1997). Mechanistic model of G-protein signal transduction. Determinants of efficacy and effect of precoupled receptors. *Biochemical Pharmacology*, *53*, 519–530.

Somogyi, P., Tamas, G., Lujan, R., et al. (1998). Salient features of synaptic organisation in the cerebral cortex. *Brain Research. Brain Research Reviews*, *26*, 113–135.

Steiger, J. L., Bandyopadhyay, S., Farb, D. H., et al. (2004). CAMP response element-binding protein, activating transcription factor-4, and upstream stimulatory factor differentially control hippocampal $GABA_{BR1a}$ and $GABA_{BR1b}$ subunit gene expression through alternative promoters. *Journal of Neuroscience*, *24*, 6115–6126.

Stuart, G., Spruston, N., Sakmann, B., et al. (1997). Action potential initiation and backpropagation in neurons of the mammalian CNS. *Trends in Neurosciences*, *20*, 125–131.

Sugita, S., Tanaka, E., & North, R. A. (1993). Membrane properties and synaptic potentials of three types of neurone in rat lateral amygdala. *Journal of Physiology*, *460*, 705–718.

Thompson, S. M., & Gahwiler, B. H. (1992). Comparison of the actions of baclofen at pre- and postsynaptic receptors in the rat hippocampus in vitro. *Journal of Physiology*, *451*, 329–345.

Tiao, J. Y., Bradaia, A., Biermann, B., et al. (2008). The sushi domains of secreted GABA(B1) isoforms selectively impair GABA(B) heteroreceptor function. *Journal of Biological Chemistry*, *283*, 31005–31011.

Tobin, A. B., Butcher, A. J., & Kong, K. C. (2008). Location, location, location... Site-specific GPCR phosphorylation offers a mechanism for cell-type-specific signalling. *Trends in Pharmacological Sciences*, *29*, 413–420.

Tosetti, P., Bakels, R., Colin-Le Brun, I., et al. (2004). Acute desensitization of presynaptic $GABA_B$-mediated inhibition and induction of epileptiform discharges in the neonatal rat hippocampus. *European Journal of Neuroscience*, *19*, 3227–3234.

Tsao, P., & von Zastrow, M. (2000). Downregulation of G protein-coupled receptors. *Current Opinion in Neurobiology*, *10*, 365–369.

Ulrich, D., Besseyrias, V., & Bettler, B. (2007). Functional mapping of GABA(B)-receptor subtypes in the thalamus. *Journal of Neurophysiology*, *98*, 3791–3795.

Ulrich, D., & Bettler, B. (2007). GABA(B) receptors: Synaptic functions and mechanisms of diversity. *Current Opinion in Neurobiology*, *17*, 298–303.

Vigot, R., Barbieri, S., Brauner-Osborne, H., et al. (2006). Differential compartmentalization and distinct functions of $GABA_B$ receptor variants. *Neuron*, *50*, 589–601.

Vogt, K. E., & Nicoll, R. A. (1999). Glutamate and gamma-aminobutyric acid mediate a heterosynaptic depression at mossy fiber synapses in the hippocampus. *Proceedings of the National Academy of Sciences of the United States of America*, *96*, 1118–1122.

Wagner, P. G., & Dekin, M. S. (1993). $GABA_b$ receptors are coupled to a barium-insensitive outward rectifying potassium conductance in premotor respiratory neurons. *Journal of Neurophysiology*, *69*, 286–289.

Waldmeier, P. C., Wicki, P., Feldtrauer, J. J., et al. (1994). GABA and glutamate release affected by $GABA_B$ receptor antagonists with similar potency: No evidence for pharmacologically different presynaptic receptors. *British Journal of Pharmacology*, *113*, 1515–1521.

Xu, J., & Wojcik, W. J. (1986). Gamma aminobutyric acid B receptor-mediated inhibition of adenylate cyclase in cultured cerebellar granule cells: Blockade by islet-activating protein. *Journal of Pharmacology and Experimental Therapeutics*, *239*, 568–573.

Yamada, K., Yu, B., & Gallagher, J. P. (1999). Different subtypes of $GABA_B$ receptors are present at pre- and postsynaptic sites within the rat dorsolateral septal nucleus. *Journal of Neurophysiology*, *81*, 2875–2883.

Yoon, K. W., & Rothman, S. M. (1991). The modulation of rat hippocampal synaptic conductances by baclofen and gamma-aminobutyric acid. *Journal of Physiology*, *442*, 377–390.

Yu, B., Yamada, K., & Gallagher, J. P. (1999). GABA(B) auto- versus hetero-receptor sensitivity: Implications for novel pharmacotherapy. *Neuropharmacology*, *38*, 1805–1809.

Zilberter, Y., Kaiser, K. M., & Sakmann, B. (1999). Dendritic GABA release depresses excitatory transmission between layer 2/3 pyramidal and bitufted neurons in rat neocortex. *Neuron*, *24*, 979–988.

De-Pei Li[*] and Hui-Lin Pan[†]

[*]Department of Critical Care, The University of Texas, M. D. Anderson Cancer Center, Houston, Texas, USA

[†]Department of Anesthesiology and Perioperative Medicine, The University of Texas, M. D. Anderson Cancer Center, Houston, Texas, USA

Role of $GABA_B$ Receptors in Autonomic Control of Systemic Blood Pressure

Abstract

$GABA_B$ receptors belong to family III G protein-coupled receptors (GPCRs) and are widely distributed in the peripheral and central nervous systems. The $GABA_B$ receptor is one of the most important therapeutic targets in the treatment for spasticity. $GABA_B$ agonists, such as baclofen, are used as muscle relaxants clinically and are effective for the treatment of anxiety, depression, epilepsy, and cognitive disorders (Caddick & Hosford, 1996; Dichter, 1997; Enna & Bowery, 1997). In addition, $GABA_B$ receptors regulate neurotransmitter release and neuronal excitability in the brain

Advances in Pharmacology, Volume 58

1054-3589/10 $35.00
10.1016/S1054-3589(10)58011-6

regions involved in the autonomic nervous system. Recent studies have led to a better understanding of the role of $GABA_B$ in the regulation of the autonomic nervous system, especially in disease conditions such as hypertension. Here, we provide an overview of the recent progress, a discussion of disparate and contradictory findings, and a description of theories used to explain various cardiovascular effects of $GABA_B$ receptor drugs. Particular emphasis is placed on the role of $GABA_B$ receptors in the neural plasticity of brain regions related to the control of sympathetic outflow in cardiovascular disorders.

I. Introduction

γ-Aminobutyric acid (GABA) is a ubiquitous inhibitory neurotransmitter and is widely distributed throughout the central nervous system (CNS). The responses to GABA are mediated by the activation of membrane receptors and by specific uptake mechanisms that control the spatial and temporal pattern of GABAergic transmission. GABA activates three pharmacologically distinct types of receptors: ionotropic $GABA_A$ and $GABA_C$ receptors and G protein-coupled $GABA_B$ receptors (Table I) (Bowery, 1993; Bowery & Enna, 2000). The early component of GABA inhibition is mediated by $GABA_A$ receptors (Macdonald & Olsen, 1994). These receptors belong to the ion channel superfamily and typically produce a rapid, chloride-mediated membrane hyperpolarization in response to GABA and $GABA_A$ receptor agonists. The late component of inhibitory transmission, first identified in sympathetic neurons, is slower and not affected by $GABA_A$ receptor antagonists (Hill & Bowery, 1981; Mott & Lewis, 1994). Three decades ago, the receptors responsible for the late component of inhibition were named $GABA_B$ (Hill & Bowery, 1981). Upon being activated, $GABA_B$ receptors mediate hyperpolarization of postsynaptic membranes and inhibition of presynaptic neurotransmitter release (Misgeld et al., 1995). Also, $GABA_B$ receptors play important roles in regulating long-term potentiation (Davies et al., 1991), neuroblast migration (Behar et al., 1995),

TABLE I GABA Receptor Subtypes

Receptor subtypes	*$GABA_A$*	*$GABA_B$*	*$GABA_C$*
Effector	Chloride channel	$G_{i/o}$, GIRK current (K^+ channel)	Chloride channel
Agonists	Muscimol	Baclofen	CACA
Antagonists	Bicuculline, gabazine	Saclofen, CGP55845, CGP52432, etc.	TPMPA

and rhythmic activity in the hippocampus (Scanziani, 2000). $GABA_B$ receptors are family III G protein-coupled receptors (GPCRs) and activation of these receptors results in GDP/GTP exchange in the associated G proteins and diffusion of Gα and G$\beta\gamma$ subunits for the activation of a wide variety of intracellular effector systems (Kubo & Tateyama, 2005; Mott & Lewis, 1994). Activation of $GABA_B$ receptors leads to either an increase or a decrease in the intracellular level of cyclic AMP, depending upon the types of adenylyl cyclases in the cell (Bowery & Enna, 2000).

Numerous $GABA_B$ receptor subunit isoforms have been identified. Initial studies indicated that fully functional $GABA_B$ receptors must be composed of a $GABA_{B1}$ and a $GABA_{B2}$ protein (Bowery & Enna, 2000; Chronwall et al., 2001; Jones et al., 1998; Kaupmann et al., 1998). The cDNA for $GABA_{B1}$ has been cloned (Kaupmann et al., 1997) and this gene produces two predominant amino-terminal splice variants: $GABA_{B1a}$ and $GABA_{B1b}$. Expression of either $GABA_{B1a}$ or $GABA_{B1b}$ alone does not possess a fully functional $GABA_B$ receptor. In addition, the $GABA_{B2}$ cDNA was cloned and characterized (Jones et al., 1998; Kaupmann et al., 1998; Kuner et al., 1999; White et al., 1998). Coexpression of $GABA_{B1}$ and $GABA_{B2}$ proteins represents the similar form of heteromeric dimers of the wild-type $GABA_B$ receptors (Jones et al., 1998; Kaupmann et al., 1998; Kuner et al., 1999; White et al., 1998). The dimers result from a physical interaction between the $GABA_{B1}$ and $GABA_{B2}$ proteins within their carboxyl-terminal cytoplasmic tails (Kuner et al., 1999). $GABA_{B2}$ receptors act as chaperons for the trafficking and insertion of $GABA_{B1}$ proteins into the cell membrane (White et al., 1998). However, it is essential that $GABA_{B2}$ attach to $GABA_{B1}$ after membrane insertion for the complete expression of $GABA_B$ receptor function. However, more recent findings suggest that $GABA_{B1}$ alone or $GABA_{B1}$ homodimers can display some activity as well (Gassmann et al., 2004).

In the past decades, increasing evidence demonstrated that $GABA_B$ receptors play important roles in the regulation of the autonomic nervous system including sympathetic output, cardiovascular reflexes (e.g., baro- and chemoreflexes), and energy balance (Amano & Kubo, 1993; Avanzino et al., 1994; Lemus et al., 2008; Persson, 1981; Suzuki et al., 1999; Sved & Sved, 1989; Sved & Sved, 1990). Furthermore, alterations of $GABA_B$ receptor expression and function in autonomic centers have been reported in cardiovascular disease conditions such as hypertension (Durgam et al., 1999; Li & Pan, 2006, 2007a; Li et al., 2008a; Sved & Sved, 1989; Sved & Tsukamoto, 1992; Takenaka et al., 1996; Tsukamoto & Sved, 1993; Zhang & Mifflin, 2010b). However, it is not clear that the plasticity of $GABA_B$ receptors in hypertension is a cause of elevated sympathetic activity or an adaptive change to high level of blood pressure. This chapter summarizes the latest findings about the role of $GABA_B$ receptors in the regulation of cardiovascular function.

II. Overview of Autonomic Nervous System

The autonomic nervous system controls essential physiological functions such as circulation, respiration, body temperature, visceral functions, metabolism, water and electrolyte balance, and body fluid balance (Dampney, 1994). The autonomic nervous system has two main divisions: the sympathetic and the parasympathetic. The afferent and efferent nerves linked by autonomic neurons in the CNS regulate many physiological functions such as the stability of blood pressure (Guyenet, 2006; Spyer, 1990). The neurons involved in the regulation of cardiovascular functions are located mainly in the spinal cord, brainstem, and hypothalamus. Generally, the primary afferent nerves that sense blood pressure and visceral signals project to the nucleus tractus solitarii (NTS) (Andresen et al., 2001; Andresen et al., 2004; Dampney, 1994). The second-order NTS neurons convey this information to other nuclei such as rostral ventrolateral medulla (RVLM), caudal ventrolateral medulla (CVLM), and hypothalamus. The autonomic neurons with a cardiovascular function (controlling the heart, blood vessels, and adrenal medulla) project to and control the activity of the preganglionic neurons in the intermediolateral cell column in the spinal cord (Dampney, 1994). The baroreceptor reflex is the principal neural mechanism involved in blood pressure regulation. Baroreceptors in the carotid arteries, cardiac chambers, and the aortic arch are activated by beat-to-beat fluctuations in systemic blood pressure. The baroreceptors relay information to the NTS and the ventrolateral medulla and is further processed in the insula, medial prefrontal cortex, cingulate cortex, amygdala, hypothalamus, thalamus, and cerebellum (Dampney, 1994; Henderson et al., 2004). Activation of baroreceptors by a rise in blood pressure leads to an increase in cardioinhibitory vagal outflow and a decrease in sympathetic vasomotor tone, leading to a decrease in peripheral vascular and cardiac tone. This results in a lower heart rate, decreased cardiac contractility, decreased vascular resistance, and venous return (Dampney, 1994). It has been shown that increased sympathetic outflow is an important pathophysiological characteristics of hypertension and chronic heart failure (Allen, 2002; Anderson et al., 1989; Cohn et al., 1984; Judy et al., 1976; Zucker et al., 1995). Sympathetic nerve activity is elevated in animal models of hypertension including spontaneously hypertensive rats (SHRs), renin transgenic rats, Dahl salt-sensitive rats, and deoxycorticosterone acetate (DOCA)-salt rats (Cabassi et al., 2002; Judy et al., 1976; Takeda & Bunag, 1980). There is also evidence of elevated sympathetic nerve activity in hypertensive patients (Anderson et al., 1989; Grassi, 1998; Greenwood et al., 1999; Mancia et al., 1999). The central alterations in hypertensive animals have focused on the hypothalamus and brainstem (Allen, 2002; Ciriello et al., 1984; de Wardener, 2001; Esler & Kaye, 2000; Ito et al., 2002).

The three central control regions—the RVLM, NTS, and hypothalamus—that regulate the barosensitive sympathetic efferents and consequently BP are described, together with their potential contributions to various forms of hypertension.

III. Distribution of $GABA_B$ Receptors in CNS Autonomic Centers

$GABA_B$ receptors are widely distributed in the CNS including regions related to autonomic functions (Bowery et al., 1987; Hill & Bowery, 1981). $GABA_B$ receptor-binding sites, measured by $[^3H]$GABA quantitative autoradiography, are uniformly distributed in several brain nuclei related to autonomic functions such as the hypothalamus, RVLM, NTS, periaqueductal gray, parabrachial nucleus, and medullary raphe nucleus (Bowery et al., 1987; Chu et al., 1990; Ichida & Kuriyama, 1998; Ichida et al., 1995). Furthermore, immunochemical labeling study using the antibody against $GABA_{B1}$ reveals that $GABA_{B1}$ immunoreactivity distributes throughout the CNS, with distinct distribution patterns (Margeta-Mitrovic et al., 1999). The distribution of $GABA_{B1}$ immunoreactivity is overlapped with $GABA_B$ receptor-binding sites measured by $[^3H]$GABA quantitative autoradiography, especially in the autonomic center such as hypothalamus, periaqueductal gray, RVLM, NTS, and medullary raphe nucleus. A high density of $GABA_{B1}$ is present in the brainstem monoaminergic neurons and cholinergic regions in the CNS (Margeta-Mitrovic et al., 1999).

IV. $GABA_B$ Receptor Function in the RVLM

The putative pressor area RVLM is a pivotal brainstem region in maintaining tonic sympathetic nerve activity and basal arterial blood pressure (Madden & Sved, 2003; Sapru, 2002; Willette et al., 1983a). Stimulation of the RVLM by microinjection of glutamate or kainic acid produces massive sympathoexcitation in both anesthetized and conscious rats (Bachelard et al., 1990; Dampney & Moon, 1980; McAllen, 1986; Ross et al., 1984b). On the other hand, bilateral destruction or inhibition of the RVLM results in the virtual elimination of sympathetic vasomotor tone (Feldberg & Guertzenstein, 1976; Guertzenstein & Silver, 1974; Willette et al., 1983b). Not only is this region of critical importance in the maintenance of baseline sympathetic vasomotor activity, it is also involved in mediating many cardiovascular reflexes such as baroreceptor reflex, chemoreceptor reflex, cardiopulmonary reflex, and somato-sympathetic reflex (Dampney, 1994; Gordon, 1995; Guyenet et al., 1996; Madden & Sved, 2003; Sapru, 2002). The RVLM is a heterogeneous brain region that contains many projection neurons and has reciprocal connections with other areas of the CNS that are involved in the regulation of autonomic functions. Electrophysiological and anatomic evidence has shown that the RVLM premotor neurons send monosynaptic projections to the intermediolateral cell column of the spinal cord (Brown & Guyenet, 1985; Dampney et al., 1987; Guyenet et al., 1996; Milner et al., 1988; Ross et al., 1984a). The vasomotor neurons in the RVLM provide a major source of tonic excitatory drive to the preganglionic neurons in the spinal cord and control sympathetic nerve discharges to the heart and blood vessels (Dampney, 1994; McAllen & Dampney, 1990; Morrison et al., 1991; Ross et al., 1984b). Many of these RVLM vasomotor neurons use glutamate as a transmitter in the spinal cord (Deuchars et al., 1995; Guyenet et al., 1996) and most of them are C1 adrenergic neurons containing the enzymes tyrosine hydroxylase and phenylethanolamine *N*-methyltransferase (Milner et al., 1988). These RVLM neurons are tonically active and are powerfully inhibited by GABAergic innervation upon activation of baroreceptors (Brown, 1984; Dampney, 1994; Guyenet, 2006; Verberne et al., 1999).

The involvement of GABAergic mechanisms in the central control of cardivovascular function has been known for many years (Antonaccio & Taylor, 1977; DiMicco et al., 1987). Particularly, the role of GABA in mediating inhibition of RVLM vasopressor neurons has been shown in several studies (Blessing, 1988; Lovick, 1988; Ruggiero et al., 1985; Willette et al., 1984). In addition to $GABA_A$ receptor-mediated inhibition in the RVLM (Meeley et al., 1985; Ruggiero et al., 1985), GABA influences neuronal functions by activating $GABA_B$ receptors (Amano & Kubo, 1993; Avanzino et al., 1994). For example, microinjection of the $GABA_B$ receptor agonist baclofen into the RVLM produces a profound drop in blood pressure in anesthetized rats, whereas injection of the $GABA_B$ receptor antagonists 2-hydroxysaclofen and CGP35348 produces opposite effects (Amano & Kubo, 1993; Avanzino et al., 1994). These data clearly suggest that functional $GABA_B$ receptors are present in the RVLM and play an important physiological role in controlling blood pressure. Also, baclofen produces hyperpolarization and inhibitory effects on the majority of RVLM neurons in brain slices from neonatal rats (Li & Guyenet, 1995). The $GABA_B$ antagonists CGP55845 or 2-hydroxysaclofen abolish the inhibition induced by baclofen but has no effect on inhibition caused by the $GABA_A$ receptor agonist muscimol (Li & Guyenet, 1995). Furthermore, baclofen induces outward currents in retrogradely labeled spinally projecting RVLM vasomotor neurons. The baclofen-induced currents exhibit a reversal potential close to potassium current equilibrium potential and is abolished by Ba^{2+} and the $GABA_B$ receptor antagonist CGP55845 (Li & Guyenet, 1996a). Interestingly, all histologically verified tyrosine hydroxylase immunoreactive neurons are inhibited by baclofen. These data provide information about ionic mechanisms of activation of $GABA_B$ receptors in the RVLM. Baclofen could exert its effect by a combination of pre and postsynaptic actions. Since baclofen also inhibits the discharge activity of RVLM vasomotor neurons when it is administered through iontophoresis (Li & Guyenet, 1996b), it is likely that baclofen acts on the postsynaptic site in the RVLM vasomotor neurons.

V. $GABA_B$ Function in the NTS

The NTS, located within the dorsomedial medulla oblongata, is a critical region involved in the integration of visceral sensory information. Thus, the NTS is an important site in the regulation of multiple autonomic reflexes related to cardiovascular, respiratory, gastrointestinal, hepatic, and renal functions (Aicher et al., 1995; Dias et al., 2003; Kannan & Yamashita, 1985; Lawrence & Jarrott, 1996; Wilson et al., 1996; Zhang & Mifflin, 1993). Stimulation of the NTS with the excitatory amino acid L-glutamate produces decreases in sympathetic outflow and blood pressure, an effect mimicking the baroreflex (Reis et al., 1981; Talman et al., 1980; Yin et al., 1994). On the other hand, lesions of the NTS cause elevation of sympathetic vasomotor tone, fulminating hypertension, and loss of baroreflex control of blood pressure and heart rate in both humans (Montgomery, 1961) and experimental animals (Doba & Reis, 1973, 1974). Furthermore, the importance of the NTS in the baroreflex in humans is demonstrated in a case report of a patient with baroreflex damage evidenced by a complete lack of reflex tachycardia following a hypotensive dose of nitroprusside. Postmortem examination of the patient's brain revealed that he had bilateral infarctions and gliosis of the NTS, but no damage to other medullar cardiovascular control such as the RVLM (Biaggioni et al., 1994). The baroreceptor afferent fibers terminate in the intermediate portion of the NTS (Ciriello, 1983; Donoghue et al., 1984; Housley et al., 1987) and form asymmetric (excitatory) synaptic contacts with second-order NTS neurons (Aicher et al., 1999). GABAergic mechanisms are responsible for monosynaptic connections from the CVLM to the RVLM and for the inhibition of vasomotor neurons in the RVLM (Agarwal & Calaresu, 1991; Agarwal et al., 1989).

The baroreceptor reflex plays a critical role in the normal regulation of arterial blood pressure (Spyer, 1990). The baroreceptive region of the NTS contains a high density of GABA-containing nerve terminals (Blessing, 1990; Hwang & Wu, 1984; Lasiter & Kachele, 1988; Maley & Newton, 1985; Maqbool et al., 1991; Meeley et al., 1985; Pickel et al., 1989) and a high density of both $GABA_A$ and $GABA_B$ receptors (Bowery et al., 1987; Singh & Ticku, 1985). GABAergic neurons, terminals, and GABA receptors, including $GABA_A$ and $GABA_B$ receptors, are highly expressed in the NTS (Bowery et al., 1987; Izzo et al., 1992; Torrealba & Muller, 1999).

Stimulation of $GABA_B$ receptors with baclofen in the NTS elicits a marked pressor response (Bousquet et al., 1982; Florentino et al., 1990; Persson, 1981; Sved & Sved, 1989; Sved & Sved, 1990). The cardiovascular response elicited by microinjections of nipecotic acid, a GABA uptake blocker, into the NTS of rats is antagonized by the $GABA_B$ receptor antagonist phaclofen (Sved & Sved, 1989; Sved & Sved, 1990), but not by the $GABA_A$ receptor antagonist bicuculline (Catelli et al., 1987). However, controversies exist as to which GABA receptor subtype is involved in the cardiovascular regulation in the NTS (van Giersbergen et al., 1992). For example, microinjection of $GABA_A$ receptor agonists and antagonists into the NTS produces depressor and pressor responses, respectively (Bousquet et al., 1982; Catelli et al., 1987). On the other hand, injection of baclofen into the NTS elicits pressor and tachycardic responses (Florentino et al., 1990; Sved & Sved, 1989). However, it is possible that both $GABA_A$ and $GABA_B$ receptors are involved in the inhibition of the baroreceptor reflex (Suzuki et al., 1993) and spontaneous discharges of the NTS baroreceptive neurons (Ruggeri et al., 1996). In addition to influencing baroreflex, $GABA_B$ receptors are also involved in modulation of other reflexes. In this regard, microinjection of baclofen into the commissural subnucleus of the NTS attenuates pressor response elicited by carotid chemoreceptor stimulation, an effect blocked by the $GABA_B$ antagonist 2-OH-saclofen (Suzuki et al., 1999). Also, $GABA_B$ receptor activation is involved in the inhibition of NTS neurons induced by stimulating vagal afferents (Wang et al., 2010). In addition to being involved in the regulation of cardiovascular reflexes, $GABA_B$ receptors in the NTS is important in the regulation of blood glucose concentration (Lemus et al., 2008). Microinjection of baclofen into the NTS elicits significant decreases in blood glucose concentrations, while $GABA_B$ receptor antagonists phaclofen or CGP55845A have an opposite effect. However, microinjection of $GABA_A$ agonist muscimol or $GABA_A$ receptor antagonist bicuculline fails to elicit significant changes in blood glucose concentrations (Lemus et al., 2008).

Increasing line of evidence suggests that the $GABA_B$ receptor expression and function in control of blood pressure are altered in pathological states such as chronic hypertension. The pressor response to activation of $GABA_B$ receptors with baclofen in the NTS is enhanced in SHR (Catelli & Sved, 1988; Sved & Sved, 1989; Sved & Tsukamoto, 1992; Tsukamoto & Sved, 1993) and DOCA-salt hypertension rats (Tsukamoto & Sved, 1993). In this regard, microinjection of baclofen into the NTS elicits an exaggerated depressor response in SHR compared with Wistar Kyoto (WKY) rats, whereas the pressor response to stimulation of $GABA_A$ receptors by microinjection of muscimol is not different between WKY and SHR (Catelli & Sved, 1988). Furthermore, in a rat model of chronic renal-wrap hypertension, microinjection of baclofen into the NTS causes an enhanced depressor response compared with the response in sham-operated normotensive rats (Durgam et al., 1999). These data suggest that the $GABA_B$ receptor in the

NTS is upregulated in chronic hypertensive state. One would expect that blocking GABA$_B$ receptors in this region will produce opposing responses to activation of GABA$_B$ receptors. However, blocking GABA$_B$ receptors in the NTS produces inconsistent responses in different models of hypertension. For instance, microinjection of the GABA$_B$ receptor antagonist CGP35348 into the NTS produces a greater depressor response in SHR and DOCA-salt hypertensive rats than in their normotensive controls (Tsukamoto & Sved, 1993). However, in the renal-wrap hypertensive rats, the cardiovascular response to microinjection of the GABA$_B$ antagonist CGP55845 and GABA reuptake inhibitor nipecotic acid are not different compared with normotensive sham-operated rats (Durgam et al., 1999). Thus, it seems that GABA$_B$ receptors are tonically activated in certain models of hypertension but not in the renal hypertension.

In supporting the notion that GABA$_B$ receptor function is increased in the NTS in hypertension, Singh and Ticku (1985reported the number of GABA$_B$ receptors in the NTS is increased by comparison of the [^{3}H]-baclofen-binding sites in the NTS in WKY and SHR. Consistently, the GABA$_B$ receptor mRNA levels in the caudal NTS shows a 3-fold increase in chronic renal-wrap hypertension compared with that in sham-operated normotensive rats (Durgam et al., 1999). It is unclear whether the increase in GABA$_B$ mRNA levels occurs in all NTS neurons or in a subpopulation of neurons with either distinct function or specific biochemical markers. However, it has been shown that in normotensive rats, NTS neurons receiving monosynaptic inputs from aortic depressor nerve are only slightly inhibited by baclofen, whereas neurons receiving polysynaptic inputs are markedly inhibited by baclofen (Zhang & Mifflin, 1998). Furthermore, in renal-wrap hypertensive rats, baclofen-induced inhibition of NTS neurons receiving monosynaptic inputs is greater than that in NTS neurons receiving polysynaptic aortic nerve input at 4 weeks of hypertension (Mei et al., 2003). These data suggest that the GABA$_B$ receptor is predominantly upregulated in second-order NTS neurons that receives monosynaptic inputs from the aortic depressor nerve in hypertension.

Baclofen can act both presynaptically to inhibit glutamate release from synaptic terminals and postsynaptically to induce outward potassium currents to reduce neuronal excitability in the NTS (Brooks et al., 1992; Zhang & Mifflin, 2010a). It is difficult to determine the role of presynaptic and postsynaptic mechanisms underlying baclofen-induced pressor responses in hypertension using whole-animal preparation. By taking advantage of electrophysiological recording in anatomically identified NTS neurons receiving aortic nerve (baroreceptor) inputs, Zhang et al. (2007) reported that baclofen-induced postsynaptic outward currents in the NTS second-order baroreceptor neurons are significantly greater in renal-wrap hypertensive rats than in sham-operated rats. This finding is consistent with in vivo data showing that bacofen-induced pressor response in the NTS is exaggerated in renal-wrap hypertension (Durgam et al., 1999). Also, the EC_{50} for baclofen-induced outward currents is significantly lower in hypertensive rats than in normotensive rats, suggesting that the response to activation of postsynaptic $GABA_B$ receptors is enhanced in NTS second-order baroreceptor neurons in renal-wrap hypertension. These data are consistent with previous data showing that $GABA_B$ receptor mRNA is increased in the NTS in renal-wrap hypertensive rats (Durgam et al., 1999). Also, the $[^3H]$-baclofen-binding sites in the NTS are increased in SHR compared with the control strain (Singh & Ticku, 1985). Therefore, the increased baclofen-induced currents could be due to increased $GABA_B$ receptor numbers in NTS neurons in hypertension. Furthermore, presynaptic $GABA_B$ receptor inhibition of glutamate and GABA release is also altered in hypertension. After blocking postsynaptic action of $GABA_B$ receptors with including cesium in the recording pipette solution, baclofen-induced inhibition of synaptic glutamate release is significantly greater in NTS second-order baroreceptor neurons in hypertensive rats than normotensive rats (Zhang & Mifflin, 2010b).

Little is known about the mechanisms underlying the alteration of $GABA_B$ receptor function in the NTS in hypertension. It has been shown that activation of $GABA_A$ receptors in cerebellar granule cells increases the number of binding sites for $[^3H]$-baclofen (Kardos et al., 1994; Schousboe, 1999). Furthermore, recent studies have shown that angiotensin II increases $GABA_B$ receptor function and $GABA_{B1}$ expression in the NTS (Yao et al., 2008; Zhang et al., 2009). In this regard, intracerebroventricular infusion of angiotensin II causes a significant increase in blood pressure and an increase in $GABA_{B1}$, but not $GABA_A$, receptor expression in the NTS (Zhang et al., 2009). However, the $GABA_B$ receptor expression does not differ after treatment with the nitric oxide synthase inhibitor NG-nitro-L-arginine methyl ester (L-NAME) (Zhang et al., 2009). Because treatment of L-NAME or angiotensin II produces similar increases in blood pressure, the increased expression of $GABA_B$ receptors is not due to elevation of blood pressure, but specifically attributes to angiotensin II treatment. Furthermore, $GABA_{B1}$ mRNA is increased by chronic treatment of NTS neuronal cultures with angiotensin II, and the inhibitory effect of baclofen on the excitability of the NTS neurons is exaggerated after chronic treatment of angiotensin II (Yao et al., 2008). These results provide evidence that increased angiotensin II may be responsible for the enhanced pressor response to microinjection of baclofen into the NTS in hypertensive rats (Durgam et al., 1999; Mei et al., 2003; Tsukamoto & Sved, 1993; Vitela & Mifflin, 2001; Zhang et al., 2007). Angiotensin II is a potent effector of the renin–angiotensin system and plays an important role in the control of hydromineral and fluid volume, and autonomic function (Swanson & Sawchenko, 1983). It is well known that increased angiotensin II activity contributes to the pathogenesis of hypertension (Raizada et al., 2000; Tan et al., 2005). Angiotensin II activity is increased in the central regions regulating cardiovascular function in hypertensive animals (Brooks et al., 2001; Katsunuma et al., 2003), which could consequently increase $GABA_{B1}$ expression and enhance $GABA_B$ receptor function in hypertension. Increased $GABA_B$ receptor function in the NTS may elicit an inhibitory action on baroreflex input signals and result in an increase in sympathetic outflow and elevation of blood pressure. It has been reported that the inducible $GABA_B$ receptors have functional and pharmacological properties, which seem to differ from those exhibited by the constitutively expressed receptors (Kardos et al., 1994; Vale et al., 1998). It is not clear whether there is a differential change of expression of the two isoforms of the $GABA_B$ receptors ($GABA_{B1}$ and $GABA_{B2}$), because the functional properties of $GABA_B$ receptors are dependent on the formation of heterooligomeric complexes of $GABA_{B1}$ and $GABA_{B2}$ (Jones et al., 1998; Kaupmann et al., 1998; White et al., 1998).

What is the physiological role of upregulation of $GABA_B$ receptors in the NTS in hypertension? Upregulation of $GABA_B$ receptors may enhance the GABAergic inhibition and serve as a protective mechanism in the NTS in hypertension. Activation of baroreceptor afferent inputs elicits a negative feedback inhibitory pathway that limits the initial excitatory inputs (Mifflin, 1996; Mifflin et al., 1988). It has been suggested that this feedback inhibition is primarily mediated by activation of both $GABA_A$ and $GABA_B$ receptors (Edwards et al., 2009), although other mechanisms could also be involved. Alternatively, increased $GABA_B$ receptor function in the NTS in hypertension might serve to dampen excitatory inputs and maintain normal baroreflex buffering upon increased arterial pressure.

VI. $GABA_B$ Receptor Function in the Hypothalamus

The hypothalamic paraventricular nucleus (PVN) is closely involved in the regulation of various neuroendocrine and autonomic functions (Cui et al., 2001; Imaki et al., 1998; Pyner & Coote, 1999, 2000; Swanson & Sawchenko, 1980, 1983). Previous anatomic and functional studies support the view that the PVN is an important source of excitatory drive for sympathetic vasomotor tone (Allen, 2002; Dampney et al., 2005; Griffiths et al., 1998; Kannan et al., 1989; Martin & Haywood, 1993; Reddy et al., 2005; Swanson & Sawchenko, 1983). The PVN is a heterogenous nucleus, and the PVN neurons can be differentiated based on both morphological and electrophysiological criteria. Furthermore, the PVN is reciprocally connected with other areas of the CNS that are involved in the regulation of autonomic nervous system. The PVN is composed of at least two distinct populations of output neurons: the magnocellular and parvocellular neurons. The magnocellular neurons project to the posterior pituitary and regulate vasopressin and oxytocin secretion. The parvocellular neurons project to brainstem autonomic centers such as the RVLM (Coote et al.,

1998; Hardy, 2001), NTS (Pyner & Coote, 2000; Ricardo & Koh, 1978; Yang & Coote, 1998), and the sympathetic preganglionic neurons located in the intermediolateral cell column (IML) of the spinal cord (Hardy, 2001; Pyner & Coote, 1999, 2000). The PVN presympatheic neurons may contribute to elevated sympathetic outflow in some pathophysiological conditions, such as heart failure (Patel, 2000) and hypertension (Allen, 2002; Li & Pan, 2006). Electrical stimulation of subregions of hypothalamus produces pronounced increases in sympathetic nerve activity and elevates blood pressure. For example, stimulation of posterior or ventromedial hypothalamus or the PVN elicits sympathetic excitatory response (Swanson & Sawchenko, 1980, 1983).

GABAergic inputs make up about 50% of the synaptic innervation of PVN neurons (Decavel & Van den Pol, 1990). Electrophysiological studies have further demonstrated that the majority of the local synaptic inputs to PVN neurons are GABAergic (Boudaba et al., 1996; Tasker & Dudek, 1993). In addition to activation of $GABA_A$ receptors to produce postsynaptic inhibition, synaptically released GABA also stimulates $GABA_B$ receptors located at both presynaptic and postsynaptic sites (Misgeld et al., 1995). It has been shown that $GABA_B$ receptors are widely distributed in the CNS including the PVN (Margeta-Mitrovic et al., 1999). Activation of $GABA_B$ receptors contribute to the regulation of neuronal activity and sympathetic outflow. In this regard, microinjection of baclofen into one of the hypothalamic pressor areas, the ventromedial hypothalamus, decreases sympathetic nerve activity, blood pressure, and heart rate in both normotensive rats and SHR (Takenaka et al., 1996). Also, activation of $GABA_B$ receptors with baclofen in the PVN decreases the lumbar sympathetic nerve activity and arterial blood pressure in anesthetized hypertensive SHR and normotensive control strains (Li & Pan, 2007a) (Fig. 1). Furthermore, these depressor responses induced by baclofen are significantly larger in SHR and Dahl salt-sensitive rats than in their respective normotensive controls (Bunag et al., 1983; Takeda & Bunag, 1978).

Similar to the enhancement of $GABA_B$ receptor function in the NTS in hypertensive rats, activation of $GABA_B$ receptors in the PVN also produces an augmented inhibitory effect on sympathetic vasomotor tone in SHR compared with WKY rats (Li & Pan, 2007a). In fact, activation of $GABA_B$ receptors with baclofen in the PVN decreases lumbar sympathetic nerve activity and arterial blood pressure only at the highest dose in anesthetized normotensive WKY and SD rats (Li & Pan, 2007a). However, microinjection of baclofen into the PVN significantly decreases lumbar sympathetic activity and arterial blood pressure in all of the doses tested in the SHR (Li & Pan, 2007a). Furthermore, blockade of $GABA_B$ receptors with CGP52432 significantly increases lumbar sympathetic nerve activity and arterial blood pressure in SHR but not in normotensive control rats (Li & Pan, 2007a). These in vivo data provide strong evidence that $GABA_B$

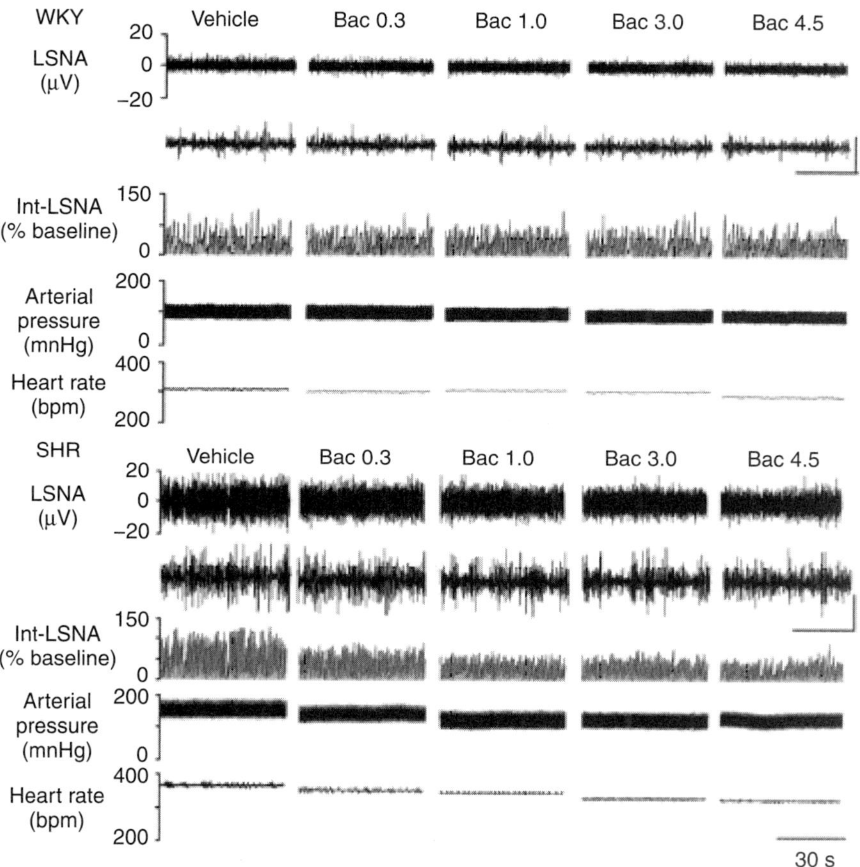

FIGURE 1 Microinjection of baclofen into the PVN produced greater reduction of lumbar sympathetic nerve activity (LSNA), arterial blood pressure (ABP), and heart rate (HR) in spontaneously hypertensive rats (SHR) than in Wistar Kyoto (WKY) rats. The raw tracings below LSNA show the nerve activity on an expanded time scale (horizontal bar, 1 s; vertical bar, 10 μV). Bac, baclofen. Reproduced from Li & Pan (2007a) with permission.

receptors in the PVN are not involved in tonic GABAergic inhibition of sympathetic outflow in normotensive rats. However, in SHRs, tonic GABAergic inhibition of sympathetic vasomotor tone is mediated by both GABA$_A$ and GABA$_B$ receptors, because blockade of GABA$_A$ receptors with bicuculline or GABA$_B$ receptors with CGP52432 increases sympathetic vasomotor tone (Li & Pan, 2007a). Consistent with the finding that GABA$_B$ function is upregulated in the PVN in SHR, blockade of GABA$_B$ receptors with CGP55845 significantly increased the excitability of 75% of PVN neurons projecting to the RVLM tested in hypertensive 13-week-old SHR (Li & Pan, 2006). However, CGP55845 has no effect on the excitability of the PVN–RVLM output neurons in normotensive controls in a brain slice preparation (Li & Pan, 2006) (Fig. 2). These electrophysiological data suggest that $GABA_B$ receptors play a greater role in GABAergic inhibition of PVN–RVLM projection neurons in hypertension.

On the contrary to enhanced baclofen-induced depressor response in the PVN in hypertensive rats, microinjection of baclofen into the PVN in a

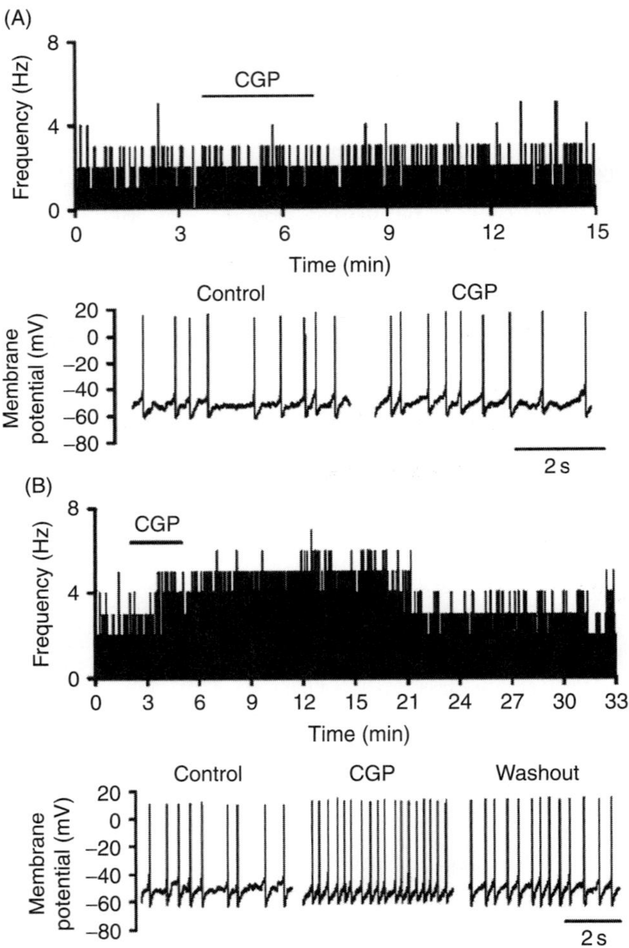

FIGURE 2 Bath application of $GABA_B$ receptor antagonist CGP55845 increases firing activity of spinally projecting PVN neurons in hypertensive SHR but not in prehypertensive SHR rats. (A) Top: effect of 1 μM CGP on firing activity of labeled PVN neuron in 6-week-old SHR. (A) Bottom: firing activity of same cell during control, application of CGP, and washout. (B) Top: effect of 1 μM CGP on firing activity of labeled PVN neuron in 13-week-old SHR. (B) Bottom: spontaneous firing activity of same cell during control, application of CGP, and washout. Reproduced from Li & Pan (2006) with permission.

rat model of chronic heart failure induced by coronary artery ligation produces a smaller inhibitory effect on renal sympathetic nerve activity, arterial blood pressure, and heart rate than in sham-operated rats (Wang et al., 2009). Furthermore, the mRNA levels of $GABA_{B1a}$ and $GABA_{B1b}$ are significantly reduced in the PVN in this chronic heart failure model (Wang et al., 2009). This downregulation of $GABA_B$ receptors in the PVN may contribute to the elevated sympathetic outflow in chronic heart failure.

$GABA_B$ receptors are expressed on both presynaptic terminals and postsynaptic soma, and activation of presynaptic $GABA_B$ receptors reduces neurotransmitter release at many central synapses (Misgeld et al., 1995). For instance, baclofen reduces GABA and glutamate release in PVN neurons with projection to the suprachiasmatic nucleus in the hypothalamus (Cui et al., 2000; Wang et al., 2003). To determine the effect of activation of $GABA_B$ receptors on the neurotransmitter release in the PVN presympathetic neurons, we retrogradely labeled the PVN neurons projecting to the RVLM and spinal cord by preinjection of fluorescence tracer into the RVLM or spinal cord (Li & Pan, 2006; Li et al., 2008a, 2008b). We found that baclofen dose-dependently decreases the frequency of GABAergic inhibitory postsynaptic currents (IPSCs) and glutamatergic excitatory postsynaptic currents (EPSCs) in the spinally projecting and RVLM-projecting PVN neurons in rat hypothalamic slices (Chen & Pan, 2006; Li et al., 2008a). Furthermore, baclofen significantly reduced the frequency, but not the amplitude and decay time constant, of miniature IPSCs and miniature EPSCs. Because bath application of a Na^+ channel blocker tetrodotoxin has little effect on spontaneous EPSCs and IPSCs, these spontaneous EPSCs and IPSCs may represent local release of glutamate and GABA, respectively. These data suggest that $GABA_B$ receptors in the PVN directly act on the presynaptic terminals to decrease glutamate and GABA release to the spinally projecting PVN neurons (Chen & Pan, 2006). However, blocking $GABA_B$ receptors did not alter the frequency of spontaneous EPSCs and IPSCs, suggesting that under the basal condition, the endogenous GABA level in the synapse is not sufficient to stimulate the presynaptic $GABA_B$ receptors. However, we have demonstrated that increased endogenous GABA release can activate $GABA_B$ receptors to inhibit glutamatergic and glycinergic inputs to spinal dorsal horn neurons (Li et al., 2002; Wang et al., 2006).

Also, the intracellular signaling pathways involved in the effect of activation of $GABA_B$ receptors have been revealed. It has been shown that $GABA_B$ receptors are functionally coupled to *N*-ethylmaleimide-sensitive G proteins, especially $G_{i\alpha\text{-}2}$ proteins, in rat cerebral cortical membranes (Odagaki et al., 2000). *N*-Ethylmaleimide can uncouple pertussis toxin-sensitive $G_{i/o}$ subunits by modifying cysteine residues (Jakobs et al., 1982; Shapiro et al., 1994). In hypothalamic slices, the inhibitory effects of baclofen on the frequency of mIPSCs and mEPSCs are completely blocked by bath application of *N*-ethylmaleimide (Chen & Pan, 2006). Therefore, $G_{i/o}$ proteins are probably involved in the signal transduction of the presynaptic action of $GABA_B$ receptor agonists. At the postsynaptic site, $GABA_B$ receptors can activate $G_{i/o}$ proteins, thereby enhancing G protein-coupled inwardly rectifying K^+ channel (GIRK) currents (Harayama et al., 1998; Sodickson & Bean, 1996). Voltage-dependent K^+ channels is one of the effectors of activation of $GABA_B$ receptors (Misgeld et al., 1995) and presynaptic voltage-dependent K^+ channels are considered important in the regulation of neurotransmitter release (Ishikawa et al., 2003). Therefore, it is possible that the presynaptic voltage-dependent K^+ channels are involved in the effect of $GABA_B$ receptor agonists on glutamate and GABA release in the PVN. However, 4-aminopyridine has no significant effect on baclofen-induced inhibition of mIPSCs and mEPSCs in the PVN spinally projecting neurons (Chen & Pan, 2006). Consistently, Ba^{2+}, another voltage-dependent K^+ current blocker, also has no effect on $GABA_B$ receptor-mediated presynaptic inhibition in the calyx of Held (Takahashi et al., 1998). Therefore, it is less likely that voltage-dependent K^+ currents are involved in the presynaptic $GABA_B$ receptor-mediated inhibition on both GABA and glutamate release to PVN presympathetic neurons.

The function of presynaptic $GABA_B$ receptors in inhibiting GABAergic and glutamatergic synaptic inputs in the PVN is altered in hypertension. In this regard, baclofen produces a greater inhibitory effect on the frequency of glutamatergic spontaneous EPSCs in SHR than in WKY rats (Li et al., 2008a). In contrast, baclofen caused a smaller inhibitory effect on the frequency of spontaneous IPSCs in SHR than in WKY rats (Li et al., 2008a). These findings suggest that the presynaptic $GABA_B$ receptor function in the control of glutamatergic and GABAergic synaptic inputs to the PVN presympathetic neurons is differentially affected in SHR. The increased $GABA_B$ receptor function on glutamatergic terminals and reduced $GABA_B$ receptor function on GABAergic terminals could contribute to the enhanced inhibitory effect of baclofen on the firing activity of PVN neurons and sympathetic vasomotor tone in SHR. It is still not clear whether different $GABA_B$ receptor subunits (i.e., $GABA_{B1}$ and $GABA_{B2}$) are differentially expressed at glutamatergic and GABAergic terminals in the PVN in SHR. In the hippocampus, both $GABA_{B1}$ and $GABA_{B2}$ subtypes are mainly present at glutamatergic axon terminals (Kulik et al., 2003). However, in the dorsal cochlear nucleus, $GABA_{B1}$ is primarily located at glutamatergic synapse but not GABAergic terminals (Lujan et al., 2004). Alternatively, changes in the downstream signaling also can preferentially alter the $GABA_B$ receptor function in glutamatergic and GABAergic terminals in the PVN in SHR.

To assess whether the postsynaptic $GABA_B$ receptor function is altered in spinally projecting PVN neurons in hypertension, we compared postsynaptic $GABA_B$ receptor currents evoked by puff application of baclofen directly to labeled PVN neurons in WKY rats and SHR. We found that the amplitude of baclofen-elicited outward currents in labeled PVN neurons was significantly greater in SHR than in WKY rats (Fig. 3). Bath application of the GABA$_B$ receptor antagonist CGP55845 or inhibition of G proteins with intracellular dialysis of GDP-β-S completely blocked baclofen-elicited outward currents in spinally projecting PVN neurons (Li et al., 2008a). These data suggest that the postsynaptic GABA$_B$ receptor function is upregulated in the PVN presympathetic neurons in SHR. It is likely that this upregulation of postsynaptic GABA$_B$ receptor function contributes to the potentiated inhibitory effect of baclofen on the firing activity of spinally projecting PVN neurons and sympathetic vasomotor tone in SHR (Li & Pan, 2006, 2007a). At this time, little is known about the mechanisms

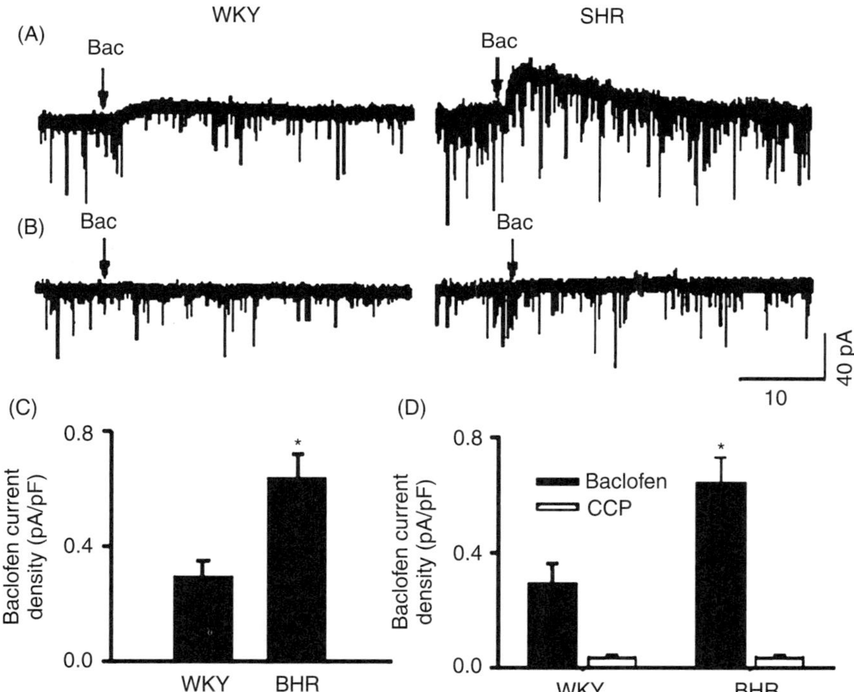

FIGURE 3 Baclofen-sensitive currents in labeled PVN neurons in WKY rats and SHR. (A) Original tracings showing currents elicited by puff application of 100 μM baclofen to labeled PVN neurons in WKY rat and SHR. Arrows indicate puff application of baclofen. (B) Raw tracings showing undetectable current by puff application of 100 μM baclofen when 1 mM GDP-β-S was included in the recording pipette solution in WKY rats and SHR. (C) Summary data showing the baclofen-sensitive current elicited by puff application of 100 μM baclofen in labeled PVN neurons in WKY rats and SHR. (D) Summary data showing baclofen-sensitive currents in the absence and presence of 2 μM CGP55845 (CGP) in labeled PVN neurons in WKY rats and SHR. Data are means ± SE. $^{*}p < 0.05$ compared with the corresponding value in WKY rats. Reproduced from Li et al. (2008a) with permission.

underlying enhanced postsynaptic $GABA_B$ receptor function in the PVN in hypertension. Studies have shown that *N*-methyl-D-aspartic acid (NMDA) receptor activation can increase $GABA_B$ receptor gene expression and postsynaptic $GABA_B$ receptor function in the spinal cord and hippocampus (Benardo, 1995; Cole et al., 1989; McCarson & Enna, 1999). Since the glutamatergic input in the PVN is enhanced in SHR (Li & Pan, 2007b; Li et al., 2008b), it is possible that the enhanced glutamatergic input in SHR causes more Ca^{2+} influx into postsynaptic PVN neurons, which could upregulate postsynaptic $GABA_B$ receptors through Ca^{2+}-dependent phosphorylation or dephosphorylation mechanisms. Acutely blocking NMDA receptors with AP5 or MK-801 had no significant effect on the baclofen-elicited currents in spinally projecting PVN neurons in SHR (Li et al., 2008a). Because acute blockade of NMDA receptors may not be sufficient to inhibit the downstream signaling cascade leading to increased expression of $GABA_B$ receptors by enhanced glutamatergic inputs in SHR, we cannot rule out the possibility that chronic blockade of NMDA receptor is required to affect $GABA_B$ receptors in SHR. Because angiotensin II has been shown to increase $GABA_B$ receptor expression and function in the NTS (Li et al., 2008a), and because angiotensin II concentration and angiotensin receptors are increased in the PVN in SHR (Gutkind et al., 1988; Phillips & Kimura, 1988), increased angiotensin II could play a role in the upregulation of $GABA_B$ receptors. We have shown previously that the $GABA_A$ receptor function in the PVN is attenuated, which contributes to increased firing activity of PVN presympathetic neurons in SHR (Li & Pan, 2006). Hence, upregulation of postsynaptic $GABA_B$ receptors in the PVN may represent a compensatory response to the reduced $GABA_A$ receptor function and enhanced glutamatergic input to these neurons in SHR and serve to dampen the increased sympathetic outflow in hypertension.

VII. Conclusion

The role of $GABA_B$ receptors in the regulation of autonomic functions has been recognized (Fig. 4). The cellular and molecular mechanisms underlying altered $GABA_B$ receptor function in hypertension are being studied by using advanced molecular and electrophysiological techniques. Enhanced $GABA_B$ receptor function in the autonomic centers in hypertension suggests that $GABA_B$ receptor plays an increasingly significant role in GABAergic control of the excitability of the autonomic neurons in hypertension (Table II). Thus, the $GABA_B$ receptor may represent new target for treatment of neurogenic hypertension. Indeed, baclofen appears effective in reducing blood pressure in severely hypertensive patients (Becker et al., 2000). However, it remains unclear that whether altered $GABA_B$ receptor function in the autonomic centers contributes to development and/or

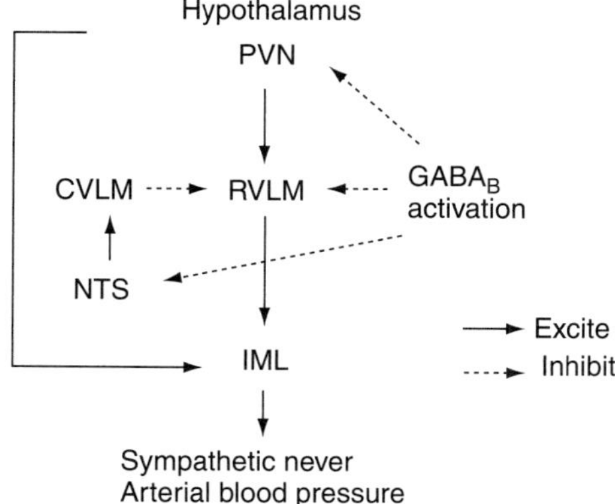

FIGURE 4 Diagram illustrates overall functions of GABA$_B$ receptors in regulating arterial blood pressure. Activation of GABA$_B$ receptors inhibits neurons in the PVN, RVLM, and NTS. Because the barosensitive neurons in the NTS inhibit RVLM vasomotor neurons through exciting neurons in the CVLM, GABA$_B$ receptor-mediated inhibition of NTS neurons may attenuate the inhibition of RVLM neurons. PVN: paraventricular nucleus of hypothalamus; RVLM: rostral ventrolateral medulla; NTS: nucleus tractus solitarii; CVLM: caudal ventrolateral medulla; IML: intermediolateral cell column in the spinal cord.

TABLE II Alteration of GABA$_B$ Receptor Function in Animal Models of Hypertension

Hypertension model	PVN	NTS
SHR	↑ Li & Pan, 2007a, 2006; Li et al., 2008a; Takenaka et al., 1996;	↑ Catelli & Sved, 1988; Sved & Sved, 1989; Sved & Tsukamoto, 1992; Tsukamoto & Sved, 1993; Singh & Ticku, 1985
DOCA-salt hypertensive rat		↑ Tsukamoto & Sved, 1993
Renal-wrap hypertension		↑ Durgam et al., 1999; Mei et al., 2003; Zhang et al., 2007; Zhang & Mifflin, 2010b

maintenance of hypertension or just simply an adaptive response to elevated blood pressure. Further studies are warranted to define the physiological role of GABA$_B$ receptors in the neural control of autonomic functions in various disease conditions.

Acknowledgments

Work conducted in the authors' laboratory was supported by grants GM64830 and NS45602 from the National Institutes of Health.

Conflict of Interest Statement: The authors have no conflicts of interest to declare.

References

Agarwal, S. K., & Calaresu, F. R. (1991). Monosynaptic connection from caudal to rostral ventrolateral medulla in the baroreceptor reflex pathway. *Brain Research*, *555*, 70–74.

Agarwal, S. K., Gelsema, A. J., & Calaresu, F. R. (1989). Neurons in rostral VLM are inhibited by chemical stimulation of caudal VLM in rats. *American Journal of Physiology*, *257*, R265–R270.

Aicher, S. A., Kurucz, O. S., Reis, D. J., & Milner, T. A. (1995). Nucleus tractus solitarius efferent terminals synapse on neurons in the caudal ventrolateral medulla that project to the rostral ventrolateral medulla. *Brain Research*, *693*, 51–63.

Aicher, S. A., Sharma, S., & Pickel, V. M. (1999). *N*-methyl-D-aspartate receptors are present in vagal afferents and their dendritic targets in the nucleus tractus solitarius. *Neuroscience*, *91*, 119–132.

Allen, A. M. (2002). Inhibition of the hypothalamic paraventricular nucleus in spontaneously hypertensive rats dramatically reduces sympathetic vasomotor tone. *Hypertension*, *39*, 275–280.

Amano, M., & Kubo, T. (1993). Involvement of both $GABA_A$ and $GABA_B$ receptors in tonic inhibitory control of blood pressure at the rostral ventrolateral medulla of the rat. *Naunyn-Schmiedeberg's Archives of Pharmacology*, *348*, 146–153.

Anderson, E. A., Sinkey, C. A., Lawton, W. J., & Mark, A. L. (1989). Elevated sympathetic nerve activity in borderline hypertensive humans. Evidence from direct intraneural recordings. *Hypertension*, *14*, 177–183.

Andresen, M. C., Doyle, M. W., Bailey, T. W., & Jin, Y. H. (2004). Differentiation of autonomic reflex control begins with cellular mechanisms at the first synapse within the nucleus tractus solitarius. *Brazilian Journal of Medical and Biological Research*, *37*, 549–558.

Andresen, M. C., Doyle, M. W., Jin, Y. H., & Bailey, T. W. (2001). Cellular mechanisms of baroreceptor integration at the nucleus tractus solitarius. *Annals of the New York Academy of Sciences*, *940*, 132–141.

Antonaccio, M. J., & Taylor, D. G. (1977). Involvement of central GABA receptors in the regulation of blood pressure and heart rate of anesthetized cats. *European Journal of Pharmacology*, *46*, 283–287.

Avanzino, G. L., Ruggeri, P., Blanchi, D., Cogo, C. E., Ermirio, R., & Weaver, L. C. (1994). $GABA_B$ receptor-mediated mechanisms in the RVLM studied by microinjections of two $GABA_B$ receptor antagonists. *American Journal of Physiology*, *266*, H1722–H1728.

Bachelard, H., Gardiner, S. M., & Bennett, T. (1990). Cardiovascular responses elicited by chemical stimulation of the rostral ventrolateral medulla in conscious, unrestrained rats. *Journal of the Autonomic Nervous System*, *31*, 185–190.

Becker, R., Benes, L., Sure, U., Hellwig, D., & Bertalanffy, H. (2000). Intrathecal baclofen alleviates autonomic dysfunction in severe brain injury. *Journal of Clinical Neuroscience*, *7*, 316–319.

Behar, T. N., Schaffner, A. E., Tran, H. T., & Barker, J. L. (1995). GABA-induced motility of spinal neuroblasts develops along a ventrodorsal gradient and can be mimicked by agonists of $GABA_A$ and $GABA_B$ receptors. *Journal of Neuroscience Research*, *42*, 97–108.

Benardo, L. S. (1995). *N*-methyl-D-aspartate transmission modulates $GABA_B$-mediated inhibition of rat hippocampal pyramidal neurons in vitro. *Neuroscience*, *68*, 637–643.

Biaggioni, I., Whetsell, W. O., Jobe, J., & Nadeau, J. H. (1994). Baroreflex failure in a patient with central nervous system lesions involving the nucleus tractus solitarii. *Hypertension*, *23*, 491–495.

Blessing, W. W. (1988). Depressor neurons in rabbit caudal medulla act via GABA receptors in rostral medulla. *American Journal of Physiology*, *254*, H686–H692.

Blessing, W. W. (1990). Distribution of glutamate decarboxylase-containing neurons in rabbit medulla oblongata with attention to intramedullary and spinal projections. *Neuroscience*, *37*, 171–185.

Blessing, W. W. (1997). Arterial pressure and blood flow to the tissues. In *The lower brainstem and bodily homeostasis* (pp. 165–268). New York: Oxford University Press.

Boudaba, C., Szabo, K., & Tasker, J. G. (1996). Physiological mapping of local inhibitory inputs to the hypothalamic paraventricular nucleus. *Journal of Neuroscience*, *16*, 7151–7160.

Bousquet, P., Feldman, J., Bloch, R., & Schwartz, J. (1982). Evidence for a neuromodulatory role of GABA at the first synapse of the baroreceptor reflex pathway. Effects of GABA derivatives injected into the NTS. Naunyn schmiedebergs. *Archives of Pharmacology*, *319*, 168–171.

Bowery, N. G. (1993). $GABA_B$ receptor pharmacology. *Annual Review of Pharmacology and Toxicology*, *33*, 109–147.

Bowery, N. G., & Enna, S. J. (2000). Gamma-aminobutyric acid(B) receptors: First of the functional metabotropic heterodimers. *Journal of Pharmacology and Experimental Therapeutics*, *292*, 2–7.

Bowery, N. G., Hudson, A. L., & Price, G. W. (1987). $GABA_A$ and $GABA_B$ receptor site distribution in the rat central nervous system. *Neuroscience*, *20*, 365–383.

Brooks, P. A., Glaum, S. R., Miller, R. J., & Spyer, K. M. (1992). The actions of baclofen on neurones and synaptic transmission in the nucleus tractus solitarii of the rat in vitro. *Journal of Physiology (Paris)*, *457*, 115–129.

Brooks, V. L., Scrogin, K. E., & McKeogh, D. F. (2001). The interaction of angiotensin II and osmolality in the generation of sympathetic tone during changes in dietary salt intake. An hypothesis. *Annals of the New York Academy of Sciences*, *940*, 380–394.

Brown, D. D. (1984). The role of stable complexes that repress and activate eucaryotic genes. *Cell*, *37*, 359–365.

Brown, D. L., & Guyenet, P. G. (1985). Electrophysiological study of cardiovascular neurons in the rostral ventrolateral medulla in rats. *Circulation Research*, *56*, 359–369.

Bunag, R. D., Butterfield, J., & Sasaki, S. (1983). Hypothalamic pressor responses and salt-induced hypertension in Dahl rats. *Hypertension*, *5*, 460–467.

Cabassi, A., Vinci, S., Cantoni, A. M., Quartieri, F., Moschini, L., Cavazzini, S., et al. (2002). Sympathetic activation in adipose tissue and skeletal muscle of hypertensive rats. *Hypertension*, *39*, 656–661.

Caddick, S. J., & Hosford, D. A. (1996). The role of $GABA_B$ mechanisms in animal models of absence seizures. *Molecular Neurobiology*, *13*, 23–32.

Catelli, J. M., Giakas, W. J., & Sved, A. F. (1987). GABAergic mechanisms in nucleus tractus solitarius alter blood pressure and vasopressin release. *Brain Research*, *403*, 279–289.

Catelli, J. M., & Sved, A. F. (1988). Enhanced pressor response to GABA in the nucleus tractus solitarii of the spontaneously hypertensive rat. *European Journal of Pharmacology*, *151*, 243–248.

Chen, Q., & Pan, H. L. (2006). Regulation of synaptic input to hypothalamic presympathetic neurons by GABA(B) receptors. *Neuroscience*, *142*, 595–606.

Chronwall, B. M., Davis, T. D., Severidt, M. W., Wolfe, S. E., McCarson, K. E., Beatty, D. M., et al. (2001). Constitutive expression of functional GABA(B) receptors in mIL-tsA58 cells requires both GABA(B(1)) and GABA(B(2)) genes. *Journal of Neurochemistry*, *77*, 1237–1247.

Chu, D. C., Albin, R. L., Young, A. B., & Penney, J. B. (1990). Distribution and kinetics of $GABA_B$ binding sites in rat central nervous system: A quantitative autoradiographic study. *Neuroscience*, *34*, 341–357.

Ciriello, J. (1983). Brainstem projections of aortic baroreceptor afferent fibers in the rat. *Neuroscience Letters*, *36*, 37–42.

Ciriello, J., Kline, R. L., Zhang, T. X., & Caverson, M. M. (1984). Lesions of the paraventricular nucleus alter the development of spontaneous hypertension in the rat. *Brain Research*, *310*, 355–359.

Cohn, J. N., Levine, T. B., Olivari, M. T., Garberg, V., Lura, D., Francis, G. S., et al. (1984). Plasma norepinephrine as a guide to prognosis in patients with chronic congestive heart failure. *The New England Journal of Medicine*, *311*, 819–823.

Cole, A. J., Saffen, D. W., Baraban, J. M., & Worley, P. F. (1989). Rapid increase of an immediate early gene messenger RNA in hippocampal neurons by synaptic NMDA receptor activation. *Nature*, *340*, 474–476.

Coote, J. H., Yang, Z., Pyner, S., & Deering, J. (1998). Control of sympathetic outflows by the hypothalamic paraventricular nucleus. *Clinical and Experimental Pharmacology and Physiology*, *25*, 461–463.

Cui, L. N., Coderre, E., & Renaud, L. P. (2000). GABA(B) presynaptically modulates suprachiasmatic input to hypothalamic paraventricular magnocellular neurons. *American Journal of Physiology – Regulatory, Integrative and Comparative Physiology*, *278*, R1210–R1216.

Cui, L. N., Coderre, E., & Renaud, L. P. (2001). Glutamate and GABA mediate suprachiasmatic nucleus inputs to spinal-projecting paraventricular neurons. *American Journal of Physiology – Regulatory, Integrative and Comparative Physiology*, *281*, R1283–R1289.

Dampney, R. A. (1994). Functional organization of central pathways regulating the cardiovascular system. *Physiological Reviews*, *74*, 323–364.

Dampney, R. A., Czachurski, J., Dembowsky, K., Goodchild, A. K., & Seller, H. (1987). Afferent connections and spinal projections of the pressor region in the rostral ventrolateral medulla of the cat. *Journal of the Autonomic Nervous System*, *20*, 73–86.

Dampney, R. A., Horiuchi, J., Killinger, S., Sheriff, M. J., Tan, P. S., & McDowall, L. M. (2005). Long-term regulation of arterial blood pressure by hypothalamic nuclei: Some critical questions. *Clinical and Experimental Pharmacology and Physiology*, *32*, 419–425.

Dampney, R. A., & Moon, E. A. (1980). Role of ventrolateral medulla in vasomotor response to cerebral ischemia. *American Journal of Physiology*, *239*, H349–H358.

Davies, C. H., Starkey, S. J., Pozza, M. F., & Collingridge, G. L. (1991). GABA autoreceptors regulate the induction of LTP. *Nature*, *349*, 609–611.

Decavel, C., & Van den Pol, A. N. (1990). GABA: A dominant neurotransmitter in the hypothalamus. *Journal of Comparative Neurology*, *302*, 1019–1037.

Deuchars, S. A., Morrison, S. F., & Gilbey, M. P. (1995). Medullary-evoked EPSPs in neonatal rat sympathetic preganglionic neurones in vitro. *Journal of Physiology (Paris)*, *487*(Pt 2), 453–463.

de Wardener, H. E. (2001). The hypothalamus and hypertension. *Physiological Reviews*, *81*, 1599–1658.

DiMicco, J. A., Abshire, V. M., Shekhar, A., & Wilbe, J. H. Jr. (1987). Role of GABAergic mechanisms in the central regulation of arterial pressure. *European Heart Journal*, *8*(Suppl. B), 133–138.

Dias, A. C., Colombari, E., & Mifflin, S. W. (2003). Effect of nitric oxide on excitatory amino acid-evoked discharge of neurons in NTS. *American Journal of Physiology. Heart and Circulatory Physiology*, *284*, H234–H240.

Dichter, M. A. (1997). Basic mechanisms of epilepsy: Targets for therapeutic intervention. *Epilepsia*, *38*(Suppl. 9), S2–S6.

Doba, N., & Reis, D. J. (1973). Acute fulminating neurogenic hypertension produced by brainstem lesions in the rat. *Circulation Research*, *32*, 584–593.

Doba, N., & Reis, D. J. (1974). Role of central and peripheral adrenergic mechanisms in neurogenic hypertension produced by brainstem lesions in rat. *Circulation Research*, *34*, 293–301.

Donoghue, S., Felder, R. B., Jordan, D., & Spyer, K. M. (1984). The central projections of carotid baroreceptors and chemoreceptors in the cat: A neurophysiological study. *Journal of Physiology (Paris)*, *347*, 397–409.

Durgam, V. R., Vitela, M., & Mifflin, S. W. (1999). Enhanced gamma-aminobutyric acid-B receptor agonist responses and mRNA within the nucleus of the solitary tract in hypertension. *Hypertension*, *33*, 530–536.

Edwards, I. J., Deuchars, S. A., & Deuchars, J. (2009). The intermedius nucleus of the medulla: A potential site for the integration of cervical information and the generation of autonomic responses. *Journal of Chemical Neuroanatomy*, *38*, 166–175.

Enna, S. J., & , Bowery, N. G. (1997). *The GABA receptors* (2nd ed.), Clifton, New Jersey: Humana Press.

Esler, M., & Kaye, D. (2000). Sympathetic nervous system activation in essential hypertension, cardiac failure and psychosomatic heart disease. *Journal of Cardiovascular Pharmacology*, *35*, S1–S7.

Feldberg, W., & Guertzenstein, P. G. (1976). Vasodepressor effects obtained by drugs acting on the ventral surface of the brain stem. *Journal of Physiology (Paris)*, *258*, 337–355.

Florentino, A., Varga, K., & Kunos, G. (1990). Mechanism of the cardiovascular effects of $GABA_B$ receptor activation in the nucleus tractus solitarii of the rat. *Brain Research*, *535*, 264–270.

Gassmann, M., Shaban, H., Vigot, R., Sansig, G., Haller, C., Barbieri, S., et al. (2004). Redistribution of $GABA_{B(1)}$ protein and atypical $GABA_B$ responses in $GABA_{B(2)}$-deficient mice. *Journal of Neuroscience*, *24*, 6086–6097.

Gordon, F. J. (1995). Excitatory amino acid receptors in central cardiovascular regulation. *Clinical and Experimental Hypertension*, *17*, 81–90.

Grassi, G. (1998). Role of the sympathetic nervous system in human hypertension. *Journal of Hypertension*, *16*, 1979–1987.

Greenwood, J. P., Stoker, J. B., & Mary, D. A. (1999). Single-unit sympathetic discharge: Quantitative assessment in human hypertensive disease. *Circulation*, *100*, 1305–1310.

Griffiths, R. I., Bleecker, G. C., Jabs, D. A., Dieterich, D. T., Coleson, L., Winters, D., et al. (1998). Pharmacoeconomic analysis of 3 treatment strategies for cytomegalovirus retinitis in patients with AIDS. *PharmacoEconomics*, *13*, 461–474.

Guertzenstein, P. G., & Silver, A. (1974). Fall in blood pressure produced from discrete regions of the ventral surface of the medulla by glycine and lesions. *Journal of Physiology (Paris)*, *242*, 489–503.

Gutkind, J. S., Kurihara, M., Castren, E., & Saavedra, J. M. (1988). Increased concentration of angiotensin II binding sites in selected brain areas of spontaneously hypertensive rats. *Journal of Hypertension*, *6*, 79–84.

Guyenet, P. G. (2006). The sympathetic control of blood pressure. *Nature Reviews. Neuroscience*, *7*, 335–346.

Guyenet, P. G., Koshiya, N., Huangfu, D., Baraban, S. C., Stornetta, R. L., & Li, Y. W. (1996). Role of medulla oblongata in generation of sympathetic and vagal outflows. *Progress in Brain Research*, *107*, 127–144.

Harayama, N., Shibuya, I., Tanaka, K., Kabashima, N., Ueta, Y., & Yamashita, H. (1998). Inhibition of N- and P/Q-type calcium channels by postsynaptic $GABA_B$ receptor activation in rat supraoptic neurones. *Journal of Physiology (Paris)*, *509*(Pt 2), 371–383.

Hardy, S. G. (2001). Hypothalamic projections to cardiovascular centers of the medulla. *Brain Research*, *894*, 233–240.

Henderson, L. A., Richard, C. A., Macey, P. M., Runquist, M. L., Yu, P. L., Galons, J. P., et al. (2004). Functional magnetic resonance signal changes in neural structures to baroreceptor reflex activation. *Journal of Applied Physiology*, *96*, 693–703.

Hill, D. R., & Bowery, N. G. (1981). 3H-baclofen and 3h-GABA bind to bicuculline-insensitive GABA B sites in rat brain. *Nature*, *290*, 149–152.

Housley, G. D., Martin-Body, R. L., Dawson, N. J., & Sinclair, J. D. (1987). Brain stem projections of the glossopharyngeal nerve and its carotid sinus branch in the rat. *Neuroscience*, *22*, 237–250.

Hwang, B. H., & Wu, J. Y. (1984). Ultrastructural studies on catecholaminergic terminals and GABAergic neurons in nucleus tractus solitarius of the rat medulla oblongata. *Brain Research*, *302*, 57–67.

Ichida, T., & Kuriyama, K. (1998). Age-related development of gamma-aminobutyric acid (GABA)B receptor functions in various brain regions of spontaneously hypertensive rats. *Neurochemical Research*, *23*, 89–95.

Ichida, T., Sasaki, S., Takeda, K., Kuriyama, K., & Nakagawa, M. (1995). Functional alteration of the $GABA_B$ receptor in the brain of spontaneously hypertensive rats. *Clinical and Experimental Pharmacology & Physiology. Supplement*, *22*, S51–S53.

Imaki, T., Naruse, M., Harada, S., Chikada, N., Nakajima, K., Yoshimoto, T., et al. (1998). Stress-induced changes of gene expression in the paraventricular nucleus are enhanced in spontaneously hypertensive rats. *Journal of Neuroendocrinology*, *10*, 635–643.

Ishikawa, T., Nakamura, Y., Saitoh, N., Li, W. B., Iwasaki, S., & Takahashi, T. (2003). Distinct roles of kv1 and kv3 potassium channels at the calyx of held presynaptic terminal. *Journal of Neuroscience*, *23*, 10445–10453.

Ito, S., Komatsu, K., Tsukamoto, K., Kanmatsuse, K., & Sved, A. F. (2002). Ventrolateral medulla AT1 receptors support blood pressure in hypertensive rats. *Hypertension*, *40*, 552–559.

Izzo, P. N., Sykes, R. M., & Spyer, K. M. (1992). Gamma-aminobutyric acid immunoreactive structures in the nucleus tractus solitarius: A light and electron microscopic study. *Brain Research*, *591*, 69–78.

Jakobs, K. H., Lasch, P., Minuth, M., Aktories, K., & Schultz, G. (1982). Uncoupling of alpha-adrenoceptor-mediated inhibition of human platelet adenylate cyclase by *N*-ethylmaleimide. *Journal of Biological Chemistry*, *257*, 2829–2833.

Jones, K. A., Borowsky, B., Tamm, J. A., Craig, D. A., Durkin, M. M., Dai, M., et al. (1998). GABA(B) receptors function as a heteromeric assembly of the subunits GABA(B)R1 and GABA(B)R2. *Nature*, *396*, 674–679.

Judy, W. V., Watanabe, A. M., Henry, D. P., Besch, H. R. Jr., Murphy, W. R., & Hockel, G. M. (1976). Sympathetic nerve activity: Role in regulation of blood pressure in the spontaenously hypertensive rat. *Circulation Research*, *38*, 21–29.

Kannan, H., Hayashida, Y., & Yamashita, H. (1989). Increase in sympathetic outflow by paraventricular nucleus stimulation in awake rats. *American Journal of Physiology*, *256*, R1325–R1330.

Kannan, H., & Yamashita, H. (1985). Connections of neurons in the region of the nucleus tractus solitarius with the hypothalamic paraventricular nucleus: Their possible involvement in neural control of the cardiovascular system in rats. *Brain Research*, *329*, 205–212.

Kardos, J., Elster, L., Damgaard, I., Krogsgaard-Larsen, P., & Schousboe, A. (1994). Role of $GABA_B$ receptors in intracellular Ca^{2+} homeostasis and possible interaction between $GABA_A$ and $GABA_B$ receptors in regulation of transmitter release in cerebellar granule neurons. *Journal of Neuroscience Research*, *39*, 646–655.

Katsunuma, N., Tsukamoto, K., Ito, S., & Kanmatsuse, K. (2003). Enhanced angiotensin-mediated responses in the nucleus tractus solitarii of spontaneously hypertensive rats. *Brain Research Bulletin*, *60*, 209–214.

Kaupmann, K., Huggel, K., Heid, J., Flor, P. J., Bischoff, S., Mickel, S. J., et al. (1997). Expression cloning of GABA(B) receptors uncovers similarity to metabotropic glutamate receptors. *Nature*, *386*, 239–246.

Kaupmann, K., Malitschek, B., Schuler, V., Heid, J., Froestl, W., Beck, P., et al. (1998). GABA (B)-receptor subtypes assemble into functional heteromeric complexes. *Nature*, *396*, 683–687.

Kubo, Y., & Tateyama, M. (2005). Towards a view of functioning dimeric metabotropic receptors. *Current Opinion in Neurobiology*, *15*, 289–295.

Kulik, A., Vida, I., Lujan, R., Haas, C. A., Lopez-Bendito, G., Shigemoto, R., et al. (2003). Subcellular localization of metabotropic GABA(B) receptor subunits GABA(B1a/b) and GABA(B2) in the rat hippocampus. *Journal of Neuroscience*, *23*, 11026–11035.

Kuner, R., Kohr, G., Grunewald, S., Eisenhardt, G., Bach, A., & Kornau, H. C. (1999). Role of heteromer formation in GABA$_B$ receptor function. *Science*, *283*, 74–77.

Lasiter, P. S., & Kachele, D. L. (1988). Organization of GABA and GABA-transaminase containing neurons in the gustatory zone of the nucleus of the solitary tract. *Brain Research Bulletin*, *21*, 623–636.

Lawrence, A. J., & Jarrott, B. (1996). Neurochemical modulation of cardiovascular control in the nucleus tractus solitarius. *Progress in Neurobiology*, *48*, 21–53.

Lemus, M., Montero, S., Cadenas, J. L., Lara, J. J., Tejeda-Chavez, H. R., Alvarez-Buylla, R., et al. (2008). GabaB receptors activation in the NTS blocks the glycemic responses induced by carotid body receptor stimulation. *Autonomic Neuroscience*, *141*, 73–82.

Li, D. P., Chen, S. R., Pan, Y. Z., Levey, A. I., & Pan, H. L. (2002). Role of presynaptic muscarinic and GABA(B) receptors in spinal glutamate release and cholinergic analgesia in rats. *Journal of Physiology (Paris)*, *543*, 807–818.

Li, Y. W., & Guyenet, P. G. (1995). Neuronal inhibition by a GABA$_B$ receptor agonist in the rostral ventrolateral medulla of the rat. *American Journal of Physiology*, *268*, R428–R437.

Li, Y. W., & Guyenet, P. G. (1996a). Activation of GABA$_B$ receptors increases a potassium conductance in rat bulbospinal neurons of the C1 area. *American Journal of Physiology*, *271*, R1304–R1310.

Li, Y. W., & Guyenet, P. G. (1996b). Angiotensin II decreases a resting K^+ conductance in rat bulbospinal neurons of the C1 area. *Circulation Research*, *78*, 274–282.

Li, D. P., & Pan, H. L. (2006). Plasticity of GABAergic control of hypothalamic presympathetic neurons in hypertension. *American Journal of Physiology. Heart and Circulatory Physiology*, *290*, H1110–H1119.

Li, D. P., & Pan, H. L. (2007a). Role of gamma-aminobutyric acid (GABA)A and GABAB receptors in paraventricular nucleus in control of sympathetic vasomotor tone in hypertension. *Journal of Pharmacology and Experimental Therapeutics*, *320*, 615–626.

Li, D. P., & Pan, H. L. (2007b). Glutamatergic inputs in the hypothalamic paraventricular nucleus maintain sympathetic vasomotor tone in hypertension. *Hypertension*, *49*, 916–925.

Li, D. P., Yang, Q., Pan, H. M., & Pan, H. L. (2008a). Plasticity of pre- and postsynaptic GABA$_B$ receptor function in the paraventricular nucleus in spontaneously hypertensive rats. *American Journal of Physiology. Heart and Circulatory Physiology*, *295*, H807–H815.

Li, D. P., Yang, Q., Pan, H. M., & Pan, H. L. (2008b). Pre- and postsynaptic plasticity underlying augmented glutamatergic inputs to hypothalamic presympathetic neurons in spontaneously hypertensive rats. *Journal of Physiology (Paris)*, *586*, 1637–1647.

Lovick, T. A. (1988). GABA-mediated inhibition in nucleus paragigantocellularis lateralis in the cat. *Neuroscience Letters*, *92*, 182–186.

Lujan, R., Shigemoto, R., Kulik, A., & Juiz, J. M. (2004). Localization of the GABA$_B$ receptor 1a/b subunit relative to glutamatergic synapses in the dorsal cochlear nucleus of the rat. *Journal of Comparative Neurology*, *475*, 36–46.

Macdonald, R. L., & Olsen, R. W. (1994). GABA$_A$ receptor channels. *Annual Review of Neuroscience*, *17*, 569–602.

Madden, C. J., & Sved, A. F. (2003). Rostral ventrolateral medulla C1 neurons and cardiovascular regulation. *Cellular and Molecular Neurobiology*, *23*, 739–749.

Maley, B., & Newton, B. W. (1985). Immunohistochemistry of gamma-aminobutyric acid in the cat nucleus tractus solitarius. *Brain Research*, *330*, 364–368.

Mancia, G., Grassi, G., Giannattasio, C., & Seravalle, G. (1999). Sympathetic activation in the pathogenesis of hypertension and progression of organ damage. *Hypertension*, *34*, 724–728.

Maqbool, A., Batten, T. F., & McWilliam, P. N. (1991). Ultrastructural relationships between GABAergic terminals and cardiac vagal preganglionic motoneurons and vagal afferents in the cat: A combined HRP tracing and immunogold labelling study. *European Journal of Neuroscience*, *3*, 501–513.

Margeta-Mitrovic, M., Mitrovic, I., Riley, R. C., Jan, L. Y., & Basbaum, A. I. (1999). Immunohistochemical localization of GABA(B) receptors in the rat central nervous system. *Journal of Comparative Neurology*, *405*, 299–321.

Martin, D. S., & Haywood, J. R. (1993). Hemodynamic responses to paraventricular nucleus disinhibition with bicuculline in conscious rats. *American Journal of Physiology*, *265*, H1727–H1733.

McAllen, R. M. (1986). Action and specificity of ventral medullary vasopressor neurones in the cat. *Neuroscience*, *18*, 51–59.

McAllen, R. M., & Dampney, R. A. (1990). Vasomotor neurons in the rostral ventrolateral medulla are organized topographically with respect to type of vascular bed but not body region. *Neuroscience Letters*, *110*, 91–96.

McCarson, K. E., & Enna, S. J. (1999). Nociceptive regulation of GABA(B) receptor gene expression in rat spinal cord. *Neuropharmacology*, *38*, 1767–1773.

Meeley, M. P., Ruggiero, D. A., Ishitsuka, T., & Reis, D. J. (1985). Intrinsic gamma-aminobutyric acid neurons in the nucleus of the solitary tract and the rostral ventrolateral medulla of the rat: An immunocytochemical and biochemical study. *Neuroscience Letters*, *58*, 83–89.

Mei, L., Zhang, J., & Mifflin, S. (2003). Hypertension alters GABA receptor-mediated inhibition of neurons in the nucleus of the solitary tract. *American Journal of Physiology – Regulatory, Integrative and Comparative Physiology*, *285*, R1276–R1286.

Mifflin, S. W. (1996). Convergent carotid sinus nerve and superior laryngeal nerve afferent inputs to neurons in the NTS. *American Journal of Physiology*, *271*, R870–R880.

Mifflin, S. W., Spyer, K. M., & Withington-Wray, D. J. (1988). Baroreceptor inputs to the nucleus tractus solitarius in the cat: Postsynaptic actions and the influence of respiration. *Journal of Physiology (Paris)*, *399*, 349–367.

Milner, T. A., Morrison, S. F., Abate, C., & Reis, D. J. (1988). Phenylethanolamine *N*-methyltransferase-containing terminals synapse directly on sympathetic preganglionic neurons in the rat. *Brain Research*, *448*, 205–222.

Misgeld, U., Bijak, M., & Jarolimek, W. (1995). A physiological role for $GABA_B$ receptors and the effects of baclofen in the mammalian central nervous system. *Progress in Neurobiology*, *46*, 423–462.

Montgomery, B. M. (1961). The basilar artery hypertensive syndrome. *Archives of Internal Medicine*, *108*, 559–569.

Morrison, S. F., Callaway, J., Milner, T. A., & Reis, D. J. (1991). Rostral ventrolateral medulla: A source of the glutamatergic innervation of the sympathetic intermediolateral nucleus. *Brain Research*, *562*, 126–135.

Mott, D. D., & Lewis, D. V. (1994). The pharmacology and function of central $GABA_B$ receptors. *International Review of Neurobiology*, *36*, 97–223.

Odagaki, Y., Nishi, N., & Koyama, T. (2000). Functional coupling of GABA(B) receptors with G proteins that are sensitive to *N*-ethylmaleimide treatment, suramin, and benzalkonium chloride in rat cerebral cortical membranes. *Journal of Neural Transmission*, *107*, 1101–1116.

Patel, K. P. (2000). Role of paraventricular nucleus in mediating sympathetic outflow in heart failure. *Heart Failure Reviews*, *5*, 73–86.

Persson, B. (1981). A hypertensive response to baclofen in the nucleus tractus solitarii in rats. *Journal of Pharmacy and Pharmacology*, *33*, 226–231.

Phillips, M. I., & Kimura, B. (1988). Brain angiotensin in the developing spontaneously hypertensive rat. *Journal of Hypertension*, *6*, 607–612.

Pickel, V. M., Chan, J., & Milner, T. A. (1989). Cellular substrates for interactions between neurons containing phenylethanolamine *N*-methyltransferase and GABA in the nuclei of the solitary tracts. *Journal of Comparative Neurology*, *286*, 243–259.

Pyner, S., & Coote, J. H. (1999). Identification of an efferent projection from the paraventricular nucleus of the hypothalamus terminating close to spinally projecting rostral ventrolateral medullary neurons. *Neuroscience*, *88*, 949–957.

Pyner, S., & Coote, J. H. (2000). Identification of branching paraventricular neurons of the hypothalamus that project to the rostroventrolateral medulla and spinal cord. *Neuroscience*, *100*, 549–556.

Raizada, M. K., Francis, S. C., Wang, H., Gelband, C. H., Reaves, P. Y., & Katovich, M. J. (2000). Targeting of the renin-angiotensin system by antisense gene therapy: A possible strategy for the long-term control of hypertension. *Journal of Hypertension*, *18*, 353–362.

Reddy, M. K., Patel, K. P., & Schultz, H. D. (2005). Differential role of the paraventricular nucleus of the hypothalamus in modulating the sympathoexcitatory component of peripheral and central chemoreflexes. *American Journal of Physiology – Regulatory, Integrative and Comparative Physiology*, *289*, R789–R797.

Reis, D. J., Granata, A. R., Perrone, M. H., & Talman, W. T. (1981). Evidence that glutamic acid is the neurotransmitter of baroreceptor afferent terminating in the nucleus tractus solitarius (NTS). *Journal of the Autonomic Nervous System*, *3*, 321–334.

Ricardo, J. A., & Koh, E. T. (1978). Anatomical evidence of direct projections from the nucleus of the solitary tract to the hypothalamus, amygdala, and other forebrain structures in the rat. *Brain Research*, *153*, 1–26.

Ross, C. A., Ruggiero, D. A., Joh, T. H., Park, D. H., & Reis, D. J. (1984a). Rostral ventrolateral medulla: Selective projections to the thoracic autonomic cell column from the region containing C1 adrenaline neurons. *Journal of Comparative Neurology*, *228*, 168–185.

Ross, C. A., Ruggiero, D. A., Park, D. H., Joh, T. H., Sved, A. F., Fernandez-Pardal, J., et al. (1984b). Tonic vasomotor control by the rostral ventrolateral medulla: Effect of electrical or chemical stimulation of the area containing C1 adrenaline neurons on arterial pressure, heart rate, and plasma catecholamines and vasopressin. *Journal of Neuroscience*, *4*, 474–494.

Ruggeri, P., Cogo, C. E., Picchio, V., Molinari, C., Ermirio, R., & Calaresu, F. R. (1996). Influence of GABAergic mechanisms on baroreceptor inputs to nucleus tractus solitarii of rats. *American Journal of Physiology*, *271*, H931–H936.

Ruggiero, D. A., Meeley, M. P., Anwar, M., & Reis, D. J. (1985). Newly identified GABAergic neurons in regions of the ventrolateral medulla which regulate blood pressure. *Brain Research*, *339*, 171–177.

Sapru, H. N. (2002). Glutamate circuits in selected medullo-spinal areas regulating cardiovascular function. *Clinical and Experimental Pharmacology and Physiology*, *29*, 491–496.

Scanziani, M. (2000). GABA spillover activates postsynaptic GABA(B) receptors to control rhythmic hippocampal activity. *Neuron*, *25*, 673–681.

Schousboe, A. (1999). Pharmacologic and therapeutic aspects of the developmentally regulated expression of GABA(A) and GABA(B) receptors: Cerebellar granule cells as a model system. *Neurochemistry International*, *34*, 373–377.

Shapiro, M. S., Wollmuth, L. P., & Hille, B. (1994). Modulation of Ca^{2+} channels by PTX-sensitive G-proteins is blocked by *N*-ethylmaleimide in rat sympathetic neurons. *Journal of Neuroscience*, *14*, 7109–7116.

Singh, R., & Ticku, M. K. (1985). Comparison of [3h]baclofen binding to $GABA_B$ receptors in spontaneously hypertensive and normotensive rats. *Brain Research*, *358*, 1–9.

Sodickson, D. L., & Bean, B. P. (1996). $GABA_B$ receptor-activated inwardly rectifying potassium current in dissociated hippocampal CA3 neurons. *Journal of Neuroscience*, *16*, 6374–6385.
Spyer, K. M. (1990). *Central regulation of autonomic functions*. New York: Oxford University Press.
Suzuki, M., Kuramochi, T., & Suga, T. (1993). GABA receptor subtypes involved in the neuronal mechanisms of baroreceptor reflex in the nucleus tractus solitarii of rabbits. *Journal of the Autonomic Nervous System*, *43*, 27–35.
Suzuki, M., Tetsuka, M., & Endo, M. (1999). GABA(B) receptors in the nucleus tractus solitarii modulate the carotid chemoreceptor reflex in rats. *Neuroscience Letters*, *260*, 21–24.
Sved, J. C., & Sved, A. F. (1989). Cardiovascular responses elicited by gamma-aminobutyric acid in the nucleus tractus solitarius: Evidence for action at the $GABA_B$ receptor. *Neuropharmacology*, *28*, 515–520.
Sved, A. F., & Sved, J. C. (1990). Endogenous GABA acts on $GABA_B$ receptors in nucleus tractus solitarius to increase blood pressure. *Brain Research*, *526*, 235–240.
Sved, A. F., & Tsukamoto, K. (1992). Tonic stimulation of $GABA_B$ receptors in the nucleus tractus solitarius modulates the baroreceptor reflex. *Brain Research*, *592*, 37–43.
Swanson, L. W., & Sawchenko, P. E. (1980). Paraventricular nucleus: A site for the integration of neuroendocrine and autonomic mechanisms. *Neuroendocrinology*, *31*, 410–417.
Swanson, L. W., & Sawchenko, P. E. (1983). Hypothalamic integration: Organization of the paraventricular and supraoptic nuclei. *Annual Review of Neuroscience*, *6*, 269–324.
Takahashi, T., Kajikawa, Y., & Tsujimoto, T. (1998). G-protein-coupled modulation of presynaptic calcium currents and transmitter release by a $GABA_B$ receptor. *Journal of Neuroscience*, *18*, 3138–3146.
Takeda, K., & Bunag, R. D. (1978). Sympathetic hyperactivity during hypothalamic stimulation in spontaneously hypertensive rats. *Journal of Clinical Investigation*, *62*, 642–648.
Takeda, K., & Bunag, R. D. (1980). Augmented sympathetic nerve activity and pressor responsiveness in DOCA hypertensive rats. *Hypertension*, *2*, 97–101.
Takenaka, K., Sasaki, S., Uchida, A., Fujita, H., Nakamura, K., Ichida, T., et al. (1996). GABAB-ergic stimulation in hypothalamic pressor area induces larger sympathetic and cardiovascular depression in spontaneously hypertensive rats. *American Journal of Hypertension*, *9*, 964–972.
Talman, W. T., Perrone, M. H., & Reis, D. J. (1980). Evidence for L-glutamate as the neurotransmitter of baroreceptor afferent nerve fibers. *Science*, *209*, 813–815.
Tan, P. S., Potas, J. R., Killinger, S., Horiuchi, J., Goodchild, A. K., Pilowsky, P. M., et al. (2005). Angiotensin II evokes hypotension and renal sympathoinhibition from a highly restricted region in the nucleus tractus solitarii. *Brain Research*, *1036*, 70–76.
Tasker, J. G., & Dudek, F. E. (1993). Local inhibitory synaptic inputs to neurones of the paraventricular nucleus in slices of rat hypothalamus. *Journal of Physiology (Paris)*, *469*, 179–192.
Torrealba, F., & Muller, C. (1999). Ultrastructure of glutamate and GABA immunoreactive axon terminals of the rat nucleus tractus solitarius, with a note on infralimbic cortex afferents. *Brain Research*, *820*, 20–30.
Tsukamoto, K., & Sved, A. F. (1993). Enhanced gamma-aminobutyric acid-mediated responses in nucleus tractus solitarius of hypertensive rats. *Hypertension*, *22*, 819–825.
Vale, C., Damgaard, I., Sunol, C., Rodriguez-Farre, E., & Schousboe, A. (1998). Cytotoxic action of lindane in cerebellar granule neurons is mediated by interaction with inducible GABA(B) receptors. *Journal of Neuroscience Research*, *52*, 286–294.
van Giersbergen, P. L., Palkovits, M., & De Jong, W. (1992). Involvement of neurotransmitters in the nucleus tractus solitarii in cardiovascular regulation. *Physiological Reviews*, *72*, 789–824.
Verberne, A. J., Stornetta, R. L., & Guyenet, P. G. (1999). Properties of C1 and other ventrolateral medullary neurones with hypothalamic projections in the rat. *Journal of Physiology (Paris)*, *517*(Pt 2), 477–494.

Vitela, M., & Mifflin, S. W. (2001). Gamma-aminobutyric acid(B) receptor-mediated responses in the nucleus tractus solitarius are altered in acute and chronic hypertension. *Hypertension*, *37*, 619–622.

Wang, D., Cui, L. N., & Renaud, L. P. (2003). Pre- and postsynaptic GABA(B) receptors modulate rapid neurotransmission from suprachiasmatic nucleus to parvocellular hypothalamic paraventricular nucleus neurons. *Neuroscience*, *118*, 49–58.

Wang, Y., Jordan, D., & Ramage, A. G. (2010). Both GABA$_A$ and GABA$_B$ receptors mediate vagal inhibition in nucleus tractus solitarii neurones in anaesthetized rats. *Autonomic Neuroscience*, *152*, 75–83.

Wang, R. J., Zeng, Q. H., Wang, W. Z., & Wang, W. (2009). GABA(A) and GABA(B) receptor-mediated inhibition of sympathetic outflow in the paraventricular nucleus is blunted in chronic heart failure. *Clinical and Experimental Pharmacology and Physiology*, *36*, 516–522.

Wang, X. L., Zhang, H. M., Li, D. P., Chen, S. R., & Pan, H. L. (2006). Dynamic regulation of glycinergic input to spinal dorsal horn neurones by muscarinic receptor subtypes in rats. *Journal of Physiology (Paris)*, *571*, 403–413.

White, J. H., Wise, A., Main, M. J., Green, A., Fraser, N. J., Disney, G. H., et al. (1998). Heterodimerization is required for the formation of a functional GABA(B) receptor. *Nature*, *396*, 679–682.

Willette, R. N., Barcas, P. P., Krieger, A. J., & Sapru, H. N. (1983a). Vasopressor and depressor areas in the rat medulla. Identification by microinjection of L-glutamate. *Neuropharmacology*, *22*, 1071–1079.

Willette, R. N., Barcas, P. P., Krieger, A. J., & Sapru, H. N. (1984). Endogenous GABAergic mechanisms in the medulla and the regulation of blood pressure. *Journal of Pharmacology and Experimental Therapeutics*, *230*, 34–39.

Willette, R. N., Krieger, A. J., Barcas, P. P., & Sapru, H. N. (1983b). Medullary gamma-aminobutyric acid (GABA) receptors and the regulation of blood pressure in the rat. *Journal of Pharmacology and Experimental Therapeutics*, *226*, 893–899.

Wilson, C. G., Zhang, Z., & Bonham, A. C. (1996). Non-NMDA receptors transmit cardiopulmonary C fibre input in nucleus tractus solitarii in rats. *Journal of Physiology (Paris)*, *496*(Pt 3), 773–785.

Yang, Z., & Coote, J. H. (1998). Influence of the hypothalamic paraventricular nucleus on cardiovascular neurones in the rostral ventrolateral medulla of the rat. *Journal of Physiology (Paris)*, *513*(Pt 2), 521–530.

Yao, F., Sumners, C., O'Rourke, S. T., & Sun, C. (2008). Angiotensin II increases GABA$_B$ receptor expression in nucleus tractus solitarii of rats. *American Journal of Physiology. Heart and Circulatory Physiology*, *294*, H2712–H2720.

Yin, M., Lee, C. C., Ohta, H., & Talman, W. T. (1994). Hemodynamic effects elicited by stimulation of the nucleus tractus solitarii. *Hypertension*, *23*, I73–I77.

Zhang, W., Herrera-Rosales, M., & Mifflin, S. (2007). Chronic hypertension enhances the postsynaptic effect of baclofen in the nucleus tractus solitarius. *Hypertension*, *49*, 659–663.

Zhang, W., & Mifflin, S. W. (1993). Excitatory amino acid receptors within NTS mediate arterial chemoreceptor reflexes in rats. *American Journal of Physiology*, *265*, H770–H773.

Zhang, J., & Mifflin, S. W. (1998). Receptor subtype specific effects of GABA agonists on neurons receiving aortic depressor nerve inputs within the nucleus of the solitary tract. *Journal of the Autonomic Nervous System*, *73*, 170–181.

Zhang, W., & Mifflin, S. (2010a). Plasticity of GABAergic mechanisms within the nucleus of the solitary tract in hypertension. *Hypertension*, *55*, 201–206.

Zhang, W., & Mifflin, S. (2010b). Chronic hypertension enhances presynaptic inhibition by baclofen in the nucleus of the solitary tract. *Hypertension*, *55*, 481–486.

Zhang, Q., Yao, F., O'Rourke, S. T., Qian, S. Y., & Sun, C. (2009). Angiotensin II enhances GABA(B) receptor-mediated responses and expression in nucleus tractus solitarii of rats. *American Journal of Physiology. Heart and Circulatory Physiology*, *297*, H1837–H1844.
Zucker, I. H., Wang, W., & Schultz, H. D. (1995). Cardiac receptor activity in heart failure: Implications for the control of sympathetic nervous outflow. *Advances in Experimental Medicine & Biology*, *381*, 109–124.

Anders Lehmann[*], Jörgen M. Jensen[*], and Guy E. Boeckxstaens[†]

[*]AstraZeneca R&D, Mölndal, Sweden

[†]Department of Gastroenterology, University Hospital Leuven, Catholic University of Leuven, Belgium; Department of Gastroenterology & Hepatology, Academic Medical Center, Amsterdam, The Netherlands

$GABA_B$ Receptor Agonism as a Novel Therapeutic Modality in the Treatment of Gastroesophageal Reflux Disease

Abstract

Defined pharmacologically by its insensitivity to the $GABA_A$ antagonist bicuculline and sensitivity to the GABA analogue baclofen, the G protein-linked γ-aminobutyric acid type B ($GABA_B$) receptor couples to adenylyl cyclase, voltage-gated calcium channels, and inwardly-rectifying potassium channels. On the basis of a wealth of preclinical data in conjunction with early clinical observations that baclofen improves symptoms of

Advances in Pharmacology, Volume 58

1054-3589/10 $35.00
10.1016/S1054-3589(10)58012-8

gastroesophageal reflux disease (GERD), the $GABA_B$ receptor has been proposed as a therapeutic target for a number of diseases including GERD. Subsequently, there has been a significant effort to develop a peripherally-restricted $GABA_B$ agonist that is devoid of the central nervous system side effects that are observed with baclofen. In this article we review the *in vitro* and *in vivo* pharmacology of the peripherally-restricted $GABA_B$ receptor agonists and the preclinical and clinical development of lesogaberan (AZD3355, (R)-(3-amino-2-fluoropropyl) phosphinic acid), a potent and predominately peripherally-restricted $GABA_B$ receptor agonist with a preclinical therapeutic window superior to baclofen.

I. Introduction

The clinical benefits of γ-aminobutyric acid (GABA) receptor agonism have been recognized for a long time. For example, the GABA receptor type B ($GABA_B$) agonist baclofen was introduced as a treatment for spasticity in 1966 (Hudgson & Weightman, 1971). A further unanticipated clinical benefit of baclofen that emerged many years later was a reduction in gastroesophageal reflux events and subsequent improvement in the measures of gastroesophageal reflux disease (GERD) (Beaumont & Boeckxstaens, 2009; Cange et al., 2002; Ciccaglione & Marzio, 2003; Koek et al., 2003; Zhang et al., 2002). GERD is a common disorder that affects up to 20% of the Western population (Dent et al., 2005; Vakil et al., 2006) and the classic symptoms of GERD (heartburn and regurgitation) have a significant negative impact upon patients' health-related quality of life (Wiklund, 2004; Wiklund et al., 2006). Furthermore, GERD may give rise to esophageal complications such as esophagitis (Vakil et al., 2006) and Barrett's esophagus (Lieberman et al., 1997), and non-esophageal manifestations such as cough and asthma (Canning & Mazzone, 2003).

The recommended approach for the treatment of GERD involves the suppression of gastric acid production, and proton pump inhibitors (PPIs) are currently the cornerstone of drug therapy (DeVault & Castell, 2005). However, even with PPI treatment, the management of GERD is often suboptimal: 20–30% of patients with GERD continue to experience symptoms despite treatment with a PPI (Fass et al., 2005), and almost one-quarter of patients take over-the-counter support medication for additional symptom control (Jones et al., 2006).

Although acid-suppressive therapy provides good GERD symptom relief (DeVault & Castell, 2005), the rate of persistent symptoms despite therapy highlights the need for additional treatments with novel therapeutic targets. As the majority of reflux episodes occur during transient relaxations of the lower esophageal sphincter (LES) (Dodds et al., 1982; Lehmann, 2006), inhibition of these relaxations is a valid therapeutic strategy for the management of GERD. In this review we will discuss the potential modality of GABA$_B$ receptor agonism in this regard.

II. Underlying Pathophysiological Mechanisms of GERD

Taken together, the LES and the crural diaphragm (CD) form the gastroesophageal junction (Fig. 1) and act as the major physiological pressure barrier to gastroesophageal reflux. Whether the LES exists as an anatomically distinct structure or a region of high pressure is currently a topic of debate, but it is clear that the anatomy and physiology of the LES is much more complicated than originally believed (Brasseur et al., 2007). One previous description of the LES is an approximately 4 cm long thickening of the esophageal wall, the distal 2 cm of which is encircled by the CD as it passes through the crural hiatus (Liebermann-Meffert et al., 1986). Pressure at the gastroesophageal junction is therefore governed by dual sphincter mechanisms, comprised of LES smooth muscle and the extrinsic skeletal muscle of the CD. The high pressure (>10 mmHg) of the LES is maintained through neural, myogenic, and hormonal mechanisms (Lipan et al., 2006).

Transient lower esophageal sphincter relaxation (TLESR) is a normal physiological response to post-prandial gastric distension that allows venting of gases from the stomach into the esophagus. However, numerous studies have also implicated TLESRs as a major mechanism of reflux of

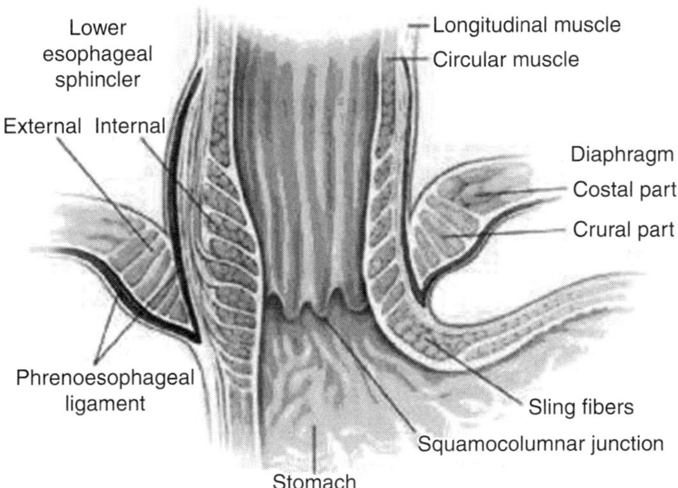

FIGURE 1 Anatomy of the esophagogastric junction (Mittal & Balaban, 1997). *Reproduced with permission from* Mittal & Balaban, 1997. *Copyright © 1997 Massachusetts Medical Society. All rights reserved.*

gastric contents in both healthy volunteers and patients with reflux symptoms (Dent et al., 1980; Dodds et al., 1982; Holloway et al., 1991; Iwakiri et al., 2005; Mittal & McCallum, 1988). TLESRs precede 94 and 65% of reflux events in healthy volunteers and patients with reflux esophagitis, respectively (Dent et al., 1980), but it should be noted that other additional mechanisms may be important causes of reflux in certain cases, especially those involving hiatal hernia (van Herwaarden et al., 2000). It may be these additional mechanisms that account for the wider variability in the percentage of reflux events preceded by TLESRs in patients with GERD, compared with healthy volunteers.

TLESRs are controlled by neural pathways with both efferent and afferent neurones and a pattern generator located within the brainstem (Mittal et al., 1995). There is significant evidence to suggest that the neuronal control of TLESRs is a vago-vagal response as cold blockade of the vagal nerve trunks of conscious dogs abolishes air-induced TLESRs (Martin et al., 1986). Moreover, sectioning of the splanchnic nerve in ferrets was shown to neither affect the triggering of TLESRs nor the number of reflux episodes, further supporting the exclusive vago-vagal control of TLESRs (Staunton et al., 2000b). This response is initiated by gastric distension that presumably is detected primarily by intraganglionic laminar endings (specialized transduction sites of vagal mechanoreceptors). The signal is transmitted along vagal afferent neurones to the nucleus of the solitary tract within the brainstem (Fig. 2), relayed through the dorsal motor nucleus of the vagus (DMV), and then vagal efferent neurones relay an effector signal to the LES via inhibitory esophageal enteric neurones (Lehmann, 2009; Mittal et al., 1995). Such neural circuitry contains several neuropharmacological targets that can be modified to affect TLESR frequency, including $GABA_B$, metabotropic glutamate 5, cannabinoid 1, cholecystokinin 1, muscarinic, and μ-opioid receptors which have been extensively reported elsewhere (Lehmann, 2006). Of these, the target that appears to be the most promising is the $GABA_B$ receptor. Agonism of this receptor has been shown to reduce the frequency of TLESRs in all species studied to date, including humans (Blackshaw et al., 1999; Lehmann et al., 1999; Liu et al., 2002; Zhang et al., 2002). The $GABA_B$ receptor, its actions, and development of $GABA_B$ receptor agonists are discussed below in further detail.

III. The $GABA_B$ Receptor

The $GABA_B$ receptor was initially identified by Norman Bowery's pioneering work demonstrating that the effects of baclofen are insensitive to bicuculline (Bowery et al., 1980). Almost 20 years passed before the $GABA_B$ receptor became the last major neurotransmitter receptor to be

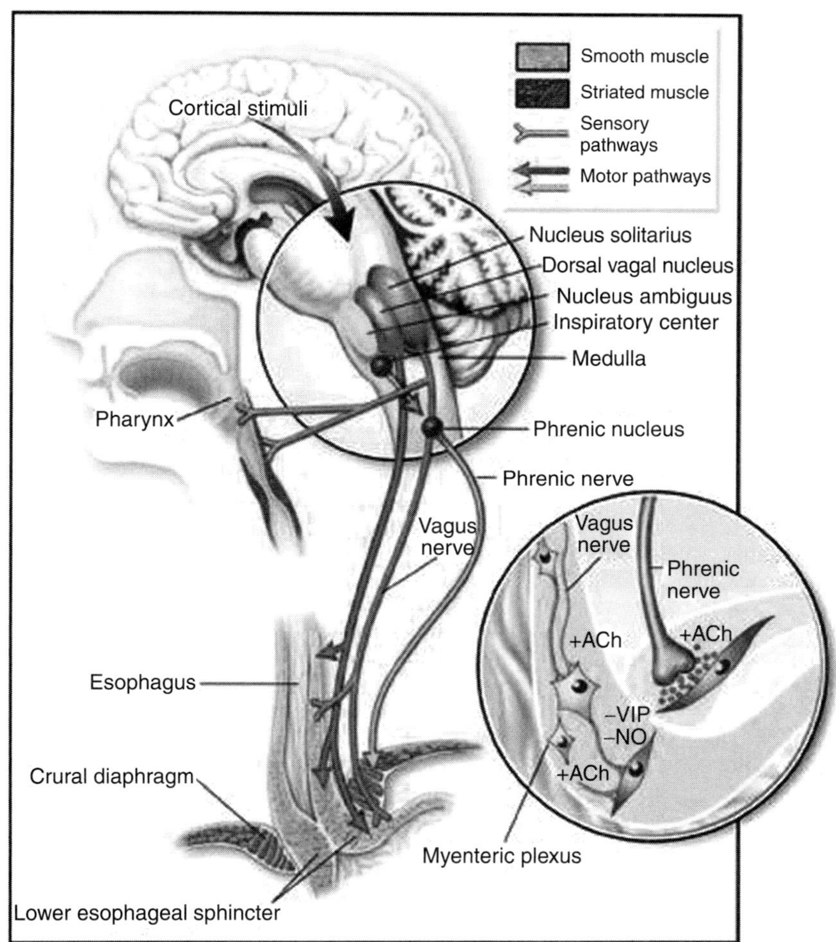

FIGURE 2 Neural pathways to the lower esophageal sphincter (LES) and crural diaphragm (CD). Esophageal peristalsis and relaxation of the LES induced by swallow result from the excitation of receptors in the pharynx. The afferent stimulus travels to the sensory nucleus, the nucleus solitarius (small insert). A programmed set of events from the dorsal vagal nucleus and the nucleus ambiguus mediates esophageal peristalsis and sphincter relaxation. The vagal efferent fibers communicate with myenteric neurons that mediate LES relaxation (large inset). The postganglionic transmitters are nitric oxide (NO) and vasoactive intestinal peptide (VIP). Transient lower esophageal sphincter relaxation (TLESR), the principal mechanism of reflux, appears to use the same efferent neural pathway as the swallow reflex. The afferent signals for TLESR may originate in the pharynx, larynx or the stomach. The efferent pathway is in the vagus nerve, and nitric oxide is the postganglionic neurotransmitter responsible for LES relaxation. Contraction of the crural diaphragm is controlled by the inspiratory center in the brainstem and the nucleus of the phrenic nerve. The crural diaphragm is innervated by right and left phrenic nerves through nicotinic cholinergic receptor acetylcholine (Ach). +, excitatory effects; –, inhibitory effects. (Mittal & Balaban, 1997). *Reproduced with permission from* Mittal & Balaban, 1997.

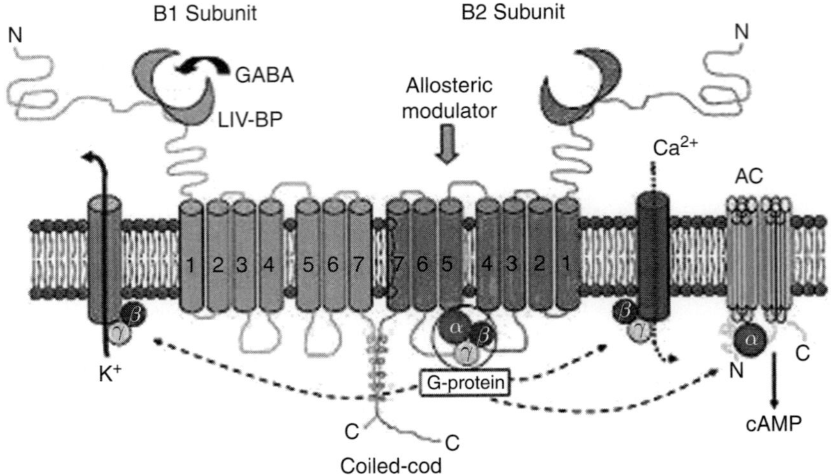

FIGURE 3 Heterodimeric structure of the $GABA_B$ receptor. The two seven-transmembrane receptor subunits are coupled via their intracellular C-termini. The binding domain for GABA is located in the extracellular domain (ECD) of $GABA_{B(1)}$. The heptahelical domain of $GABA_{B(2)}$ contains an allosteric modulator site as well as the G-protein-coupling site. Neither of these appears to be present in the $GABA_{B(1)}$ subunit. The receptor is coupled indirectly to potassium and calcium channels, the former of which predominates postsynaptically while the latter is mainly presynaptic. Reproduced with permission from Smart, T. G. (2004). Gamma-aminobutyric acid (GABA) receptors. In G. Adelman, B. H. Smith (Eds.), *Encyclopedia of Neuroscience* (3rd ed., CD-Rom). Amsterdam: Elsevier.

cloned (Kaupmann et al., 1997). In addition to an absence of high-affinity, irreversibly binding radioligands and modern cloning techniques, one explanation as to why there was such delay between the pharmacological and genetic identification of the $GABA_B$ receptor is related to the nature of the receptor itself. The $GABA_B$ receptor exists as an obligate heterodimer with two distinct seven-transmembrane subunits ($GABA_{B(1)}$ and $GABA_{B(2)}$) (Fig. 3). The discovery of the $GABA_{B(2)}$ subunit (Jones et al., 1998; Kaupmann et al., 1998; Kuner et al., 1999; White et al., 1998) placed the $GABA_B$ receptor as the first G protein-coupled receptor heterodimer to be identified. As with other Family 3 G protein-linked receptors, $GABA_B$ agonists and competitive antagonists bind to the extracellular domain (ECD) of the $GABA_{B(1)}$ subunit (Malitschek et al., 1999).

A. $GABA_B$ Receptor Ligand Binding Site

X-ray crystallography of the $GABA_B$ receptor has shown that the agonist binding pocket consists of two globular lobes ($GABA_{B(1)}$ and $GABA_{B(2)}$) that are separated by a C-terminal hinge region. Within this pocket, several amino acid residues that are key for ligand binding have been identified, including Ser246, Ser269, Asp417, Glu464, and Tyr366

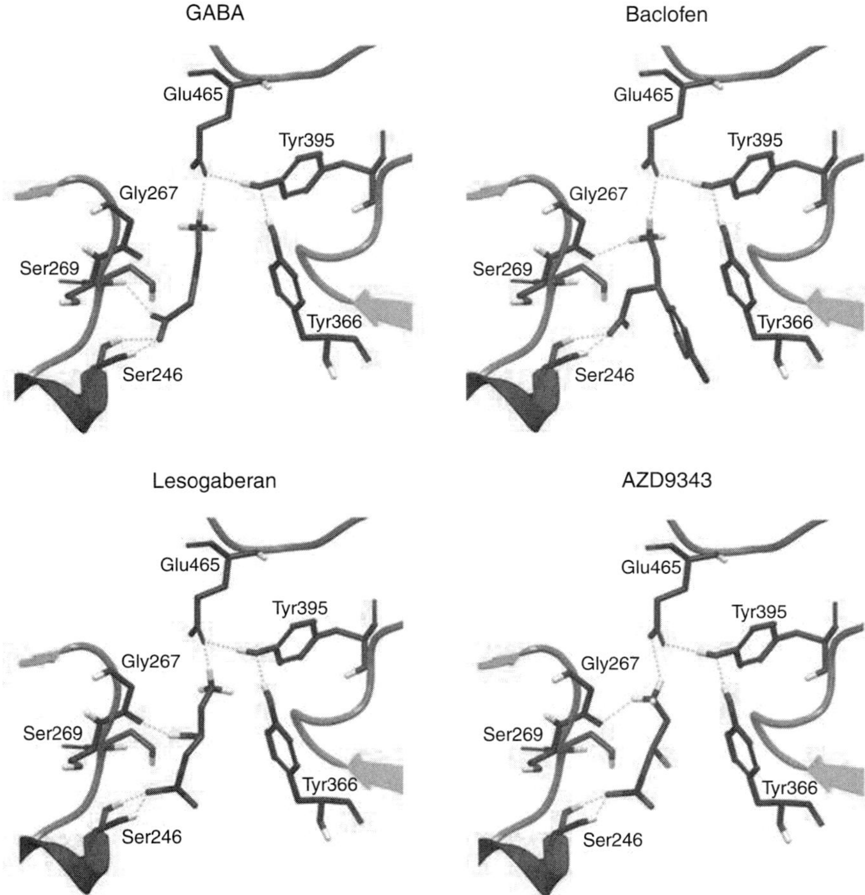

FIGURE 4 Three-dimensional model of the Venus Fly Trap module of GABA$_{B(1)}$ with GABA, baclofen, lesogaberan, and AZD9343. Copyright AstraZeneca 2010.

(Fig. 4) (KniazeffLipan et al., 2002). It is suggested that upon ligand binding, the two lobes close together in a manner analogous to a Venus Fly Trap (Quiocho & Ledvina, 1996) and that such "closing" induces a stabilized conformational change that activates the receptor (Galvez et al., 2000a). The "closing" of the Venus Fly Trap model is further stabilized by calcium-mediated allosteric modulation. It is thought that Ser269 is a primary calcium-sensing amino acid residue within the GABA$_{B(1)}$ subunit (Galvez et al., 2000a) that when associated with calcium is capable of compensating for the lack of an α-amino acid group in GABA and optimising the position of the GABA carboxylic acid group for contact with Ser246 (Galvez et al., 2000b).

In silico molecular docking studies of the $GABA_B$ receptor with GABA have revealed several potential points of interaction, and a low-energy interaction between the receptor subunit and the ligand. One such investigation (Costantino et al., 2001) has indicated that when GABA and calcium are simultaneously docked onto the $GABA_B$ receptor, the carboxylic moiety of GABA forms a hydrogen bond network with Ser246 and Tyr366. In addition, a strong interaction with the GABA carboxylic group and these amino acid residues plus Ser269 also occurs. Using the sequence homology data presented by Galvez et al. (2000a), the interactions between the $GABA_B$ receptor and GABA were modeled *in silico*, and are presented in Fig. 4. The binding characteristics of baclofen to the $GABA_B$ receptor share some common features with the binding of GABA (Fig. 4).

B. $GABA_B$ Receptor Signal Transduction

Signal transduction from $GABA_B$ receptor agonism and subsequent G protein activation is primarily provided through the heptahelical ECD of the $GABA_{B(2)}$ subunit (Bettler et al., 2004) by coupling to voltage-gated calcium channels, adenylyl cyclase, and inwardly-rectifying potassium channels (Bettler et al., 2004; Billinton et al., 2001; Bowery et al., 2002).

In general, the consequences of presynaptic $GABA_B$ agonism are mediated through voltage-dependent inhibition of N-, P-, and Q-type calcium channels (Amico et al., 1995; Cardozo & Bean, 1995; Menon-Johansson et al., 1993; Mintz & Bean, 1993; Takahashi et al., 1998). Inhibition of these channels following postsynaptic agonism has also been observed (Harayama et al., 1998). Furthermore, $GABA_B$ receptor agonism has been shown to inhibit or facilitate L- (Amico et al., 1995; Maguire et al., 1989; Marchetti et al., 1991; Shen & Slaughter, 1999) and T-type calcium channels (Crunelli & Leresche, 1991; Matsushima et al., 1993; Scott et al., 1990).

In addition to calcium channels, the $GABA_B$ receptor interacts with numerous isoforms of neuronal adenylyl cyclase. These interactions include $G_i\alpha$- and $G_o\alpha$-mediated inhibition of type I, III, V, and VI adenylyl cyclase, and $G\beta\gamma$-mediated stimulation of type II, IV, and VII adenylyl cyclase. The inhibitory and stimulatory effects of $GABA_B$ agonism on cAMP levels has been confirmed *in vivo* by microdialysis techniques (Hashimoto & Kuriyama, 1997).

A further mechanism of $GABA_B$ signal transduction of particular importance in vagal afferent neurones is the opening of potassium channels (Page et al., 2006). $GABA_B$ channels induce a slow inhibitory postsynaptic current (IPSC) via inwardly-rectifying G protein-coupled potassium channels known as either GIRK or Kir3 channels (Luscher et al., 1997; Schuler et al., 2001). The normal physiological effect of Kir3 channel opening is potassium efflux with resulting neuronal hyperpolarisation with a time scale that clearly differs from that of GABA$_A$ receptor-mediated fast IPSC (Lacaille, 1991; Otis et al., 1993).

IV. GABA$_B$ Receptor Agonists as Reflux Inhibitors

The major early evidence for the utility of a GABA$_B$ agonist in the treatment of GERD came from observations of healthy volunteers and patients treated with the skeletal muscle antispastic agent, baclofen. Initial preclinical investigations into the usefulness of baclofen reported that this selective agonist inhibits TLESRs in a dose-dependent manner in both dogs and ferrets (Blackshaw & Partosoedarso, 1999; Blackshaw et al., 1999; Lehmann et al., 1999). These early preclinical findings were subsequently supported by several investigations in humans. Indeed, activation of the GABA$_B$ receptor by baclofen produces several beneficial effects in relation to the treatment of GERD. These benefits include a reduction in TLESR frequency (Lidums et al., 2000; van Herwaarden et al., 2002; Zhang et al., 2002), reduced esophageal acid exposure (van Herwaarden et al., 2002), reduced acidic and weakly/non-acidic reflux (Vela et al., 2003), and reduced biliary reflux (Koek et al., 2003). In turn, patients with GERD report relief of their symptoms (Ciccaglione & Marzio, 2003; Koek et al., 2003; Vela et al., 2003), and such benefits extend to those already treated with a PPI (Koek et al., 2003) and with hiatal hernia (Beaumont & Boeckxstaens, 2009).

The primary complication associated with baclofen as a therapeutic agent for GERD is its propensity to cause central nervous system (CNS) side effects such as drowsiness and daytime sedation with occasional reports of lassitude, exhaustion, light-headedness, confusion, dizziness, headache, insomnia, and hypothermia (Baclofen Summary of Product Characteristics, 2009).

A. Peripherally-Restricted GABA$_B$ Agonists

To achieve the maximal clinical utility of a GABA$_B$ agonist for the treatment of GERD, the agonist itself must be peripherally restricted, or selectively target a pharmacologically distinct subpopulation of GABA$_B$ receptors to avoid the CNS side effects seen with baclofen. With the exception of arbaclofen placarbil (AP), the new GABA$_B$ agonists under current development are peripherally restricted because of unique properties which limit their availability in the CNS, compared with baclofen.

When the drug discovery project to identify peripherally-restricted GABA$_B$ agonists was initiated at Astra (now AstraZeneca) during the 1990s, it was already known that GABA$_B$ agonism inhibits TLESRs (Blackshaw & Partosoedarso, 1999; Lehmann et al., 1999). Despite this, there were many significant challenges. Firstly, there was no indication as to the site of action of the agonists evaluated, hence any distinctions between peripheral versus central sites of action were speculative. Additional critical questions were as follows: could peripheral $GABA_B$ receptors be stimulated selectively based on drug distribution properties; and could new agonists be discovered given the very narrow structure–activity relationships for agonists when even small variations of the GABA pharmacophore mostly render the compounds inactive or even transform them into antagonists (Froestl et al., 1995a, 1995b)? Furthermore, there was also the question of whether there were any differences in the pharmacological properties of central and peripheral $GABA_B$ receptors.

The finding that there are a number of different splice variants of both $GABA_{B(1)}$ and $GABA_{B(2)}$ offered a theoretical possibility for selectivity of agonists between splice variants but this has not yet been verified (Bettler et al., 2004; Lehmann et al., 2009). Reports claiming that the antiepileptic gabapentin has a selective action on presynaptic $GABA_{B(1a)}$ receptors (Ng et al., 2001) raised considerable initial interest, but other groups could not confirm such a selectivity or even an action on $GABA_B$ receptors in general (Cheng et al., 2004; Jensen et al., 2002). Furthermore, gabapentin does not affect TLESR in dogs (Jensen et al., 2002). Interestingly, there are pronounced disparities in the potency of $GABA_B$ agonists in brain and peripheral tissues, but potency in the former may be 1–3 orders of magnitude higher. These differences may simply relate to a mismatch between the expression of $GABA_{B(1)}$ and $GABA_{B(2)}$ (Calver et al., 2000), the stoichiometry of which determines the efficiency of the heterodimer. However, one agonist described in the literature was of particular interest. The properties of 3-aminopropylphosphinic acid were reported in the late 1980s, and it was noted that while this compound is a very potent full agonist at the $GABA_B$ receptor (Hills et al., 1989), it has no measurable effects on CNS function when administered *in vivo*. Since most scientific interest in the $GABA_B$ receptor in academia and pharmaceutical industry alike was and continues to be CNS-centric, the reason for the lack of correlation between *in vitro* and *in vivo* effects of 3-aminopropylphosphinic acid was largely unexplored. However, the finding that 3-aminopropylphosphinic acid did attenuate the number of TLESRs in dogs (Lehmann et al, manuscript in preparation), even though it was ineffective at the doses tested in ferrets (Blackshaw et al., 1999), did provide a starting point to discover novel agents of that structural class with improved *in vivo* activity. These efforts eventually led to the identification of lesogaberan (AZD3355) and AZD9343, which were selected for further clinical development as described below.

I. Lesogaberan and AZD9343

Lesogaberan ((2R)-3-amino-2-fluoropropylphosphinic acid) and AZD9343 ((2S)-3-amino-2-hydroxypropyl phosphinic acid) are structural analogues of GABA. As would be expected, the binding of both lesogaberan and AZD9343 (Fig. 4) to the GABA$_B$ receptor shares many common characteristics with the binding profile of GABA. For instance, the phosphinic acid moiety of both lesogaberan and AZD9343 forms a hydrogen bond network with Ser246, and the amino acid groups form a series of hydrogen bonds with Glu465, Tyr395, and possibly Tyr366.

Structure–activity investigations have demonstrated that for 2-substituted 3-aminopropylphosphinic acids, the properties of the substituent govern receptor binding selectivity. Thus, 2(S)-3-amino-2-fluoropropylphosphinic acid is a considerably weaker agonist than lesogaberan, while a 2(R)-hydroxyl group yields a less potent compound compared with AZD9343. The same relationship is seen for the corresponding enantiomers of 2-substituted 3-aminopropyl(methyl)phosphinic acids (Alstermark et al., 2008; Froestl et al., 1995a).

2. *Preclinical Studies with Lesogaberan and AZD9343*

A range of *in vitro* and *in vivo* experiments were devised to characterize the preclinical pharmacology of lesogaberan and AZD9343, in relation to both GABA and baclofen. These included receptor affinity/intrinsic activity studies; *in vivo* distribution in the rat brain; GABA transporter affinity studies; ferret gastric mechanosensitive vagal afferent sensitivity; effects on TLESR as the primary pharmacodynamic endpoint; and hypothermia as a measure of side effects (Lehmann et al., 2009).

a. Receptor Affinity Studies The affinities of lesogaberan, AZD9343, GABA, and baclofen for the GABA$_B$ receptor were assessed in rat brain tissue. Overall, lesogaberan showed greater inhibition of ^{3}H-GABA binding to the rat brain GABA$_B$ receptor than baclofen (lesogaberan K_i, 5.1 ± 1.2 nM; baclofen K_i, 220 ± 50 nM). The K_i of GABA and AZD9343 were 110 ± 21 nM (Lehmann et al., 2009) and 50 ± 8 nM (Lehmann et al., 2008), respectively. Furthermore, the K_i of lesogaberan at the GABA$_B$ receptor was almost 300 times lower than that for the GABA$_A$ receptor indicating a high degree of GABA$_B$ receptor selectivity.

b. In vitro Potency on GABA$_B$ Receptor-Mediated Intracellular Calcium Release The potency of lesogaberan, AZD9343, baclofen, and GABA on GABA$_B$ receptor-mediated intracellular calcium release was determined *in vitro* in cells stably expressing functional recombinant human GABA$_B$ receptors (Lehmann et al., 2005). The EC_{50} of GABA in this assessment of potency was 160 ± 10 nM and both lesogaberan and AZD9343 were more potent than baclofen (8.6 ± 0.77 nM, 130 ± 16, and 750 ± 50 nM, respectively). These data also showed that lesogaberan and AZD9343 are full agonists at the GABA$_B$ receptor.

c. Ferret Gastric Mechanosensitive Vagal Afferent Sensitivity Studies As a vago-vagal mechanism of control of TLESRs has been suggested (Martin et al., 1986; Smid et al., 2001; Staunton et al., 2000a), one important observation made by Blackshaw and his associates was that baclofen inhibits the peripheral endings of gastric vagal mechanosensitive fibers (Blackshaw et al., 2000). Baclofen also appears to have an additional central effect on the vago-vagal loop (Partosoedarso et al., 2001). In a sense, these effects are reminiscent of the mode of action of baclofen in spasticity with the important distinction that the site of action of baclofen on the skeletal muscle afferent–efferent loop seems to reside exclusively within the spinal cord (Skoog, 1996). Consequently, these findings suggested that there is an anatomical rationale for the suppression of TLESR by activation of peripheral $GABA_B$ receptors. The nodose ganglion is the origin of vagal afferents, and the presence of $GABA_B$ receptors in the nodose neurons has been demonstrated electrophysiologically and immunohistochemically. While baclofen has been shown to have a very low potency in nodose neurons (Zagorodnyuk et al., 2002), nothing is known about potential differences between central and nodose ganglion $GABA_B$ receptors in terms of ligand binding or signaling mechanisms. The effect of racemic lesogaberan on single vagal gastric afferent mucosal and tension receptors *in vitro* was studied to confirm the notion that those fibers are sensitive to $GABA_B$ receptor stimulation with agonists other than baclofen (Lehmann et al., 2009). Overall, racemic lesogaberan inhibited mechanically-stimulated firing of muscular vagal afferent neurones at all concentrations and tensions tested in a concentration-dependent manner. However, mucosal receptor fibers, incidentally not believed to trigger TLESR, were largely unaffected.

In order to further explore the significance of these *in vitro* data in relation to the *in vivo* setting, inhibition of TLESRs in dogs was assessed.

d. Inhibition of TLESRs in Dogs and Murine Hypothermia The quantitation of gastric distension-induced TLESR was investigated in Labrador Retrievers. TLESRs were stimulated by infusion of an acidified liquid nutrient followed by insufflation of air. Lesogaberan and baclofen were administered intravenously or intragastrically and GABA was infused intravenously at previously described doses (Thirlby et al., 1988).

Results showed that lesogaberan dose-dependently inhibited TLESRs with an ED_{50} for TLESRs at an oral dose of approximately 7 μmol/kg (Fig. 5). The dose–response curve was biphasic, and the initial phase was similar to that of GABA. In contrast, baclofen displayed a monophasic dose–response curve. Almost complete inhibition (90%) of TLESRs was seen with lesogaberan at a dose of 300 μmol/kg, which was close to the threshold for induction of CNS side effects (500 μmol/kg). The approximate ED_{50} and ED_{90} values for baclofen were 1 and 10 μmol/kg. Typical $GABA_B$ side effects in dogs are seen with baclofen at a dose of approximately 10 μmol/kg (Lehmann et al., unpublished observations). Moreover, racemic lesogaberan was shown to reduce TLESR in ferrets (Lehmann et al., 2009).

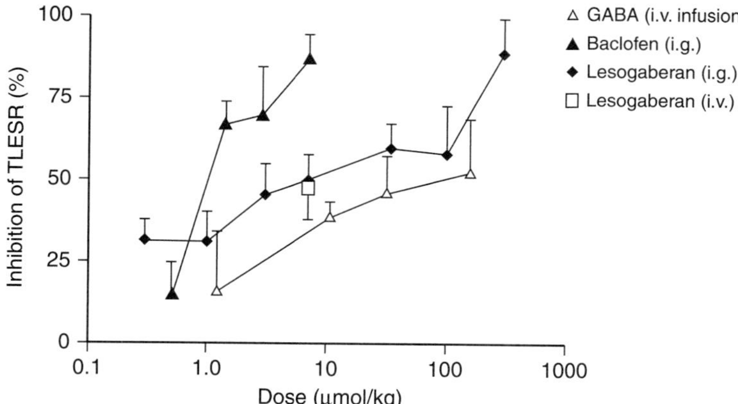

FIGURE 5 Inhibition of TLESR in dogs following administration of GABA, baclofen, and lesogaberan. Reproduced with permission from Lehmann A, Antonsson M, Holmberg AA, Blackshaw LA, Brändén L, Bräuner-Osborne H, Christiansen B, Dent J, Elebring T, Jacobson BM, Jensen J, Mattsson JP, Nilsson K, Oja SS, Page AJ, Saransaari P, von Unge S.(R)-(3-amino-2-fluoropropyl) phosphinic acid (AZD3355), a novel GABA$_B$ receptor agonist, inhibits transient lower esophageal sphincter relaxation through a peripheral mode of action. J Pharmacol Exp Ther. 2009;331:504–12. Copyright American Society for Pharmacology and Experimental Therapeutics 2009.

It is also known that the administration of GABA$_B$ agonists in mice can induce a state of hypothermia. This model can therefore be used to test the propensity of a GABA$_B$ agonist to induce unwanted CNS effects. In relevant experiments, lesogaberan and baclofen were administered to mice and body temperature was monitored using thermosensitive chips. Comparisons between lesogaberan and baclofen were assessed by the calculation of the dose that resulted in a 2°C drop in body temperature (ED_2). Results showed that baclofen had a pronouncedly lower ED_2 value compared with lesogaberan despite being less active at the GABA$_B$ receptor (~10 μmol/kg and ~1,000 μmol/kg, respectively) (Lehmann et al., 2009). To ascertain that the hypothermic effects of lesogaberan were mediated by the GABA$_B$ receptor, a selective GABA$_B$ receptor antagonist was administered before injection of lesogaberan. The antagonist did not have any effects on body temperature in its own right, but it prevented hypothermia induced by lesogaberan. Furthermore, the hypothermic effect of lesogaberan was abolished in GABA$_B$ receptor knockout mice. Taken together these data demonstrate that while lesogaberan is a much more potent GABA$_B$ agonist than baclofen *in vitro*, the dose required to induce GABA$_B$-mediated CNS side effects is approximately two orders of magnitude higher than baclofen.

Even if baclofen is more potent than lesogaberan in terms of inhibition of TLESR, the therapeutic index of lesogaberan is nevertheless clearly superior to that of baclofen. The caveat in this comparison is that it is based on data from two different species.

The differences in the therapeutic windows of lesogaberan and baclofen were hypothesized to arise from different cellular accumulation and transport characteristics. It was postulated that GABA and lesogaberan but not baclofen may be efficiently transported into neuronal cells via a GABA transporter (GAT), and that such transport would prevent extracellular accumulation in the CNS and therefore minimize adverse central effects compared with baclofen (Lehmann et al., 2009). This hypothesis was based on the observations that the electrophysiological effects of 3-aminopropylphosphinic acid but not baclofen in rat brain slices were markedly enhanced by concomitant addition of a GAT inhibitor (Ong & Kerr, 1998). To test this hypothesis the accumulation and transport of GABA, lesogaberan, AZD9343, and baclofen were investigated.

e. *Accumulation and Transport* The first indication that lesogaberan is transported into CNS cells came from autoradiographic studies *in vivo*. It was then demonstrated that there is a significant difference between lesogaberan and baclofen with respect to distribution within and around the circumventricular organs in the rat brain. Similar differences in distribution have been noted between GABA and baclofen in the mouse brain (Kuroda et al., 2000). A possible explanation for this difference is that lesogaberan is a substrate for active GABA uptake mechanisms, whereas baclofen is not.

The hypothesis that lesogaberan is sequestered by neural cells was investigated directly using rat brain cortical slices. As expected, GABA was efficiently taken up by the cerebrocortical slices (68.8 ± 4.4 μmol/kg h). The accumulation of lesogaberan, AZD9343, and baclofen was 19.4 ± 0.5 μmol/kgh, 10.9 ± 0.7 μmol/kgh, and 1.64 ± 0.1 μmol/kgh, respectively. These data support the hypothesis that lesogaberan and AZD9343 do accumulate into neural cells while there is negligible uptake of baclofen.

The mechanism of lesogaberan and baclofen accumulation into glial cells was further investigated by the method of competitive GAT binding studies in rat cerebrocortical membranes. Overall, lesogaberan inhibited GABA binding with an IC_{50} of 0.67 mM. However, baclofen at concentrations of 1mM inhibited less than 25% of GABA binding. These data support the view that lesogaberan is a substrate for native GATs, whereas baclofen is not (Rosenstein et al., 1990).

Because selective accumulation of lesogaberan in neural cells via GATs was used to explain the low propensity of lesogaberan to cause CNS side effects in experimental animals, compared with baclofen, the transport of lesogaberan by human GAT was also investigated for translational purposes and to ultimately prove that GAT mediates the cellular uptake. Indeed, lesogaberan displayed pronounced competition of ^{3}H-GABA transport at all four human GATs (Table I). In contrast, baclofen displayed only minor competition of ^{3}H-GABA transport at all four GATs (Table 1) (Lehmann et al., 2009).

TABLE 1 Effects of GABA, lesogaberan and baclofen on ^{3}H-GABA uptake and membrane potential in tsA201 cells transiently expressing the four human GATs (Lehmann et al., 2009)

	[^{3}H]GABA uptake assay IC_{50} (pIC_{50} ± S.E.M.) (μM)				*FLIPR® membrane potential assay EC_{50} (pEC_{50} ± S.E.M.) (μM)*			
Compound	*hGAT-1*	*hBGT-1*	*hGAT-2*	*hGAT-3*	*hGAT-1*	*hBGT-1*	*hGAT-2*	*hGAT-3*
GABA	24 (4.6 ± 0.02)	30 (4.5 ± 0.07)	15 (4.8 ± 0.02)	12 (5.0 ± 0.1)	11 (4.9 ± 0.07)	31 (4.5 ± 0.05)	6.6 (5.2±0.08)	10 (5.0±0.05)
AZD3355	>1000[a]	110 (4.0 ± 0.06)	>1000[a]	>1000[a]	440 (3.4 ± 0.06)[b]	110 (4.0 ± 0.01)	600 (3.2±0.07)[b]	770 (3.1±0.07)[b]
Baclofen	>3000[a]	>3000[a]	>3000[a]	>3000[a]	>3000	>3000	>3000	>3000

[a] The compounds displayed less than 75% inhibition of the indicated GABA transporters at the maximal tested concentration and it was therefore not possible to generate concentration-inhibition curves. These compounds exhibited the following maximal inhibition at 3 mM (mean ± S.E.M.): lesogaberan at hGAT-1, 54 ± 1.1%; lesogaberan at hGAT-2, 46 ± 1.2%; lesogaberan at hGAT-3, 46 ± 5.3%; Baclofen at hGAT-1, 4.9 ± 1.2%; Baclofen at hBGT-1, 11 ± 1.1%; Baclofen at hGAT-2, 10 ± 2.8%; Baclofen at hGAT-3, 5.1 ± 3.0%.

[b] It was not possible to generate fully completed concentration–response curves and the curves were therefore fitted to the maximal response of GABA (3 mM). Reproduced with permission from Lehmann A, Antonsson M, Holmberg AA, Blackshaw LA, Brändén L, Bräuner-Osborne H, Christiansen B, Dent J, Elebring T, Jacobson BM, Jensen J, Mattsson JP, Nilsson K, Oja SS, Page AJ, Saransaari P, von Unge S.(R)-(3-amino-2-fluoropropyl) phosphinic acid (AZD3355), a novel $GABA_B$ receptor agonist, inhibits transient lower esophageal sphincter relaxation through a peripheral mode of action. J Pharmacol Exp Ther. 2009;331:504–12.

Taken together, these distribution, accumulation, and transport data provide the first clear support for the hypothesis that lesogaberan, as with GABA but not baclofen, is accumulated into neuronal/glial cells by specific GABA transporters, and that these serve to control extracellular levels of lesogaberan *in vivo*.

f. Translational Validity of the Dog Model The finding that baclofen reduces TLESR in dogs and ferrets has repeatedly been reproduced in both healthy volunteers and GERD patients. Indeed, it is rare in drug discovery research to identify an animal model that uses a surrogate endpoint (TLESR) that so precisely translates into the human setting with such a convincing relationship to the clinically relevant endpoint (GERD symptoms). In order to analyse the predictive resolution of the dog model to humans, plots of dose administered *versus* inhibition of TLESRs in dogs and humans were constructed for lesogaberan, AZD9343, and baclofen (AstraZeneca, unpublished data; Fig. 6). In healthy human volunteers the dose 5.7 μmol/kg of lesogaberan resulted in 36% inhibition of TLESRs; the inhibition of TLESRs in dogs at a similar dose level was approximately 45%. Similar correlations were also found for AZD9343 and baclofen and these findings testify as to the validity of the canine model. Plotting plasma concentrations *versus* effect yielded similar results.

3. *Clinical Pharmacodynamics of $GABA_B$ Agonists*

Baclofen is active in the CNS, resulting in side effects such as dizziness, sleepiness, and tiredness. Furthermore, pharmacokinetic characteristics such as a short half-life limit the clinical utility of baclofen as a treatment for patients with GERD. The development of AP, a prodrug of R-baclofen with improved uptake from the lower GI tract, has attempted to redress some of

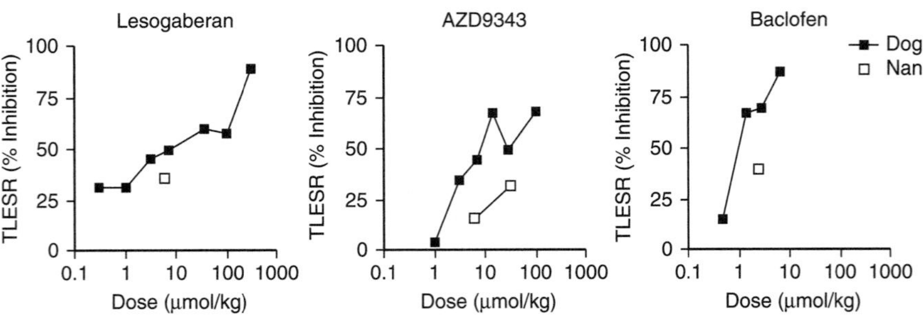

FIGURE 6 The mean inhibitory effect of lesogaberan, AZD9343 and baclofen on TLESR in dogs compared to the mean effect in healthy volunteers. Dog data are from Lehmann et al., 2009 (lesobaberan and baclofen) and AstraZeneca unpublished information (AZD9343). Data from healthy volunteers are compiled from Boeckxstaens et al., 2009c (lesogaberan) and Beaumont et al., 2009 (AZD9343 and baclofen), and recalculated to μmol/kg based on a body weight of 75 kg.

these limitations. In contrast, the mechanism of action of lesogaberan at peripheral GABA$_B$ receptors, and its pharmacodynamic effects demonstrated in Phase I and Phase II clinical trials, should translate into an effective add-on therapy with an acceptable side-effect profile for patients with persistent symptoms of GERD while on a PPI. Clinical findings with lesogaberan, AZD9343, and AP are summarized below.

In a single-blind, placebo-controlled, randomized, single-centre Phase I study in healthy subjects, in which baclofen was used as a positive control, a single oral dose of lesogaberan (0.8 mg/kg) reduced the number of TLESRs and reflux episodes and increased LES pressure compared with placebo, and had a favorable safety and tolerability profile (Boeckxstaens et al., 2009c). The number of TLESRs 0–3 h after intake of a standardized meal (completed 1 h after drug intake) were reduced by an average of 36% and 47% for lesogaberan and baclofen (40 mg), respectively, as compared to placebo. During the same recording period, an increase in LES pressure of 39% was seen after administration of lesogaberan compared with placebo. This effect was also seen for baclofen and confirmed earlier studies on baclofen in humans (Lidums et al., 2000; Zhang et al., 2002). Such an effect may be beneficial in the minority of GERD patients with a pathologically low LES resting pressure. The estimated number of acid reflux episodes per hour for the 3-h recording period (as measured by esophageal pH) was significantly reduced by both lesogaberan (-1.6; ~ 44%) and baclofen (-2.2) compared with placebo. The number of swallows was reduced with baclofen but not with lesogaberan compared with placebo. The number of subjects who experienced CNS adverse events was lower with lesogaberan (7/21) and placebo (6/22) than with baclofen (14/22). These findings underscored the translational value of the preclinical models and provided the impetus for the further development of lesogaberan as a treatment option for GERD.

In addition to lesogaberan, the effects of AZD9343 and baclofen on TLESR and reflux episodes in healthy volunteers have also been studied. For example, the effects of AZD9343 (60 mg or 320 mg) and baclofen (40 mg) once daily on TLESR and reflux episodes, compared with placebo, in 27 healthy volunteers were studied in a dual-centre, crossover study using esophageal manometry and pH-metry measurements. Tolerability assessments were also performed (Beaumont et al., 2009). Pre-prandial TLESRs were significantly reduced in number by AZD9343 320 mg, compared with placebo (1.6 *versus* 4.0, $p < 0.03$). In addition, meal-induced TLESRs were significantly reduced in number by both AZD9343 320 mg (3.6 ± 0.6 per hour) and baclofen (2.7 ± 0.2 per hour), compared with placebo (4.7 ± 0.3 per hour). Furthermore, the increase in post-prandial reflux episodes was also inhibited with both doses of AZD9343 (60 mg, 6.1 ± 0.9 episodes per 3 hours; 320 mg, 2.8 ± 0.5 episodes per 3 hours) and baclofen (3.8 ± 0.6 episodes per 3 hours, compared with placebo (8.9 ± 1.4 episodes per 3 hours). Both AZD9343 320 mg and baclofen significantly ($p < 0.05$)

reduced swallowing rate. With respect to tolerability, the profile of the high dose ofAZD9343 was comparable to that of baclofen, with somnolence and short-lasting paraesthesia being the most frequently reported adverse events. It was concluded that whilst AZD9343 significantly improves reflux outcomes, investigations into other agents with improved efficacy and tolerability profiles are warranted.

The efficacy of lesogaberan has also been shown in patients with GERD. In patients with reflux symptoms despite PPI treatment, add-on treatment with lesogaberan decreased the number of TLESRs and reflux episodes, and increased LES pressure compared with placebo in a randomized, double-blind, crossover study in which patients (n = 27) received lesogaberan 65 mg or placebo twice on day 1 (morning/evening) and once on day 2 (morning), in addition to existing PPI treatment (Boeckxstaens et al., 2009b). After a standardized meal, treatment with lesogaberan reduced the mean number of reflux events, measured by ambulatory pH-impedance monitoring, by ~35% compared with placebo during the first 24-h period, with the acid component being the most reduced (Table II). In addition, esophageal acid exposure and the number of proximal reflux events over 24 h were also lower with lesogaberan. On day 2, patients underwent stationary manometry and pH-impedance monitoring for 4 hours after the third dose, including 3 h following a standardized meal, and it was found that lesogaberan reduced the mean number of TLESRs by 25%, and increased the mean LES pressure by 28%, compared with placebo. Overall, lesogaberan was well tolerated with a low incidence of CNS-related side effects.

A further proof-of-concept study has demonstrated that add-on treatment with lesogaberan significantly improved heartburn and regurgitation symptoms, and was well tolerated in patients with GERD symptoms despite PPI therapy (Boeckxstaens et al., 2009a). In this double-blind, multicenter Phase IIa study, patients with continued GERD symptoms despite ≥ 6 weeks PPI treatment were randomized to 4 weeks treatment with lesogaberan 65 mg twice daily (n = 114) or placebo (n = 118) as add-on to existing PPI therapy. The patients included in this trial had a heavy GERD symptom load at baseline and the stringent response definition for the primary

TABLE II Arithmetic mean number of reflux events with lesogaberan and placebo over 24 h on day 1 in patients with reflux symptoms despite PPI treatment (n = 21) (Boeckxstaens et al., 2009b)

	Placebo	*Lesogaberan 65 mg*	*Mean difference (95% CI)*
Total reflux events	62.7	41.0	−22 (−23, −15)
Acid reflux	29.1	13.3	−16 (−23, −8.3)
Weakly acidic reflux	31.2	24.7	−6.5 (−12, −0.53)
Weakly alkaline reflux	2.3	2.9	0.64 (−2.1, 3.34)

endpoint was at most one 24-h period with heartburn or regurgitation of not more than mild intensity during the last week of treatment. Treatment with lesogaberan, compared with placebo, resulted in a statistically significantly larger proportion of responders to treatment (16% *versus* 8% of patients; $p = 0.026$), a significant treatment effect on the cumulative proportion of responders over time (log-rank $p = 0.0195$), and a higher proportion of symptom-free days (Table III). These findings suggested that lesogaberan can potentially be a new therapeutic agent to meet the needs of this patient population with persistent GERD symptoms despite PPI therapy. Further studies are planned to confirm these findings.

The clinical utility of AP as a treatment for patients with GERD has also been studied in a small multicenter, randomized crossover cohort study (Gerson et al., 2010). In this study single doses of AP (10, 20, 40, and 60 mg) and placebo were followed by high-fat meals 2 and 6 h after dosing. Reflux events were assessed for 12 h post-treatment by esophageal manometry and impedance monitoring. The combined data from all the AP dose cohorts (n = 44) showed a statistically significant reduction in reflux episodes, compared with placebo. The mean ($\pm$SD) number of reflux episodes during 12 h in the total cohort was 50.5 ± 27.2, compared with 60.9 ± 35.3 events in placebo-treated patients. For the 60-mg dose cohort alone, the mean (SD) number of reflux episodes after AP treatment was 39.3 ± 15.0 compared with 53.0 ± 15.3 for placebo. Acidic reflux events (detected by impedance recording of retrograde movement in the esophagus and defined as reflux episodes with an esophageal pH of <4 for at least 4 s) were significantly reduced for all dose groups combined, compared with placebo. The number of acidic reflux episodes in the 60 mg AP group was 27.1 ± 13.8 over 12 h, compared with 41.4 ± 9.5 for placebo. The

TABLE III Rates of treatment response and symptom-free days following therapy with lesogaberan as an add-on treatment to a proton pump inhibitor in patients who experience GERD symptoms despite PPI treatment.(Boeckxstaens et al., 2009a)

	Treatment response (Percentage of patients)[a]	*Proportion of symptom-free days (Percentage of patients)*[b]		
	(Proportion of patients, %)	*Heartburn*	*Regurgitation*	*Heartburn and regurgitation*
Placebo	8	21	23	10
Lesogaberan	16	36	37	19
p-value	0.026	<0.05	<0.05	<0.05
95% C.I.	0–17	7–23	6–23	4–15

[a] Placebo $n = 105$, lesogaberan $n = 104$.
[b] placebo $n = 114$, lesogaberan $n = 114$.

corresponding comparisons for non-acidic reflux were generally not statistically significant. Furthermore the incidence of heartburn associated with reflux episodes (either acid or non-acid) was significantly reduced in the total cohort (12.9 ± 11.9 over 12 h) compared with placebo (16.7 ± 18.2 over 12 h). The rate of heartburn events in the 60 mg AP group was significantly lower than placebo: 7.8 ± 6.7 *versus* 15.3 ± 9.3. The tolerability of AP was generally comparable to placebo and within the known safety profile of baclofen.

V. Conclusions

The original effort in the 1960s to discover novel GABA analogues led to the discovery of baclofen for the treatment of spasticity (to reverse the inhibition of skeletal muscle relaxation characteristic of this symptom). Although the seminal work of Bowery et al. (1980) did provide evidence for peripheral $GABA_B$ receptors, the overwhelming majority of work in the years to come focused on the role of central $GABA_B$ receptors. It is therefore somewhat ironic that the second most clinically validated indication (GERD) for the next generation of $GABA_B$ receptor agonists encompasses smooth muscle relaxation as the major pathophysiologic mechanism.

Investigations into the mechanism of action and efficacy of these next-generation $GABA_B$ agonists have revealed interesting differentiating characteristics of a pharmacophore that is somewhat intolerant of chemical modification. In particular, it seems that the chirality of the substituent in the 2-position of 3-aminopropylphosphinic acids may seriously impact the efficacy of both 2-fluoro and 2-hydroxy analogues, i.e. lesogaberan and AZD9343. In addition, differences in the accumulation of lesogaberan, AZD9343, and baclofen into neuronal/glial cells and transport by GATs appear to go some way to explaining the differing clinical profiles of these agents. For example, it has been demonstrated that lesogaberan is a substrate for native GATs, whereas baclofen is not. Selective accumulation of lesogaberan into neural cells via GATs has been used to explain the low propensity of lesogaberan to cause CNS side effects in experimental animals, healthy volunteers, and patients with GERD, in comparison with baclofen. Furthermore, lesogaberan has been shown to be transported into neural cells to a greater degree than AZD9343, which may explain the superior clinical profile of lesogaberan compared with AZD9343. In addition, the lower level of transport of AZD9343 into neuronal cells may also explain why this analogue of GABA shares a similar adverse event profile to baclofen, and why therefore its clinical development has been halted.

The encouraging results in lesogaberan Phase II studies coincide not only with the 30th anniversary of the discovery of the $GABA_B$ receptor, but also with the retirement of the eminent scientist who not only lay the foundation to the field of GABA$_B$ receptor pharmacology but also has continued to be a key player in developing it: Norman Bowery.

Acknowledgments

The authors thank Andrew Stead and Simon Lancaster, from Wolters Kluwer (Chester, UK), who provided medical writing and editing support funded by AstraZeneca.

Conflict of Interest Statement Anders Lehmann and Jörgen M. Jensen are employees of AstraZeneca. Guy E. Boeckxstaens has received financial support in the form of consultation fees and grant/research funding from AstraZeneca, and in the form of consultation fees from Movetis.

Abbreviations

AE	adverse event
AP	arbaclofen placarbil
CD	crural diaphragm
CNS	central nervous system
DMV	dorsal motor nucleus of the vagus
ECD	extracellular domain
ED	effective dose
GABA	γ-aminobutyric acid
GAT	GABA transporter
GERD	gastroesophageal reflux disease
IC	inhibitory concentration
IPSC	inhibitory postsynaptic current
LES	lower esophageal sphincter
PPI	proton pump inhibitor
TLESR	transient lower esophageal sphincter relaxation

References

Alstermark, C., Amin, K., Dinn, S. R., Elebring, T., Fjellström, O., Fitzpatrick, K., et al. (2008). Synthesis and pharmacological evaluation of novel gamma-aminobutyric acid type B (GABA$_B$) receptor agonists as gastroesophageal reflux inhibitors. *Journal of Medicinal Chemistry*, *51*(14), 4315–4320.

Amico, C., Marchetti, C., Nobile, M., & Usai, C. (1995). Pharmacological types of calcium channels and their modulation by baclofen in cerebellar granules. *Journal of Neuroscience*, *15*(4), 2839–2848.

Baclofen Tablets BP 10 mg Summary of Product Characteristics (27 October 2009). Retrieved 12 January 2010, from http://emc.medicines.org.uk/medicine/22404/SPC/Baclofen+Tablets+BP +10 mg

Beaumont, H., & Boeckxstaens, G. E. (2009). Does the presence of a hiatal hernia affect the efficacy of the reflux inhibitor baclofen during add-on therapy?. *American Journal of Gastroenterology*, *104*(7), 1764–1771.

Beaumont, H., Smout, A., Aanen, M., Rydholm, H., Lei, A., Lehmann, A., et al. (2009). The GABA(B) receptor agonist AZD9343 inhibits transient lower oesophageal sphincter relaxations and acid reflux in healthy volunteers: A phase I study. *Alimentary Pharmacology & Therapeutics*, *30*(9), 937–946.

Bettler, B., Kaupmann, K., Mosbacher, J., & Gassmann, M. (2004). Molecular structure and physiological functions of $GABA_B$ receptors. *Physiological Reviews*, *84*(3), 835–867.

Billinton, A., Ige, A. O., Bolam, J. P., White, J. H., Marshall, F. H., & Emson, P. C. (2001). Advances in the molecular understanding of $GABA_B$ receptors. *Trends in Neurosciences*, *24*(5), 277–282.

Blackshaw, L. A., & Partosoedarso, E. R. (1999). Baclofen inhibits mechanical sensitivity of gastrointestinal vagal afferents in vivo. *Gastroenterology*, *116*(4 Pt 2), A959.

Blackshaw, L. A., Smid, S. D., O'Donnell, T. A., & Dent, J. (2000). $GABA_B$ receptor-mediated effects on vagal pathways to the lower oesophageal sphincter and heart. *British Journal of Pharmacology*, *130*(2), 279–288.

Blackshaw, L. A., Staunton, E., Lehmann, A., & Dent, J. (1999). Inhibition of transient LES relaxations and reflux in ferrets by GABA receptor agonists. *American Journal of Physiology*, *277*(4 Pt 1), G867–G874.

Boeckxstaens, G., Beaumont, H., Hatlebakk, J., Silberg, D., Adler, J., & Denison, H. (2009b). Efficacy and tolerability of the novel reflux inhibitor, AZD3355, as add-on treatment in GERD patients with symptoms despite proton pump inhibitor therapy [abstract]. *Gastroenterology*, *136*(Suppl. 1), M1875.

Boeckxstaens, G., Denison, H., Ruth, M., Adler, J., Silberg, D., & Sifrim, D. (2009c). Effect of AZD3355, a novel $GABA_B$ agonist, on reflux and lower esophageal sphincter function in patients with GERD with symptoms despite proton pump inhibitor treatment [abstract]. *Gastroenterology*, *136*(Suppl. 1), M1861.

Boeckxstaens, G., Rydholm, H., Adler, J., & Ruth, M. (2009). Effect of AZD3355, a novel $GABA_B$ receptor agonist, on transient lower esophageal sphincter relaxations in healthy subjects [abstract]. *Gastroenterology*, *136*(Suppl. 1), 772.

Bowery, N. G., Bettler, B., Froestl, W., Gallagher, J. P., Marshall, F., Raiteri, M., et al. (2002). International Union of Pharmacology. XXXIII. Mammalian gamma-aminobutyric acid(B) receptors: Structure and function. *Pharmacological Reviews*, *54*(2), 247–264.

Bowery, N. G., Hill, D. R., Hudson, A. L., Doble, A., Middlemiss, D. N., Shaw, J., et al. (1980). (-)Baclofen decreases neurotransmitter release in the mammalian CNS by an action at a novel GABA receptor. *Nature*, *283*(5742), 92–94.

Brasseur, J. G., Ulerich, R., Dai, Q., Patel, D. K., Soliman, A. M., & Miller, L. S. (2007). Pharmacological dissection of the human gastro-oesophageal segment into three sphincteric components. *Journal of Physiology (Paris)*, *580*(Pt 3), 961–975.

Calver, A. R., Medhurst, A. D., Robbins, M. J., Charles, K. J., Evans, M. L., Harrison, D. C., et al. (2000). The expression of GABA(B1) and GABA(B2) receptor subunits in the CNS differs from that in peripheral tissues. *Neuroscience*, *100*(1), 155–170.

Cange, L., Johnsson, E., Rydholm, H., Lehmann, A., Finizia, C., Lundell, L., et al. (2002). Baclofen-mediated gastro-oesophageal acid reflux control in patients with established reflux disease. *Alimentary Pharmacology & Therapeutics*, *16*(5), 869–873.

Canning, B. J., & Mazzone, S. B. (2003). Reflex mechanisms in gastroesophageal reflux disease and asthma. *American Journal of Medicine*, *115*(Suppl. 3A), 45S–48S.

Cardozo, D. L., & Bean, B. P. (1995). Voltage-dependent calcium channels in rat midbrain dopamine neurons: Modulation by dopamine and GABAB receptors. *Journal of Neurophysiology*, *74*(3), 1137–1148.

Cheng, J. K., Lee, S. Z., Yang, J. R., Wang, C. H., Liao, Y. Y., Chen, C. C., et al. (2004). Does gabapentin act as an agonist at native GABA(B) receptors?. *Journal of Biomedical Science*, *11*(3), 346–355.

Ciccaglione, A. F., & Marzio, L. (2003). Effect of acute and chronic administration of the $GABA_B$ agonist baclofen on 24 hour pH metry and symptoms in control subjects and in patients with gastro-oesophageal reflux disease. *Gut*, *52*(4), 464–470.

Costantino, G., Macchiarulo, A., Entrena Guadix, A., & Pellicciari, R. (2001). QSAR and molecular modeling studies of baclofen analogues as GABA(B) agonists. Insights into the role of the aromatic moiety in GABA(B) binding and activation. *Journal of Medicinal Chemistry*, *44*(11), 1827–1832.

Crunelli, V., & Leresche, N. (1991). A role for GABAB receptors in excitation and inhibition of thalamocortical cells. *Trends in Neurosciences*, *14*(1), 16–21.

Dent, J., Dodds, W. J., Friedman, R. H., Sekiguchi, T., Hogan, W. J., Arndorfer, R. C., et al. (1980). Mechanism of gastroesophageal reflux in recumbent asymptomatic human subjects. *Journal of Clinical Investigation*, *65*(2), 256–267.

Dent, J., El-Serag, H. B., Wallander, M. A., & Johansson, S. (2005). Epidemiology of gastro-oesophageal reflux disease: A systematic review. *Gut*, *54*(5), 710–717.

DeVault, K. R., & Castell, D. O. (2005). Updated guidelines for the diagnosis and treatment of gastroesophageal reflux disease. *American Journal of Gastroenterology*, *100*(1), 190–200.

Dodds, W. J., Dent, J., Hogan, W. J., Helm, J. F., Hauser, R., Patel, G. K., et al. (1982). Mechanisms of gastroesophageal reflux in patients with reflux esophagitis. *New England Journal of Medicine*, *307*(25), 1547–1552.

Fass, R., Shapiro, M., Dekel, R., & Sewell, J. (2005). Systematic review: Proton-pump inhibitor failure in gastro-oesophageal reflux disease – where next? *Alimentary Pharmacology & Therapeutics*, *22*(2), 79–94.

Froestl, W., Mickel, S. J., Hall, R. G., von Sprecher, G., Strub, D., Baumann, P. A., et al. (1995a). Phosphinic acid analogues of GABA. 1. New potent and selective GABAB agonists. *Journal of Medicinal Chemistry*, *38*(17), 3297–3312.

Froestl, W., Mickel, S. J., von Sprecher, G., Diel, P. J., Hall, R. G., Maier, L., et al. (1995b). Phosphinic acid analogues of GABA. 2. Selective, orally active GABAB antagonists. *Journal of Medicinal Chemistry*, *38*(17), 3313–3331.

Galvez, T., Prezeau, L., Milioti, G., Franek, M., Joly, C., Froestl, W., et al. (2000a). Mapping the agonist-binding site of GABAB type 1 subunit sheds light on the activation process of GABAB receptors. *Journal of Biological Chemistry*, *275*(52), 41166–41174.

Galvez, T., Urwyler, S., Prezeau, L., Mosbacher, J., Joly, C., Malitschek, B., et al. (2000b). Ca(2+) requirement for high-affinity gamma-aminobutyric acid (GABA) binding at GABA (B) receptors: Involvement of serine 269 of the GABA(B)R1 subunit. *Molecular Pharmacology*, *57*(3), 419–426.

Gerson, L. B., Huff, F. J., Hila, A., Hirota, W. K., Reilley, S., Agrawal, A., et al. (2010). Arbaclofen placarbil decreases postprandial reflux in patients with gastroesophageal reflux disease. *American Journal of Gastroenterology*, *105*, 1266–1275.

Harayama, N., Shibuya, I., Tanaka, K., Kabashima, N., Ueta, Y., & Yamashita, H. (1998). Inhibition of N- and P/Q-type calcium channels by postsynaptic GABAB receptor activation in rat supraoptic neurones. *Journal of Physiology (Paris)*, *509*(Pt 2), 371–383.

Hashimoto, T., & Kuriyama, K. (1997). In vivo evidence that GABA(B) receptors are negatively coupled to adenylate cyclase in rat striatum. *Journal of Neurochemistry*, *69* (1), 365–370.

Hills, J. M., Dingsdale, R. A., Parsons, M. E., Dolle, R. E., & Howson, W. (1989). 3-Aminopropylphosphinic acid–a potent, selective GABAB receptor agonist in the guinea-pig ileum and rat anococcygeus muscle. *British Journal of Pharmacology*, *97*(4), 1292–1296.

Holloway, R. H., Kocyan, P., & Dent, J. (1991). Provocation of transient lower esophageal sphincter relaxations by meals in patients with symptomatic gastroesophageal reflux. *Digestive Diseases and Sciences*, *36*(8), 1034–1039.

Hudgson, P., & Weightman, D. (1971). Baclofen in the treatment of spasticity. *British Medical Journal*, *4*(5778), 15–17.

Iwakiri, K., Hayashi, Y., Kotoyori, M., Tanaka, Y., Kawakami, A., Sakamoto, C., et al. (2005). Transient lower esophageal sphincter relaxations (TLESRs) are the major mechanism of gastroesophageal reflux but are not the cause of reflux disease. *Digestive Diseases and Sciences*, *50*(6), 1072–1077.

Jensen, A. A., Mosbacher, J., Elg, S., Lingenhoehl, K., Lohmann, T., Johansen, T. N., et al. (2002). The anticonvulsant gabapentin (neurontin) does not act through gamma-aminobutyric acid-B receptors. *Molecular Pharmacology*, *61*(6), 1377–1384.

Jones, R., Armstrong, D., Malfertheiner, P., & Ducrotte, P. (2006). Does the treatment of gastroesophageal reflux disease (GERD) meet patients' needs? A survey-based study. *Current Medical Research and Opinion*, *22*(4), 657–662.

Jones, K. A., Borowsky, B., Tamm, J. A., Craig, D. A., Durkin, M. M., Dai, M., et al. (1998). GABA(B) receptors function as a heteromeric assembly of the subunits GABA(B)R1 and GABA(B)R2. *Nature*, *396*(6712), 674–679.

Kaupmann, K., Huggel, K., Heid, J., Flor, P. J., Bischoff, S., Mickel, S. J., et al. (1997). Expression cloning of $GABA_B$ receptors uncovers similarity to metabotropic glutamate receptors. *Nature*, *386*, 239–246.

Kaupmann, K., Malitschek, B., Schuler, V., Heid, J., Froestl, W., Beck, P., et al. (1998). GABA (B)-receptor subtypes assemble into functional heteromeric complexes. *Nature*, *396* (6712), 683–687.

Kniazeff, J., Galvez, T., Labesse, G., & Pin, J. P. (2002). No ligand binding in the GB2 subunit of the GABA(B) receptor is required for activation and allosteric interaction between the subunits. *Journal of Neuroscience*, *22*(17), 7352–7361.

Koek, G. H., Sifrim, D., Lerut, T., Janssens, J., & Tack, J. (2003). Effect of the $GABA_B$ agonist baclofen in patients with symptoms and duodeno-gastro-oesophageal reflux refractory to proton pump inhibitors. *Gut*, *52*(10), 1397–1402.

Kuner, R., Kohr, G., Grunewald, S., Eisenhardt, G., Bach, A., & Kornau, H. C. (1999). Role of heteromer formation in GABAB receptor function. *Science*, *283*(5398), 74–77.

Kuroda, E., Watanabe, M., Tamayama, T., & Shimada, M. (2000). Autoradiographic distribution of radioactivity from ^{14}C-GABA in the mouse. *Microscopy Research and Technique*, *48*(2), 116–126.

Lacaille, J. C. (1991). Postsynaptic potentials mediated by excitatory and inhibitory amino acids in interneurons of stratum pyramidale of the CA1 region of rat hippocampal slices in vitro. *Journal of Neurophysiology*, *66*(5), 1441–1454.

Lehmann, A. (2006). Inhibitors of transient lower esophageal sphincter relaxations ("reflux inhibitors") in the future treatment of GERD. *Gastroenterology & Hepatology – Annual Review*, *1*, 109–117.

Lehmann, A. (2009). $GABA_B$ receptors as drug targets to treat gastroesophageal reflux disease. *Pharmacology & Therapeutics*, *122*(3), 239–245.

Lehmann, A., Antonsson, M., Bremner-Danielsen, M., Flärdh, M., Hansson-Brändén, L., & Kärrberg, L. (1999). Activation of the $GABA_B$ receptor inhibits transient lower esophageal sphincter relaxations in dogs. *Gastroenterology*, *117*(5), 1147–1154.

Lehmann, A., Antonsson, M., Holmberg, A. A., Blackshaw, L. A., Brändén, L., Bräuner-Osborne, H., et al. (2009). (R)-(3-amino-2-fluoropropyl) phosphinic acid (AZD3355), a novel GABAB receptor agonist, inhibits transient lower esophageal sphincter relaxation through a peripheral mode of action. *Journal of Pharmacology and Experimental Therapeutics*, *331*(2), 504–512.

Lehmann, A., Elebring, T., Jensen, J., Mattsson, J. P., Nilsson, K. A., Saransaari, P., et al. (2008). In vivo and in vitro characterization of AZD9343, a novel GABA$_B$ receptor agonist and reflux inhibitor [abstract]. *Gastroenterology*, *134*(4), A131–A132.

Lehmann, A., Holmberg, A. A., Bhatt, U., Bremner-Danielsen, M., Brändén, L., Elg, S., et al. (2005). Effects of (2r)-(3-amino-2-fluoropropyl)sulphinic acid (AFPSiA) on transient lower oesophageal sphincter relaxation in dogs and mechanism of hypothermic effects in mice. *British Journal of Pharmacology*, *146*, 89–97.

Lidums, I., Lehmann, A., Checklin, H., Dent, J., & Holloway, R. H. (2000). Control of transient lower esophageal sphincter relaxations and reflux by the GABA$_B$ agonist baclofen in normal subjects. *Gastroenterology*, *118*(1), 7–13.

Lieberman, D. A., Oehlke, M., & Helfand, M. (1997). Risk factors for Barrett's esophagus in community-based practice. GORGE consortium. Gastroenterology outcomes research group in endoscopy. *American Journal of Gastroenterology*, *92*(8), 1293–1297.

Liebermann-Meffert, D., Riedo, I., & Allgöwer, M. (1986). Interrelation between functional and muscular lower esophageal sphincter. *Digestive Surgery*, *3*, 101.

Lipan, M. J., Reidenberg, J. S., & Laitman, J. T. (2006). Anatomy of reflux: A growing health problem affecting structures of the head and neck. *Anatomical Record. Part B, New Anatomist*, *289*(6), 261–270.

Liu, J., Pehlivanov, N., & Mittal, R. K. (2002). Baclofen blocks LES relaxation and crural diaphragm inhibition by esophageal and gastric distension in cats. *American Journal of Physiology*, *283*(6), G1276–G1281.

Luscher, C., Jan, L. Y., Stoffel, M., Malenka, R. C., & Nicoll, R. A. (1997). G protein-coupled inwardly rectifying K+ channels (GIRKs) mediate postsynaptic but not presynaptic transmitter actions in hippocampal neurons. *Neuron*, *19*(3), 687–695.

Maguire, G., Maple, B., Lukasiewicz, P., & Werblin, F. (1989). Gamma-aminobutyrate type B receptor modulation of L-type calcium channel current at bipolar cell terminals in the retina of the tiger salamander. *Proceedings of the National Academy of Sciences of the United States of America*, *86*(24), 10144–10147.

Malitschek, B., Schweizer, C., Keir, M., Heid, J., Froestl, W., Mosbacher, J., et al. (1999). The N-terminal domain of gamma-aminobutyric acid(B) receptors is sufficient to specify agonist and antagonist binding. *Molecular Pharmacology*, *56*(2), 448–454.

Marchetti, C., Carignani, C., & Robello, M. (1991). Voltage-dependent calcium currents in dissociated granule cells from rat cerebellum. *Neuroscience*, *43*(1), 121–133.

Martin, C. J., Patrikios, J., & Dent, J. (1986). Abolition of gas reflux and transient lower esophageal sphincter relaxation by vagal blockade in the dog. *Gastroenterology*, *91*(4), 890–896.

Matsushima, T., Tegner, J., Hill, R. H., & Grillner, S. (1993). GABAB receptor activation causes a depression of low- and high-voltage-activated Ca2+ currents, postinhibitory rebound, and postspike afterhyperpolarization in lamprey neurons. *Journal of Neurophysiology*, *70*(6), 2606–2619.

Menon-Johansson, A. S., Berrow, N., & Dolphin, A. C. (1993). G(o) transduces GABAB-receptor modulation of N-type calcium channels in cultured dorsal root ganglion neurons. *Pflugers Archiv*, *425*(3–4), 335–343.

Mintz, I. M., & Bean, B. P. (1993). GABAB receptor inhibition of P-type Ca2+ channels in central neurons. *Neuron*, *10*(5), 889–898.

Mittal, R. K., & Balaban, D. H. (1997). The esophagogastric junction. *New England Journal of Medicine*, *336*(13), 924–932.

Mittal, R. K., Holloway, R. H., Penagini, R., Blackshaw, L. A., & Dent, J. (1995). Transient lower esophageal sphincter relaxation. *Gastroenterology*, *109*(2), 601–610.

Mittal, R. K., & McCallum, R. W. (1988). Characteristics and frequency of transient relaxations of the lower esophageal sphincter in patients with reflux esophagitis. *Gastroenterology, 95*(3), *593–599.*

Ng, G. Y., Bertrand, S., Sullivan, R., Ethier, N., Wang, J., Yergey, J., et al. (2001). Gamma-aminobutyric acid type B receptors with specific heterodimer composition and postsynaptic actions in hippocampal neurons are targets of anticonvulsant gabapentin action. *Molecular Pharmacology, 59*(1), 144–152.

Ong, J., & Kerr, D.I.B. (1998). The gamma-aminobutyric acid uptake inhibitor NO-711 potentiates 3-aminopropylphosphinic acid-induced actions in rat neocortical slices. *European Journal of Pharmacology, 347*, 197–200.

Otis, T. S., De Koninck, Y., & Mody, I. (1993). Characterization of synaptically elicited GABAB responses using patch-clamp recordings in rat hippocampal slices. *Journal of Physiology (Paris), 463*, 391–407.

Page, A. J., O'Donnell, T. A., & Blackshaw, L. A. (2006). Inhibition of mechanosensitivity in visceral primary afferents by $GABA_B$ receptors involves calcium and potassium channels. *Neuroscience, 137*(2), 627–636.

Partosoedarso, E. R., Young, R. L., & Blackshaw, L. A. (2001). GABA(B) receptors on vagal afferent pathways: Peripheral and central inhibition. *American Journal of Physiology. Gastrointestinal and Liver Physiology, 280*(4), G658–G668.

Quiocho, F. A., & Ledvina, P. S. (1996). Atomic structure and specificity of bacterial periplasmic receptors for active transport and chemotaxis: Variation of common themes. *Molecular Microbiology, 20*(1), 17–25.

Rosenstein, R. E., Chuluyan, H. E., Diaz, M. C., & Cardinali, D. P. (1990). GABA as a presumptive paracrine signal in the pineal gland. Evidence on an intrapineal GABAergic system. *Brain Research Bulletin, 25*, 339–344.

Schuler, V., Luscher, C., Blanchet, C., Klix, N., Sansig, G., Klebs, K., et al. (2001). Epilepsy, hyperalgesia, impaired memory, and loss of pre- and postsynaptic GABA(B) responses in mice lacking GABA(B(1)). *Neuron, 31*(1), 47–58.

Scott, R. H., Wootton, J. F., & Dolphin, A. C. (1990). Modulation of neuronal T-type calcium channel currents by photoactivation of intracellular guanosine 5′-O(3-thio) triphosphate. *Neuroscience, 38*(2), 285–294.

Shen, W., & Slaughter, M. M. (1999). Metabotropic GABA receptors facilitate L-type and inhibit N-type calcium channels in single salamander retinal neurons. *Journal of Physiology (Paris), 516*(Pt 3), 711–718.

Skoog, B. (1996). A comparison of the effects of two antispastic drugs, tizanidine and baclofen, on synaptic transmission from muscle spindle afferents to spinal interneurones in cats. *Acta Physiologica Scandinavica, 156*(1), 81–90.

Smart, T. G. (2004). Gamma-aminobutyric acid (GABA) receptors. In G. Adelman, B. H. Smith (Eds.), *Encyclopedia of Neuroscience* (3rd ed., CD-Rom). Amsterdam: Elsevier

Smid, S. D., Young, R. L., Cooper, N. J., & Blackshaw, L. A. (2001). $GABA_BR$ expressed on vagal afferent neurones inhibit gastric mechanosensitivity in ferret proximal stomach. *Am. J. Physiology, 281*(6), G1494–G1501.

Staunton, E., Smid, S. D., Dent, J., & Blackshaw, L. A. (2000a). Triggering of transient LES relaxations in ferrets: Role of sympathetic pathways and effects of baclofen. *American Journal of Physiology, 279*(1), G157–G162.

Staunton, E., Smid, S. D., Dent, J., & Blackshaw, L. A. (2000b). Triggering of transient LES relaxations in ferrets: Role of sympathetic pathways and effects of baclofen. *American Journal of Physiology. Gastrointestinal and Liver Physiology, 279*(1), G157–G162.

Takahashi, T., Kajikawa, Y., & Tsujimoto, T. (1998). G-Protein-coupled modulation of presynaptic calcium currents and transmitter release by a GABAB receptor. *Journal of Neuroscience, 18*(9), 3138–3146.

Thirlby, R. C., Stevens, M. H., Blair, A. J., Petty, F., Crawford, I. L., Taylor, I. L., et al. (1988). Effect of GABA on basal and vagally mediated gastric acid secretion and hormone release in dogs. *American Journal of Physiology*, *254*(5 Pt 1), G723–G731.

Vakil, N., van Zanten, S. V., Kahrilas, P., Dent, J., & Jones, R. (2006). The Montreal definition and classification of gastroesophageal reflux disease: A global evidence-based consensus. *American Journal of Gastroenterology*, *101*(8), 1900–1920.

van Herwaarden, M. A., Samsom, M., Rydholm, H., & Smout, A. J. (2002). The effect of baclofen on gastro-oesophageal reflux, lower oesophageal sphincter function and reflux symptoms in patients with reflux disease. *Aliment. Pharmacol. Therapy*, *16*(9), 1655–1662.

van Herwaarden, M. A., Samsom, M., & Smout, A. J. (2000). Excess gastroesophageal reflux in patients with hiatus hernia is caused by mechanisms other than transient LES relaxations. *Gastroenterology*, *119*(6), 1439–1446.

Vela, M. F., Tutuian, R., Katz, P. O., & Castell, D. O. (2003). Baclofen decreases acid and non-acid post-prandial gastro-oesophageal reflux measured by combined multichannel intraluminal impedance and pH. *Aliment. Pharmacol. Therapy*, *17*(2), 243–251.

White, J. H., Wise, A., Main, M. J., Green, A., Fraser, N. J., Disney, G. H., et al. (1998). Heterodimerization is required for the formation of a functional GABA(B) receptor. *Nature*, *396*(6712), 679–682.

Wiklund, I. (2004). Review of the quality of life and burden of illness in gastroesophageal reflux disease. *Digestive Diseases*, *22*(2), 108–114.

Wiklund, I., Carlsson, J., & Vakil, N. (2006). Gastroesophageal reflux symptoms and well-being in a random sample of the general population of a Swedish community. *American Journal of Gastroenterology*, *101*(1), 18–28.

Zagorodnyuk, V. P., D'Antona, G., Brookes, S. J., & Costa, M. (2002). Functional GABA$_B$ receptors are present in guinea pig nodose ganglion cell bodies but not in peripheral mechanosensitive endings. *Autonomic Neuroscience*, *102*(1–2), 20–29.

Zhang, Q., Lehmann, A., Rigda, R., Dent, J., & Holloway, R. H. (2002). Control of transient lower oesophageal sphincter relaxations and reflux by the GABA$_B$ agonist baclofen in patients with gastro-oesophageal reflux disease. *Gut*, *50*(1), 19–24.

Styliani Vlachou and Athina Markou

Department of Psychiatry, School of Medicine,
University of California San Diego, La Jolla, California, USA

$GABA_B$ Receptors in Reward Processes

Abstract

γ-aminobutyric acid (GABA) is the predominant inhibitory neurotransmitter in the brain which acts through different receptor subtypes. Metabotropic $GABA_B$ receptors are widely distributed throughout the brain. Alterations in GABA signaling through pharmacological activation or deactivation of the $GABA_B$ receptor regulate behavior and brain reward processes. $GABA_B$ receptor agonists and, most recently, positive modulators have been found to inhibit the reinforcing effects of drugs of abuse, such as cocaine, amphetamine, nicotine, ethanol, and opiates. This converging evidence of the effects of $GABA_B$ compounds on the reinforcing properties of addictive drugs is based on behavioral studies that used a variety of

Advances in Pharmacology, Volume 58

1054-3589/10 $35.00
10.1016/S1054-3589(10)58013-X

procedures with relevance to reward processes and drug abuse liability, including intracranial self-stimulation, intravenous self-administration under both fixed- and progressive-ratio schedules of reinforcement, reinstatement, and conditioned place preference. $GABA_B$ receptor agonists and positive modulators block the reinforcing effects of drugs of abuse in these animal models. However, $GABA_B$ receptor agonists also have undesirable side-effects. $GABA_B$ receptor modulators have potential advantages as medications for drug addiction. These compounds have a better side-effect profile than $GABA_B$ agonists because they are devoid of intrinsic agonistic activity in the absence of GABA. They only exert their modulatory actions in concert with endogenous GABAergic activity. Thus, $GABA_B$ receptor positive modulators are promising therapeutics for the treatment of various aspects of dependence (e.g., initiation, maintenance, and relapse) on various drugs of abuse, such as cocaine, nicotine, heroin, and alcohol.

I. Introduction

Studies in laboratory animals strongly suggest that the neurotransmitter γ-aminobutyric acid (GABA) is critically involved in brain reward processes. Drugs of abuse powerfully stimulate the brain's reward pathways and provide the means to study how GABA transmission modulates circuits involved in reward processes. Alterations in reward processes and hedonic homeostasis have been implicated in the development of drug addictions in humans (for review, see Koob & Volkow, 2010). The purpose of this review is to primarily discuss findings from studies in laboratory animals that explored the role of GABA transmission in reward processes, particularly in the context of the effects of drugs of abuse on reward. Accruing evidence from preclinical studies support the hypothesis that $GABA_B$ receptor agonists or positive modulators attenuate the rewarding effects of various drugs of abuse, such as cocaine, amphetamine, heroin, nicotine, and ethanol (see below; Table I). Accordingly, compounds modulating GABA transmission through actions at $GABA_B$ receptors have been proposed as useful therapeutics for the treatment of drug addictions.

A. The GABAergic System in Brain Reward Circuits

Dopaminergic neurons projecting from the ventral tegmental area (VTA) to the nucleus accumbens (NAc) (Fallon & Moore, 1978) and amygdala (Fallon et al., 1978) have been implicated in reward processes (Koob & Volkow, 2010; Schultz, 2001, 2007; Tobler et al., 2005). These dopaminergic neurons receive descending GABAergic inputs from the ventral pallidum and NAc (Sugita et al., 1992; Walaas & Fonnum, 1979) which have an inhibitory effect on dopaminergic tone (Engberg et al., 1993; Klitenick et al., 1992) at

TABLE I Effects of $GABA_B$ receptor compounds in reward processes and the rewarding effects of addictive drugs measured by behavioral procedures

Drug	*Drug Dose*	*$GABA_B$ compound*	*Dose/Dose range*	*Species*	*Method*	*Effects of $GABA_B$ receptor compounds on the rewarding/reinforcing effects of drugs of abuse*	*References*
		Baclofen	0, 3, 5.4, 10 mg/kg, i.p.	Fisher rats	ICSS	↓ response decrement pattern (i.e., decrease in responding to receive electrical stimulation)	(Fenton & Liebman, 1982)
		Baclofen	0, 0.064, 0.128, 0.26, 0.52 μg/kg, intra-VTA	Wistar rats	ICSS	↑ ICSS thresholds by high doses of baclofen (0.128, 0.26, and 0.52 μg/kg)	(Willick & Kokkinidis, 1995)
		CGP44532	0, 0.125, 0.25, 0.5, 1 mg/kg, s.c.	Wistar rats	ICSS	↑ ICSS thresholds by all compounds. Additive effects after co-administration of CGP44532 with either of the two antagonists	(Macey et al., 2001)
		CGP56433A	0, 2.5, 5, 7.5, 10 mg/kg, s.c.				
		CGP51176	0, 0.3, 3, 30, 100, 300 mg/kg, s.c.				
		Baclofen	0.12, 0.48 μg/kg, intra-VTA	Sprague–Dawley rats	ICSS	↑ ICSS thresholds	(Panagis & Kastellakis, 2002)
		Baclofen	0, 0.1, 0.5, 2.5 mM into the MR and the DR	Wistar rats	Self-administration (FR1 schedule) CPP	Self-administration of baclofen in the MR Injections in the DR induced CPP	(Shin & Ikemoto, 2010)

(*Continued*)

TABLE I (*Continued*)

Drug	*Drug Dose*	*$GABA_B$ compound*	*Dose/Dose range*	*Species*	*Method*	*Effects of $GABA_B$ receptor compounds on the rewarding/reinforcing effects of drugs of abuse*	*References*
Psychostimulants							
Cocaine	1.5 mg/kg/inj	Baclofen	0, 1.25, 2.5, 5 mg/kg, i.p.	Wistar rats	Self-administration (FR1 and PR schedules)	↓ breakpoints for cocaine self-administration (dose-dependent effect) No effect of baclofen under FR1	(Roberts et al., 1996)
	1.5 mg/kg/inj	Baclofen	0, 1.25, 2.5, 5 mg/kg, i.p.	Wistar rats	Self-administration (discrete trials schedule)	↓ responding for cocaine self-administration	(Roberts & Andrews, 1997)
	0.2, 0.4 mg/kg/inj Priming dose: 3.2 mg/kg, i.v.	Baclofen	0, 1.25, 2.5, 5 mg/kg, i.p.	Wistar rats	Self-administration (FR1 schedule) Drug-induced reinstatement	↓ responding for cocaine self-administration (dose-dependent effect) in both cocaine groups (0.2, 0.4 mg/kg) Baclofen dose-dependently blocked cocaine-induced responding	(Campbell et al., 1999)
	0.3, 0.6 mg/kg/inj	GVG	0, 100, 200, 300, 400 mg/kg, i.p.	Wistar rats	Self-administration (FR5 and PR schedules)	↓ responding for cocaine self-administration under both FR5 and PR (dose-dependent effect)	(Kushner et al., 1999)
	20 mg/kg, i.p.	GVG	0, 75, 112, 150, 300 mg/kg, i.p.	Sprague–Dawley rats	CPP	GVG (112, 150, 300 mg/kg) blocked the acquisition and expression of cocaine-induced CPP	(Dewey et al., 1998)

2.5, 5 mg/kg, i.p.	GVG	0, 200, 300, 400 mg/kg, i.p.	F-344 rats	ICSS	↓ thresholds by cocaine, blocking effect by GVG (400 mg/kg)	(Kushner et al., 1997)
0.66 mg/kg/inj	Baclofen	0, 2.5, 10 mg/kg, i.p. 100–300 ng (intra-NAc) 300 ng (intra-VTA) 2.5, 5 mg/kg, i.p. (for 3 or 5 days, respectively)	Sprague–Dawley rats	Self-administration (FR5 and FR5 multiple schedule)	↓ responding for cocaine self-administration (dose-dependent effect) after systemic, intra-NAc, or intra-VTA administration or after subchronic administration for 3 or 5 days	(Shoaib et al., 1998)
0.75 mg/kg/inj or 0.18, 0.37, 0.75, 1.5 mg/kg/inj	CGP44532	0.063–0.5 mg/kg, i.p. 0.125 mg/kg, i.p.	Wistar rats	Self-administration (PR and discrete-trials schedule)	↓ breakpoints or responding for cocaine self-administration under both PR and discrete-trials schedules (dose-dependent effect)	(Brebner et al., 1999)
0.19, 0.38, 0.75, 1.5 mg/kg/inj	Baclofen	0, 1.8, 3.2, 5.6 mg/kg, i.p.	Wistar rats	Self-administration (FR1 and PR schedules)	↓ cocaine self-administration under both FR1 and PR (dose-dependent effect) Larger intra-NAc doses of baclofen (180, 320 ng/kg) were required to produce a similar effect compared with intra-VTA and intra-striatum doses (56, 100 ng/kg)	(Brebner et al., 2000a)
1.5 mg/kg/inj	Baclofen	0, 32, 56, 100, 178, 320 ng/kg/inf, intra-VTA, intra-NAc, intra-striatum	Wistar rats	Self-administration (PR schedule)	↓ breakpoints for cocaine self-administration (dose-dependent effect)	(Brebner et al., 2000b)

(*Continued*)

TABLE I *(Continued)*

Drug	*Drug Dose*	*$GABA_B$ compound*	*Dose/Dose range*	*Species*	*Method*	*Effects of $GABA_B$ receptor compounds on the rewarding/reinforcing effects of drugs of abuse*	*References*
	0.3 mg/kg/inj	Baclofen	0, 30, 60 ng/kg/inf, intra-VTA	Long–Evans rats	Self-administration (FR5 schedule)	↓ cocaine self-administration (nonsignificant)	(Corrigall et al., 2000)
	3 mg/ml	Baclofen	0, 30, 60 ng/kg/inf, intra-pedunculo-pontine tegmental nucleus	Long–Evans rats	Self-administration (FR5 schedule)	No effect on cocaine self-administration	(Corrigall et al., 2001)
	20 mg/kg, i.p.	vigabatrin	300 mg/kg, i.p.	Sprague–Dawley rats	CPP	↑ NAc dopamine in response to cocaine-paired cues Vigabatrin completely abolished this increase	(Gerasimov et al., 2001)
	1.5 mg/kg/inj	CGP56433A	0, 0.6, 1, 1.8 mg/kg, i.p.	Wistar rats	Self-administration (FR1 and PR schedules)	No effect on cocaine self-admnistration under either of the schedules tested Blocked the effect of baclofen on cocaine self-administration	(Brebner et al., 2002b)
	10 mg/kg, i.p.	CGP44532	0.063, 0.125, 0.25 mg/kg, i.p.	Sprague–Dawley rats	ICSS	No effect of CGP44532 on brain stimulation reward CGP44532 dose-dependently reduced cocaine-induced reward enhancement	(Dobrovitsky et al., 2002)

1.5 mg/kg/inj (PR) 0.19, 0.38, 0.75, 1.5 mg/kg/inj (FR1)	CGP7930 GS39783 Baclofen	0, 1, 10, 30 mg/kg, i.p. 0, 1, 10, 30 mg/kg, i.p. 2.5 mg/kg, i.p.	Sprague–Dawley rats	Self-administration (FR1, PR, and discrete-trials schedule)	↓ breakpoints for cocaine self-administration by CGP7930 (10, 30 mg/kg) and GS39783 (30 mg/kg) ↓ low doses of cocaine self-administration under FR1 schedule by both modulators ↓ cocaine self-administration in discrete-trials schedule by baclofen (2.5 mg/kg), CGP7930 (30 mg/kg), and GS39783 (30 mg/kg)	(Smith et al., 2004)
0, 0.03, 0.1, 0.3, 1, 3.2 mg/kg/inj	GVG Baclofen	0, 180, 320, 560 mg/kg, i.p. 0, 1.8, 3.2, 5.6 mg/kg, i.p.	Sprague–Dawley rats	Self-administration (FR5 multiple schedule)	↓ cocaine self-administration by baclofen and GVG at doses that also affected food-maintained responding	(Barrett et al., 2005)
10 mg/kg, i.p.	Baclofen GS39783	0, 2.5, 5 mg/kg, p.o. 0, 10, 30, 100 mg/kg, p.o.	Sprague–Dawley rats	ICSS	No effects of GS39783 on ICSS ↑ thresholds by baclofen (dose-dependent effect) Co-administration of baclofen with GS39783 attenuated the lowering of thresholds induced by cocaine (dose-dependent effect)	(Slattery et al., 2005)

(*Continued*)

TABLE I (*Continued*)

Drug	*Drug Dose*	*$GABA_B$ compound*	*Dose/Dose range*	*Species*	*Method*	*Effects of $GABA_B$ receptor compounds on the rewarding/reinforcing effects of drugs of abuse*	*References*
	0.5 mg/kg/inj	SCH50911 Baclofen SKF97541 CGP7930	0, 3, 10 mg/kg, i.p. 0, 1.25, 2.5, 5 mg/kg, i.p. 0, 0.03, 0.1, 0.3 mg/kg, i.p. 0, 10, 30, 100 mg/kg, i.p.	Wistar rats	Self-administration (FR5 schedule)	↓ cocaine self-administration by baclofen (2.5, 5 mg/kg), SKF97541 (0.1, 0.3 mg/kg), and CGP7930 (30, 100 mg/kg) No effect of SCH50911	(Filip et al., 2007a)
	0.5 mg/kg/inj Priming dose: 10 mg/kg, i.p.	Gabapentine Tiagabine GVG	0, 10, 30 mg/kg, i.p. 0, 5, 10 mg/kg, i.p. 0, 150, 250, i.p.	Wistar rats	Self-administration (FR5 schedule) Drug-induced reinstatement	↓ cocaine self-administration by GVG and tiagabine (10 mg/kg) GVG blocked cocaine-induced reinstatement Tiagabine nonsignificantly attenuated cocaine-induced reinstatement No effect of gabapentin in either of the procedures	(Filip et al., 2007b)
	0.5 mg/kg/inj Priming dose: 10 mg/kg, i.p.	Baclofen SKF97541 CGP7930 SCH50911	1.25–5 mg/kg, i.p. 0.03–0.3 mg/kg, i.p. 10–30 mg/kg, i.p. 3–10 mg/kg, i.p.	Wistar rats	Cue- and drug-induced reinstatement	All compounds dose-dependently blocked both cue- and cocaine induced reinstatement Baclofen and SKF97541 also affected reinstatement of food-seeking behavior	(Filip & Frankowska, 2007)

	0.032 mg/kg/inj Priming dose: 0.1–1.8 mg/kg, i.v.	Baclofen CGP44532 Tiagabine	0.32 mg/kg, i.m. (for all three compounds)	Baboons (*Papio Anubis*)	Drug-induced reinstatement	Baclofen and CGP44532 attenuated cocaine-induced reinstatement of cocaine-seeking behavior Tiagabine did not have an effect	(Weerts et al., 2007)
	0.5 mg/kg/inj	Baclofen SKF97541 CGP7930 SCH50911	0, 0.125 mg/kg, i.p. 0, 0.005 mg/kg, i.p. 0, 0.3 mg/kg, i.p. 0, 0.3 mg/kg, i.p.	Wistar rats	Self-administration (FR5 schedule) Chronic mild stress (forced swim)	All compounds counteracted the increased immobility time in the forced-swim test produced by discontinuation of cocaine self-administration	(Frankowska et al., 2010)
D-Amphetamine	0.5 mg/kg, i.p.	Baclofen	0, 1.5, 2.5, 5 mg/kg, i.p.	Wistar rats	CPP	Baclofen attenuated the development (2.5 and 5 mg/kg) and expression (1.5, 2.5, 5 mg/kg) of D-methamphetamine-induced CPP	(Li et al., 2001)
	0.1, 0.2 mg/kg/inj	Baclofen	0, 1.8, 3.2, 5.6 mg/kg, i.p.	Sprague–Dawley rats	Self-administration (FR1 and PR schedules)	↓ D-amphetamine self-administration by baclofen (3.2, 5.8 mg/kg) in rats that were trained to self-administer 0.1 mg/kg/injection of D-amphetamine No effect on self-administration of 0.2 mg/kg/injection of D-amphetamine ↓ breakpoints for D-amphetamine self-administration by baclofen (dose-dependent effect)	(Brebner et al., 2005)

(*Continued*)

TABLE I *(Continued)*

Drug	*Drug Dose*	*GABA$_B$ compound*	*Dose/Dose range*	*Species*	*Method*	*Effects of GABA$_B$ receptor compounds on the rewarding/reinforcing effects of drugs of abuse*	*References*
Methamphetamine	0, 0.0625, 0.125, 0.25 mg/kg/inj	Baclofen	0, 2.5, 5 mg/kg, i.p.	Long–Evans rats	Self-administration (PR schedule)	↓ breakpoints for all doses of D-amphetamine self-administration	(Ranaldi & Poeggel, 2002)
	0, 1, 2.5, 5, 10mg/kg, i.p.	GVG	0, 300 mg/kg, i.p.	Sprague–Dawley rats	CPP	GVG blocked methamphetamine-induced reinstatement of CPP	(DeMarco et al., 2009)
Nicotine	0.4 mg/kg, s.c.	GVG	18.75, 37.5, 75, 150 mg/kg, i.p.	Sprague–Dawley rats	CPP	All doses abolished the expression of nicotine-induced CPP 75 mg/kg also blocked the acquisition phase	(Dewey et al., 1999)
	30 μg/kg/inj	Baclofen	0, 30, 60 ng/kg/inj, intra-VTA	Long–Evans rats	Self-administration (FR5 schedule)	↓ nicotine self-administration	(Corrigall et al., 2000)
	0.4 mg/kg, s.c.	ACC	0, 300 mg/kg, i.p.	Sprague–Dawley rats	CPP	ACC attenuated the expression of nicotine-induced CPP	(Ashby et al., 2002)
	0.075 mg/kg/inj	*R*(+) baclofen *S*(-) baclofen	0, 0.625, 1.25, 2.5 mg/kg, i.p.	Mice Long–Evans rats	Self-administration (FR1 schedule)	↓ nicotine self-administration by *R*,*S* baclofen (1.25, 2.5 mg/kg) in mice ↓ nicotine self-administration by *R*,*S* baclofen (5 mg/kg) in rats	(Fattore et al., 2002)

0.01, 0.03 mg/kg/inj	GVG	0, 75, 150, 300 mg/kg, i.p.	Wistar rats	Self-administration (FR5 schedule)	↓ nicotine (0.03 mg/kg) and food self-administration by the two highest doses of GVG ↓ nicotine (0.01 mg/kg) self-administration by the highest dose of GVG	(Paterson & Markou, 2002)
0.03 mg/kg/inj	Baclofen CGP44532	0.9, 1.6, 2.8 mg/kg, i.p. 0, 0.125, 0.5, 1 mg/kg, s.c.	Wistar rats	Self-administration (FR5 and PR schedules) Cue-induced reinstatement	↓ nicotine self-administration by both compounds (dose-dependent effect) Baclofen also decreased food-maintained responding ↓ breakpoints for both nicotine and food self-administration by CGP44532	(Paterson et al., 2004)
0.03 mg/kg/inj	CGP44532	0, 0.25, 0.5 mg/kg, s.c. (acute or repeated administration for 14 days)	Wistar rats	Self-administration (FR5 and PR schedules)	↓ nicotine self-administration in the first 7 days (0.25 mg/kg) ↓ nicotine and food-maintained responding (0.5 mg/kg) ↓ cue-induced reinstatement of nicotine-seeking behavior (0.125, 0.25 mg/kg)	(Paterson et al., 2005b)

(Continued)

TABLE I (*Continued*)

Drug	*Drug Dose*	*GABA$_B$ compound*	*Dose/Dose range*	*Species*	*Method*	*Effects of GABA$_B$ receptor compounds on the rewarding/reinforcing effects of drugs of abuse*	*References*
	3.16 mg/kg/day, free base, s.c.	GVG CGP44532	0, 75, 150, 300 mg/kg, i.p. 0, 065, 0.125, 0.25, 0.5 mg/kg, s.c.	Wistar rats	ICSS	No effect of GVG or CGP44532 on brain reward thresholds in chronically saline- or nicotine-treated rats ↑ thresholds by the highest dose of each drug ↑ thresholds by intra-VTA CGP44532 in both saline- and nicotine-treated groups	(Paterson et al., 2005a)
	0.06 mg/kg, s.c.	GS39783	0, 10, 30, 100 mg/kg, p.o.	Wistar rats	CPP	GS39783 (30–100 mg/kg) blocked the development, but not expression, of nicotine-induced CPP	(Mombereau et al., 2007)
	0.4 mg/kg, s.c.	Baclofen	0, 0.3, 1, 3 mg/kg, i.p.	Sprague–Dawley rats	CPP	Baclofen blocked the expression of nicotine-induced CPP at the highest dose (3 mg/kg)	(Le Foll et al., 2008)

	0.03 mg/kg/inj	CGP7930	0, 5, 10, 30 mg/kg, i.p.	Wistar rats	Self-administration (FR5 and PR schedules)	↓ nicotine self-administration (CGP7930, GS39783, CGP44532, BHF177)	(Paterson et al., 2008)
		GS39783	0, 10, 20, 40 mg/kg, p.o.		ICSS	Co-administration of GS39783 and CGP44532 had additive effects	
		CGP44532	0, 0.125, 0.25 mg/kg, s.c. (FR5); 0, 0.125, 0.25, 0.375, 0.5 mg/kg, s.c. (ICSS)				
		BHF177	0, 10, 20, 40 mg/kg, p.o. (FR5 and PR); 0, 3.75, 7.5, 15, 30 mg/kg, i.p. (ICSS)			↓ breakpoints for nicotine self-administration (BHF177) Nicotine-induced enhancement of brain reward function was blocked by BHF177 and CGP44532	
	30 μg/kg/inj Priming dose: 0.15 mg/kg, i.p.	Baclofen	0,0.612, 1.25, 2.5 mg/kg, i.p.	Sprague–Dawley rats C57/BL6 mice	Drug-induced reinstatement CPP	Baclofen blocked nicotine-induced (0.15 mg/kg) reinstatement of nicotine-seeking behavior in rats (dose-dependent effect) Baclofen also abolished nicotine-induced (0.3 mg/kg) reinstatement of CPP in mice	(Fattore et al., 2009)
Ethanol							
	12%, v/v	Balofen	3 mg/kg, i.p. for 14 days	Long–Evans rats	Self-administration (two-bottle choice)	↓ ethanol intake	(Daoust et al., 1987)

(Continued)

TABLE I (*Continued*)

Drug	*Drug Dose*	*GABA$_B$ compound*	*Dose/Dose range*	*Species*	*Method*	*Effects of GABA$_B$ receptor compounds on the rewarding/reinforcing effects of drugs of abuse*	*References*
	20%, v/v, i.g.	Baclofen	0, 10, 20, 40 mg/kg, i.p. (acute) 0, 2.5, 5, 10 mg/kg, i.p. (chronic)	Wistar rats sP rats	Self-administration (two-bottle choice)	Baclofen decreased the intensity of ethanol withdrawal (dose-dependent effect) Baclofen selectively and dose-dependently reduced voluntary ethanol intake	(Colombo et al., 2000)
	10%, v/v	Baclofen CGP44532	0, 1, 3 mg/kg, i.p. 0, 0.1, 0.3, 1 mg/kg, i.p.	sP rats	Self-administration (FR4, two-bottle choice)	Repeated administration of baclofen or CGP44532 suppressed the acquisition of alcohol-drinking behavior	(Colombo et al., 2002)
	10%, v/v	Baclofen	0, 1.8, 3.2, 5.6 mg/kg, i.p.	Long–Evans rats	Self-administration (FR1, two-bottle choice)	↓ alcohol and sucrose self-administration similarly (dose-dependent effect)	(Anstrom et al., 2003)
	10%, v/v	Baclofen	0, 1, 1.7, 3 mg/kg, i.p.	sP rats	Self-administration (FR4, two-bottle choice)	Baclofen (0, 1, 1.7, 3 mg/kg, i.p.) completely suppressed the extra amount of alcohol consumed during the first hour of re-exposure to alcohol after 7 days of deprivation	(Colombo et al., 2003a)

10%, v/v	Baclofen	0, 1, 2, 3 mg/kg, i.p.	sP rats	Self-administration (FR4, two-bottle choice)	All doses of baclofen produced a marked suppression of extinction responding for alcohol	(Colombo et al., 2003b)
10%, v/v	Baclofen SKF97541	0, 1, 3, 10, 17 mg/kg, i.p. 0, 0.01, 0.03, 0.1, 0.3, 1 mg/kg, i.p.	C57BL/6J mice	Self-administration (FR1 schedule)	↓ ethanol self-administration by both baclofen (10 mg/kg) and SKF97541 (0.3 mg/kg)	(Besheer et al., 2004)
10%, v/v	Baclofen	0, 0.5, 1 mg/kg, i.p.	sP rats	Self-administration (FR4, two-bottle choice)	No effect of baclofen on alcohol intake Baclofen dose-dependently suppressed morphine- (1 mg/kg) and WIN 55,212-2- (2 mg/kg) induced increase in alcohol drinking	(Colombo et al., 2004)
2 g/kg, i.p.	Baclofen	0, 25, 50 ng/kg, intra-VTA	DBA/2J mice	CPP	↓ ethanol-induced CPP	(Bechtholt & Cunningham, 2005)
15%, v/v	Baclofen	0, 1.7, 3 mg/kg, i.p.	sP rats	Self-administration (FR4, two-bottle choice)	↓ ethanol- and sucrose-reinforced responding	(Maccioni et al., 2005)
10%, v/v	CGP7930 GS39783	0, 25, 50, 100 mg/kg, i.p. 0, 6.25, 12.5, 25 mg/kg, i.p.	sP rats	Self-administration (FR4, two-bottle choice)	Both CGP7930 and GS39783 dose-dependently suppressed the acquisition of alcohol drinking behavior ↓ alcohol intake only at the highest dose tested	(Orru et al., 2005)

(*Continued*)

TABLE I *(Continued)*

Drug	*Drug Dose*	*$GABA_B$ compound*	*Dose/Dose range*	*Species*	*Method*	*Effects of $GABA_B$ receptor compounds on the rewarding/reinforcing effects of drugs of abuse*	*References*
	10%, 20%, 30% v/v	Baclofen	0, 1, 2, 3 mg/kg, i.p.	sP rats	Self-administration (FR4, two-bottle choice)	Baclofen suppressed both aspects of the alcohol deprivation effect (i.e., the extra intake of alcohol and the selection of the highest alcohol concentration solution)	(Colombo et al., 2006)
	10%, v/v	CGP7930 Baclofen	0, 10, 20 mg/kg, i.p. 0, 2, 3 mg/kg, i.p.	Inbred alcohol-preferring rats	Self-administration (FR3, two-bottle choice)	↓ ethanol self-administration	(Liang et al., 2006)
	10%, v/v	Baclofen	0, 0.5, 1, 2, 4 mg/kg, i.p. (FR1) 0, 2 mg/kg, i.p. (PR)	Wistar rats	Self-administration (FR1 and PR schedules)	↓ ethanol self-administration in both dependent and nondependent rats (dose-dependent effect) under the FR1 schedule ↓ breakpoints for ethanol self-administration in both groups	(Walker & Koob, 2007)
	15%, v/v	GS39783	0, 25, 50, 100 mg/kg, i.p.	sP rats	Self-administration (FR4, two-bottle choice)	↓ ethanol self-administration (dose-dependent effect)	(Maccioni et al., 2007)

	15%, v/v	Baclofen	0, 3 mg/kg, i.p.	sP rats	Self-administration (FR4, two-bottle choice) Cue-induced reinstatement	Baclofen attenuated cue-induced reinstatement of alcohol-seeking behavior	(Maccioni et al., 2008a)
	15%, v/v	GS39783 Baclofen	0, 25, 50, 100 mg/kg, i.p. 0, 1, 3 mg/kg, i.p.	sP rats	Self-administration (PR schedule, two-bottle choice)	↓ breakpoints for ethanol self-administration (dose-dependent effect) by GS39783 Baclofen also reduced the breakpoints for sucrose self-administration	(Maccioni et al., 2008b)
	15%, v/v	GHB	0, 25, 50, 100 mg/kg, i.p.	sP rats	Self-administration (PR schedule, two-bottle choice, single-session extinction responding)	↓ breakpoints and single-session extinction responding for alcohol	(Maccioni et al., 2008c)
	15%, v/v	BHF177	0, 12.5, 25, 50 mg/kg, i.g.	sP rats	Self-administration (FR4 and PR schedules, two-bottle choice)	↓ ethanol self-administration (25 and 50 mg/kg) ↓ breakpoints for ethanol self-administration (50 mg/kg)	(Maccioni et al., 2009)
	15%, v/v	*rac*-BHFF	0, 50, 100, 200 mg/kg, i.g.	sP rats	Self-administration (FR4, two-bottle choice)	↓ ethanol self-administration (dose-dependent effect)	(Maccioni et al., 2010)
Opiates							
Heroin	0.06 mg/kg/inj	Baclofen	0, 0.5, 1 mg/kg/inj	Sprague–Dawley rats	Self-administration (FR1 schedule)	↓ acquisition and maintenance of heroin self-administration	(Xi & Stein, 1999)

(*Continued*)

TABLE I (*Continued*)

Drug	*Drug Dose*	*$GABA_B$ compound*	*Dose/Dose range*	*Species*	*Method*	*Effects of $GABA_B$ receptor compounds on the rewarding/reinforcing effects of drugs of abuse*	*References*
	0.06 mg/kg/inj	GVG	0, 20, 50 μg/kg, intra-VTA, intra-ventral pallidum, i.c.v.	Sprague–Dawley rats	Self-administration (FR1 schedule)	↓ heroin self-administration by i.c.v., intra-VTA, and intra-ventral pallidum, but not intra-NAc GVG (dose-dependent effect), an effect blocked by 2-hydroxysaclofen	(Xi & Stein, 2000)
		2-OH-saclofen	2 mg/kg, i.v.			↓ acquisition of heroin self-administration by inta-VTA GVG in drug-naive rats	
	1.5 mg/kg, i.p.	GVG	0, 300 mg/kg, i.p.	Sprague–Dawley rats	CPP	GVG attenuated the expression of heroin-induced CPP	(Paul et al., 2001)
	30 μg/kg/inj Priming dose: 0.25 mg/kg, s.c.	Baclofen	0, 0.625, 1.25, 2.5 mg/kg, i.p.	Sprague–Dawley rats	Drug-induced reinstatement	Baclofen blocked heroin-induced reinstatement of heroin-seeking behavior (dose-dependent effect)	(Xi & Stein, 1999)
Morphine	8 mg/kg, s.c.	Baclofen	0.1–1 nmol/side, intra-VTA	Sprague–Dawley rats	CPP	Baclofen suppressed morphine-induced CPP	(Tsuji et al., 1996)
	0, 5 mg/kg, s.c.	Baclofen	0, 10 mg/kg, i.p.	C57BL/6 mice	CPP	Baclofen blocked the acquisition of morphine-induced CPP	(Kaplan et al., 2003)

	0, 1, 3 mg/kg, s.c.	Baclofen	0, 1, 2 μg/kg, intra-dorsal hippocampus	Wistar rats	CPP	Baclofen decreased the acquisition of morphine-induced CPP	(Zarrindast et al., 2006)
		Phaclofen	0, 2, 3 μg/kg			Phaclofen increased the acquisition of morphine-induced CPP	
	0.25, 0.5, 1 mg/kg, s.c.	Baclofen	0, 1.5, 6 and 12 μg/kg/inj, intra-VTA	Female Wistar rats	CPP	Both baclofen (1.5 and 12 μg/kg) and CGP35348 (12 μg/kg) reduced the expression of morphine-induced CPP, whereas the 6 μg/kg dose of baclofen increased it	(Sahraei et al., 2009)
		CGP35348	0, 1.5, 6 and 12 μg/kg/inj			Both compounds significantly reduced the acquisition of morphine-induced CPP in morphine-sensitized animals	
Morphine **Fentanyl**	8 mg/kg, s.c. 56 μg/kg, s.c.	Baclofen	0, 1.5, 3 mg/kg, s.c.	Sprague–Dawley rats	CPP	Baclofen blocked both morphine- and fentanyl-induced CPP	(Suzuki et al., 2005)

↑ elevation/increase; ↓ lowering/decrease; s.c., subcutaneous; inj., injection; i.c.v., intracerebroventricular; i.g., intragastric; i.m., intramuscular; i.p., intraperitoneal; i.v., intravenous; p.o., *per os*; see text for more abbreviations.

the level of both the VTA and NAc (Dewey et al., 1992; Heimer et al., 1991; Kalivas, 1993; Kalivas et al., 1992). There are GABA inhibitory afferents to dopaminergic VTA neurons (e.g., (Walaas & Fonnum, 1980; Yim & Mogenson, 1980a, 1980b), inhibitory GABA interneurons within the VTA, and medium spiny GABA neurons in the NAc that also inhibit mesolimbic dopamine release (for review, see (Kalivas, 1993; Kalivas et al., 1992).

Other non-dopaminergic systems also play a key role in mediating reward processes (Hernandez et al., 2006; Nestler, 2005). For example, the dorsal and median raphe nuclei (DR and MR, respectively) contain serotonergic neurons projecting to the forebrain. These neurons are involved in inhibitory control over reward processes. These nuclei also contain $GABA_B$ receptors, which may be involved in reward processes (Wirtshafter & Sheppard, 2001). Furthermore, the dorsal striatum does not appear to play a major role in the reinforcing effects of addictive drugs. However, it seems to play an important role in the development and maintenance of compulsive drug-seeking behavior (Everitt et al., 2008; Yin et al., 2005). Specifically, there is activation of dopaminergic neurons in the ventral striatum after acute administration of drugs of abuse, decreased glutamatergic activity during withdrawal, and increased glutamate release during drug-primed and cue-induced drug seeking. These actions of drugs of abuse affect the functioning of the dorsal striatum where compulsivity occurs and the frontal cortex where deficits in executive function and poor decision-making contribute to increased incentive motivation for drugs over natural reinforcers (Koob & Volkow, 2010). Moreover, the extended amygdala, including the central nucleus of the amygdala, bed nucleus of the stria terminalis, and NAc shell, and its projections to the hypothalamus and brainstem are hypothesized to be a substrate for negative reinforcement processes. Considering that mesolimbic dopaminergic and dopamine-independent transmission contributes to aspects of reward anticipation, reward expectation, and the rewarding properties of drugs of abuse (e.g., cocaine, nicotine, ethanol, and opiates; for review, see Feltenstein & See, 2008; Hernandez et al., 2006; Kalivas & Volkow, 2005; Nestler, 2005; Weiss, 2005) and that the GABA system is the major inhibitory system in the brain altering the function of many other neurotransmitter systems, including those critically involved in reward processes, GABAergic pharmacological manipulations modulate the reinforcing and motivational properties of a variety of drugs of abuse. Here, we review preclinical studies that support the hypothesis that alterations in GABA neurotransmission, particularly through the $GABA_B$ receptor, affect brain reward system function and, consequently, the rewarding and motivational effects of drugs of abuse. Table I summarizes the effects of $GABA_B$ receptor compounds in reward processes and in the reinforcing effects of drugs of abuse. Table II provides descriptions of the main behavioral procedures used to study the reinforcing effects of addictive drugs. Table III presents the chemical name and description of all of the $GABA_B$ compounds referenced in this review.

TABLE II Definitions of behavioral procedures

Behavioral procedure	*Description*
Intracranial self-stimulation (ICSS)	A behavioral operant paradigm in which experimental animals learn to deliver brief electrical pulses into specific regions of their own brains that are considered to be part of the brain's reward pathways mediating both natural and ICSS reward. Acute administration of most drugs of abuse, including cocaine, amphetamine, nicotine, morphine, and heroin, lowers ICSS thresholds in experimental animals, indicating a reward-facilitating effect. By contrast, withdrawal from chronic administration of these compounds induces elevations in ICSS thresholds, indicating an anhedonic state that resembles the negative affective state of the drug withdrawal syndrome experienced by humans. Lowering of the ICSS threshold indicates an increase in the reward value of the stimulation because less electrical stimulation is required for the subject to perceive the stimulation as rewarding. Conversely, elevations in thresholds indicate a decrease in the reward value of the self-stimulation because higher frequencies or current-intensity values are required before the subject perceives the stimulation as rewarding.
Self-administration	A procedure based on operant conditioning, in which animals learn to make a response to self-administer a reinforcer (e.g., a drug of abuse) after the completion of the reinforcement schedule requirement. A reinforcer is an event that follows a response and increases the probability of a response to reoccur.
Fixed-ratio schedule of reinforcement	Under this schedule, the reinforcer is delivered every time a predetermined number of responses is completed, and the delivery of a reinforcer is usually followed by a timeout period in self-administration studies to prevent the subjects from overdosing (e.g., FR1 or continuous reinforcement schedule, FR4, FR5, etc.). Data obtained from a fixed-ratio schedule provide a measure of drug intake and reinforcement efficacy.
Progressive-ratio schedule of reinforcement	Under this schedule, the response requirements are progressively increased after the delivery of each reinforcer, according to a predetermined progression. For example, the number of responses required to earn a nicotine infusion or food pellet on the progressive-ratio can be determined by the exponential progression $[5e^{(0.25 \times \text{(infusion number +3)}} - 5]$, with the first two values replaced by 5 and 10, so that the response requirements for successive reinforcers are 5, 10, 17, 24, 32, 42, 56, 73, 95, 124, 161, 208, etc. Breakpoints in this schedule are typically defined as the highest response rate achieved to obtain a single reinforcer before an animal fails to complete the next ratio requirement within a predetermined time period (e.g., 60 min). Data obtained from a progressive-ratio schedule provide a measure of the motivation (i.e., incentive value) to obtain a reinforcer.

(Continued)

TABLE II (*Continued*)

Behavioral procedure	*Description*
Discrete-trials schedule of reinforcement	A procedure in which only a single injection of the drug is delivered during individual trials. The intertrial interval (ITI) can be adjusted to manipulate the influence of one injection on subsequent trials. When short ITIs are used, animals continuously self-administer a drug for long periods of time (hours or even days). When long ITIs are used, a regular circadian pattern of self-administration occurs (i.e., periods of abstinence during the light phase of the cycle alternate with periods of self-administration during the dark phase). Data obtained from a discrete-trials schedule provide a measure of the motivation to initiate drug-taking behavior.
Conditioned place preference	A procedure in which drug administration is paired with a specific environment and vehicle administration with a different environment. Subsequently, in a drug-free state, animals are tested for their preference for the drug-paired or non-paired environment.
Reinstatement procedure (cue-, context-, drug-, or stress-induced reinstatement of drug-seeking behavior)	A procedure used to study cue-, context-, drug-, or stress-induced reinstatement of drug seeking, hypothesized to be a putative model of relapse to drug seeking in humans. Animals learn to self-administer a drug for a period of time. Drug-reinforced lever responding is then extinguished, and reinstatement of drug-seeking behavior is subsequently triggered by a priming injection of a compound (drug-induced), a cue (or context) previously associated with the self-administration of the drug (cue- or context- induced), or a stressor (stress-induced reinstatement).

II. $GABA_B$ Receptor Agonists, $GABA_B$ Receptor Positive Modulators, and Reward: Effects on the Rewarding Properties of Food, Intracranial Self-Stimulation, and Drugs of Abuse

GABA acts through $GABA_A$, $GABA_B$, **and** $GABA_C$ receptors. $GABA_B$ receptors consist of two subunits, $GABA_{B1}$ and $GABA_{B2}$. The $GABA_{B1}$ subunit exists in two isoforms, $GABA_{B1a}$ and $GABA_{B1b}$. Either of these two isoforms can combine with the $GABA_{B2}$ subunit to form a functional receptor (Mohler & Fritschy, 1999; White et al., 1998). These two types of receptors are differentially expressed presynaptically (both as heteroreceptors and autoreceptors) and postsynaptically on GABA and glutamate-releasing neurons (for details, see Chapter 3 in this volume).

Numerous studies have investigated the effects of compounds acting on the $GABA_B$ receptor in reward processes *per se* (e.g., Fenton & Liebman,

TABLE III Definitions of GABA$_B$ receptor compounds used in the text

GABA$_B$ receptor active compound	*Chemical name/structure*	*Description*
GVG (γ-vinyl GABA, vigabatrin, antiepileptic drug)	(*R*,*S*)-4-aminohex-5-enoic acid	Indirect GABA$_B$ receptor agonist; irreversible inhibitor of GABA transaminase
Baclofen (Kemstro, Lioresal; treatment for spasticity)	(*R*,*S*)-4-amino-3-(4-chlorophenyl) butanoic acid	GABA$_B$ receptor agonist
CGP44532	(3-amino-2[*S*]-hydroxypropyl)-methylphosphinic acid	GABA$_B$ receptor agonist
CGP7930	2,6-Di-*tert*-butyl-4-(3-hydroxy-2,2-dimethyl-propyl)-phenol	GABA$_B$ receptor positive modulator
BHF177	*N*-[(1*R*,2*R*,4*S*)-bicyclo[2.2.1]hept-2-yl]-2-methyl-5-[4-(trifluoromethyl) phenyl]-4-pyrimidinamine	GABA$_B$ receptor positive modulator
GS39783	*N*,*N'*-dicyclopentyl-2-methylsulfanyl-5-nitro-pyrimidine-4,6-diamine	GABA$_B$ receptor positive modulator
SKF97541	3-aminopropyl(methyl)phoshinic acid	GABA$_B$ receptor agonist
SCH50911	(2*S*)-(+)-5,5-dimethyl-2-morpholineacetic acid	GABA$_B$ receptor antagonist
GHB (Xyrem; treatment for cataplexy and narcolepsy)	γ-hydroxybutyric acid	Weak GABA$_B$ receptor agonist; GHB receptor (GPR172A) agonist
ACC	1*R*,4*S*-4-amino-cyclopent-2-ene-carboxylic acid	Reversible inhibitor of GABA transaminase
rac-BHFF	(*R*,*S*)-5,7-di-*tert*-butyl-3-hydroxy-3-trifluoromethyl-3*H*-benzofuran-2-one	GABA$_B$ receptor positive modulator
CGP35348	3-Aminopropyl(diethoxymethyl) phosphinic acid	GABA$_B$ receptor antagonist
CGP56433A	[3-{1-(*S*)-[{3-(cyclohexylmethyl) hydroxy phosphinyl}-2-(*S*) hydroxypropyl]amino}ethyl] benzoic acid	GABA$_B$ receptor antagonist
CGP51176	3-amino-2-(*R*)-hydroxypropyl-cyclohexylmethyl-phosphinic acid	GABA$_B$ receptor antagonist
Tiagabine (Gabitril; antiepileptic drug)	(*R*)-1-[4,4-bis(3-methylthiophen-2-yl)but-3-enyl] piperidine-3-carboxylic acid	GABA reuptake inhibitor

(Continued)

TABLE III (*Continued*)

$GABA_B$ receptor active compound	*Chemical name/structure*	*Description*
Gabapentin (Neurontin; antiepileptic drug, treatment for neuralgia)	2-[1-(aminomethyl)cyclohexyl]acetic acid	Cyclic GABA analog
2-hydroxysaclofen (2-OH-saclofen)	(*R*,*S*)-3-Amino-2-(4-chlorophenyl)-2-hydroxypropyl-sulfonic acid	$GABA_B$ receptor antagonist
Phaclofen (phosphonobaclofen)	[3-amino-2-(4-chlorophenyl)propyl] phosphonic acid	$GABA_B$ receptor antagonist

1982; Macey et al., 2001; Panagis & Kastellakis, 2002; Shin & Ikemoto, 2010; Willick & Kokkinidis, 1995). The results of these studies showed that $GABA_B$ receptors differentially affect reward processes in different brain regions, and $GABA_B$ receptor activation has different effects in behavioral procedures that assess reward processes.

The intracranial self-stimulation (ICSS) procedure is a behavioral operant paradigm in which experimental animals learn to deliver brief electrical pulses into specific regions of their own brains that are considered to be part of the brain's reward pathways mediating both natural and ICSS reward (Table II). Acute administration of most drugs of abuse, including cocaine, D-amphetamine, nicotine, morphine, and heroin, lower ICSS thresholds in experimental animals, indicating a reward-facilitating effect (Vlachou & Markou, 2010). By contrast, withdrawal from chronic administration of these compounds induces elevations in ICSS thresholds, indicating an anhedonic state that resembles the negative affective state of the early drug withdrawal syndrome experienced by humans.

In the ICSS procedure, the $GABA_B$ receptor agonist baclofen negatively regulated brain reward function, reflected by elevated reward thresholds for medial forebrain bundle stimulation (Fenton & Liebman, 1982; Macey et al., 2001; Panagis & Kastellakis, 2002; Willick & Kokkinidis, 1995). Similarly, intra-VTA injections of baclofen also reduced the reward value of the stimulation, reflected in threshold elevations, when the electrode was placed in the lateral hypothalamus or ventral pallidum (Panagis & Kastellakis, 2002; Willick & Kokkinidis, 1995). Interestingly, both agonists and antagonists induce a reward decrement in the ICSS procedure. The $GABA_B$ receptor agonist (3-amino-2[*S*]-hydroxypropyl)-methylphosphinic acid (CGP44532) and the $GABA_B$ receptor antagonists [3-{1-(*S*)-[{3-(cyclohexylmethyl)hydroxy phosphinyl}-2-(*S*) hydroxypropyl]amino}ethyl]benzoic acid (CGP56433A) and 3-amino-2-(*R*)-hydroxypropyl-cyclohexylmethyl-phosphinic acid (CGP51176) elevated ICSS thresholds, reflecting diminished reward, when administered separately, whereas co-administration of either of the two antagonists with the agonist induced an additive effect on threshold elevations rather than blockade of the effects of the agonist (Macey et al., 2001). These findings are unexpected and reveal a complex pattern of action for the GABA$_B$ compounds, possibly through differential effects on pre- versus postsynaptic GABA$_B$ receptors.

In a recent study, Shin and Ikemoto (Shin & Ikemoto, 2010) examined the effects of direct injections of baclofen into the MR and DR nuclei in reward processes, assessed by the self-administration and conditioned place preference (CPP) procedures. As mentioned previously, the raphe nuclei contain GABA$_B$ receptors that negatively modulate their function. The results indicated that rats trained under a fixed-ratio 1 (FR1) schedule of reinforcement (Table II) more readily self-administered baclofen (0–2.5 mM) in the MR than in the DR. By contrast, injections of baclofen (0, 0.1, and 0.5 mM) in the DR were more effective than injections into the MR at inducing CPP. These effects of baclofen were attenuated by co-administration of the GABA$_B$ receptor antagonist (2*S*)-(+)-5,5-dimethyl-2-morpholineacetic acid (SCH50911) or the dopamine D$_1$ receptor antagonist 7-chloro-3-methyl-1-phenyl-1,2,4,5-tetrahydro-3-benzazepin-8-ol (SCH23390), indicating that GABA$_B$ receptors located on MR and DR serotonergic neurons are involved in reward processes. GABA$_B$ receptors can be found on GABAergic afferents, which inhibit GABAergic inputs to serotonergic neurons by disinhibiting them. Thus, baclofen administration in MR and DR nuclei could differentially affect reward processes by exerting opposing effects on serotonergic neurons (Shin & Ikemoto, 2010).

The aforementioned results clearly indicate a critical role of the GABA system, particularly the GABA$_B$ receptor, in reward processes. Partly because of these strong indications about the role of the GABA system in reinforcement processes, multiple studies in laboratory animals have explored the effects of GABA$_B$ compounds on the rewarding effects of drugs of abuse from a variety of pharmacological classes. These findings are reviewed below, organized by drug of abuse.

A. The Psychomotor Stimulant Cocaine

1. GABA$_B$ Receptor Agonists

γ-vinyl GABA (GVG), also referred to as vigabatrin, is an indirect GABA receptor agonist, whose administration leads to increased GABA levels (Jung et al., 1977) (Table III). GVG is not an addictive drug and does not appear to produce tolerance or show withdrawal effects (Takada & Yanagita, 1997). GVG is used clinically for the treatment of partial complex seizures. Systemic injections of GVG (i) decreased cocaine self-administration in rats (Kushner et al., 1999), (ii) decreased cocaine self-administration in a preliminary open clinical trial in human cocaine addicts (Ling et al., 1998), (iii) attenuated cocaine-induced lowering of brain reward thresholds (i.e., attenuated the reward-enhancing effects of cocaine on ICSS reward; Kushner et al., 1997), and (iv) abolished the expression and acquisition of cocaine- and nicotine-induced CPP (Dewey et al., 1998, 1999). However, although no toxicity was observed after administration of GVG or co-administration of high doses of GVG and cocaine in rats (Molina et al., 1999), GVG has severe side-effects, including visual field constriction and blurring after chronic therapy, which limit its clinical use (Ascaso et al., 2003; Schmitz et al., 2002; Wohlrab et al., 1999; but see Brodie, 2005).

Baclofen, a direct $GABA_B$ receptor agonist with fewer side effects than GVG (Table III), decreases cocaine self-administration in rats under several different schedules of reinforcement (Brebner et al., 2000a, 2000b; Campbell et al., 1999; Roberts et al., 1996; Shoaib et al., 1998; for review, see Roberts & Brebner, 2000; Roberts, 2005). Specifically, baclofen decreased cocaine self-administration under both FR1 and progressive-ratio (PR) schedules of reinforcement (Brebner et al., 2000a) (Table II), suggesting that baclofen decreases both the rewarding and incentive properties of cocaine (Fig. 1A). Importantly, in this study, the effect of baclofen depended on the unit injection dose of cocaine (Brebner et al., 2000a). Baclofen had no effect on cocaine self-administration with high unit injection doses (1.5 mg/kg/injection). In another study, baclofen dose-dependently decreased breakpoints for cocaine self-administration under a PR schedule, indicating that baclofen decreases the motivation for cocaine (Roberts et al., 1996) (Fig. 2A). Importantly, baclofen had small effects on food-maintained responding and only at the highest dose tested, indicating that cocaine reinforcement is more readily disrupted by baclofen administration than responding for food. Interestingly, when a discrete-trials schedule of reinforcement was used (Table II), acute baclofen treatment suppressed cocaine intake in two groups of rats self-administering cocaine for 4 h under different 12 h light/dark cycles (10:00 AM–10:00 PM or 3:00 AM–3:00 PM; Roberts & Andrews, 1997). Baclofen also blocked cocaine- and cue-induced reinstatement of cocaine-seeking behavior in both rats (Campbell et al., 1999; Filip & Frankowska, 2007) and baboons (Weerts et al., 2007), indicating potential anti-relapse properties of baclofen in psychostimulant addiction. The reinstatement procedure is an animal model commonly used to study addictive drug-seeking behavior (Epstein et al., 2006; Schmidt et al., 2005; See, 2005) (Table II). Furthermore, the $GABA_B$ receptor agonist CGP44532 dose-dependently decreased cocaine self-administration on a PR schedule of reinforcement (Brebner et al., 1999). Additionally, CGP44532 reduced cocaine-induced enhancement of brain stimulation reward (Dobrovitsky et al., 2002).

Various other GABA-mimetic drugs had diverse effects on cocaine self-administration (Filip et al., 2007b). Both GVG and the GABA reuptake inhibitor tiagabine (Table III) significantly decreased cocaine-maintained responding (0.5 mg/kg/injection) on an FR5 schedule of reinforcement. GVG blocked reinstatement of cocaine seeking after a cocaine priming injection, whereas tiagabine non-significantly attenuated cocaine priming-induced cocaine-seeking behavior. However, gabapentin (Table III) did not alter cocaine self-administration or cocaine-induced reinstatement of cocaine-seeking behavior (Filip et al., 2007b). Moreover, administration of the $GABA_A$ receptor agonists muscimol, baclofen, and GVG, among other $GABA_B$ receptor agonists and $GABA_A$ receptor agonists and modulators, decreased cocaine self-administration, although at doses that also affected food-maintained responding (Barrett et al., 2005).

Furthermore, microinjections of baclofen or CGP44532 directly into the NAc shell, VTA, or peduncular pontine nucleus, but not caudate-putamen (Brebner et al., 2000b; Corrigall et al., 2000, 2001; Shoaib et al., 1998), decreased the reinforcing effects of cocaine, heroin, alcohol, and nicotine (Colombo et al., 2002; Fattore et al., 2002; Paterson & Markou, 2002; Xi & Stein, 1999) and cocaine-seeking behavior in a second-order schedule of reinforcement which assesses the motivation to self-administer drugs (Di Ciano & Everitt, 2002). By contrast, 4,5,6,7-tetrahydroisoxazolo(5,4-c)pyridin-3-ol hydrochloride, a $GABA_A$ receptor agonist, or the inactive (+) baclofen enantiomer did not modify cocaine (Shoaib et al., 1998) or nicotine (Paterson & Markou, 2002) self-administration. Consistent with the above data, GVG administration abolished the dopamine increase in the NAc triggered by exposure of animals to environmental cues previously paired with cocaine administration (Gerasimov et al., 2001).

These effects of $GABA_B$ receptor agonists on drug self-administration in experimental animals are consistent with clinical studies showing baclofen to be an effective treatment in cocaine-abstinent humans (Ling et al., 1998; Shoptaw et al., 2003) and in preventing cue-induced craving and concurrent activation of relevant brain areas in cocaine-abstinent humans (Brebner et al., 2002a; but see Kahn et al., 2009). In this latter study, individuals with severe cocaine dependence received either placebo or 60 mg/kg/day baclofen for 8 weeks. The results showed no significant differences between the two groups on self-reported cocaine use and urine benzoylecgonine levels. The discrepancy between the results of this study and other clinical studies using the same dose of baclofen may be attributed to the focus of this study on heavy cocaine-dependent users (Kahn et al., 2009) who may require higher baclofen doses for a therapeutic effect to emerge. The most promising results on the effects of baclofen on cocaine- and cue-induced craving come from brain imaging studies in cocaine patients. These studies showed that patients receiving baclofen twice daily for 7–10 days showed substantial inhibition of cue-induced cocaine craving and no activation of

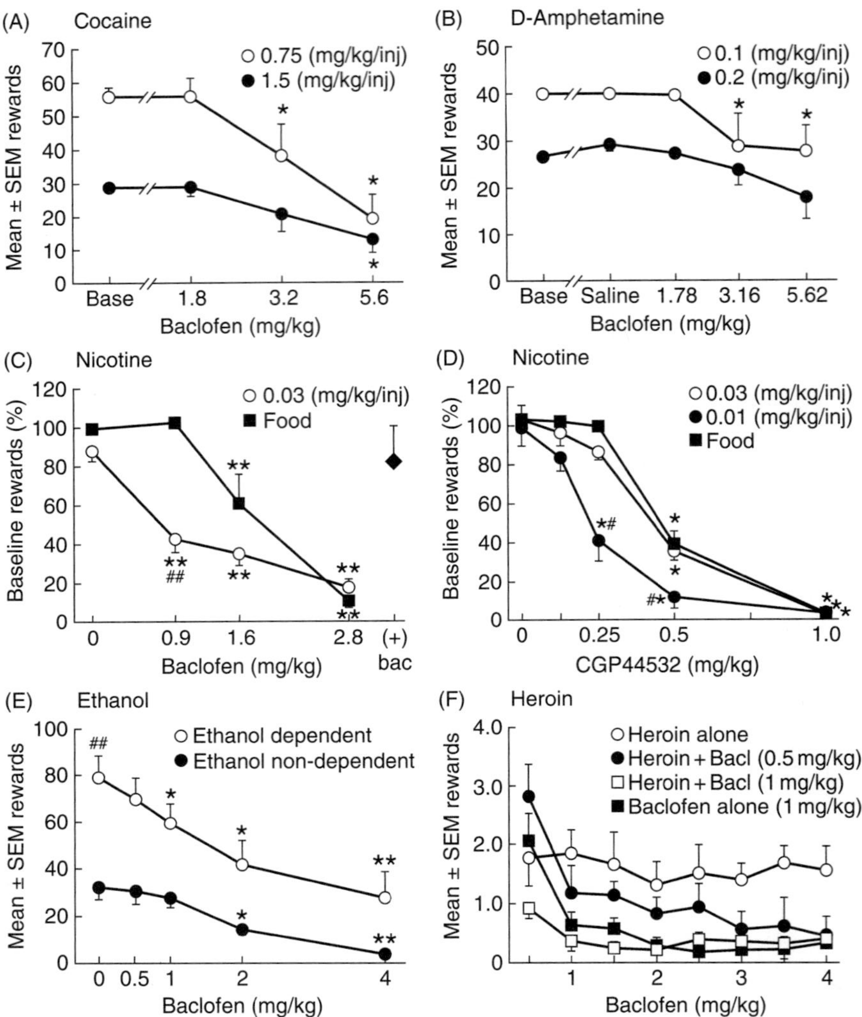

FIGURE 1 Effects of $GABA_B$ receptor agonists on the reinforcing effects of psychoactive drugs. (A) Effect of baclofen (1.8, 3.2, or 5.6 mg/kg, i.p.) on cocaine self-administration under an FR1 schedule of reinforcement in Wistar rats. Data are expressed as mean (± SEM) number of injections self-administered during a 3 h baseline (Base) or test session. Separate groups of animals were trained to respond for either 0.75 or 1.5 mg/kg/injection of cocaine and were pretreated with baclofen 30 min before the session. Self-administration of the lower unit dose (0.75 mg/kg/injection) was significantly affected by both 3.2 and 5.6 mg/kg baclofen ($p < 0.05$), whereas the high unit injection dose (1.5 mg/kg) was significantly reduced only by 5.6 mg/kg baclofen ($p < 0.05$). (B) Effect of baclofen (1.78, 3.16, and 5.62 mg/kg, i.p.) on D-amphetamine self-administration under an FR1 schedule of reinforcement in Sprague–Dawley rats. Data are expressed as mean (± SEM) number of injections self-administered during a 5 h baseline (Base) or test session. Two separate groups of rats were trained to respond for either 0.2 or 0.1 mg/kg/injection of D-amphetamine and were pretreated with baclofen

the limbic anterior cingulate or amygdala during cocaine versus non-drug video presentation (for review, see Brebner et al., 2002a). However, administration of baclofen resulted in side-effects in a number of studies, often attributable to the high doses required to achieve effective or desirable effects (Adams & Lawrence, 2007). Some of these side-effects include sedation, nausea, dizziness, drowsiness, mental confusion, and motor impairment. These side-effects may potentially limit the usefulness of baclofen in the treatment of drug addiction. GABA$_B$ receptor positive modulators, therefore, may provide a more attractive therapeutic approach than direct GABA$_B$ agonists for the treatment of drug dependence.

High-affinity GABA$_B$ receptor antagonists that rapidly cross the blood-brain barrier have also been developed (Froestl et al., 1995b). As expected from the effects of the GABA$_B$ receptor agonists, the GABA$_B$ receptor antagonist CGP56433A had no effect on cocaine self-administration (1.5 mg/kg/injection) under either an FR1 or PR schedule of reinforcement, whereas at the highest dose tested (1.8 mg/kg) it attenuated the effect of baclofen on cocaine, but not heroin, self-administration (Brebner et al., 2002b). Moreover, the selective GABA$_B$ receptor antagonist SCH50911 abolished GVG's attenuation of cocaine-induced increases in NAc dopamine, suggesting that the mechanisms underlying this attenuation are partly,

---

(saline, 1.78, 3.16, and 5.62 mg/kg, i.p.) 30 min before the session. Self-administration of the lower dose of D-amphetamine was significantly affected by both the 3.2 and 5.6 mg/kg doses of baclofen ($*p < 0.05$), whereas baclofen did not affect responding for the high dose of D-amphetamine. (C) Effects of (-)baclofen (0, 0.9, 1.6, and 2.8 mg/kg, i.p.) on nicotine- and food-maintained responding under an FR5 schedule of reinforcement in Wistar rats. Data are expressed as a percentage of baseline (mean $\pm$ SEM) across a range of (-)baclofen (active enantiomer) doses and 0.09 mg/kg (+)baclofen (inactive enantiomer; nicotine group only). $*p < 0.05$, $**p < 0.01$, significant difference from the vehicle pretreatment condition; $^{##}p < 0.01$, significant difference from food-maintained responding. (D) Effects of CGP44532 (0, 0.25, 0.5, and 1 mg/kg, s.c.) on nicotine- and food-maintained responding under an FR5 schedule of reinforcement in Wistar rats. Data are expressed as a percentage of baseline (mean $\pm$ SEM). $*p < 0.05$, $**p < 0.01$, significant difference from the vehicle pretreatment condition; $^{#}p < 0.05$, significant difference from food-maintained responding. (E) Effects of baclofen (0, 0.5, 1, 2, and 4 mg/kg, i.p.) on ethanol administration in nondependent and ethanol-dependent Wistar rats under an FR1 schedule of reinforcement. Data are expressed as mean ($\pm$ SEM) lever-presses for ethanol. $^{##}p < 0.01$, significant difference between control and ethanol-dependent groups; $*p < 0.05$, $**p < 0.01$, compared with the appropriate ethanol nondependent or ethanol-dependent vehicle control. (F) Within-session effects of heroin (0.06 mg/kg) plus baclofen co-administration on heroin self-administration behavior under an FR1 schedule of reinforcement in Sprague–Dawley rats. Data are expressed as mean ($\pm$ SEM) lever-presses for heroin self-administration. Baclofen dose-dependently decreased the acquisition of heroin self-administration but did not support self-administration behavior *per se*. No direct comparisons of effect sizes can be made among the figures presented here because these studies used different methodologies, GABA$_B$ receptor compounds, dose ranges, and data analyses. [Figures taken with permission from: Brebner et al., 2000a (cocaine; Fig. 1A), Brebner et al., 2005 (D-amphetamine; Fig. 1B), Paterson et al., 2004 (nicotine; Fig. 1C and 1D), Walker & Koob, 2007 (ethanol; Fig. 1E), Xi & Stein, 1999 (heroin; Fig. 1F)].

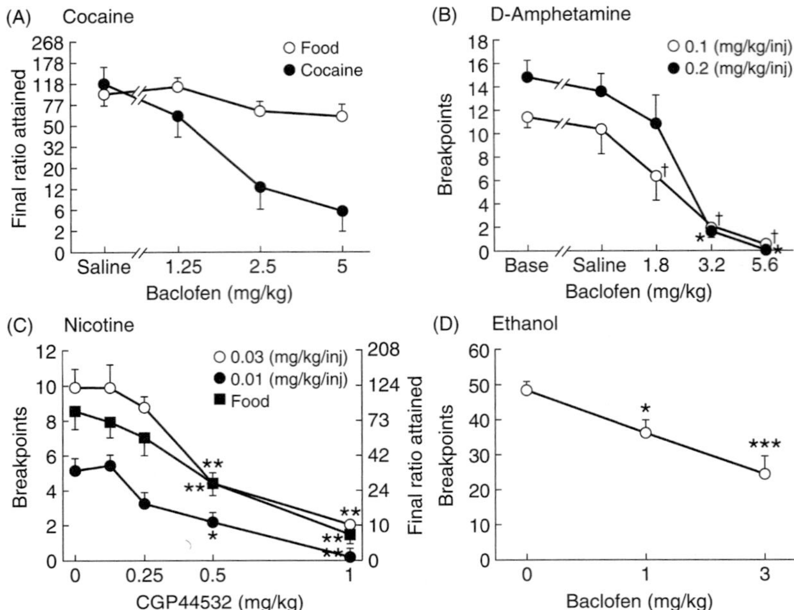

FIGURE 2 Effects of $GABA_B$ receptor agonists on the motivational effects of psychoactive drugs. (A) Effects of baclofen on responding for either cocaine or food reinforcement in Wistar rats. Data are expressed as mean ($\pm$ SEM) breakpoints on a PR schedule of reinforcement. Animals received various doses of baclofen or vehicle 30 min before the test session. Baclofen pretreatment dose-dependently decreased breakpoints for cocaine and food reinforcement, with a relatively small effect on food intake. (B) Effects of baclofen on D-amphetamine self-administration on a PR schedule of reinforcement in Sprague–Dawley rats. Two separate groups of rats were treated with baclofen (saline, 1.8, 3.2, and 5.6 mg/kg, i.p.) 30 min prior to the test session. Data are expressed as mean ($\pm$ SEM) breakpoints on a PR schedule of D-amphetamine reinforcement (0.1 or 0.2 mg/kg/injection) under baseline (Base) conditions or after baclofen administration. Baclofen dose-dependently decreased breakpoints for D-amphetamine. All three doses of baclofen significantly reduced breakpoints in rats responding for the low dose of D-amphetamine ($^{\dagger}p < 0.05$), whereas only the two highest doses of baclofen reduced responding for 0.2 mg/kg D-amphetamine ($*p < 0.05$). (C) Effects of CGP44532 administration on nicotine- and food-maintained responding under a PR schedule of reinforcement in Wistar rats. Data are expressed as mean ($\pm$ SEM) number of infusions/food pellets earned after CGP44532 pretreatment. Ordinates reflect nicotine-reinforced breakpoints. $*p < 0.05$, $**p < 0.01$, significant difference from vehicle pretreatment for each group. (D) Effect of baclofen on breakpoints for alcohol in selectively bred Sardinian alcohol-preferring (sP) rats. Saline and baclofen (1 and 3 mg/kg, i.p.) were injected 30 min before the start of the test session. Each bar represents the mean ($\pm$ SEM) of breakpoints. $*p < 0.05$, $**p < 0.01$, $***p < 0.001$, compared with saline-treated rats. No direct comparisons can be made among the findings presented here because different studies used different methodologies, $GABA_B$ receptor compounds, dose ranges, and data analyses. [Figures taken with permission from: Roberts et al., 1996 (cocaine; Fig. 2A), Brebner et al., 2005 (D-amphetamine; Fig. 2B), Paterson et al., 2004 (nicotine; Fig. 2C), Maccioni et al., 2008b (ethanol; Fig. 2D)].

if not completely, mediated by activation of GABA$_B$ receptors (Ashby et al., 1999).

2. *GABA$_B$ Receptor Positive Allosteric Modulators*

The GABA$_B$ receptor positive modulator 2,6-di-*tert*-butyl-4-(3-hydroxy-2,2-dimethyl-propyl)-phenol (CGP7930) (Urwyler et al., 2001) reduced cocaine self-administration (1.5 mg/kg/injection) under a PR schedule of reinforcement, indicating reduced motivation to self-administer cocaine (Smith et al., 2004) (Fig. 3A). This effect of CGP7930 was more potent than the effect of another GABA$_B$ receptor positive modulator, *N,N′*-dicyclopentyl-2-methylsulfanyl-5-nitro-pyrimidine-4,6-diamine (GS39783) (Urwyler et al., 2003), which only decreased cocaine self-administration at the highest dose tested (Smith et al., 2004) (Fig. 3B). The effects of both CGP7930 and GS39783 were also tested on cocaine self-administration under FR1 and discrete-trials schedules of reinforcement in rats trained to self-administer different doses of cocaine (Smith et al., 2004). Both GABA$_B$ receptor positive modulators decreased responding for the lowest dose of cocaine (0.19 mg/kg) (Fig. 4B) but did not have an effect at higher cocaine doses (0.38–1.5 mg/kg). In the discrete-trials procedure, baclofen, CGP7930, and GS39783 decreased cocaine self-administration, although at different time-points after administration for each compound (Smith et al., 2004). In another recent study, the GABA$_B$ receptor agonists baclofen and 3-aminopropyl(methyl)phoshinic acid (SKF 97541) and the GABA$_B$ receptor positive modulator CGP7930 decreased cocaine self-administration under an FR5 schedule of reinforcement (Fig. 4A), whereas the GABA$_B$ receptor antagonist SCH50911 did not alter cocaine self-administration (Filip et al., 2007a). Baclofen or SKF97541, but not CGP7930 or SCH50911, also attenuated food-maintained responding, indicating a selective effect of the GABA$_B$ receptor positive modulator compared with the GABA$_B$ receptor agonist on self-administration of cocaine versus the natural reinforcer food. The inhibitory effects of the GABA$_B$ receptor agonists and positive modulators were blocked by SCH50911 (Filip et al., 2007a). GS39783 selectively decreased cocaine-, but not food-, maintained responding, whereas after a period of extinction, the GABA$_B$ receptor antagonist SCH50911 also dose-dependently attenuated responding on the previously cocaine-paired lever during both cue- and cocaine priming-induced reinstatement conditions. SCH50911 failed to alter reinstatement of food-seeking behavior. The GABA$_B$ receptor agonists baclofen and SKF97541 attenuated cocaine- and food-seeking behavior, whereas the GABA$_B$ receptor positive modulator CGP7930 reduced cocaine seeking without affecting food-induced reinstatement of food-seeking behavior (Filip & Frankowska, 2007). These results indicate that tonic activation of GABA$_B$ receptors is required for cocaine-seeking behavior in rats.

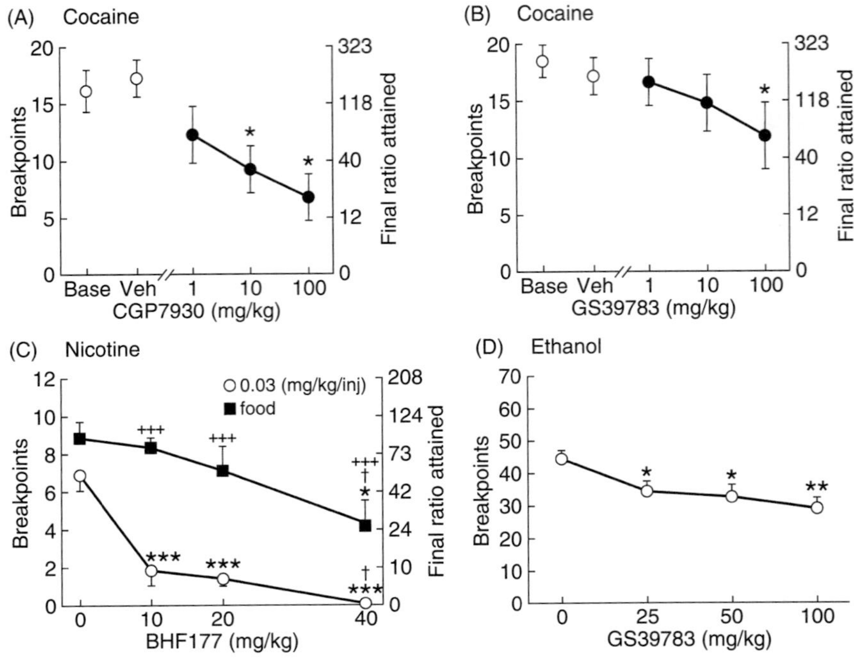

FIGURE 3 Effects of $GABA_B$ receptor positive modulators on the motivational effects of psychoactive drugs. (A and B) Effects of various doses of CGP7930 (left panel) and GS39783 (right panel) on cocaine self-administration under a PR schedule of reinforcement in Sprague–Dawley rats. Ordinates reflect cocaine-reinforced breakpoints ($\pm$ SEM) on the left and final ratio values on the right; abscissa shows doses of CGP7930 and GS39783 in mg/kg body weight. Data above "Base" represent the baseline values; data above "Veh" represent the effects of vehicle. $*p < 0.05$, significant difference from baseline values. (C) Effects of BHF177 on nicotine- and food-maintained responding under a PR schedule of reinforcement in Wistar rats. Data are expressed as mean ($\pm$ SEM) number of reinforcers earned (left ordinal axis) and corresponding final ratio attained (i.e., breakpoints; right ordinal axis). $*p < 0.05$, $***p < 0.001$, significant difference compared with nicotine-maintained responding after administration of vehicle; $^{\dagger}p < 0.05$, significant difference compared with food-maintained responding after administration of vehicle. $^{+++}p < 0.001$, significant difference between nicotine- and food-maintained responding within specific doses of BHF177. (D) Effect of GS39783 on breakpoints (mean $\pm$ SEM) for alcohol in selectively bred Sardinian alcohol-preferring (sP) rats. Rats were trained to lever-press for oral alcohol (15% v/v). Breakpoint was defined as the lowest response requirement not completed by each rat. Water or GS39783 (25, 50, and 100 mg/kg, i.g.) were administered 60 min before the start of the test session. $*p < 0.05$, $**p < 0.01$, compared with vehicle-treated rats. No direct comparisons can be made among the findings presented here because different studies used different methodologies, $GABA_B$ receptor compounds, dose ranges, and data analyses. [Figures taken with permission from: Smith et al., 2004 (cocaine; Fig. 3A and 3B), Paterson et al., 2008 (nicotine; Fig. 3C), Maccioni et al., 2008b (ethanol; Fig. 3D)].

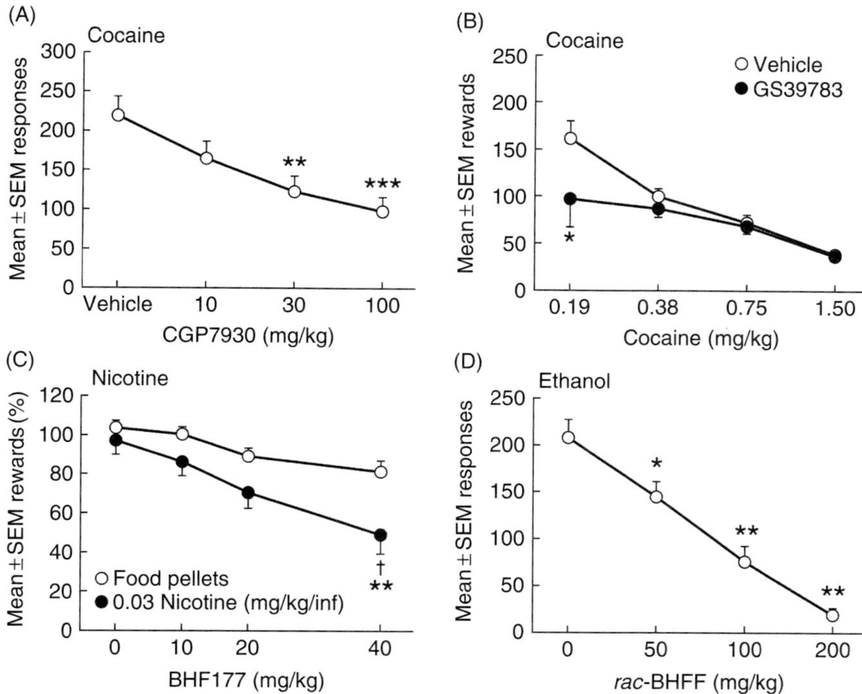

FIGURE 4 Effects of GABA$_B$ receptor positive modulators on the reinforcing effects of psychoactive drugs. (A) Effects of CGP7930 on cocaine self-administration under an FR5 schedule of reinforcement in Wistar rats. Number of active lever presses during cocaine self-administration (mean of the last 3 maintenance days ± SEM). During self-administration, active lever responses resulted in the delivery of a cocaine infusion (0.5 mg/kg/injection) and simultaneous presentation of a light+tone compound stimulus. $**p < 0.01$, $***p < 0.001$, compared with vehicle group. (B) Effects of GS39783 (10 mg/kg) on cocaine self-administration under an FR1 schedule of reinforcement in Sprague–Dawley rats. Ordinates reflect the number of responses during the 4 h session (mean ± SEM); abscissa shows unit doses of cocaine in mg/kg/injection. $*p < 0.05$, significant difference from vehicle control values. (C) Effects of BHF177 on nicotine- and food-maintained responding under an FR5 schedule of reinforcement in Wistar rats. The graph shows nicotine- and food-maintained responding expressed as a percentage of baseline number of rewards earned (mean ± SEM). Asterisks and daggers indicate significant differences compared with nicotine- or food-maintained responding after administration of either vehicle ($**p < 0.01$) or 10 mg/kg BHF177 ($^{\dagger}p < 0.05$), respectively. (D) Effect of *rac*-BHFF under an FR4 schedule of reinforcement on the total number of responses for alcohol (mean ± SEM) in selectively bred Sardinian alcohol-preferring (sP) rats. $*p < 0.01$, $**p < 0.001$, compared with vehicle-treated rats. No direct comparisons can be made among the findings presented here because different studies used different methodologies, GABA$_B$ receptor compounds, dose ranges, and data analyses. [Figures taken with permission from: Filip et al., 2007a (cocaine; Fig. 4A), Smith et al., 2004 (cocaine; Fig. 4B), Paterson et al., 2008 (nicotine; Fig. 4C), Maccioni et al., 2010 (ethanol; Fig. 4D)].

Importantly, although the $GABA_B$ receptor positive modulator GS39783 had no effects in the ICSS procedure *per se*, baclofen dose-dependently elevated thresholds. Co-administration of baclofen with GS39783 dose-dependently attenuated the lowering of thresholds induced by cocaine in rats (Slattery et al., 2005). Interestingly, acute administration of baclofen, SKF97541, the $GABA_B$ receptor positive allosteric modulator CGP7930, and the $GABA_B$ receptor antagonist SCH50911 counteracted the increased immobility time in the forced-swim test produced by discontinuation of cocaine self-administration in rats trained to self-administer cocaine for 14 days (Frankowska et al., 2010). This finding indicates that either blockade or enhancement of GABA transmission through administration of $GABA_B$ receptor ligands can reduce cocaine craving.

B. The Psychomotor Stimulants Amphetamine and Methamphetamine

In addition to cocaine, studies have also been conducted investigating the effects of $GABA_B$ receptor compounds on other psychostimulants, such as D-amphetamine and methamphetamine. Methamphetamine induced CPP after 20 days of conditioning (10 alternating pairing days of conditioning with either saline or methamphetamine). After subsequent extinction sessions, acute GVG administration blocked methamphetamine-induced reinstatement of CPP when administered 2.5 h prior to a methamphetamine challenge dose (DeMarco et al., 2009). This effect indicated a possible therapeutic effect of GVG in the treatment of methamphetamine dependence in terms of relapse prevention. Furthermore, GVG administration dose-dependently inhibited methamphetamine-induced increases in NAc dopamine (Gerasimov et al., 1999).

Furthermore, baclofen attenuated the development and expression of D-methamphetamine-induced CPP in rats when it was administered 30 min prior to the test session (Li et al., 2001). Baclofen also decreased self-administration of D-amphetamine under an FR schedule of reinforcement in rats trained to self-administer 0.1 mg/kg/injection of D-amphetamine, but it had no effect on self-administration supported by 0.2 mg/kg/injection of D-amphetamine (Brebner et al., 2005) (Fig. 1B). When a PR schedule of reinforcement was used, baclofen induced a dose-dependent decrease in breakpoints for D-amphetamine (Brebner et al., 2005) (Fig. 2B) and methamphetamine (Ranaldi & Poeggel, 2002), indicating attenuation of the motivation for the drug. Baclofen also attenuated the D-amphetamine-induced increases in dopamine efflux in the NAc (Brebner et al., 2005).

C. Nicotine

1. GABA$_B$ Receptor Direct Agonists

Administration of either of two GABA$_B$ receptor agonists, CGP44532 or baclofen (Bowery et al., 1985; Froestl et al., 1995a), decreased intravenous nicotine self-administration (Corrigall et al., 2000; Fattore et al., 2002; Paterson et al., 2004). Specifically, baclofen administration prevented acute nicotine self-administration in drug-naive mice (Fattore et al., 2002). Baclofen also decreased nicotine self-administration at the highest dose tested (5 mg/kg) in rats trained to chronically self-administer nicotine under an FR1 schedule of reinforcement (Fattore et al., 2002). Moreover, administration of either CGP44532 or baclofen dose-dependently reduced nicotine self-administration under an FR schedule, whereas the highest doses of baclofen also decreased food-maintained responding (Paterson et al., 2004) (Fig. 1C, D). Furthermore, CGP44532 decreased breakpoints for both nicotine and food under a PR schedule of reinforcement (Paterson et al., 2004) (Fig. 2C). Baclofen also reduced nicotine self-administration (30 μg/kg/injection) when it was administered directly into the VTA, and this effect was more potent for nicotine than for cocaine (Corrigall et al., 2000). Importantly, repeated 14-day administration of CGP44532 (0.25 mg/kg) selectively decreased nicotine self-administration, with little evidence of tolerance to this effect and without affecting food-maintained responding under an FR schedule of reinforcement. The 0.5 mg/kg dose of CGP44532 non-selectively decreased both nicotine- and food-maintained responding (Paterson et al., 2005b). Moreover, acute administration of CGP44532 decreased cue-induced reinstatement of nicotine-seeking behavior in rats (Paterson et al., 2005b). Most recently, baclofen dose-dependently blocked nicotine-induced reinstatement of nicotine-seeking behavior in rats previously trained to self-administer nicotine under an FR1 schedule of reinforcement (Fattore et al., 2009). This effect was selective for nicotine because it did not affect responding for natural reinforcers, such as food and water, and it did not affect locomotor activity, indicating that baclofen's effects on nicotine intake were not attributable to general disruption of activity, even with doses up to 2.5 mg/kg (Fattore et al., 2009). Baclofen also abolished nicotine-induced reinstatement of CPP in mice (Fattore et al., 2009).

Administration of GVG dose- and time-dependently (at 2.5 h, but not at 12 or 24 h, prior to nicotine administration) lowered the nicotine-induced increases in NAc dopamine in both naive and chronically nicotine-treated rats, measured by *in vivo* microdialysis. In two studies using positron emission tomography, administration of the *S*(+) enantiomer of GVG completely inhibited the nicotine-induced dopamine increase in the corpus striatum at half the dose of the racemic mixture (*R*,*S*-GVG, 300 mg/kg; Schiffer et al., 2000). GVG also abolished nicotine-induced increases in dopamine in the striatum of primates (Dewey et al., 1999). Accordingly, microinjections of baclofen into the VTA decreased, whereas the $GABA_B$ receptor antagonist 2-hydroxy-saclofen increased, NAc dopamine levels measured by fast-cyclic voltammetry (Xi & Stein, 1998). Furthermore, baclofen dose-dependently decreased the nicotine-, cocaine-, and morphine-induced increase in dopamine levels in the NAc shell (Fadda et al., 2003).

Systemic administration of low doses of either GVG or CGP44532 did not affect brain reward thresholds in the ICSS procedure in chronically saline- or nicotine-treated rats, whereas the highest dose of each drug elevated thresholds in both groups of animals (Paterson et al., 2005a). Additionally, microinfusion of CGP44532 into the VTA elevated thresholds equally in both saline- and nicotine-treated groups, indicating that prolonged nicotine exposure did not alter $GABA_B$ receptor-mediated regulation of brain reward function (Paterson et al., 2005a). Thus, based on these findings, changes in $GABA_B$ receptor activity do not appear to play a role in brain reward deficits associated with spontaneous nicotine withdrawal.

In studies investigating the effects of $GABA_B$ compounds on CPP for nicotine, baclofen blocked the expression of nicotine-induced CPP in rats (Le Foll et al., 2008). Furthermore, pretreatment with the reversible inhibitor of GABA transaminase 1*R*,4*S*-4-amino-cyclopent-2-ene-carboxylic acid (ACC) (Table III) attenuated the expression of nicotine- and cocaine-induced CPP in rats, respectively (Ashby et al., 2002). Interestingly, a lower dose of ACC was sufficient to block nicotine-induced CPP compared with the dose used to block cocaine-induced CPP, yielding a dose-effect function that was also observed with the effects of GVG on nicotine- and cocaine-induced CPP (Dewey et al., 1998, 1999).

Importantly, in a clinical study by Cousins and colleagues, a single acute dose of baclofen (20 mg/kg) negatively impacted cigarette pleasure and increased feelings of relaxation, although it did not reduce cigarette craving or smoking, indicating a possible important effect of baclofen in smoking abstinence and relapse prevention (Cousins et al., 2001). In a more recent study, baclofen was administered chronically for 9 weeks to humans at doses ranging from 10 mg to 80 mg per day. In this smoking-reduction study, baclofen significantly reduced the number of cigarettes smoked per day (Franklin et al., 2009). These preliminary clinical findings add to the promising results from preclinical studies on the effects of $GABA_B$ receptor agonists on nicotine dependence.

2. $GABA_B$ Receptor Positive Allosteric Modulators

Administration of either GS39783 or the newly synthesized $GABA_B$ receptor positive modulator *N*-([1*R*,2*R*,4*S*]-bicyclo[2.2.1]hept-2-yl)-2-methyl-5-(4-[trifluoromethyl]phenyl)-4-pyrimidinamine (BHF177) decreased nicotine intake at doses that had no effect on responding for food under an FR5 schedule of reinforcement in rats (Fig. 4C). Additionally, administration of intermediate doses of BHF177 attenuated the motivational properties of nicotine without decreasing breakpoints for food, whereas only the highest dose of BHF177 significantly decreased breakpoints for both nicotine and food (Paterson et al., 2008) (Fig. 3C). Furthermore, BHF administration dose-dependently and selectively blocked cue-induced reinstatement of nicotine-, but not food-, seeking behavior in rats (Vlachou and Markou, unpublished observations). Additionally, repeated administration of 20 mg/kg BHF177 for 14 days selectively blocked nicotine self-administration but not food-maintained responding in rats (Vlachou and Markou, unpublished observations). Administration of the GABA$_B$ receptor agonist CGP44532 selectively decreased nicotine versus food self-administration under an FR5 schedule of reinforcement, whereas co-administration of GS39783 with CGP44532 had additive effects in decreasing nicotine self-administration.

In the ICSS procedure, only the highest BHF177 dose significantly elevated brain reward thresholds when administered alone, whereas lower doses had no effect, indicating that relatively low doses of BHF177 do not affect reward processes under baseline conditions. Interestingly, the intermediate doses of BHF177 blocked the ICSS threshold-lowering effect of non-contingent nicotine (Fig. 5A). These doses of BHF177 are within the same range as the doses of BHF177 (10 and 20 mg/kg) that selectively decreased nicotine-, but not food-, maintained responding under FR5 and PR schedules of reinforcement. In contrast to the effects of BHF177, only the highest dose of the GABA$_B$ receptor agonist CGP44532 blocked the threshold-lowering effect of nicotine at a dose that significantly impaired brain reward function when administered alone (0.5 mg/kg) (Fig. 5B). A lower dose of CGP44532 (0.25 mg/kg), which elevated ICSS thresholds when administered alone but did not block the reward-enhancing effect of nicotine, decreased nicotine self-administration under an FR5 schedule. The behavioral effects of the GABA$_B$ receptor positive modulators used in this study are consistent with the GS39783-induced blockade of the development, but not expression, of nicotine-induced CPP (Mombereau et al., 2007) (Fig. 6A). Administration of GS39783 before the CPP acquisition phase also prevented nicotine-induced accumulation of ΔFosB in the NAc, and this effect correlated with the degree of CPP exhibited by each rat (Mombereau et al., 2007) (Fig. 6B, C). ΔFosB is a Fos family protein that appears to play a unique role in the addiction process. Specifically, chronic administration of drugs of abuse, including nicotine, induces accumulation of ΔFosB in specific brain regions, most prominently in the NAc and dorsal striatum (Nestler, 2008). Thus, the reversal of the effects of nicotine by GS39783 can also be detected at the molecular level. In conclusion, findings from both behavioral and molecular studies provide converging evidence about the blockade of the effects of nicotine by GABA$_B$ receptor positive modulators.

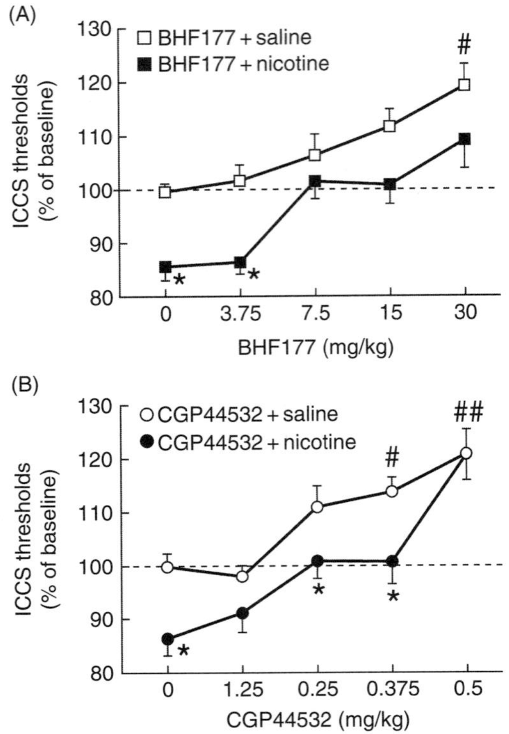

FIGURE 5 Effects of BHF177 and CGP44532 on nicotine-induced facilitation of brain reward function in Wistar rats. Data are expressed as a percentage of baseline ICSS thresholds (mean ± SEM) before each drug session. $*p < 0.05$, significant difference between saline- and nicotine-treated rats within the same dose of the $GABA_B$ receptor agonist (B) or positive modulator (A). $^{\#}p < 0.05$, $^{\#\#}p < 0.01$, significant difference in thresholds compared with values after vehicle administration in the absence of nicotine. [Figures taken with permission from Paterson et al., 2008].

Although ample evidence supports a role for GABA neurotransmission in the neurochemical, reinforcing, and reward-facilitating effects of acute or chronic nicotine, puzzling data have been generated about the potential role for GABA transmission in the non-rewarding, depressive-like aspects of nicotine withdrawal. Specifically, both receptor activation, through administration of either the $GABA_B$ receptor agonist CGP44532 or the $GABA_B$ receptor positive modulator BHF177, and blockade, through administration of the $GABA_B$ receptor antagonist CGP56433A, exacerbated the anhedonic depression-like aspects of nicotine withdrawal in nicotine-treated groups (Vlachou and Markou, unpublished observations). The effects of $GABA_B$ receptor agonists and antagonists in the same direction (i.e., elevations in

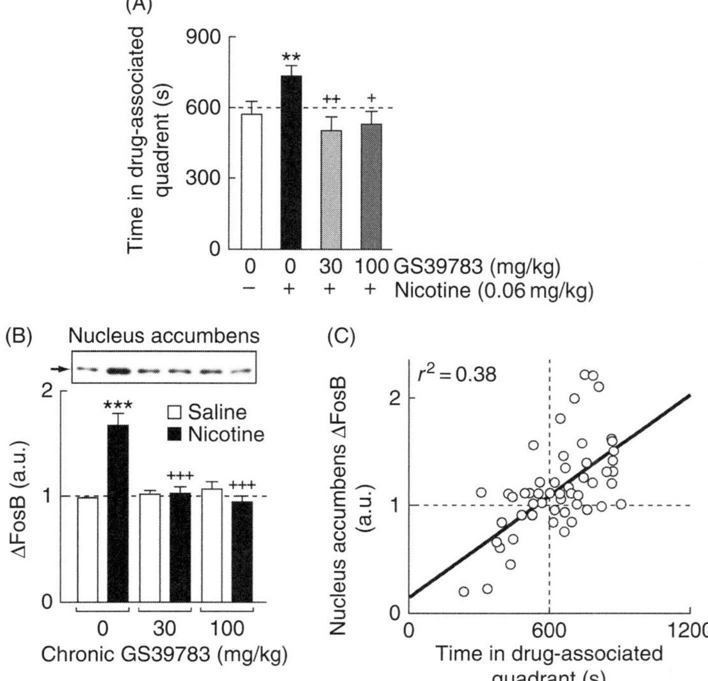

FIGURE 6 Effects of GS39783 on the acquisition of nicotine-induced conditioned place preference in Wistar rats: correlation with ΔFosB accumulation. (A) Nicotine (0.06 mg/kg, s.c.) elicited significant place preference (black bar) that was blocked by administration of GS39783 at both 30 and 100 mg/kg during the conditioning phase (gray bars). Each bar represents the mean ($\pm$ SEM) of the time spent in the drug-associated quadrants. $**p < 0.01$, compared with saline group; $^{+}p < 0.05$, $^{++}p < 0.01$, compared with nicotine group. (B) Effects of chronic GS39783 administration on ΔFosB accumulation in the nucleus accumbens (NAc). Rats were treated daily with saline (white bars) or nicotine (0.06 mg/kg, i.p., black bars) during the 5 days of pre-exposure and during the CPP acquisition phase. During conditioning, GS39783 was administered (0, 30, and 100 mg/kg, p.o.) 30 min before nicotine/saline administration. Animals were sacrificed immediately after the final test. The NAc was dissected and processed for immunoblot analysis. Representative immunoblots (top) and averaged densitometry values (bottom) are shown. ΔFosB changes are expressed in arbitrary units (a.u.). Nicotine induced ΔFosB upregulation in the NAc. GS39783 completely inhibited ΔFosB induction in the NAc at both doses. $^{+++}p < 0.001$, compared with saline group; $***p < 0.001$, compared with nicotine group. (C) ΔFosB induction correlated with nicotine reinforcement. Data from both behavioral and biochemical analyses are included. Relative nucleus accumbens ΔFosB abundance is plotted against the subject's CPP score. The solid line represents the best fit with a least-square method. Dashed lines indicate chance level (x-axis) and ΔFosB basal level (y-axis). [Figures taken with permission from Mombereau et al., 2007].

ICSS thresholds) under baseline conditions have been previously reported (Macey et al., 2001). This effect is surprising and perhaps reflects differential efficacy of these compounds at presynaptic hetero- and autoreceptors versus postsynaptic $GABA_B$ receptors which may have contributed to these findings. These data also indicate the importance of the development of selective $GABA_B$ receptor subtype compounds to further investigate the role of $GABA_B$ receptor subtypes in the anhedonic aspects of nicotine withdrawal and depression.

D. Alcohol

1. $GABA_B$ Receptor Direct Agonists

Studies focusing on the effects of $GABA_B$ receptor agonists, and most recently $GABA_B$ receptor positive modulators, on alcohol use and alcoholism have drawn much attention in the past few years as promising treatments for alcohol dependence. Preliminary clinical studies (e.g., Addolorato et al., 2000, 2002a, 2002b, 2006a, 2006b, 2007; Evans & Bisaga, 2009; Flannery et al., 2004; Ling et al., 1998) have shown that treatment with the $GABA_B$ receptor agonist baclofen, a clinically available compound, is effective at reducing alcohol consumption and craving, leads to the achievement of abstinence from alcohol drinking, and/or treats the alcohol withdrawal syndrome after either acute or chronic treatment with doses ranging from 15 to 75 mg/kg/day (for reviews, see Addolorato et al., 2009; Leggio et al., 2010; Maccioni & Colombo, 2009). Furthermore, a preliminary study showed that baclofen may also work on the hypothalamic-pituitary-adrenal axis to reduce alcohol craving, as indicated by significant correlations between hormonal levels (e.g., free T3 and thyroid-stimulating hormone) and scores on psychometric evaluation scales for alcohol craving and drinking after a 12-week treatment with 10 mg/kg/day baclofen (Leggio et al., 2008). Case reports (Addolorato et al., 2003; Agabio et al., 2007; Ameisen, 2005; Bucknam, 2007) have also shown that administration of 15–30 mg/kg/day either acutely or chronically up to 24 weeks was effective in reducing alcohol intake, craving for alcohol, or severe symptoms of the alcohol withdrawal syndrome (for details on clinical studies, see Chapter 14 in this volume).

At the preclinical level, baclofen and other $GABA_B$ receptor agonists decreased ethanol self-administration in a number of studies, some of which showed a nonselective effect of baclofen on drug and non-drug rewards (e.g., sucrose) or side-effects (Anstrom et al., 2003; Besheer et al., 2004; Maccioni et al., 2005; Walker & Koob, 2007). Baclofen, CGP44532, and SKF97541 attenuated convulsions in alcohol-withdrawing mice (Humeniuk et al., 1994), suppressed the acquisition of alcohol self-administration (Colombo et al., 2000; Daoust et al., 1987), suppressed the motivation to consume alcohol (Colombo et al., 2003b), and suppressed the alcohol-deprivation effect (Colombo et al., 2003a, 2006) in alcohol-experienced rats continuously exposed to a two-bottle choice regimen (alcohol versus water). In the latter study, acute administration of baclofen suppressed both aspects of the alcohol deprivation effect, specifically the extra intake of alcohol and the selection of the highest alcohol concentration solution among four bottles containing either water or 10, 20, and 30% (v/v) alcohol (Colombo et al., 2006). Both baclofen and CGP44532 dose-dependently and similarly suppressed daily alcohol intake in alcohol-naive rats that received the drug treatment for 10 or 14 consecutive days (Colombo et al., 2000, 2002). In one of these studies, acute administration of baclofen dose-dependently decreased the intensity of ethanol withdrawal signs in rats physically dependent on ethanol induced by repeated administration of intoxicating doses of ethanol for 6 consecutive days (Colombo et al., 2000). Additionally, acute administration of baclofen produced a marked suppression of extinction responding for alcohol, which was defined as the maximum number of lever responses reached in the absence of reinforcement. However, only the 3 mg/kg dose of baclofen significantly affected extinction responding for sucrose (Colombo et al., 2003b). In a different range of non-sedative doses, baclofen completely suppressed the extra amount of alcohol consumed during the first hour of re-exposure to alcohol after 7 days of deprivation, suggesting that baclofen may possess anti-relapse properties (Colombo et al., 2003a). Interestingly, pretreatment of Sardinian alcohol-preferring (sP) rats with baclofen dose-dependently suppressed morphine- and WIN 55,212-2-induced increases in alcohol drinking, although it failed to alter alcohol intake when administered alone (Colombo et al., 2004).

Moreover, baclofen dose-dependently decreased ethanol self-administration in both dependent (1 month vapor exposure for 14 h/day) and nondependent rats on an FR1 schedule of reinforcement (Walker & Koob, 2007) (Fig. 1E). Baclofen also decreased breakpoints in both nondependent and dependent animals, with an increased potency in dependent animals, indicating that baclofen could be a potential therapeutic target for the treatment of chronic alcoholism (Walker & Koob, 2007). Moreover, baclofen decreased breakpoints (Maccioni et al., 2008b) (Fig. 2D) and produced a dose-dependent and specific delay in the onset of alcohol-reinforced responding, suggesting that it suppressed the motivation to start drinking alcohol (Maccioni et al., 2005). Furthermore, baclofen attenuated cue-induced reinstatement of alcohol-seeking behavior in sP rats exposed to two within-session 70 min extinction/reinstatement tests (Maccioni et al., 2008a), reflected by 60% fewer responses on the active lever, an increased latency to the first response, and a decreased response rate. Administration of γ-hydroxybutyric acid (GHB) (Table III) also specifically suppressed alcohol's motivational properties in sP rats by reducing both breakpoints in a PR schedule and single-session extinction responding for alcohol (Maccioni et al., 2008c).

Moreover, an intra-VTA injection of baclofen or the nonselective opioid antagonist methylnaloxonium decreased ethanol-induced CPP, whereas intra-NAc injection of methylnaloxonium had no effect. These effects indicate that the conditioned rewarding effect of ethanol is expressed through a VTA-dependent mechanism that involves both opioid and $GABA_B$ receptors (Bechtholt & Cunningham, 2005). Accordingly, GVG completely abolished ethanol-induced increases in dopamine in the NAc (Gerasimov et al., 1999) at half the dose that was required for inhibition of heroin-induced dopamine increases in the NAc.

Importantly, in C57BL/6J (B6) ethanol-preferring mice, the $GABA_B$ receptor agonists baclofen and SKF97541 reduced ethanol-reinforced responding. Interestingly, these doses of the $GABA_B$ agonists also potentiated the sedative effects of ethanol (4 g/kg) by converting a non-sedative dose of ethanol (2 g/kg) into a fully sedative dose. These data question the safety of prescribing $GABA_B$ agonists to alcoholics because they may potentiate ethanol's sedative/hypnotic effects during relapse (Besheer et al., 2004).

2. *$GABA_B$ Receptor Positive Allosteric Modulators*

CGP7930 and GS39783 administered for 5 consecutive days dose-dependently reduced alcohol intake in alcohol-naive sP rats exposed to the two-bottle choice regimen, without altering total daily alcohol intake (Orru et al., 2005). CGP7930 had no effect on responding for water (Liang et al., 2006). The two $GABA_B$ receptor positive modulators had the same effect in alcohol-experienced sP rats (Orru et al., 2005). Similarly, administration of GS39783 dose-dependently reduced responding for alcohol self-administration. This effect of GS39783 was selective because it did not alter responding for a natural reinforcer, such as sucrose, in a different group of sP rats (Maccioni et al., 2007). Administration of either baclofen or CGP7930 reduced responding for ethanol self-administration (Liang et al., 2006) under an FR3 schedule of reinforcement. Interestingly, co-administration of subeffective doses of baclofen or CGP7930 decreased ethanol self-administration (Liang et al., 2006), similar to the effects of co-administration of CGP44532 and GS39783 on nicotine self-administration (Paterson et al., 2008). These findings indicate that the $GABA_B$ receptor positive modulators facilitated $GABA_B$ receptor-mediated signaling through administration of the $GABA_B$ receptor agonists. Finally, GS39783 administration had an effect on the motivational properties of alcohol, reflected by dose-dependent decreases in breakpoints for alcohol self-administration on a PR schedule of reinforcement (Maccioni et al., 2008b) (Fig. 3D). Importantly, the effects of the previously tested $GABA_B$ receptor positive modulators were extended by studies utilizing the newly synthesized positive modulators. Most recently, the newly synthesized GABA$_B$ receptor positive modulator (*R,S*)-5,7-di-*tert*-butyl-3-hydroxy-3-trifluoromethyl-3*H*-benzofuran-2-one (*rac*-BHFF; Malherbe et al., 2008) produced a selective and dose-dependent suppression of responding for alcohol in sP rats on an FR4 schedule of reinforcement (Maccioni et al., 2010) (Fig. 4D). The same effects were observed with the GABA$_B$ receptor positive modulator BHF177 in both the reinforcing and motivational properties of alcohol on FR and PR schedules of reinforcement, respectively, in alcohol-preferring rats (Maccioni et al., 2009). In all three experiments, the effects of BHF177 and *rac*-BHFF on alcohol self-administration were specific. Responding for sucrose was unaltered by BHF177 or *rac*-BHFF pretreatment, whereas it was reduced only by pretreatment with 200 mg/kg *rac*-BHFF (Maccioni et al., 2009, 2010).

E. Opiates

GABAergic mechanisms appear to be critically involved in opiate reinforcement (for review, see Xi & Stein, 2002). GVG administered 2.5 h prior to heroin administration attenuated the expression of heroin-induced CPP in male Sprague–Dawley rats (Paul et al., 2001). Importantly, a magnetic resonance imaging study in Sprague–Dawley rats showed that heroin increased the blood oxygen level-dependent (BOLD) signal in cortical regions, such as the prefrontal cortex, cingulate cortex, and olfactory cortex, but decreased the BOLD signal in subcortical areas, such as the caudate-putamen, thalamus, hypothalamus, and NAc. These changes were attenuated by pretreatment with GVG, independent of the dose used (Xi et al., 2002). The same research group also showed that increased mesolimbic GABAergic concentrations induced by administration of various GABA$_B$ receptor active compounds blocked heroin self-administration in rats (Xi & Stein, 2000). Specifically, GVG administered directly into the VTA, lateral ventricle, or ventral pallidum, but not NAc, dose-dependently blocked heroin self-administration, an effect that lasted for 3–5 days. This effect was blocked by administration of the GABA$_B$ receptor antagonist 2-hydroxysaclofen (Table III), but not by bicuculline, indicating a GABA$_B$ receptor-specific effect. Furthermore, intra-VTA GVG administration also blocked the acquisition of heroin self-administration in drug-naive rats (Xi & Stein, 2000). Although in contrast to the study by Xi and Stein, Gerasimov and colleagues showed that direct intra-NAc GVG administration inhibited or completely abolished (65 and 100% inhibition, respectively) the dopamine increase induced by heroin (Gerasimov et al., 1999) or by a cocaine/heroin (speedball) challenge (Gerasimov & Dewey, 1999). Baclofen also attenuated heroin-induced increases in extracellular dopamine concentrations in the NAc (Xi & Stein, 1999).

Additionally, in studies conducted with the CPP procedure, baclofen reduced morphine-induced CPP in rodents when administered systemically (Kaplan et al., 2003; Suzuki et al., 2005) or directly into the hippocampus (Zarrindast et al., 2006) or VTA (Sahraei et al., 2009) and had the same effect on fentanyl-induced CPP (Suzuki et al., 2005). Baclofen blocked the rewarding effects of morphine measured by acquisition of CPP in mice (Kaplan et al., 2003). Microinjections of baclofen or the $GABA_B$ receptor antagonist phaclofen in the dorsal hippocampus of rats decreased or increased, respectively, the acquisition of CPP induced by morphine (Zarrindast et al., 2006). Consistent with the above study, microinjections of baclofen into the VTA dose-dependently suppressed morphine-induced CPP in rats (Tsuji et al., 1996). In the study by Sahraei and colleagues (Sahraei et al., 2009), female rats sensitized to 5 mg/kg of morphine for 3 days exhibited CPP. Acute administration of baclofen (1.5 and 12 μg/kg/injection) directly into the VTA reduced the expression of morphine-induced CPP, whereas the 6 μg/kg dose increased it. All three doses were effective at reducing the acquisition of CPP (Sahraei et al., 2009). Interestingly, administration of the $GABA_B$ receptor antagonist 3-aminopropyl(diethoxymethyl) phosphinic acid (CGP35348) had the same effects as baclofen on morphine-induced CPP, with the 12 μg/kg dose reducing the expression and all three doses (1.5, 6, and 12 μg/kg) reducing the acquisition of CPP (Sahraei et al., 2009).

Moreover, baclofen decreased heroin self-administration (Xi & Stein, 1999) and drug-induced reinstatement of heroin-seeking behavior in rats (Spano et al., 2007). In the study by Xi and Stein (Xi & Stein, 1999), when baclofen was co-administered with heroin, it suppressed both the development of heroin self-administration in drug-naive rats and the maintenance of heroin self-administration in heroin-treated rats under an FR1 schedule of reinforcement (Fig. 1F). Additionally, direct intra-VTA, but not NAc, injections of baclofen reduced heroin reinforcement, reflected by an increase in responding for heroin self-administration, and dose-dependently decreased heroin-induced dopamine release in the NAc when baclofen was administered either alone or in combination with heroin. This effect was abolished by pretreatment with the $GABA_B$ receptor antagonist 2-hydroxysaclofen into the VTA (Xi & Stein, 1999). Additionally, baclofen dose-dependently reduced heroin-seeking behavior induced by a priming injection of heroin after a period of extinction (Spano et al., 2007).

Finally, baclofen prevented naloxone-precipitated morphine withdrawal in both female and male mice, with a more marked effect in males than in females (Diaz et al., 2006), regardless of the seasonal variation (Diaz et al., 2001; Kemmling et al., 2002). Intra-locus coeruleus injections of baclofen also dose-dependently attenuated morphine withdrawal signs after chronic treatment, an effect that was reversed by the GABA$_B$ receptor antagonist CGP35348 (Riahi et al., 2009), indicating that the GABA$_B$ receptor compounds are involved in both the aversive and somatic aspects of withdrawal from addictive drugs and, specifically, in opiate withdrawal.

III. GABA$_B$ Receptor Agonists versus GABA$_B$ Receptor Positive Modulators

Although GABA$_B$ receptor agonists exhibit promising effects on different aspects of drug dependence, these compounds have undesirable side-effects, including impaired working memory (DeSousa et al., 1994), impaired spatial learning (McNamara & Skelton, 1996), disruption of performance on the rotarod test (a measure of locomotor impairment; Cryan et al., 2004), and decreased responding for non-drug rewards, such as food and electrical brain stimulation (Macey et al., 2001; Paterson et al., 2005b; Slattery et al., 2005). For example, baclofen, which is the only clinically used GABA$_B$ receptor agonist used to treat spasticity, exerts some serious side-effects, such as sedation, nausea, dizziness, drowsiness, mental confusion, and motor impairment, at high doses (Ong & Kerr, 2005). Tolerance to baclofen's therapeutic effects develops after prolonged use. Its short duration of action (3–4 h) is another challenge in its potential use in the treatment of drug addiction (Brebner et al., 2002a).

Positive allosteric modulators of the GABA$_B$ receptor increase both the potency and efficacy of GABA in activating GABA$_B$ receptors. GABA$_B$ receptor positive modulator compounds, such as GS39783, are devoid of substantial intrinsic agonistic activity in the absence of GABA (Urwyler et al., 2003). These compounds only act synergistically with GABA to enhance its effects. Thus, positive modulators offer more physiological means to activate receptors *in vivo* than agonists. The modulators do not perturb receptor signaling on their own but potentiate the effect of GABA only where and when it is endogenously released to activate GABA$_B$ receptors. Furthermore, allosteric modulators bind to a site that is distinct from the agonist binding site; therefore, co-applying positive modulators with agonists is possible to potentiate the effects of the agonists, which was reflected in some of the studies presented above (e.g., Liang et al., 2006; Paterson et al., 2008). In this case, lower doses of agonists than the ones leading to side-effects could be used to achieve efficacious results (Christopoulos, 2002; Conn et al., 2009). Positive allosteric modulators can also be administered at relatively high doses without concerns about side-effects resulting from overstimulation of the receptors, overdosage, or the development of tolerance (Ong & Kerr, 2005).

Overall, the potential advantages of $GABA_B$ receptor modulators are that these compounds may be more effective with a better side-effect profile than $GABA_B$ agonists because the modulatory actions occur in concert with endogenous GABAergic activity. Finally, based on the discussion above, $GABA_B$ modulators may be effective clinical treatments for addiction to various drugs of abuse, such as nicotine, cocaine, heroin, and alcohol.

IV. Conclusion

The results from the experimental studies reviewed here suggest that enhancement of GABA transmission through activation of $GABA_B$ receptors blocks reward in general and the rewarding effects of various drugs of abuse. Drugs of abuse exert their effects by producing changes in the brain reward pathways and affecting reward processes at both molecular and behavioral levels in both humans and animals. Agonists and positive allosteric modulators acting at $GABA_B$ receptors blocked the reinforcing and reward-facilitating effects of addictive drugs, which was shown in self-administration studies using an FR schedule of reinforcement and in ICSS studies, respectively. Both agonists and positive modulators were also effective at attenuating the motivation to self-administer drug, which was shown in the self-administration studies using the PR schedule, and effective at blocking the reinstatement of drug-seeking for different drugs of abuse. The effects of the $GABA_B$ compounds in the withdrawal phase from addictive drugs were not as promising. In some studies, $GABA_B$ receptor agonists, antagonists, and positive modulators exacerbated the aversive, depression-like effects of drug-withdrawal. These findings indicate that $GABA_B$ compounds may be most effective at blocking the initiation and maintenance of drug abuse by decreasing the reinforcing effects of addictive drugs and in preventing relapse in humans, which is suggested by the blockade of reinstatement of drug-seeking behavior in animals. However, $GABA_B$ receptor agonists have also shown side-effects in clinical and preclinical studies. Thus, $GABA_B$ receptor positive modulators could be promising therapeutics for the treatment of drug dependence.

Acknowledgments

This work was supported by U.S. National Institutes of Health grants R01DA11946, R01DA232090, and U01MH69062 to Athina Markou. Dr. Styliani Vlachou was supported by individual Postdoctoral Research Fellowship 18FT-0048 from the Tobacco-Related Disease Research Program from the State of California, USA. The authors would like to thank Mike Arends for editorial assistance and Janet Hightower for computer graphics.

Disclosure Statement: AM has received contract research support from Intracellular Therapeutics, Inc., Lundbeck Research USA, Inc., Bristol-Myers Squibb Co., F. Hoffman-La Roche Inc., Pfizer, and Astra-Zeneca and honorarium/consulting fees from Abbott GmbH and Company, AstraZeneca, and Pfizer during the past 3 years. SV discloses no actual or possible conflict of interest including financial, personal, or other relationships with other people or organizations within 3 years of beginning the work submitted.

Non-standard abbreviations

ACC	1*R*,4*S*-4-amino-cyclopent-2-ene-carboxylic acid
BHF177	*N*-([1*R*,2*R*,4*S*]-bicyclo[2.2.1] hept-2-yl)-2-methyl-5-(4-[trifluoromethyl]phenyl)-4-pyrimidinamine
BOLD	blood oxygen level-dependent
CGP35348	3-aminopropyl (diethoxymethyl)phosphinic acid
CGP44532	(3-amino-2[*S*]-hydroxypropyl)-methylphosphinic acid
CGP51176	3-amino-2-(*R*)-hydroxypropyl-cyclohexylmethyl-phosphinic acid
CGP56433A	[3-{1-(*S*)-[{3-(cyclohexylmethyl) hydroxy phosphinyl}-2-(*S*) hydroxypropyl]amino}ethyl] benzoic acid
CGP7930	2,6-di-*tert*-butyl-4-(3-hydroxy-2,2-dimethyl-propyl)-phenol
CPP	conditioned place preference
DR	dorsal raphe
FR	fixed-ratio
GABA	γ-aminobutyric acid
$GABA_B$ receptor	γ-aminobutyric acid B-type receptor
GHB	γ-hydroxybutyric acid

GS39783	*N,N′*-dicyclopentyl-2-methylsulfanyl-5-nitro-pyrimidine-4,6-diamine
GVG	γ-vinyl GABA
ICSS	intracranial self-stimulation
MR	median raphe
NAc	nucleus accumbens
PR	progressive-ratio
rac-BHFF	(*R,S*)-5,7-di-*tert*-butyl-3-hydroxy-3-trifluoromethyl-3*H*-benzofuran-2-one
SCH23390	7-chloro-3-methyl-1-phenyl-1,2,4,5-tetrahydro-3-benzazepin-8-ol
SCH50911	(2*S*)-(+)-5,5-dimethyl-2-morpholineacetic acid
SKF97541	3-aminopropyl(methyl) phoshinic acid
sP	Sardinian alcohol-preferring
VTA	ventral tegmental area

References

Adams, C. L., & Lawrence, A. J. (2007). CGP7930: A positive allosteric modulator of the $GABA_B$ receptor. *CNS Drug Reviews*, *13*, 308–316.

Addolorato, G., Caputo, F., Capristo, E., et al. (2000). Ability of baclofen in reducing alcohol craving and intake: II–preliminary clinical evidence. *Alcoholism, Clinical and Experimental Research*, *24*, 67–71.

Addolorato, G., Caputo, F., Capristo, E., et al. (2002). Baclofen efficacy in reducing alcohol craving and intake: A preliminary double-blind randomized controlled study. *Alcohol and Alcoholism*, *37*, 504–508.

Addolorato, G., Caputo, F., Capristo, E., et al. (2002). Rapid suppression of alcohol withdrawal syndrome by baclofen. *American Journal of Medicine*, *112*, 226–229.

Addolorato, G., Leggio, L., Abenavoli, L., et al. (2003). Suppression of alcohol delirium tremens by baclofen administration: A case report. *Clinical Neuropharmacology*, *26*, 258–262.

Addolorato, G., Leggio, L., Abenavoli, L., et al. (2006). Baclofen in the treatment of alcohol withdrawal syndrome: A comparative study vs diazepam. *American Journal of Medicine*, *119*, 276e13–276e18.

Addolorato, G., Leggio, L., Agabio, R., et al. (2006). Baclofen: A new drug for the treatment of alcohol dependence. *International Journal of Clinical Practice*, *60*, 1003–1008.

Addolorato, G., Leggio, L., Cardone, S., et al. (2009). Role of the $GABA_B$ receptor system in alcoholism and stress: Focus on clinical studies and treatment perspectives. *Alcohol*, *43*, 559–563.

Addolorato, G., Leggio, L., Ferrulli, A., et al. (2007). Effectiveness and safety of baclofen for maintenance of alcohol abstinence in alcohol-dependent patients with liver cirrhosis: Randomised, double-blind controlled study. *Lancet*, *370*, 1915–1922.

Agabio, R., Marras, P., Addolorato, G., et al. (2007). Baclofen suppresses alcohol intake and craving for alcohol in a schizophrenic alcohol-dependent patient: A case report. *Journal of Clinical Psychopharmacology*, *27*, 319–320.

Ameisen, O. (2005). Complete and prolonged suppression of symptoms and consequences of alcohol-dependence using high-dose baclofen: A self-case report of a physician. *Alcohol and Alcoholism*, *40*, 147–150.

Anstrom, K. K., Cromwell, H. C., Markowski, T., et al. (2003). Effect of baclofen on alcohol and sucrose self-administration in rats. *Alcoholism, Clinical and Experimental Research*, *27*, 900–908.

Ascaso, F. J., Lopez, M. J., Mauri, J. A., et al. (2003). Visual field defects in pediatric patients on vigabatrin monotherapy. *Documenta Ophthalmologica*, *107*, 127–130.

Ashby, C. R. Jr., Paul, M., Gardner, E. L., et al. (2002). Systemic administration of 1*r*,4*s*-4-amino-cyclopent-2-ene-carboxylic acid, a reversible inhibitor of GABA transaminase, blocks expression of conditioned place preference to cocaine and nicotine in rats. *Synapse*, *44*, 61–63.

Ashby, C. R. Jr., Rohatgi, R., Ngosuwan, J., et al. (1999). Implication of the GABA$_B$ receptor in gamma vinyl-GABA's inhibition of cocaine-induced increases in nucleus accumbens dopamine. *Synapse*, *31*, 151–153.

Barrett, A. C., Negus, S. S., Mello, N. K., et al. (2005). Effect of GABA agonists and GABA-A receptor modulators on cocaine- and food-maintained responding and cocaine discrimination in rats. *Journal of Pharmacology and Experimental Therapeutics*, *315*, 858–871.

Bechtholt, A. J., & Cunningham, C. L. (2005). Ethanol-induced conditioned place preference is expressed through a ventral tegmental area dependent mechanism. *Behavioral Neuroscience*, *119*, 213–223.

Besheer, J., Lepoutre, V., & Hodge, C. W. (2004). GABA$_B$ receptor agonists reduce operant ethanol self-administration and enhance ethanol sedation in C57BL/6j mice. *Psychopharmacology*, *174*, 358–366.

Bowery, N. G., Hill, D. R., & Hudson, A. L. (1985). [3H](-)baclofen: An improved ligand for GABA$_B$ sites. *Neuropharmacology*, *24*, 207–210.

Brebner, K., Ahn, S., & Phillips, A. G. (2005). Attenuation of d-amphetamine self-administration by baclofen in the rat: Behavioral and neurochemical correlates. *Psychopharmacology*, *177*, 409–417.

Brebner, K., Childress, A. R., & Roberts, D. C. S. (2002a). A potential role for GABA$_B$ agonists in the treatment of psychostimulant addiction. *Alcohol and Alcoholism*, *37*, 478–484.

Brebner, K., Froestl, W., Andrews, M., et al. (1999). The GABA$_B$ agonist CGP 44532 decreases cocaine self-administration in rats: Demonstration using a progressive ratio and a discrete trials procedure. *Neuropharmacology*, *38*, 1797–1804.

Brebner, K., Froestl, W., & Roberts, D. C. S. (2002b). The GABA$_B$ antagonist CGP56433A attenuates the effect of baclofen on cocaine but not heroin self-administration in the rat. *Psychopharmacology*, *160*, 49–55.

Brebner, K., Phelan, R., & Roberts, D. C. S. (2000a). Effect of baclofen on cocaine self-administration in rats reinforced under fixed-ratio 1 and progressive-ratio schedules. *Psychopharmacology*, *148*, 314–321.

Brebner, K., Phelan, R., & Roberts, D. C. S. (2000b). Intra-VTA baclofen attenuates cocaine self-administration on a progressive ratio schedule of reinforcement. *Pharmacology Biochemistry and Behavior*, *66*, 857–862.

Brodie, J. D., Figueroa, E., Laska, E. M., et al. (2005). Safety and efficacy of γ-vinyl GABA (GVG) for the treatment of methamphetamine and/or cocaine addiction. *Synapse*, *55*, 122–125.

Bucknam, W. (2007). Suppression of symptoms of alcohol dependence and craving using high-dose baclofen. *Alcohol and Alcoholism*, *42*, 158–160.

Campbell, U. C., Lac, S. T., & Carroll, M. E. (1999). Effects of baclofen on maintenance and reinstatement of intravenous cocaine self-administration in rats. *Psychopharmacology*, *143*, 209–214.

Christopoulos, A. (2002). Allosteric binding sites on cell-surface receptors: Novel targets for drug discovery. *Nature reviews. Drug discovery*, *1*, 198–210.

Colombo, G., Agabio, R., Carai, M. A., et al. (2000). Ability of baclofen in reducing alcohol intake and withdrawal severity: I–preclinical evidence. *Alcoholism, Clinical and Experimental Research*, *24*, 58–66.

Colombo, G., Serra, S., Brunetti, G., et al. (2002). The $GABA_B$ receptor agonists baclofen and CGP 44532 prevent acquisition of alcohol drinking behaviour in alcohol-preferring rats. *Alcohol and Alcoholism*, *37*, 499–503.

Colombo, G., Serra, S., Brunetti, G., et al. (2003a). Suppression by baclofen of alcohol deprivation effect in Sardinian alcohol-preferring (sP) rats. *Drug and Alcohol Dependence*, *70*, 105–108.

Colombo, G., Serra, S., Vacca, G., et al. (2004). Suppression by baclofen of the stimulation of alcohol intake induced by morphine and WIN 55,212-2 in alcohol-preferring rats. *European Journal of Pharmacology*, *492*, 189–193.

Colombo, G., Serra, S., Vacca, G., et al. (2006). Baclofen-induced suppression of alcohol deprivation effect in Sardinian alcohol-preferring (sP) rats exposed to different alcohol concentrations. *European Journal of Pharmacology*, *550*, 123–126.

Colombo, G., Vacca, G., Serra, S., et al. (2003b). Baclofen suppresses motivation to consume alcohol in rats. *Psychopharmacology*, *167*, 221–224.

Conn, P. J., Christopoulos, A., & Lindsley, C. W. (2009). Allosteric modulators of GPCRs: A novel approach for the treatment of CNS disorders. *Nature reviews. Drug discovery*, *8*, 41–54.

Corrigall, W. A., Coen, K. M., Adamson, K. L., et al. (2000). Response of nicotine self-administration in the rat to manipulations of mu-opioid and γ-aminobutyric acid receptors in the ventral tegmental area. *Psychopharmacology*, *149*, 107–114.

Corrigall, W. A., Coen, K. M., Zhang, J., et al. (2001). GABA mechanisms in the pedunculopontine tegmental nucleus influence particular aspects of nicotine self-administration selectively in the rat. *Psychopharmacology*, *158*, 190–197.

Cousins, M. S., Stamat, H. M., & de Wit, H. (2001). Effects of a single dose of baclofen on self-reported subjective effects and tobacco smoking. *Nicotine & Tobacco Research*, *3*, 123–129.

Cryan, J. F., Kelly, P. H., Chaperon, F., et al. (2004). Behavioral characterization of the novel $GABA_B$ receptor-positive modulator GS39783 (*N,n'*-dicyclopentyl-2-methylsulfanyl-5-nitro-pyrimidine-4,6-diamine): Anxiolytic-like activity without side effects associated with baclofen or benzodiazepines. *Journal of Pharmacology and Experimental Therapeutics*, *310*, 952–963.

Daoust, M., Saligaut, C., Lhuintre, J. P., et al. (1987). GABA transmission, but not benzodiazepine receptor stimulation, modulates ethanol intake by rats. *Alcohol*, *4*, 469–472.

DeMarco, A., Dalal, R. M., Pai, J., et al. (2009). Racemic gamma vinyl-GABA (R,S-GVG) blocks methamphetamine-triggered reinstatement of conditioned place preference. *Synapse*, *63*, 87–94.

DeSousa, N. J., Beninger, R. J., Jhamandas, K., et al. (1994). Stimulation of $GABA_B$ receptors in the basal forebrain selectively impairs working memory of rats in the double Y-maze. *Brain Research*, *641*, 29–38.

Dewey, S. L., Brodie, J. D., Gerasimov, M., et al. (1999). A pharmacologic strategy for the treatment of nicotine addiction. *Synapse*, *31*, 76–86.

Dewey, S. L., Morgan, A. E., Ashby, C. R.Jr., et al.(1998). A novel strategy for the treatment of cocaine addiction. *Synapse*, *30*, 119–129.

Dewey, S. L., Smith, G. S., Logan, J., et al. (1992). GABAergic inhibition of endogenous dopamine release measured *in vivo* with ^{11}C-raclopride and positron emission tomography. *Journal of Neuroscience*, *12*, 3773–3780.

Di Ciano, P., & Everitt, B. J. (2002). Reinstatement and spontaneous recovery of cocaine-seeking following extinction and different durations of withdrawal. *Behavioural Pharmacology*, *13*, 397–405.

Diaz, S. L., Barros, V. G., Antonelli, M. C., et al. (2006). Morphine withdrawal syndrome and its prevention with baclofen: Autoradiographic study of μ-opioid receptors in prepubertal male and female mice. *Synapse*, *60*, 132–140.

Diaz, S. L., Kemmling, A. K., Rubio, M. C., et al. (2001). Lack of sex-related differences in the prevention by baclofen of the morphine withdrawal syndrome in mice. *Behavioural Pharmacology*, *12*, 75–79.

Dobrovitsky, V., Pimentel, P., Duarte, A., et al. (2002). CGP 44532, a $GABA_B$ receptor agonist, is hedonically neutral and reduces cocaine-induced enhancement of reward. *Neuropharmacology*, *42*, 626–632.

Engberg, G., Kling-Petersen, T., & Nissbrandt, H. (1993). $GABA_B$-receptor activation alters the firing pattern of dopamine neurons in the rat substantia nigra. *Synapse*, *15*, 229–238.

Epstein, D. H., Preston, K. L., Stewart, J., et al. (2006). Toward a model of drug relapse: An assessment of the validity of the reinstatement procedure. *Psychopharmacology*, *189*, 1–16.

Evans, S. M., & Bisaga, A. (2009). Acute interaction of baclofen in combination with alcohol in heavy social drinkers. *Alcoholism, Clinical and Experimental Research*, *33*, 19–30.

Everitt, B. J., Belin, D., Economidou, D., et al. (2008). Neural mechanisms underlying the vulnerability to develop compulsive drug-seeking habits and addiction. *Philosophical Transactions of the Royal Society of London. Series B, Biological Sciences*, *363*, 3125–3135.

Fadda, P., Scherma, M., Fresu, A., et al. (2003). Baclofen antagonizes nicotine-, cocaine-, and morphine-induced dopamine release in the nucleus accumbens of rat. *Synapse*, *50*, 1–6.

Fallon, J. H., Koziell, D. A., & Moore, R. Y. (1978). Catecholamine innervation of the basal forebrain. II. Amygdala, suprarhinal cortex and entorhinal cortex. *Journal of Comparative Neurology*, *180*, 509–532.

Fallon, J. H., & Moore, R. Y. (1978). Catecholamine innervation of the basal forebrain. IV. Topography of the dopamine projection to the basal forebrain and neostriatum. *Journal of Comparative Neurology*, *180*, 545–580.

Fattore, L., Cossu, G., Martellotta, M. C., et al. (2002). Baclofen antagonizes intravenous self-administration of nicotine in mice and rats. *Alcohol and Alcoholism*, *37*, 495–498.

Fattore, L., Spano, M. S., Cossu, G., et al. (2009). Baclofen prevents drug-induced reinstatement of extinguished nicotine-seeking behaviour and nicotine place preference in rodents. *European Neuropsychopharmacology*, *19*, 487–498.

Feltenstein, M. W., & See, R. E. (2008). The neurocircuitry of addiction: An overview. *British Journal of Pharmacology*, *154*, 261–274.

Fenton, H. M., & Liebman, J. M. (1982). Self-stimulation response decrement patterns differentiate clonidine, baclofen and dopamine antagonists from drugs causing performance deficit. *Pharmacology Biochemistry and Behavior*, *17*, 1207–1212.

Filip, M., & Frankowska, M. (2007). Effects of $GABA_B$ receptor agents on cocaine priming, discrete contextual cue and food induced relapses. *European Journal of Pharmacology*, *571*, 166–173.

Filip, M., Frankowska, M., & Przegalinski, E. (2007a). Effects of $GABA_B$ receptor antagonist, agonists and allosteric positive modulator on the cocaine-induced self-administration and drug discrimination. *European Journal of Pharmacology*, *574*, 148–157.

Filip, M., Frankowska, M., Zaniewska, M., et al. (2007b). Diverse effects of GABA-mimetic drugs on cocaine-evoked self-administration and discriminative stimulus effects in rats. *Psychopharmacology*, *192*, 17–26.

Flannery, B. A., Garbutt, J. C., Cody, M. W., et al. (2004). Baclofen for alcohol dependence: A preliminary open-label study. *Alcoholism, Clinical and Experimental Research*, *28*, 1517–1523.

Franklin, T. R., Harper, D., Kampman, K., et al. (2009). The $GABA_B$ agonist baclofen reduces cigarette consumption in a preliminary double-blind placebo-controlled smoking reduction study. *Drug and Alcohol Dependence*, *103*, 30–36.

Frankowska, M., Golda, A., Wydra, K., et al. (2010). Effects of imipramine or $GABA_B$ receptor ligands on the immobility, swimming and climbing in the forced swim test in rats following discontinuation of cocaine self-administration. *European Journal of Pharmacology*, *627*, 142–149.

Froestl, W., Mickel, S. J., Hall, R. G., et al. (1995a). Phosphinic acid analogues of GABA: 1. New potent and selective $GABA_B$ agonists. *Journal of Medicinal Chemistry*, *38*, 3297–3312.

Froestl, W., Mickel, S. J., von Sprecher, G., et al. (1995b). Phosphinic acid analogues of GABA; 2. Selective, orally active $GABA_B$ antagonists. *Journal of Medicinal Chemistry*, *38*, 3313–3331.

Gerasimov, M. R., Ashby, C. R. Jr., Gardner, E. L., et al. (1999). γ-vinyl GABA inhibits methamphetamine, heroin, or ethanol-induced increases in nucleus accumbens dopamine. *Synapse*, *34*, 11–19.

Gerasimov, M. R., & Dewey, S. L. (1999). Gamma-vinyl γ-aminobutyric acid attenuates the synergistic elevations of nucleus accumbens dopamine produced by a cocaine/heroin (speedball) challenge. *European Journal of Pharmacology*, *380*, 1–4.

Gerasimov, M. R., Schiffer, W. K., Gardner, E. L., et al. (2001). GABAergic blockade of cocaine-associated cue-induced increases in nucleus accumbens dopamine. *European Journal of Pharmacology*, *414*, 205–209.

Heimer, L., Zahm, D. S., Churchill, L., et al. (1991). Specificity in the projection patterns of accumbal core and shell in the rat. *Neuroscience*, *41*, 89–125.

Hernandez, G., Hamdani, S., Rajabi, H., et al. (2006). Prolonged rewarding stimulation of the rat medial forebrain bundle: Neurochemical and behavioral consequences. *Behavioral Neuroscience*, *120*, 888–904.

Humeniuk, R. E., White, J. M., & Ong, J. (1994). The effects of $GABA_B$ ligands on alcohol withdrawal in mice. *Pharmacology Biochemistry and Behavior*, *49*, 561–566.

Jung, M. J., Lippert, B., Metcalf, B. W., et al. (1977). γ-Vinyl GABA (4-amino-hex-5-enoic acid), a new selective irreversible inhibitor of GABA-T: Effects on brain GABA metabolism in mice. *Journal of neurochemistry*, *29*, 797–802.

Kahn, R., Biswas, K., Childress, A. R., et al. (2009). Multi-center trial of baclofen for abstinence initiation in severe cocaine-dependent individuals. *Drug and Alcohol Dependence*, *103*, 59–64.

Kalivas, P. W. (1993). Neurotransmitter regulation of dopamine neurons in the ventral tegmental area. *Brain Research. Brain Research Reviews*, *18*, 75–113.

Kalivas, P. W., Striplin, C. D., Steketee, J. D., et al. (1992). Cellular mechanisms of behavioral sensitization to drugs of abuse. *Annals of the New York Academy of Sciences*, *654*, 128–135.

Kalivas, P. W., & Volkow, N. D. (2005). The neural basis of addiction: A pathology of motivation and choice. *American Journal of Psychiatry*, *162*, 1403–1413.

Kaplan, G. B., Leite-Morris, K. A., Joshi, M., et al. (2003). Baclofen inhibits opiate-induced conditioned place preference and associated induction of fos in cortical and limbic regions. *Brain Research*, *987*, 122–125.

Kemmling, A. K., Rubio, M. C., & Balerio, G. N. (2002). Baclofen prevents morphine withdrawal irrespective of seasonal variation. *Behavioural Pharmacology*, *13*, 87–92.

Klitenick, M. A., DeWitte, P., & Kalivas, P. W. (1992). Regulation of somatodendritic dopamine release in the ventral tegmental area by opioids and GABA: An *in vivo* microdialysis study. *Journal of Neuroscience*, *12*, 2623–2632.

Koob, G. F., & Volkow, N. D. (2010). Neurocircuitry of addiction. *Neuropsychopharmacology*, *35*, 217–238.

Kushner, S. A., Dewey, S. L., & Kornetsky, C. (1997). γ-Vinyl GABA attenuates cocaine-induced lowering of brain stimulation reward thresholds. *Psychopharmacology*, *133*, 383–388.

Kushner, S. A., Dewey, S. L., & Kornetsky, C. (1999). The irreversible γ-aminobutyric acid (GABA) transaminase inhibitor γ-vinyl-GABA blocks cocaine self-administration in rats. *Journal of Pharmacology and Experimental Therapeutics*, *290*, 797–802.

Le Foll, B., Wertheim, C. E., & Goldberg, S. R. (2008). Effects of baclofen on conditioned rewarding and discriminative stimulus effects of nicotine in rats. *Neuroscience Letters*, *443*, 236–240.

Leggio, L., Ferrulli, A., Cardone, S., et al. (2008). Relationship between the hypothalamic-pituitary-thyroid axis and alcohol craving in alcohol-dependent patients: A longitudinal study. *Alcoholism, Clinical and Experimental Research*, *32*, 2047–2053.

Leggio, L., Garbutt, J. C., & Addolorato, G. (2010). Effectiveness and safety of baclofen in the treatment of alcohol dependent patients. *CNS & Neurological Disorders Drug Targets*, *9*, 33–44.

Li, S. M., Yin, L. L., Ren, Y. H., et al. (2001). GABA$_B$ receptor agonist baclofen attenuates the development and expression of d-methamphetamine-induced place preference in rats. *Life Sciences*, *70*, 349–356.

Liang, J. H., Chen, F., Krstew, E., et al. (2006). The GABA$_B$ receptor allosteric modulator CGP7930, like baclofen, reduces operant self-administration of ethanol in alcohol-preferring rats. *Neuropharmacology*, *50*, 632–639.

Ling, W., Shoptaw, S., & Majewska, D. (1998). Baclofen as a cocaine anti-craving medication: A preliminary clinical study. *Neuropsychopharmacology*, *18*, 403–404.

Maccioni, P., Bienkowski, P., Carai, M. A., et al. (2008a). Baclofen attenuates cue-induced reinstatement of alcohol-seeking behavior in Sardinian alcohol-preferring (sP) rats. *Drug and Alcohol Dependence*, *95*, 284–287.

Maccioni, P., Carai, M. A., Kaupmann, K., et al. (2009). Reduction of alcohol's reinforcing and motivational properties by the positive allosteric modulator of the GABA$_B$ receptor, BHF177, in alcohol-preferring rats. *Alcoholism, Clinical and Experimental Research*, *33*, 1749–1756.

Maccioni, P., & Colombo, G. (2009). Role of the GABA$_B$ receptor in alcohol-seeking and drinking behavior. *Alcohol*, *43*, 555–558.

Maccioni, P., Fantini, N., Froestl, W., et al. (2008b). Specific reduction of alcohol's motivational properties by the positive allosteric modulator of the GABA$_B$ receptor, GS39783 – comparison with the effect of the GABA$_B$ receptor direct agonist, baclofen. *Alcoholism, Clinical and Experimental Research*, *32*, 1558–1564.

Maccioni, P., Pes, D., Fantini, N., et al. (2008c). γ-Hydroxybutyric acid (GHB) suppresses alcohol's motivational properties in alcohol-preferring rats. *Alcohol*, *42*, 107–113.

Maccioni, P., Pes, D., Orru, A., et al. (2007). Reducing effect of the positive allosteric modulator of the GABA$_B$ receptor, GS39783, on alcohol self-administration in alcohol-preferring rats. *Psychopharmacology*, *193*, 171–178.

Maccioni, P., Serra, S., Vacca, G., et al. (2005). Baclofen-induced reduction of alcohol reinforcement in alcohol-preferring rats. *Alcohol*, *36*, 161–168.

Maccioni, P., Thomas, A. W., Carai, M. A., et al. (2010). The positive allosteric modulator of the GABA$_B$ receptor, *rac*-BHFF, suppresses alcohol self-administration. *Drug and Alcohol Dependence*, *109*, 96–103.

Macey, D. J., Froestl, W., Koob, G. F., et al. (2001). Both GABA$_B$ receptor agonist and antagonists decreased brain stimulation reward in the rat. *Neuropharmacology*, *40*, 676–685.

Malherbe, P., Masciadri, R., Norcross, R. D., et al. (2008). Characterization of (*R,S*)-5,7-di-*tert*-butyl-3-hydroxy-3-trifluoromethyl-3*h*-benzofuran-2-one as a positive allosteric modulator of $GABA_B$ receptors. *British Journal of Pharmacology*, *154*, 797–811.

McNamara, R. K., & Skelton, R. W. (1996). Baclofen, a selective $GABA_B$ receptor agonist, dose-dependently impairs spatial learning in rats. *Pharmacology Biochemistry and Behavior*, *53*, 303–308.

Mohler, H., & Fritschy, J. M. (1999). $GABA_B$ receptors make it to the top - as dimers. *Trends in Pharmacological Sciences*, *20*, 87–89.

Molina, P. E., Ahmed, N., Ajmal, M., et al. (1999). Co-administration of γ-vinyl GABA and cocaine: Preclinical assessment of safety. *Life Sciences*, *65*, 1175–1182.

Mombereau, C., Lhuillier, L., Kaupmann, K., et al. (2007). $GABA_B$ receptor-positive modulation-induced blockade of the rewarding properties of nicotine is associated with a reduction in nucleus accumbens ΔFosB accumulation. *Journal of Pharmacology and Experimental Therapeutics*, *321*, 172–177.

Nestler, E. J. (2005). Is there a common molecular pathway for addiction?. *Nature Neuroscience*, *8*, 1445–1449.

Nestler, E. J. (2008). Transcriptional mechanisms of addiction: Role of ΔFosB. *Philosophical Transactions of the Royal Society of London. Series B, Biological Sciences*, *363*, 3245–3255.

Ong, J., & Kerr, D. I. (2005). Clinical potential of $GABA_B$ receptor modulators. *CNS Drug Reviews*, *11*, 317–334.

Orru, A., Lai, P., Lobina, C., et al. (2005). Reducing effect of the positive allosteric modulators of the $GABA_B$ receptor, CGP7930 and GS39783, on alcohol intake in alcohol-preferring rats. *European Journal of Pharmacology*, *525*, 105–111.

Panagis, G., & Kastellakis, A. (2002). The effects of ventral tegmental administration of $GABA_A$, $GABA_B$, NMDA and AMPA receptor agonists on ventral pallidum self-stimulation. *Behavioural Brain Research*, *131*, 115–123.

Paterson, N. E., Bruijnzeel, A. W., Kenny, P. J., et al. (2005a). Prolonged nicotine exposure does not alter $GABA_B$ receptor-mediated regulation of brain reward function. *Neuropharmacology*, *49*, 953–962.

Paterson, N. E., Froestl, W., & Markou, A. (2004). The $GABA_B$ receptor agonists baclofen and CGP44532 decreased nicotine self-administration in the rat. *Psychopharmacology*, *172*, 179–186.

Paterson, N. E., Froestl, W., & Markou, A. (2005b). Repeated administration of the $GABA_B$ receptor agonist CGP44532 decreased nicotine self-administration, and acute administration decreased cue-induced reinstatement of nicotine-seeking in rats. *Neuropsychopharmacology*, *30*, 119–128.

Paterson, N. E., & Markou, A. (2002). Increased GABA neurotransmission via administration of γ-vinyl GABA decreased nicotine self-administration in the rat. *Synapse*, *44*, 252–253.

Paterson, N. E., Vlachou, S., Guery, S., et al. (2008). Positive modulation of $GABA_B$ receptors decreased nicotine self-administration and counteracted nicotine-induced enhancement of brain reward function in rats. *Journal of Pharmacology and Experimental Therapeutics*, *326*, 306–314.

Paul, M., Dewey, S. L., Gardner, E. L., et al. (2001). γ-Vinyl GABA (GVG) blocks expression of the conditioned place preference response to heroin in rats. *Synapse*, *41*, 219–220.

Ranaldi, R., & Poeggel, K. (2002). Baclofen decreases methamphetamine self-administration in rats. *NeuroReport*, *13*, 1107–1110.

Riahi, E., Mirzaii-Dizgah, I., Karimian, S. M., et al. (2009). Attenuation of morphine withdrawal signs by a $GABA_B$ receptor agonist in the locus coeruleus of rats. *Behavioural Brain Research*, *196*, 11–14.

Roberts, D. C. S. (2005). Preclinical evidence for GABA$_B$ agonists as a pharmacotherapy for cocaine addiction. *Physiology & Behavior*, *86*, 18–20.
Roberts, D. C. S., & Andrews, M. M. (1997). Baclofen suppression of cocaine self-administration: Demonstration using a discrete trials procedure. *Psychopharmacology*, *131*, 271–277.
Roberts, D. C. S., Andrews, M. M., & Vickers, G. J. (1996). Baclofen attenuates the reinforcing effects of cocaine in rats. *Neuropsychopharmacology*, *15*, 417–423.
Roberts, D. C. S., & Brebner, K. (2000). GABA modulation of cocaine self-administration. *Annals of the New York Academy of Sciences*, *909*, 145–158.
Sahraei, H., Etemadi, L., Rostami, P., et al. (2009). GABA$_B$ receptors within the ventral tegmental area are involved in the expression and acquisition of morphine-induced place preference in morphine-sensitized rats. *Pharmacology Biochemistry and Behavior*, *91*, 409–416.
Schiffer, W. K., Gerasimov, M. R., Bermel, R. A., et al. (2000). Stereoselective inhibition of dopaminergic activity by gamma vinyl-GABA following a nicotine or cocaine challenge: A PET/microdialysis study. *Life Sciences*, *66*, PL169–PL173.
Schmidt, H. D., Anderson, S. M., Famous, K. R., et al. (2005). Anatomy and pharmacology of cocaine priming-induced reinstatement of drug seeking. *European Journal of Pharmacology*, *526*, 65–76.
Schmitz, B., Schmidt, T., Jokiel, B., et al. (2002). Visual field constriction in epilepsy patients treated with vigabatrin and other antiepileptic drugs: A prospective study. *Journal of Neurology*, *249*, 469–475.
Schultz, W. (2001). Reward signaling by dopamine neurons. *Neuroscientist*, *7*, 293–302.
Schultz, W. (2007). Multiple dopamine functions at different time courses. *Annual Review of Neuroscience*, *30*, 259–288.
See, R. E. (2005). Neural substrates of cocaine-cue associations that trigger relapse. *European Journal of Pharmacology*, *526*, 140–146.
Shin, R., & Ikemoto, S. (2010). The GABA$_B$ receptor agonist baclofen administered into the median and dorsal raphe nuclei is rewarding as shown by intracranial self-administration and conditioned place preference in rats. *Psychopharmacology*, in press.
Shoaib, M., Swanner, L. S., Beyer, C. E., et al. (1998). The GABA$_B$ agonist baclofen modifies cocaine self-administration in rats. *Behavioural Pharmacology*, *9*, 195–206.
Shoptaw, S., Yang, X., Rotheram-Fuller, E. J., et al. (2003). Randomized placebo-controlled trial of baclofen for cocaine dependence: Preliminary effects for individuals with chronic patterns of cocaine use. *Journal of Clinical Psychiatry*, *64*, 1440–1448.
Slattery, D. A., Markou, A., Froestl, W., et al. (2005). The GABA$_B$ receptor-positive modulator GS39783 and the GABA$_B$ receptor agonist baclofen attenuate the reward-facilitating effects of cocaine: Intracranial self-stimulation studies in the rat. *Neuropsychopharmacology*, *30*, 2065–2072.
Smith, M. A., Yancey, D. L., Morgan, D., et al. (2004). Effects of positive allosteric modulators of the GABA$_B$ receptor on cocaine self-administration in rats. *Psychopharmacology*, *173*, 105–111.
Spano, M. S., Fattore, L., Fratta, W., et al. (2007). The GABA$_B$ receptor agonist baclofen prevents heroin-induced reinstatement of heroin-seeking behavior in rats. *Neuropharmacology*, *52*, 1555–1562.
Sugita, S., Johnson, S. W., & North, R. A. (1992). Synaptic inputs to GABA$_A$ and GABA$_B$ receptors originate from discrete afferent neurons. *Neuroscience Letters*, *134*, 207–211.
Suzuki, T., Nurrochmad, A., Ozaki, M., et al. (2005). Effect of a selective GABA$_B$ receptor agonist baclofen on the μ-opioid receptor agonist-induced antinociceptive, emetic and rewarding effects. *Neuropharmacology*, *49*, 1121–1131.
Takada, K., & Yanagita, T. (1997). Drug dependence study on vigabatrin in rhesus monkeys and rats. *Arzneimittel-Forschung*, *47*, 1087–1092.

Tobler, P. N., Fiorillo, C. D., & Schultz, W. (2005). Adaptive coding of reward value by dopamine neurons. *Science*, *307*, 1642–1645.

Tsuji, M., Nakagawa, Y., Ishibashi, Y., et al. (1996). Activation of ventral tegmental $GABA_B$ receptors inhibits morphine-induced place preference in rats. *European Journal of Pharmacology*, *313*, 169–173.

Urwyler, S., Mosbacher, J., Lingenhoehl, K., et al. (2001). Positive allosteric modulation of native and recombinan gamma-aminobutyric $acid_B$ receptors by 2,6-di-*tert*-butyl-4-(3-hydroxy-2,2-dimethyl-propyl)-phenol (CGP7930) and its aldehyde analog CGP13501. *Molecular Pharmacology*, *60*, 963–971.

Urwyler, S., Pozza, M. F., Lingenhoehl, K., et al. (2003). *N,N′*-dicyclopentyl-2-methylsulfanyl-5-nitro-pyrimidine-4,6-diamine (GS39783) and structurally related compounds: Novel allosteric enhancers of γ-aminobutyric $acid_B$ receptor function. *Journal of Pharmacology and Experimental Therapeutics*, *307*, 322–330.

Vlachou, S., & Markou, A. (2010). The use of the intracranial self-stimulation in drug abuse research: Addictive drugs and the neurobiology of reward and motivation. In M. C. Olmstead (Ed.), *Animal models of drug addiction*. New York: Springer, in press.

Walaas, I., & Fonnum, F. (1979). The distribution and origin of glutamate decarboxylase and choline acetyltransferase in ventral pallidum and other basal forebrain regions. *Brain Research*, *177*, 325–336.

Walaas, I., & Fonnum, F. (1980). Biochemical evidence for γ-aminobutyrate containing fibres from the nucleus accumbens to the substantia nigra and ventral tegmental area in the rat. *Neuroscience*, *5*, 63–72.

Walker, B. M., & Koob, G. F. (2007). The γ-aminobutyric acid-B receptor agonist baclofen attenuates responding for ethanol in ethanol-dependent rats. *Alcoholism, Clinical and Experimental Research*, *31*, 11–18.

Weerts, E. M., Froestl, W., Kaminski, B. J., et al. (2007). Attenuation of cocaine-seeking by $GABA_B$ receptor agonists baclofen and CGP44532 but not the GABA reuptake inhibitor tiagabine in baboons. *Drug and Alcohol Dependence*, *89*, 206–213.

Weiss, F. (2005). Neurobiology of craving, conditioned reward and relapse. *Current Opinion in Pharmacology*, *5*, 9–19.

White, J. H., Wise, A., Main, M. J., et al. (1998). Heterodimerization is required for the formation of a functional $GABA_B$ receptor. *Nature*, *396*, 679–682.

Willick, M. L., & Kokkinidis, L. (1995). The effects of ventral tegmental administration of $GABA_A$, $GABA_B$ and NMDA receptor agonists on medial forebrain bundle self-stimulation. *Behavioural Brain Research*, *70*, 31–36.

Wirtshafter, D., & Sheppard, A. C. (2001). Localization of $GABA_B$ receptors in midbrain monoamine containing neurons in the rat. *Brain Research Bulletin*, *56*, 1–5.

Wohlrab, G., Boltshauser, E., Schmitt, B., et al. (1999). Visual field constriction is not limited to children treated with vigabatrin. *Neuropediatrics*, *30*, 130–132.

Xi, Z. X., & Stein, E. A. (1998). Nucleus accumbens dopamine release modulation by mesolimbic $GABA_A$ receptors – an in vivo electrochemical study. *Brain Research*, *798*, 156–165.

Xi, Z. X., & Stein, E. A. (1999). Baclofen inhibits heroin self-administration behavior and mesolimbic dopamine release. *Journal of Pharmacology and Experimental Therapeutics*, *290*, 1369–1374.

Xi, Z. X., & Stein, E. A. (2000). Increased mesolimbic GABA concentration blocks heroin self-administration in the rat. *Journal of Pharmacology and Experimental Therapeutics*, *294*, 613–619.

Xi, Z. X., & Stein, E. A. (2002). GABAergic mechanisms of opiate reinforcement. *Alcohol and Alcoholism*, *37*, 485–494.

Xi, Z. X., Wu, G., Stein, E. A., et al. (2002). GABAergic mechanisms of heroin-induced brain activation assessed with functional MRI. *Magnetic Resonance in Medicine*, *48*, 838–843.

Yim, C. Y., & Mogenson, G. J. (1980a). Effect of picrotoxin and nipecotic acid on inhibitory response of dopaminergic neurons in the ventral tegmental area to stimulation of the nucleus accumbens. *Brain Research*, *199*, 466–473.
Yim, C. Y., & Mogenson, G. J. (1980b). Electrophysiological studies of neurons in the ventral tegmental area of Tsai. *Bain Research*, *181*, 301–313.
Yin, H. H., Ostlund, S. B., Knowlton, B. J., et al. (2005). The role of the dorsomedial striatum in instrumental conditioning. *European Journal of Neuroscience*, *22*, 513–523.
Zarrindast, M. R., Massoudi, R., Sepehri, H., et al. (2006). Involvement of GABA$_B$ receptors of the dorsal hippocampus on the acquisition and expression of morphine-induced place preference in rats. *Physiology & Behavior*, *87*, 31–38.

Robin J. Tyacke, Anne Lingford-Hughes, Laurence J. Reed, and David J. Nutt

Neuropsychopharmacology Unit, Division of Experimental Medicine and Therapeutics, Imperial College London, London, UK

$GABA_B$ Receptors in Addiction and Its Treatment

Abstract

The $GABA_B$ receptor plays an important role in the control of neurotransmitter release, and experiments using preclinical models have shown that modulation of this receptor can have profound effects on the reward process. This ability to affect the reward process has led to clinical investigations into the possibility that this could be a viable target in the treatment of addiction. Presented here is an overview of a number of studies testing this hypothesis in different drug dependencies. The studies reviewed have used the $GABA_B$ receptor agonist baclofen, which is currently the only $GABA_B$ agonist for use in humans. In addition, studies using the non-specific $GABA_B$ receptor agonists vigabatrin and tiagabine have been included. In some of

Advances in Pharmacology, Volume 58

1054-3589/10 $35.00
10.1016/S1054-3589(10)58014-1

the studies these were found to have efficacy in the initiation and maintenance of abstinence, as an anti-craving treatment and alleviation of withdrawal syndromes, while in other studies showing limited effects. However, there is enough evidence to suggest that modulators of the $GABA_B$ receptor have potential as adjunct treatments to aid in the initiation of abstinence, maintenance of abstinence, and prevention of cue-related relapse in some addictions. This potential is at present poorly understood or studied and warrants further investigation.

I. Introduction

A. $GABA_B$ Receptors as Targets to Treat Substance Addictions

The central role that the $GABA_B$ receptor plays in the control of neurotransmitter release, especially dopamine, has led to extensive investigations into its role in the reward process (see previous chapter). The role the reward process plays in the reinforcement and conditioning of individuals to pleasurable experiences, for example, drugs, and the ability of the $GABA_B$ receptor to modulate this has led to interest in this system's potential to provide new ways in treating addictions to these drugs. Preclinical studies using animal models have repeatedly shown the ability of the $GABA_B$ agonists to attenuate many of the positive reinforcing effects of reward, especially to drugs and the antagonists to augment them. The potential that the $GABA_B$ receptor has to modulate the reward system has led many to test its applicability in clinical environments.

B. $GABA_B$ Agonists

1. Baclofen

Currently, there is only one selective $GABA_B$ agonist available for human use, baclofen. Baclofen was originally developed as an anti-epileptic in the 1920s but its effectiveness in epilepsy was disappointing. However, it was found to have anti-spastic effects and is currently used for the treatment of spastic movement, especially in instances of spinal cord injury, spastic diplegia, multiple sclerosis, and amyotrophic lateral sclerosis. It is an orally active γ-aminobutyric acid (GABA) derivative, *p*-chlorophenyl-gamma-aminobutyric acid, with a half-life of 2–4 h. Over its many years of use, it has proven to be a very safe drug with few side effects. The main adverse effects of baclofen are somnolence, dizziness, muscle weakness, and headache but these are not universal and can be minimized by titrating doses up over a few days allowing desensitization to the side effects and optimum therapeutic response. Baclofen has good absorption after oral administration (75%), with peak serum concentrations achieved in 2–4 h. Baclofen is weakly bound (30%) to plasma proteins. It is eliminated primarily via the kidneys, 85% as the unchanged parent compound.

2. *Non-Specific $GABA_B$ Agonists*

As baclofen is currently the only $GABA_B$ agonist or modulator available for human use, much of the research into addiction treatment has focused on its use. However, some studies have used tiagabine or vigabatrin (γ-vinyl-GABA) clinically in the treatment of different substance addictions. This again is based on strong preclinical evidence of their effectiveness in the modulation of the reward process in animal models. Tiagabine and vigabatrin are anti-convulsants that work by increasing GABA levels, but by different mechanisms. Tiagabine is a selective GABA reuptake inhibitor (Fink-Jensen et al., 1992), while vigabatrin inhibits the catabolism of GABA by irreversibly inhibiting GABA transaminase. These compounds are not direct agonists of the $GABA_B$ receptor but they do all work via the GABA system by elevating GABA levels either acutely, tiagabine, or chronically, vigabatrin. Although this elevation of GABA in the brain will affect the $GABA_A$ receptors there will also be agonist effects at the $GABA_B$ receptor. For this reason, studies utilizing these agents have also been included in this chapter. Other anti-epileptics have also been shown to have efficacy in treating addiction, for example, gabapentin and pregabalin, but despite their names, they do not have a mechanism of action directly related to GABA elevation and so will have even more confounding issues than tiagabine and vigabatrin and thus have not been included.

II. Alcohol Dependence

Alcohol dependence has received the most investigation regarding modulation of the $GABA_B$ receptor and its treatment. Building on a number of preclinical experiments showing the effect of $GABA_B$ receptor agonists and modulators to affect alcohol intake in alcohol-preferring rats (preceding chapter), the utility of baclofen to have an effect on alcohol dependence in humans has been carried out. Two open-label trials testing the effect of baclofen on alcohol reduction/abstinence and craving (Addolorato et al., 2000; Flannery et al., 2004) gave exciting preliminary results. In the first of these studies, 10 male current alcoholics were given baclofen titrated up to 30 mg/day (10 mg three times a day (t.i.d.)) over the 4 weeks of the study (Addolorato et al., 2000). The medication and its administration in this study was entrusted to a family member. The participants attended a weekly outpatient visit where their alcohol craving level was evaluated by the Alcohol Craving Scale, and abstinence from alcohol was assessed based on the individual's self-evaluation, family member interview, and the main blood biological markers of alcohol abuse. Counselling was also performed at these visits. Two participants continued to drink alcohol although they substantially reduced their daily drinks in the first week of treatment, whereas the seven remaining completers were abstinent for the entire 4 weeks. Alcohol craving was significantly reduced, as were the main biological markers of alcohol abuse: aspartate amino transferase, alanine aminotranferase, gamma glutamyl-transpeptidase, and mean cellular volume. In addition, some participants also reported that their obsessional thinking about alcohol had disappeared. In the other open-label study, 12 alcohol-dependent individuals were given baclofen titrated up to 30 mg/day (10 mg t.i.d.) for the 12 weeks of the trial (Flannery et al., 2004). Although the subjects in this trial were alcohol dependent they were not necessarily seeking to give up. Participants attended weekly clinic, visits where self-reported, drinking data were collected. The Beck Depression Inventory and the Beck Anxiety Inventory were completed, and alcohol craving was assessed using the Penn Alcohol Craving Scale. In addition, at four visits over the 12 weeks, participants received motivational enhancement therapy sessions. There was a significant reduction in the number of drinks per drinking day and the number of heavy-drinking days, and an increase in the number of abstinent days. Significant decreases in anxiety and craving were also shown. There was a much greater attrition rate in this study compared with the other with six non-completers (Flannery et al., 2004) as apposed to one in Addolorato et al. (2000). This might represent differences in the motivation and support of the different population in either study. In Addolorato et al. (2000), they were treatment-seeking with family support while in Flannery et al. (2004), they were not treatment-seeking. Nevertheless the results of these two open-label trials led both groups to follow them up with full randomized controlled trials.

In the first of these studies, 39 alcohol-dependent subjects were given either baclofen, titrated up to 30 mg/day (10 mg t.i.d., $n = 20$), or placebo ($n = 19$) in a 4-week double-blind trial to test the efficacy of baclofen in inducing and maintaining abstinence and reduce craving for alcohol (Addolorato et al., 2002a). Subjects were only included if they had been referred by a family member who could be entrusted with the study medication and its administration. The subjects had weekly outpatient visits where the study measures were obtained and counselling given. Alcohol craving was measured using the Obsessive-Compulsive Drinking Scale, abstinence was measured by self-report and family member interview, and other biological markers of alcohol dependence were also measured as well as state and trait anxiety using the Spielberg inventories and depression using the Zung Self-Rating Depression Scale. Three subjects in the baclofen group and 8 in the placebo group failed to complete the study. There were a significantly greater number of abstinent subjects treated with baclofen compared with placebo, and the duration of abstinence was also significantly greater in the baclofen group. There was also a significant reduction in overall alcohol craving in the baclofen group compared with placebo as measured with the Obsessive-Compulsive Drinking Scale with significant reductions in both the compulsive and obsessive subscales. In addition, state anxiety was also significantly reduced in the baclofen group, but there was no effect on the depression score. These results were confirmed in a larger ($n=84$), 12-week, double-blind, placebo-controlled trial of baclofen 30 mg/day (20 mg t.i.d., $n=42$) versus placebo ($n=42$) (Addolorato et al., 2007). The methods for these two studies were broadly similar with the same measures of alcohol dependence recorded, the use of a family member to keep and administer the study medication, etc. The main difference between the two studies was that this was conducted in alcohol patients with liver cirrhosis, a group usually not included in such trials. In addition to confirming the findings of the previous study that baclofen is effective at promoting and maintaining alcohol abstinence in alcohol-dependent individuals, it also found that there were no hepatic side effects or worsening of liver function tests. This showed the potential of baclofen to treat those with significant liver impairment. An important consideration, as liver disease is often a key precipitant for alcoholics to seek sobriety. Many of the other current treatments for alcohol dependence, for example, disulfiram, naltrexone, and acamprosate carry specific warnings against use in patients with marked liver impairment so the opportunity to use baclofen in these patients is very welcome.

However, the promising and exciting effects of baclofen on alcohol dependence were not seen in the follow-up trial to the other open-label study (Flannery et al., 2004). In this trial they saw no effect of baclofen on either heavy-drinking days or abstinent days compared with placebo (Garbutt, 2009). As with the differences in the open-label trial, the lack of significant effect in this study may be due to the differing motivations in the two populations. In Addolorato et al. (2002a, 2007), the participants were more severely dependent, required medication for detoxification, were more anxious, and all wanted sobriety. In contrast, those in the Garbutt et al. study (Garbutt, 2009) were recruited by advert, did not require detoxification medication, wanted "controlled drinking," and were less anxious and had fewer adverse health consequences of their drinking.

The acute safety of baclofen and alcohol taken together were tested in a human laboratory-based study (Evans & Bisaga, 2009). Eighteen non-dependent, non-treatment-seeking heavy social drinkers were tested using a double-blind, double-dummy design. Acute baclofen (0, 40, and 80 mg) was given 2.5 h before alcohol (1.5 g/l body water or approximately 0.75 g/kg) or placebo drinks. They found that baclofen, alone or with alcohol, caused only modest increases in heart rate and blood pressure, did not have its own positive subjective effects (e.g. stimulant effects, drug liking), and did not alter alcohol craving or alcohol-induced positive subjective effects. However, baclofen did increase sedation and impair performance, which is unsurprising considering the known pharmacology of baclofen especially at the high doses used in this study. Such doses are usually titrated up to slowly in two or three doses per day and rarely given as one dose. Nevertheless, the study confirmed that baclofen alone has minimal abuse liability in these heavy drinkers and is safe when given in combination with intoxicating doses of alcohol.

There have been a few case studies that also describe the beneficial effects of baclofen in alcoholics (Ameisen, 2005; Bucknam, 2007) and in a schizophrenic alcohol-dependent patient (Agabio et al., 2007). These case studies showed that baclofen was very effective in suppressing alcohol craving and preventing relapse, and apparently safe when coadministered with medications for depression (Bucknam, 2007) and schizophrenia (Agabio et al., 2007). Of note is that in two of these studies the subjects used very high doses of baclofen, 100 mg/day (Bucknam, 2007) and 120 mg/day (Ameisen, 2005), increasing the dose as needed in stressful situations or periods. This presents interesting possibilities for physician-monitored treatments as opposed to generically "safe" dosing regimens.

Baclofen has also been studied for its ability to treat alcohol withdrawal syndrome in comparison with diazepam (Addolorato et al., 2006). Thirty-seven patients with alcohol withdrawal syndrome were given either baclofen 30 mg/day (10 mg t.i.d., $n = 18$) or diazepam (0.5-0.75 mg/kg/day, $n = 19$) for 10 days and the Clinical Institute Withdrawal Assessment (CIWA-Ar) used to evaluate the efficacy of the treatments. The study found that there was no significant difference between the treatments in their ability to alleviate the alcohol withdrawal symptoms. Closer inspection of the data suggested that diazepam acted slightly more rapidly to reduce the sweating, anxiety, and agitation subscale scores of the CIWA-Ar. However, these data do suggest that baclofen provides another option for treating alcohol withdrawal syndrome other than the benzodiazepines, which might be clinically important in patients who abuse benzodiazepines.

In addition to this randomized controlled trial there have been case reports showing the efficacy of baclofen to treat alcohol withdrawal syndrome (Addolorato et al., 2002b) and alcohol withdrawal syndrome complicated with delirium tremens (Addolorato et al., 2003). Both tiagabine (Myrick et al., 2005) and vigabatrin (Stuppaeck et al., 1996) have also been shown to be effective in treating alcohol withdrawal syndrome. In addition, there was a trend for the patients in the tiagabine study to have less post-detoxification drinking indicating a decreased tendency to relapse (Myrick et al., 2005).

Taken together, the evidence seems to suggest that agonism of the $GABA_B$ receptor by baclofen or a non-specific agonist (tiagabine or vigabatrin) has a great potential to treat all aspects of alcohol dependence: alleviating the acute withdrawal symptoms, initiating and maintaining abstinence, and reducing craving and craving-related relapse. Larger controlled trials are now required.

III. Cocaine

The role the GABA$_B$ receptor plays in cocaine addiction and the potential for agonists to assist in cessation and abstinence treatment have been studied in a number of laboratory experiments, open-label studies, and randomized controlled trials. In one of the early open-label studies (Ling et al., 1998), baclofen was investigated as a cocaine anti-craving treatment. Ten cocaine-abusing males were treated with baclofen titrated up to 60 mg/day (20 mg t.i.d.), in conjunction with three group counselling sessions a week. The results of this study were generally positive with patients reporting decreased craving and reduction in cocaine use (measured by urine analysis) with an average continuous abstinence of nearly 5 weeks.

This open-label trial was followed by a randomized controlled trial of 70 treatment-seeking cocaine-dependent individuals (Shoptaw et al., 2003). This study followed a similar protocol with baclofen titrated up to 60 mg/day (20 mg t.i.d.), in conjunction with counselling sessions, urine analysis to confirm reported abstinence and following the subjects over a 16-week period. As with the open-label trial, the results, while not overwhelming, were promising, showing that baclofen was significantly better than placebo at reducing cocaine use using longitudinal analysis. In addition, they found that the baclofen group had a significant reduction in positive urine tests at weeks 3 and 8 using post hoc analysis. Their analysis also showed that those subjects with higher cocaine use at baseline were more likely to respond favorably to baclofen than those with lower cocaine use. However, there was no significant difference in self-reported cocaine craving. The authors did admit that the data provided "only a suggestion of a treatment effect." However, also as they observed, the tests they used in their analysis were particularly sensitive to attrition of participants, especially with the small numbers that they had; attrition in this study was high with only nine in the baclofen group and eight in the placebo group reaching the end of the study.

A subsequent multicenter randomized controlled trial failed to replicate the promising results of these two trials (Kahn et al., 2009). The study was conducted over eight centers with 160 participants, 80 per group. These subjects were cocaine dependent and seeking treatment for their addiction. The dose of baclofen was again titrated up to 60 mg/day (20 mg t.i.d.) in conjunction with weekly cognitive behavioral therapy. However, the length of the study was only 8 weeks. They found no significant effect of baclofen over placebo on their primary outcome measure of self-reported cocaine use substantiated by urine analysis. However, both of these studies only investigated the ability of baclofen to initiate and maintain cocaine abstinence compared with the placebo group and it may be that baclofen is more effective in preventing relapse in people who have already achieved a period of abstinence (Brebner et al., 2002).

The ability of baclofen to attenuate the direct effects of cocaine use has been tested in human laboratory-based studies. In one such study, Lile et al. (2004a) investigated the ability of acute administration of baclofen to modify the reinforcing, subject-rated, and cardiovascular effects of a moderate dose of cocaine in current cocaine users under placebo-controlled, double-blind crossover conditions. Seven non-treatment-seeking participants were admitted to a clinical research center and given baclofen (placebo, 10, 20, and 30 mg) 1.5 h prior to intranasal cocaine (45 mg or placebo 4 mg). Disappointingly, baclofen was found to have no significant effect on any of the parameters tested.

In a more recent study, baclofen seemed more promising at modifying cocaine administration (Haney et al., 2006). In this study, the effect of baclofen on smoked cocaine self-administration, cocaine craving, subjective-effects ratings, and cardiovascular effects were studied in two populations of cocaine-dependent volunteers; those maintained on methadone and those who were not. Again, these subjects were not seeking treatment for their cocaine use. The design of this study was also a placebo-controlled, double-blind crossover for the effect of baclofen (0, 10, and 20 mg t.i.d.) on smoked cocaine base (0, 12, 25, and 50 mg) in both groups. Participants were admitted to a clinical research center for the entire study period, 21 days. The participants were maintained on various baclofen treatments prior to smoking the cocaine. Cocaine sessions began with a sample trial, where participants smoked the cocaine dose available that session, and five choice trials, where participants chose between smoking the available cocaine dose or receiving one $5 voucher for food. The effects of baclofen were most pronounced in the non-methadone group. The 60 mg dose significantly decreased the number of times the 12 mg dose of cocaine was self-administered as compared with placebo, while having no effect in the methadone group. Baclofen (30 mg) also significantly reduced the perceived monetary value of the 50 mg dose of cocaine but only in the non-methadone group. Baclofen had no effect on the ratings of "high/stimulated" score for either group but did significantly reduce the cocaine craving caused by cocaine administration in the methadone-maintained group. There was no such effect in the non-methadone group.

The differences in the results for these two apparently similar studies may be due to the differing sample sizes and methodology. The first study (Lile et al., 2004a) looked purely at an acute effect of baclofen on subjects who could have been using cocaine up to admission, while the second study (Haney et al., 2006) was looking at the effect on baclofen-maintained participants, who due to the study design were already abstinent from cocaine for at least 2 days. In addition, as commented by the authors (Haney et al., 2006), a more prolonged initial period of baclofen treatment may have produced more robust effects. However, the small reduction in self-administered cocaine is interesting especially as these individuals were not seeking to give up. In addition, the results of this study, especially on the non-methadone group, are in agreement with those seen in preclinical models (previous chapter) where the effects of baclofen are surmountable by high cocaine doses, further indicating that baclofen may have more benefit in preventing relapse, rather than in initiating cocaine abstinence.

Further evidence for the potential of baclofen to have a positive impact on the treatment of cocaine addiction comes from imaging studies reported in Brebner et al. (2002). Initial studies using oxygen-15-labeled water PET showed that regional cerebral blood flow was increased in different limbic regions, anterior cingulate and amygdala, in unmedicated cocaine patients exposed to cocaine-related video cues. The cue-induced craving and increased limbic blood flow was hypothesized to indicate activation of the mesolimbic dopamine reward system. When a similar imaging paradigm was used with cocaine-dependent individuals who were pretreated with baclofen (10–20 mg for 7–10 days), an attenuation of this increase in blood flow was seen. Three of the cohort who received their last dose of baclofen 1–2 h prior to the scanning session showed a substantial blunting of cue-induced craving and did not show anterior cingulate or amygdala activation during the cue exposure. However, this effect seems to be time dependent as some of the cohort who received the baclofen more than 4 h prior to the scanning session showed less of an effect of the medication, a finding probably accounted for by the short half-life of baclofen (Brebner et al., 2002).

Vigabatrin has been shown to be effective in many preclinical models of cocaine addiction, reward, etc. (Dewey et al., 1997, 1998; Kushner et al., 1999) among others. These promising indications have led to investigations in humans who are dependent on cocaine. An initial outpatient, open-label, fixed-dose, time-limited trial in a setting with psychotherapeutic support and intervention studied the effect that vigabatrin had on cocaine cessation and abstinence (Brodie et al., 2003). Treatment-seeking subjects ($n=20$) were given vigabatrin (titrated up to 2 g twice daily) for the period of the study. Urine analysis was used to monitor abstinence from cocaine and subjects who were cocaine-free for 4 weeks had their vigabatrin reduced 1 g of vigabatrin per day per week. Twelve subjects failed to complete the study, with 8 leaving the study within 10 days having changed their minds regarding stopping cocaine use. The four remaining non-completers stayed in the study but failed to stop using cocaine. However, in three of them, their (self-reported) use was greatly reduced; two of the four had a >80% reduction, one out of the four had a 50% reduction, and the other did not reduce at all. The eight remaining subjects were drug-free for 46–58 days. The completers reported that the vigabatrin did not eliminate their craving until weeks 2–3 of treatment. The use of vigabatrin is limited by safety concerns, especially ocular problems (visual field constriction) that have been shown to occur after prolonged use (Hilton et al., 2004). For this reason, this and other ocular (visual acuity) and physiological measures were monitored in this study and other similar studies by this group; in all cases they found that vigabatrin was safe in the context of these treatment periods (Brodie et al., 2005; Fechtner et al., 2006). These results were extremely promising suggesting that vigabatrin could be a viable treatment for the instigation of cocaine abstinence in treatment-seeking cocaine-dependent individuals. However, as this was only an open-label study, confirmation of its effectiveness was still required by a randomized controlled trial.

Such a randomized controlled trial has recently been completed by the same group (Brodie et al., 2009). The participants in the study were treatment-seeking parolees who were actively using cocaine and had a history of cocaine dependence. Subjects were randomly assigned to a fixed titration of vigabatrin ($n = 50$, up to 3 g per day) or placebo ($n = 53$) in a 9-week double-blind trial and 4-week follow-up assessment. In addition to the vigabatrin, the subjects also received cognitive behavioral therapy during their clinic visits. Cocaine use was determined by directly observed urine toxicology testing twice weekly. The primary endpoint was full abstinence for the last 3 weeks of the trial. Thirty-one subjects completed the trial in the vigabatrin group (62%) and 22 in the placebo group (41.5%). A significantly higher number of completers on vigabatrin reached full end of trial abstinence ($n = 14$) compared with those in the placebo group ($n = 4$) and similarly more achieved partial abstinence ($n = 17$) in the vigabatrin group compared with the placebo group ($n = 5$). Twelve subjects in the vigabatrin group and two subjects in the placebo group maintained abstinence through the follow-up period. These studies demonstrate the efficacy and safety of vigabatrin in the treatment of cocaine dependence. However, more studies, preferably across different groups of patients would give more confidence in this as a viable treatment. In addition, the safety implications in using vigabatrin chronically may also limit its successfulness in long-term relapse prevention.

Another GABA modulating agent that has been investigated for its ability to affect cocaine dependence is the GABA reuptake blocker tiagabine. In a 10-week, randomized, double-blind, placebo-controlled pilot study (Gonzalez et al., 2003) 45 cocaine-dependent methadone-treated patients were randomized to either tiagabine (12 or 24 mg per day; $n = 15$ per group) or placebo ($n = 15$). The high dose of tiagabine significantly reduced the use of cocaine, as measured by the number of cocaine-free urines per week, compared with baseline in weeks 6–10 of the trial. This was also accompanied by a self-reported decrease in cocaine use. However, there was no clear evidence of an increase in abstinence rates relative to placebo.

These results prompted a second, better-powered, 10-week, randomized controlled trial, again in cocaine-dependent methadone-treated patients, comparing the effects of tiagabine (24 mg/day, $n = 25$), gabapentin (2,400 mg/day, $n = 26$), or placebo ($n = 25$) (Gonzalez et al., 2007). The main outcome of the study was the same as before, the number of cocaine-free urines per week. The urine was tested three times per week. This study confirmed the findings of the previous study that tiagabine significantly reduced the proportion of cocaine-free urine samples during weeks 6–10. In addition, they were able to show a significantly greater abstinent rate for the tiagabine-treated group compared with the other two treatment groups.

However, these results were not confirmed by two other trials of tiagabine in cocaine-dependent individuals (Winhusen et al., 2005; 2007). A preliminary 10-week study (Winhusen et al., 2005) comparing daily doses of tiagabine (20 mg, $n = 17$), sertraline (100 mg, $n = 16$), donepezil (10 mg, $n = 17$), and placebo ($n = 17$) with 1 h of manualized cognitive behavioral therapy a week showed only a non-significant trend toward cocaine-free urine in the tiagabine group using baseline-endpoint analysis. This reduction in cocaine-free urines was not significant compared with any of the other treatments or placebo. From these data the authors calculated that their study was underpowered and they would need at least 42 participants in the tiagabine group to have 80% power to detect a difference at the 0.05 level (two-sided). However, these authors felt these initial findings warranted a further properly powered study (Winhusen et al., 2007) to study the effects of tiagabine (20 mg/day, $n = 70$) versus placebo ($n = 71$). This was a 12-week, double-blind, placebo-controlled study in cocaine-dependent outpatients. As with the pilot study (Winhusen et al., 2005) the subjects also received 1 h of manualized individual cognitive behavioral therapy on a weekly basis. The outcomes of the study were not promising with there being no significant difference in the amount of cocaine non-use days or cocaine-free urine analysis between tiagabine and placebo. However, when these authors analyzed their data using the statistical analysis used in Gonzalez et al. (2007) they did find a significant effect suggesting that tiagabine, compared with placebo, significantly decreased cocaine use. The difference in significance depending on statistical test used does suggest that the effect of tiagabine at these doses, 24 mg/day (Gonzalez et al., 2003, 2007) and 20 mg/day (Winhusen et al., 2005, 2007) are not robust enough to conclude whether it is useful in treating cocaine dependence, but equally they also do not indicate conclusively that it is not. Other differences between the studies of the two groups that could account for the different results were that the subjects in Gonzalez et al. (2003, 2007) were also addicted to opiates and maintained on methodone. It is noteworthy that the tiagabine had an effect in this population when baclofen did not (Haney et al., 2006). Another difference in the two studies is that the subjects in Winhusen et al. (2005, 2007) were heavier users than those in Gonzalez et al. (2003, 2007), with nearly all smoking crack cocaine. This in combination with the lower dose of tiagabine used, 20 mg, may well account for the smaller effect.

Two human-based laboratory experiments (Lile et al., 2004b; Sofuoglu et al, 2005b) on the effect of tiagabine on cocaine have also produced conflicting evidence. Lile et al. (2004b) showed that acute tiagabine 4 and 8 mg had no effect on the behavioral or physiological effects of oral cocaine. However, Sofuoglu et al. (2005b) showed that tiagabine 8 mg had a significant effect on the subjective ratings of "stimulated" and "crave cocaine" compared with placebo following intravenous cocaine, but had no effect on any other behavioral or physiological measure. Apart from the route of cocaine administration these studies also differed in the pretreatment with tiagabine: Lile et al. (2004b) gave the tiagabine acutely 2 h before the cocaine challenge, while Sofuoglu et al. (2005b) gave the tiagabine in two doses of 4 mg one the night before and the other 2 h prior to testing. Based on these limited data we think that tiagabine clearly warrants further study as a potential for the treatment of cocaine dependence.

It seems that in line with the preclinical models of cocaine addiction, modulation of the $GABA_B$ receptor either directly with baclofen or indirectly with vigabatrin or tiagabine does have promise as a treatment target. Of the possible treatments mentioned here vigabatrin appears to be the most promising. This may reflect its mechanism of GABA elevation and therefore $GABA_B$ modulation. Vigabatrin is a mechanism-based irreversible inhibitor of GABA transaminase and so results in a prolonged elevation in GABA. This is in comparison to the GABA reuptake blocker tiagabine that causes an increase in synaptic GABA levels or baclofen that agonizes the $GABA_B$ receptor directly. These sites of action in conjunction with short pharmacokinetic profile of these two drugs mean that the dose schedule and level may be of much more importance in obtaining and maintaining their therapeutic effect.

IV. Nicotine

Modification of the $GABA_B$ receptor in nicotine addiction has received relatively little investigation in humans with only a few studies. In one such study, the acute effect of baclofen (20 mg) on cigarette smoking, craving for nicotine, cigarette taste, and smoking satisfaction was investigated in a human laboratory test (Cousins et al., 2001). Sixteen smokers (>15 cigarettes/day) were given baclofen or placebo after overnight abstinence. The subjects were then allowed to smoke their preferred cigarettes during a 3-h ad libitum smoking period, during which the self-reported ratings of craving for nicotine and the number of cigarettes smoked by the subjects were measured. Acute baclofen treatment had no effect on the number of cigarettes smoked nor did it affect the ratings of nicotine craving. However it did significantly change the "sensory" properties of the cigarettes, increasing ratings of "harsh" and decreasing ratings of "like cigarette's effects." This acute alteration in the smoker's perception of the cigarettes might indicate potential for a longer-term alteration in the pleasurable rewards associated with smoking.

In a similar study, tiagabine was tested to see its effects on intravenous nicotine (Sofuoglu et al., 2005a). Like the previous study this was a double-blind, placebo-controlled crossover design where 12 subjects were treated with tiagabine (4 or 8 mg) or placebo after overnight abstinence. Two hours posttreatment the subjects received i.v. saline, followed 30 min later by nicotine (1.5 mg/70 kg i.v.). Tiagabine had no effects on the physiological effects of nicotine on heart rate or blood pressure. However, it did significantly attenuate the ratings of "good effects" and "drug liking" compared with placebo and the 8 mg dose attenuated the craving for cigarettes compared with placebo or 4 mg tiagabine. These data indicate that it too may have the potential to reduce the rewarding effects of nicotine.

Recently, a preliminary double-blind, placebo-controlled smoking reduction study has been completed with promising results (Franklin et al., 2009). This was a 9-week pilot study of baclofen (80 mg/day), with the primary outcome of the number of cigarettes smoked per day. Participants were smokers contemplating, but not quite ready to quit and were given baclofen (titrated up to 20 mg four times a day (q.i.d.), $n = 30$) or placebo ($n = 30$). Baclofen was found to be significantly better at reducing the number of cigarettes smoked per day compared with placebo by the end of 9 weeks. These results do seem to imply that baclofen will be of use in smoking cessation and abstinence. Of particular note regarding this study is the dose and dose regimen of baclofen of 80 mg (20 mg q.i.d.). This is a higher dose than that used in any other study, other than case reports, investigating baclofen as an addiction treatment and given it was tolerated in this study might be a target dosage for future work.

V. Opiates

There are very few studies so far examining the use of $GABA_B$ modulation in opiate addiction in humans, even though the preclinical models show that there is potential for its efficacy (previous chapter). One possible reason for this is the wide acceptance and use of methadone and buprenorphine in the treatment of heroin addiction. However, these treatments are not available in all countries; therefore, other possible treatments could be of use. For this reason, a study of baclofen as such a treatment was carried out (Assadi et al., 2003). In this study, opioid-dependent patients were detoxified, over 2 weeks using a combination of clonidine, clonazepam, and thioridazine, and randomly assigned to receive baclofen (60 mg/day; 20 mg t.i.d.) or placebo in a 12-week, double-blind, parallel-group trial. The primary outcome measure for this study was retention in treatment and they found that subjects were significantly more likely to remain in treatment in the baclofen group compared with placebo. Baclofen was also significantly better compared with placebo in terms of opiate withdrawal syndrome as measured using the Short Opiate Withdrawal Scale and depression using the HAM-D. There was no significant effect on opioid craving, self-reported opioid use, and rates of opioid-positive urine tests compared with placebo, but the trends were toward a better outcome with baclofen. The main difficulty faced by the authors in this study was the small number of initial participants compounded by a very high dropout rate making accurate statistical analysis difficult. However, the results do indicate the possibility that baclofen may be of use as an abstinence-promoting maintenance treatment in opiate withdraw.

Baclofen has also been looked at with regard to its potential in treating opiate withdrawal. An initial open-label pilot study in five opiate-dependent individuals who were given baclofen (80 mg/day) after abrupt discontinuation of methadone (Krystal et al., 1992) found that while all of the subjects reported that the treatment reduced their discomfort it was not sufficient for three of the five who had to switch to clonidine for the remainder of their detoxification. The main complaints regarding baclofen's ineffectiveness were that there was not enough suppression of vomiting, myalgias, and headache. It is possible though that some of these problems were confounded by such a high dose of baclofen (80 mg/day), as vomiting and headaches are known side effects. Normally, baclofen is titrated up to this dose over a number of days to reduce these actions. Such a slow titration is impractical in rapid detoxification and may limit either the use of higher doses or even its usefulness in detoxification.

A more promising result was shown in another study where 62 addicts were treated with a lower dose of baclofen (40 mg/day, $n = 29$) or clonidine (0.8 mg/day, $n = 33$) during a 14-day, randomized, double-blind trial of their effectiveness in alleviating opiate withdrawal symptoms (Ahmadi-Abhari et al., 2001; Akhondzadeh et al., 2000). Treatment side effects (Ahmadi-Abhari et al., 2001) and subjective (Akhondzadeh et al., 2000) measures of opiate withdrawal were measured. The Short Opiate Withdrawal Scale with additional five items, dysphoria, anxiety, agitation, irritability, and craving for substances, to probe the psychological aspects was the main outcome. The results showed that baclofen and clonidine were equally effective at treating the physical symptoms. However, baclofen was significantly better at attenuating the psychological items studied compared with clonidine and this effect was visible from day 2 (Akhondzadeh et al., 2000). Generally there was no significant difference in retention in the trial due to the "side effects" of the two medications. Not unsurprisingly, it was found that there was significantly more hypotension and dry mouth with clonidine and more headaches, nausea, and vomiting with baclofen (Ahmadi-Abhari et al., 2001).

VI. Methamphetamine

A 16-week, randomized, placebo-controlled, double-blind trial was performed to test baclofen (20 mg t.i.d., $n = 25$) and gabapentin (800 mg t.i.d., $n = 26$) compared with placebo ($n = 37$), for the treatment of methamphetamine dependence (Heinzerling et al., 2006). In addition to the study medication, the participants had psychosocial counselling three times a week when they visited the clinic. At these clinic visits, study assessments and urine monitoring was carried out. The hypothesis for this study was that participants taking baclofen or gabapentin would have greater reductions in methamphetamine use, longer retention in treatment, greater reductions in depressive symptoms, and greater reductions in methamphetamine craving compared with placebo. Unfortunately, there were no significant effects of treatment compared with placebo for any of these measures. However, after post hoc analysis there was a significant association of methamphetamine-free urine in those participants in the baclofen group who took a higher percentage of the prescribed study medication compared with placebo. These data though preliminary suggest that in patients showing high compliance, baclofen might have some utility in treating methamphetamine addiction.

Vigabatrin has also been investigated for its ability to treat methamphetamine addiction in a 9-week open-label safety and efficacy pilot (Brodie et al., 2005). Thirty subjects with methamphetamine and/or cocaine dependency were treated with vigabatrin, titrated up to 3 g/day. Eighteen of the original 30 completed the trial and of those 18, 16 gave methamphetamine- and cocaine-free urine for the last 6 weeks. In addition, there were no changes in any of the vital signs or ocular measures (Fechtner et al., 2006).

The safety of vigabatrin was also confirmed in a human laboratory-based, double-blind, placebo-controlled, between-subjects study in non-treatment-seeking methamphetamine-dependent volunteers (De La Garza et al., 2009). Participants were admitted to the study center and given vigabatrin, titrated up to 5 g/day. They were then exposed to different doses of methamphetamine (15 and 30 mg i.v.) and cardiovascular and subjective measures taken. There was no significant effect of vigabatrin treatment on any cardiovascular measures or in attenuating the positive subjective effects produced by methamphetamine.

VII. GHB

Clinical experience with the GABA precursor compounds GHB (γ-hydroxybutyrate), GBL (γ-butyrolactone), and 1,4-butandiol has proved illuminating in considering the role of GABA in general in addictive processes. These relatively simple compounds are used in a variety of areas including as solvents or fertilizers, and experience has shown these to possess both abuse liability and dependence liability. The compounds act by gross elevation of the metabolic precursors of GABA, enhancing supply of GABA as the intermediate enzymatic pathway is not involved in the regulation of GABA production. The resultant enhanced GABA neurotransmission gives rise to the abuse-liable properties of these drugs, producing effects such as relaxation, intoxication, and euphoria. In these latter respects, it is very likely that the GABA receptor profile involved in exerting effects includes the $GABA_B$ as well as $GABA_A$ receptors (Lingenhoehl et al., 1999). There is also evidence for a distinct GHB receptor; however, the role of this particular receptor is at present uncertain. There are now a series of case reports detailing GHB dependence, which identify certain distinct qualities of the withdrawal syndrome—compared to those of alcohol or benzodiazepines—with increased incidence of paranoia, autonomic instability, and acute confusional states (Craig et al., 2000; Miotto et al., 2001; van Noorden et al., 2009; Wojtowicz et al., 2008). Further evidence for $GABA_B$ involvement in the action of GHB and withdrawal is that the withdrawal state is somewhat refractory to high-dose benzodiazepines (Addolorato et al., 1999) yet not entirely alleviated by baclofen administration (LeTourneau et al., 2008).

GHB itself has a very circumscribed clinical indication in the treatment of narcolepsy (Wong et al., 2003). Further, there are a series of studies detailing the efficacy of this compound in the treatment of alcohol withdrawal (Leone et al., 2010). In these studies there is undoubtedly evidence of alleviation of withdrawal symptoms with similar efficacy to benzodiazepine schedules. Studies detailing efficacy in maintaining abstinence have been less clear (Caputo et al., 2009; Leone et al., 2010) and should be set against the overwhelming evidence of the abuse liability of these compounds.

VIII. Discussion

The evidence of the literature seems to indicate that the $GABA_B$ receptor and its modulation have potential as a target for the treatment of addiction. There are two areas where it has been shown to be useful. One is the maintenance of abstinence and as an anti-craving treatment and the other is the alleviation of withdrawal syndromes. However, the amount of human clinical evidence of this is limited and not conclusive. There are a number of well-performed trials that indicate its usefulness and equally there are some that show it is of limited value. Therefore, there is a definite need for more of such trials to try and show conclusively one way or another if modulation of the $GABA_B$ receptor is going to be of use. These new studies need to build on the results of the previous trials and seek to avoid some of the problems of the earlier studies such as dosing regime, optimized compliance, and comorbid anxiety. Another interesting point that should be addressed is whether the modulation of the GABA$_B$ receptor by these preclinical and clinical studies is affecting the reward circuits in general and so has a true utility in any addiction, or whether it is only useful in certain narcotic dependencies.

A. Dose

The first obvious possibility for the discrepancies in the results of the various trials is do we have the right dose of the drugs? The doses used have generally been those that are considered normal for the indications that the drugs are being used before, for example, spasticity (baclofen) and epilepsy (tiagabine and vigabatrin). However, there is no reason that the effective doses in these conditions should also be effective in addiction. In addition, baclofen and tiagabine both have relatively short *in vivo* half-lives necessitating regular dosing to maintain a therapeutic level. The possibility that at these doses we are fluctuating around therapeutic level would be one way of explaining the conflicting results of the different trials. The evidence for this is currently anecdotal and comes from the various successful case studies that have been reported (Agabio et al., 2007; Ameisen, 2005; Brebner et al., 2002; Bucknam, 2007). All these reports cite higher doses of baclofen being used than most of the trials completed so far. In addition, they comment that usually the subjects were able to further increase their dose as needed if events became stressful and this helped to further suppress their anxiety and control cue-related craving maintaining their abstinence. Only one controlled trial used a dose of baclofen approaching those used in the case reports (Franklin et al., 2009). This was a successful study showing a significant effect on their primary and secondary outcome measures in subjects who were only interested in quitting but were not quite ready. The dose of baclofen was 80 mg/day refracted over four equal dose of 20 mg at 8:00 AM, 12:00 PM, 4:00 PM, and 8:00 PM. Another piece of evidence toward the idea that tiagabine and baclofen are not at an optimum dose is the relative success of vigabatrin compared with tiagabine in cocaine (Brodie et al., 2003, 2009) and methamphetamine (Brodie et al., 2005) dependency. The mechanism by which vigabatrin elevates GABA is by permanently and covalently inhibiting GABA transaminase, the main catabolic enzyme in the metabolism of GABA. This results in a chronic and long-lived elevation of GABA as compared with the usually short-lived elevation in synaptic GABA with the GABA reuptake inhibitor tiagabine.

Another consideration regarding dose that would be interesting to investigate is whether the "anti-addiction" dose differs for the narcotic of choice. Baclofen has shown most success in the treatment of alcohol addiction (Addolorato et al., 2000, 2002a; Addolorato et al., 2002b; Flannery et al., 2004) where they used the lowest dose, 30 mg/day (10 mg t.i.d.). Alcohol has more similarities to GABA-acting drugs than either stimulants or opioids.

While the less convincing studies into baclofen's effects on cocaine (Brebner et al., 2002; Haney et al., 2006; Kahn et al., 2009; Ling et al., 1998; Shoptaw et al., 2003) or methamphetamine (Heinzerling et al., 2006) addiction routinely used 60 mg/day (20 mg t.i.d.), do these differing efficacies represent the nature and mechanism of the addictive drugs? For example, cocaine and methamphetamine act directly on dopamine reuptake and release and therefore reward pathways while alcohol acts less so.

B. Compliance

A related question to dose is whether compliance in taking the medication is having a particularly detrimental effect of the results of many of the studies. Motivation, participant attrition, and compliance of any addict to keep taking the treatment are a problem in all studies in such addiction medications. However, as mentioned, the short half-life of baclofen *in vivo* means that three or four separate doses are required to be taken over the day. Failure to do this regularly may well result in fluctuations in baclofen concentrations below the optimum therapeutic levels. In addition, the motivation of the subject to stop taking the drug will also impact on compliance. Evidence that this may be the case are the particularly successful studies with alcohol where highly motivated patients with good social and family support were used (Addolorato et al., 2000, 2002a; Addolorato et al., 2002b). A family member was responsible for keeping and administering the baclofen and attending the clinic visits with them. This level of family support in conjunction with their own motivation to stop drinking may have resulted in a much higher level of compliance taking the baclofen. These factors probably explain why studies were so positive compared to any other despite the lowest daily doses of baclofen, 30 mg.

C. Comorbid Affective Disorders

Addiction and its treatment is often complicated by comorbid anxiety disorders (Lingford-Hughes et al., 2004). Therefore, the use of modulators of the $GABA_B$ receptor to treat addiction may come with the added benefit of addressing the treatment of the comorbid anxiety, as the $GABA_B$ receptor has been linked with anxiety and agonists have been shown to be anxiolytic (Cryan & Kaupmann, 2005). In addition, some of the studies using baclofen have also indicated improvements in the subject's anxiety (Addolorato et al., 2002a; Akhondzadeh et al., 2000; Ameisen, 2005; Flannery et al., 2004).

D. Future Directions

The results of these human trials investigating the $GABA_B$ receptor's modulation in the various treatments of addiction show this is an area warranting further study. The results are conflicting but show promise. There is a definite need for more research into the currently available $GABA_B$ modulators available for human use, baclofen, tiagabine and vigabatrin, to prove or disprove convincingly if this is a real target for the treatment of addiction, and in the case of the latter two drugs that they do affect the $GABA_B$ system. This research needs to address adequately the problems of dose and dose regimens, especially with regard to therapeutic effect. In addition, the possibilities of different doses for initiation of abstinence, and longer-term maintenance doses to suppress craving so preventing relapse and promoting abstinence in conjunction with psychosocial treatment should be investigated. In addictions where withdrawal syndromes occur, the treatment of these with $GABA_B$ drugs also needs clarification. The safety concerns surrounding vigabatrin's long-term use may limit its utility but it is used long-term in refractory epilepsy and so should not be ruled out in addicts whose drug use is life-threatening, provided it was showing clear benefits. The long-term safety of baclofen and tiagabine is widely accepted. Another area that requires resolution is that of compliance with medication especially if three or four daily doses are required. One way to resolve this would be the development of longer-acting agonists or positive allosteric modulators for human use. Another strategy would be the development of slow-release formulations of baclofen, some of which are already in development for spasticity (Abdelkader et al., 2008; Lagarce et al., 2005). More research in humans is also needed in confirming the actual role the $GABA_B$ receptor system is playing in the modulation of the reward pathways: Is it specific to the dopamine system or is it due to a more generic modulation? What is happening to the $GABA_B$ receptor density or in chronic substance misuse? Are there differential effects in the regulation of $GABA_{B1}$ and $GABA_{B2}$? Is this different depending on the narcotic? There is limited information from preclinical studies to answer these questions. There is a study suggesting that there is no change in [^{3}H]baclofen binding density after chronic ethanol treatment in rats (Peris et al., 1997) while there is another that suggests that chronic nicotine exposure may differentially affect the $GABA_{B1}$ and $GABA_{B2}$ subunit regulation (Li et al., 2004). In addition, there is also evidence that alcohol (Castelli et al., 2005) and nicotine (Amantea et al., 2004) may alter the G-protein coupling. Understanding this both preclinically and clinically will be invaluable. Human imaging would be one way forward in this area, using either functional or pharmacological MRI, or PET. The labeling of baclofen with carbon-11 has recently been reported (Kato et al., 2009) but its usefulness *in vivo* is still yet to be evaluated. However, this may well be an invaluable tool, and the development of other similar tracers, in the study of the $GABA_B$ system and its role in humans.

IX. Conclusion

In conclusion, while modulators of $GABA_B$ receptors, such as baclofen, may not be "cures" for addiction, they certainly seem to have potential as adjunct treatments to aid in the initiation of abstinence, maintenance of abstinence, and prevention of cue-related relapse. This potential is at present poorly understood or studied and warrants further investigation with either higher dosing or different formulations of the shorter-acting agents such as baclofen or tiagabine or the development of other $GABA_B$ receptor agonists or positive allosteric modulators, with better pharmacokinetic profiles, for use in humans.

Conflict of Interest Statement: The authors are all members of the Neuropsychopharmacology Unit, Imperial College London and Professors D. J. Nutt and A. Lingford-Hughes and Dr L. J. Reed also hold clinical posts with Central and North West London NHS Trust.

Professor D. J. Nutt has provided consultancy services to Pfizer, GSK, Novartis, Organon, Cypress, Lilly, Janssen, Lundbeck, BMS, Astra Zeneca, Servier, Hythiam, and Sepracor; received honoraria from Wyeth, Reckitt-Benkiser, and Cephalon; and received grants or clinical trials payments from MSD, GSK, Novartis, Servier, Janssen, Lundbeck, Pfizer, Wyeth, Organon, and P1Vital.

Professor A. Lingford-Hughes has received honoraria from Servier and Janssen-Cilag and consultancy fees from NET Device Corp, is member of BMS Core Faculty, and has research collaborations with GSK and unrestricted research grants from Wyeth, Interneuron, and Merck.

Dr L. J. Reed has accepted honoraria for attendance at discussion meetings held by Eli Lilly and Company and General Electric Healthcare, and has accepted grant funding in the form of experimental medicine grants from Pfizer Global Medicine and Amylin Pharmaceuticals.

Dr R. J. Tyacke has research collaborations with GSK.

References

Abdelkader, H., Youssef Abdalla, O., & Salem, H. (2008). Formulation of controlled-release baclofen matrix tablets. II. Influence of some hydrophobic excipients on the release rate and in vitro evaluation. *AAPS PharmSciTech*, *9*(2), 675–683.

Addolorato, G., Caputo, F., Capristo, E., et al. (1999). A case of gamma-hydroxybutyric acid withdrawal syndrome during alcohol addiction treatment: Utility of diazepam administration. *Clinical Neuropharmacology*, *22*(1), 60–62.

Addolorato, G., Caputo, F., Capristo, E., et al. (2000). Ability of baclofen in reducing alcohol craving and intake: II–Preliminary clinical evidence. *Alcoholism, Clinical and Experimental Research*, *24*(1), 67–71.

Addolorato, G., Caputo, F., Capristo, E., et al. (2002a). Baclofen efficacy in reducing alcohol craving and intake: A preliminary double-blind randomized controlled study. *Alcohol and Alcoholism*, *37*(5), 504–508.

Addolorato, G., Caputo, F., Capristo, E., et al. (2002b). Rapid suppression of alcohol withdrawal syndrome by baclofen. *American Journal of Medicine*, *112*(3), 226–229.

Addolorato, G., Leggio, L., Abenavoli, L., et al. (2003). Suppression of alcohol delirium tremens by baclofen administration: A case report. *Clinical Neuropharmacology*, *26*(5), 258–262.

Addolorato, G., Leggio, L., Abenavoli, L., et al. (2006). Baclofen in the treatment of alcohol withdrawal syndrome: A comparative study vs diazepam. *American Journal of Medicine*, *119*(3), 276, e213–e278.

Addolorato, G., Leggio, L., Ferrulli, A., et al. (2007). Effectiveness and safety of baclofen for maintenance of alcohol abstinence in alcohol-dependent patients with liver cirrhosis: Randomised, double-blind controlled study. *Lancet*, *370*(9603), 1915–1922.

Agabio, R., Marras, P., Addolorato, G., et al. (2007). Baclofen suppresses alcohol intake and craving for alcohol in a schizophrenic alcohol-dependent patient: A case report. *Journal of Clinical Psychopharmacology*, *27*(3), 319–320.

Ahmadi-Abhari, S. A., Akhondzadeh, S., Assadi, S. M., et al. (2001). Baclofen versus clonidine in the treatment of opiates withdrawal, side-effects aspect: A double-blind randomized controlled trial. *Journal of Clinical Pharmacy and Therapeutics*, *26*(1), 67–71.

Akhondzadeh, S., Ahmadi-Abhari, S. A., Assadi, S. M., et al. (2000). Double-blind randomized controlled trial of baclofen vs. clonidine in the treatment of opiates withdrawal. *Journal of Clinical Pharmacy and Therapeutics*, *25*(5), 347–353.

Amantea, D., Tessari, M., & Bowery, N. G. (2004). Reduced G-protein coupling to the GABA$_B$ receptor in the nucleus accumbens and the medial prefrontal cortex of the rat after chronic treatment with nicotine. *Neuroscience Letters*, *355*(3), 161–164.

Ameisen, O. (2005). Complete and prolonged suppression of symptoms and consequences of alcohol-dependence using high-dose baclofen: A self-case report of a physician. *Alcohol and Alcoholism*, *40*(2), 147–150.

Assadi, S. M., Radgoodarzi, R., & Ahmadi-Abhari, S. A. (2003). Baclofen for maintenance treatment of opioid dependence: A randomized double-blind placebo-controlled clinical trial [ISRCTN32121581]. *BMC Psychiatry*, *3*, 16.

Brebner, K., Childress, A. R., & Roberts, D. C. (2002). A potential role for GABA(B) agonists in the treatment of psychostimulant addiction. *Alcohol and Alcoholism*, *37*(5), 478–484.

Brodie, J. D., Case, B. G., Figueroa, E., et al. (2009). Randomized, double-blind, placebo-controlled trial of vigabatrin for the treatment of cocaine dependence in Mexican parolees. *American Journal of Psychiatry*, *166*(11), 1269–1277.

Brodie, J. D., Figueroa, E., & Dewey, S. L. (2003). Treating cocaine addiction: From preclinical to clinical trial experience with gamma-vinyl GABA. *Synapse*, *50*(3), 261–265.

Brodie, J. D., Figueroa, E., Laska, E. M., et al. (2005). Safety and efficacy of gamma-vinyl GABA (GVG) for the treatment of methamphetamine and/or cocaine addiction. *Synapse*, *55*(2), 122–125.

Bucknam, W. (2007). Suppression of symptoms of alcohol dependence and craving using high-dose baclofen. *Alcohol and Alcoholism*, *42*(2), 158–160.

Caputo, F., Vignoli, T., Maremmani, I., et al. (2009). Gamma hydroxybutyric acid (GHB) for the treatment of alcohol dependence: A review. *International Journal of Environmental Research and Public Health*, *6*(6), 1917–1929.

Castelli, M. P., Pibiri, F., Piras, A. P., et al. (2005). Differential G-protein coupling to GABA$_B$ receptor in limbic areas of alcohol-preferring and -nonpreferring rats. *European Journal of Pharmacology*, *523*(1–3), 67–70.

Cousins, M. S., Stamat, H. M., & de Wit, H. (2001). Effects of a single dose of baclofen on self-reported subjective effects and tobacco smoking. *Nicotine & Tobacco Research*, *3*(2), 123–129.

Craig, K., Gomez, H. F., McManus, J. L., et al. (2000). Severe gamma-hydroxybutyrate withdrawal: A case report and literature review. *Journal of Emergency Medicine*, *18*(1), 65–70.

Cryan, J. F., & Kaupmann, K. (2005). Don't worry 'b' happy!: A role for GABA(B) receptors in anxiety and depression. *Trends in Pharmacological Sciences*, *26*(1), 36–43.

De La Garza, R. 2nd., Zorick, T., Heinzerling, K. G., et al. (2009). The cardiovascular and subjective effects of methamphetamine combined with gamma-vinyl-gamma-aminobutyric acid (GVG) in non-treatment seeking methamphetamine-dependent volunteers. *Pharmacology Biochemistry and Behavior*, *94*(1), 186–193.

Dewey, S. L., Chaurasia, C. S., Chen, C. E., et al. (1997). GABAergic attenuation of cocaine-induced dopamine release and locomotor activity. *Synapse*, *25*(4), 393–398.

Dewey, S. L., Morgan, A. E., Ashby, C. R. Jr., et al. (1998). A novel strategy for the treatment of cocaine addiction. *Synapse*, *30*(2), 119–129.

Evans, S. M., & Bisaga, A. (2009). Acute interaction of baclofen in combination with alcohol in heavy social drinkers. *Alcoholism, Clinical and Experimental Research*, *33*(1), 19–30.

Fechtner, R. D., Khouri, A. S., Figueroa, E., et al. (2006). Short-term treatment of cocaine and/or methamphetamine abuse with vigabatrin: Ocular safety pilot results. *Archives of Ophthalmology*, *124*(9), 1257–1262.

Fink-Jensen, A., Suzdak, P. D., Swedberg, M. D., et al. (1992). The gamma-aminobutyric acid (GABA) uptake inhibitor, tiagabine, increases extracellular brain levels of GABA in awake rats. *European Journal of Pharmacology*, *220*(2–3), 197–201.

Flannery, B. A., Garbutt, J. C., Cody, M. W., et al. (2004). Baclofen for alcohol dependence: A preliminary open-label study. *Alcoholism, Clinical and Experimental Research*, *28*(10), 1517–1523.

Franklin, T. R., Harper, D., Kampman, K., et al. (2009). The GABA B agonist baclofen reduces cigarette consumption in a preliminary double-blind placebo-controlled smoking reduction study. *Drug and Alcohol Dependence*, *103*(1–2), 30–36.

Garbutt, J. C. (2009). The state of pharmacotherapy for the treatment of alcohol dependence. *Journal of Substance Abuse Treatment*, *36*(1), S15–S23.

Gonzalez, G., Desai, R., Sofuoglu, M., et al. (2007). Clinical efficacy of gabapentin versus tiagabine for reducing cocaine use among cocaine dependent methadone-treated patients. *Drug and Alcohol Dependence*, *87*(1), 1–9.

Gonzalez, G., Sevarino, K., Sofuoglu, M., et al. (2003). Tiagabine increases cocaine-free urines in cocaine-dependent methadone-treated patients: Results of a randomized pilot study. *Addiction*, *98*(11), 1625–1632.

Haney, M., Hart, C. L., & Foltin, R. W. (2006). Effects of baclofen on cocaine self-administration: Opioid- and nonopioid-dependent volunteers. *Neuropsychopharmacology*, *31*(8), 1814–1821.

Heinzerling, K. G., Shoptaw, S., Peck, J. A., et al. (2006). Randomized, placebo-controlled trial of baclofen and gabapentin for the treatment of methamphetamine dependence. *Drug and Alcohol Dependence*, *85*(3), 177–184.

Hilton, E. J., Hosking, S. L., & Betts, T. (2004). The effect of antiepileptic drugs on visual performance. *Seizure*, *13*(2), 113–128.

Kahn, R., Biswas, K., Childress, A. R., et al. (2009). Multi-center trial of baclofen for abstinence initiation in severe cocaine-dependent individuals. *Drug and Alcohol Dependence*, *103*(1–2), 59–64.

Kato, K., Zhang, M. R., & Suzuki, K. (2009). Synthesis of (R,S)-[4-11c]baclofen via Michael addition of nitromethane labeled with short-lived 11c. *Bioorganic & Medicinal Chemistry Letters*, *19*(21), 6222–6224.

Krystal, J. H., McDougle, C. J., Kosten, T. R., et al. (1992). Baclofen-assisted detoxification from opiates. A pilot study. *Journal of Substance Abuse Treatment*, *9*(2), 139–142.

Kushner, S. A., Dewey, S. L., & Kornetsky, C. (1999). The irreversible gamma-aminobutyric acid (GABA) transaminase inhibitor gamma-vinyl-GABA blocks cocaine self-administration in rats. *Journal of Pharmacology and Experimental Therapeutics*, *290*(2), 797–802.

Lagarce, F., Faisant, N., Desfontis, J. C., et al. (2005). Baclofen-loaded microspheres in gel suspensions for intrathecal drug delivery: In vitro and in vivo evaluation. *European Journal of Pharmaceutics and Biopharmaceutics*, *61*(3), 171–180.

Leone, M. A., Vigna-Taglianti, F., Avanzi, G., et al. (2010). Gamma-hydroxybutyrate (GHB) for treatment of alcohol withdrawal and prevention of relapses. *Cochrane Database of Systematic Reviews*, *2*, CD006266.

LeTourneau, J. L., Hagg, D. S., & Smith, S. M. (2008). Baclofen and gamma-hydroxybutyrate withdrawal. *Neurocritical Care*, *8*(3), 430–433.

Li, S. P., Park, M. S., Kim, J. H., et al. (2004). Chronic nicotine and smoke treatment modulate dopaminergic activities in ventral tegmental area and nucleus accumbens and the gamma-aminobutyric acid type B receptor expression of the rat prefrontal cortex. *Journal of Neuroscience Research*, *78*(6), 868–879.

Lile, J. A., Stoops, W. W., Allen, T. S., et al. (2004a). Baclofen does not alter the reinforcing, subject-rated or cardiovascular effects of intranasal cocaine in humans. *Psychopharmacology (Berl)*, *171*(4), 441–449.

Lile, J. A., Stoops, W. W., Glaser, P. E., et al. (2004b). Acute administration of the GABA reuptake inhibitor tiagabine does not alter the effects of oral cocaine in humans. *Drug and Alcohol Dependence*, *76*(1), 81–91.

Ling, W., Shoptaw, S., & Majewska, D. (1998). Baclofen as a cocaine anti-craving medication: A preliminary clinical study. *Neuropsychopharmacology*, *18*(5), 403–404.

Lingenhoehl, K., Brom, R., Heid, J., et al. (1999). Gamma-hydroxybutyrate is a weak agonist at recombinant GABA(B) receptors. *Neuropharmacology*, *38*(11), 1667–1673.

Lingford-Hughes, A. R., Welch, S., & Nutt, D. J. (2004). Evidence-based guidelines for the pharmacological management of substance misuse, addiction and comorbidity: Recommendations from the British Association for Psychopharmacology. *Journal of Psychopharmacology*, *18*(3), 293–335.

Miotto, K., Darakjian, J., Basch, J., et al. (2001). Gamma-hydroxybutyric acid: Patterns of use, effects and withdrawal. *American Journal on Addictions*, *10*(3), 232–241.

Myrick, H., Taylor, B., LaRowe, S., et al. (2005). A retrospective chart review comparing tiagabine and benzodiazepines for the treatment of alcohol withdrawal. *Journal of Psychoactive Drugs*, *37*(4), 409–414.

Peris, J., Eppler, B., Hu, M., et al. (1997). Effects of chronic ethanol exposure on GABA receptors and GABAB receptor modulation of 3h-GABA release in the hippocampus. *Alcoholism, Clinical and Experimental Research*, *21*(6), 1047–1052.

Shoptaw, S., Yang, X., Rotheram-Fuller, E. J., et al. (2003). Randomized placebo-controlled trial of baclofen for cocaine dependence: Preliminary effects for individuals with chronic patterns of cocaine use. *Journal of Clinical Psychiatry*, *64*(12), 1440–1448.

Sofuoglu, M., Mouratidis, M., Yoo, S., et al. (2005a). Effects of tiagabine in combination with intravenous nicotine in overnight abstinent smokers. *Psychopharmacology (Berl)*, *181*(3), 504–510.

Sofuoglu, M., Poling, J., Mitchell, E., et al. (2005b). Tiagabine affects the subjective responses to cocaine in humans. *Pharmacology Biochemistry and Behavior*, *82*(3), 569–573.

Stuppaeck, C. H., Deisenhammer, E. A., Kurz, M., et al. (1996). The irreversible gamma-aminobutyrate transaminase inhibitor vigabatrin in the treatment of the alcohol withdrawal syndrome. *Alcohol and Alcoholism*, *31*(1), 109–111.

van Noorden, M. S., van Dongen, L. C., Zitman, F. G., et al. (2009). Gamma-hydroxybutyrate withdrawal syndrome: Dangerous but not well-known. *General Hospital Psychiatry*, *31*(4), 394–396.

Winhusen, T., Somoza, E., Ciraulo, D. A., et al. (2007). A double-blind, placebo-controlled trial of tiagabine for the treatment of cocaine dependence. *Drug and Alcohol Dependence*, *91*(2–3), 141–148.

Winhusen, T. M., Somoza, E. C., Harrer, J. M., et al. (2005). A placebo-controlled screening trial of tiagabine, sertraline and donepezil as cocaine dependence treatments. *Addiction*, *100*(Suppl. 1), 68–77.

Wojtowicz, J. M., Yarema, M. C., & Wax, P. M. (2008). Withdrawal from gamma-hydroxybutyrate, 1,4-butanediol and gamma-butyrolactone: A case report and systematic review. *Canadian Journal of Emergency Medical Care*, *10*(1), 69–74.

Wong, C. G., Bottiglieri, T., & Snead, O. C. 3rd. (2003). GABA, gamma-hydroxybutyric acid, and neurological disease. *Annals of Neurology*, *54*(Suppl. 6), S3–S12.

Nathan P. Cramer[*], Tyler K. Best[*], Marcus Stoffel[‡], Richard J. Siarey[*], and Zygmunt Galdzicki[*]

[*]Department of Anatomy, Physiology, and Genetics, School of Medicine, Uniformed Services University of the Health Sciences, Bethesda, Maryland, USA

[‡]ETH Zurich, Institute of Molecular Systems Biology, Zurich, Switzerland

GABA$_B$–GIRK2-Mediated Signaling in Down Syndrome

Abstract

Down syndrome (DS) results from the presence of an extra copy of genes on the long-arm of chromosome 21. Aberrant expression of these trisomic genes leads to widespread neurological changes that vary in their severity. However, how the presence of extra genes affects the physiological and behavioral phenotypes associated with DS is not well understood. The most likely cause of the complex DS phenotypes is the overexpression of dosage-sensitive genes. However, other factors, such as the complex interactions between gene products as proteins and noncoding RNAs, certainly play significant roles contributing to the spectrum of severity. Here we will

Advances in Pharmacology, Volume 58

1054-3589/10 $35.00
10.1016/S1054-3589(10)58015-3

review evidence regarding how the overexpression of one particular gene encoding for G-protein-activated inward rectifying potassium type 2 (GIRK2) channel subunit and its coupling to $GABA_B$ receptors may contribute to a range of mental and functional disabilities in DS.

I. Introduction

Down syndrome (DS) is a genetic disorder that results from the presence of an extra copy of the long arm of chromosome 21. It is characterized by a wide range of physiological and behavioral abnormalities that vary in their severity between individuals (reviewed in Lejeune, 1959; Neri & Opitz, 2009). Phenotypes common to all individuals with DS are some degree of cognitive impairments as well as specific craniofacial abnormalities, muscle hypotonia, and the histopathology of Alzheimer's disease after 35 years of age (Antonarakis et al., 2004; Prandini et al., 2007). It is widely accepted that the expression of genes and nonprotein coding functional sequences (e.g. microRNA (miRNA)) on the extra copy of chromosome 21 is causally related to the appearance of DS phenotypes but the exact mechanisms through which the phenotypes are conferred remain uncertain (Patterson, 2009). However, a variation of the gene-dosage hypothesis is the most likely explanation of many of DS phenotypes.

According to the gene-dosage hypothesis, the increased expression of a specific or small set of dosage-sensitive genes results in a specific DS phenotype (Rachidi & Lopes, 2007). Studies looking at the expression levels of trisomic genes in both humans (Mao et al., 2003) and mouse models (Amano et al., 2004; Kahlem et al., 2004; Lyle et al., 2004) provide some support for a gene-dosage based hypothesis (reviewed in Gardiner, 2004). These studies found increases in the level of trisomic gene expression in a variety of tissues including the central nervous system (CNS). However, where examined, there was a wide variation in the expression levels between individuals and the expression levels of some nontrisomic genes were also altered (Amano et al., 2004; Mao et al., 2003). Such variable DS cognitive phenotypes likely result from the complex interactions of multiple genes suppressing or enhancing cognitive performance (Korbel et al., 2009; Olson et al., 2007). Nevertheless, given the observed increase in expression levels of trisomic genes, it is reasonable to conclude that gene dosage levels are causally related to DS phenotypes.

Consistent with the hypothesis of gene-dosage, it has been proposed that there exists a Down syndrome critical region (DSCR) chromosomal segment that, when triplicated, is sufficient to produce many of the DS phenotypes (Roper & Reeves, 2006). This concept had its origins in the observations of shared features between partial and full trisomic individuals. This hypothesis was tested directly by Olson et al. (2004, 2007) using a mouse model trisomic for the murine analogs of the human DSCR. The investigators found that mice trisomic only for this segment did not display the physical and behavioral changes normally seen in DS mice. On the other hand, mice that were trisomic for murine analogs of human chromosome 21 genes with the exception of the DSCR (produced by breeding general trisomic mice with mice monosomic for the DSCR) displayed normal performance on the Morris water maze test indicating that this region is necessary but not sufficient for the DS phenotype. In a more recent study, Belichenko et al. reexamined the DSCR hypothesis using a different strain of mice and found that the DSCR was sufficient for producing some DS phenotypes in facia dentate (Belichenko et al., 2009). These and other similar results suggest that, although changes in the dosage levels of specific genes may contribute to a phenotype, there is no single gene or region of Hsa21 responsible for all features of DS (Korbel et al., 2009; Olson et al., 2004, 2007; Pereira et al., 2009). The expression of different alleles, interactions between different genes, and the additive effects of altered expression over time also contribute to the presence or severity of any given phenotype (Roper & Reeves, 2006). The relative importance of each permutation, however, will need to be determined for each characteristic of DS.

Previously, we have shown that the Kcnj6 gene that encodes the type 2 G-protein-activated inward rectifying potassium (GIRK2) channel is overexpressed in a mouse model of DS (Harashima et al., 2006a). We also found that this led to an exaggerated $GABA_B$ receptor-mediated inhibitory response in neurons from the hippocampus, a brain region critically involved in learning and memory (Best et al., 2007). In this review, we will focus on the putative role of increased inhibition in the DS brain in conferring aberrant cognition with a special emphasis on $GABA_B$–GIRK signaling. We will also speculate as to how the overexpression of GIRK may contribute to other DS pathologies through its interactions with $GABA_B$ receptors.

II. Mouse Models of DS

To better understand how triplication of a subset of genes, such as Kcnj6/Girk2, leads to the pathologies associated with DS it is necessary to have representative animal models in which detailed behavioral, genetic, and neurophysiological experiments can be performed and then pharmacological intervention tested. Multiple mouse models for DS currently exist and are made possible by the substantial homology between the long-arm of human chromosome 21 and the distal portion of mouse chromosome 16 (Patterson, 2009). This suggests that the trisomy 16 (Ts16) mouse (one extra "full" chromosome 16) would be a useful genetic animal model for human DS (Galdzicki et al., 2001; Lacey-Casem & Oster-Granite, 1994). Indeed, Ts16 mouse fetuses have a number of phenotypic characteristics similar to those of DS subjects, including cardiac defects and anasarca (Epstein et al.,

1985). Unfortunately, Ts16 fetuses die *in utero*, so research on them has been limited to early developmental stages, cell culture systems, and neuronal Ts16 tissue grafted into brain ventricles (Rapoport & Galdzicki, 1994). In addition, although segments of mouse chromosome 16 have significant homology with portions of human chromosome 21, most of the genes on chromosome 16 are not triplicated in humans with DS (Mural et al., 2002).

Mouse models that are trisomic for different segments of chromosome 16 (Ts65Dn, Ts1Cje, and Ts1Rhr) have been produced to offer a better match with the genes triplicated in humans (Fig. 1; Patterson, 2009). These segmental trisomic mice survive into adulthood and show a range of phenotypes similar to those observed in DS humans, depending on the specific genes that have been triplicated. Importantly, all three mouse models show differing degrees of impairment on spatial learning tasks and hippocampal synaptic plasticity with an apparent correlation between performance and the size of the trisomic gene segment (Belichenko et al., 2009; Siarey et al., 2005). Tc1 mice, a more recent model, carry an almost complete copy of human chromosome 21 (Hsa21) thereby overexpressing genes not triplicated in other mouse models (O'Doherty et al., 2005). The varying levels of trisomy offered by these multiple mouse models provide a powerful tool for investigating how clusters of genes contribute to DS phenotypes.

III. Evidence for Cognitive Impairment in Mouse Models of DS

In the Morris water maze test, the Ts65Dn mouse is impaired at finding both visible and hidden platforms; on removal of the platform, the Ts65Dn mouse has a decreased search time in the correct (platform) quadrant compared with the diploid control mouse (Escorihuela et al., 1995; Holtzman et al., 1996; Reeves et al., 1995). Experiments using the radial arm maze demonstrate that the Ts65Dn mouse makes fewer correct choices, performing at or near chance levels, than the diploid control mouse (Demas et al., 1996). In passive avoidance tests, both control and Ts65Dn mice learned the task (Coussons-Read & Crnic, 1996; Holtzman et al., 1996), but more variability was found in the Ts65Dn data (Holtzman et al., 1996). Impaired spatial working and reference memory was reported in the Ts65Dn mouse in a 12-arm radial maze (Demas et al., 1996). Between 4 and 8 months of age, the Ts65Dn mouse shows a decrease in performance in spatial learning and reversal, but not visual discrimination learning and reversal (Granholm et al., 2000; Hunter et al., 2003).

The Ts1Cje mouse possesses a smaller extra segment of mouse chromosome 16 than that of the Ts65Dn mouse (Sago et al., 1998). These mice also show behavioral deficits in the Morris water maze similar to the Ts65Dn but the reverse probe test indicates that impairments are less dramatic than the ones reported for the Ts65Dn mouse (Sago et al., 1998, 2000). The locomotor activity of the two mouse models also differs with Ts1Cje mice

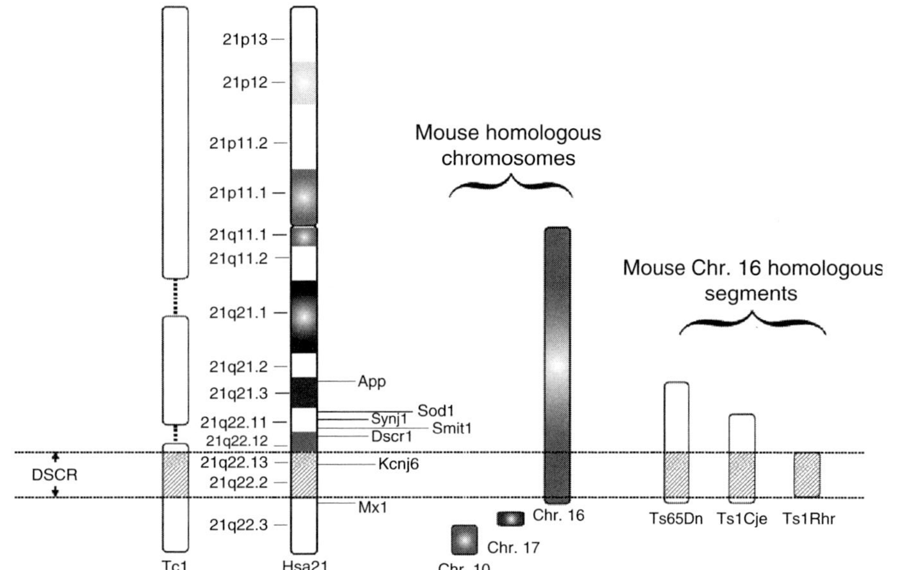

FIGURE 1 Human chromosome 21 (Hsa21) map showing the Down syndrome critical region (DSCR) and the relationship of chromosomes from three common mouse models for DS. Additional segments from mouse chromosomes 10 and 17 are shown along with their homologous portions of Hsa21. Noted genes: amyloid precursor protein (App), superoxide dismutase-1 (Sod1), Synaptojanin 1 (Synj1), solute carrier family 5 (inositol transporters) member 3 (Smit1/Slc5a3), Down syndrome candidate region-1 (Dscr1), G-protein-coupled inward rectifying potassium channel subunit-2 (Girk2), and myxovirus (influenza virus) resistance 1 (Mx1).

tending to be hypoactive and Ts65Dn more hyperactive. The Ts65Rhr mouse may show cognitive deficiencies depending on the strain (Belichenko et al., 2009; Olson et al., 2004, 2007). As outlined in the following sections, it is likely that dysfunctional information processing within the hippocampus is causally related to these cognitive impairments.

IV. Hippocampal Deficit in DS and in DS Mouse Models

Neuropsychological evidences point to hippocampal dysfunction in DS individuals (Nadel, 2003; Pennington et al., 2003), a feature well represented in mouse models. The hippocampus is part of the limbic system and plays an important role in learning and memory. It is a site for a long-term synaptic plasticity that appears to be critical to memory formation, consolidation, and retrieval, and, therefore, has been extensively studied in the modeling of learning and memory (Neves et al., 2008). Brief high-frequency activation of specific inputs causes a persistent increase in synaptic responsiveness (an increase in the excitatory postsynaptic potentials termed long-term potentiation (LTP)) that under certain circumstances can last for hours, days, or weeks (Bliss & Collingridge, 1993). The hippocampi from Ts65Dn, Ts1Cje, Ts65Rhr and Ts1 mice show a reduced capacity for LTP compared with age-matched controls (Costa & Grybko, 2005; Kleschevnikov et al., 2004; Siarey et al., 1997b; Siarey et al., 2005; O'Doherty et al., 2005; Morice et al., 2008; however, see Olson et al., 2007 concerning LTP in Ts65Rhr mice). In contrast, in the CA1, long-term depression (LTD), a decrease in the magnitude of excitatory postsynaptic potentials following low-frequency stimulation, is enhanced in Ts65Dn mice (Siarey et al., 1999) and in Ts1Cje mice (Siarey et al., 2005). Reductions in both excitatory and inhibitory inputs to pyramidal neurons in CA3 of the Ts65Dn hippocampus were also reported (Hanson et al., 2007). Thus, critical mechanisms for selectively strengthening or weakening specific synapses within the hippocampi are aberrant in DS mice (Fig. 2).

Several lines of evidence suggest that the cognitive deficits and LTP abnormalities observed in mouse models of DS arise, at least in part, from overinhibition in the hippocampal circuitries. Within the Ts65Dn hippocampus there is a widespread and significant decrease in the presumed excitatory synapse-to-neuron ratio while the ratio of presumed inhibitory synapse-to-neuron ratio was not significantly different from controls (Kurt et al., 2004). Consequently, the overall ratio of inhibition to excitation in DS hippocampi is likely increased. Additionally, the observed deficiencies in LTP of synapses in all three mouse models of DS can be rescued by reducing the magnitude of $GABA_A$ receptor-mediated inhibition (Belichenko et al., 2007; Costa & Grybko, 2005; Fernandez et al., 2007; Kleschevnikov et al., 2004). In behavioral tests, administration of $GABA_A$ receptor antagonists improves the performance of Ts65Dn mice (Fernandez et al., 2007; Rueda

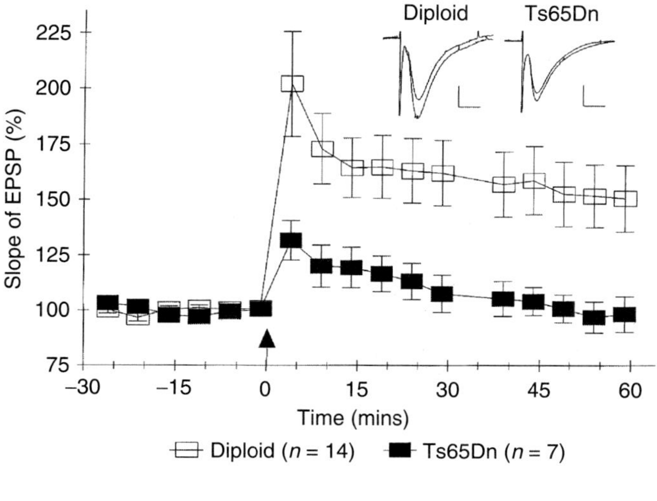

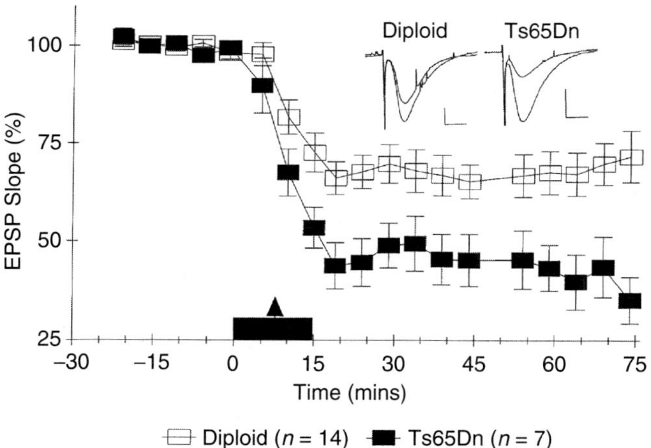

FIGURE 2 Altered synaptic plasticity of Schaffer collateral inputs to CA1 pyramidal cells in Ts65Dn mice hippocampi. Long-term potentiation (LTP; top) is significantly reduced while long-term depression (LTD; bottom) is enhanced compared to diploid litter mates. Mean ± SEM (Siarey et al., 1997, 1999).

et al., 2008) establishing a causal link between increased inhibitory tone in the hippocampus and diminished cognitive capabilities.

The roles of GABA$_B$ receptor-mediated inhibition on the cognitive deficits associated with DS remain to be investigated. However, evidence from other mouse models where cognitive impairments are present suggest that the activity level of GABA$_B$ receptors can impact learning and memory (Cadet & Krasnova, 2009). Abuse of methamphetamines causes cognitive difficulties. In a mouse model of methamphetamine abuse, Arai et al. (2009) found that the $GABA_B$ agonist baclofen helped restore performance on a novel object recognition task. Similarly, the $GABA_B$ receptor antagonist CGP35348 reversed cognitive deficits in a rat model of atypical absence seizures (Chan et al., 2006).

Even under normal conditions, the $GABA_B$ receptor appears to play a role in suppressing cognitive performance. The $GABA_B$ receptor antagonist SGS742, 3-aminopropyl-*n*-butyl phosphinic acid, is frequently called a "cognitive enhancer" for its ability to improve performance on learning and memory tasks (Froestl et al., 2004). The drug works in part by post-synaptically suppressing the activity of CREB2 that interferes with long-term memory formation by inhibiting cAMP response element-binding (CREB)-mediated gene transcription (Helm et al., 2005). In addition, SGS742 treatment in mice is associated with increases in synapsin isoform levels in the hippocampus, which are suggested to play an important role in the synaptic plasticity underlying hippocampal LTP (John et al., 2009). At least one isoform, synapsin I, is expressed at lower levels in neurospheres obtained from human DS embryonic tissue (Bahn et al., 2002) further supporting a role for $GABA_B$ in DS pathologies. In addition to its effects on the expression levels of proteins important for LTP, there is also a possible link between enhanced $GABA_B$ receptor-mediated inhibition of hippocampal LTP through its coupling to GIRK channels. The mechanisms through which this connection may meditate cognitive impairments are discussed in further detail below.

V. A Role of $GABA_B$–GIRK Coupling in Hippocampal Overinhibition in DS

As outlined above, evidence from mouse models of DS suggests that overinhibition within the hippocampus may underlie the phenotypic cognitive deficits. To date, the main focuses of investigations into the role of inhibition in the DS brain have been on $GABA_A$ receptor-mediated phenomena. However, strong genetic and physiological evidence suggests that $GABA_B$ receptor-mediated inhibition should also play a prominent role through its interactions with GIRK channels (Fig. 3) that are overexpressed in DS mouse model. Our previous results (Best et al., 2007; Harashima et al., 2006a, 2006b) and the results presented here indicate that the increased expression of the gene encoding GIRK channels resulting from the presence of the GIRK2 gene within the trisomic region has a dosage-sensitive effect on features of CA1 pyramidal neurons and Ts65Dn mice that have one of the three copies of Girk2 deleted (disomic for Girk2) show a rescue of some DS phenotypes. In particular, the intrinsic properties (cell size and resting membrane potential) of hippocampal neurons resembled those of normal euploid mice (Fig. 4). Thus, regulation of the magnitude of GIRK2

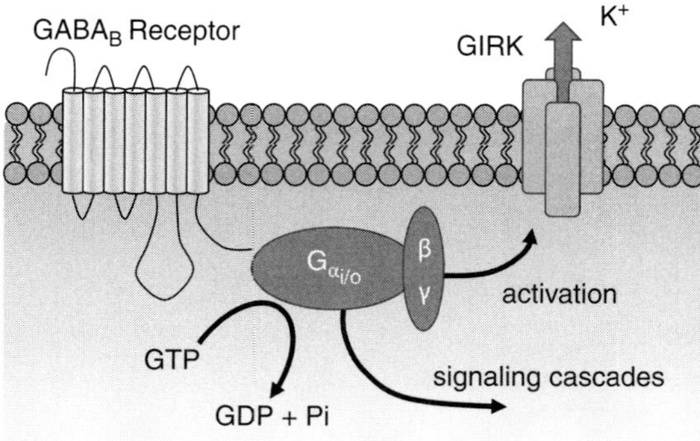

FIGURE 3 Coupling of metabotropic GABA$_B$ receptors to GIRK channels. The G-protein βγ subunit contributes to GIRK channel activation while the α subunit is available to participate in additional intracellular signaling cascades.

expression confers dosage-sensitive changes in neuronal properties. The potential contributions of GABA$_B$-mediated exaggeration of GIRK currents to DS phenotypes will be postulated in the remainder of the chapter.

A. Properties of Mammalian GIRK Channels

Four genes encoding mammalian GIRK channels have been isolated and are designated GIRK1/KIR3.1/KCNJ1, GIRK2/KIR3.2/KCNJ6, GIRK3/KIR3.3/KCNJ9, and GIRK4/KIR3.4/KCNJ5 (reviewed by Hibino et al., 2010). Typically, functional GIRK channels are heterotetrameric complex consisting of GIRK1 and one of the other three GIRK subunits (reviewed by Hibino et al., 2010). Within the CNS, channels containing GIRK1 and GIRK2 subunits are most common with GIRK2 subunits playing critical roles in the translocation of channel complexes to the cell membrane (GIRK4 plays a similar role although it is less widely expressed in the CNS). GIRK3 subunits contain a lysosomal targeting signal and are important for recycling or degrading channel complexes (reviewed by Hibino et al., 2010).

As their name suggests, GIRK channels allow a flux of K^+ ions across the cell membrane and conduct more efficiently at voltages that allow K^+ influx. Under physiological conditions, the inward rectification results from a voltage-dependent block of the channel pore by intracellular Mg^{2+} or polyamines such as spermine (Fakler et al., 1995; Nishida & MacKinnon, 2002). GIRK channels are activated by $G_{\beta\gamma}$ heterodimers released from pertussis toxin-sensitive G-proteins by a variety of metabotropic neurotransmitter receptors; however, other intracellular messenger systems make additional contributions to GIRK channel activation (Luján et al., 2009).

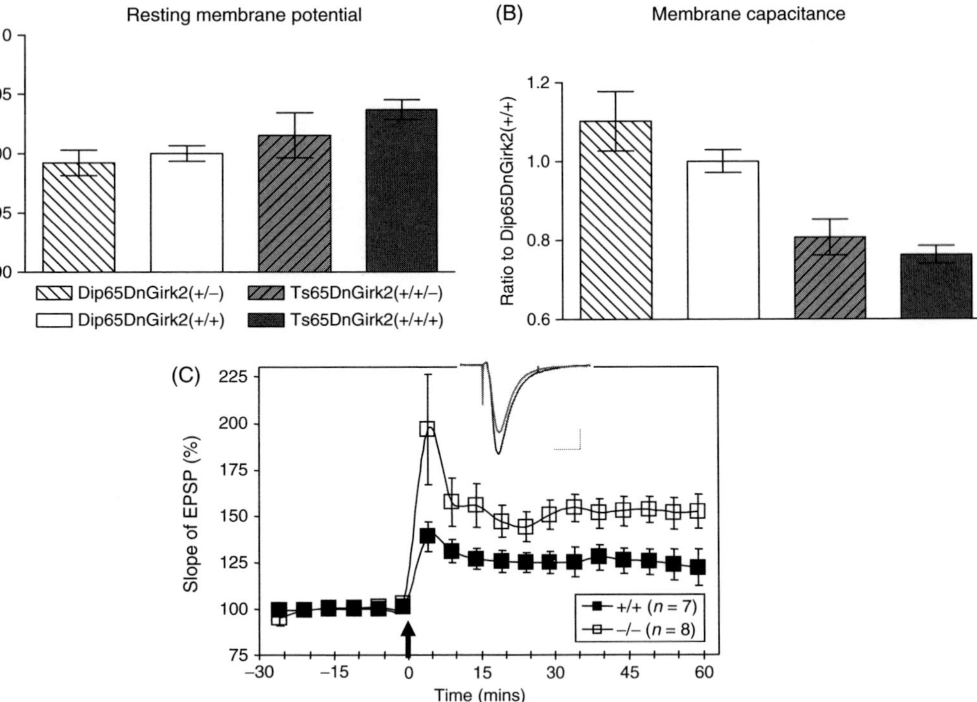

FIGURE 4 Gene dosage effect of intrinsic properties of CA1 pyramidal neurons. Whole-cell recordings from CA1 pyramidal neurons from acute hippocampal slices show significance in (A) resting membrane potential (ANOVA $p < 0.005$; post hoc trend $p < 0.005$) and on (B) membrane capacitance (ANOVA $p < 0.001$; post hoc trend $p < 0.001$) from mice with varying levels of GIRK2 gene dosage and Ts65Dn background. Hippocampal slices were derived from offsprings of Ts65Dn–GIRK2 knockout mice mating scheme. Ts65Dn females were crossed with GIRK2$^{+/-}$ males to produce homozygous-diploid/*Girk2*$^{(+/+)}$, heterozygous-diploid/*Girk2*$^{(+/-)}$, disomic-Ts65Dn/*Girk2*$^{(+/+/-)}$, and Ts65Dn/*Girk2*$^{(+/+/+)}$ such that mice had one, two, and three copies of *Girk2* gene, respectively. Recordings and analysis were done without prior knowledge of mouse genotype. (C) Complete (homozygous) Girk2 knockout mice show enhanced hippocampal LTP. Inset depicts evoked field potential following potentiation for diploid (red) and Girk2-knockout (blue) mice. Recording conditions and recording setup were as in Siarey et al. (1997, 1999).

The ability of neurotransmitters to activate these channels constitutes a potent mechanism for the regulation of neuronal excitability. Presumably, alterations in the dosage of GIRK2 expressions, such as those that occur in DS, would confer a coincident change in the excitability of the affected neurons.

B. $GABA_B$–GIRK2 Signaling and Interference from Other GPCRs and Signaling Pathways: Gene Dosage Effect Gets Complicated

$GABA_B$–GIRK2 signaling is not only regulated by expression levels of $GABA_B$ receptors and GIRK2-containing channel mRNAs. Their assembly is dynamically regulated at translational level as well. The $GIRK2^{(-/-)}$ mouse displays a decrease in GIRK1 protein expression (Signorini et al., 1997), which suggests that GIRK2 controls GIRK1 translational expression. A similar result was reported in the *weaver* mouse that has a mutation in the GIRK2 gene (Liao et al., 1996). A linkage also exists in Ts65Dn mice where GIRK2 overexpression is accompanied by higher expression of GIRK1 protein (Fig. 5; Harashima et al., 2006a). These observations underscore the need to consider additional interactions between gene products beyond the direct dosage-sensitive effects of trisomic genes.

In addition to interactions at the translational level, interactions between other activators and modulators of GIRK channels add to the complexity of the $GABA_B$–GIRK response. Various neurotransmitters such as acetylcholine, serotonin, adenosine, glutamate, somatostatin, and neuropeptide Y (Paredes et al., 2003; Sodickson & Bean, 1998; Svoboda & Lupica, 1998) converge upon and are capable of activating GIRK channels. In addition, other G-protein-coupled receptors (GPCRs) such as M_2-muscarinic, 5-HT_{1A}-serotonergic, μ-,δ-,κ-opioid, β_2-adrenergic, D_2-dopaminergic, somatostatin, and galanin receptors may modulate performance of GIRK channels through the family of G-proteins setting the stage for a potentially significant degree of cross talk between $GABA_B$–GIRK assemblies and other signal transduction pathways (Dascal, 1997; Roberson et al., 2002; Yamada et al., 1999). The potential for occlusive and/or supra-additive effects of simultaneously acting neurotransmitters may reshape dynamic properties of DS neuronal networks in both hippocampus and cortex (Sodickson & Bean, 1998).

Beyond the direct effects of G-proteins on GIRK channels, the overall GIRK response is modulated by a variety of intracellular mechanisms (Fig. 6). For example, the phospholipid phosphatidylinositol 4,5-bisphosphate (PtdIns(4,5)P2) is essential for sustained normal GIRK channel activity (reviewed by Hibino et al., 2010). By facilitating the interaction between GIRK channels and PtdIns(4,5)P2, intracellular Na^+ ions also play a role

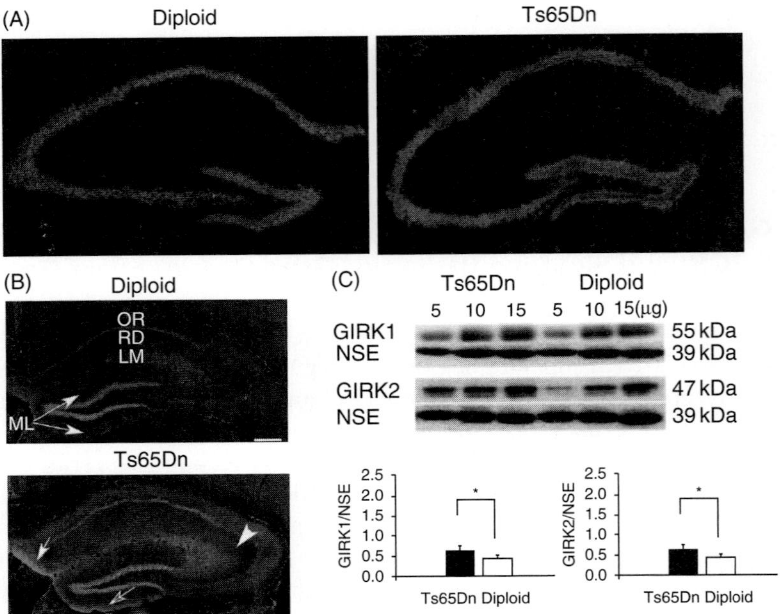

FIGURE 5 Increased expression of GIRK channel subunits in Ts65Dn hippocampus demonstrated by *in situ* hybridization (A) and GIRK2 immunoreactivity (B). Western blot analysis (C) revealed a significant increase in GIRK2 protein from Ts65Dn hippocampal homogenates compared to controls. Neuron-specific enolase (NSE) used as a standard. Abbreviations: LM, stratum lacunosum moleculare; ML, molecular layer; OR, stratum oriens; RD, stratum radiatum (Harashima et al., 2006a).

in regulating GIRK channel activity (reviewed by Hibino et al., 2010). In addition to the activity of individual channels, the overall GIRK-mediated response in a cell is also a function of the density of channels in the cell membrane. Agents such as protein phosphatase-1 (PP1) that dephosphorylates a serine residue on GIRK2 subunits to promote channel recycling and insertion into the cell membrane can thus also powerfully regulate the magnitude of GIRK-mediated currents (Chung et al., 2009; Luján et al., 2009). These mechanisms of potassium channel activation may attenuate or enhance gene dosage effects related to the overexpression of GIRK2 gene.

C. Interactions between Intracellular Signaling Mechanisms in DS

Given the ability of PtdIns(4,5)P2 and PP1 to modulate GIRK activity, changes in the activity or expression levels of these two agents could either help mitigate or exacerbate the aberrant $GABA_B$–GIRK signaling observed in DS mouse models. PtdIns(4,5)P2 levels in Ts65Dn brain and Ts65Dn

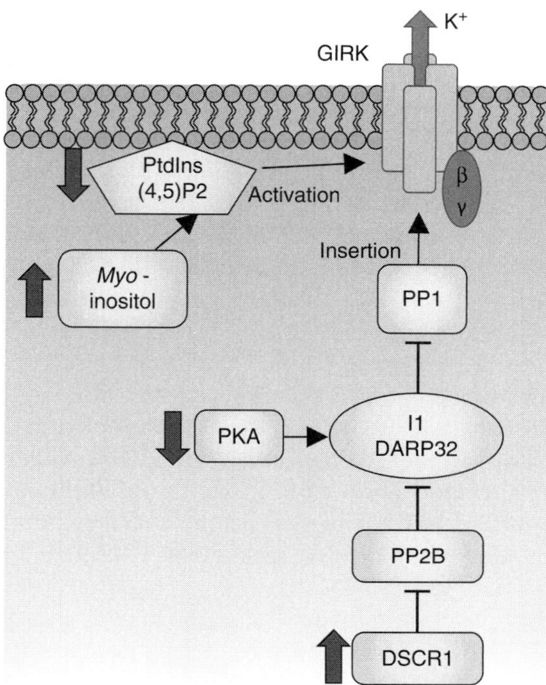

FIGURE 6 Gene dosage induced alterations in signaling pathways responsible for GIRK channel trafficking and activation in DS. The expression of DSCR1 in DS can alter the insertion of GIRK channels into the cell membrane by interfering indirectly with PP1. A reduction in PKA activity could help compensate for increased PP1 inhibition. Observed changes in the PtdIns (4,5)P2 pathway necessary for GIRK channel activation underscore the complexity of the interactions resulting from DS gene dosage effects. Block arrows indicate changes in signaling molecules reported in DS.

cortical synaptosomes are slightly lower than those found in control diploid mice (Voronov et al., 2008) suggesting reduced PtdIns(4,5)P2-mediated activity in DS neurons. This decrease is linked to increased expression of the gene (*Synj1*) encoding Synaptojanin 1, which dephosphorylates PtdIns (4,5)P2 and is located within the trisomic region in both mouse models and humans (Voronov et al., 2008). Chang and Min (2009) also observed an increase in Synaptojanin in a *Drosophila* model of *Synj* overexpression. Interestingly, the co-overexpression of another Hsa21 homolog, *Dap160*, resulted in a differential expression pattern of Synaptojanin such that levels were increased in the soma but were normal in synaptic terminals. In addition, *myo*-inositol, a PtdIns(4,5)P2 precursor, is elevated in the CNS of DS humans (Alexander et al., 1997; Berry et al., 1999; Huang et al., 1999; Shonk & Ross, 1995) and DS mouse models (Shetty et al., 1996; Yao et al., 2000) most likely due to overexpression of Chr. 21 Smit1 gene

(McVeigh et al., 2000), encoding for Na^+/*myo*-inositol cotransporter (Smit1 knockout mice show decrease in level of *myo*-inositol in a gene dosage manner (Berry et al., 2003)). Thus, shifts in gene dosages for functionally related genes may have interactions at the compartmental level that may be difficult to detect with global assays.

As outlined above, PP1 also plays an important role in $GABA_B$–GIRK signaling by promoting the recycling of GIRK channel complexes from endosomes to the cell membrane (Chung et al., 2009; Luján et al., 2009). Recent evidence suggests that PP1 insertion of GIRK channels is enhanced following *N*-methyl-D-aspartate (NMDA) receptor activation (Chung et al., 2009) by a mechanism that links strong excitatory inputs, such as those capable of inducing LTP, to increased inhibition. Such a mechanism may normally act as a brake on overexcitation but may also suppress induction of LTP in situations like DS where the balance between inhibition and excitation is abnormal. Indeed, interfering with PP1 activity in the forebrain of mice enhances memory formation (Peters et al., 2009) possibly by reducing the transport of GIRK channels to the cell membrane and thereby reducing the potential for $GABA_B$ and other G-protein-linked receptor-mediated suppression of LTP.

The overexpression of individual genes, such as those encoding Synaptojanin 1 and GIRK2, is likely to have direct dosage-related effects on affected neurons such as those described above. However, the extensive interactions between signaling mechanisms means that the overall effect of increased gene dosage is likely to be quite complex. This is illustrated by the potential impact of trisomy on PP1 in Ts65Dn mice (see Fig. 6). The activity of PP1 is suppressed by phosphorylated inhibitor 1 (I1) and dopamine and cAMP-regulated phosphoprotein of 32 kDa (DARP32; D'Alcantara et al., 2003) resulting in reduced transport of GIRK channel complexes to the cell membrane (Luján et al., 2009). These endogenous inhibitors of PP1 are themselves inhibited by PP2B (calcineurin; D'Alcantara et al., 2003). PP2B activity is suppressed by its own endogenous inhibitor, DSCR1, that, due to gene dosage effects, is overexpressed in DS (Arron et al., 2006; Baek et al., 2009). Thus, the inhibitors of PP1 are under less inhibition and would be expected to more completely suppress PP1 activity (see Fig. 6). This effect may potentially be counteracted by low protein kinase A (PKA) activity also observed in Ts65Dn mice (Siarey et al., 2006), which normally increases inhibition of PP1 by I1 and DARP32 activation (see Fig. 6). Shifts in the activity of proteins from nontrisomic genes, like that observed for PKA, may reflect attempts by the neurons to maintain homeostasis. This possibility, however, remains to be examined.

Regardless of whether an observed change in gene expression reflects dysfunction or attempts to restore homeostasis, it is clear that it is difficult to interpret single gene dosage effects in isolation from other cellular responses. However, given the capability of enhanced $GABA_B$–GIRK to cause excessive inhibition in neurons, we can speculate how the effects of GIRK overexpression may contribute to DS phenotypes, such as cognitive deficits and altered pain responses, in a dosage-dependent manner.

D. Impact of GIRK2 Dosage on Hippocampal Function in DS

Within the hippocampus, both $GABA_B$ and GIRK2 proteins are expressed in high densities and in close proximity to each other supporting a potentially significant role for GABAergic modulation of neuronal excitability and regulation of synaptic function in this brain region (Koyrakh et al., 2005; Kulik et al., 2006). In Ts65Dn mice, we have found that GIRK2 is significantly overexpressed in the hippocampus compared to diploid controls (Fig. 5; Harashima et al., 2006a). The overexpression is particularly evident in the stratum lacunosum moleculare (SLM) in CA1. Neural circuitry within this lamina, including the apical dendrites of CA1 pyramidal cells and, most likely, GABAergic interneurons, receives direct inputs from the temporoammonic (TA) pathway (Dvorak-Carbone & Schuman, 1999). In addition, the apical dendrites in this stratum are the target of feedforward inhibition from interneurons in stratum oriens alveus interneurons (OAIs). Thus, the increased inhibitory tone in this area would be expected to have a significant impact on the processing of information in hippocampal circuitry that involves SLM.

Major inputs to the hippocampus arise from the entorhinal cortex (EC) and converge on pyramidal neurons in CA1 through two principal pathways: (1) a trisynaptic connection via the perforant pathway (PP) and (2) directly through the TA pathway (Kajiwara et al., 2008). In the normal hippocampus, these two pathways are configured to separately process inputs arriving at different frequencies such that higher-frequency inputs will travel through the trisynaptic PP–CA3–CA1 pathway while lower-frequency inputs will traverse directly to CA1 pyramidal cells via the TA pathway (Maccaferri & McBain, 1995). As outlined by Maccaferri and McBain (1995), the two mechanisms that allow this to occur are frequency-based plasticity at the CA3–CA1 synapse coupled with a feedforward inhibition loop driven by CA1 pyramidal cells themselves (Fig. 7).

In the normal hippocampus, the synapse from Schaffer collaterals arising from CA3 pyramidal neurons and impinging upon proximal CA1 pyramidal dendrites is susceptible to long-lasting plastic changes (Pöschel & Stanton, 2007). During high-frequency events, the synapse potentiates causing succeeding events to more strongly depolarize and more effectively drive CA1 pyramidal cells. This increased excitation is transmitted to interneurons in OA through CA1 axon collaterals, which in turn excite inhibitory interneurons in SLM to hyperpolarize distal CA1 dendrites (Maccaferri

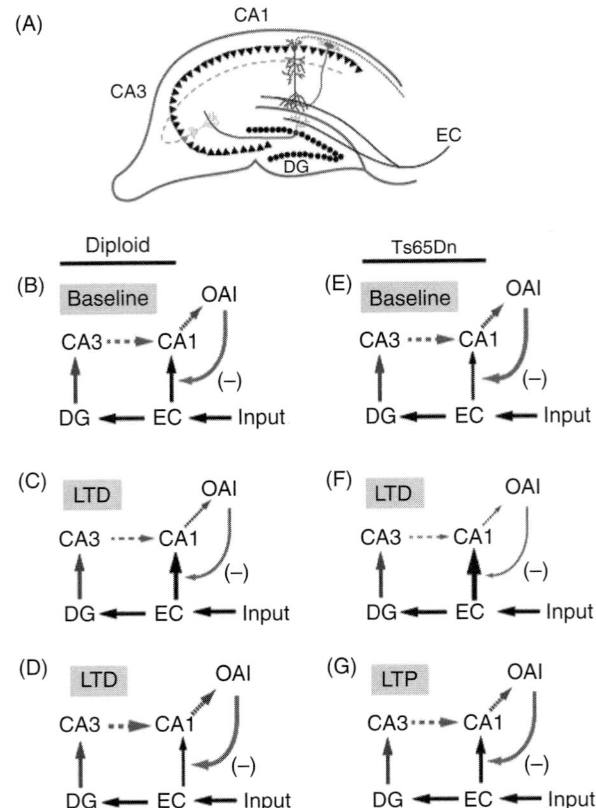

FIGURE 7 Model for disrupted hippocampal circuitry in Ts65Dn. Basic hippocampal circuitry depiction (A) including entorhinal cortex (EC), dentate gyrus (DG), CA3, CA1, and stratum oriens alveus interneuron (OAI) projections. Diploid (B–D) and Ts65Dn (E–G) circuit diagrams depicting the putative relative strength (size of arrow) of synaptic connections under baseline (B, E), LTD (C, F), and LTP (D, G) conditions. Adapted from Maccaferri & McBain (1995) in Best 2007. Under baseline conditions, the inhibitory input onto CA1 pyramidal neurons is relatively greater in Ts65Dn than in diploid. In diploid mice, when LTD is induced at CA3–CA1 synapses in Ts65Dn, the drive by CA1 pyramidal collaterals onto OAIs is diminished resulting in less feedforward inhibition by OAIs, resulting in enhanced influence of direct EC input onto CA1 pyramidal neurons. Since in Ts65Dn, LTD protocols induce more depression of CA3–CA1 synapses, the inhibition by OAIs is less and as a result the EC influence is greater. The converse is true for LTP, where in Ts65Dn, the level of potentiation is less than in diploid, again resulting in greater influence of direct EC inputs onto CA1 pyramidal neurons than diploid.

& McBain, 1995). Since this region is the target of direct TA inputs, this feedforward inhibition loop minimizes the impact of TA inputs on the hippocampus during the processing of high-frequency information. On the other hand, during low-frequency events, the Schaffer collateral synapses undergo LTD and excitation of CA1 pyramidal cells by this pathway becomes less effective (Maccaferri & McBain, 1995). This reduction in excitation in the trisynaptic pathway means that feedforward inhibition from CA1 axon collaterals driving interneurons in OA is also reduced, relieving TA inputs to the distal dendrites of CA1 pyramidal cells from inhibition. Consequently, TA inputs to CA1 pyramidal cells are better able to impact hippocampal circuitry. Thus, high- and low-frequency events can be segregated into two distinct pathways in the hippocampus and interference between the two is minimized.

As outlined above, the hippocampi of Ts65Dn mice overexpress GIRK channels, particularly in SLM (Harashima et al., 2006a). This is coupled with a high level of $GABA_B$ receptors known to be present in the normal mouse (Kulik et al., 2006) and presumed to be present in the DS mouse as well. Indeed, we find that CA1 pyramidal cells from Ts65Dn mice are more strongly hyperpolarized by baclofen, a $GABA_B$ receptor agonist, and show a larger $GABA_B$-mediated response to direct stimulation of SLM than control mice (Fig. 8; Best, 2007). These exaggerated responses to $GABA_B$ receptor activation are mediated largely by GIRK channels since pharmacological blockade of these channels reverses the effects (Best, 2007). Thus the overexpression of GIRK2 channels in SLM of Ts65Dn mice translates into a functional increase in inhibitory input to distal CA1 dendrites.

A second significant abnormality is present at the CA3–CA1 synapse. Here, LTP normally induced by high-frequency stimulation is significantly attenuated (Belichenko et al., 2009; Costa & Grybko, 2005; Kleschevnikov et al., 2004; Siarey et al., 1997a; Siarey et al., 2005; O'Doherty et al., 2005; Morice et al., 2008). Conversely, the magnitude of LTD evoked by low-frequency inputs at these synapses is significantly enhanced (Siarey et al., 1999, 2005). As a result, the CA1–OAI feedforward inhibition of the TA pathway that relies on potentiation of the CA3–CA1 synapse is likely to be less efficacious in the DS hippocampus. This suggests that TA inputs that would normally be suppressed during the processing of high-frequency events within the hippocampus now become superimposed upon and interfere with the normal flow of high-frequency information processing. In contrast, the flow of low-frequency information that is processed by the TA–CA1 pathway would most likely remain relatively intact since the enhanced LTD in the trisynaptic pathway would help prevent interference.

Electroencephalogram (EEG) recordings from individuals with DS suggest indirectly that such an aberrant processing of information may exist in the hippocampus. During eyes-closed resting conditions, humans with DS show increased power at low EEG frequencies with a reduction in power at higher frequencies when compared to controls (Babiloni et al., 2009). Similar shifts in EEG frequency content were observed in Ts65Dn mice during sleep and, although they did not show a significant difference, there was a similar trend in the EEGs recorded during wakefulness (see Fig. 2A in

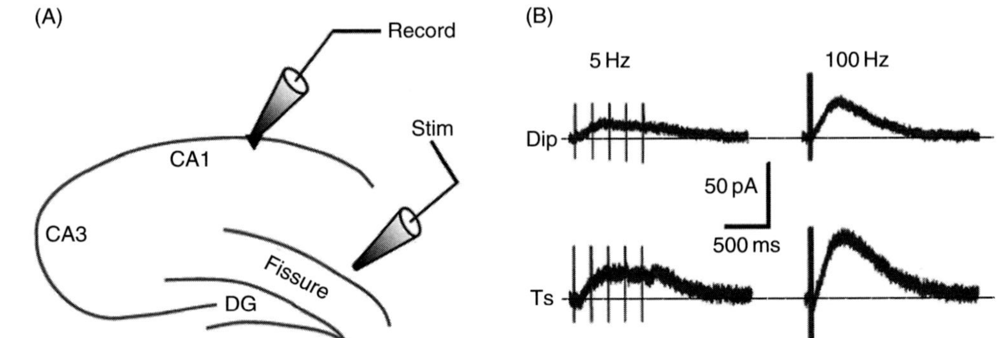

FIGURE 8 $GABA_B$-mediated GIRK currents after stimulation of the temporoammonic pathway. (A) Diagram showing whole-cell recording electrode on CA1 pyramidal neuron and stimulating electrode placed just opposite the hippocampal fissure in the SLM. (B) Whole-cell recordings from diploid (top) or Ts65Dn (bottom) hippocampal slices after stimulating the TA path with five pulses at 5 or 100 Hz.

Colas et al., 2008). While EEGs do not necessarily reflect oscillations of deeper brain structures such as the hippocampus, these findings are consistent with enhanced processing of low-frequency information and aberrant handling of higher-frequency content. While there is evidence to suggest that the magnitude of EEG abnormalities correlates with cognitive function (Partanen et al., 1996), exactly how the putative aberrant flow of information through the hippocampus resulting from gene dosage-enhanced $GABA_B$-mediated GIRK currents translates into cognitive deficits remains to be determined.

E. Effect of $GABA_B$–GIRK2 Localization at Postsynaptic Active Zones

While $GABA_B$ receptors and GIRK2-containing channels are found distributed along dendritic shafts and spines of pyramidal cells within the hippocampus, they are in closest proximity to each other near glutamatergic synapses on dendritic spines (Kulik et al., 2006). At this location, GIRK channels may also strongly influence the gating of NMDA receptors in Ts65Dn neurons by enhancing their voltage-dependent Mg^{2+} block thereby contributing to the decrease in LTP that we previously reported in the Ts65Dn hippocampus (Siarey et al., 1997a). Opposite effects related to attenuation of voltage-dependent Mg^{2+} block should increase LTP in $GIRK2^{-/-}$ knockout mice. In fact we found that LTP is increased in the CA1 region of hippocampus, which is consistent with a weaker Mg^{2+} block (Fig. 4c (bottom)). In addition, the inhibitory currents mediated by these clusters can be potentiated by the same intracellular signaling cascades that are responsible for LTP of glutamatergic EPSCs (Huang et al., 2005). This suggests that GABAergic signaling can also closely regulate the magnitude of excitatory transmission at a fundamental level by narrowing the window during which glutamatergic inputs can summate to produce LTP. The overexpression of GIRK2 channels in the DS hippocampus and the exaggerated $GABA_B$ response seen in DS hippocampal cells might thus actively suppress excitatory inputs at the synaptic level and constitute an additional brake on the induction of LTP. Given the importance of hippocampal LTP in memory formation and consolidation, this $GABA_B$-mediated inhibition of LTP resulting from a dosage-sensitive increase in GIRK expression may also contribute to the cognitive deficits associated with DS.

VI. Potential Evidence for $GABA_B$-Mediated Developmental and Morphological Changes in DS

Individuals with DS have a reduction in absolute brain volume (Pinter et al., 2001) that likely results from a combination of early developmental problems and neurodegeneration later in life. In Ts65Dn mice, longer cell cycle durations and reduced neurogenesis during the prenatal period contribute to a delay in the growth of the cerebral cortex and hippocampus (Contestabile et al., 2007). An apparent increased vulnerability to apoptosis in DS (Sawa, 1999) may also contribute to the hypocellularity. Apoptosis-related genes present on Hsa21 likely contribute to the increased susceptibility of DS neurons to the initiation of apoptosis (Sawa, 1999; Wolvetang et al., 2003).

Evidence that fluoxetine, a selective serotonin reuptake inhibitor (SSRI) and GIRK channel antagonist (Kobayashi et al., 2003), can rescue deficits in neurogenesis in Ts65Dn mice (Clark et al., 2006) suggests that $GABA_B$–GIRK coupling may also play a role in adult neurogenesis. In fact, observations in the mutuant *weaver* mouse support a role for GIRK2 in neurogenesis. These mice have a single amino acid mutation in the pore-forming region of the GIRK2 subunit, which causes a loss of K^+ selectivity and permits a tonic Na^+ current through the channel (Slesinger et al., 1996). *Weaver* mice show enhanced neuronal death during development, an effect that can be prevented with fluoxetine treatment, which effectively blocks mutant GIRK2 channels (Takahashi et al., 2006). While the effects of fluoxetine on serotonergic pathways must also be considered, these findings support an important role for GIRK channel activity in development. Similarly, ethanol was reported to open GIRK channel (Lewohl et al., 1999) and is a cause of fetal alcohol syndrome leading to mental retardation and numerous developmental disabilities.

During the embryonic stage, cortical lamination is delayed and disorganized in DS individuals and in Ts65Dn mice and may indicate disruption of axonal and dendritic connections (Chakrabarti et al., 2007; Golden & Hyman, 1994). In addition to reduced neuronal populations, individuals with DS show changes in neuronal morphology and dendritic spine density (Takashima et al., 1981; Weitzdoerfer et al., 2001). Initially, dendritic branching is greater in DS infants but, by 2 years of age, it declines below that observed in normal children (Becker et al., 1986). Ts65Dn mice show similar changes in dendritic malformations and have a preferential loss of presumably excitatory synapses (Ayberk et al., 2004). While the exact mechanisms underlying these changes remain to be determined, a potential enhancement in the magnitude of hyperpolarization in forming synapses in DS brains conferred by exaggerated $GABA_B$–GIRK signaling could make significant contributions. The formation of synapses is believed to be use-dependent requiring depolarizing events (Jontes & Smith, 2000). As structural changes in dendritic spines involve calcium (Schubert & Dotti, 2007), a $GABA_B$–GIRK suppression of depolarizing events that drive calcium influx would be expected to disrupt this important spine regulation pathway. However, such a role for $GABA_B$ in spine regulation remains to be determined.

VII. Role of GIRK in DS Pain and Cerebellar Phenotypes

In addition to the cognitive and morphological alterations highlighted above, individuals with DS show a range of phenotypes including an altered response to nociceptive stimuli and other phenotypes that can be caused by abnormal cerebellar and/or sensory nuclei function. Evidence from mouse models suggests that DS is associated with an increase in pain thresholds. Following injection of formalin into the paw, Ts65Dn mice spend less time licking the injured paw than their control litter mates (Martinez-Cue et al., 1999). Similarly, in a noxious heat test, Ts65Dn mice show longer latencies to remove their tail from the stimulus than their control litter mates (Martinez-Cue et al., 1999). Thus, it appears that DS confers a reduction in pain sensitivity. Investigations with GIRK knockout mice suggest a potential mechanism for these observations.

Using a targeted mutation technique, GIRK2 knockout mice were generated that show normal presynaptic response to neurotransmitters known to act through GPCRs but lack typical postsynaptic effects (Lüscher et al., 1997). These mice survive into adulthood and perform equally well as control mice on many behavioral tests (Blednov et al., 2001). They do, however, show a slight thermal hyperalgesia. Additionally, the antinociceptive effects of drugs that target GIRK2 channels were reduced or completely eliminated (Blednov et al., 2003). This included the GABA$_B$ agonist baclofen. In investigating the observation that male mice tend to have higher pain thresholds than female mice, Mitrovic et al. (2003) noted that this difference is lost in GIRK2 knockout strains. The knockout mice also showed less antinociceptive benefits from clonidine (an alpha2 agonist) and morphine (μ-receptor agonist). Thus, the increased pain thresholds observed in Ts65Dn mice may be related to the overexpression of GIRK2 channels. It also suggests that drugs that activate these channels, such as GABA$_B$ agonists, may convey a stronger analgesic effect in Ts65Dn mice. This increased analgesic potency, however, remains to be investigated.

Early reports in humans, based primarily on behavioral observations, suggested that individuals with DS also have higher pain thresholds than individuals from the general population (Biersdorff, 1994). In a study by Hennequin et al. (2000), testing the response of individuals to cold pain, it was found that DS individuals had a longer latency until the expression of discomfort and were less successful in localizing the site of discomfort than normal controls. While the results seem to support the notion that individuals with DS have increased pain thresholds, the authors noted that other factors such as aberrant processing of sensory information may underlie the slow response latencies. The sensations are clearly felt but the recognition of the input as painful is delayed. In contrast, Defrin et al. (2004) found that when a test that did not rely on reaction times was used, individuals with DS had a lower threshold for heat-induced pain. They also found evidence for slower conduction velocities of sensory information with corresponding slower reaction times in DS individuals. These latter observations, combined with cognitive delays in processing sensory information, may be sufficient to explain the perception of higher pain thresholds in individuals with DS. In either case, it is noted that DS individuals are not insensitive to pain and the proper use of anesthetics and analgesics is still essential. More research is required to resolve the apparent discrepancies between animal model and human observations regarding pain perception in DS.

One of the possible approaches would be to use a special mating scheme where crossing Ts65Dn mice and heterozygous knockout for a single gene from chromosome 21 mice produces small percentage offsprings that have all triplicated genes encoded by Ts65Dn chromosomal segment beyond this selected gene. We applied this strategy to produce Girk2-disomic Ts65Dn mice using Girk2 knockout (Harashima et al., 2006b; we have already discussed the results of this strategy for hippocampal neurons in Fig. 5). In the Ts65Dn cerebellum we found dramatically elevated levels of GIRK2 in unipolar brush cells (UBCs; Harashima et al., 2006b). UBCs are unique interneurons with a brush-like dendrite structure that are found in the granule cell layer of the vestibular cerebellum and in the dorsal cochlear nucleus (DCn; Abbott & Jacobowitz, 1995; Floris et al., 1994; Mugnaini et al., 1997). UBCs are excitatory granule layer interneurons that are capable of intrinsic firing that at least in part is regulated by GIRK channels (Jaarsma et al., 1996; Russo et al., 2008). The Girk2 disomic Ts65Dn mice (Ts65DnGirk2$^{-/+/+}$) show restoration of cerebellar phenotype related to overexpression of GIRK2 in UBCs (Harashima et al., 2006b). The behavioral consequences of this restoration were not addressed because of poor efficiency of Ts65Dn breeding scheme. The increased expression levels of the GIRK2 subunit may contribute to the motor and sensorimotor cognitive sequelae seen in the Ts65Dn mouse and DS individuals by impacting physiological properties of cerebellar and auditory neuronal circuits and may contribute early in development to hypotonia that is a prominent feature of DS during the time of birth.

VIII. Conclusions

It is clear that the complexity of Down syndrome cannot be linked to the overexpression of a single gene or even the limited segment of genes on chromosome 21 (Korenberg, 2009; Olson et al., 2007; Pereira et al., 2009). However, examining the effects of increased dosage in particular genes offers a tractable starting point for investigating the complex interactions that occur in DS. Using this approach we presented evidence that the presence of a gene encoding GIRK2 subunits on a triplicated region of mouse chromosome 16 (and also present on human chromosome 21) leads to dosage-sensitive changes in the properties of neurons within the CNS (Harashima et al., 2006a). Based on the properties of these channels and their functional coupling to $GABA_B$ receptors, we postulated how enhanced inhibitory tone can affect proper functioning at both the cellular and behavioral levels.

In postulating the effects of a dosage-sensitive increase in GIRK expression, we also elaborated how alterations in the expression of functionally related genes and their resulting products may interact to either ameliorate or exacerbate DS phenotypes. As signaling pathways do not act in isolation and overlap extensively, untangling the functional impact of these alterations becomes quite complex. This complexity can occur at different hierarchical levels of brain anatomy (in set of networks through single microcircuit and/or in a single neuron) potentially leading to attenuating or enhancing consequences of the chromosome 21 gene overexpression in not easily predictable manner. As therapies targeting specific DS phenotypes are developed it will be critically important to keep the interactions of gene products and miRNAs in mind. In some instances, like the ability to rescue cognitive deficits in DS mice by interfering with GABAergic signaling (Fernandez et al., 2007; Rueda et al., 2008), interactions between signaling pathways and $GABA_A$ receptors may be of less concern although cross talk between pathways that are abnormal in DS and inhibitory $GABA_A$ activity likely takes place and may provide explanation why this approach is successful (Jacob et al., 2008; Wang et al., 2003). Therefore, targeting specific pathways altered by one or a few overexpressed genes provides a strong foundation to begin to understand the complexities of DS. In this light, $GABA_B$–GIRK2 signaling emerges as a potentially important and promising pharmacological target to reverse some of the DS cognitive phenotypes.

Acknowledgments

This work was supported by grants from the National Institutes of Health, J. Lejeune Foundation, and USUHS (ZG). The authors wish to thank Mrs. Angelina KlineBurgess and Mrs. Madelaine Clark for assistance with the care and genotyping of the GIRK2 and Ts65Dn mice.

Conflicts of Interest: The authors declare no conflicts of interest.

References

Abbott, L. C., & Jacobowitz, D. M. (1995). Development of calretinin-immunoreactive unipolar brush-like cells and an afferent pathway to the embryonic and early postnatal mouse cerebellum. *Anatomy and Embryology (Berl)*, *191*, 541–559.

Alexander, G. E., Huang, W., Krasuski, J. S., Szczepanik, J., Pietrini, P., Furey, M. L., et al. (1997). Elevated brain myo-inositol correlates with cognition in non-demented Down syndrome (DS) adults. *Society for Neuroscience Abstract*, *23*, 859.

Amano, K., Sago, H., Uchikawa, C., Suzuki, T., Kotliarova, S. E., Nukina, N., et al. (2004). Dosage-dependent over-expression of genes in the trisomic region of Ts1Cje mouse model for Down syndrome. *Human Molecular Genetics*, *13*, 1333–1340.

Antonarakis, S. E., Lyle, R., Dermitzakis, E. T., Reymond, A., & Deutsch, S. (2004). Chromosome 21 and Down syndrome: From genomics to pathophysiology. *Nature Reviews. Genetics*, *5*, 725–738.

Arai, S., Takuma, K., Mizoguchi, H., Ibi, D., Nagai, T., Kamei, H., et al. (2009). $GABA_B$ receptor agonist baclofen improves methamphetamine-induced cognitive deficit in mice. *European Journal of Pharmacology*, *602*, 101–104.

Arron, J. R., Winslow, M. M., Polleri, A., Chang, C. P., Wu, H., Gao, X., et al. (2006). NFAT dysregulation by increased dosage of DSCR1 and DYRK1A on chromosome 21. *Nature*, *441*, 595–600.

Ayberk, K. M., Ilker, K. M., Dierssen, M., & Ceri, D. D. (2004). Deficits of neuronal density in CA1 and synaptic density in the dentate gyrus, CA3 and CA1, in a mouse model of Down syndrome. *Brain Research*, *1022*, 101–109.

Babiloni, C., Albertini, G., Onorati, P., Vecchio, F., Buffo, P., Sarà, M., et al. (2009). Inter-hemispheric functional coupling of eyes-closed resting EEG rhythms in adolescents with Down syndrome. *Clinical Neurophysiology*, *120*, 1619–1627.

Baek, K., Zaslavsky, A., Lynch, R., Britt, C., Okada, Y., Siarey, R., et al. (2009). Down's syndrome suppression of tumour growth and the role of the calcineurin inhibitor DSCR1. *Nature*, *459*, 1126–1130.

Bahn, S., Mimmack, M., Ryan, M., Caldwell, M., Jauniaux, E., Starkey, M., et al. (2002). Neuronal target genes of the neuron-restrictive silencer factor in neurospheres derived from fetuses with Down's syndrome: A gene expression study. *Lancet*, *359*, 310–315.

Becker, L. E., Armstrong, D. L., & Chan, F. (1986). Dendritic atrophy in children with Down's syndrome. *Annals of Neurology*, *20*, 520–526.

Belichenko, N., Belichenko, P., Kleschevnikov, A., Salehi, A., Reeves, R., & Mobley, W. (2009). The "Down syndrome critical region" is sufficient in the mouse model to confer behavioral, neurophysiological, and synaptic phenotypes characteristic of Down syndrome. *Journal of Neuroscience*, *29*, 5938–5948.

Belichenko, P., Kleschevnikov, A., Salehi, A., Epstein, C., & Mobley, W. (2007). Synaptic and cognitive abnormalities in mouse models of Down syndrome: Exploring genotype-phenotype relationships. *Journal of Comparative Neurology*, *504*, 329–345.

Berry, G. T., Wang, Z. Y. J., Dreha, S. F., Finucane, B. M., & Zimmerman, R. A. (1999). In vivo brain myo-inositol levels in children with Down syndrome. *Journal of Pediatrics*, *135*, 94–97.

Berry, G. T., Wu, S., Buccafusca, R., Ren, J., Gonzales, L. W., Ballard, P. L., et al. (2003). Loss of murine Na^+/myo-inositol cotransporter leads to brain myo-inositol depletion and central apnea. *Journal of Biological Chemistry*, *278*, 18297–18302.

Best, T. K. (2007) Disruption of inhibitory function in the Ts65Dn mouse hippocampus through overexpression of GIRK2. PhD Thesis USUHS, SoM.

Best, T., Siarey, R., & Galdzicki, Z. (2007). Ts65dn, a mouse model of Down syndrome, exhibits increased $GABA_B$-induced potassium current. *Journal of Neurophysiology*, *97*, 892–900.

Biersdorff, K. K. (1994). Incidence of significantly altered pain experience among individuals with developmental disabilities. *American Journal of Mental Retardation*, *98*, 619–631.

Blednov, Y., Stoffel, M., Alva, H., & Harris, R. (2003). A pervasive mechanism for analgesia: Activation of GIRK2 channels. *Proceedings of the National Academy of Sciences of the United States of America*, *100*, 277–282.

Blednov, Y., Stoffel, M., Chang, S., & Harris, R. (2001). GIRK2 deficient mice. Evidence for hyperactivity and reduced anxiety. *Physiology & Behavior*, *74*, 109–117.

Bliss, T. V., & Collingridge, G. L. (1993). A synaptic model of memory: Long-term potentiation in the hippocampus. *Nature*, *361*, 31–39.

Cadet, J., & Krasnova, I. (2009). Molecular bases of methamphetamine-induced neurodegeneration. *International Review of Neurobiology*, *88*, 101–119.

Chakrabarti, L., Galdzicki, Z., & Haydar, T. (2007). Defects in embryonic neurogenesis and initial synapse formation in the forebrain of the Ts65Dn mouse model of Down syndrome. *Journal of Neuroscience*, *27*, 11483–11495.

Chan, K., Burnham, W., Jia, Z., Cortez, M., & Or, S. (2006). GABA$_B$ receptor antagonism abolishes the learning impairments in rats with chronic atypical absence seizures. *European Journal of Pharmacology*, *541*, 64–72.

Chang, K., & Min, K. (2009). Upregulation of three *Drosophila* homologs of human chromosome 21 genes alters synaptic function: Implications for Down syndrome. *Proceedings of the National Academy of Sciences of the United States of America*, *106*, 17117–17122.

Chung, H., Ge, W., Qian, X., Wiser, O., Jan, Y., & Jan, L. (2009). G protein-activated inwardly rectifying potassium channels mediate depotentiation of long-term potentiation. *Proceedings of the National Academy of Sciences of the United States of America*, *106*, 635–640.

Clark, S., Schwalbe, J., Stasko, M., Yarowsky, P., & Costa, A. (2006). Fluoxetine rescues deficient neurogenesis in hippocampus of the Ts65Dn mouse model for Down syndrome. *Experimental Neurology*, *200*, 256–261.

Colas, D., Valletta, J. S., Takimoto-Kimura, R., Nishino, S., Fujiki, N., Mobley, W. C., et al. (2008). Sleep and EEG features in genetic models of Down syndrome. *Neurobiology of Disease*, *30*, 1–7.

Contestabile, A., Fila, T., Ceccarelli, C., Bonasoni, P., Bonapace, L., Santini, D., et al. (2007). Cell cycle alteration and decreased cell proliferation in the hippocampal dentate gyrus and in the neocortical germinal matrix of fetuses with Down syndrome and in Ts65Dn mice. *Hippocampus*, *17*, 665–678.

Costa, A., & Grybko, M. (2005). Deficits in hippocampal CA1 LTP induced by TBS but not HFS in the Ts65Dn mouse: A model of Down syndrome. *Neuroscience Letters*, *382*, 317–322.

Coussons-Read, M. E., & Crnic, L. S. (1996). Behavioral assessment of the Ts65Dn mouse, a model for Down syndrome: Altered behavior in the elevated plus maze and open field. *Behavior Genetics*, *26*, 7–13.

D'Alcantara, P., Schiffmann, S., & Swillens, S. (2003). Bidirectional synaptic plasticity as a consequence of interdependent Ca^{2+}-controlled phosphorylation and dephosphorylation pathways. *European Journal of Neuroscience*, *17*, 2521–2528.

Dascal, N. (1997). Signalling via the G protein-activated K^+ channels. *Cellular Signalling*, *9*, 551–573.

Defrin, R., Pick, C. G., Peretz, C., & Carmeli, E. (2004). A quantitative somatosensory testing of pain threshold in individuals with mental retardation. *Pain*, *108*, 58–66.

Demas, G. E., Nelson, R. J., Krueger, B. K., & Yarowsky, P. J. (1996). Spatial memory deficits in segmental trisomic Ts65Dn mice. *Behavioural Brain Research*, *82*, 85–92.

Dvorak-Carbone, H., & Schuman, E. M. (1999). Patterned activity in stratum lacunosum moleculare inhibits CA1 pyramidal neuron firing. *Journal of Neurophysiology*, *82*, 3213–3222.

Epstein, C., Cox, D., & Epstein, L. (1985). Mouse trisomy 16: An animal model of human trisomy 21 (Down syndrome). *Annals of the New York Academy of Sciences*, *450*, 157–168.

Escorihuela, R. M., Fernandez-Teruel, A., Vallina, I. F., Baamonde, C., Lumbreras, M. A., Dierssen, M., et al. (1995). A behavioral assessment of Ts65Dn mice: A putative Down syndrome model. *Neuroscience Letters*, *199*, 143–146.

Fakler, B., Brandle, U., Glowatzki, E., Weidemann, S., Zenner, H. P., & Ruppersberg, J. P. (1995). Strong voltage-dependent inward rectification of inward rectifier K^+ channels is caused by intracellular spermine. *Cell*, *80*, 149–154.

Fernandez, F., Morishita, W., Zuniga, E., Nguyen, J., Blank, M., Malenka, R. C., et al. (2007). Pharmacotherapy for cognitive impairment in a mouse model of Down syndrome. *Nature Neuroscience*, *10*, 411–413.

Floris, A., Dino, M., Jacobowitz, D. M., & Mugnaini, E. (1994). The unipolar brush cells of the rat cerebellar cortex and cochlear nucleus are calretinin-positive: A study by light and electron microscopic immunocytochemistry. *Anatomy and Embryology (Berl)*, *189*, 495–520.

Froestl, W., Gallagher, M., Jenkins, H., Madrid, A., Melcher, T., Teichman, S., et al. (2004). SGS742: The first GABA(B) receptor antagonist in clinical trials. *Biochemical Pharmacology*, *68*, 1479–1487.

Galdzicki, Z., Siarey, R., Pearce, R., Stoll, J., & Rapoport, S. (2001). On the cause of mental retardation in Down syndrome: Extrapolation from full and segmental trisomy 16 mouse models. *Brain Research. Brain Research Reviews*, *35*, 115–145.

Gardiner, K. (2004). Gene-dosage effects in Down syndrome and trisomic mouse models. *Genome Biology*, *5*, 244.

Golden, J. A., & Hyman, B. T. (1994). Development of the superior temporal neocortex is anomalous in trisomy 21. *Journal of Neuropathology and Experimental Neurology*, *53*, 513–520.

Granholm, A. C., Sanders, L. A., & Crnic, L. S. (2000). Loss of cholinergic phenotype in basal forebrain coincides with cognitive decline in a mouse model of Down's syndrome. *Experimental Neurology*, *161*, 647–663.

Hanson, J. E., Blank, M., Valenzuela, R. A., Garner, C. C., & Madison, D. V. (2007). The functional nature of synaptic circuitry is altered in area CA3 of the hippocampus in a mouse model of Down's syndrome. *Journal of Physiology*, *579*, 53–67.

Harashima, C., Jacobowitz, D. M., Stoffel, M., Chakrabarti, L., Haydar, T. F., Siarey, R. J., et al. (2006b). Elevated expression of the G-protein activated inwardly rectifying potassium channel 2 (GIRK2) in cerebellar unipolar brush cells of Down syndrome mouse model. *Cellular and Molecular Neurobiology*, *26*, 719–734.

Harashima, C., Jacobowitz, D., Witta, J., Borke, R., Best, T., Siarey, R., et al. (2006a). Abnormal expression of the G-protein-activated inwardly rectifying potassium channel 2 (GIRK2) in hippocampus, frontal cortex, and substantia nigra of Ts65Dn mouse: A model of Down syndrome. *Journal of Comparative Neurology*, *494*, 815–833.

Helm, K., Haberman, R., Dean, S., Hoyt, E., Melcher, T., Lund, P., et al. (2005). $GABA_B$ receptor antagonist SGS742 improves spatial memory and reduces protein binding to the cAMP response element (CRE) in the hippocampus. *Neuropharmacology*, *48*, 956–964.

Hennequin, M., Morin, C., & Feine, J. S. (2000). Pain expression and stimulus localisation in individuals with Down's syndrome. *Lancet*, *356*, 1882–1887.

Hibino, H., Inanobe, A., Furutani, K., Murakami, S., Findlay, I., & Kurachi, Y. (2010). Inwardly rectifying potassium channels: Their structure, function, and physiological roles. *Physiological Reviews*, *90*, 291–366.

Holtzman, D. M., Santucci, D., Kilbridge, J., Chua-Couzens, J., Fontana, D. J., Daniels, S. E., et al. (1996). Developmental abnormalities and age-related neurodegeneration in a mouse model of Down syndrome. *Proceedings of the National Academy of Sciences of the United States of America*, *93*, 13333–13338.

Huang, W., Alexander, G. E., Daly, E. M., Shetty, H. U., Krasuski, J. S., Rapoport, S. I., et al. (1999). High brain myo-inositol levels in the predementia phase of Alzheimer's disease in adults with Down's syndrome: A 1H MRS study. *American Journal of Psychiatry*, *156*, 1879–1886.

Huang, C., Shi, S., Ule, J., Ruggiu, M., Barker, L., Darnell, R., et al. (2005). Common molecular pathways mediate long-term potentiation of synaptic excitation and slow synaptic inhibition. *Cell*, *123*, 105–118.

Hunter, C. L., Bimonte, H. A., & Granholm, A. C. (2003). Behavioral comparison of 4 and 6 month-old Ts65Dn mice: Age-related impairments in working and reference memory. *Behavioural Brain Research*, *138*, 121–131.

Jaarsma, D., Dino, M. R., Cozzari, C., & Mugnaini, E. (1996). Cerebellar choline acetyltransferase positive mossy fibres and their granule and unipolar brush cell targets: A model for central cholinergic nicotinic neurotransmission. *Journal of Neurocytology*, *25*, 829–842.

Jacob, T. C., Moss, S. J., & Jurd, R. (2008). GABA(A) receptor trafficking and its role in the dynamic modulation of neuronal inhibition. *Nature Review Neuroscience*, *9*, 331–343.

John, J., Sunyer, B., Höger, H., Pollak, A., & Lubec, G. (2009). Hippocampal synapsin isoform levels are linked to spatial memory enhancement by SGS742. *Hippocampus*, *19*, 731–738.

Jontes, J., & Smith, S. (2000). Filopodia, spines, and the generation of synaptic diversity. *Neuron*, *27*, 11–14.

Kahlem, P., Sultan, M., Herwig, R., Steinfath, M., Balzereit, D., Eppens, B., et al. (2004). Transcript level alterations reflect gene dosage effects across multiple tissues in a mouse model of Down syndrome. *Genome Research*, *14*, 1258–1267.

Kajiwara, R., Wouterlood, F., Sah, A., Boekel, A., Baks-te Bulte, L., & Witter, M. (2008). Convergence of entorhinal and CA3 inputs onto pyramidal neurons and interneurons in hippocampal area CA1—an anatomical study in the rat. *Hippocampus*, *18*, 266–280.

Kleschevnikov, A., Belichenko, P., Villar, A., Epstein, C., Malenka, R., & Mobley, W. (2004). Hippocampal long-term potentiation suppressed by increased inhibition in the Ts65Dn mouse, a genetic model of Down syndrome. *Journal of Neuroscience*, *24*, 8153–8160.

Kobayashi, T., Washiyama, K., & Ikeda, K. (2003). Inhibition of G protein-activated inwardly rectifying K^+ channels by fluoxetine (Prozac). *British Journal of Pharmacology*, *138*, 1119–1128.

Korbel, J., et al. (2009). The genetic architecture of Down syndrome phenotypes revealed by high-resolution analysis of human segmental trisomies. *Proceedings of the National Academy of Sciences of the United States of America*, *106*, 12031–12036.

Korenberg, J. (2009). Down syndrome: The crucible for treating genomic imbalance. *Genetics in Medicine*, *11*, 617–619.

Koyrakh, L., Lujan, R., Colon, J., Karschin, C., Kurachi, Y., Karschin, A., et al. (2005). Molecular and cellular diversity of neuronal G-protein-gated potassium channels. *Journal of Neuroscience*, *25*, 11468–11478.

Kulik, A., Vida, I., Fukazawa, Y., Guetg, N., Kasugai, Y., Marker, C. L., et al. (2006). Compartment-dependent colocalization of kir3.2-containing K^+ channels and GABA$_B$ receptors in hippocampal pyramidal cells. *Journal of Neuroscience*, *26*, 4289–4297.

Kurt, M. A., Kafa, M. I., Dierssen, M., & Davies, D. C. (2004). Deficits of neuronal density in CA1 and synaptic density in the dentate gyrus, CA3 and CA1, in a mouse model of Down syndrome. *Brain Research*, *1022*, 101–109.

Lacey-Casem, M., & Oster-Granite, M. (1994). The neuropathology of the trisomy 16 mouse. *Critical Reviews in Neurobiology*, *8*, 293–322.

Lejeune, J. T. R. G. M. (1959). Le mogolisme, premier exemple d'aberration autosomique humanie. *Annales de génétique*, *1*, 41–49.

Lewohl, J., Wilson, W., Mayfield, R., Brozowski, S., Morrisett, R., & Harris, R. (1999). G-protein-coupled inwardly rectifying potassium channels are targets of alcohol action. *Nature Neuroscience*, *2*, 1084–1090.

Liao, Y. J., Jan, Y. N., & Jan, L. Y. (1996). Heteromultimerization of G-protein-gated inwardly rectifying K^+ channel proteins GIRK1 and GIRK2 and their altered expression in weaver brain. *Journal of Neuroscience*, *16*, 7137–7150.

Luján, R., Maylie, J., & Adelman, J. (2009). New sites of action for GIRK and SK channels. *Nature Reviews. Neuroscience*, *10*, 475–480.

Lüscher, C., Jan, L., Stoffel, M., Malenka, R., & Nicoll, R. (1997). G protein-coupled inwardly rectifying K^+ channels (GIRKs) mediate postsynaptic but not presynaptic transmitter actions in hippocampal neurons. *Neuron*, *19*, 687–695.

Lyle, R., Gehrig, C., Neergaard-Henrichsen, C., Deutsch, S., & Antonarakis, S. (2004). Gene expression from the aneuploid chromosome in a trisomy mouse model of Down syndrome. *Genome Research*, *14*, 1268–1274.

Maccaferri, G., & McBain, C. (1995). Passive propagation of LTD to stratum oriens-alveus inhibitory neurons modulates the temporoammonic input to the hippocampal CA1 region. *Neuron*, *15*, 137–145.

Mao, R., Zielke, C., Zielke, H., & Pevsner, J. (2003). Global up-regulation of chromosome 21 gene expression in the developing Down syndrome brain. *Genomics*, *81*, 457–467.

Martinez-Cue, C., Baamonde, C., Lumbreras, M. A., Vallina, I. F., Dierssen, M., & Florez, J. (1999). A murine model for Down syndrome shows reduced responsiveness to pain. *NeuroReport*, *10*, 1119–1122.

McVeigh, K. E., Mallee, J. J., Lucente, A., Barnoski, B. L., Wu, S., & Berry, G. T. (2000). Murine chromosome 16 telomeric region, homologous with human chromosome 21q22, contains the osmoregulatory Na^{+}/myo-inositol cotransporter (SLC5A3) gene. *Cytogenetics and Cell Genetics*, *88*, 153–158.

Mitrovic, I., Margeta-Mitrovic, M., Bader, S., Stoffel, M., Jan, L., & Basbaum, A. (2003). Contribution of GIRK2-mediated postsynaptic signaling to opiate and alpha 2-adrenergic analgesia and analgesic sex differences. *Proceedings of the National Academy of Sciences of the United States of America*, *100*, 271–276.

Morice, E., Andreae, L. C., Cooke, S. F., Vanes, L., Fisher, E. M., Tybulewicz, V. L., et al. (2008). Preservation of long-term memory and synaptic plasticity despite short-term impairments in the Tc1 mouse model of Down syndrome. *Learning and Memory*, *15*, 492–500.

Mugnaini, E., Dino, M. R., & Jaarsma, D. (1997). The unipolar brush cells of the mammalian cerebellum and cochlear nucleus: Cytology and microcircuitry. *Progress in Brain Research*, *114*, 131–150.

Mural, R., et al. (2002). A comparison of whole-genome shotgun-derived mouse chromosome 16 and the human genome. *Science*, *296*, 1661–1671.

Nadel, L. (2003). Down's syndrome: A genetic disorder in biobehavioral perspective. *Genes, Brain, and Behavior*, *2*, 156–166.

Neri, G., & Opitz, J. (2009). Down syndrome: Comments and reflections on the 50th anniversary of Lejeune's discovery. *American Journal of Medical Genetics. Part A*, *149A*, 2647–2654.

Neves, G., Cooke, S., & Bliss, T. (2008). Synaptic plasticity, memory and the hippocampus: A neural network approach to causality. *Nature Reviews. Neuroscience*, *9*, 65–75.

Nishida, M., & MacKinnon, R. (2002). Structural basis of inward rectification: Cytoplasmic pore of the G protein-gated inward rectifier GIRK1 at 1.8 A resolution. *Cell*, *111*, 957–965.

O'Doherty, A., Ruf, S., Mulligan, C., Hildreth, V., Errington, M. L., Cooke, S., et al. (2005). An aneuploid mouse strain carrying human chromosome 21 with Down syndrome phenotypes. *Science*, *309*, 2033–2037.

Olson, L., Roper, R., Baxter, L., Carlson, E., Epstein, C., & Reeves, R. (2004). Down syndrome mouse models Ts65Dn, Ts1Cje, and Ms1Cje/Ts65Dn exhibit variable severity of cerebellar phenotypes. *Developmental Dynamics*, *230*, 581–589.

Olson, L. E., Roper, R. J., Sengstaken, C. L., Peterson, E. A., Aquino, V., Galdzicki, Z., et al. (2007). Trisomy for the Down syndrome "critical region" is necessary but not sufficient for brain phenotypes of trisomic mice. *Human Molecular Genetics*, *16*, 774–782.

Paredes, M. F., Greenwood, J., & Baraban, S. C. (2003). Neuropeptide Y modulates a G protein-coupled inwardly rectifying potassium current in the mouse hippocampus. *Neuroscience Letters*, *340*, 9–12.

Partanen, J., Soininen, H., Könönen, M., Kilpeläinen, R., Helkala, E., & Riekkinen, P. S. (1996). EEG reactivity correlates with neuropsychological test scores in Down's syndrome. *Acta Neurologica Scandinavica*, *94*, 242–246.

Patterson, D. (2009). Molecular genetic analysis of Down syndrome. *Human Genetics*, *126*, 195–214.

Pennington, B. F., Moon, J., Edgin, J., Stedron, J., & Nadel, L. (2003). The neuropsychology of Down syndrome: Evidence for hippocampal dysfunction. *Child Development*, *74*, 75–93.

Pereira, P., Magnol, L., Sahún, I., Brault, V., Duchon, A., Prandini, P., et al. (2009). A new mouse model for the trisomy of the Abcg1-U2af1 region reveals the complexity of the combinatorial genetic code of Down syndrome. *Human Molecular Genetics*, *18*, 4756–4769.

Peters, M., Bletsch, M., Catapano, R., Zhang, X., Tully, T., & Bourtchouladze, R. (2009). RNA interference in hippocampus demonstrates opposing roles for CREB and PP1alpha in contextual and temporal long-term memory. *Genes, Brain, and Behavior*, *8*, 320–329.

Pinter, J. D., Brown, W. E., Eliez, S., Schmitt, J. E., Capone, G. T., & Reiss, A. L. (2001). Amygdala and hippocampal volumes in children with Down syndrome: A high- resolution MRI study. *Neurology*, *56*, 972–974.

Prandini, P., Deutsch, S., Lyle, R., Gagnebin, M., Delucinge Vivier, C., Delorenzi, M., et al. (2007). Natural gene-expression variation in Down syndrome modulates the outcome of gene-dosage imbalance. *American Journal of Human Genetics*, *81*, 252–263.

Pöschel, B., & Stanton, P. (2007). Comparison of cellular mechanisms of long-term depression of synaptic strength at perforant path-granule cell and Schaffer collateral-CA1 synapses. *Progress in Brain Research*, *163*, 473–500.

Rachidi, M., & Lopes, C. (2007). Mental retardation in Down syndrome: From gene dosage imbalance to molecular and cellular mechanisms. *Neuroscience Research*, *59*, 349–369.

Rapoport, S. I., & Galdzicki, Z. (1994). Electrical studies of cultured fetal human trisomy 16 and mouse trisomy 16 neurons identify functional deficits that may lead to mental retardation in Down syndrome. *Developmental Brain Dysfunction*, *7*, 265–288.

Reeves, R. H. (2006). Down syndrome mouse models are looking up. *Trends in Molecular Medicine*, *12*, 237–240.

Reeves, R. H., Irving, N. G., Moran, T. H., Wohn, A., Kitt, C., Sisodia, S. S., et al. (1995). A mouse model for Down syndrome exhibits learning and behaviour deficits [see comments]. *Nature Genetics*, *11*, 177–184.

Roberson, C., Clapham, D. E., Pangalos, M. N., & Davies, C. H. (2002). GPCR signalling through ion channels. In R. W. Davies, G. L. Collingridge, S. P. Hunt, (Eds.), *Understanding G protein-coupled receptors and their role in the CNS* (pp. 108–123.). Oxford: Oxford University Press.

Roper, R., & Reeves, R. (2006). Understanding the basis for Down syndrome phenotypes. *PLoS Genetics*, *2*, e50.

Rueda, N., Florez, J., & Martinez-Cue, C. (2008). Chronic pentylenetetrazole but not donepezil treatment rescues spatial cognition in Ts65Dn mice, a model for Down syndrome. *Neuroscience Letters*, *433*, 22–27.

Russo, M., Yau, H., Nunzi, M., Mugnaini, E., & Martina, M. (2008). Dynamic metabotropic control of intrinsic firing in cerebellar unipolar brush cells. *Journal of Neurophysiology*, *100*, 3351–3360.

Sago, H., Carlson, E. J., Smith, D. J., Kilbridge, J., Rubin, E. M., Mobley, W. C., et al. (1998). Ts1Cje, a partial trisomy 16 mouse model for Down syndrome, exhibits learning and behavioral abnormalities. *Proceedings of the National Academy of Sciences of the United States of America*, *95*, 6256–6261.

Sago, H., Carlson, E. J., Smith, D. J., Rubin, E. M., Crnic, L. S., Huang, T. T., et al. (2000). Genetic dissection of region associated with behavioral abnormalities in mouse models for Down syndrome. *Pediatric Research*, *48*, 606–613.

Sawa, A. (1999). Neuronal cell death in Down's syndrome. *Journal of Neural Transmission-Supplement*, *57*, 87–97.

Schubert, V., & Dotti, C. (2007). Transmitting on actin: Synaptic control of dendritic architecture. *Journal of Cell Science*, *120*, 205–212.

Shetty, H., Holloway, H., Acevedo, L., & Galdzicki, Z. (1996). Brain accumulation of myo-inositol in the trisomy 16 mouse, an animal model of Down's syndrome. *Biochemical Journal*, *313*(Pt 1), 31–33.

Shonk, T., & Ross, B. D. (1995). Role of increased cerebral myo-inositol in the dementia of Down syndrome. *Magnetic Resonance in Medicine*, *33*, 858–861.

Siarey, R. J., Carlson, E. J., Epstein, C. J., Balbo, A., Rapoport, S. I., & Galdzicki, Z. (1999). Increased synaptic depression in the Ts65Dn mouse, a model for mental retardation in Down syndrome. *Neuropharmacology*, *38*, 1917–1920.

Siarey, R., Kline-Burgess, A., Cho, M., Balbo, A., Best, T., Harashima, C., et al. (2006). Altered signaling pathways underlying abnormal hippocampal synaptic plasticity in the Ts65Dn mouse model of Down syndrome. *Journal of Neurochemistry*, *98*, 1266–1277.

Siarey, R., Stoll, J., Rapoport, S., & Galdzicki, Z. (1997a). Altered long-term potentiation in the young and old Ts65Dn mouse, a model for Down syndrome. *Neuropharmacology*, *36*, 1549–1554.

Siarey, R., Villar, A., Epstein, C., & Galdzicki, Z. (2005). Abnormal synaptic plasticity in the Ts1Cje segmental trisomy 16 mouse model of Down syndrome. *Neuropharmacology*, *49*, 122–128.

Signorini, S., Liao, Y. J., Duncan, S. A., Jan, L. Y., & Stoffel, M. (1997). Normal cerebellar development but susceptibility to seizures in mice lacking G protein-coupled, inwardly rectifying K^+ channel GIRK2. *Proceedings of the National Academy of Sciences of the United States of America*, *94*, 923–927.

Slesinger, P., Patil, N., Liao, Y., Jan, Y., Jan, L., & Cox, D. (1996). Functional effects of the mouse weaver mutation on G protein-gated inwardly rectifying K^+ channels. *Neuron*, *16*, 321–331.

Sodickson, D. L., & Bean, B. P. (1998). Neurotransmitter activation of inwardly rectifying potassium current in dissociated hippocampal CA3 neurons: Interactions among multiple receptors. *Journal of Neuroscience*, *18*, 8153–8162.

Svoboda, K. R., & Lupica, C. R. (1998). Opioid inhibition of hippocampal interneurons via modulation of potassium and hyperpolarization-activated cation (1h) currents. *Journal of Neuroscience*, *18*, 7084–7098.

Takahashi, T., Kobayashi, T., Ozaki, M., Takamatsu, Y., Ogai, Y., Ohta, M., et al. (2006). G protein-activated inwardly rectifying K^+ channel inhibition and rescue of weaver mouse motor functions by antidepressants. *Neuroscience Research*, *54*, 104–111.

Takashima, S., Becker, L. E., Armstrong, D. L., & Chan, F. (1981). Abnormal neuronal development in the visual cortex of the human fetus and infant with Down's syndrome. A quantitative and qualitative Golgi study. *Brain Research*, *225*, 1–21.

Voronov, S. V., Frere, S. G., Giovedi, S., Pollina, E. A., Borel, C., Zhang, H., et al. (2008). Synaptojanin 1-linked phosphoinositide dyshomeostasis and cognitive deficits in mouse models of Down's syndrome. *Proceedings of the National Academy of Sciences of the United States of America*, *105*, 9415–9420.

Wang, J., Liu, S., Haditsch, U., Tu, W., Cochrane, K., Ahmadian, G., et al. (2003). Interaction of calcineurin and type-A GABA receptor gamma 2 subunits produces long-term depression at CA1 inhibitory synapses. *Journal of Neuroscience*, *23*, 826–836.

Weitzdoerfer, R., Dierssen, M., Fountoulakis, M., & Lubec, G. (2001). Fetal life in Down syndrome starts with normal neuronal density but impaired dendritic spines and synaptosomal structure. *Journal of Neural Transmission. Supplementum*, *61*, 59–70.

Wolvetang, E., Bradfield, O., Hatzistavrou, T., Crack, P., Busciglio, J., Kola, I., et al. (2003). Overexpression of the chromosome 21 transcription factor Ets2 induces neuronal apoptosis. *Neurobiology of Disease*, *14*, 349–356.

Yamada, K., Yu, B., & Gallagher, J. P. (1999). Different subtypes of $GABA_B$ receptors are present at pre- and postsynaptic sites within the rat dorsolateral septal nucleus. *Journal of Neurophysiology*, *81*, 2875–2883.

Yao, F. S., Caserta, M. T., & Wyrwicz, A. M. (2000). In vitro 1H and 31P NMR spectroscopic evidence of multiple aberrant biochemical pathways in murine trisomy 16 brain development. *International Journal of Developmental Neuroscience*, *18*, 833–841.

John F. Cryan[*] **and David A. Slattery**[†]

[*]School of Pharmacy, Department of Pharmacology & Therapeutics, Alimentary Pharmabiotic Centre, University College Cork, Cork, Ireland

[†]Institute of Zoology, University of Regensburg, Regensburg, Germany

GABA$_B$ Receptors and Depression: Current Status

Abstract

Dysfunction of the gamma amino butyric acid (GABA)-ergic system has been purported to play a role in psychiatric disorders, including anxiety and major depression. A clear link between GABA$_A$ receptors and anxiety has long been established. However, despite the GABA system being the prominent inhibitory neurotransmitter in the brain, a role in depression has been less well validated. GABA$_B$ receptors, first characterized by Bowery and colleagues 30 years ago, have been long postulated to be involved in the etiology of depression, but a lack of selective, orally active, pharmacological compounds slowed down their assessment. From the mid-1990s, more selective pharmacological and genetic tools for examining the GABA$_B$

Advances in Pharmacology, Volume 58

1054-3589/10 $35.00
10.1016/S1054-3589(10)58016-5

system have provided greater insight into their role in complex behavioral and molecular characteristics. The resulting literature garnered from recent behavioral studies employing selective $GABA_B$ receptor antagonists and knockout mice suggests in the main that the blockade, or loss of function, of $GABA_B$ receptors causes an antidepressant-like phenotype. The $GABA_B$ receptor system has been shown to have substantial interactions with the serotonergic system and neurotrophic factors, such as BDNF. We argue that the potential antidepressant properties of $GABA_B$ receptor antagonists may be, at least in part, mediated via these interactions. Clinical studies have repeatedly reported alterations in GABA levels, particularly in cortical areas, but studies to specifically assess $GABA_B$ receptors are lacking. Those available have documented differential gene expression of the $GABA_{B(2)}$ subunit but require replication. Therefore, while further research is necessary, it is suggested that $GABA_B$ receptor antagonists may represent a new class of antidepressant compounds.

I. Introduction

Despite a biological basis for mood disorders been described as early as the fifth century BC, by Hippocrates, and substantial research efforts, the underlying etiology of major depressive disorder (MDD) remains poorly understood. The World Health Organization positions depression as the second largest global health burden (Licinio & Wong, 2005; Murray & Lopez, 1997), and current lifetime prevalence rates are estimated at 16.2% in the United States (Kessler et al., 2003). Moreover, the average cost burden on society in the United States is approximately $80 billion per year and in Europe as high as €106 billion (Andlin-Sobocki & Wittchen, 2005). All currently available antidepressants have a delayed onset of action (Slattery et al., 2004) and are ineffective in up to 30–40% of patients (Kirsch et al., 2008; Matthews et al., 2005). Moreover, all currently available antidepressants are believed to exert their therapeutic benefits by modulation of monoamine neurotransmission. Thus, efforts to develop more efficacious antidepressants have more recently focused on other neurotransmitter systems and molecular pathways (Berton & Nestler, 2006; Cryan et al., 2008; McKernan et al., 2009). Therefore, attention has been focused on the GABAergic (gamma amino butyric acid) system (Brambilla et al., 2003; Slattery & Cryan, 2006). GABA is the main inhibitory transmitter in the central nervous system (CNS), and there is accumulating evidence purporting a specific GABAergic dysfunction in mood disorders (Slattery & Cryan, 2006). While there are two main receptor classes, ionotropic $GABA_A$ and $GABA_C$ receptors and metabotropic $GABA_B$ receptors, via which GABA acts (Cryan & Kaupmann, 2005), the following review focuses on $GABA_B$ receptors.

Metabotropic $GABA_B$ receptors were first implicated to play a role in MDD more than 25 years ago (Pilc & Lloyd, 1984). However, the lack of pharmacological tools with which to study this system prevented detailed assessment of their role, particularly as the prototypical receptor agonist, baclofen (Brogden et al., 1974), causes sedation, motor incoordination, and hypothermia (Jacobson & Cryan, 2005). However, within the past 15 years, novel receptor antagonists, positive modulators, and subunit-specific knockout mice have become available to study with more specificity the role of these receptors (Bettler et al., 2004; Cryan & Kaupmann, 2005).

II. $GABA_B$ Receptors

It is appropriate to mention in this special issue that it is now 30 years since $GABA_B$ receptors were first identified by Norman Bowery and coworkers. Their seminal work showed that a GABA-sensitive blockade of noradrenaline release from rat atria was not altered by application of the $GABA_A$ receptor antagonist bicuculline but by baclofen, the prototypical $GABA_B$ receptor agonist (Bowery et al., 1980; Hill & Bowery, 1981). It was then almost 20 years before the $GABA_B$ receptor was cloned revealing it as the first heteromeric metabotropic receptor (Kaupmann et al., 1997). There are two major subunits, $GABA_{B(1)}$ and $GABA_{B(2)}$, both of which are required to generate functional receptors with the former providing the ligand-binding domain and the latter coupling to the intracellular signaling cascade (Couve et al., 2000; Kaupmann et al., 1998 but see Gassmann et al., 2004). There are also two major isoforms of the $GABA_{B(1)}$ subunit, namely, the $GABA_{B(1a)}$ and $GABA_{B(1b)}$ (although $GABA_{B1c\text{–}g}$ splice variants have been identified, their physiological relevance remains to be elucidated). While the $GABA_{B(1a)}$ is the predominant isoform in the developing brain, $GABA_{B(1b)}$ is the main isoform in the adult brain (Cryan & Kaupmann, 2005). The cloning of $GABA_B$ receptor cDNAs in 1997 (Kaupmann et al., 1997) also paved the way for the generation of mice lacking either the $GABA_{B(1)}$ (both $GABA_{B(1a)}$ and $GABA_{B(1b)}$) or the $GABA_{B(2)}$ subunits (Gassmann et al., 2004; Prosser et al., 2001; Queva et al., 2003; Schuler et al., 2001). More recently, mice lacking either the $GABA_{B(1a)}$ or $GABA_{B(1b)}$ receptor isoform have been generated (Vigot et al., 2006) and should be very useful in understanding the role of these receptor isoforms in brain function and behavior (Jacobson et al., 2006a, 2006b, 2007; Vigot et al., 2006).

In situ hybridization and autoradiography studies have demonstrated that $GABA_B$ receptors are abundantly and widely expressed in the CNS. Presynaptic $GABA_B$ receptors inhibit neurotransmitter release via inhibition of voltage-gated Ca^{2+} channels while postsynaptic receptors are coupled to K^+ channels to generate inhibitory postsynaptic potentials (IPSPs) (see Bettler et al., 2004; Bowery et al., 2002 for reviews).

High-affinity phosphinic acid-derived $GABA_B$ receptor antagonists, such as CGP36742, CGP55845, and CGP56433A have recently been described (Froestl et al., 2003). These compounds have largely replaced the $GABA_B$ receptor antagonists phaclofen, saclofen, and 2-hydroxysaclofen that had emerged in the late 1980s as tools for assessing $GABA_B$ function *in vitro* and *in vivo* (Bowery et al., 2002; Kerr & Ong, 1992). Following the characterization of these novel antagonists, $GABA_B$ receptor-positive modulators, GS39783 and CGP7930, were designed, which have been shown to bind to the transmembrane region of the $GABA_{B(2)}$ subunit (Binet et al., 2004; Dupuis et al., 2006), a site distinct from where GABA binds (Jensen & Spalding, 2004; Soudijn et al., 2002). These ligands provide a powerful tool to assess $GABA_B$ receptors, as they have been shown to potentiate $GABA_B$ receptor activity following stimulation by an agonist but retain no intrinsic activity (Urwyler et al., 2003). Thus, positive modulators have been shown to be devoid of the side effects such as motor incoordination and hypothermia, which limit *in vivo* use of full agonists, such as baclofen (Cryan et al., 2004; Slattery et al., 2005a).

Therefore, the combination of the new pharmacological and genetic tools has provided highly selective mechanisms by which to study the function of $GABA_B$ receptors using both neurochemical and behavioral paradigms.

III. Role of $GABA_B$ Receptors in Animal Models of Antidepressant Action

Animal models of depression are responsive to acute and/or chronic administration of currently available antidepressants and can thus be used to investigate the therapeutic potential of novel compounds (Cryan et al., 2002). The following section reviews the literature to date that has examined the $GABA_B$ receptor system in different animal models.

A. Learned Helplessness

In this model of depression, an animal is subjected to inescapable shocks, which result in a reduced number of escape attempts when they are subsequently placed in an escapable situation (Cryan et al., 2002). This behavior is reversed by a wide range of antidepressants and the GABAergic system was first shown to be involved in helpless behavior by Sherman and Petty (1980). They showed that microinjection of GABA into the frontal cortex, hippocampus, or lateral geniculate nucleus prevented or reversed learned helplessness and that hippocampal GABA release was attenuated in animals that exhibited helpless behavior (Petty & Sherman, 1981). However, Nakagawa and colleagues provided evidence in a series of experiments that $GABA_B$ receptor blockade displayed antidepressant-like activity in the learned helplessness paradigm. Thus, systemic baclofen, but not bicuculline (a $GABA_A$ receptor agonist), administration exacerbated learned helplessness and also abolished the antidepressant-like effect of the tricyclic antidepressant desipramine in this paradigm (Nakagawa et al., 1996a, 1996b), while the $GABA_B$ receptor antagonist CGP36742 dose-dependently improved the number of escapes (Nakagawa et al., 1999).

It has also been demonstrated that animals that did not develop helpless behavior (typically 30%) had decreased $GABA_B$ receptor but increased $GABA_A$ receptor levels in the medial and lateral septum (Kram et al., 2000), a region that has been implicated in depressive behavior (Duncan et al., 1996; Estrada-Camarena et al., 2002; Price et al., 2002; Sheehan et al., 2004). This suggests that decreased septal levels of $GABA_B$ receptors confer the animal with an antidepressant-like coping strategy, at least in this model. However, it is noteworthy that a prior study revealed that $GABA_B$ receptor levels were decreased in the frontal cortex of learned helpless rats (Martin et al., 1989). A number of differences in protocol may account for this discrepancy, including rat strain (Sprague Dawley vs. Wistar), shock protocol (100 shocks lasting 5 s at an intensity of 1.0 mA vs. 60 15 s 0.8 mA inescapable shocks), and the technique to assess $GABA_B$ receptor levels (quantitative autoradiography vs. membrane ligand binding). However, it may also help to explain the conflicting pharmacological findings as $GABA_B$ receptor levels in different brain regions may be differentially regulated in order to provide an antidepressant-like effect.

B. Olfactory Bulbectomy

The removal of the olfactory bulbs results in numerous behavioral, neuroendocrine, neurochemical, and immune alterations, which are also observed in clinical studies (Harkin et al., 2003; Song & Leonard, 2005). The olfactory bulbectomy model of depression is solely responsive to chronic administration of antidepressants and as such represents an animal model that is responsive in a clinical time frame (Cryan et al., 2002; Song & Leonard, 2005; Willner & Mitchell, 2002). The most consistent behavioral alterations caused by bulbectomy are hyperactivity in a novel brightly lit open field arena and deficits in passive avoidance learning, both of which are responsive to chronic antidepressant administration (Cryan et al., 1998, 1999; Kelly et al., 1997; Song & Leonard, 2005). Baclofen was reported to reverse the bulbectomy-induced hyperactivity (Leonard & Tuite, 1981) but it is problematic to separate an antidepressant-like response to the known sedative properties of baclofen. The GABA mimetic progabide has also been demonstrated to show antidepressant-like behavior in the olfactory bulbectomy model by reversing the passive avoidance deficits (Lloyd et al., 1983). The generation of novel $GABA_B$ receptor antagonists has enabled a more selective examination of the $GABA_B$ system's involvement in bulbectomy-induced deficits. Thus, CGP36742 and CGP51176 were shown to reverse the passive avoidance deficits in bulbectomized rats (Nowak et al., 2006). $GABA_B$ receptor expression has been assessed in a number of studies following bulbectomy and has consistently been shown to be decreased in the frontal cortices (Lloyd & Pichat, 1986; cited in Dennis et al., 1993, 1994; Lloyd et al., 1989). Therefore, when these findings are coupled with those showing a decreased level of $GABA_B$ receptors in the frontal cortex of learned helpless rats (Martin et al., 1989), it appears that low expression of $GABA_B$ receptors in this region may confer an increased risk of depression following stress exposure.

Thus, results obtained from these studies collectively demonstrate that $GABA_B$ receptors are likely to play a role in the olfactory bulbectomy animal model of depression. However, further research is required to determine whether $GABA_B$ receptor activation is also beneficial as the initial baclofen reports suggest. The assessment of the behavioral effects of the novel allosteric positive modulators GS39783 or CGP7930, which lack the sedative effects of the full agonist, would help to address this question (Cryan et al., 2004; Jacobson & Cryan, 2008).

C. Forced Swim Test

The most widely utilized animal model of antidepressant action is the forced swim test (FST) (see Cryan et al., 2002, 2005b; Lucki, 1997 for reviews). When a rat is exposed to an inescapable cylinder filled with water, after initial escape-directed behavior the animals quickly adopt an immobile posture, which is believed to reflect either a failure to persist with escape-directed behavior or a passive behavior to cease active forms of coping to the stressful stimuli (Cryan et al., 2002; Lucki, 1997). While the traditional rat version (Porsolt et al., 1977, 1978) comprises of an initial pre-exposure followed 24 h later by the test exposure, the mouse swim test only requires a single exposure to the swim cylinder. Nevertheless, in both the rat and mouse versions, antidepressants have been selectively shown to increase the time that rodents spend in active behaviors (Borsini, 1995; Borsini & Meli, 1988; Cryan & Mombereau, 2004; Petit-Demouliere et al., 2005).

Baclofen administration has been shown to attenuate the antidepressant-like effect induced by desipramine, mianserin, and buspirone in the rat FST (Nakagawa et al., 1996c) and to be without intrinsic effect itself in the test (Nakagawa et al., 1996c) at doses that do not impair motoric behavior. These findings are similar to those observed in the learned helpless paradigm, which suggest that $GABA_B$ receptor activation blocks the antidepressant-like effects of clinically efficacious drugs. These effects were not due to any sedative effect of baclofen as it was without any effect itself on immobility behavior at the doses tested (Nakagawa et al., 1996c).

In agreement with the above findings, both $GABA_B$ receptor antagonists (Mombereau et al., 2004; Nowak et al., 2006) and genetic ablation of either $GABA_{B(1)}$ or $GABA_{B(2)}$ receptor subunits (Mombereau et al., 2004, 2005) induce an antidepressant-like phenotype in the mouse FST. Interestingly, the antidepressant-like effect of CGP51176 was prevented by coadministration of CGP44532, a $GABA_B$ receptor agonist, further supporting a $GABA_B$-mediated effect (Nowak et al., 2006).

A modified version of the rat FST has emerged over the past decade, championed by Lucki and colleagues (Cryan et al., 2005; Lucki, 1997) in part as a response to the fact that the traditional rat FST has proven to be somewhat unreliable in detecting the antidepressant-like efficacy of selective serotonin reuptake inhibitors (SSRIs) (Borsini, 1995). The modified version enabled the scoring of two active behaviors, swimming, which is increased by serotonergic antidepressants, and climbing, which is increased by catecholaminergic drugs (Cryan et al., 2002, 2005; Lucki, 1997). We have examined $GABA_B$ receptor antagonists in this version of the FST and the results confirm those observed in the traditional and mouse versions of the test (Slattery et al., 2005a). However, the ability to distinguish between the different active behaviors revealed that $GABA_B$ receptor antagonists selectively increased swimming behavior, as observed by SSRI administration (Cryan et al., 2002). Interestingly, and confirming the link to the 5-hydroxytryptamine (5-HT) system, depletion of central 5-HT levels with *para*-chlorophenylalanine blocked the antidepressant-like effect of $GABA_B$ receptor blockade (Slattery et al., 2005a). Interestingly, the $GABA_B$ receptor-positive modulator GS39783 did not alter any behavior in this test, which is similar to that shown for the full agonist baclofen in the traditional rat swim test (Nakagawa et al., 1996c). However, it remains unknown at present whether GS39783 would attenuate the effect of antidepressants in a similar vein to that reported by baclofen. The data described above supports a role of the $GABA_B$ receptor in the depression-related behaviors as assessed by the FST and provides evidence that antagonism of these receptors display antidepressant-like effects in this model. More recently, Frankowska et al. (2010) demonstrated that the $GABA_B$ receptor antagonist SCH 50911 (0.3 mg/kg) reduced cocaine withdrawal-induced elevations in immobility in the FST at doses lower than that which induces antidepressant-like effects (Frankowska et al., 2007). However, the same studies also show that activation of $GABA_B$ receptors with agonists and positive modulators has antidepressant-like effects. This conundrum cannot be easily explained currently and reinforces the difficulty in dissecting out the behavioral role of a receptor that has both pre- and postsynaptic effects. Interestingly, both $GABA_B$ receptor and antagonists also behave similarly in intracranial self-stimulation procedures, in cocaine discrimination, or cocaine-induced seeking behavior in self-administration procedures in rats (Filip & Frankowska, 2007; Filip et al., 2007; Macey et al., 2001). Theoretically, $GABA_B$ receptor agonists (used in low doses) might cause the stimulation of presynaptic $GABA_B$ receptors leading to decreased GABA release, thus mimicking the effects of antagonists at $GABA_B$ postsynaptic sites. On the other hand, blockade of presynaptic $GABA_B$ receptors may result in the enhanced GABA release, which may stimulate postsynaptically localized GABA receptors. Interestingly, there is no evidence to date of such effects in other paradigms used to assess antidepressant activity.

D. Tail Suspension Test

When a mouse is suspended by its tail after initial struggling behavior, the animal will quickly adopt an immobile posture. Antidepressant administration increases the duration of struggling (Cryan et al., 2005a; Steru et al., 1985) in the mouse. Similar to the results obtained in the olfactory bulbectomy model and FST, baclofen was shown to have a prodepressant-like effect in this model but as was the case in the former these effects are likely to be due to the motor impairing, muscle relaxing, and sedating effects induced by $GABA_B$ receptor activation (Cryan et al., 2005; Jacobson & Cryan, 2005). Interestingly, it was shown that the $GABA_B$ receptor antagonist CGP56433A did not have any antidepressant-like effect in this test (Mombereau et al., 2004). Similar $GABA_{B(1)}$ receptor subunit-deficient mice also failed to demonstrate altered behavior in the tail suspension test (TST) (Mombereau et al., 2004). Interestingly, GABA transporter subtype 1 knockout mice display reduced immobility in the TST suggesting that an enhanced GABAergic drive leads to an antidepressant-like phenotype (Liu et al., 2007). However, it should also be noted that these animals displayed greater locomotor activity, as measured in the open field and that a general increase in activity may underlie the decreased immobility in the TST. Despite the similarities between the TST and FST it is becoming increasing apparent that different biological substrates underlie the tests (Cryan & Mombereau, 2004; Cryan et al., 2005a, b), which may have a $GABA_B$ receptor component to them.

E. Chronic Mild Stress

The chronic mild stress (CMS) paradigm is based on the initial discovery by Katz (1982) that exposure to severe predictable stressors leads to an anhedonic phenotype in rats. The severity of the stressors was reduced (e.g. change of cage-mate, overnight lighting, novel housing, restraint, and deprivation of food or water) and was also extended to over a period of weeks to months (Willner, 1997, 2005). Similar to the original paradigm, CMS has been shown to decrease consumption of a sweet (sucrose or saccharin) solution, i.e. an anhedonic-like effect, and that this effect is long-lasting and reversible by numerous antidepressants when administered chronically but not acutely (Willner, 1997, 2005). It should be noted that there is much controversy in the literature surrounding the reliability and robustness of the CMS paradigm (Reid et al., 1997; Willner, 1997).

Studies of $GABA_B$ receptor ligands in this model have been limited but it has been demonstrated that the $GABA_B$ receptor antagonist CGP51176A fully normalized sucrose consumption in rats following 4 weeks treatment at doses of 3 and 10 mg/kg p.o. but not at 0.3 mg/kg (Bittiger et al., 1996). The antidepressant imipramine (10 mg/kg p.o.) was included in this study as a positive control and also fully restored sucrose consumption in rats that had undergone CMS (Bittiger et al., 1996). This finding has recently been recapitulated for CGP51176 (Nowak et al., 2006). Moreover, the time-course of treatment was assessed and it demonstrated that following 3–4 weeks of treatment sucrose intake was increased compared with vehicle-treated CMS rats and after 5 weeks there was no difference between the CMS CGP51176 group and controls (Nowak et al., 2006). Therefore, in agreement with data outlined above in other models, these data demonstrate that $GABA_B$ receptor antagonists display antidepressant-like behavioral effects. Furthermore, it has been demonstrated that CMS resulted in decreased sucrose preference as well as decreased GABA levels in the hippocampus (Gronli et al., 2007). However, there was no correlation between sucrose intake and hippocampal GABA levels and the authors did not attempt to reverse these deficits with GABAergic ligands. Contrastingly, it has been demonstrated that baclofen administration blocked the reduced social interaction of rats repeatedly exposed to ethanol or restraint stress (Knapp et al., 2007). While the authors suggest that this effect is related to anxiety, previous studies have shown that repeated exposure to restraint stress or ethanol causes depression-like behavior and that chronic antidepressant administration can reverse the social interaction deficits caused by such stressors (Berton et al., 2006; Krishnan et al., 2007; Tsankova et al., 2006). Thus, it is possible that the reversal of the effects of stress and alcohol exposure by baclofen may also be related to an antidepressant-like response but further studies are warranted to test this hypothesis.

IV. $GABA_B$ Receptors and Cognition

Cognitive impairments, including decreased ability to concentrate, decreased learning and memory, and deficits in executive function (the ability to plan and participate in goal-directed behavior) are observed in depressed patients (Beck, 2008; Elliott et al., 1996; Wang et al., 2008). The deficits in cognition can be partially explained from neuroimaging studies demonstrating decreased activity in many regions of the frontal cortex (Phillips et al., 2003), as executive function has been shown to be dependent on these regions (Dalley et al., 2004; Rogers et al., 2004). There is preclinical data showing that administration of the $GABA_B$ receptor antagonist SGS742 (or CGP36742) improves cognition, or more specifically spatial working memory, using the Morris water maze radial maze (Helm et al., 2005; Sunyer et al., 2007, 2008). Interestingly, this observation has been replicated in a small clinical sample (Bullock, 2005; Froestl et al., 2004). A potential mechanism whereby $GABA_B$ receptor antagonism improves cognitive functioning is by enhancing hippocampal theta and gamma rhythms (Leung & Shen, 2007). This study demonstrated that icv administration of CGP35348 increased gamma waves in the hippocampus while medial septal infusion increased the power and frequency of theta waves in the hippocampus (Leung & Shen, 2007). Thus, $GABA_B$ receptor antagonism may enhance cognition by elevating neural activity in the hippocampus. Adding further credence to this hypothesis is the recent finding by Pilc and colleagues that administration of CGP51176 increased $GABA_B$ receptor density in the mouse hippocampus (Nowak et al., 2006).

Another potential mechanism of action for this improved cognition is via modulation of activating transcription factor 4 (ATF4, or CREB2) as SGS742 was shown to reduce basal CRE binding in the hippocampus, which was mainly associated with ATF4 binding (Helm et al., 2005). ATF4 acts to suppress gene transcription (Cardin & Abel, 1999) and $GABA_B$ receptors are functionally linked to this transcription factor (Nehring et al., 2000; Vernon et al., 2001; White et al., 2000). Thus, together with improving spatial memory, this mechanism may be responsible, at least in part, for the antidepressant-like properties of $GABA_B$ receptor antagonists. Such a mechanism is also in line with the recent neurotrophic hypotheses of depression (Berton & Nestler, 2006). One neurotrophic factor that has received much interest is brain-derived neurotropic factor (BDNF), which leads to neuronal plasticity (Duman, 2002, 2004; Duman et al., 1997). A number of studies have shown that a variety of antidepressants increase levels of BDNF in rodents and intracerebral administration of BDNF has been shown to result in antidepressant-like behavior (Duman et al., 1997; Eisch et al., 2003; Hoshaw et al., 2005; Shirayama et al., 2002; Siuciak et al., 1997), although there are also some negative studies (for example, Altar et al., 2003). In humans, depression has been associated with a decreased level of postmortem BDNF levels in untreated depressed patients compared with control and treated groups (Shimizu et al., 2003). Heese et al. (2000) demonstrated that a single administration of four different $GABA_B$ receptor antagonists increased the concentration of BDNF in the hippocampus, cortex, and spinal cord, which was confirmed by Enna and colleagues in the striatum (Enna et al., 2006).

Therefore, this represents a possible mechanism that may contribute to the antidepressant-like effects of GABA$_B$ receptor antagonists.

V. Chronic Antidepressants and GABA$_B$ Receptor Function

There has been a substantial research effort over the past three decades into the effects of chronic antidepressant administration on GABA$_B$ receptor function and density and how this may correlate with their delayed onset of action (for review see Enna & Bowery, 2004). This research has focused on the premise that secondary alterations on neurotransmitter receptors caused by chronic administration of antidepressants may underlie or contribute to their overall efficacy (Duman, 2002, 2004; Nestler et al., 2002). However, in the context of GABA$_B$, this research has been largely inconsistent with findings of no change or increases in GABA$_B$ receptor density, mRNA, or function following chronic antidepressant administration (see Enna & Bowery, 2004; Slattery & Cryan, 2006 for comprehensive reviews of the literature). Reasons for the variability may be due to discrete differences in the methodologies used between various laboratories.

VI. GABA$_B$ Receptors and the Reward System

Anhedonia, the inability to experience pleasure from normally rewarding stimuli, is a core symptom of depression (Cryan & Slattery, 2007). Thus, a number of studies have been performed examining the influence of the GABA$_B$ receptor ligands on rewarding-drug administration and on the dopamine system, the key neurotransmitter in the brain reward circuitry (Wise, 2002). Thus, direct effects on dopamine release have been observed following intracerebral administration of GABA$_B$ receptor ligands. While intra-ventral tegmental area (VTA) baclofen decreased dopamine release in the prefrontal cortex (Westerink et al., 1998) and nucleus accumbens (Westerink et al., 1996; Xi & Stein, 1999), intracerebral phaclofen was shown to increase dopamine release in numerous regions associated with the reward system (Giorgetti et al., 2002; Gong et al., 1998; Santiago et al., 1993; Smolders et al., 1995). Therefore, the antidepressant-like potential of GABA$_B$ receptor antagonists may be due in part to an increase of dopaminergic transmission in the brain reward circuitry, which may help to alleviate the anhedonic-like symptoms.

In agreement with this hypothesis a number of behavioral studies have been performed, which reveal that GABA$_B$ receptor activation, either with baclofen or the positive modulator GS39783, attenuate the rewarding properties of cocaine and nicotine (Lhuillier et al., 2007; Mombereau et al., 2007; Slattery et al., 2005b; Willick & Kokkinidis, 1995) in addition to the neurochemical (Fadda et al., 2003) and molecular consequences (Lhuillier et al., 2007; Mombereau et al., 2007).

VII. $GABA_B$–5-HT Interactions

As alluded to above, a link between $GABA_B$ receptors and the serotonergic system has been demonstrated in behavioral studies (Slattery et al., 2005a). In addition there is substantial evidence from neurochemical and electrophysiological studies. Varga et al. (2002) demonstrated that over 95% of 5-HT immunoreactive cell bodies in the raphe complex are copositive with $GABA_B$ receptors. Systemic administration of baclofen was shown to increase 5-HT release with the dorsal raphe nucleus (DRN) while intra-DRN application of baclofen resulted in dose-specific alterations in 5-HT release (Abellan et al., 2000). In another study, intra-DRN administration of phaclofen (a $GABA_B$ receptor antagonist) did not alter basal 5-HT levels, which suggests that activation of $GABA_B$ receptors is required before affecting the 5-HT system (Tao et al., 1996). Administration of baclofen into the DRN has also been demonstrated to decrease 5-HT release in the nucleus accumbens (Tao et al., 1996), a region that is implicated in the reward circuitry of the brain (see Wise, 2002 for review). Therefore, it is possible that $GABA_B$ receptors in the DRN could be overactive in major depression and lead to the decrease in 5-HT neurotransmission that is seen in patients. However, as yet there is no direct evidence for this.

A link between the $GABA_B$ receptor and 5-HT system has also been supported in studies employing genetically-modified mice. Thus, in mice lacking the 5-HT transporter, which also have a desensitization of DRN $5\text{-}HT_{1A}$ autoreceptors, baclofen was shown to have a reduced potency in inhibiting 5-HT cell firing (Fabre et al., 2000; Mannoury la Cour et al., 2004). Given that no deficits in $GABA_B$ receptor function were observed in $5\text{-}HT_{1A}$ receptor knockout mice and that $GABA_A$ receptors are unaffected in both knockouts, the link between $GABA_B$ receptors and $5\text{-}HT_{1A}$ appears to be specific (Mannoury la Cour et al., 2004). Furthermore, $GABA_B$ and $5\text{-}HT_{1A}$ receptors share the same pool of G-proteins (Andrade et al., 1986) and are coupled to the same G-protein-gated potassium (GirK) channels in the DRN (Innis et al., 1988). Therefore, it is plausible to assume that downstream effects are responsible for these observations. In support of this, targeted deletion of GIRK2 has been shown to attenuate the hypothermic response elicited by $5\text{-}HT_{1A}$ and $GABA_B$ receptor agonists (Costa et al., 2005). Moreover, a recent study revealed that chronic treatment with the SSRI antidepressant fluoxetine reduced $5\text{-}HT_{1A}$ and $GABA_B$ receptor-mediated GirK currents in the dorsal raphe in control rats or following a social stressor, daily defeat for 5 consecutive days (Cornelisse et al., 2007). These findings suggest that fluoxetine desensitizes a shared downstream component of the receptors (Cornelisse et al., 2007). It was also recently shown that a member of regulator of G-protein signaling (RGS) family, RGS2, was responsible for the coupling of GABA$_B$ receptors and GirK channels in dopaminergic VTA neurones (Labouebe et al., 2007). It is intriguingly possible that a similar mechanism exists within the DRN. Of interest, increased RGS2 immunoreactivity in the prefrontal cortex and amygdala of suicide victims has been reported (Cui et al., 2007) and RGS2 knockout mice display increased emotionality and anxiety levels (Oliveira-Dos-Santos et al., 2000). Given that desensitization of 5-HT$_{1A}$ autoreceptors is hypothesized to occur before the onset of antidepressant efficacy following SSRI treatment (Artigas et al., 2006; Blier & Ward, 2003; Hensler, 2003), this may in turn lead to a decrease in GABA$_B$ergic function in the DRN. Therefore, it is plausible that this link between GABA$_B$ receptors and the 5-HT system may contribute to the clinical benefit of SSRIs and provide an explanation for the antidepressant-like potential of GABA$_B$ receptor antagonists.

VIII. Clinical Evidence for a Role of GABA$_B$ Receptors in Depression

The recent preclinical interest in GABA$_B$ receptors in the etiology of depression has not been replicated in clinical studies, mainly for the same reason that hindered earlier preclinical research, a lack of appropriate tools and therapeutic agents. Therefore, despite GABAergic dysfunction being reported in patients, a specific role of GABA$_B$ receptors has not yet been comprehensively studied (Brambilla et al., 2003; Cryan & Kaupmann, 2005; Krystal et al., 2002). Emrich et al. (1980) were among the first to ascribe a role for GABAergic dysfunction in bipolar depression based on the fact that that valproate (an indirect GABA agonist) displayed antimanic effects. The GABA mimetics, progabide and fengabine, were reported to result in a clinical improvement in depressed patients to a similar extent as classical tricyclic antidepressants (Morselli et al., 1988). However, regarding the GABA$_B$ system, only one small study with five patients, to date, has been performed, which employed baclofen. In agreement with the recent preclinical evidence, baclofen was reported in three out of the five patients to worsen their symptoms but clearly such a small sample size means that the findings are questionable (Post et al., 1991).

While studies examining the direct effect of baclofen on depressive symptomology have not been extensively performed, a number of studies have assessed the impact of acute baclofen administration on endocrine parameters in depressed patients. Both growth hormone (Cavagnini et al., 1977 but see also Markar et al., 1989) and cortisol (Morosini et al., 1980) concentrations are increased following acute baclofen administration. However, the majority of these studies employed low numbers and nonuniform patient populations and thus it may not be surprising that blunted (Marchesi et al., 1991; O'Flynn & Dinan, 1993) or no effect (Davis et al., 1997; Monteleone et al., 1990) on growth hormone levels have been reported.

Further assessments of the GABAergic system have been performed utilizing cerebrospinal fluid (CSF) measurements and neuroimaging in depressed patients. These studies have shown, in the main, that GABA levels are decreased in the CSF of depressed patients (see Brambilla et al., 2003 for review) and within the occipital and dorsal medial/dorsal anterolateral cortices of patients with depression (Bhagwagar et al., 2004; Hasler et al., 2007; Sanacora et al., 2002, 2003, 2004). Interestingly, the neuroimaging studies demonstrated that fluoxetine, citalopram, or electroconvulsive therapy (ECT) normalized the GABA levels in patients (Sanacora et al., 2002, 2003, 2004). Correspondingly, increased GABA levels in the occipital cortex were reported following acute administration of citalopram in healthy volunteers (Bhagwagar et al., 2004). In agreement with these findings a reduction in glutamic acid decarboxylase (GAD—the enzyme that catalyzes the synthesis of GABA from glutmate) activity has been reported in several cortical and subcortical brain regions in depressed patients (Perry et al., 1977), as well as a reduction in plasma GAD activity (see Brambilla et al., 2003 for review). A recent postmortem study demonstrated significantly lower expression of GAD65 and GAD67 mRNA (GAD67 is the predominant isoform in humans whereas GAD65 is the most prominent in rodents) in the cerebellum of unipolar depressed patients (Fatemi et al., 2005 but see Bielau et al., 2007). Similiarly, a reduction in the density and size of GABAergic interneurones in the prefrontal cortex, and density in the occipital, of depressed patients compared with control subjects was recently revealed (Maciag et al., 2009; Rajkowska et al., 2007). These findings clearly indicate a disruption of the GABAergic system in the frontal cortices of patients with depression but need to be extended to additional brain regions implicated in mood disorders and also elaborated upon to distinguish the role of $GABA_A$ and $GABA_B$ receptors downstream of the changes in GABA levels.

A microarray analysis of suicide and control subjects also revealed many alterations in relation to $GABA_A$-relevant genes but also when assessed using a global analysis paradigm found that $GABA_{B(2)}$ subunit was differentially expressed between controls and suicide victims (Sequeira et al., 2009). This finding was also demonstrated in postmortem samples taken from BA44–BA47 in a different cohort of suicide victims (Klempan et al., 2009). These findings require additional replication and verification (i.e. by RT-PCR) but further support a role of the $GABA_B$ system in MDD. Transcranial magnetic stimulus (TMS) can be used to assess cortical inhibition, a process mediated partly via GABA interneurones. The paradigm employed to assess cortical silent period (CSP) is believed to be more dependent on $GABA_B$ receptors than $GABA_A$ based on studies with baclofen, which lengthen this period. This technique provided more direct evidence for an alteration in $GABA_B$ receptor functioning in MDD as the CSP was longer in MDD patients compared with controls (Levinson et al., 2009). This suggests that MDD patients display deficits in neurophysiological indexes of $GABA_B$ function. Moreover, while a deficit in both $GABA_A$ (short-interval cortical inhibition (SICI)) and $GABA_B$ function was found in severe treatment-resistant patients, only $GABA_B$ function was altered in unmedicated MDD and euthymic MDD patients. Therefore, deficits in $GABA_B$ receptor functioning, at least in relation to cortical inhibition, appear to be common across different modalities (Levinson et al., 2009).

IX. Conclusion

The recent preclinical findings regarding the $GABA_B$ receptor system utilizing novel ligands and genetic tools have shown that antagonists may represent a novel therapeutic strategy for the treatment of depression. The molecular basis of such effects remains elusive although increases in both BDNF (Enna et al., 2006; Froestl et al., 2004; Heese et al., 2000) and serotonergic neurotransmission (Cornelisse et al., 2007; Costa et al., 2005; Mannoury la Cour et al., 2004; Slattery et al., 2005a) have been implicated. On the clinical front, deficits in GABAergic neurotransmission have been consistently reported, particularly in the prefrontal cortex of MDD patients; however, a more causal role of the $GABA_B$ receptor system from clinical studies is lacking. Finally, a conundrum exists that activation of $GABA_B$ receptor reduces anxiety in a variety of paradigms, whereas antagonists have an antidepressant potential (Cryan & Kaupmann, 2005) especially given the extensive comorbidity of such disorders clinically. Further studies are needed to understand the functional interactions of $GABA_B$ receptors with other neurotransmitter systems and how these might contribute to the manifestation of differential anxiolytic and antidepressant-like effects of $GABA_B$ receptor-positive allosteric modulators and antagonists, respectively. Therefore, despite much more research required, the data gathered to date support a role of the $GABA_B$ receptor system in MDD and that $GABA_B$ receptor antagonists may represent a viable approach to a novel therapeutic intervention strategy.

Acknowledgments

JFC is funded by European Community's Seventh Framework Programme, Grant Number FP7/2007-2013, Grant Agreement 201714.

Conflict of Interest: The authors do not have any conflict of interests to report.

References

Abellan, M. T., Martin-Ruiz, R., & Artigas, F. (2000). Local modulation of the 5-HT release in the dorsal striatum of the rat: An in vivo microdialysis study. *European Neuropsychopharmacology*, *10*(6), 455–462.

Altar, C. A., Whitehead, R. E., Chen, R., Wortwein, G., & Madsen, T. M. (2003). Effects of electroconvulsive seizures and antidepressant drugs on brain-derived neurotrophic factor protein in rat brain. *Biological Psychiatry*, *54*(7), 703–709.

Andlin-Sobocki, P., & Wittchen, H. U. (2005). Cost of affective disorders in Europe. *European Journal of Neurology*, *12*(Suppl. 1), 34–38.

Andrade, R., Malenka, R. C., & Nicoll, R. A. (1986). A G protein couples serotonin and $GABA_B$ receptors to the same channels in hippocampus. *Science*, *234*(4781), 1261–1265.

Artigas, F., Adell, A., & Celada, P. (2006). Pindolol augmentation of antidepressant response. *Current Drug Targets*, *7*(2), 139–147.

Beck, A. T. (2008). The evolution of the cognitive model of depression and its neurobiological correlates. *American Journal of Psychiatry*, *165*(8), 969–977.

Berton, O., McClung, C. A., Dileone, R. J., Krishnan, V., Renthal, W., Russo, S. J., et al. (2006). Essential role of BDNF in the mesolimbic dopamine pathway in social defeat stress. *Science*, *311*(5762), 864–868.

Berton, O., & Nestler, E. J. (2006). New approaches to antidepressant drug discovery: Beyond monoamines. *Nature Reviews Neuroscience*, *7*(2), 137–151.

Bettler, B., Kaupmann, K., Mosbacher, J., & Gassmann, M. (2004). Molecular structure and physiological functions of $GABA_B$ receptors. *Physiological Reviews*, *84*(3), 835–867.

Bhagwagar, Z., Wylezinska, M., Taylor, M., Jezzard, P., Matthews, P. M., & Cowen, P. J. (2004). Increased brain GABA concentrations following acute administration of a selective serotonin reuptake inhibitor. *American Journal of Psychiatry*, *161*(2), 368–370.

Bielau, H., Steiner, J., Mawrin, C., Trubner, K., Brisch, R., Meyer-Lotz, G., et al. (2007). Dysregulation of GABAergic neurotransmission in mood disorders: A postmortem study. *Annals of the New York Academy of Sciences*, *1096*, 157–169.

Binet, V., Brajon, C., Le Corre, L., Acher, F., Pin, J. P., & Prezeau, L. (2004). The heptahelical domain of $GABA_{(B2)}$ is activated directly by CGP7930, a positive allosteric modulator of the $GABA_{(B)}$ receptor. *Journal of Biological Chemistry*, *279*(28), 29085–29091.

Bittiger, H., Froestl, W., Gentsch, C., Jaekel, J., Mickel, S. J., Mondadori, C. G., et al. (1996). $GABA_B$ receptor antagonists: Potential therapeutic applications. In C. Tanaka & N. G. Bowery (Eds.), *GABA: Receptors, transporters and metabolism* (pp. 297–305). Basel: Birkhäuser Verlag.

Blier, P., & Ward, N. M. (2003). Is there a role for 5-HT1A agonists in the treatment of depression? *Biological Psychiatry*, *53*(3), 193–203.

Borsini, F. (1995). Role of the serotonergic system in the forced swimming test. *Neuroscience & Biobehavioral Reviews*, *19*(3), 377–395.

Borsini, F., & Meli, A. (1988). Is the forced swimming test a suitable model for revealing antidepressant activity? *Psychopharmacology (Berl)*, *94*(2), 147–160.

Bowery, N. G., Bettler, B., Froestl, W., Gallagher, J. P., Marshall, F., Raiteri, M., et al. (2002). International Union of Pharmacology. XXXIII. Mammalian gamma-aminobutyric $acid_{(B)}$ receptors: Structure and function. *Pharmacological Reviews*, *54*(2), 247–264.

Bowery, N. G., Hill, D. R., Hudson, A. L., Doble, A., Middlemiss, D. N., Shaw, J., et al. (1980). (-)Baclofen decreases neurotransmitter release in the mammalian CNS by an action at a novel GABA receptor. *Nature*, *283*(5742), 92–94.

Brambilla, P., Perez, J., Barale, F., Schettini, G., & Soares, J. C. (2003). GABAergic dysfunction in mood disorders. *Molecular Psychiatry*, *8*(8), 721–737.

Brogden, R. N., Speight, T. M., & Avery, G. S. (1974). Baclofen: A preliminary report of its pharmacological properties and therapeutic efficacy in spasticity. *Drugs*, *8*(1), 1–14.

Bullock, R. (2005). SGS-742 novartis. *Current Opinion in Investigational Drugs*, *6*(1), 108–113.

Cardin, J. A., & Abel, T. (1999). Memory suppressor genes: Enhancing the relationship between synaptic plasticity and memory storage. *Journal of Neuroscience Research*, *58* (1), 10–23.

Cavagnini, F., Invitti, C., Di Landro, A., Tenconi, L., Maraschini, C., & Girotti, G. (1977). Effects of a gamma aminobutyric acid (GABA) derivative, baclofen, on growth hormone and prolactin secretion in man. *Journal of Clinical Endocrinology and Metabolism*, *45*(3), 579–584.

Cornelisse, L. N., Van der Harst, J. E., Lodder, J. C., Baarendse, P. J., Timmerman, A. J., Mansvelder, H. D., et al. (2007). Reduced 5-HT1A- and GABA$_B$ receptor function in dorsal raphe neurons upon chronic fluoxetine treatment of socially stressed rats. *Journal of Neurophysiology*, *98*(1), 196–204.

Costa, A. C., Stasko, M. R., Stoffel, M., & Scott-McKean, J. J. (2005). G-protein-gated potassium (GIRK) channels containing the GIRK2 subunit are control hubs for pharmacologically induced hypothermic responses. *Journal of Neuroscience*, *25*(34), 7801–7804.

Couve, A., Moss, S. J., & Pangalos, M. N. (2000). GABA$_B$ receptors: A new paradigm in G protein signaling. *Molecular and Cellular Neuroscience*, *16*(4), 296–312.

Cryan, J. F., & Kaupmann, K. (2005). Don't worry 'B' happy!: A role for GABA$_B$ receptors in anxiety and depression. *Trends in Pharmacological Sciences*, *26*(1), 36–43.

Cryan, J. F., Kelly, P. H., Chaperon, F., Gentsch, C., Mombereau, C., Lingenhoehl, K., et al. (2004). Behavioral characterization of the novel GABA$_B$ receptor-positive modulator GS39783 (*N,N'*-dicyclopentyl-2-methylsulfanyl-5-nitro-pyrimidine-4,6-diamine): Anxiolytic-like activity without side effects associated with baclofen or benzodiazepines. *Journal of Pharmacology and Experimental Therapeutics*, *310*(3), 952–963.

Cryan, J. F., Markou, A., & Lucki, I. (2002). Assessing antidepressant activity in rodents: Recent developments and future needs. *Trends in Pharmacological Sciences*, *23*(5), 238–245.

Cryan, J. F., McGrath, C., Leonard, B. E., & Norman, T. R. (1998). Combining pindolol and paroxetine in an animal model of chronic antidepressant action—can early onset of action be detected? *European Journal of Pharmacology*, *352*(1), 23–28.

Cryan, J. F., McGrath, C., Leonard, B. E., & Norman, T. R. (1999). Onset of the effects of the 5-HT1A antagonist, WAY-100635, alone, and in combination with paroxetine, on olfactory bulbectomy and 8-OH-DPAT-induced changes in the rat. *Pharmacology Biochemistry and Behavior*, *63*(2), 333–338.

Cryan, J. F., & Mombereau, C. (2004). In search of a depressed mouse: Utility of models for studying depression-related behavior in genetically modified mice. *Molecular Psychiatry*, *9*(4), 326–357.

Cryan, J. F., Mombereau, C., & Vassout, A. (2005a). The tail suspension test as a model for assessing antidepressant activity: Review of pharmacological and genetic studies in mice. *Neuroscience and Biobehavioral Reviews*, *29*(4–5), 571–625.

Cryan, J. F., & Slattery, D. A. (2007). Animal models of mood disorders: Recent developments. *Current Opinion in Psychiatry*, *20*(1), 1–7.

Cryan, J. F., Sánchez, C., Dinan, T. G., & Borsini, F. (2008). Developing more efficacious antidepressant medications: Improving and aligning preclinical and clinical assessment tools. In R. A. McArthur & F. Borsini (Eds.), *Animal and translational models for drug discovery: Psychiatric disorders* (pp. 165–197). New York: Elsevier.

Cryan, J. F., Valentino, R. J., & Lucki, I. (2005b). Assessing substrates underlying the behavioral effects of antidepressants using the modified rat forced swimming test. *Neuroscience and Biobehavioral Reviews*, *29*(4–5), 547–569.

Cui, H., Nishiguchi, N., Ivleva, E., Yanagi, M., Fukutake, M., Nushida, H., et al. (2007). Association of RGS2 gene polymorphisms with suicide and increased RGS2 immunoreactivity in the postmortem brain of suicide victims. *Neuropsychopharmacology*, *33*(7), 1537–1544.

Dalley, J. W., Cardinal, R. N., & Robbins, T. W. (2004). Prefrontal executive and cognitive functions in rodents: Neural and neurochemical substrates. *Neuroscience and Biobehavioral Reviews*, *28*(7), 771–784.

Davis, L. L., Trivedi, M., Choate, A., Kramer, G. L., & Petty, F. (1997). Growth hormone response to the $GABA_B$ agonist baclofen in major depressive disorder. *Psychoneuroendocrinology*, *22*(3), 129–140.

Dennis, T., Beauchemin, V., & Lavoie, N. (1993). Differential effects of olfactory bulbectomy on $GABA_A$ and $GABA_B$ receptors in the rat brain. *Pharmacology Biochemistry and Behavior*, *46*(1), 77–82.

Dennis, T., Beauchemin, V., & Lavoie, N. (1994). Antidepressant-induced modulation of $GABA_A$ receptors and beta-adrenoceptors but not $GABA_B$ receptors in the frontal cortex of olfactory bulbectomised rats. *European Journal of Pharmacology*, *262*(1–2), 143–148.

Duman, R. S. (2002). Synaptic plasticity and mood disorders. *Molecular Psychiatry*, *7*(Suppl. 1), S29–S34.

Duman, R. S. (2004). Depression: A case of neuronal life and death? *Biological Psychiatry*, *56*(3), 140–145.

Duman, R. S., Heninger, G. R., & Nestler, E. J. (1997). A molecular and cellular theory of depression. *Archives of General Psychiatry*, *54*(7), 597–606.

Duncan, G. E., Knapp, D. J., Johnson, K. B., & Breese, G. R. (1996). Functional classification of antidepressants based on antagonism of swim stress-induced fos-like immunoreactivity. *Journal of Pharmacology and Experimental Therapeutics*, *277*, 1076–1089.

Dupuis, D. S., Relkovic, D., Lhuillier, L., Mosbacher, J., & Kaupmann, K. (2006). Point mutations in the transmembrane region of $GABA_{B2}$ facilitate activation by the positive modulator *N,N′*-dicyclopentyl-2-methylsulfanyl-5-nitro-pyrimidine-4,6-diamine (GS39783) in the absence of the $GABA_{B1}$ subunit. *Molecular Pharmacology*, *70*(6), 2027–2036.

Eisch, A. J., Bolanos, C. A., de Wit, J., Simonak, R. D., Pudiak, C. M., Barrot, M., et al. (2003). Brain-derived neurotrophic factor in the ventral midbrain-nucleus accumbens pathway: A role in depression. *Biological Psychiatry*, *54*(10), 994–1005.

Elliott, R., Sahakian, B. J., McKay, A. P., Herrod, J. J., Robbins, T. W., & Paykel, E. S. (1996). Neuropsychological impairments in unipolar depression: The influence of perceived failure on subsequent performance. *Psychological Medicine*, *26*(5), 975–989.

Emrich, H. M., von Zerssen, D., Kissling, W., Moller, H. J., & Windorfer, A. (1980). Effect of sodium valproate on mania. The GABA-hypothesis of affective disorders. *Archiv für Psychiatrie und Nervenkrankheiten*, *229*(1), 1–16.

Enna, S. J., & Bowery, N. G. (2004). $GABA_{(B)}$ receptor alterations as indicators of physiological and pharmacological function. *Biochemical Pharmacology*, *68*(8), 1541–1548.

Enna, S. J., Reisman, S. A., & Stanford, J. A. (2006). CGP 56999a, a $GABA_{(B)}$ receptor antagonist, enhances expression of brain-derived neurotrophic factor and attenuates dopamine depletion in the rat corpus striatum following a 6-hydroxydopamine lesion of the nigrostriatal pathway. *Neuroscience Letters*, *406*(1–2), 102–106.

Estrada-Camarena, E., Contreras, C. M., Saavedra, M., Luna-Baltazar, I., & Lopez-Rubalcava, C. (2002). Participation of the lateral septal nuclei (LSN) in the antidepressant-like actions of progesterone in the forced swimming test (FST). *Behavioral and Brain Sciences*, *134*(1–2), 175–183.

Fabre, V., Beaufour, C., Evrard, A., Rioux, A., Hanoun, N., Lesch, K. P., et al. (2000). Altered expression and functions of serotonin $5\text{-}HT_{1A}$ and $5\text{-}HT_{1B}$ receptors in knock-out mice lacking the 5-HT transporter. *European Journal of Neuroscience*, *12*(7), 2299–2310.

Fadda, P., Scherma, M., Fresu, A., Collu, M., & Fratta, W. (2003). Baclofen antagonizes nicotine-, cocaine-, and morphine-induced dopamine release in the nucleus accumbens of rat. *Synapse*, *50*(1), 1–6.

Fatemi, S. H., Hossein Fatemi, S., Stary, J. M., Earle, J. A., Araghi-Niknam, M., & Eagan, E. (2005). GABAergic dysfunction in schizophrenia and mood disorders as reflected by decreased levels of glutamic acid decarboxylase 65 and 67 kDa and reelin proteins in cerebellum. *Schizophrenia Research*, *72*(2–3), 109–122.

Filip, M., & Frankowska, M. (2007). Effects of $GABA_{(B)}$ receptor agents on cocaine priming, discrete contextual cue and food induced relapses. *European Journal of Pharmacology*, *571*(2–3), 166–173.

Filip, M., Frankowska, M., & Przegalinski, E. (2007). Effects of $GABA_{(B)}$ receptor antagonist, agonists and allosteric positive modulator on the cocaine-induced self-administration and drug discrimination. *European Journal of Pharmacology*, *574*(2–3), 148–157.

Frankowska, M., Filip, M., & Przegalinski, E. (2007). Effects of $GABA_B$ receptor ligands in animal tests of depression and anxiety. *Pharmacological Reports*, *59*(6), 645–655.

Frankowska, M., Golda, A., Wydra, K., Gruca, P., Papp, M., & Filip, M. (2010). Effects of imipramine or $GABA_{(B)}$ receptor ligands on the immobility, swimming and climbing in the forced swim test in rats following discontinuation of cocaine self-administration. *European Journal of Pharmacology*, *627*(1–3), 142–149.

Froestl, W., Bettler, B., Bittiger, H., Heid, J., Kaupmann, K., Mickel, S. J., et al. (2003). Ligands for expression cloning and isolation of $GABA_{(B)}$ receptors. *Farmaco*, *58*(3), 173–183.

Froestl, W., Gallagher, M., Jenkins, H., Madrid, A., Melcher, T., Teichman, S., et al. (2004). SGS742: The first $GABA_{(B)}$ receptor antagonist in clinical trials. *Biochemical Pharmacology*, *68*(8), 1479–1487.

Gassmann, M., Shaban, H., Vigot, R., Sansig, G., Haller, C., Barbieri, S., et al. (2004). Redistribution of $GABA_{B(1)}$ protein and atypical $GABA_B$ responses in $GABA_{B(2)}$-deficient mice. *Journal of Neuroscience*, *24*(27), 6086–6097.

Giorgetti, M., Hotsenpiller, G., Froestl, W., & Wolf, M. E. (2002). In vivo modulation of ventral tegmental area dopamine and glutamate efflux by local $GABA_B$ receptors is altered after repeated amphetamine treatment. *Neuroscience*, *109*(3), 585–595.

Gong, W., Neill, D. B., & Justice, J. B. Jr. (1998). GABAergic modulation of ventral pallidal dopamine release studied by in vivo microdialysis in the freely moving rat. *Synapse*, *29*(4), 406–412.

Gronli, J., Fiske, E., Murison, R., Bjorvatn, B., Sorensen, E., Ursin, R., et al. (2007). Extracellular levels of serotonin and GABA in the hippocampus after chronic mild stress in rats. A microdialysis study in an animal model of depression. *Behavioural Brain Research*, *181*(1), 42–51.

Harkin, A., Kelly, J. P., & Leonard, B. E. (2003). A review of the relevance and validity of olfactory bulbectomy as a model of depression. *Clinical Neuroscience Research*, *3*, 253–262.

Hasler, G., van der Veen, J. W., Tumonis, T., Meyers, N., Shen, J., & Drevets, W. C. (2007). Reduced prefrontal glutamate/glutamine and gamma-aminobutyric acid levels in major depression determined using proton magnetic resonance spectroscopy. *Archives of General Psychiatry*, *64*(2), 193–200.

Heese, K., Otten, U., Mathivet, P., Raiteri, M., Marescaux, C., & Bernasconi, R. (2000). $GABA_B$ receptor antagonists elevate both mRNA and protein levels of the neurotrophins nerve growth factor (NGF) and brain-derived neurotrophic factor (BDNF) but not neurotrophin-3 (NT-3) in brain and spinal cord of rats. *Neuropharmacology*, *39*(3), 449–462.

Helm, K. A., Haberman, R. P., Dean, S. L., Hoyt, E. C., Melcher, T., Lund, P. K., et al. (2005). $GABA_B$ receptor antagonist SGS742 improves spatial memory and reduces protein binding to the cAMP response element (CRE) in the hippocampus. *Neuropharmacology*, *48*(7), 956–964.

Hensler, J. G. (2003). Regulation of 5-HT_{1A} receptor function in brain following agonist or antidepressant administration. *Life Sciences*, *72*(15), 1665–1682.

Hill, D. R., & Bowery, N. G. (1981). 3H-baclofen and 3H-GABA bind to bicuculline-insensitive GABA B sites in rat brain. *Nature*, *290*(5802), 149–152.

Hoshaw, B. A., Malberg, J. E., & Lucki, I. (2005). Central administration of IGF-I and BDNF leads to long-lasting antidepressant-like effects. *Brain Research*, *1037*(1–2), 204–208.

Innis, R. B., Nestler, E. J., & Aghajanian, G. K. (1988). Evidence for G protein mediation of serotonin- and $GABA_B$-induced hyperpolarization of rat dorsal raphe neurons. *Brain Research*, *459*(1), 27–36.

Jacobson, L. H., Bettler, B., Kaupmann, K., & Cryan, J. F. (2006a). $GABA_{B1}$ receptor subunit isoforms exert a differential influence on baseline but not $GABA_B$ receptor agonist-induced changes in mice. *Journal of Pharmacology and Experimental Therapeutics*, *319*(3), 1317–1326.

Jacobson, L. H., & Cryan, J. F. (2005). Differential sensitivity to the motor and hypothermic effects of the $GABA_B$ receptor agonist baclofen in various mouse strains. *Psychopharmacology (Berl)*, *179*(3), 688–699.

Jacobson, L. H., & Cryan, J. F. (2008). Evaluation of the anxiolytic-like profile of the $GABA_B$ receptor positive modulator CGP7930 in rodents. *Neuropharmacology*, *54*(5), 854–862.

Jacobson, L. H., Kelly, P. H., Bettler, B., Kaupmann, K., & Cryan, J. F. (2006b). $GABA_{(B(1))}$ receptor isoforms differentially mediate the acquisition and extinction of aversive taste memories. *Journal of Neuroscience*, *26*(34), 8800–8803.

Jacobson, L. H., Kelly, P. H., Bettler, B., Kaupmann, K., & Cryan, J. F. (2007). Specific roles of $GABA_{(B(1))}$ receptor isoforms in cognition. *Behavioural Brain Research*, *181*(1), 158–162.

Jensen, A. A., & Spalding, T. A. (2004). Allosteric modulation of G-protein coupled receptors. *European Journal of Pharmaceutical Sciences*, *21*(4), 407–420.

Katz, R. J. (1982). Animal model of depression: Pharmacological sensitivity of a hedonic deficit. *Pharmacology Biochemistry and Behavior*, *16*(6), 965–968.

Kaupmann, K., Huggel, K., Heid, J., Flor, P. J., Bischoff, S., Mickel, S. J., et al. (1997). Expression cloning of $GABA_{(B)}$ receptors uncovers similarity to metabotropic glutamate receptors. *Nature*, *386*(6622), 239–246.

Kaupmann, K., Malitschek, B., Schuler, V., Heid, J., Froestl, W., Beck, P., et al. (1998). $GABA_{(B)}$-receptor subtypes assemble into functional heteromeric complexes. *Nature*, *396*(6712), 683–687.

Kelly, J. P., Wrynn, A. S., & Leonard, B. E. (1997). The olfactory bulbectomized rat as a model of depression: An update. *Pharmacology & Therapeutics*, *74*(3), 299–316.

Kerr, D. I., & Ong, J. (1992). GABA agonists and antagonists. *Medicinal Research Reviews*, *12*(6), 593–636.

Kessler, R. C., Berglund, P., Demler, O., Jin, R., Koretz, D., Merikangas, K. R., et al. (2003). The epidemiology of major depressive disorder: Results from the national comorbidity survey replication (NCS-R). *JAMA*, *289*(23), 3095–3105.

Kirsch, I., Deacon, B. J., Huedo-Medina, T. B., Scoboria, A., Moore, T. J., & Johnson, B. T. (2008). Initial severity and antidepressant benefits: A meta-analysis of data submitted to the food and drug administration. *PLoS Medicine*, *5*, e45.

Klempan, T. A., Sequeira, A., Canetti, L., Lalovic, A., Ernst, C., ffrench-Mullen, J., et al. (2009). Altered expression of genes involved in ATP biosynthesis and GABAergic neurotransmission in the ventral prefrontal cortex of suicides with and without major depression. *Molecular Psychiatry*, *14*(2), 175–189.

Knapp, D. J., Overstreet, D. H., & Breese, G. R. (2007). Baclofen blocks expression and sensitization of anxiety-like behavior in an animal model of repeated stress and ethanol withdrawal. *Alcoholism, Clinical and Experimental Research*, *31*(4), 582–595.

Kram, M. L., Kramer, G. L., Steciuk, M., Ronan, P. J., & Petty, F. (2000). Effects of learned helplessness on brain GABA receptors. *Neuroscience Research*, *38*(2), 193–198.

Krishnan, V., Han, M. H., Graham, D. L., Berton, O., Renthal, W., Russo, S. J., et al. (2007). Molecular adaptations underlying susceptibility and resistance to social defeat in brain reward regions. *Cell*, *131*(2), 391–404.

Krystal, J. H., Sanacora, G., Blumberg, H., Anand, A., Charney, D. S., Marek, G., et al. (2002). Glutamate and GABA systems as targets for novel antidepressant and mood-stabilizing treatments. *Molecular Psychiatry*, *7*(Suppl. 1), S71–S80.

Labouebe, G., Lomazzi, M., Cruz, H. G., Creton, C., Lujan, R., Li, M., et al. (2007). RGS2 modulates coupling between $GABA_B$ receptors and GIRK channels in dopamine neurons of the ventral tegmental area. *Nature Neuroscience*, *10*(12), 1559–1568.

Leonard, B. E., & Tuite, M. (1981). Anatomical, physiological, and behavioral aspects of olfactory bulbectomy in the rat. *International Review of Neurobiology*, *22*, 251–286.

Leung, L. S., & Shen, B. (2007). $GABA_B$ receptor blockade enhances theta and gamma rhythms in the hippocampus of behaving rats. *Hippocampus*, *17*(4), 281–291.

Levinson, A. J., Fitzgerald, P. B., Favalli, G., Blumberger, D. M., Daigle, M., & Daskalakis, Z. J. (2009). Evidence of cortical inhibitory deficits in major depressive disorder. *Biological Psychiatry*, *67*(5), 458–464.

Lhuillier, L., Mombereau, C., Cryan, J. F., & Kaupmann, K. (2007). $GABA_{(B)}$ receptor-positive modulation decreases selective molecular and behavioral effects of cocaine. *Neuropsychopharmacology*, *32*(2), 388–398.

Licinio, J., & Wong, M. L. (2005). Depression, antidepressants and suicidality: A critical appraisal. *Nature Reviews Drug Discovery*, *4*(2), 165–171.

Liu, G. X., Cai, G. Q., Cai, Y. Q., Sheng, Z. J., Jiang, J., Mei, Z., et al. (2007). Reduced anxiety and depression-like behaviors in mice lacking GABA transporter subtype 1. *Neuropsychopharmacology*, *32*(7), 1531–1539.

Lloyd, K. G., Morselli, P. L., Depoortere, H., Fournier, V., Zivkovic, B., Scatton, B., et al. (1983). The potential use of GABA agonists in psychiatric disorders: Evidence from studies with progabide in animal models and clinical trials. *Pharmacology Biochemistry and Behavior*, *18*(6), 957–966.

Lloyd, K. G., & Pichat, P. (1986). Decrease in $GABA_B$ binding to the frontal cortex of olfactory bulbectomized rats. *British Journal of Pharmacology*, *87*, 36P.

Lloyd, K. G., Zivkovic, B., Scatton, B., Morselli, P. L., & Bartholini, G. (1989). The gabaergic hypothesis of depression. *Progress in Neuro-Psychopharmacology & Biological Psychiatry*, *13*(3–4), 341–351.

Lucki, I. (1997). The forced swimming test as a model for core and component behavioral effects of antidepressant drugs. *Behavioural Pharmacology*, *8*(6–7), 523–532.

Macey, D. J., Froestl, W., Koob, G. F., & Markou, A. (2001). Both $GABA_{(B)}$ receptor agonist and antagonists decreased brain stimulation reward in the rat. *Neuropharmacology*, *40* (5), 676–685.

Maciag, D., Hughes, J., O'Dwyer, G., Pride, Y., Stockmeier, C. A., Sanacora, G., et al. (2009). Reduced density of calbindin immunoreactive GABAergic neurons in the occipital cortex in major depression: Relevance to neuroimaging studies. *Biological Psychiatry*, *67*(5), 465–470.

Mannoury la Cour, C., Hanoun, N., Melfort, M., Hen, R., Lesch, K. P., Hamon, M., et al. (2004). $GABA_B$ receptors in 5-HT transporter- and $5\text{-}HT_{1A}$ receptor-knock-out mice: Further evidence of a transduction pathway shared with $5\text{-}HT_{1A}$ receptors. *Journal of Neurochemistry*, *89*(4), 886–896.

Marchesi, C., Chiodera, P., De Ferri, A., De Risio, C., Dasso, L., Menozzi, P., et al. (1991). Reduction of GH response to the $GABA_{\text{-}B}$ agonist baclofen in patients with major depression. *Psychoneuroendocrinology*, *16*(6), 475–479.

Markar, H., Bennie, J., & Goodwin, G. M. (1989). Stimulation of growth hormone secretion by baclofen in man: Failure to confirm published findings. *Human Psychopharmacology*, *4*(2), 135–138.

Martin, P., Pichat, P., Massol, J., Soubrie, P., Lloyd, K. G., & Puech, A. J. (1989). Decreased $GABA_B$ receptors in helpless rats: Reversal by tricyclic antidepressants. *Neuropsychobiology*, *22*(4), 220–224.

Matthews, K., Christmas, D., Swan, J., & Sorrell, E. (2005). Animal models of depression: Navigating through the clinical fog. *Neuroscience and Biobehavioral Reviews*, *29*(4–5), 503–513.

McKernan, D. P., Dinan, T. G., & Cryan, J. F. (2009). "Killing the blues": A role for cellular suicide (apoptosis) in depression and the antidepressant response? *Progress in Neurobiology, 88*(4), 246–263.

Mombereau, C., Kaupmann, K., Froestl, W., Sansig, G., van der Putten, H., & Cryan, J. F. (2004). Genetic and pharmacological evidence of a role for $GABA_{(B)}$ receptors in the modulation of anxiety- and antidepressant-like behavior. *Neuropsychopharmacology, 29*(6), 1050–1062.

Mombereau, C., Kaupmann, K., Gassmann, M., Bettler, B., van der Putten, H., & Cryan, J. F. (2005). Altered anxiety and depression-related behaviour in mice lacking $GABA_{B(2)}$ receptor subunits. *NeuroReport, 16*(3), 307–310.

Mombereau, C., Lhuillier, L., Kaupmann, K., & Cryan, J. F. (2007). $GABA_B$ receptor-positive modulation-induced blockade of the rewarding properties of nicotine is associated with a reduction in nucleus accumbens DeltaFosB accumulation. *Journal of Pharmacology and Experimental Therapeutics, 321*(1), 172–177.

Monteleone, P., Maj, M., Iovino, M., & Steardo, L. (1990). GABA, depression and the mechanism of action of antidepressant drugs: A neuroendocrine approach. *Journal of Affective Disorders, 20*(1), 1–5.

Morosini, P. P., Carletti, P., Ferretti, G. F., Campanella, N., & Governatori, D. (1980). [Plasma cortisol levels after acute administration of baclofen]. *Bollettino della Società Italiana di Bbiologia Sperimentale, 56*(23), 2428–2431.

Morselli, P. L., Priore, P., Loeb, C., Albano, C., Nielsen, N. P., & Serrati, C. (1988). Antidepressant activity of progabide and fengabine. In B. Lerer & S. Gershon (Eds.), *New directions in affective disorders* (pp. 660–664). New York: Springer-Verlag.

Murray, C. J., & Lopez, A. D. (1997). Alternative projections of mortality and disability by cause 1990–2020: Global burden of disease study. *Lancet, 349*(9064), 1498–1504.

Nakagawa, Y., Ishima, T., Ishibashi, Y., Tsuji, M., & Takashima, T. (1996). Involvement of $GABA_B$ receptor systems in action of antidepressants. II: Baclofen attenuates the effect of desipramine whereas muscimol has no effect in learned helplessness paradigm in rats. *Brain Research, 728*(2), 225–230.

Nakagawa, Y., Ishima, T., Ishibashi, Y., Tsuji, M., & Takashima, T. (1996a). Involvement of $GABA_B$ receptor systems in experimental depression: Baclofen but not bicuculline exacerbates helplessness in rats. *Brain Research, 741*(1–2), 240–245.

Nakagawa, Y., Ishima, T., Ishibashi, Y., Yoshii, T., & Takashima, T. (1996b). Involvement of $GABA_{(B)}$ receptor systems in action of antidepressants: Baclofen but not bicuculline attenuates the effects of antidepressants on the forced swim test in rats. *Brain Research, 709*(2), 215–220.

Nakagawa, Y., Sasaki, A., & Takashima, T. (1999c). The $GABA_B$ receptor antagonist CGP36742 improves learned helplessness in rats. *European Journal of Pharmacology, 381*(1), 1–7.

Nehring, R. B., Horikawa, H. P., El Far, O., Kneussel, M., Brandstatter, J. H., Stamm, S., et al. (2000). The metabotropic $GABA_B$ receptor directly interacts with the activating transcription factor 4. *Journal of Biological Chemistry, 275*(45), 35185–35191.

Nestler, E. J., Barrot, M., DiLeone, R. J., Eisch, A. J., Gold, S. J., & Monteggia, L. M. (2002). Neurobiology of depression. *Neuron, 34*(1), 13–25.

Nowak, G., Partyka, A., Palucha, A., Szewczyk, B., Wieronska, J. M., Dybala, M., et al. (2006). Antidepressant-like activity of CGP 36742 and CGP 51176, selective $GABA_B$ receptor antagonists, in rodents. *British Journal of Pharmacology, 149*(5), 581–590.

O'Flynn, K., & Dinan, T. G. (1993). Baclofen-induced growth hormone release in major depression: Relationship to dexamethasone suppression test result. *American Journal of Psychiatry, 150*(11), 1728–1730.

Oliveira-Dos-Santos, A. J., Matsumoto, G., Snow, B. E., Bai, D., Houston, F. P., Whishaw, I. Q., et al. (2000). Regulation of T cell activation, anxiety, and male aggression by RGS2.

Proceedings of the National Academy of Sciences of the United States of America, *97*(22), 12272–12277.

Perry, E. K., Gibson, P. H., Blessed, G., Perry, R. H., & Tomlinson, B. E. (1977). Neurotransmitter enzyme abnormalities in senile dementia. Choline acetyltransferase and glutamic acid decarboxylase activities in necropsy brain tissue. *Journal of the Neurological Sciences*, *34*(2), 247–265.

Petit-Demouliere, B., Chenu, F., & Bourin, M. (2005). Forced swimming test in mice: A review of antidepressant activity. *Psychopharmacology (Berl)*, *177*(3), 245–255.

Petty, F., & Sherman, A. D. (1981). GABAergic modulation of learned helplessness. *Pharmacology Biochemistry and Behavior*, *15*(4), 567–570.

Phillips, M. L., Drevets, W. C., Rauch, S. L., & Lane, R. (2003). Neurobiology of emotion perception II: Implications for major psychiatric disorders. *Biological Psychiatry*, *54*(5), 515–528.

Pilc, A., & Lloyd, K. G. (1984). Chronic antidepressants and GABA "B" receptors: A GABA hypothesis of antidepressant drug action. *Life Sciences*, *35*(21), 2149–2154.

Porsolt, R. D., Bertin, A., & Jalfre, M. (1978). "Behavioural despair" in rats and mice: Strain differences and the effects of imipramine. *European Journal of Pharmacology*, *51*(3), 291–294.

Porsolt, R. D., Le Pichon, M., & Jalfre, M. (1977). Depression: A new animal model sensitive to antidepressant treatments. *Nature*, *266*(5604), 730–732.

Post, R. M., Ketter, T. A., Joffe, R. T., & Kramlinger, K. L. (1991). Lack of beneficial effects of L-baclofen in affective disorder. *International Clinical Psychopharmacology*, *6*(4), 197–207.

Price, M. L., Kirby, L. G., Valentino, R. J., & Lucki, I. (2002). Evidence for corticotropin-releasing factor regulation of serotonin in the lateral septum during acute swim stress: Adaptation produced by repeated swimming. *Psychopharmacology (Berl)*, *162*(4), 406–414.

Prosser, H. M., Gill, C. H., Hirst, W. D., Grau, E., Robbins, M., Calver, A., et al. (2001). Epileptogenesis and enhanced prepulse inhibition in $GABA_{(B1)}$-deficient mice. *Molecular and Cellular Neuroscience*, *17*(6), 1059–1070.

Queva, C., Bremner-Danielsen, M., Edlund, A., Ekstrand, A. J., Elg, S., Erickson, S., et al. (2003). Effects of GABA agonists on body temperature regulation in $GABA_{(B(1))}{}^{-/-}$ mice. *British Journal of Pharmacology*, *140*(2), 315–322.

Rajkowska, G., O'Dwyer, G., Teleki, Z., Stockmeier, C. A., & Miguel-Hidalgo, J. J. (2007). GABAergic neurons immunoreactive for calcium binding proteins are reduced in the prefrontal cortex in major depression. *Neuropsychopharmacology*, *32*(2), 471–482.

Reid, I., Forbes, N., Stewart, C., & Matthews, K. (1997). Chronic mild stress and depressive disorder: A useful new model? *Psychopharmacology (Berl)*, *134*(4), 365–367, and see pg 371–377, for further discussion.

Rogers, M. A., Kasai, K., Koji, M., Fukuda, R., Iwanami, A., Nakagome, K., et al. (2004). Executive and prefrontal dysfunction in unipolar depression: A review of neuropsychological and imaging evidence. *Neuroscience Research*, *50*(1), 1–11.

Sanacora, G., Gueorguieva, R., Epperson, C. N., Wu, Y.-T., Appel, M., Rothman, D. L., et al. (2004). Subtype-specific alterations of {gamma}-aminobutyric acid and glutamate in patients with major depression. *Archives of General Psychiatry*, *61*(7), 705–713.

Sanacora, G., Mason, G. F., Rothman, D. L., Hyder, F., Ciarcia, J. J., Ostroff, R. B., et al. (2003). Increased cortical GABA concentrations in depressed patients receiving ECT. *American Journal of Psychiatry*, *160*(3), 577–579.

Sanacora, G., Mason, G. F., Rothman, D. L., & Krystal, J. H. (2002). Increased occipital cortex GABA concentrations in depressed patients after therapy with selective serotonin reuptake inhibitors. *American Journal of Psychiatry*, *159*(4), 663–665.

Santiago, M., Machado, A., & Cano, J. (1993). Regulation of the prefrontal cortical dopamine release by $GABA_A$ and $GABA_B$ receptor agonists and antagonists. *Brain Research*, *630*(1–2), 28–31.

Schuler, V., Luscher, C., Blanchet, C., Klix, N., Sansig, G., Klebs, K., et al. (2001). Epilepsy, hyperalgesia, impaired memory, and loss of pre- and postsynaptic $GABA_{(B)}$ responses in mice lacking $GABA_{(B(1))}$. *Neuron, 31*(1), 47–58.

Sequeira, A., Mamdani, F., Ernst, C., Vawter, M. P., Bunney, W. E., Lebel, V., et al. (2009). Global brain gene expression analysis links glutamatergic and GABAergic alterations to suicide and major depression. *PLoS One, 4*(8), e6585.

Sheehan, T. P., Chambers, R. A., & Russell, D. S. (2004). Regulation of affect by the lateral septum: Implications for neuropsychiatry. *Brain Research. Brain Research Reviews, 46*(1), 71–117.

Sherman, A. D., & Petty, F. (1980). Neurochemical basis of the action of antidepressants on learned helplessness. *Behavioral and Neural Biology, 30*(2), 119–134.

Shimizu, E., Hashimoto, K., Okamura, N., Koike, K., Komatsu, N., Kumakiri, C., et al. (2003). Alterations of serum levels of brain-derived neurotrophic factor (BDNF) in depressed patients with or without antidepressants. *Biological Psychiatry, 54*(1), 70–75.

Shirayama, Y., Chen, A. C., Nakagawa, S., Russell, D. S., & Duman, R. S. (2002). Brain-derived neurotrophic factor produces antidepressant effects in behavioral models of depression. *Journal of Neuroscience, 22*(8), 3251–3261.

Siuciak, J. A., Lewis, D. R., Wiegand, S. J., & Lindsay, R. M. (1997). Antidepressant-like effect of brain-derived neurotrophic factor (BDNF). *Pharmacology Biochemistry and Behavior, 56*(1), 131–137.

Slattery, D. A., & Cryan, J. F. (2006). The role of $GABA_B$ receptors in depression and antidepressant-related behavioural responses. *Drug Development Research, 67*(6), 477–494.

Slattery, D. A., Desrayaud, S., & Cryan, J. F. (2005a). $GABA_B$ receptor antagonist-mediated antidepressant-like behavior is serotonin-dependent. *Journal of Pharmacology and Experimental Therapeutics, 312*(1), 290–296.

Slattery, D. A., Hudson, A. L., & Nutt, D. J. (2004). Invited review: The evolution of antidepressant mechanisms. *Fundamental & Clinical Pharmacology, 18*(1), 1–21.

Slattery, D. A., Markou, A., Froestl, W., & Cryan, J. F. (2005b). The $GABA_{(B)}$ receptor-positive modulator GS39783 and the GABA(B) receptor agonist baclofen attenuate the reward-facilitating effects of cocaine: Intracranial self-stimulation studies in the rat. *Neuropsychopharmacology, 30*(11), 2065–2072.

Smolders, I., De Klippel, N., Sarre, S., Ebinger, G., & Michotte, Y. (1995). Tonic GABA-ergic modulation of striatal dopamine release studied by in vivo microdialysis in the freely moving rat. *European Journal of Pharmacology, 284*(1–2), 83–91.

Song, C., & Leonard, B. E. (2005). The olfactory bulbectomised rat as a model of depression. *Neuroscience and Biobehavioral Reviews, 29*(4–5), 627–647.

Soudijn, W., van Wijngaarden, I., & AP, I. J. (2002). Allosteric modulation of G protein-coupled receptors. *Current Opinion in Drug Discovery & Development, 5*(5), 749–755.

Steru, L., Chermat, R., Thierry, B., & Simon, P. (1985). The tail suspension test: A new method for screening antidepressants in mice. *Psychopharmacology (Berl), 85*(3), 367–370.

Sunyer, B., Patil, S., Frischer, C., Hoger, H., Selcher, J., Brannath, W., et al. (2007). Strain-dependent effects of SGS742 in the mouse. *Behavioural Brain Research, 181*(1), 64–75.

Sunyer, B., Shim, K. S., Hoger, H., & Lubec, G. (2008). The cognitive enhancer SGS742 does not involve major known signaling cascades in OF1 mice. *Neurochemical Research, 33*(7), 1384–1392.

Tao, R., Ma, Z., & Auerbach, S. B. (1996). Differential regulation of 5-hydroxytryptamine release by $GABA_A$ and $GABA_B$ receptors in midbrain raphe nuclei and forebrain of rats. *British Journal of Pharmacology, 119*(7), 1375–1384.

Tsankova, N. M., Berton, O., Renthal, W., Kumar, A., Neve, R. L., & Nestler, E. J. (2006). Sustained hippocampal chromatin regulation in a mouse model of depression and antidepressant action. *Nature Neuroscience, 9*(4), 519–525.

Urwyler, S., Pozza, M. F., Lingenhoehl, K., Mosbacher, J., Lampert, C., Froestl, W., et al. (2003). *N,N'*-dicyclopentyl-2-methylsulfanyl-5-nitro-pyrimidine-4,6-diamine (GS39783) and structurally related compounds: Novel allosteric enhancers of gamma-aminobutyric acidB receptor function. *Journal of Pharmacology and Experimental Therapeutics*, *307*(1), 322–330.

Varga, V., Sik, A., Freund, T. F., & Kocsis, B. (2002). $GABA_B$ receptors in the median raphe nucleus: Distribution and role in the serotonergic control of hippocampal activity. *Neuroscience*, *109*(1), 119–132.

Vernon, E., Meyer, G., Pickard, L., Dev, K., Molnar, E., Collingridge, G. L., et al. (2001). $GABA_{(B)}$ receptors couple directly to the transcription factor ATF4. *Molecular and Cellular Neuroscience*, *17*(4), 637–645.

Vigot, R., Barbieri, S., Brauner-Osborne, H., Turecek, R., Shigemoto, R., Zhang, Y. P., et al. (2006). Differential compartmentalization and distinct functions of $GABA_B$ receptor variants. *Neuron*, *50*(4), 589–601.

Wang, L., LaBar, K. S., Smoski, M., Rosenthal, M. Z., Dolcos, F., Lynch, T. R., et al. (2008). Prefrontal mechanisms for executive control over emotional distraction are altered in major depression. *Psychiatry Research*, *163*(2), 143–155.

Westerink, B. H., Enrico, P., Feimann, J., & De Vries, J. B. (1998). The pharmacology of mesocortical dopamine neurons: A dual-probe microdialysis study in the ventral tegmental area and prefrontal cortex of the rat brain. *Journal of Pharmacology and Experimental Therapeutics*, *285*(1), 143–154.

Westerink, B., Kwint, H., & de Vries, J. (1996). The pharmacology of mesolimbic dopamine neurons: A dual-probe microdialysis study in the ventral tegmental area and nucleus accumbens of the rat brain. *Journal of Neuroscience*, *16*(8), 2605–2611.

White, J. H., McIllhinney, R. A., Wise, A., Ciruela, F., Chan, W. Y., Emson, P. C., et al. (2000). The $GABA_B$ receptor interacts directly with the related transcription factors CREB2 and ATFx. *Proceedings of the National Academy of Sciences of the United States of America*, *97*(25), 13967–13972.

Willick, M. L., & Kokkinidis, L. (1995). The effects of ventral tegmental administration of $GABA_A$, $GABA_B$ and NMDA receptor agonists on medial forebrain bundle self-stimulation. *Behavioural Brain Research*, *70*(1), 31–36.

Willner, P. (1997). Validity, reliability and utility of the chronic mild stress model of depression: A 10-year review and evaluation. *Psychopharmacology (Berl)*, *134*(4), 319–329.

Willner, P. (2005). Chronic mild stress (CMS) revisited: Consistency and behavioural-neurobiological concordance in the effects of CMS. *Neuropsychobiology*, *52*(2), 90–110.

Willner, P., & Mitchell, P. J. (2002). The validity of animal models of predisposition to depression. *Behavioural Pharmacology*, *13*(3), 169–188.

Wise, R. A. (2002). Brain reward circuitry: Insights from unsensed incentives. *Neuron*, *36*(2), 229–240.

Xi, Z. X., & Stein, E. A. (1999). Baclofen inhibits heroin self-administration behavior and mesolimbic dopamine release. *Journal of Pharmacology and Experimental Therapeutics*, *290*(3), 1369–1374.

Hari Manev and Svetlana Dzitoyeva

The Psychiatric Institute, Department of Psychiatry, University of Illinois, Chicago, Illinois, USA

GABA-B Receptors in *Drosophila*

Abstract

Drosophila melanogaster, the "fruit fly," is being increasingly used as an experimental model in neurosciences, including neuropharmacology. The advantages of *Drosophila* over typical mammalian models in neuropharmacology include better access to genetic manipulation and the availability of almost unlimited numbers of experimental subjects at relatively low cost and with minimal regulatory restrictions. Nevertheless, one should remain cognizant of the substantial differences between insects and mammals. Insects, including *Drosophila*, utilize γ-aminobutyric acid (GABA) as a neurotransmitter and express both ionotropic GABA receptors and metabotropic GABA-B receptors. Before cloning of the *Drosophila* GABA-B receptors (subunits 1–3), it had been assumed that flies did not express these receptors since baclofen, a typical agonist for mammalian GABA-B receptors,

Advances in Pharmacology, Volume 58

1054-3589/10 $35.00
10.1016/S1054-3589(10)58017-7

does not produce any effects in insects. Subsequently, it was confirmed that cloned *Drosophila* GABA-B receptors exhibit a unique pharmacology. Using *Drosophila* as a model, it has been shown that GABA-B receptors are involved in the behavioral actions of alcohol and γ-hydroxybutyric acid, and possibly in pain. Furthermore, recent research suggests that in flies these receptors may play an important developmental role and that they participate in olfaction and in regulation of circadian rhythms.

I. Introduction

Drosophila melanogaster, the "fruit fly," is a preeminent model organism in experimental genetics and developmental biology. Its usefulness for unraveling the mysteries of neuroscience became evident with the pioneering work of Seymour Benzer, who introduced *Drosophila* in studies of genetic basis of behavior (for review, see Vosshall, 2007). In one of the first papers in this field, he reported on the discovery of a specific gene mutation that led to a deficit in the locomotor response of these flies to light (i.e., a defect in the natural affinity of flies to move toward light—phototaxis). Subsequent studies have identified *Drosophila* genes responsible for complex behaviors and nervous system functions, such as sexual behavior and even learning and memory. In spite of major differences between insects and mammals, striking resemblances exist with respect to behavior, including the effects of social environment on their behavior. For example, social isolation increases aggression as well in *Drosophila* as it does in mammals (Wang et al., 2008).

II. *Drosophila* Model in Pharmacology

In pharmacology, insects including fruit flies have typically been considered in experiments designed to discover drugs that selectively kill them—insecticides (Raymond-Delpech et al., 2005). Hence, dieldrin, one of the now obsolete major insecticides (its extensive use in the period after World War II has been blamed for development of resistance to insecticides), is an antagonist of insect-specific ionotropic GABA (γ-aminobutyric acid) receptors (Buckingham et al., 2005). Only recently has the fruit fly model become of interest for broader neuropharmacological research and possibly for therapeutic drug discovery (Atkinson, 2009; Manev et al., 2003; Nichols, 2006). Surprisingly, neuropharmacological studies of the mechanisms involved in a psychiatric disorder, that is, drug addiction, appear to have benefited substantially from the use of the *Drosophila* model (Atkinson, 2009; Wolf & Heberlein, 2003). For example, in flies, psychostimulants such as cocaine produce similar behavioral effects as in mammals, that is, repeated/intermittent cocaine administration to

Drosophila triggers locomotor sensitization (Dimitrijevic et al., 2004; McClung & Hirsh, 1998). Using a method for *in vivo* electrochemistry in the *Drosophila* central nervous system (CNS) to measure dopamine uptake (a carbon fiber microelectrode was inserted into the protocerebral anterior medial area of an adult *Drosophila* brain), Makos et al. (2010) demonstrated that cocaine potently decreases dopamine uptake and that this action of cocaine requires the presence of an intact dopamine transporter. Thus, in spite of major anatomical differences between the fly and the mammalian CNS, the shared molecular architecture of the insect and mammalian nervous system appears to be similar enough to allow for the use of fruit flies in neuropharmacology.

The advantages of the *Drosophila* model over typical mammalian models in neuropharmacology include better access to genetic manipulation and the availability of almost unlimited numbers of experimental subjects at relatively low cost and with minimal if any regulatory (e.g., animal protection) restrictions. Nevertheless, one should remain cognizant of the substantial differences between insects and mammals that sometimes go beyond the typical species differences, such as the species-specific sequence of genes encoding the same functional protein structures, for example, receptors (Elphick & Egertová, 2005).

One peculiarity of the fruit fly model is the characteristic of the *Drosophila* life cycle; fruit flies hatch as larvae, which behave as fully functional organisms expressing a wide array of behaviors before they develop into pupae and subsequently into flies. Hence, using *Drosophila*, pharmacological and behavioral experiments (e.g., in pain research) can be performed with two dramatically different organisms—the wingless and legless larvae (Tracey et al., 2003) and the fully developed fruit fly (Manev & Dimitrijevic, 2004; Xu et al., 2006). With respect to GABA, the larva model has been used to demonstrate developmental involvement of GABA-B receptors (Dzitoyeva et al., 2005).

In taking advantage of the *Drosophila* model for pharmacological research, an important technical aspect relates to the methods of drug delivery. These include exposing large numbers of flies to pharmacologically active compounds by mixing drugs with food, volatilizing drugs and delivering them as vapor, or injecting flies or larvae using specially designed microinjecting procedures (Manev et al., 2003). With these techniques at hand, one can fully tackle the unique methodological advances offered by the *Drosophila* model. Not only are fruit flies amenable to large-scale mutagenesis screens, but they are also easily modifiable into transgenic flies that can be used in combination with pharmacological challenges (Nichols, 2006). Also useful is the method of specific gene silencing via RNA interference (RNAi), which can be performed both in *Drosophila* cells in culture (Bai et al., 2009) and in adult flies (Dzitoyeva et al., 2001; Goto et al., 2003).

III. *Drosophila* GABA System

A. GABA Synthesis

In mammals, GABA is synthesized in brain via at least two molecular forms of glutamic acid decarboxylase (GAD), the principal synthetic enzyme for GABA, and in at least two compartments, the transmitter and the metabolic compartment (Martin & Rimvall, 1993). The two GAD forms, GAD65 and GAD67, are the products of two different genes. GAD65 appears to be localized in nerve terminals and responds to short-term changes in demand for transmitter GABA, whereas GAD67 appears to be more uniformly distributed throughout the cell. In *Drosophila*, GABA is synthesized by two enzymes, GAD1 encoded by *Drosophila GAD* (*Dgad1*) and GAD2 (*Dgad2*). GAD1 is expressed exclusively within the nervous system during embryogenesis and in adult flies (Featherstone et al., 2000; Jackson et al., 1990). On the other hand, GAD2 appears to be expressed in glia (Phillips et al., 1993). In adult flies, the cells expressing GAD1 form clusters in the rind (cortex) around the neuropils, whereas GAD2 expression was detected preferentially in the lamina (Okada et al., 2009). Mutant flies heterozygous for *Dgad2* deletions show wild-type levels of GAD activity but *Dgad1* mutants lack GAD activity. The latter experimental model of GAD1-deficient flies demonstrated that GAD1 is necessary for synaptogenesis at the glutamatergic *Drosophila* neuromuscular junction. Hence, it was shown that GAD function is required in the presynaptic motor neurons for the proper development of a functional postsynaptic glutamate receptor field; these authors proposed that GAD functions at this synapse to locally regulate glutamate levels, which in turn regulate levels of postsynaptic receptors (Featherstone et al., 2000).

B. GABA Transport

A key component in terminating the neurotransmitter action of GABA in a GABAergic synapse and in recycling GABA molecules is the sequestration of GABA into presynaptic neurons and glia via a GABA transporter. The distribution of immunoreactivity to the putative *Drosophila* vesicular GABA transporter (vGAT) (CG8394) has been mapped in the fly brain (Enell et al., 2007). These authors observed that the vGAT antiserum did not label cell bodies or axonal projections but the calyces of the mushroom bodies were labeled, as were distinct layers in the optic lobe neuropils and many nonglomerular neuropil areas in the brain and subesophageal ganglia. The antennal lobe glomeruli were also densely immunolabeled. Pharmacological experiments using systemic treatment of *Drosophila* larvae with the GABA transport inhibitor D,L-2,4-diaminobutyric acid (DABA) revealed that blocking the transporter leads to inhibition of motor-controlled body wall and mouth hook contractions and impaired rollover activity and contractile responses to touch stimulation (Leal et al., 2004). On the other hand, DABA-treated larvae responded normally in olfaction and phototaxis assays. Recovery of DABA-impaired behaviors was achieved by a GABA-A receptor antagonist picrotoxin. These pharmacological studies suggested shared conservation of GABA transport mechanisms between *Drosophila* and mammals (Leal et al., 2004).

C. GABA Receptors

Insects including *Drosophila* express both ionotropic and metabotropic GABA receptors (Table I). *Drosophila* expresses three genes that encode ionotropic GABA receptors, RDL, LCCH3, and GRD. The corresponding three *Drosophila* genes are *Rdl* (resistance to dieldrin), *Lcch3* (ligand-gated chloride channel homolog 3), and *Grd* (GABA and glycine-like receptor of *Drosophila*) (Hosie et al., 1997). The majority of mammalian ionotropic GABA receptors are classified as GABA-A (bicuculline-sensitive and regulated by allosteric modulators) whereas a smaller proportion is known as GABA-A-ρ receptors (bicuculline- and allosteric-modulator-insensitive). In contrast, insect ionotropic GABA receptors are generally insensitive to bicuculline (like GABA-A-ρ) but respond to GABA-A allosteric modulators (Buckingham et al., 2005; Hosie et al., 1997). Nevertheless, both mammalian and *Drosophila* ionotropic GABA receptors are blocked by picrotoxin. It has been established that RDL receptors are widely expressed throughout the insect CNS and are important in inhibitory neurotransmission. For these reasons, RDL receptors are a major insecticidal target site (McGonigle & Lummis, 2009). Based on common sensitivity of mammalian and *Drosophila* ionotropic GABA receptors to picrotoxin it was proposed that a pharmacological model in *Drosophila* (i.e., feeding flies picrotoxin) could be used as a seizure model for *in vivo* high-throughput anticonvulsant drug screening (Stilwell et al., 2006).

Since baclofen, a typical agonist for metabotropic GABA receptors used to pharmacologically characterize the mammalian GABA-B receptors, does

TABLE I *Drosophila* GABA Receptors

Receptor type	Drosophila *name*	*Mammalian ortholog*
Ionotropic	RDL	GABA-A/A-ρ
Ionotropic	LCCH3	GABA-A/A-ρ
Ionotropic	GRD	GABA-A/A-ρ
Metabotropic	D-GABA-B-R1	GABA-B
Metabotropic	D-GABA-B-R2	GABA-B
Metabotropic	D-GABA-B-R3	None

TABLE II Effects of Various GABA Receptor Ligands (Concentrations up to 100 μM) on *Drosophila* GABA-B Receptors (Tested in Stably Transfected HEK293 Cells with D-GABA-B-R1 + D-GABA-B-R2)

Drug	*Effect (yes/no)*
GABA-B agonist	
GABA	Yes
Baclofen	No
Saclofen	No
3-APMPA	Yes
GABA-B antagonist	
CGP52432	No
CGP46381	No
CGP35348	No
CGP54626	Yes
CGP55845	Yes
GABA-B positive modulator	
GS39783	No
GABA-A antagonist	
Picrotoxin	No
Bicuculline	No

Adapted from Mezler et al. (2001) and Dupuis et al. (2006).

not produce any biochemical and electrophysiological effects in insects (Table II), it had been assumed that *Drosophila* does not express GABA-B receptors. This was disproved by Mezler et al. (2001) who cloned three GABA-B receptors isolated from *Drosophila*: D-GABA-B-R1, D-GABA-B-R2, and D-GABA-B-R3 (Table I).

IV. *Drosophila* GABA-B Receptors

A. Physiology

As in other species, GABA-B receptors also belong to the family of G-protein-coupled receptors in *Drosophila*. D-GABA-B-R1 and D-GABA-B-R2 but not the D-GABA-B-R3 are coupled to the Gαi signaling pathway so that they inhibit cAMP formation when stimulated (Mezler et al., 2001). Hence, an expression of D-GABA-B-R1 or D-GABA-B-R2 alone in mammalian cell lines did not induce measurable activation of second messenger cascades, but coexpression (in a ratio of 1:1) of these *Drosophila* subunits produced a functional receptor. On the other hand, a coexpression of D-GABA-B-R3 with either D-GABA-B-R1 or R2 or D-GABA-B-R3 alone did not result in the expression of a functional receptor. Furthermore, coexpression of all three receptors failed to modulate the GABA response compared to the coexpression of D-GABA-B-R1/2.

Mezler et al. (2001) described the distribution of the mRNA of all three *Drosophila* GABA-B subunit mRNA in the embryonic CNS. They found that *GABA-B-R1* and *GABA-B-R2* were expressed in similar regions, whereas *GABA-B-R3* demonstrated a unique expression pattern. In the adult brain, all three genes were expressed in overlapping areas, although *GABA-B-R3* showed a slightly different pattern (Okada et al., 2009). Furthermore, it was found that the expression pattern of GABA-B receptors differs from the pattern of GABA-A (e.g., *Rdl*) receptor expression. The distribution of GABA-B-R protein has been analyzed for GABA-B-R2. Its presence was found in pacemaker neurons of the circadian clock of *Drosophila* (Hamasaka et al., 2005), as well as in other areas of the CNS (Enell et al., 2007). The latter studies found a selective distribution of GABA-B-R2 immunoreactivity in the CNS of larval and adult *Drosophila*. In the adult brain, the most prominently labeled structures were neuropils of the antennal lobes, mushroom body calyces, ellipsoid body, and specific layers of the optic lobes. No GABA-B-R2 immunoreactivity was found in peripheral tissues.

A developmental role for *Drosophila* GABA-B-R1 receptors was demonstrated using a method for GABA-B-R1 knockdown based on injectable RNAi (Dzitoyeva et al., 2005). This method was originally developed for selective gene suppression in adult flies (Dzitoyeva et al., 2001; Goto et al., 2003) and adapted for embryo injections. For embryo collection, female flies were placed on agar/grape juice plates and syncytial blastoderm-stage embryos were collected every 30–40 min and injected with double-stranded RNA oligonucleotides against D-GABA-B-R1 or with a GABA-B antagonist CGP54626 (Dzitoyeva et al., 2005). Ultimately, these treatments were lethal by the time of the third instar larvae. The affected larvae were smaller than controls, and compared to controls with symmetrical and straight tracheae, the tracheae of GABA-B-blocked larvae were compressed and folded. The authors suggested that inhibition of GABA-B receptors during development impairs the larvae molting process.

It has been reported that in the adult *Drosophila* the expression of GABA-B receptors by the olfactory receptor neurons (ORNs) innervating different glomeruli is heterogeneous (Root et al., 2008), and that their expression is greater by ORNs innervating the VM2 glomerulus, compared to the DM2 glomerulus (Dacks et al., 2009). In these studies, serotonin enhanced the responses of inhibitory local interneurons, resulting in a reduction of neurotransmitter release from the olfactory sensory neurons via GABA-B receptor-dependent presynaptic inhibition (Dacks et al., 2009). These authors proposed a role for GABA-B receptors in a mechanism underlying the odorant-specific modulation of projection neurons.

It appears that circadian rhythms influence the pharmacology and physiology of all living systems and that G-protein-coupled receptors participate in these mechanisms (Manev & Uz, 2009). In *Drosophila*, GABA-B receptors play a role in the regulation of circadian rhythms via the clock neurons and neuronal circuits of the circadian system. Hamasaka et al. (2005) showed that the master clock neurons, s-LN(v)s, in *Drosophila* utilize GABA as a slow inhibitory neurotransmitter and that the response to GABA can be blocked by a GABA-B antagonist, CGP54626. Furthermore, they demonstrated the presence of GABA-B-R2 immunoreactivity in the dendritic regions of the s-LN(v)s in both adults and larvae, as well as in the dissociated s-LN(v)s.

B. Pharmacology

In their original studies of cloned *Drosophila* GABA-B receptors, Mezler et al. (2001) reported that the coexpressed D-GABA-B-R1/2 receptor exhibited a unique pharmacology. Even prior to their work, it was known that GABA and the potent GABA-B agonist 3-APMPA stimulate insect receptors at low concentrations, but the insect receptors did not respond to the widely used GABA-B agonist baclofen. This phenomenon has been observed in insect preparations, based on biochemical as well as electrophysiological experiments (for review, see Mezler et al., 2001). In preparations using the cloned *Drosophila* GABA-B receptor, it was shown that neither the antagonist saclofen nor CGP35348, which is highly effective on mammalian GABA-B receptors, inhibits D-GABA-B-R1/2, and that picrotoxin and bicuculline are ineffective. Table II summarizes the pharmacology of cloned *Drosophila* GABA-B receptors.

Dupuis et al. (2006) used interspecies combinations of GABA-B receptor subunits to characterize the interaction of the positive modulator *N,N′*-dicyclopentyl-2-methylsulfanyl-5-nitro-pyrimidine-4,6-diamine (GS39783) with the GABA-B receptor heterodimer. Using functional guanosine 5′-*O*-(3-[^{35}S]thio)triphosphate binding assays, these authors observed positive modulation by GS39783 in different vertebrate species but not in *Drosophila*. However, coexpression of *Drosophila* GABA-B-R1 with rat GABA-B-R2 yielded functional receptors positively modulated by GS39783. These studies helped to identify a critical role of the GABA-B-R2 transmembrane region for positive modulation (Dupuis et al., 2006).

C. Modeling System

Although GABA-B ligands have not reached a wide clinical use, and many are still at the level of preclinical development, a number of clinical conditions have been identified as putative targets for GABA-B pharmacology. These include spasticity, drug addiction, pain, cognition, and depression and anxiety (Bowery, 2006).

In drug addiction research, GABA-B receptors have been studied in preclinical models and at least one GABA-B receptor agonist, baclofen, has been tested clinically (Filip & Frankowska, 2008). Thus, baclofen reduced cocaine and alcohol use and was able to reduce some aspects of alcohol withdrawal. In *Drosophila*, it was found that a reduced expression of GABA-B-R1 via RNAi, that is, injection of double-stranded RNA complementary to GABA-B-R1 into adult *Drosophila* attenuated the behavioral actions of the GABA-B agonist 3-APMPA. In addition, both GABA-B-R1 RNAi and the GABA-B antagonist CGP54626 reduced the behavior-impairing effects of ethanol (Dzitoyeva et al., 2003). Thus, it appears that *Drosophila* could be used not only to model the general aspects of alcohol addiction (Devineni & Heberlein, 2009) but also to explore the putative role of GABA-B receptors in behavioral effects of alcohol.

Another drug of abuse, gamma-hydroxybutyric acid (GHB), has been studied in *Drosophila* (Dimitrijevic et al., 2005). In addition to its pharmacological use, GHB occurs endogenously in the brain, including in flies (Satta et al., 2003). Although putative GHB receptors have been cloned, it has been proposed that, similar to the behavior-impairing effects of ethanol, the *in vivo* effects of pharmacological GHB may involve GABA-B receptors. Injecting GHB into flies produced a dose-dependent motor impairment, which was greater in ethanol-sensitive cheapdate mutant *Drosophila* than in wild-type flies. This effect of GHB was attenuated by the GABA-B antagonist CGP54626 and by GABA-B-R1 RNAi as was previously observed with the behavioral effects of alcohol. Interestingly, GHB pretreatment diminished the behavioral response to subsequent GHB injections, that is, it triggered GHB tolerance, but did not produce alcohol tolerance. In contrast, alcohol pretreatment produced both alcohol and GHB tolerance, suggesting that in spite of many similarities between alcohol and GHB, the primary sites of their action may differ (Dimitrijevic et al., 2005).

There is evidence for a key function of GABA-B receptors in the modulation of pain (Bettler et al., 2004), which could involve both the peripherally and centrally expressed GABA-B receptors. Gangadharan et al. (2009) addressed contribution of GABA-B receptors expressed on primary afferent nociceptive fibers to the modulation of pain, and found that neither the development of pain nor the analgesic effects of GABA-B agonists required peripherally expressed GABA-B receptors. These authors concluded that GABA-B receptors in the peripheral nervous system play a less important role than those in the CNS in the regulation of pain. Currently, it is believed that in *Drosophila* GABA-B receptors are expressed selectively in the CNS. An apparatus has been designed to assess pain threshold in freely moving adult flies (Manev & Dimitrijevic, 2004). It consists of a spiral plastic tube that forms a tunnel for flies to negotiate—hot water is pumped through this tube to produce a heat barrier, which prevents naive or control flies from crossing. Flies treated with a GABA-B agonist 3-APMPA cross this heat barrier suggesting that, similar to a rat model in which a GABA-B receptor agonist produced antinociception in a hot-plate test, the *Drosophila* model can be used to explore a role for GABA-B receptors in pain (Manev & Dimitrijevic, 2004).

V. Conclusion

Since the cloning of *Drosophila* GABA-B receptors, fruit flies are increasingly being used as a model system to investigate the molecular mechanisms of GABAergic neurotransmission and physiological pathways modulated by GABA-B receptors. Notwithstanding significant differences between the mammalian and *Drosophila* GABA-B receptors, *Drosophila* could also prove useful in research related to development of GABA-B receptor pharmacology.

Conflict of Interest: The authors do not have any conflict of interest.

References

Atkinson, N. S. (2009). Tolerance in *Drosophila*. *Journal of Neurogenetics*, *23*, 293–302.

Bai, J., Sepp, K. J., & Perrimon, N. (2009). Culture of *Drosophila* primary cells dissociated from gastrula embryos and their use in RNAi screening. *Nature Protocols*, *4*, 1502–1512.

Bettler, B., Kaupmann, K., Mosbacher, J., & Gassmann, M. (2004). Molecular structure and physiological functions of $GABA_B$ receptors. *Physiology Reviews*, *84*, 835–867.

Bowery, N. G. (2006). $GABA_B$ receptor: A site of therapeutic benefit. *Current Opinion in Pharmacology*, *6*, 37–43.

Buckingham, S. D., Biggin, P. C., Sattelle, B. M., Brown, L. A., & Sattelle, D. B. (2005). Insect GABA receptors: Splicing, editing, and targeting by antiparasitics and insecticides. *Molecular Pharmacology*, *68*, 942–951.

Dacks, A. M., Green, D. S., Root, C. M., Nighorn, A. J., & Wang, J. W. (2009). Serotonin modulates olfactory processing in the antennal lobe of *Drosophila*. *Journal of Neurogenetics*, *23*, 366–377.

Devineni, A. V., & Heberlein, U. (2009). Preferential ethanol consumption in *Drosophila* models features of addiction. *Current Biology*, *19*, 2126–2132.

Dimitrijevic, N., Dzitoyeva, S., & Manev, H. (2004). An automated assay of the behavioral effects of cocaine injections in adult *Drosophila*. *Journal of Neuroscience Methods*, *137*, 181–184.

Dimitrijevic, N., Dzitoyeva, S., Satta, R., Imbesi, M., Yildiz, S., & Manev, H. (2005). *Drosophila* $GABA_B$ receptors are involved in behavioral effects of gamma-hydroxybutyric acid (GHB). *European Journal of Pharmacology*, *519*, 246–252.

Dupuis, D. S., Relkovic, D., Lhuillier, L., Mosbacher, J., & Kaupmann, K. (2006). Point mutations in the transmembrane region of GABA-B-2 facilitate activation by the positive modulator *N,N′*-dicyclopentyl-2-methylsulfanyl-5-nitro-pyrimidine-4,6-diamine (GS39783) in the absence of the GABA-B-1 subunit. *Molecular Pharmacology*, *70*, 2027–2036.

Dzitoyeva, S., Dimitrijevic, N., & Manev, H. (2001). Intra-abdominal injection of double-stranded RNA into anesthetized adult *Drosophila* triggers RNA interference in the central nervous system. *Molecular Psychiatry*, *6*, 665–670.

Dzitoyeva, S., Dimitrijevic, N., & Manev, H. (2003). Gamma-aminobutyric acid B receptor 1 mediates behavior-impairing actions of alcohol in *Drosophila*: Adult RNA interference and pharmacological evidence. *Proceedings of the National Academy of Sciences of the United States of America*, *100*, 5485–5490.

Dzitoyeva, S., Gutnov, A., Imbesi, M., Dimitrijevic, N., & Manev, H. (2005). Developmental role of $GABA_{B1}$ receptors in *Drosophila*. *Brain Research Developmental Brain Research*, *158*, 111–114.

Elphick, M. R., & Egertová, M. (2005). The phylogenetic distribution and evolutionary origins of endocannabinoid signalling. *Handbook of Experimental Pharmacology*, *168*, 283–297.

Enell, L., Hamasaka, Y., Kolodziejczyk, A., & Nässel, D. R. (2007). γ-Aminobutyric acid (GABA) signaling components in *Drosophila*: Immunocytochemical localization of $GABA_B$ receptors in relation to the $GABA_A$ receptor subunit RDL and a vesicular GABA transporter. *Journal of Comparative Neurology*, *505*, 18–31.

Featherstone, D. E., Rushton, E. M., Hilderbrand-Chae, M., Phillips, A. M., Jackson, F. R., & Broadie, K. (2000). Presynaptic glutamic acid decarboxylase is required for induction of the postsynaptic receptor field at a glutamatergic synapse. *Neuron*, *27*, 71–84.

Filip, M., & Frankowska, M. (2008). $GABA_B$ receptors in drug addiction. *Pharmacological Reports*, *60*, 755–770.

Gangadharan, V., Agarwal, N., Brugger, S., Tegeder, I., Bettler, B., Kuner, R., et al. (2009). Conditional gene deletion reveals functional redundancy of $GABA_B$ receptors in peripheral nociceptors in vivo. *Molecular Pain*, *5*, 68.

Goto, A., Blandin, S., Royet, J., Reichhart, J. M., & Levashina, E. A. (2003). Silencing of toll pathway components by direct injection of double-stranded RNA into *Drosophila* adult flies. *Nucleic Acids Research*, *31*, 6619–6623.

Hamasaka, Y., Wegener, C., & Nässel, D. R. (2005). GABA modulates *Drosophila* circadian clock neurons via $GABA_B$ receptors and decreases in calcium. *Journal of Neurobiology*, *65*, 225–240.

Hosie, A. M., Aronstein, K., Sattelle, D. B., & ffrench-Constant, R. H. (1997). Molecular biology of insect neuronal GABA receptors. *Trends in Neurosciences*, *20*, 578–583.

Jackson, F. R., Newby, L. M., & Kulkarni, S. J. (1990). *Drosophila* GABAergic systems: Sequence and expression of glutamic acid decarboxylase. *Journal of Neurochemistry*, *54*, 1068–1078.

Leal, S. M., Kumar, N., & Neckameyer, W. S. (2004). GABAergic modulation of motor-driven behaviors in juvenile *Drosophila* and evidence for a nonbehavioral role for GABA transport. *Journal of Neurobiology*, *61*, 189–208.

Makos, M. A. et al., Han, K.-A., Heien, M. L., & Ewing, A. G. (2010). Using *in vivo* electrochemistry to study the physiological effects of cocaine and other stimulants on the *Drosophila melanogaster* dopamine transporter. *ACS Chemical Neuroscience*, *20*, 74–83.

Manev, H., & Dimitrijevic, N. (2004). *Drosophila* model for in vivo pharmacological analgesia research. *European Journal of Pharmacology*, *491*, 207–208.

Manev, H., Dimitrijevic, N., & Dzitoyeva, S. (2003). Techniques: Fruit flies as models for neuropharmacological research. *Trends in Pharmacological Sciences*, *24*, 41–43.

Manev, H., & Uz, T. (2009). Dosing time-dependent actions of psychostimulants. *International Reviews in Neurobiology*, *88*, 25–41.

Martin, D. L., & Rimvall, K. (1993). Regulation of gamma-aminobutyric acid synthesis in the brain. *Journal of Neurochemistry*, *60*, 395–407.

McClung, C., & Hirsh, J. (1998). Stereotypic behavioral responses to free-base cocaine and the development of behavioral sensitization in *Drosophila*. *Current Biology*, *8*, 109–112.

McGonigle, I., & Lummis, S. C. (2009). RDL receptors. *Biochemical Society Transactions*, *37*(Pt 6), 1404–1406.

Mezler, M., Müller, T., & Raming, K. (2001). Cloning and functional expression of $GABA_B$ receptors from *Drosophila*. *European Journal of Neuroscience*, *13*, 477–486.

Nichols, C. D. (2006). *Drosophila melanogaster* neurobiology, neuropharmacology, and how the fly can inform central nervous system drug discovery. *Pharmacology & Therapeutics*, *112*, 677–700.

Okada, R., Awasaki, T., & Ito, K. (2009). Gamma-aminobuyric acid (GABA)-mediated neural connections in the *Drosophila* antennal lobe. *Journal of Comparative Neurology, 514*, 74–91.

Phillips, A. M., Salkoff, L. B., & Kelly, L. E. (1993). A neural gene from *Drosophila melanogaster* with homology to vertebrate and invertebrate glutamate decarboxylases. *Journal of Neurochemistry, 61*, 1291–1301.

Raymond-Delpech, V., Matsuda, K., Sattelle, B. M., Rauh, J. J., & Sattelle, D. B. (2005). Ion channels: Molecular targets of neuroactive insecticides. *Invertebrate Neuroscience, 5*, 119–133.

Root, C. M., Masuyama, K., Green, D. S., Enell, L. E., Nässel, D. R., Lee, C.-H., et al. (2008). A presynaptic gain control mechanism fine-tunes olfactory behavior. *Neuron, 59*, 311–321.

Satta, R., Dimitrijevic, N., & Manev, H. (2003). *Drosophila* metabolize 1,4-butanediol into gamma-hydroxybutyric acid in vivo. *European Journal of Pharmacology, 473*, 149–152.

Stilwell, G. E., Saraswati, S., Littleton, J. T., & Chouinard, S. W. (2006). Development of a *Drosophila* seizure model for in vivo high-throughput drug screening. *European Journal of Neuroscience, 24*, 2211–2222.

Tracey, W. D. Jr., Wilson, R. I., Laurent, G., & Benzer, S. (2003). Painless, a *Drosophila* gene essential for nociception. *Cell, 113*, 261–273.

Vosshall, L. B. (2007). Into the mind of a fly. *Nature, 450*, 193–197.

Wang, L., Dankert, H., Perona, P., & Anderson, D. J. (2008). A common genetic target for environmental and heritable influences on aggressiveness in *Drosophila*. *Proceedings of the National Academy of Sciences of the United States of America, 105*, 5657–5663.

Wolf, F. W., & Heberlein, U. (2003). Invertebrate models of drug abuse. *Journal of Neurobiology, 54*, 161–178.

Xu, S. Y., Cang, C. L., Liu, X. F., Peng, Y. Q., Ye, Y. Z., Zhao, Z. Q., et al. (2006). Thermal nociception in adult *Drosophila*: Behavioral characterization and the role of the painless gene. *Genes Brain and Behavior, 5*, 602–613.

Index

Note: The letters 'f' and 't' following locators denote figures and tables respectively.

Addiction, 118–119
 alcohol dependence, 375–378
  Alcohol Craving Scale, 375–376
  alcohol withdrawal syndrome, 378
  baclofen on alcohol abstinence/craving/ reduction, 375
  Beck Anxiety Inventory, 376
  Clinical Institute Withdrawal Assessment (CIWA-Ar), 378
  controlled drinking, 377
  Obsessive-Compulsive Drinking Scale, 376
  Penn Alcohol Craving Scale, 376
  Zung Self-Rating Depression Scale, 376
 cocaine
  addiction treatment, baclofen on, 380
  administration, 119
  baclofen on direct effects of, 380
  GABA reuptake blocker tiagabine, 382–384
  vigabatrin, effect of, 380–381
 discussion
  anti-addiction dose, 389
  in chronic substance, 391–392
  comorbid affective disorders, 390
  compliance, 390
  dose, 388–390
  future directions, 391–392
 $GABA_B$ agonists, 374–375
  baclofen, 374–375
  non-specific $GABA_B$ agonists, 375
 GHB, 387–388
  abuse-liable properties, 388
  treatment of narcolepsy, 388
 immunoblotting, 119
 immunoprecipitation assay, 119
 methamphetamine, 387
  safety of vigabatrin, 387
 nicotine, 384–385
  double-blind, placebo-controlled smoking reduction study, 385
  sensory properties of cigarettes, 384
  tiagabine, 384
 opiates, 385–386
  Short Opiate Withdrawal Scale and depression using HAM-D, 386
  side effects, 386
 targets to treat substance addictions, 374
Affective disorders
 antidepressants, 9
 anxiolytic activity, 9
 swim test and learned helplessness models, 9
Age-related learning impairment, 36–37
A-kinase anchoring protein (AKAP), 105
Alcohol
 dependence, 375–378
  Alcohol Craving Scale, 375–376
  baclofen on alcohol abstinence/craving/ reduction, 375

Alcohol *(Continued)*
baclofen to treat alcohol withdrawal syndrome, 378
Beck Anxiety Inventory, 376
Clinical Institute Withdrawal Assessment (CIWA-Ar), 378
"controlled drinking," 377
Obsessive-Compulsive Drinking Scale, 376
Penn Alcohol Craving Scale, 376
Zung Self-Rating Depression Scale, 376
fetal alcohol syndrome, 416
reducing craving for
baclofen, 23
withdrawal syndrome, 354–355, 378
Allosteric modulation/modulators, 7
CGP7930/GS39783 and BHF177, positive modulators, 7
location, 7
mechanism
activation of $GABA_BR$, 79
CGP7930 and GS39783, 77–79
chemical structures of $GABA_BR$ PAM, 78*f*
effect of PAM CGP7390, 78*f*
Kaupmann's group, 78
pharmacokinetics, 77
positive (PAM) or negative (NAM) manner, 77
Alzheimer's disease, 33
Ambient GABA, 152–154, 156*f*, 158*f*, 162, 166–167, 179, 187, 190
Amino-3-hydroxy-5-methyl-4-isoxazolepropionic acid-type ionotropic glutamate receptor (AMPAR), 116, 151, 154–155, 163, 166
3-Aminopropyl-methylphosphinic acid (3-APMPA), 6, 458*t*, 460–461
3-Aminopropyl-phosphinic acid (3-APPA), 6
γ−Aminopropyl-phosphinic acids, 26–27, 36, 40
baclofen in Rhesus monkeys, 26
CGP44532
act as antagonists to $GABA_C$ receptors, 27
in animal models of psychosis, 27
lesogaberan, 27
phosphonous acid derivative AZD3355, 27
reduces craving for alcohol/cocaine/nicotine, 27
$GABA_C$ receptors, 26
histone [^{3}H]CGP27492, 26
methyl-phosphinic acid derivative CGP35024, 26
phosphonous acid analogue of GABA (CGP27492), 26
AMP-activated protein kinase (AMPK), 117
anoxia and ischemia, 117
Edman degradation analysis, 117
mass spectrometry, 117
serine/threonine protein kinase, 117
Analgesia, 8–9
baclofen, 8
$GABA_BR$, 9
activation, 8
significance of, 9
GAT-1, 9
positive allosteric modulator CGP7930, 8
Antidepressant(s)
actions of CGP36742 and CGP51176, 34
CGP7930, 30
chronic administration of, 431
effect of, 431
imipramine, 435
-like coping strategy, 431
tricyclic, 431, 439
Antidepressant action, $GABA_BR$ in animal models
chronic mild stress, 434–435
anhedonic-like effect, 435
chronic mild stress (CMS)
anhedonic-like effect, 435
GABAergic ligands, 435
social interaction deficits, 435
forced swim test (FST), 432–434
baclofen administration, 432
catecholaminergic drugs, 433
cocaine-induced seeking behavior, 434
escape-directed behavior or passive behavior, 432
$GABA_BR$ antagonist SCH 50911, 433
$GABA_BR$-positive modulator GS39783, 433
5-hydroxytryptamine (5-HT) system, 433
modified version of rat FST, 433
selective serotonin reuptake inhibitors (SSRI), 433
learned helplessness, 430–431
antidepressant-like coping strategy, 431
systemic baclofen, 431
olfactory bulbectomy, 431–432

allosteric positive modulators GS39783 or CGP7930, 432
bulbectomy-induced hyperactivity, 431
$GABA_BR$ expression, 432
GABA mimetic progabine, 431
hyperactivity, 431
tail suspension test (TST), 434
Arrestins, 115
β-arrestin, 82
β-arrestin 2, 82
2-EGFP or 3-EGFP, 104
Asthma, 12*t*, 13, 288
A-type ionotropic GABA receptor ($GABA_AR$), 116, 152, 160, 176, 182–195, 208*f*, 219*f*
Augmentation, $GABA_BR$-mediated mGluR1 signaling, 155–157
A1 adenosine receptor (A1R), 155
baclofen-induced augmentation, 156
CGP62349, $GABA_BR$-selective antagonist, 155
R,S-3,5-dihydroxyphenylglycine (DHPG), 156
N-ethylmaleimide, 156
excitatory postsynatptic current (EPSC), 155
$GABA_BR$-mediated mGluR1 signaling augmentation, 156*f*
$G_{i/o}$ protein, 156
βγ subunit complex, 157
AZD3355, 24, 27, 288, 296, 299*f*, 301*t*
AZD9343, 293*f*, 296–297, 300, 302–304, 306

Baby hamster kidney (BHK) cells, 118
Baclofen
Ba-34647, 20
R-(−)-baclofen, 6
clinical use, 21–24
benzodiazepine withdrawal, symptoms of, 24
bronchoconstriction, 24
on chronic hiccup, 24
chronic posttraumatic stress disorder, 24
cluster headache, 22–23
craving for alcohol, reducing, 23
delirium tremens, 23
fragile X syndrome, 24
gastroesophageal reflux disease (GERD), 24
idiopathic chronic hiccup, 24
intrathecal baclofen, 22
morphine withdrawal signs, 23
muscle relaxant in spasticity, 21–22
opiate withdrawal syndrome, 23
reduced craving for nicotine/cocaine, 23
reduced panic attacks, 24
side effects, 22
Stiffman syndrome, 22
tetanus, 22
treatment of cough, 24
trigeminal neuralgia, 22
IC_{50} values, 21
prodrugs of, 25
(*R*)-(−)-baclofen (CGP11973A), 21, 21*f*
(*S*)-(+)-baclofen (CGP11974A), 21
Bactofen, 302*f*
Basal "ambient" GABA, 152
GABA- $GABA_BR$ affinity, 152
immuno-electron microscopy studies, 152
microdialysis studies, 152
BHF177 and CGP44532, 352*f*
Bicuculline, *See* Baclofen
Binding of GABA to $GABA_{B1}$ subunit, 73–74
antagonist binding, 73
$GABA_{B1}/GABA_{B2}$ subunit, role of, 73–74
ligand-binding mutagenesis of Tyr^{366}, 73
Ser^{246} and Asp^{471}, agonist binding, 73
site-directed mutagenesis studies, 73
Blood oxygen level-dependent signal (BOLD), 357
Brain-derived neurotrophic factor (BDNF), 34, 37, 436, 441
Breakpoint, 318*t*, 319*t*, 321*t*, 323*t*–325*t*, 327*t*, 330*t*–331*t*, 335*t*, 340, 344*f*, 346, 346*f*, 348–349, 351, 355–356
B-type GABA receptor ($GABA_BR$), 69*f*, 150
Buspirone, 432

cAMP-responsive element binding protein (CREB), 30, 38, 133, 404
CREB1, 33
CREB2, 33, 404, 436
CCAAT/enhancer-binding protein homologous protein (CHOP), 107
Cell-surface trafficking
of $GABA_BR$, 96*f*
$GABA_{B1}$ transport along dendritic ER, 98–99
colocalization of $GABA_{B1}$ and $GABA_{B2}$, 98
ER for trafficking, 98

Cell-surface trafficking *(Continued)*
kinesin-I, 99
protein synthesis, 98
regulation of ER exit, 96–97
arginine-rich ER retention signal (RSRR), 96
coat protein complex I (COPI), 97
ER retention signal of $GABA_{B1}$, 97
regulation of forward trafficking by C-terminal, 98
regulation of trans-Golgi to plasma membrane, 97–98
dileucine (LL) motif, 97
GEF msec7-1, 97
Cellular and subcellular localization
of $GABA_BR$, 180–181
dendritic spines and shafts, 181
in plasma membrane of nerve terminals at GABAergic/symmetric and glutamatergic/asymmetric synapses, 180
postsynaptic membrane of GABAergic/ symmetric synapses, 180
of GAT1, 181–182
diaminobenzidine peroxidation for detection of GAT1-positive structures, 181
Central nervous system (CNS), 2, 6, 11, 12*t*, 13, 23, 26, 42–43, 77, 94, 124, 128, 133, 180–181, 232, 258, 260–262, 267–268, 295–296, 298–300, 302–304, 306, 398, 405, 409, 419, 428–429, 455, 457, 459, 461
Cerebellar LTD, facilitation of, 163–166
$GABA_BR$-mediated mGluR1 sensitization, 163–164, 164*f*
2μM of CGP55845A, 164
in-vitro LTD, 163–164
postsynaptic mGluR1 modulation, 163
presynaptic inhibition, 163
$GABA_BR$-mediated mGluR1 signaling augmentation, 164–166, 165*f*
3μM of baclofen, 165
fluorometry, 166
3μM of GABA or baclofen, 165
GABA spillover, 166
in-vitro LTD, 165–166
molecular basis, 163
cerebellar motor learning, 163
pharmacological inhibition or genetic knock-out of mGluR1, 163
protein kinase C (PKC), 163
voltage-gated Ca^{2+} channels, 163
CGP7390, 78*f*, 79
CGP7930, 7–8, 28–31, 35, 38, 77–79, 321*t*–322*t*, 323*t*, 327*t*, 329*t*–330*t*, 336*t*, 345, 346*f*, 347*f*, 348, 356, 430, 432
CGP34629, 167
CGP35024, 24*f*, 26–27, 27*f*
CGP35348, 25, 32*f*, 33–35, 37, 218, 236*t*, 237, 262, 265, 333*t*, 337*t*, 358–359, 404, 436, 458*t*, 460
CGP36216, 27, 27*f*
CGP36742, 32*f*, 33–35, 236*t*, 430–432, 436
CGP44532, 24*f*, 26–27, 37, 317*t*, 319*t*–320*t*, 323*t*, 325*t*–328*t*, 336, 338, 340–342, 342*f*–343*f*, 349–351, 352*f*, 355–356, 433
CGP44533, 27, 237
CGP46381, 32*f*, 33, 35, 458*t*
CGP47656, 27–28
act as antagonist on presynaptic $GABA_B$ autoreceptors, 28
γ-aminopropyl-ethyl-phosphinic acid CGP36216
$GABA_BR$ antagonist, 27–28
partial $GABA_BR$ agonist, 28
positive allosteric modulators CGP7930 or GS39783, 28
radioligand [^{3}H]CGP62349, 28
CGP51176, 32*f*, 33–34, 337*t*, 339, 432–433, 435–436
CGP54626, 25, 36, 36*f*, 38, 458*t*, 459–461
CGP61334, 36, 36*f*
CGP62349, 155
[^{11}C]CGP62349, 36*f*, 38
[^{3}H]CGP62349, 28, 36*f*, 38, 155
CGP52432A, 37
CGP56433A, 36–37, 430
effect of baclofen on cocaine, 37
improved learning and memory retention, 37
increased swimming time, 37
CGP56999A, 36–38
CGP62349A, 38
CGP63360A, 38
$GABA_BR$ agonist/antagonist ligand, 38
learning and memory-improving effects, 38
CGP55845/CGP55845A, 36, 160, 164, 167, 270*f*, 430
age-related learning impairment, 36–37
beneficial in animal model of Down's syndrome, 37

concentration-dependent increase in dopamine levels, 37
increases swimming time, 37
CGP36742 (SGS742), 33
BBB penetration, 35
blocks $GABA_C$ receptors, 35
Chemistry and pharmacology of $GABA_BR$ ligands, 19–42
$GABA_BR$ agonists, 20–21
γ–aminopropyl-phosphinic acids, 26–27
clinical use of baclofen, 21–24
gabapentin, 25–26
phenibut, 25
prodrugs of baclofen, 25
structures of, 24*f*
$GABA_BR$ antagonists
first generation, 32–35
second generation, 36–40
third generation, 40–42
$GABA_BR$ partial agonists, 27*f*
CGP47656, 27–28
γ-hydroxy-butyric acid, 28–29
positive modulators of $GABA_BR$, 29–32
Chloride ion channels, 2
β–[4–Chlorophenyl] GABA (baclofen), 5–6
Chronic hiccup, 24
cis-Golgi network, 94, 96*f*, 97–98
Clinical Institute Withdrawal Assessment (CIWA-Ar), 378
Cloned $GABA_BR$ subtypes, functions of
distribution of $GABA_{B1}$ subunit isoforms to neuronal compartments, 245*f*
$GABA_{B1}$ and $GABA_{B2}$ subunits for classical $GABA_B$ responses, 240
localization and functions of $GABA_{B1a}$ and $GABA_{B1b}$ subunits
mechanisms involved in selective axonal targeting of $GABA_{B1a}$, 245–246
in response to pharmacological activation, 241–243
in response to physiological activation, 243–245
Cloning, expression
of $GABA_{B1}$ receptor, 65–66
in *Xenopus* oocytes, 65
Cloning techniques (modern), 292
Cluster headache, 22–23
Coat protein complex II (COPII), 95
Cognitive behavioral therapy, 379, 382–383
Cognitive enhancer, 404
Co-immunoprecipitation, 79, 154, 160
Complement protein (CP) module, 65
Conditioned place preference (CPP) procedures, 317*t*–318*t*, 320*t*, 323*t*–324*t*, 326*t*–327*t*, 329*t*, 332*t*–333*t*, 336*t*, 339–340, 348–351, 353*f*, 356–358
Control of network activity by $GABA_BR$, 210–222
-mediated direct control of slow network activity
cortex, 210–212
mediation of slow oscillations, 212*f*
thalamus, 213–214
moderate spike timing during hippocampal theta activity, 219*f*
neocortex, 217–219
theta oscillations, 217
modulate fast oscillations
gamma oscillations, 214, 216*f*
hippocampus, 215
neocortex, 215–217
in pathological network activity, 219–221
biophysical models of thalamic and cortical neurons, 220
modulating gamma oscillations, 221
severe disruption of thalamic oscillator, 220
Cortical network activity, role of $GABA_BR$, 210–221
localization, 207–209
functional role, 209–210
cholinergically-induced theta rhythm, 210
conditions leading to presynaptic activation, 209
Cortical silent period (CSP), 440–441
Cyclic adenosine monophosphate (cAMP), 5, 30
-dependent protein kinase (PKA), 115–116
$GABA_BR$ desensitization, 115
-mediated phosphorylation, role of, 116
serine 892 (S892), 115–116
"Date-rape drug," 28

Degradation of $GABA_BR$, 104–105
AKAP, scaffold proteins, 105
chloroquine, 105
ESCRT, 105
GISP, discovery of, 105
late endosomes to lysosomes, 104
leupeptin and pepstatin A, 105
lysosomal degradation, 105
proteasome inhibitor MG132, 105
TSG101, 105

Dendritic processing, 213
Depolarization, 3, 131, 150, 163–164, 166
Depression, 83
 antidepressant action in animal models
  chronic mild stress (CMS), 434–435
  forced swim test, 432–434
  learned helplessness, 430–431
  olfactory bulbectomy, 431–432
  tail suspension test, 434
 chronic antidepressants and $GABA_BR$ function, 437
 cognition, 435–437
  administration of CGP51176, 436
  brain-derived neurotropic factor (BDNF), 436
 evidence for role of $GABA_BR$ in, 439–441
  cerebrospinal fluid (CSF) measurements and neuroimaging, 440
  effect of baclofen on depressive symptomology, 439
  GABA mimetics, 439
  suicide and control subjects, 440
 $GABA_B$–5-HT interactions, 438–439
  desensitization of DRN 5- HT1A autoreceptors, 438
  G-protein-gated potassium (GirK) channels in DRN, 438
  5-HT release with dorsal raphe nucleus (DRN), 438
  regulator of G-protein signaling (RGS) family, RGS2, 439
 $GABA_BR$
  $GABA_{B(1)}$ and $GABA_{B(2)}$, 429
  high-affinity phosphinic acid-derived antagonists, 430
  inhibitory postsynaptic potentials (IPSP), 429
  metabotropic/presynaptic $GABA_BR$, 429
  positive modulators, 430
 reward system, 437–438
  dopaminergic transmission in brain reward circuitry, 437–438
  intra-ventral tegmental area (VTA), 437
Desipramine, 432
Dimerization
 on GPCR signaling, effect of, 81–82
  dopamine D1 and D2 receptors, 82
  heterodimerization of two functional 7-TM units, 81
  luteinizing hormone (LH) receptor, 82
  μ and δ oligomeric complex, 81–82
  opioid receptor heterodimerization, 81
  sea squirt, *C. intestinalis*, 82
 GPCR trafficking, role of, 80–81
  adrenergic receptor α_{1D}-AR, 80
  bioluminescent resonance energy transfer (BRET) studies, 80
  calcitonin gene-related peptide (CGRP) receptor, 81
  chemokine receptor pair CXCR1/CXCR2, 80
  homodimerization of β_2AR/CaSR, 80
  RAMP1, 81
  taste receptors TIR2/TIR3 or TIR1/TIR3, 80
Divalent cations (Ca^{2+} and Mg^{2+}), 4
Diversity, physiological functions and mechanisms of
 additional mechanisms of diversity, 246–248
  desensitization of GPCRs, 246–247
  regulators of G-protein signaling (RGS), 247
 functions of cloned $GABA_BR$ subtypes
  $GABA_{B1}/GABA_{B2}$ subunits for classical $GABA_B$ responses, 240
  localization and functions of $GABA_{B1a}/GABA_{B1b}$ subunits, 241–246
 $GABA_{B(1a,2)}$ heterodimer and its downstream effectors, 239, 239*f*
 heterogeneity of native $GABA_B$ responses, 233–234
  of postsynaptic $GABA_B$ responses, 237–238
  potency of antagonists and baclofen at pre/post-synaptic $GABA_BR$, 236*t*
  between pre- and postsynaptic $GABA_B$ responses, 234–235
  between presynaptic $GABA_B$ responses, 235–237
 physiological functions, 233–234
  electrophysiological detection of activity, 234
  $GABA_B$ -mediated activation of Kir3 channels, 233
  $GABA_BR$ activation, 233
Dorsal motor nucleus of the vagus (DMV), 290
Dorsal root ganglion (DRG), 124, 130–131
Down's syndrome, 37, 43, 397–419
 See also $GABA_B$–GIRK2-mediated signaling in Down syndrome (DS)
Down syndrome critical region (DSCR), 398–399, 401*f*, 409*f*

Drosophila GABA System, 456–458
effects of various GABAR ligands, 458*t*
GABAR, 457–458, 457*t*
baclofen, 457
RDL/LCCH3 and GRD (encode ionotropic GABAR), 457
GABA synthesis, 456
GABA transport, 456–457
Drosophila vesicular GABA transporter (vGAT), 456
inhibitor D,L-2,4-diaminobutyric acid (DABA), 456
recovery of DABA-impaired behaviors, 457
Drug addiction, 10, 12*t*, 79, 316, 344, 359, 454, 460
Dual sphincter mechanisms, 289
Dysfunctional information process, 402

Electroconvulsive therapy (ECT), 440
Endocytosis of $GABA_BR$
agonist-induced, 103–104
calcium channels in chick sensory neurons, 104
coexpression of arrestin 2-EGFP, 104
effects of receptor activation, 104
interaction of NSF protein, 104
reactivating desensitized receptors, 103
constitutive internalization, 100–101
BBS-tagged $GABA_BR$, 101
dimeric/monomeric receptor complex, 100
hemagglutinin (HA)-tagged receptors, 100
internalization of recombinant $GABA_{B1b,2}$ receptors, 100
constitutive recycling, 102–103
cell surface biotinylation experiments, 103
endocytic recycling, 103
immunofluorescence assay, 103
inhibition of recycling with monensin, 103
sorting nexin 4 (SNX4), 103
pathway and endosomal sorting, 101–102
caveolin-dependent endocytosis, 102
classical dynamin and clathrin-dependent pathway, 101
clathrin-coated vesicles, 102
colocalization studies, 102
dynamin dependent/independent mechanisms, 101
endocytic pathways, 101*f*
fast recycling endosomes, 102
Endoplasmic reticulum (ER), 74, 80–81, 94–99, 96*f*, 106–107, 125, 126*f*, 128, 161
Endosomal sorting complex required for transport (ESCRT), 101*f*, 105
Enteric nervous system, 6
Epilepsy-absences, 11–12
absence (petit-mal) epilepsy, 34
benzodiazepine, 11
seizure production, $GABA_B$ mechanisms in, 11
Exocytosis/endocytosis/degradation, mechanisms, 93–107
cell-surface trafficking
$GABA_{B1}$ transport along dendritic ER, 98–99
regulation of ER exit, 96–97
regulation of forward trafficking by C-terminal, 98
regulation of *trans*-Golgi to plasma membrane, 97–98
degradation, 104–105
endocytosis
agonist-induced, 103–104
constitutive internalization, 100–101
constitutive recycling, 102–103
pathway and endosomal sorting, 101–102
Extracellular Ca^{2+}, 153–154
Ca^{2+}-sensing receptor (CaR), 153
G protein-coupled receptors (GPCR), 153
point mutation study, 153
Extracellular signal-regulated protein kinases (1/2) (ERK(1/2)), 30, 133

Family C GPCR
evolution of, 67
mGluR group, 67
(mya) duplication, 67
sensing receptors (CaSR and T1R), 67
sequence alignments, 69*f*
First gene duplication, 67
First generation $GABA_BR$, definition, 32
Fluorescence resonance energy transfer (FRET) techniques, 75–77, 79, 135–136, 248
Fragile X syndrome, 24

GABA$_A$
receptor structure, 5
-ρ receptors (bicuculline- and allosteric-modulator insensitive), 457
GABA and glycine-like receptor of *Drosophila* (Grd), 457, 457*t*
GABA$_B$ agonists
baclofen, 374–375
effects of, 374
non-specific GABA$_B$ agonists, 375
anti-epileptics, 375
tiagabine, 375
vigabatrin, 375
GABA$_B$–GIRK2-mediated signaling in Down syndrome (DS)
developmental and morphological changes, 415–416
apoptosis related genes present on Hsa21, 416
exaggerated GABA$_B$–GIRK signaling, 416
fetal alcohol syndrome, 416
effect of GABA$_B$–GIRK2 localization at postsynaptic active zones, 415
evidence for cognitive impairment in mouse models, 400–402
behavioral deficits in Morris water maze, 400
human chromosome 21 (Hsa21) map showing DSCR, 401*f*
GABA$_B$–GIRK coupling to hippocampal overinhibition, 404–405
coupling of metabotropic GABA$_B$R to GIRK channels, 405
interactions with GIRK channels, 404
gene dosage hypothesis, 398
hippocampal deficit
abuse of methamphetamines, 403–404
administration of GABA$_A$ receptor antagonists, 402
altered synaptic plasticity, 403*f*
cognitive enhancer, 404
cognitive performance suppression, 404
CREB- mediated gene transcription, 404
long-term depression (LTD), 402
long-term potentiation (LTP), 402
reduced capacity for LTP, 402
roles of GABA$_B$R-mediated inhibition, 403
interactions between intracellular signaling mechanisms, 408–411
Dap160, 409
Drosophila model of Synj overexpression, 409
GABA$_B$–GIRK signaling, 408
N-methyl-D-aspartate (NMDA) receptor activation, 410
PtdIns(4,5)P2 levels, 408–409
synaptojanin 1 and GIRK2, 410
interference from other GPCR and signaling pathways, 407–408
gene dosage induced alterations, GIRK channel trafficking, 409*f*
GIRK channel subunits in Ts65Dn hippocampus, 408*f*
G-protein-coupled receptors (GPCR), 407
overall GIRK response, 407
mouse models, 399–400
KCNJ6/GIRK2, 399
Tc1 mice, 400
trisomic (Ts65Dn/Ts1Cje and Ts1Rhr), 400
potential impact of GIRK2 dosage on hippocampal function, 411–415
CA3–CA1 synapse, 413
electroencephalogram (EEG) recordings, 413–415
hippocampi of Ts65Dn mice, 413
model for disrupted hippocampal circuitry in Ts65Dn, 412
principal pathways, 411
stimulation of temporoammonic pathway, 414*f*
trisynaptic PP–CA3–CA1 pathway, 411
properties of mammalian GIRK channels, 405–407
activated by Gbg heterodimers, 405
gene dosage effect of CA1 pyramidal neurons, 406
role of GIRK in pain and cerebellar phenotypes, 417–418
antinociceptive benefits from clonidine, GIRK2 knockout mice, 417
behavioral consequences of restoration of cerebellar phenotype, 418
Girk2-disomic Ts65Dn mice using Girk2 knockout, 418
mating scheme, crossing Ts65Dn mice and heterozygous knockout, 418
GABA$_B$R activation by GABA from endogenous sources
GABA$_B$R-IPSP, 182–183
interneuron subtypes, 183

stimulation of dendritic-targeting synapses, 183
stronger stimulus intensities, 182–183
presynaptic $GABA_BR$ effects, 186–188
activation mechanisms, 187
GABA release regulation, 187
localized in glutamate nerve terminals, 188
unitary synaptic transmission, 183–186
fast/slow time-course of $GABA_AR$-IPSP, comparison, 184*f*
neurogliaform cells, 186
non-fast-spiking basket cells, 185
single-axon or unitary IPSC/IPSP, 183
$GABA_BR$ agonists/positive modulators and reward, 336–358
alcohol, 354–356
cocaine, effects of $GABA_BR$
agonists on effects of psychoactive drugs, 342*f*, 344–348
agonists on motivational effects of psycho drugs, 344*f*
$GABA_BR$ agonists, 339–345
$GABA_BR$-positive allosteric modulators, 345–348
GABA-mimetic drugs, negative effects, 341
high-affinity $GABA_BR$ antagonists, 343
microinjections of baclofen or CGP44532, 341–343
positive modulators on effects of psychoactive drugs, 346*f*, 347*f*
intracranial self-stimulation (ICSS) procedure, 338
intra-VTA injections of baclofen, 338
nicotine, 349–354
effects of BHF177 and CGP44532, 352*f*
effects of GS39783, 353*f*
$GABA_BR$ direct agonists, 349–350
$GABA_BR$-positive allosteric modulators, 350–354
opiates, 357–359
baclofen co-administered with heroin, 358
naloxone-precipitated morphine withdrawal, 358
psychomotor stimulants amphetamine and methamphetamine, 348
$GABA_BR$ antagonists
first generation, 32–35
administration of CGP35348, 34
antidepressant, 34
CGP46381, 35
CGP36742 (SGS742), 33
definition, 32
first $GABA_BR$ antagonists, 33
γ-hydroxybutyrate (GHB), 34–35
2-hydroxy-saclofen, second $GABA_BR$ antagonist, 33
phaclofen, 32
SCH50911, 35
structures of, 32*f*
ligands
structures of fluorescent, 39*f*
partial agonists of positive modulators, 35
CGP35348 and 2-hydroxy-saclofen, 35
CGP7930 or GS39783, 35
second generation, 36–40
benzyl substituents, 36
$[^{11}C]$CGP62349, 38
CGP55845A, 36
CGP56433A, 36–37
CGP56999A, 37
CGP62349A, 38
CGP63360A, 38
definition, 36
$[^{3}H]$CGP62349, 38
structures of, 36*f*
third generation, 40–42
class 3 G-protein-coupled receptor, 42
$[^{125}I]$CGP64213, 40
$[^{125}I]$CGP84963, 40
$[^{125}I]$CGP71872, photo affinity ligand, 40
structures of, 41*f*
$GABA_BR$ coupling to G-proteins and ion channels, 123–139
formation of macromolecular signaling heterocomplex, 134–136
$GABA_BR$-dependent desensitization, 136–138
$GABA_BR$ effectors
enzymes, 133–134
G-proteins, 129–130
G-protein independent activation, 134
potassium channels, 132–133
voltage-gated calcium channels, 130–131
$GABA_BR$ protein expression, 128–129
combined genetic/physiological and morphological approach, 129
electron microscopic studies, 128–129
electrophysiological studies, 129
fluorescence microscopy/biochemistry and quantitative colocalization, 128

GABA$_B$R coupling to G-proteins and ion channels *(Continued)*
high-resolution immunocytochemistry, 128
in situ hybridization studies, 129
light microscopy level, 128
regions of GABA$_{B1}$ and GABA$_{B2}$ expression, 128
GABA$_B$R-dependent desensitization, 136–138
baclofen stimulation, 137
changes in G-protein turnover, 137
CHO cells, 137
coexpression of Gα_q-Q209L-ΔN, 138
desensitization, 136
function of RGS proteins, 138
in oocyte, 138
GIRK channels in mammalian cells or *Xenopus* oocytes, 137
G-protein turnover affects GABA$_B$R function, 138*f*
GRK4 with HEK-293 cells, 137
hippocampal pyramidal neurons, 137
mu-opioid receptor and GABA$_B$R, 138
peri-aquaductal gray neurons, 136
phosphorylation of receptor, 137
stimulation of β2-adrenergic receptors, 137
PKA phosphorylation, 137
GABA$_B$R effectors
enzymes, 133–134
adenylyl cyclase affinity, 133
cAMP response element binding (CREB), 133
Gs-coupled GPCR, 134
modulation of neuronal excitability, 133
protein kinase A (PKA), 133
tyrosine hydroxylase and neuropeptides, 133
G-protein, 129–130
activation of downstream effectors, 129
agonist-induced receptor phosphorylation, 134
β-arrestins, 134
cerebellar Purkinje cells, 134
chemotaxis and cell survival, 134
Gαi and Gαo family, 130
Gi subtype, 130
G-protein anti-sera and toxins, 130
heterotrimeric (α, β, and γ) G-proteins, 129
hydrolysis-resistant analogs, 130
independent activation, 134
potassium channels, 132–133
GABA$_B$-GIRK currents, 132
GIRK channel activation, 132
primary ionic conductance, 132
regulator of G-protein signaling (RGS) 2 protein, 133
primary GABA$_B$, 130*f*
voltage-gated calcium channels, 130–131
calcium currents, transient and sustained, 130
electrophysiological relationship of GABA$_B$-CaV, 131
long-term potentiation (LTP) in CA1, 131
neuronal regulation, 131
presynaptic GABA$_B$ autoreceptors, 131
reduced neurotransmitter release, 131
six types of CaV channel (L-/T-/N-/P-/Q- and R-type), 130–131
GABA$_B$R in different species, 66–68
calcium receptor (CaSR), 66
evolution of Family C GPCR, 67
mGluR group, 67
(mya) duplication, 67
sensing receptors (CaSR and T1R), 67
GABA/glutamate receptor termed GrlE, 68
GrlJ, *D. discoideum* life cycle, 68
metabotropic glutamate receptors (mGluR), 66
odor receptors, 66
pheromone receptors (V2R), 66
placozoan *Trichoplax adhaerens,* 68
seven transmembrane (7TM) domain, 66
taste receptors (T1R), 66
venus fly trap domain or VFTD, 66
Xenopus laevis melanophores, 69
GABA$_B$ receptors in *Drosophila*
GABA system, *See Drosophila* GABA system
modeling system, 460–461
GABA$_B$ receptors in modulation of pain, 461
gamma-hydroxybutyric acid (GHB), 461
pharmacology, 454–455, 460
advantages, 455
coexpressed D-GABA-B-R1/2 receptor, 460
locomotor sensitization, 455
peculiarity of fruit fly model, 455
psychostimulants, cocaine, 454
RNA interference (RNAi), 455

physiology, 458–460
circadian rhythms, 459
developmental role for *Drosophila* GABA-B-R1 receptors, 459
distribution of mRNA of GABA-B subunit mRNA, 458
expression of D-GABA-B-R1 or D-GABA-B-R2, 458
expression of GABA-B receptors by olfactory receptor neurons (ORN), 459
GABA-B receptors and GABA-A, pattern of, 459
$GABA_BR$ in reward processes
agonists *vs.* positive modulators, 359–360
definitions of behavioral procedures, 335*t*–336*t*
definitions of $GABA_BR$ compounds, 336*t*–337*t*
$GABA_BR$ agonists/positive modulators/reward, 336–359
alcohol, 354–357
intracranial self-stimulation (ICSS) procedure, 338
intra-VTA injections of baclofen, 338
nicotine, 349–354
opiates, 357–359
psychomotor stimulants, 339–345, 348
GABAergic system in brain reward circuits, 316–336
dopaminergic neurons, 316
dorsal and median raphe nuclei (DR and MR), 334
dorsal striatum and compulsive drug-seeking behavior, 334
VTA neurons, 334
rewarding effects of addictive drugs, 317*t*–333*t*
$GABA_BR$ structure, 126*f*
CD spectroscopic analysis, 126
classical GPCR subunit, domains in, 125
cloning, 125
$GABA_{B1}$ and $GABA_{B2}$, 125
GABA R1 and R2, 125
coiled-coil domain, 126
critical tyrosine in lobe II, 127
$GABA_{B1a}$ subunit
"sushi" domains, 125
$GABA_B$ electrophysiological responses, 128
$GABA_B$–GPCR heterodimer relationships, 128
$GABA_{B1}$ N-terminus, 126
$GABA_{B1}$ subunit
N-terminal domain, 125
heptahelical domain, 127
hydrogen bonds and salt bridges, lobe I and GABA, 127
ligand binding, 126
M2 muscarinic receptor, 128
point mutations, 127
refined replacements, 127
sequence homology and site-directed mutagenesis, 126
venus flytrap model, 127
wild-type signaling, 127
yeast two hybrid analysis, 125–126
$GABA_BR$-mediated inhibitory postsynaptic potentials ($GABA_BR$-IPSP)
interneuron subtypes, 183
$GABA_BR$-mediated mGluR1
sensitization, 163–164, 164*f*
signaling augmentation, 164–166, 165*f*
$GABA_BR$-mediated modulation of mGluR signaling and synaptic plasticity in central neurons, 149–168
functions of $GABA_BR$ near excitatory synapses
function exerted through GIRK channels, 154–155
$GABA_BR$-mediated mGluR1 sensitization, 157–162
$GABA_BR$-mediated mGluR1 signaling augmentation, 155–157
ligands for $GABA_BR$ in cerebellar Purkinje cells
basal "ambient" GABA, 152
extracellular Ca^{2+}, 153–154
GABA spilt over from synapses, 153
physiological significance
dynamic range of slow PF-PC EPSP, 162
facilitation of cerebellar LTD, 163–166
pre- and post-synaptic $GABA_BR$, 166–167
synaptic organization around cerebellar Purkinje cells, 151–152
$GABA_BR$, protein kinases and receptor trafficking, 113–119
endocytic sorting and receptor cell surface stability, 118
biotinylation, 118
clathrin- and dynamin-dependent mechanisms, 118
glutamate treatment, 118
phosphorylation and diseases

$GABA_BR$, protein kinases and receptor trafficking *(Continued)*
addiction, 118–119
ischemia, 119
phosphorylation and functional modulation
AMPK, 117
PKA, 115–116
PKC, 116–117
phosphorylation-independent desensitization, 117–118
agonist-induced desensitization, 117
arrestin- and dynamin-dependent receptor internalization, 117
expression of $GABA_BR$ and GIRK, 118
$GABA_{B2}$ subunit
heterodimerization with, 66
in ligand binding, role of, 74–75
agonist binding to $GABA_{B1}$, 74
binding of agonists to VFTD of $GABA_{B1}$, 74
binding site within $GABA_{B2}$ VFTD, 74
chimeric receptor, 75
fluorescence energy transfer (FRET) studies, 75
increase in affinity, effects, 75
negative allosteric effects (NAM), 75
radioligand-binding studies, 74
single transmembrane domains (TMD) or GPI anchor domains, 75
$GABA_C$ receptors, 26–27, 35, 258, 336, 428
GABA (γ–amino-*n*-butyric acid), *Passim*
Gabapentin, 25–26
structures of, 21*f*
GABA synthesis, 456
Dgad2 deletions, 456
GAD65 and GAD67, 456
GABA transporter GAT1
cellular and subcellular localization of $GABA_BR$ and GAT1
localization of $GABA_BR$, 180–181
localization of GAT1, 181–182
$GABA_BR$ activation by GABA released from endogenous sources
$GABA_BR$-mediated inhibitory postsynaptic potentials, 182–183
presynaptic $GABA_BR$ effects, 186–188
unitary synaptic transmission, 183–186
GAT1 activity and $GABA_BR$ activation
GAT1 activity and $GABA_BR$-mediated IPSP, 191–194
GAT1 activity and presynaptic $GABA_BR$ activation, 194
GAT1-mediated uptake prevents GABA spillover onto synaptic $GABA_ARs$, 189–191
plasma membrane GABA transporter 1, 176–177
properties of GAT1 protein and translocation of GABA by GAT1, 177–179
regulation of GAT1-mediated GABA transport, 179–180
Gαi- and Gαo-type heteromeric type G proteins, 114
Ganglion neurons, effects, 2–3
Gastroesophageal reflux disease (GERD)
$GABA_BR$, 290–292
'closing' of venus fly trap model, 293
GABA/baclofen/lesogaberan and AZD9343, 293
$GABA_BR$ ligand binding site, 292–294
heterodimeric structure, 292
inhibitory postsynaptic current (IPSC), 294
neural pathways to LES and crural diaphragm (CD), 291*f*
seven-transmembrane subunits ($GABA_{B(1)}$ and $GABA_{B(2)}$), 292
T-type calcium channels, 294
$GABA_BR$ agonists as reflux inhibitors
complication associated with baclofen, 295
lesogaberan AZD3355 and AZD9343, 296
peripherally-restricted agonists, 295–306
underlying physiological mechanisms of GERD, 289–290
anatomy of esophagogastric junction, 289
GAT1 activity and $GABA_BR$ activation -mediated IPSP, 191–194
electrophysiological studies, suggestions, 192
neurogliaform cells, 193
NO711 application, 192
production of fast $GABA_AR$-IPSC, 192
mediated uptake prevents GABA spillover, 189–191, 190*f*–191*f*
absence of GAT1 effects on single-synapse responses, 189
prolongation by GAT1 blockade, 189
and presynaptic $GABA_BR$ activation, 194

Gene dosage hypothesis, 398
Genetic absence epilepsy rats from Strasbourg (GAERS), 11, 34–35
γ-hydroxybutyrate (GHB)
  "date-rape drug," 28
  high-affinity GHB ligands, 29
  registered drug (Xyrem), 28
  weak $GABA_BR$ partial agonist, 28
$G_{i/o}$ protein
  -coupled receptor, 150
  -independent action of GPCR in cerebellar Purkinje cells, 161–162
    A1R agonists, 161–162
    mouse Purkinje cells, 161
    [R]-N6-(1-methyl-2-phenylethyl) adenosine (R-PIA), 161
GIRK1/KIR3.1/KCNJ1, 405
GIRK2/KIR3.2/KCNJ6, 405
GIRK3/KIR3.3/KCNJ9, 405
GIRK4/KIR3.4/KCNJ5, 405
Glutamic acid decarboxylase (GAD), 440, 456
  GAD65 and GAD67, 456
Gonadotropin releasing hormone receptors (GnRHR), 82
G protein-coupled inwardly rectifying K^+ (GIRK) channels, 150, 154–155
  application of baclofen, 154
  current-clamp recordings, 154
  ectopic localization, 154
  $GABA_BR$-mediated GIRK conductance, 155
  GIRK or Kir3 channels, 294
  hyperpolarization, 155
  immuno-electron microscopy and co-immunoprecipitation study, 154
  K^+ conductance, 154
  light microscopic immunohistology studies, 155
  in mammalian cells or *Xenopus* oocytes, 137
  voltage ramp-evoked whole-cell, 154
  See also $GABA_B$–GIRK2-mediated signaling in Down syndrome (DS)
G-protein-coupled receptor interacting scaffolding protein (GISP), 105
G protein-coupled receptor kinases (GRK), 114–115, 117–118, 137, 246
G protein-coupled receptors (GPCR), 64, 231, 259
  drug discovery, relevance to, 82–83
    allosteric ligands, 83
    allosteric modulators, 83
    calcium-sensing receptor, 83
    $GABA_B$ heterodimer, 83
    G-protein-mediated signaling pathway, 82–83
    heterodimerization, 83
    heterodimer-specific drugs, developing, 83
    PAM at $GABA_BR$, 83
  Family A (rhodopsin family), 64
  Family B (secretin family), 64
  Family C (metabotropic family), 64
  signaling and drug discovery, implications for, 63–84
    See also Heterodimerization of $GABA_BR$
  signaling, effect of dimerization on, 81–82
  trafficking, role of dimerization on, 80–81
G-protein signaling
  agonist-induced activation of, 75–77
    chimeric G protein ($G_{\alpha q19}$) in HEK293 cells, coupling, 76
    coupling to GIRK channels, 76
    $GABA_{B2}$ G-protein coupling, 75
    mutation of Leu^{686}, 76
    mutations of residues Lys^{586}/Lys^{590} and Met^{587}, 76
    TMD of $GABA_{B1}$, role of, 76–77
GS39783, 7, 77–78, 432
  effects of, 353*f*
  mechanism of action of, 77–78
  positive modulators, 7
GTPase-accelerating proteins (GAP), 247
Guanine-nucleotideexchange factor (GEF), 96*f*, 97–98

Hamster β-adrenergic receptor, 65
^{3}H-baclofen, 4
[^{3}H]CGP5699, 38
[^{3}H]CGP27492, 24*f*, 26, 28, 36, 38, 40
[^{3}H]CGP54626, 25
[^{3}H]CGP62349, 38
Heptahelical transmembrane (7TM) domain
  phylogeny of $GABA_BR$, 68–69
    $GABA_BR$-like receptor ($GABA_B$ RL), 68
    six "truncated" Family C orphans (RAIG1/GPRC5A), 68
    Xenopus laevis melanophores, 69
  structure of $GABA_BR$, role of rhodopsin, 70–73
    agonist binding, 71
    Arg of Family A, 71

Heptahelical transmembrane (7TM) domain *(Continued)*
G protein transducin, 70
TMD of Family C, 71
Trp of Family A and C receptors, 71
Heterodimerization of $GABA_BR$, 63–84
functions
agonist-induced activation of G-protein signaling, 75–77
binding of GABA to $GABA_{B1}$ subunit, 73–74
mechanism of action of allosteric modulators, 77–79
role of $GABA_{B2}$ subunit in ligand binding, 74–75
for GPCR dimerization
effect of dimerization on GPCR signaling, 81–82
relevance to GPCR drug discovery, 82–83
role of dimerization on GPCR trafficking, 80–81
identification
expression cloning of $GABA_{B1}$ receptor, 65–66
heterodimerization with $GABA_{B2}$ subunit, 66
phylogeny
$GABA_BR$ in different species, 66–68
heptahelical (7TM) domain, 68–69
venus flytrap domain, 69–70
structure
heptahelical transmembrane domain, 70–72
venus flytrap domain, 72–73
Hiccup, chronic, 24
High-affinity radio labeled binding probes, 3
High-frequency information process, 413
Hill coefficient, 158–160, 158*f*
Histamine H4 receptor, 65
Histone [^{3}H]
acetylcholine, 20
-baclofen, 4, 21, 25–26, 152, 265–266, 391
CGP5699, 38
CGP27492, 24*f*, 26, 28, 36, 38, 40
CGP54626, 25
CGP62349, 28, 38
-noradrenaline, 20
Historical perspective and emergence of $GABA_BR$, 1–13
allosteric modulation, 7
$GABA_BR$ ligands, 5–6
potential therapeutic significance
affective disorders, 9
analgesia, 8–9
applications outside brain, 12–13
drug addiction, 10
epilepsy-absences, 11–12
memory, 10
skeletal muscle relaxation, 9
receptor distribution, 6–7
receptor function, 8
Human embryonic kidney (HEK) 293 cells, 78*f*, 100, 102, 104–105, 115, 137
2-Hydroxy saclofen, 6, 32–33, 35, 42, 350
Hyperparathyroidism, 83
Hyperpolarization
chloride-mediated membrane, 258
neuronal, 5
small membrane, 131
of thalamocortical neurons, 220
[^{125}I]CGP64213, 38, 40, 41*f*, 42, 73, 152
[^{125}I]CGP71872, 40, 65
[^{125}I]CGP84963, 40
[^{125}I]CGP71872, photoaffinity ligand, 40

Idiopathic chronic hiccup, 24
Immunogold particle methods, 181
Inhibitory postsynaptic currents (IPSC), 37, 124, 131–133, 183–192, 209, 235, 236*t*, 242, 271–272, 294–295
Inhibitory postsynaptic potential (IPSP), 32, 152, 182–183, 184*f*–185*f*, 186, 190*f*–191*f*, 192, 194, 196, 207, 208*f*, 209, 211, 212*f*, 214–215, 216*f*, 233, 235, 429
International Union of Basic and Clinical Pharmacology (IUPHAR) terminology, 4
$GABA_B$ and $GABA_A$, classification, 4
Intracranial self-stimulation (ICSS) procedure, 317*t*, 319*t*–321*t*, 326*t*–327*t*, 335*t*, 338–340, 348, 350–351, 352*f*, 354, 360, 433–434
Ionic lock, 71
Ionotropic $GABA_A$, 232, 258, 428
Ischemia, 119
AMPK activation, 119
model of anoxia (oxygen–glucose deprivation), 119
phosphorylation of S783 in R2 subunit, 119
Isoguvacine, 3–4

Kreb's physiological solution, 4

Larvae molting process, 459
Lesogaberan, 6, 12, 24*f*, 27, 43, 293*f*, 296–306
Ligand-gated chloride channel homolog 3 (Lcch3), 457, 457*t*
Ligand-gated chloride selective receptors ($GABA_A$ and $GABA\rho$), 124
Localization of $GABA_BR$, 207–209
  different mechanisms, 208*f*
  GIRK channels, 207
  inhibitory effects, 207
  inhibitory postsynaptic potential (IPSP), 207
  presynaptic and postsynaptic inhibition, 209
Long-term depression (LTD), 152, 163–168, 402, 403*f*, 412*f*, 413
  See also Cerebellar LTD, facilitation of
Long-term potentiation (LTP), 10, 103, 131, 233–234, 242, 244, 258, 402, 403*f*, 404, 406*f*, 410, 412*f*, 413, 415
Lower esophageal sphincter (LES), transient, 27, 288–290, 291*f*, 299*f*, 301*t*, 303–304

Macromolecular signaling heterocomplex, formation of, 134–136
  fluorescence resonance energy transfer (FRET) techniques, 136
  $GABA_{B1a}$ and $GABA_{B1b}$ subunit, 134
  hetero-oligomerization, 135
  high-resolution immuno-electronmicroscopy, 135
  kinesin-I (molecular motor), 135
  Marlin 1 (microtubule binding G-protein), 135
  membrane-associated protein CD8α, 134
  microscopic and spectroscopic techniques, 135
  PDZ proteins, 136
  PSD-95, 136
  time-resolved FRET and snap-tag technology, 135
Major depressive disorder (MDD), 428–429, 440–441
Mesolimbic system, 10, 12*t*
Metabotropic ($GABA_B$) and ionotropic ($GABA_A$) receptors, 5
Metabotropic glutamate receptor (mGluR) signaling, 42, 66–68, 149–168, 232
  mGluR1, 157–162, 158*f*
    $GABA_BR$-mediated mGluR1 sensitization, 163–164, 164*f*
    $GABA_BR$-mediated mGluR1 signaling augmentation, 164–166, 165*f*
    pharmacological inhibition or genetic knock-out of mGluR1, 163
    See also MGluR1 sensitization, $GABA_BR$-mediated
  See also $GABA_BR$-mediated modulation of mGluR signaling and synaptic plasticity in central neurons
MGluR1 sensitization, $GABA_BR$-mediated, 157–162, 158*f*
  $Ca^{2+}{}_o$
    absence of, 159
    addition of, 159
    interaction with mGluR1, 157
    physiological concentration (2mM) of, 159
    in Purkinje cells, 157
  dose–response curve with Hill coefficient, 158
  $G_{i/o}$ protein-independent action of GPCR in cerebellar Purkinje cells, 161–162
  glutamate-binding sites, 159
  ligand-binding sites, 158
  mechanisms underlying, 159–161
    $Ca^{2+}{}_o$- and baclofen, 160
    $Ca^{2+}{}_o$-sensitive GPCR, 159–160
    CGP55845, $GABA_BR$-selective antagonist, 160
    co-immunoprecipitation, 160
    co-localization in dendritic spines, 161
    direct communication or indirect communication, 160
    freeze-fracture immunoelectron microscopy, 160
    $GABA_BR$ and mGluR1, formation between, 161
    immuno-electron microscopic studies, 161
    multivalent cations including Ca^{2+}o, 159
    rat cerebellar Purkinje cells, 161
Mianserin, 432
Microcircuits
  cortical, 207–210

Microcircuits *(Continued)*
synaptic function of, 206
Microdialysis techniques, 35, 152–153, 166, 294, 349
Middle cerebral artery occlusion (MCAO), 119
Muscle hypotonia, 398
(mya) duplication in two distinct groups
Family C receptors, 67
$GABA_B$-like receptors, 67

Nerve growth factor (NGF), 34, 37
N-ethylmaleimide-sensitive fusion (NSF) protein, 104, 116–117
Neuropathic hyperalgesia, 26
Noradrenaline, 3, 20, 429

Oligomerization, 77, 135
Opiate withdrawal syndrome, 23, 386
Oriens alveus interneurons (OAI), 411, 412*f*, 413
Orphan receptors, 65, 82

Paired-pulse depression, 37
Pancreatic cancer, 12
Parkinson's disease, 37, 83
Peripherally-restricted $GABA_B$ agonists, 295–306
antiepileptic gabapentin, 296
clinical pharmacodynamics, 302–306
effects of AZD9343 and baclofen on TLESR, 303
efficacy of lesogaberan in patients with GERD, 304
mean inhibitory effect of lesogaberan, 302
rates of treatment response and symptom-free days, 305*f*
utility of AP as treatment for patients with GERD, 305
lesogaberan and AZD9343, 296–297
preclinical studies with lesogaberan and AZD9343, 297–302
accumulation and transport, 300
Ferret Gastric Mechanosensitive Vagal Afferent Sensitivity Studies, 297–298
on 3H-GABA uptake and membrane potential in tsA201 cells, 301, 301*t*
inhibition of TLESR, 299*f*
in vitro potency on intracellular calcium release, 297
murine hypothermia and inhibition of TIESR, 298–300
receptor affinity studies, 297
Pertussis toxin sensitive G-proteins ($G_{i/o}$ family), 124, 126*f*, 130*f*, 131–132, 150–151, 153–157, 158*f*, 160–162, 164–168, 258*t*, 271–272, 405
PF-Purkinje cell excitatory postsynaptic potential (PF-PC EPSP), 151
$GABA_BR$-mediated modulatory effects, 162
mGluR1-mediated, 162
slow PF-PC EPSP, 162
Phaclofen, 6, 32, 32*f*, 208*f*, 235, 236*t*, 264, 333*t*, 337*t*, 358, 430, 437–438
Phenibut, 25
IC_{50} values, 25
neuropsychotropic drug, 25
racemic (*R*)-(–)-phenibut, 25
structures of (*R*)-(–)-phenibut, 21*f*
Phosphorylation
cAMP-mediated, 116
CREB, 30
and diseases, 118–119
and functional modulation, 117
$GABA_B$ receptor-dependent desensitization, 137
G-protein agonist-induced receptor, 134
-independent desensitization, 117–118
PKA, 137
of S783 in R2 subunit, ischemia, 119
Photo-affinity radioligand [^{125}I]-CGP71872, 65
Pilc and Lloyd hypothesis, 34
Placozoans (*Trichoplax adhaerens*), 68
Plasma membrane GABA transporter 1, 176–177
properties and translocation of GABA by GAT1, 177–179, 178*f*
GAT1 slc6a1 gene, 177
localization of GAT1 at ultrastructural level, model of, 178*f*
pharmacological inhibition of GAT1 activity, 177
removal of extracellular GABA by GAT1-mediated uptake, conditions, 179
through membrane via process that requires Na^+ and Cl^-, 178*f*

regulation of GAT1-mediated GABA transport, 179–180
controlling membrane surface expression, 179
genetically-engineered GAT1 deficiency, 180
^{32}P-orthophosphate, 115
Positive modulators of GABA$_B$R, 29–32
allosteric modulators, 29
structures of positive, 30*f*
cAMP, 30
CGP7930, 30
CREB phosphorylation, 30
dose dependent effects, 29
ERK(1/2), 30
GS39783, 31
GTPγ[^{35}S] assay, 29
NVP-BHF177, 31
point mutations G706T and A708P, 31
potency and maximal efficacy, increased, 29
racemic BHFF reversed stress-induced hypothermia, 32
Postsynaptic receptors, 178*f*, 195, 235, 429, 456
Posttraumatic stress disorder, chronic, 24
Pre- and post-synaptic GABA$_B$R, 166–167
addition of CGP55845, 167
AMPAR-mediated fast PF-PC EPSC, 166
application of CGP34629, 167
augmentation at doses, 167
in vivo, ambient GABA, 166
microdialysis studies, 166
study in rat cerebellar slices, 166
Presynaptic receptors, 5, 194, 209
Probenecid-sensitive organic anion transporter, 21
Protein kinase C (PKC), 116–117
α-amino-3-hydroxy-5-methyl-4-isoxazolepropionic acid receptor (AMPAR), 116
Chinese hamster ovary (CHO) cells, 116
γ-aminobutyric acid (GABA) type A receptor (GABA$_A$R), 116
N-ethylmaleimide-sensitive fusion (NSF) protein, 116
^{35}S-GTP-γS binding, 117
Protein purification, 65
Psychomotor stimulant cocaine
GABA$_B$R agonists, 339–345
baclofen, effects of, 340
γ-vinyl GABA (GVG)/(vigabatrin), 339
GABA$_B$R-positive allosteric modulators, 345–348
Purkinje cells, 150
ligands for GABA$_B$R
basal "ambient" GABA, 152
extracellular Ca^{2+}, 153–154
GABA spilt over from synapses, 153
synaptic organization, 151–152
AMPAR and mGluR1, 151
GABA$_A$R, 152
cerebellar LTD, 152
climbing fiber (CF), 151
glutamatergic transmission, 152
IPSP, 152
parallel fibers (PF), 151
PF-PC EPSP, 151
TRPC3 cation channels, 151

Radioligands, 26, 28, 38, 40, 65, 74, 292
Rat cerebellar Purkinje cells, 161
full-length variant of mGluR1 (mGluR1a), 161
splice variant with short C-termini (mGluR1b), 161
Reflux episodes, 288, 290, 303–306
Resistance to dieldrin (Rdl), 457, 459
Reward processes, 315–360

Saclofen, 6, 258*t*, 430, 458*t*, 460
2-hydroxy saclofen, 6, 32*f*, 33, 35, 42, 350
SCH50911, 6, 32*f*, 33, 35, 38, 322*t*–323*t*, 337*t*, 339, 345, 348
Schizophrenia, 27, 43, 83, 180, 206, 221, 378
Sea squirt, *Ciona intestinalis,* 67, 82
Sensory Family C receptors, 67, 69–71, 76–77, 83
Sensory processing, 214
Serotonin reuptake inhibitor (SSRI), 416, 433, 438–439
Short consensus repeats, 238
Short-interval cortical inhibition (SICI), 441
Single-axon or unitary IPSC/IPSP, 183
Skeletal muscle relaxation, 9
properties of baclofen, 9
adverse effects, 9
side effects, 9
treatment of spasticity, 9
Slime mould *Dictyostelium discoideum,* 67
Slow wave sleep, effects on, 35, 206, 213
Snap-tag technology, 135

State-of-the-art method, 183
Stiffman syndrome, 22
Stratum lacunosum moleculare (SLM), 408*f*, 411, 413, 414*f*
βγ Subunit complex, 157
  diacylglycerol (DAG), 157
  inositol trisphosphate (IP3), 157
  phospholipase C (PLC), 157
Succinate semialdehyde dehydrogenase, 34
Suprachiasmatic nucleus (SCN), 125, 131, 271
Supraoptic nucleus (SON), 125, 131–132
Sushi domains, *See* Complement protein (CP) module
Synaptic plasticity processes, 242
Systemic blood pressure, autonomic control of
  alteration of $GABA_BR$ function in animal models of hypertension, 275*t*
  autonomic nervous system, overview, 260–261
    baroreceptor reflex, activation of, 260
    caudal ventrolateral medulla (CVLM), 260
    central control regions, RVLM/NTS/hypothalamus, 261
    nucleus tractus solitarii (NTS), 260
    rostral ventrolateral medulla (RVLM), 260
    sympathetic and parasympathetic, 260
  distribution of $GABA_BR$ in CNS autonomic centers, 261
    antibody against $GABA_{B1}$, 261
  functions of $GABA_BR$ in regulating arterial blood pressure, 275*f*
  $GABA_B$ function in NTS, 263–267
    angiotensin II, renin–angiotensin system, 266
    baroreceptor reflex, 263
    chronic hypertension, 264–265
    GABAergic mechanisms, 263
    intracerebroventricular infusion of angiotensin II, 265
    microinjection of baclofen, 264
    presynaptic/postsynaptic mechanisms, 265
    regulation of multiple autonomic reflexes, 263
    stimulation of $GABA_BR$ with baclofen, 264
    upregulation of $GABA_BR$ in NTS in hypertension, 267
  $GABA_BR$ function in hypothalamus, 267–274
    activation of $GABA_BR$ in PVN, 268
    baclofen-sensitive currents in labeled PVN neurons, 273*f*
    bath application of $GABA_BR$ antagonist CGP55845, 270*f*
    bath application of *N*-ethylmaleimide, 271–272
    GABAergic inhibitory postsynaptic currents (IPSC), 271
    glutamatergic excitatory postsynaptic currents (EPSC), 271
    hypothalamic paraventricular nucleus (PVN), 267
    microinjection of baclofen, 269*f*, 270
    N-methyl-D-aspartic acid (NMDA) receptor activation, 274
  $GABA_BR$ function in RVLM, 261–263
    neuronal functions by activating $GABA_BR$, 262
    stimulation of RVLM by microinjection of glutamate or kainic acid, 261
  GABAergic mechanisms in central control of cardivovascular function, 262
  GABAR subtypes, 258*t*

Targeted mutation technique, 417
Temporoammonic (TA) pathway, 219*f*, 411–413, 414*f*
Trafficking pathways, 95*f*
trans-Golgi network, 94–95, 97–98
Transient lower esophageal sphincter relaxation (TLESR), 27, 289–291, 291*f*, 295–299, 299*f*, 301*t*, 302–304
Transient receptor potential C3 (TRPC3) cation channels, 151, 155, 156*f*, 157, 158*f*, 162
Tumor susceptibility gene 101 (TSG101), 105

Unipolar brush cells (UBC), 418

Venus flytrap domain (VFTD)
  phylogeny of $GABA_BR$, 66–70
    $GABA_{B2}$ subunit binding domain, 70
    periplasmic amino acid-binding proteins (PBP), 69
    recapitulate ancestral receptor, 70

sensory Family C receptors, 70
structure of $GABA_BR$, 70–73
of antagonist S-MCPG ((S)-(α) methyl-4-carboxyphenylglycine), 72
of $GABA_B$ VFTD, 73
glutamate-bound structure, 72
of heterodimer in active and inactive states, 72*f*
of mGluR3 structure, 73
of mGluR1 VFT, 72
Vigabatrin, 10–11, 320*t*, 336*t*, 339, 375, 378, 381–382, 384, 387, 389, 391

Contents of Previous Volumes

Volume 40

Advances in Understanding the Pharmacological Properties of Antisense Oligonucleotides
Stanley T. Crooke

Targeted Tumor Cytotoxicity Mediated by Intracellular Single-Chain Anti-Oncogene Antibodies
David T. Curiel

In Vivo Gene Therapy with Adeno-Associated Virus Vectors for Cystic Fibrosis
Terence R. Flotte and Barrie J. Carter

Engineering Herpes Simplex Virus Vectors for Human Gene Therapy
Joseph C. Glorioso, William F. Goins, Martin C. Schmidt, Tom Oligino, Dave Krisky, Peggy Marconi, James D. Cavalcoli, Ramesh Ramakrishnan, P. Luigi Poliani, and David J. Fink

Human Adenovirus Vectors for Gene Transfer into Mammalian Cells
Mary M. Hitt, Christina L. Addison, and Frank L. Graham

Anti-Oncogene Ribozymes for Cancer Gene Therapy
Akira Irie, Hiroshi Kijima, Tsukasa Ohkawa, David Y. Bouffard, Toshiya Suzuki, Lisa D. Curcio, Per Sonne Holm, Alex Sassani, and Kevin J. Scanlon

Cytokine Gene Transduction in the Immunotherapy of Cancer
Giorgio Parmiani, Mario P. Colombo, Cecilia Melani, and Flavio Arienti

Gene Therapy Approaches to Enhance Anti-Tumor Immunity
Daniel L. Shawler, Habib Fakhrai, Charles Van Beveren, Dan Mercoa, Daniel P. Gold, Richard M. Bartholomew, Ivor Royston, and Robert E. Sobol

Modified Steroid Receptors and Steroid-Inducible Promoters as Genetic Switches for Gene Therapy
John H. White

Strategies for Approaching Retinoblastoma Tumor Suppressor Gene Therapy
Hong-Ji Xu

Immunoliposomes for Cancer Treatment
John W. Park, Keelung Hong, Dmitri B. Kirpotin, Demetrios Papahadjopoulos, and Christopher C. Benz

Antisense Inhibition of Virus Infection
R. E. Kilkuskie and A. K. Field

Volume 41

Apoptosis: An Overview of the Process and Its Relevance in Disease
Stephanie Johnson Webb, David J. Harrison, and Andrew H. Wyllie

Genetics of Apoptosis
Serge Desnoyers and Michael O. Hengartner

Methods Utilized in the Study of Apoptosis
Peter W. Mesner and Scott H. Kaufmann

In Vitro Systems for the Study of Apoptosis
Atsushi Takahashi and William C. Earnshaw

The Fas Pathway in Apoptosis
Christine M. Eischen and Paul J. Leibson

Ceramide: A Novel Lipid Mediator of Apoptosis
Miriam J. Smyth, Lina M. Obeid, and Yusuf A. Hannun

Control of Apoptosis by Proteases
Nancy A. Thornberry, Antony Rosen, and Donald W. Nicholson

Death and Dying in the Immune System
David S. Ucker

Control of Apoptosis by Cytokines
W. Stratford May, Jr.

Glucocorticoid-Induced Apoptosis
Clark W. Distelhorst

Apoptosis in AIDS
Andrew D. Badley, David Dockrell, and Carlos V. Paya

Virus-Induced Apoptosis
J. Marie Hardwick

Apoptosis in Neurodegenerative Diseases
Ikuo Nishimoto, Takashi Okamoto, Ugo Giambarella, and Takeshi Iwatsubo

Apoptosis in the Mammalian Kidney: Incidence, Effectors, and Molecular Control in Normal Development and Disease States
Ralph E. Buttyan and Glenda Gobé

Apoptosis in the Heart
Samuil R. Umansky and L. David Tomei

Apoptosis and the Gastrointestinal System
Florencia Que and Gregory J. Gores

Role of p53 in Apoptosis
Christine E. Canman and Michael B. Kastan

Chemotherapy-Induced Apoptosis
Peter W. Mesner, Jr., I. Imawati Budihardjo, and Scott H. Kaufmann

Bcl-2 Family Proteins: Strategies for Overcoming Chemoresistance in Cancer
John C. Reed

Role of Bcr-Abl Kinase in Resistance to Apoptosis
Afshin Samali, Adrienne M. Gorman, and Thomas G. Cotter

Apoptosis in Hormone-Responsive Malignancies
Samuel R. Denmeade, Diane E. McCloskey, Ingrid B. J. K. Joseph, Hillary A. Hahm, John T. Isaacs, and Nancy E. Davidson

Volume 42

Catecholamine: Bridging Basic Science
Edited by David S. Goldstein, Graeme Eisenhofer, and Richard McCarty

Part A. Catecholamine Synthesis and Release

Part B. Catecholamine Reuptake and Storage

Part C. Catecholamine Metabolism

Part D. Catecholamine Receptors and Signal Transduction

Part E. Catecholamine in the Periphery

Part F. Catecholamine in the Central Nervous System

Part G. Novel Catecholaminergic Systems

Part H. Development and Plasticity

Part I. Drug Abuse and Alcoholism

Volume 43

Overview: Pharmacokinetic Drug–Drug Interactions
Albert P. Li and Malle Jurima-Romet

Role of Cytochrome P450 Enzymes in Drug–Drug Interactions
F. Peter Guengerich

The Liver as a Target for Chemical–Chemical Interactions
John-Michael Sauer, Eric R. Stine, Lhanoo Gunawardhana, Dwayne A. Hill, and I. Glenn Sipes

Application of Human Liver Microsomes in Metabolism-Based Drug–Drug Interactions: *In Vitro–in Vivo* Correlations and the Abbott Laboratories Experience
A. David Rodrigues and Shekman L. Wong

Primary Hepatocyte Cultures as an in Vitro Experimental Model for the Evaluation of Pharmacokinetic Drug–Drug Interactions
Albert P. Li

Liver Slices as a Model in Drug Metabolism
James L. Ferrero and Klaus Brendel

Use of cDNA-Expressed Human Cytochrome P450 Enzymes to Study Potential Drug–Drug Interactions
Charles L. Crespi and Bruce W. Penman

Pharmacokinetics of Drug Interactions
Gregory L. Kedderis

Experimental Models for Evaluating Enzyme Induction Potential of New Drug Candidates in Animals and Humans and a Strategy for Their Use
Thomas N. Thompson

Metabolic Drug–Drug Interactions: Perspective from FDA Medical and Clinical Pharmacology Reviewers
John Dikran Balian and Atiqur Rahman

Drug Interactions: Perspectives of the Canadian Drugs Directorate
Malle Jurima-Romet

Overview of Experimental Approaches for Study of Drug Metabolism and Drug–Drug Interactions
Frank J. Gonzalez

Volume 44

Drug Therapy: The Impact of Managed Care
Joseph Hopkins, Shirley Siu, Maureen Cawley, and Peter Rudd

The Role of Phosphodiesterase Enzymes in Allergy and Asthma
D. Spina, L. J. Landells, and C. P. Page

Modulating Protein Kinase C Signal Transduction
Daria Mochly-Rosen and Lawrence M. Kauvar

Preventive Role of Renal Kallikrein—Kinin System in the Early Phase of Hypertension and Development of New Antihypertensive Drugs
Makoto Kartori and Masataka Majima

The Multienzyme PDE4 Cyclic Adenosine Monophosphate-Specific Phosphodiesterase Family: Intracellular Targeting, Regulation, and Selective Inhibition by Compounds Exerting Anti-inflammatory and Antidepressant Actions
Miles D. Houslay, Michael Sullivan, and Graeme B. Bolger

Clinical Pharmacology of Systemic Antifungal Agents: A Comprehensive Review of Agents in Clinical Use, Current Investigational Compounds, and Putative Targets for Antifungal Drug Development
Andreas H. Groll, Stephen C. Piscitelli, and Thomas J. Walsh

Volume 45

Cumulative Subject Index
Volumes 25–44

Volume 46

Therapeutic Strategies Involving the Multidrug Resistance Phenotype: The MDR1 Gene as Target, Chemoprotectant, and Selectable Marker in Gene Therapy
Josep M. Aran, Ira Pastan, and Michael M. Gottesman

The Diversity of Calcium Channels and Their Regulation in Epithelial Cells
Min I. N. Zhang and Roger G. O'Neil

Gene Therapy and Vascular Disease
Melina Kibbe, Timothy Billiar, and Edith Tzeng

Heparin in Inflammation: Potential Therapeutic Applications beyond Anticoagulation
David J. Tyrrell, Angela P. Horne, Kevin R. Holme, Janet M. H. Preuss, and Clive P. Page

The Regulation of Epithelial Cell cAMP-and Calcium-Dependent Chloride Channels
Andrew P. Morris

Calcium Channel Blockers: Current Controversies and Basic Mechanisms of Action
William T. Clusin and Mark E. Anderson

Mechanisms of Antithrombotic Drugs
Perumal Thiagarajan and Kenneth K. Wu

Volume 47

Hormones and Signaling
Edited by Bert W. O'Malley

New Insights into Glucocorticoid and Mineralocorticoid Signaling: Lessons from Gene Targeting
Holger M. Reichardt, Franc¸ois Tronche, Stefan Berger, Christoph Kellendonk, and Günther Shütz

Orphan Nuclear Receptors: An Emerging Family of Metabolic Regulators
Robert Sladek and Vincent Giguère

Nuclear Receptor Coactivators
Stefan Westin, Michael G. Rosenfeld, and Christopher K. Glass

Cytokines and STAT Signaling
Christian Schindler and Inga Strehlow

Coordination of cAMP Signaling Events through PKA Anchoring
John D. Scott, Mark L. Dell'Acqua, Iain D. C. Fraser, Steven J. Tavalin, and Linda B. Lester

G Protein-Coupled Extracellular Ca^{2+} ($Ca^{2+}{}_{o}$)-Sensing Receptor (CaR): Roles in Cell Signaling and Control of Diverse Cellular Functions
Toru Yamaguchi, Naibedya Chattopadhyay, and Edward M. Brown

Pancreatic Islet Development
Debra E. Bramblett, Hsiang-Po Huang, and Ming-Jer Tsai

Genetic Analysis of Androgen Receptors in Development and Disease
A. O. Brinkmann and J. Trapman

An Antiprogestin Regulable Gene Switch for Induction of Gene Expression *in Vivo*
Yaolin Wang, Sophia Y. Tsai, and Bert W. O'Malley

Steroid Receptor Knockout Models: Phenotypes and Responses Illustrate Interactions between Receptor Signaling Pathways *in Vivo*
Sylvia Hewitt Curtis and Kenneth S. Korach

Volume 48

HIV: Molecular Biology and Pathogenesis: Viral Mechanisms Edited by Kuan-Teh Jeang

Multiple Biological Roles Associated with the Repeat (R) Region of the HIV-I RNA Genome
Ben Berkhout

HIV Accessory Proteins: Multifunctional Components of a Complex System
Stephan Bour and Klaus Strebel

Role of Chromatin in HIV-I Transcriptional Regulation
Carine Van Lint

NF-κB and HIV: Linking Viral and Immune Activation
Arnold B. Rabson and Hsin-Ching Lin

Tat as a Transcriptional Activator and a Potential Therapeutic Target for HIV-1
Anne Gatignol and Kuan-Teh Jeang

From the Outside In: Extracellular Activities of HIV Tat
Douglas Noonan and Andriana Albini

Rev Protein and Its Cellular Partners
Jørgen Kjems and Peter Askjaer

HIV-I Nef: A Critical Factor in Viral-Induced Pathogenesis
A. L. Greenway, G. Holloway, and D. A. McPhee

Nucleocapsid Protein of Human Immunodeficiency Virus as a Model Protein with Chaperoning Functions and as a Target for Antiviral Drugs
Jean-Luc Darlix, Gaël Cristofari, Michael Rau, Christine Péchoux, Lionel Berthoux, and Bernard Roques

Bioactive CD4 Ligands as Pre-and/or Postbinding Inhibitors of HIV-I
Laurence Briant and Christian Devaux

Coreceptors for Human Immunodeficiency Virus and Simian Immunodeficiency Virus
Keith W. C. Peden and Joshua M. Farber

Volume 49

HIV: Molecular Biology and Pathogenesis: Clinical Applications
Edited by Kuan-Teh Jeang

Inhibitors of HIV-I Reverse Transcriptase
Michael A. Parniak and Nicolas Sluis-Cremer

HIV-I Protease: Maturation, Enzyme Specificity, and Drug Resistance
John M. Louis, Irene T. Weber, József Tözsér, G. Marius Clore, and Angela M. Gronenborn

HIV-I Integrase Inhibitors: Past, Present, and Future
Nouri Neamati, Christophe Marchand, and Yves Pommier

Selection of HIV Replication Inhibitors: Chemistry and Biology
Seongwoo Hwang, Natarajan Tamilarasu, and Tariq M. Rana

Therapies Directed against the Rev Axis of HIV Autoregulation
Andrew I. Dayton and Ming Jie Zhang

HIV-I Gene Therapy: Promise for the Future
Ralph Dornburg and Roger J. Pomerantz

Assessment of HIV Vaccine Development: Past, Present, and Future
Michael W. Cho

HIV-I-Associated Central Nervous System Dysfunction
Fred C. Krebs, Heather Ross, John McAllister, and Brian Wigdahl

Molecular Mechanisms of Human Immunodeficiency Virus Type I Mother-Infant Transmission
Nafees Ahmad

Molecular Epidemiology of HIV-I: An Example of Asia
Mao-Yuan Chen and Chun-Nan Lee

Simian Immunodeficiency Virus Infection of Monkeys as a Model System for the Study of AIDS Pathogenesis, Treatment, and Prevention
Vanessa M. Hirsch and Jeffrey D. Lifson

Animal Models for AIDS Pathogenesis
John J. Trimble, Janelle R. Salkowitz, and Harry W. Kestler

Volume 50

General Introduction to Vasopressin and Oxytocin: Structure/ Metabolism, Evolutionary Aspects, Neural Pathway/Receptor Distribution, and Functional Aspects Relevant to Memory Processing
Barbara B. McEwen

De Wied and Colleagues I: Evidence for a VP and an OT Influence on MP: Launching the "VP/OT Central Memory Theory"
Barbara B. McEwen

De Wied and Colleagues II: Further Clarification of the Roles of Vasopressin and Oxytocin in Memory Processing
Barbara B. McEwen

De Wied and Colleagues III: Brain Sites and Transmitter Systems Involved in theVasopressin and Oxytocin Influence on Memory Processing
Barbara B. McEwen

De Wied and Colleagues IV: Research into Mechanisms of Action by Which Vasopressin and Oxytocin Influence Memory Processing
Barbara B. McEwen

Research Studies of Koob and Colleagues: The "Vasopressin Dual Action Theory"
Barbara B. McEwen

Contributions of Sahgal and Colleagues: The "Vasopression Central Arousal Theory"
Barbara B. McEwen

Role of Attentional Processing in Mediating the Influence of Vasopressin on Memory Processing
Barbara B. McEwen

Expansion of Vasopressin/Oxytocin Memory Research I: Peripheral Administration
Barbara B. McEwen

Expansion of Vasopressin/Oxytocin Memory Research II: Brain Structures and Transmitter Systems Involved in the Influence of Vasopressin and Oxytocin on Memory Processing
Barbara B. McEwen

Expansion of Vasopressin/Oxytocin Memory Research III: Research Summary and Commentary on Theoretical and Methodological Issues
Barbara B. McEwen

Research Contributions of Dantzer, Bluthe, and Colleagues to the Study of the Role of Vasopressin in Olfactory-Based Social Recognition Memory
Barbara B. McEwen

Expansion of Olfactory-Based Social Recognition Memory Research: The Roles of Vasopressin and Oxytocin in Social Recognition Memory
Barbara B. McEwen

Brain–Fluid Barriers: Relevance for Theoretical Controversies Regarding Vasopressin and Oxytocin Memory Research
Barbara B. McEwen

Closing Remarks: Review and Commentary on Selected Aspects of the Roles of Vasopressin and Oxytocin in Memory Processing
Barbara B. McEwen

Volume 51

Treatment of Leukemia and Lymphoma
Edited by David A. Scheinberg and Joseph G. Jurcic

Kinase Inhibitors in Leukemia
Mark Levis and Donald Small

Therapy of Acute Promyelocytic Leukemia
Steven Soignet and Peter Maslak

Investigational Agents in Myeloid Disorders
Farhad Ravandi and Jorge Cortes

Methodologic Issues in Investigation of Targeted Therapies in Acute Myeloid Leukemia
Elihu Estey

Purine Analogs in Leukemia
Nicole Lamanna and Mark Weiss

Monoclonal Antibody Therapy in Lymphoid Leukemias
Thomas S. Lin and John C. Byrd

Native Antibody and Antibody-Targeted Chemotherapy for Acute Myeloid Leukemia
Eric L. Sievers

Radioimmunotherapy of Leukemia
John M. Burke and Joseph G. Jurcic

Immunotoxins and Toxin Constructs in the Treatment of Leukemia and Lymphoma
Michael Rosenblum

Antibody Therapy of Lymphoma
George J. Weiner and Brian K. Link

Vaccines in Leukemia
Sijie Lu, Eric Wieder, Krishna Komanduri, Qing Ma, and Jeffrey J. Molldrem

Therapeutic Idiotype Vaccines for Non-Hodgkin's Lymphoma
John M. Timmerman

Cytokine Modulation of the Innate Immune System in the Treatment of Leukemia and Lymphoma
Sherif S. Farag and Michael A. Caligiuri

Donor Lymphocyte Infusions
Vincent T. Ho and Edwin P. Alyea

Somatic Cell Engineering and the Immunotherapy of Leukemias and Lymphomas
Renier J. Brentjens and Michel Sadelain

Volume 52

Historical Background
Andrew Young

Tissue Expression and Secretion of Amylin
Andrew Young

Receptor Pharmacology
Andrew Young

Amylin and the Integrated Control of Nutrient Influx
Andrew Young

Inhibition of Food Intake
Andrew Young

Inhibition of Gastric Emptying
Andrew Young

Effects on Digestive Secretions
Andrew Young

Inhibition of Glucagon Secretion
Andrew Young

Inhibition of Insulin Secretion
Andrew Young

Effects on Plasma Glucose and Lactate
Andrew Young

Effects in Skeletal Muscle
Andrew Young

Effects in Liver
Andrew Young

Effects in Fat
Andrew Young

Cardiovascular Effects
Andrew Young

Renal Effects
Andrew Young

Effects on Bone
Andrew Young

Central Nervous System and Other Effects
Andrew Young

Clinical Studies
Andrew Young

Volume 53

Isolation, Purification, and Analysis of Chondroitin Sulfate Proteoglycans
Fumiko Matsui and Atsuhiko Oohira

Isolation and Purification of Chondroitin Sulfate
Luiz-Claudio F. Silva

Structure of Chondroitin Sulfate
Fotini N. Lamari and Nikos K. Karamanos

Progress in the Structural Biology of Chondroitin Sulfate
Barbara Mulloy

The Biosynthesis and Catabolism of Galactosaminoglycans
Vikas Prabhakar and Ram Sasisekharan

Biosynthesis of Chondroitin Sulfate: From the Early, Precursor Discoveries to Nowadays, Genetics Approaches
Mauro S. G. Pavão, Ana Cristina Vilela-Silva, and Paulo A. S. Mourão

Advances in the Analysis of Chondroitin/Dermatan Sulfate
M. Stylianou, I.-E. Triantaphyllidou, and D. H. Vynios

Chondroitin Sulfate Lyases: Applications in Analysis and Glycobiology
Emmanuel Petit, Cedric Delattre, Dulce Papy-Garcia, and Philippe Michaud

CS Lyases: Structure, Activity, and Applications in Analysis and the Treatment of Diseases
Robert J. Linhardt, Fikri Y. Avci, Toshihiko Toida, Yeong Shik Kim, and Miroslaw Cygler

Structure, Metabolism, and Tissue Roles of Chondroitin Sulfate Proteoglycans
Christopher J. Handley, Tom Samiric, and Mirna Z. Ilic

Emergence and Structural Characteristics of Chondroitin Sulfates in the Animal Kingdom
Lucia O. Sampaio and Helena B. Nader

Role of the Sulfation Pattern of Chondroitin Sulfate in its Biological Activities and in the Binding of Growth Factors
Chilkunda D. Nandini and Kazuyuki Sugahara

Chondroitin Sulfate as a Key Molecule in the Development of Atherosclerosis and Cancer Progression
A. D. Theocharis, I. Tsolakis, G. N. Tzanakakis, and N. K. Karamanos

Chondroitin Sulfate Proteoglycans in Tumor Progression
Yanusz Wegrowski and Franc¸ois-Xavier Maquart

Chondroitin Sulfate Proteoglycans in the Brain
Sachiko Aono and Atsuhiko Oohira

Chondroitin/Dermatan Sulfates in the Central Nervous System: Their Structures and Functions in Health and Disease
Uwe Rauch and Joachim Kappler

Chondroitin Sulfate Proteoglycan and its Degradation Products in CNS Repair
Asya Rolls and Michal Schwartz

Role of Chondroitin-4-Sulfate in Pregnancy-Associated Malaria
D. Channe Gowda

Immunological Activity of Chondroitin Sulfate
Toshihiko Toida, Shinobu Sakai, Hiroshi Akiyama, and Robert J. Linhardt

Antioxidant Activity of Chondroitin Sulfate
G. M. Campo, A. Avenoso, S. Campo, A. M. Ferlazzo, and A. Calatroni

Effects of Chondroitin Sulfate on the Cellular Metabolism
N. Brandl, J. Holzmann, R. Schabus, and M. Huettinger

In Vitro Effects of Chondroitin Sulfate
A. Fioravanti, R. Marcolongo, and G. Collodel

Effect of Chondroitin Sulfate as Nutraceutical in Dogs with Arthropathies
Britta Dobenecker

Chondroitin Sulfate as a Structure-Modifying Agent
Daniel Uebelhart, Ruud Knols, Eling D de Bruin, and Gust Verbruggen

Chondroitin Sulfate in the Management of Erosive Osteoarthritis of the Interphalangeal Finger Joints
Gust Verbruggen

Chondroitin Sulfate in the Management of Hip and Knee Osteoarthritis: An Overview
Géraldine Bana, Bénédicte Jamard, Evelyne Verrouil, and Bernard Mazières

Treatment of Knee Osteoarthritis with Oral Chondroitin Sulfate
Daniel Uebelhart, Ruud Knols, Eling D de Bruin, and Gust Verbruggen

Volume 54

The Role of GABA in the Mediation and Perception of Pain
S. J. Enna and Kenneth E. McCarson

Distribution of GABA Receptors in the Thalamus and Their Involvement in Nociception
Fani L. Neto, Joana Ferreira-Gomes, and José M. Castro-Lopes

GABAA Agonists and Partial Agonists: THIP (Gaboxadol) as a Non-Opioid Analgesic and a Novel Type of Hypnotic
Povl Krogsgaard-Larsen, Bente Frølund, and Tommy Liljefors

Rat Modeling for GABA Defects in Schizophrenia
Francine M. Benes and Barbara Gisabella

Epigenetic Targets in GABAergic Neurons to Treat Schizophrenia
E. Costa, E. Dong, D. R. Grayson, W. B. Ruzicka, M. V. Simonini, M. Veldic, and A. Guidotti

GABAergic Malfunction in the Limbic System Resulting from an Aboriginal Genetic Defect in Voltage-Gated Naþ-Channel SCN5A is Proposed to Give Rise to Susceptibility to Schizophrenia
Eugene Roberts

GABAA Receptor Mutations Associated with Generalized Epilepsies
Robert L. Macdonald, Jing-Qiong Kang, Martin J. Gallagher, and Hua-Jun Feng

From Gene to Behavior and Back Again: New Perspectives on GABAA Receptor Subunit Selectivity of Alcohol Actions
Stephen L. Boehm II, Igor Ponomarev, Yuri A. Blednov, and R. Adron Harris

A Role for GABA in Alcohol Dependence
George F. Koob

Structure, Pharmacology, and Function of GABAA Receptor Subtypes
Werner Sieghart

Structure–Activity Relationship and Pharmacology of g-Aminobutyric Acid (GABA) Transport Inhibitors
Rasmus Prætorius Clausen, Karsten Madsen, Orla Miller Larsson, Bente Frølund, Povl Krogsgaard-Larsen, and Arne Schousboe

Modulation of Ionotropic GABA Receptors by Natural Products of Plant Origin
Graham A. R. Johnston, Jane R. Hanrahan, Mary Chebib, Rujee K. Duke, and Kenneth N. Mewett

Volume 55

HIV-1 RNA Packaging
Andrew M. L. Lever

Structure and Function of the HIV Envelope Glycoprotein as Entry Mediator, Vaccine Immunogen, and Target for Inhibitors
Ponraj Prabakaran, Antony S. Dimitrov, Timothy R. Fouts, and Dimiter S. Dimitrov

HIV-1 Reverse Transcription: Close Encounters Between the Viral Genome and a Cellular tRNA
Truus E. M. Abbink and Ben Berkhout

Transcription of HIV: Tat and Cellular Chromatin
Anne Gatignol

Posttranscriptional Control of HIV-1 and Other Retroviruses and Its Practical Applications
Barbara K. Felber, Andrei S. Zolotukhin, and George N. Pavlakis

HIV Accessory Genes Vif and Vpu
Klaus Strebel

Interactions of HIV-1 Viral Protein R with Host Cell Proteins
Richard Y. Zhao, Robert T. Elder, and Michael Bukrinsky

HIV-1 Protease: Structure, Dynamics, and Inhibition
John M. Louis, Rieko Ishima, Dennis A. Torchia, and Irene T. Weber

Properties, Functions, and Drug Targeting of the Multifunctional Nucleocapsid Protein of the Human Immunodeficiency Virus
Jean-Luc Darlix, José Luis Garrido, Nelly Morellet, Yves Mély, and Hugues de Rocquigny

Human Immunodeficiency Virus Type 1 Assembly, Release, and Maturation
Catherine S. Adamson and Eric O. Freed

Role of Nef in HIV-1 Replication and Pathogenesis
John L. Foster and J. Victor Garcia

Treatment Implications of the Latent Reservoir for HIV-1
Susan Peterson, Alison P. Reid, Scott Kim, and Robert F. Siliciano

RNA Interference and HIV-1
Man Lung Yeung, Yamina Bennasser, Shu-Yun Le, and Kuan-Teh Jeang

Volume 56

Global Molecular Epidemiology of HIV: Understanding the Genesis of AIDS Pandemic
Yutaka Takebe, Rie Uenishi, and Xiaojie Li

Current Clinical Treatments of AIDS
Erin-Margaret Murphy, Humberto R. Jimenez, and Stephen M. Smith

HIV-1-Specific Immune Response
Alexandre Harari and Giuseppe Pantaleo

Targeting HIV Attachment and Entry for Therapy
Julie Strizki

Inhibitors of HIV-1 Reverse Transcriptase
Tatiana Ilina and Michael A. Parniak

Development of Protease Inhibitors and the Fight with Drug-Resistant HIV-1 Variants
Hiroaki Mitsuya, Kenji Maeda, Debananda Das, and Arun K. Ghosh

HIV-1 Integrase Inhibitors: Update and Perspectives
Elena A. Semenova, Christophe Marchand, and Yves Pommier

Topical Microbicides: A Promising Approach for Controlling the AIDS Pandemic via Retroviral Zinc Finger Inhibitors
Jim A. Turpin, Marco L. Schito, Lisa M. Miller Jenkins, John K. Inman, and Ettore Appella

Viral Drug Resistance and Fitness
Miguel E. Quiñones-Mateu, Dawn M. Moore-Dudley, Oyebisi Jegede, Jan Weber, and Eric J. Arts

Gene Therapy to Induce Cellular Resistance to HIV-1 Infection: Lessons from Clinical Trials
Mauro Giacca

Identification of Potential Drug Targets Using Genomics and Proteomics: A Systems Approach
Zachary A. Klase, Rachel Van Duyne, and Fatah Kashanchi

Rapid Disease Progression to AIDS due to Simian immunodeficiency virus Infection of Macaques: Host and Viral Factors
Que Dang and Vanessa M. Hirsch

Nonprimate Models of HIV-1 Infection and Pathogenesis
Viet Hoang, Elizabeth Withers-Ward, and David Camerini

Perspectives for a Protective HIV-1 Vaccine
Marco Schiavone, Ileana Quinto, and Giuseppe Scala

Molecular Mechanisms of HIV-1 Vertical Transmission and Pathogenesis in Infants
Nafees Ahmad

The Viral Etiology of AIDS-Associated Malignancies
Peter C. Angeletti, Luwen Zhang, and Charles Wood

Volume 57

Defining the Role of Pharmacology in the Emerging World of Translational Research
S. J. Enna and M. Williams

Anti-Inflammatory Agents as Cancer Therapeutics
Khosrow Kashfi

The Development and Pharmacology of Proteasome Inhibitors for the Management and Treatment of Cancer
Bruce Ruggeri, Sheila Miknyoczki, Bruce Dorsey, and Ai-Min Hui